10th Malaysian Road Conference & Exhibition 2018

IOP Conference Series: Materials Science and Engineering
Volume 512

Selangor, Malaysia
29 - 31 October 2018

Editors:

Rizati Hamidun
Fauzan Mahd Jakarni
Yuen Choon Wah

ISBN: 978-1-5108-8633-9
ISSN: 1757-8981

Printed from e-media with permission by:

Curran Associates, Inc.
57 Morehouse Lane
Red Hook, NY 12571

proceedings.com

Some format issues inherent in the e-media version may also appear in this print version.

Printed by Curran Associates, Inc. (2019)

For permission requests, please contact the Institute of Physics at the address below.

Institute of Physics
Dirac House, Temple Back
Bristol BS1 6BE UK

Phone: 44 1 17 929 7481
Fax: 44 1 17 920 0979

techtracking@iop.org

Additional copies of this publication are available from:

Curran Associates, Inc.
57 Morehouse Lane
Red Hook, NY 12571 USA
Phone: 845-758-0400
Fax: 845-758-2633
Email: curran@proceedings.com
Web: www.proceedings.com

Table of contents

Volume 512

10th Malaysian Road Conference & Exhibition 2018

29–31 October 2018, Selangor, Malaysia

Accepted papers received: 8 March 2019
Published online: 24 April 2019

Preface

Papers

IOP Conf. Series: Materials Science and Engineering **512** (2019) 011001 doi:10.1088/1757-899X/512/1/011001

Preface

The **Malaysian Road Conference & Exhibition 2018**, currently tenth in its series, was held from 29th to 31st October 2018 at the Sunway Pyramid Convention Centre in the city of Petaling Jaya, located in the suburbs of Malaysia's capital city of Kuala Lumpur. The conference was jointly organised by the Ministry of Works Malaysia, Public Works Department of Malaysia, Malaysian Highway Authority, World Road Association (PIARC) and the Road Engineering Association of Malaysia in collaboration with the Construction Industry Development Board Malaysia, International Road Federation Global and Road Engineering Association of Asia and Australasia. A World Road Association (PIARC) International Seminar on Asset Management was simultaneously hosted.

Revolve around the main theme of Global Approaches to Advance Road Asset Management, 61 technical papers from local and international participants were presented, concerning new indicators and directions relating to the research, development, cutting-edge technologies, construction, management and maintenance of road, highway and related infrastructure assets. The conference has benefited those in the road industry, be they road concessionaires, expressway regulators, related government authorities, consultants, contractors, project managers and engineers from public and corporate sector organisations.

Technical papers have undergone a rigorous peer-review process, based on initial Editors' screening, anonymous refereeing by independent expert referees, and consequent revision by the article author(s) when required. Editors would like to thank the author(s) who have contributed to the Conference Proceedings, and also to all delegates for their support towards the successful organization of the conference. Our special thanks go to the Technical Program Committee for their devoted assistance and expertise towards the preparation of the conference proceedings. Also, the Editors would like to extend gratitude to the other Committee members as well as all the individual whom are directly and indirectly involved to make the conference possible.

The Editors:

Dr. Rizati Hamidun, Malaysian Institute of Road Safety Research (MIROS)
Dr. Fauzan Mohd Jakarni, Universiti Putra Malaysia (UPM)
Ir. Dr. Yuen Choon Wah, University of Malaya (UM)

IOP Conf. Series: Materials Science and Engineering **512** (2019) 011001 doi:10.1088/1757-899X/512/1/011001

10 Malaysian Road Conference & Exhibition 2018
Technical Programme Committee

Chairman	Ir. Dr. Lim Char Ching
Committee members	Alvin Poi Wai Hoong Lim Chi Chon Ir. Muhammad Nizan Edris Dr. Rizati Hamidun Dr. Fauzan Mohd Jakarni Ir. Dr. Yuen Choon Wah

IOP Conf. Series: Materials Science and Engineering **512** (2019) 011002 doi:10.1088/1757-899X/512/1/011002

10th Malaysian Road Conference & Exhibition 2018

Chairman of Organising Committee
Ir. Haji Zulakmal Haji Sufian
Chairman of Working Committee
Dato' Ir. Dr. Ashaari Mohamad

Secretariat Committee
Ir. Johari Emby
Ir. Janari Sekeli
Safinas Saroji
Abdull Rahman Mohamad
Nor Hafizah Tajudin

Registration Committee
Ir. Razhiah Wahab
Rohana Yaakub
Muhamad Razif Norzali
Mohd Syukri Md Hanafiah
Eriena Abas
Adly Dasril
Nur Liyana Lokman

Finance Committee
Dato' Ir. Jeffrey Choong Moh Kheng
Ahmad Farehan Dahalan
Rosmah Marhamdan
Nur Izira Mohamad Azli
Rozita Ab. Hamid
Siti Mardhiah Yusof

Exhibition Committee
Dato' Ir. Mohd Shuhaimi Hassan
Siti Atikaa Hanom Hadie
Khairul Najjib Shokhaimi
Kamarulzaman @ Idham Hassan
Mohd. Noor Mohd Ali
Ir. Yap Teong Aun

2nd MRC-IIE 2018
Dr. Nadzrol Fadzilah Ahmad
Mohd. Suhaimi Sairan
Suhayya Rofik
Prof. Ratnasamy Muniandy
Prof. Madya Ir. Salehuddin Hassim

Hall Management
Ir. Nafisah Abdul Aziz
Committee members
Abdul Halim Abd. Rahman
Syahida Aripin
Ramesh Appanah
Nurul Saffrena Musa

Technical Programme Committee
Ir. Dr. Lim Char Ching
Dr. Fauzan Mohd Jakarni
Alvin Poi Wai Hong
Dr. Rizati Hamidun
Dr. Yuen Choon Wah
Ir. Mohammad Nizan Edris
Lim Chi Chon

Partnership Committee
Datuk Nik Airina Nik Jaffar
Datuk Seri Ir. Hasnur Rabiain Ismail
Azman Yunus
Zul Hisham Hamzah
Shaharuddin Abd Samad
Azmir Hasnur Rabiain

Marketing & Publicity Committee
Ir. Haji Mohamad Zulkefly Sulaiman
Ahmad Tajuddin Mohd Yusof
Noreha Mohd Ripin
Muhammad Razif Abu Hassan
Zaharah Manap
Ir. Nor Aznita Yusof
Ir. Siti Norleha Rasid
Zainuddin Hussain

PIARC International Seminar
Wan Radhiah Wan Hanafi
YDH Toh Paduka Setia Dato' Ir. Safry Kamal Hj. Ahmad
Ir. Mohd Hizam Harun
Roziawati Razali
Wan Muhammad Ridha Wan Husain

Technical Visits Committee
Dato' Ir. B. Nitchiananthan
Ir. Sivaananthan Chelladurai
Low Kian Yung
Syed Mohd Nazeri Syed Norizan
Pream Rao
Ashwini A/P Thiruthonder

IOP Conf. Series: Materials Science and Engineering **512** (2019) 011002 doi:10.1088/1757-899X/512/1/011002

Opening & Closing Ceremonies Committee
- Mohd Nizam Ismail
- Najihah Mohd Yasin
- Mashitah Mohd Arshad
- Airol Kamarul Bahrin
- Nazren Fauzi
- Rohaizad Ahamed
- Siti Norain Md Ali

Publications Committee
- Mohamad Ismail Khan
- Rosmah Marhamdan
- Nur Izira Mohamad Azli
- Muhammad Ikhwan Zainuddin
- Nur Qistina Abdullah

Accommodation & Transportation
- Noor Mazlan Muhd Noor
- Shahril Samsudin
- Noor Azreen Arshad
- Mohamad Abdullah Ismail
- Ir. Mohd Sainuzi Abdullah

Social Programme Committee
- Dato' Ir. Hj. Che Noor Azeman Yusoff
- Munira Akhmal Hashim
- Mohd Raduan Yaakob

Protocol Committee
- Zurifah Zuhairon
- Mohd Zail Arish Mohd Yusof
- Abdul Aziz Ahmad
- Azmi Helmi Omar
- Muhd Aidil Afif Sabudin

Speeches Committee
- Ir. Haji Mat Rosly Mat Daud
- Sarawanan A/L Elumali
- Shahanema Francis
- Siti Khadijah Othman
- Hazmatul Farha Hamzah
- Muhamad Adzmeer Azhari

YEP Group
- Ir. Hamzah Hashim
- Muhammad Taufek Ismail
- Nur Khairunnisa Bakri
- Tengku Mohd Hasrizal Tengku Hashim
- Fazleen Hanim Ahmad Kamar
- Noorel Syima Yatim Ishak

Peer review statement

All papers published in this volume of *IOP Conference Series: Materials Science and Engineering* have been peer reviewed through processes administered by the proceedings Editors. Reviews were conducted by expert referees to the professional and scientific standards expected of a proceedings journal published by IOP Publishing.

Coordination and Collaboration of Information for Pan Borneo Highway (Sarawak) via Common Data Environment (CDE)

Zohari Akob[1], Mohd Zaidee Abang Hipni[2] and Ravi Koka[3*]

[1]Kementerian Kerja Raya (KKR), Tingkat 1-14, Kompleks Kerja Raya, Jalan Sultan Salahuddin, 50580 Kuala Lumpur, Malaysia

[2]Lebuhraya Borneo Utara Sdn Bhd (LBU), Level 23, Unit #23-01, Gateway Kuching, No.9, Jalan Bukit Mata, 93100 Kuching, Malaysia

[3]Reveron Consulting Sdn Bhd, Suite 12-02, Level 12. Wisma E & C No 2, Lorong Dungun Kiri Bukit Damansara 50490, Kuala Lumpur, Malaysia

Email: ravi.koka@reveronconsulting.com

Abstract. Development of the Pan Borneo Highway in the Malaysian state of Sarawak is the single largest infrastructure project ever undertaken in Malaysia's largest state. Construction works over 786km across Sarawak are underway today in 25 sections at 11 works packages. With multiple disciplinary teams working concurrently, housed in different locations including at remote construction site offices, the efficient management of up-to-date data based on a single source of truth is important for effective decision-making. There can be no compromise in the quality of input at all stages of construction. As Project Delivery Partner, Lebuhraya Borneo Utara (LBU), leverages on technology to improve project management - from planning and design, to construction and monitoring, and preparing to operate and maintain the final asset. The task is challenging at best. To meet its needs for coordination and collaboration of various information sources, LBU uses a Common Data Environment (CDE) to enable teams to work together on the same information. Such information is monitored, tracked and validates, leading to an overall improved efficiency and productivity, and reduced risks. The CDE is central to LBU's Project Information Management System and Highway Information Modelling, key components for project delivery of the Pan Borneo Highway Sarawak within time and cost.

1. Introduction

Implementation of a Common or Connected Data Environment CDE has been recommended by many international BIM-related institutions and authorities. According to the PAS 1192-2:2013 Incorporating Corrigendum No. 1 specification for information management for the capital/delivery phase of construction projects using BIM from the BSI, the CDE is a means of providing a collaborative environment for sharing work. It can be implemented in a number of ways. BS 11000-1 and -2 [1] describes the development of various forms of collaboration within organisations and across project teams.

Implementation of CDE for infrastructure projects, especially mega infrastructure projects, has been adopted here in Malaysia. A successful example is the implementation of the MRT Line 2 project. Using a CDE, MRT Corp as project operator of MRT Line 2 (SSP Line) project, was able to implement BIM standards and strategies, including digital workflows, improve design collaboration and coordination, construction feasibility, and accuracy of as-built information for the underground work on the SSP line [2] across the Klang Valley.

More recently, a CDE implementation with a similar systems approach utilizing Bentley's ProjectWise has been adopted for development and upgrading of the Pan Borneo Highway in the state of Sarawak. Development of this large-scale infrastructure project has required a holistic and integrated system to be used by its Project Delivery Partner, Lebuhraya Borneo Utara (LBU), for all phases of construction to maintenance. Today, implementation of a CDE is used by LBU to support an Integrated Project Management System (IPMS) framework.

2. Study Area and Scope

The long-awaited Pan Borneo Highway in the Malaysian state of Sarawak on the island of Borneo will span an estimated 1,060 km. Works are on-going to develop and upgrade the current 2-lane single carriageway which connects the entire state into a 4-lane dual carriageway. The project is being undertaken by the Government of Malaysia and State Government of Sarawak through a Project Delivery Partner (PDP). For this, a special purpose vehicle private entity, Lebuhraya Borneo Utara (LBU), has been appointed as the PDP.

Costing RM16.5 billion, Pan Borneo Highway Sarawak is deemed to be the single largest infrastructure project ever awarded by the Government in Sarawak. Implementation of the project involves construction of 786km of a new 4-lane highway in 11 works packages. Works began in January 2016, with the Highway targeted for completion in 2021. Operations and maintenance of the Highway will follow upon completion of construction.

As PDP, LBU is in charge of the entire project lifecycle, from planning to design and construction, to operations and maintenance of the completed highway. As PDP to the Government, LBU is committed to deliver the project on time and within budget. An associated company will then undertake operations and maintenance of the highway.

In managing the delivery of a project of such scale, LBU anticipated several challenges in communicating, monitoring and managing project progress to ensure the project will be completed within the expected timeline and cost. Challenges include:

- Consolidating project information created at various locations by various parties throughout the project lifecycle.
- Managing project information and document flows that ensures the information can be leveraged throughout the entire project lifecycle.
- Providing visibility of project progress and insight to key stakeholders.
- Enforcing standards across the entire project operation, audit trails, and control of access and use of information.
- Progress reporting across various disciplines.

A project of such scale and magnitude requires the capabilities of a CDE to be in place to support numerous Bentley suite of products that is used throughout the project's lifecycle - from design to construction to operation. This ensures a timely and cost-effective project delivery and overall efficient management of a key Government asset.

3. Common Data Environment within LBU's Integrated Project Management System (IPMS)

3.1 Common Data Environment

A Common Data Environment (CDE) acts as a central repository where construction project information is housed. Contents of the CDE are not limited to assets created in a 'BIM environment'. The CDE includes documentation, graphical modelling and non-graphical assets [3]. By using a single source of information, collaboration between project members can be enhanced, mistakes reduced and duplication avoided [3]. A typical lifecycle of a project using a CDE is depicted in Figure 1.

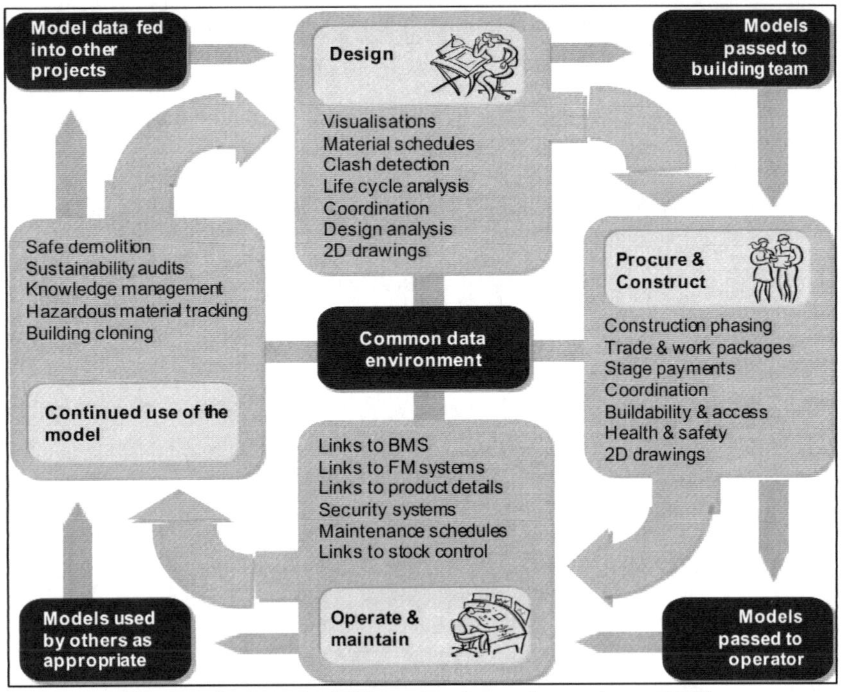

Figure 1. A typical lifecycle of a project using a CDE.

Project management of the Pan Borneo Highway Sarawak requires an increasing number of systems as the project progresses towards completion. With numerous systems, there is a need to have an end-to-end solution, such as a CDE, for project documentation and management which acts as a single platform, ready to incorporate the Highway Information Modelling (HIM) and Asset Management/Maintenance systems.

LBU decided to adopt a CDE which can be a repository of all information created. Not merely a simple document storage, its CDE is a platform which can store documents per the BS 1192 (BIM) standards, coupled with functionalities for workflows built-in for sharing and managing changes with dependency management.

A CDE has helped LBU to implement BIM, from managing the 3D design information created by consultants, to construction, through to linking this information with asset tags to develop the highway's future asset management solution. This fulfils the BIM implementation requirements of the organization. In general, a CDE enables LBU to:

- Read and consolidate information generated in different formats
- Structure and manage information created at distributed locations
- Enforce work flows and standards
- Generate audit trails and enforce controls
- Provide progress dashboards to all stakeholders
- Form an information platform as a foundation for the asset lifecycle management

3.2 LBU's Integrated Project Management System (IPMS)

LBU incorporates an Integrated Project Management System (IPMS) as a systems management roadmap which will be implemented throughout the entire project lifecycle. The IPMS is based on a technology advancement strategy and execution plan which assist LBU to monitor and provide controls for the project.

The IPMS is expected to be implemented across all phases of the Pan Borneo Highway Sarawak project. Therefore, adoption of a common systems approach is essentially important to ensure continuous effort, results and improvement.

A similar systems approach utilizing a Bentley platform has been adopted to establish a holistic and integrated system implementation under LBU for all phases.

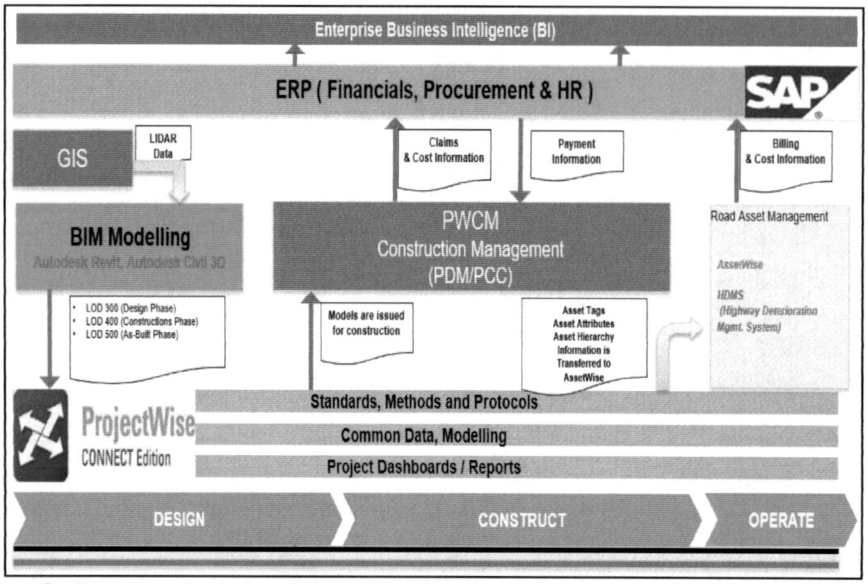

Figure 2. Conceptual solution framework diagram of an end-to-end system of the IPMS.

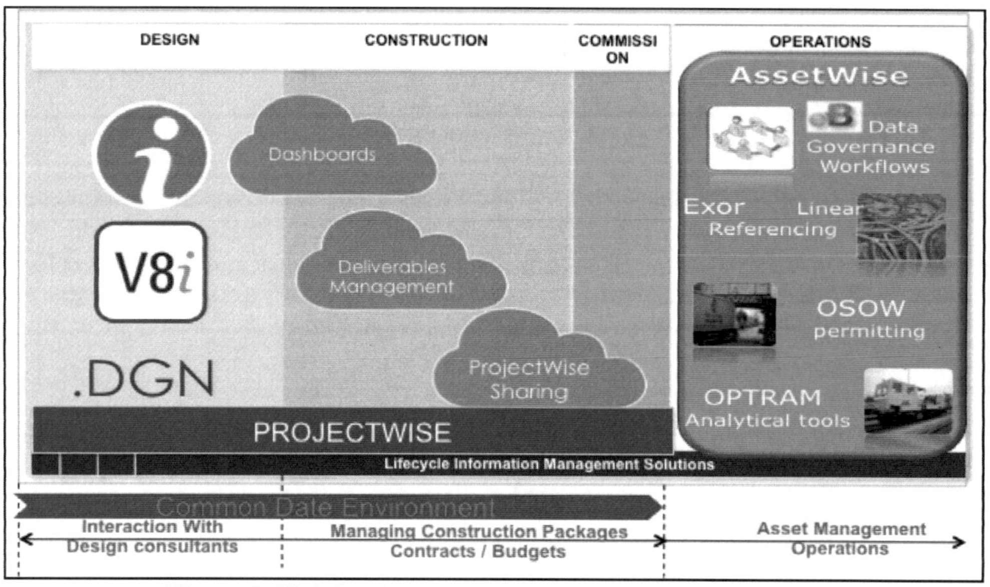

Figure 3. High-level system architecture of IPMS Phase 1A.

Based on the existing systems platform framework, the CDE platform ProjectWise supports two sub-systems , namely:

1. The Project Information Management System (PIMS); and
2. The Highway Information Modelling (HIM) system which is an integration of Building Information Modelling (BIM) and Geographical Information Systems (GIS).

The Highway Information Modelling (HIM) developed for Pan Borneo Highway Sarawak was constructed using Autodesk modelling applications, namely Autodesk Civil 3D and Autodesk Revit in the IPMS Phase 1A implementation.

The 3D models were geo-referenced using airborne LiDAR data, enabling it to have location (x and y) and height accuracy (z). Using extracted GIS features from the LiDAR data, modellers were able to have greater visibility on the existing environment surrounding the highway design. This helped the modellers conduct analysis and produce more accurate 3D models.

The HIM is useful for the construction phase. As Project Delivery Partner for the Government, LBU's ultimate goal is to evolve the models into the as-built phase of the highway where these models are expected to be used for the highway's operation and maintenance. Hence, all the models will be unified on the CDE platform for further integration with Pan Borneo Highway Sarawak's Road Asset Management System (RAMS).

Learning from implementation of Phase 1A, LBU observed that the intra-operability of the BIM modelling applications is an important element to the success of the cohesive implementation of IPMS. The models developed in individual modelling applications must be intra-operable with other modelling applications in order to create a federated modelling approach. This allows for better modelling coordination and integration with CDE and RAMS.

As a user, it is LBU's preference to use modelling applications that support a federated modelling approach for future phases of the highway project. Examples of identified solutions are Bentley OpenRoads, Bentley Open Bridge Modeller and Bentley AECOSim.

4. LBU Project Information Management System (PIMS) Implementation with Bentley's ProjectWise

LBU has identified a Project Information Management System (PIMS) designed to connect people and information across project teams no matter where they are located. The PIMS not only helps companies manage their documents, but also facilitates workflows, auditing and application integration from design to construction to asset management per BS1192 standards while improving quality, reducing rework, and helping project teams meet their deadlines.

Technically, Bentley's ProjectWise solution includes an integrated suite of desktop and server software for content management, content publishing and design review to uniquely address project needs of all connected stakeholders.

It has built-in BS 1192 folder structures, workflows for information-sharing and management, capabilities to share drawings for construction, and manage change / issue process with seamless links to source documents. With Bentley's ProjectWise, the PIMS meets the objectives in table 1.

The PIMS implementation is conducted based on industry practices of systems implementation. A waterfall methodology had been leveraged with several important milestones of the project, such as endorsement of the User Requirement Documentation by Key Users, User Acceptance Test (UAT), and others at each phase of the PIMS development to ensure the system is implemented as expected.

IOP Conf. Series: Materials Science and Engineering **512** (2019) 012001 doi:10.1088/1757-899X/512/1/012001

Table 1: PIMS implementation objectives.

IMPROVE PROJECT DELIVERY	• Improve data sharing through all project phases • Better coordination among owner – consultant - contractor teams • Produce digital plan and models through all phases
OPTIMIZE EFFICIENCY	• Reduce project delivery time and improve quality • Reduce cost while improving productivity • Accurate costing and reduced risk
CREATE BUILDING INFORMATION MODELS FOR BETTER ASSET MANEGEMENT	• Leverage on the technology platform to create & use information on the asset for better Asset Maintenance
DEVELOP LOCAL ENGINEERING INDUSTRY SKILLS	• Train local teams involved in the project about world-class design & construction processes

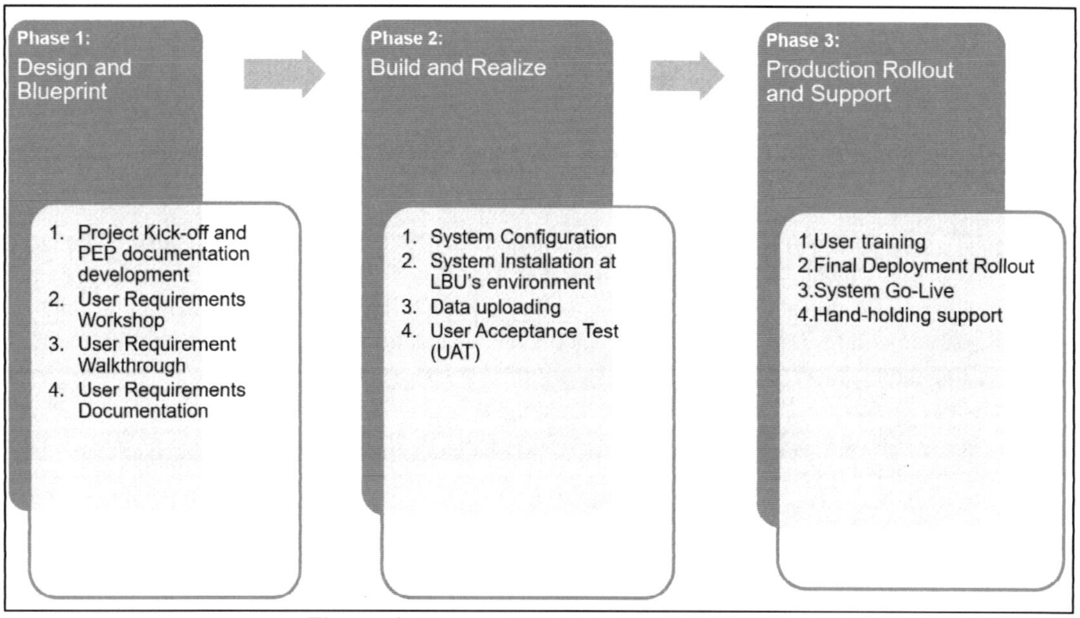

Figure 4: Phases of PIMS Implementation.

Today, a total of 74 business processes within LBU, have been incorporated into the PIMS.

These business processes involve interaction and communication between inter and intra-departments within LBU, between head office and site offices, and among the project communities which involves works package contractors (WPCs), regional engineers (REs), consultants and others. In addition, interaction and input is extended to external stakeholders, such as the project owners, i.e. relevant Government agencies.

The CDE allows project information and business processes to be incorporated into digital workflows conducted in the PIMS.

In short, a CDE, provisioned for the owners as a cloud service, now captures and intelligently shares all digital workflows, enabling participants to automate their collaboration protocols as if they were one project enterprise [4]. This capability is important for project management of the highway.

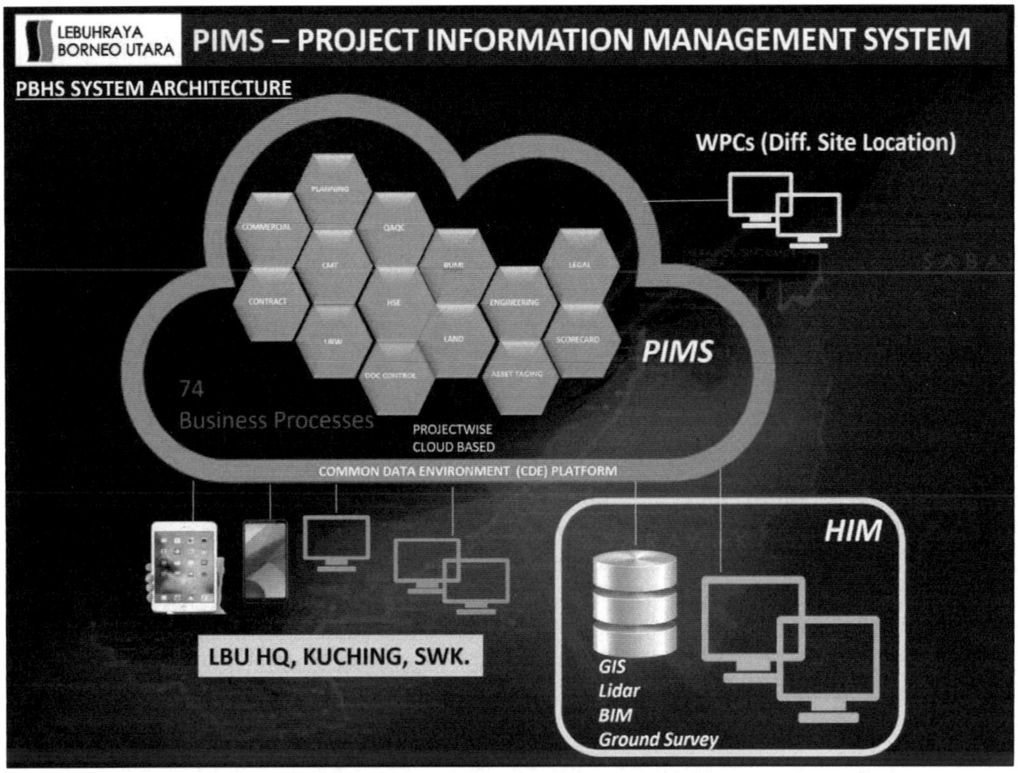

Figure 5. PIMS System Architecture.

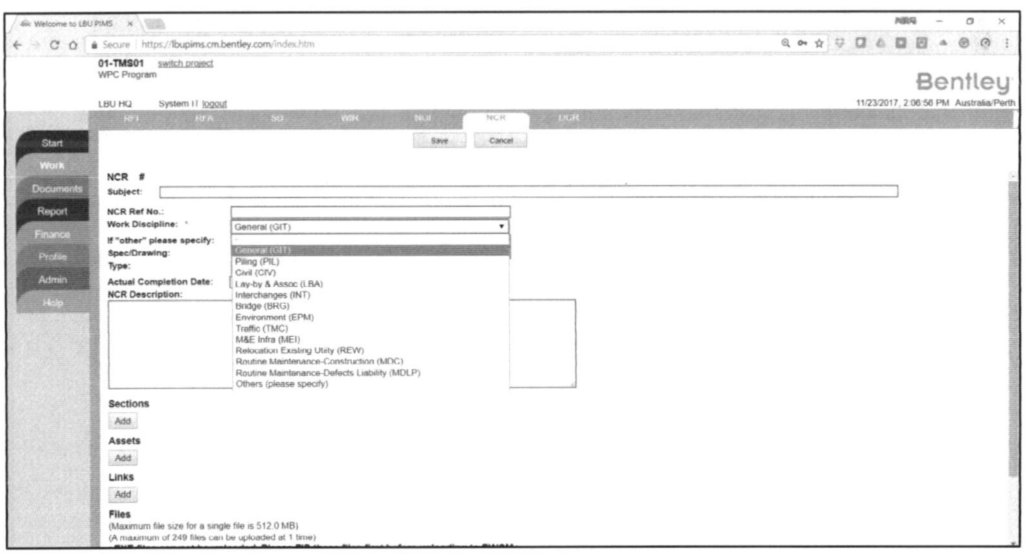

Figure 6. Online forms in the CDE.

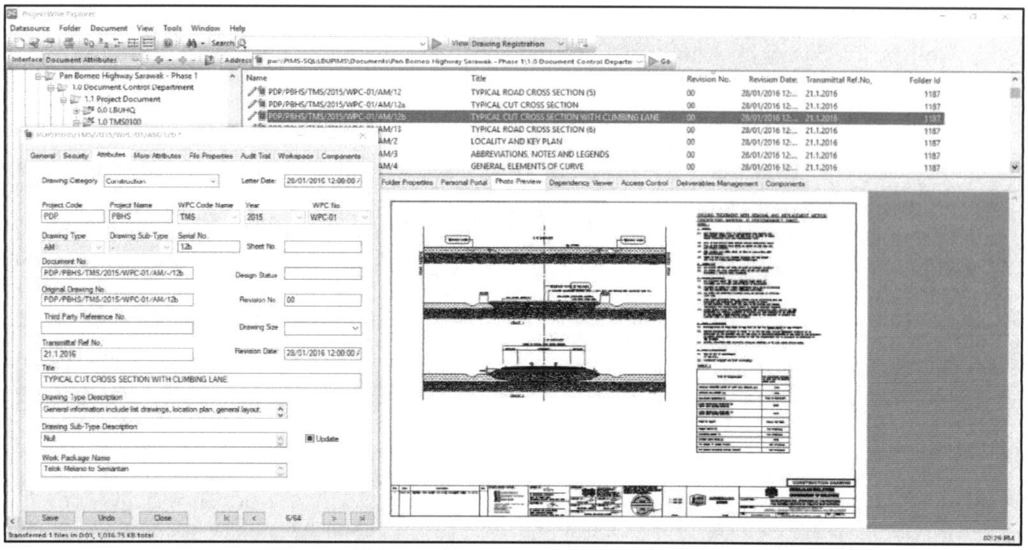

Figure 7. Drawing management in the CDE.

With project information being developed in the PIMS, it becomes a single source of information for the whole project. Information is coordinated centrally and verified before the other team members can use it within their own environment.

One of the key benefits of having centralized project information in a digital format throughout the digital workflows is the ability to extract desirable and relevant pieces of information for display. An example is summarizing project progress into project dashboards for reporting. These project dashboards provide detailed insight to the project communities to enable better monitoring and management of project progress, and ultimately better decision-making.

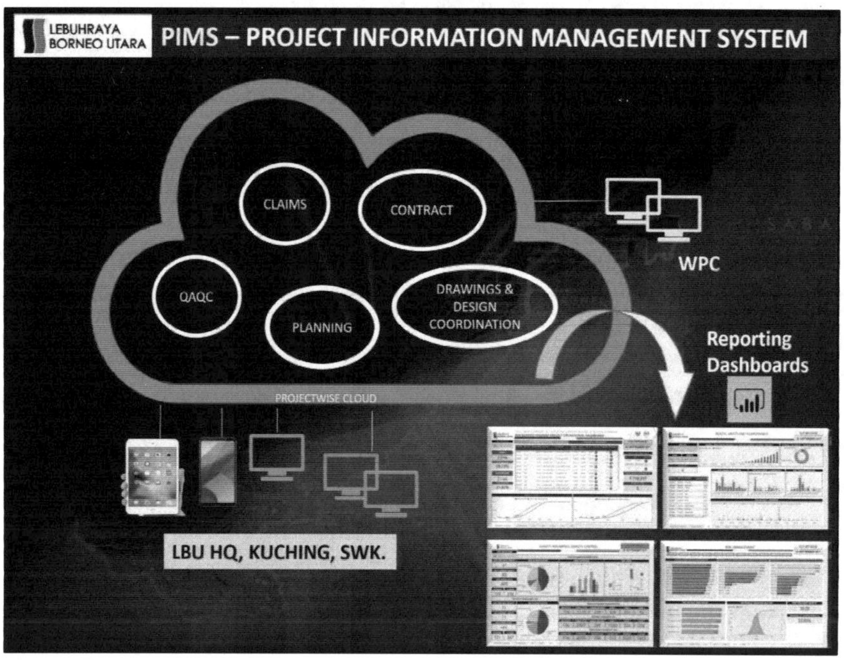

Figure 8. Illustration of project information connected to reporting dashboards.

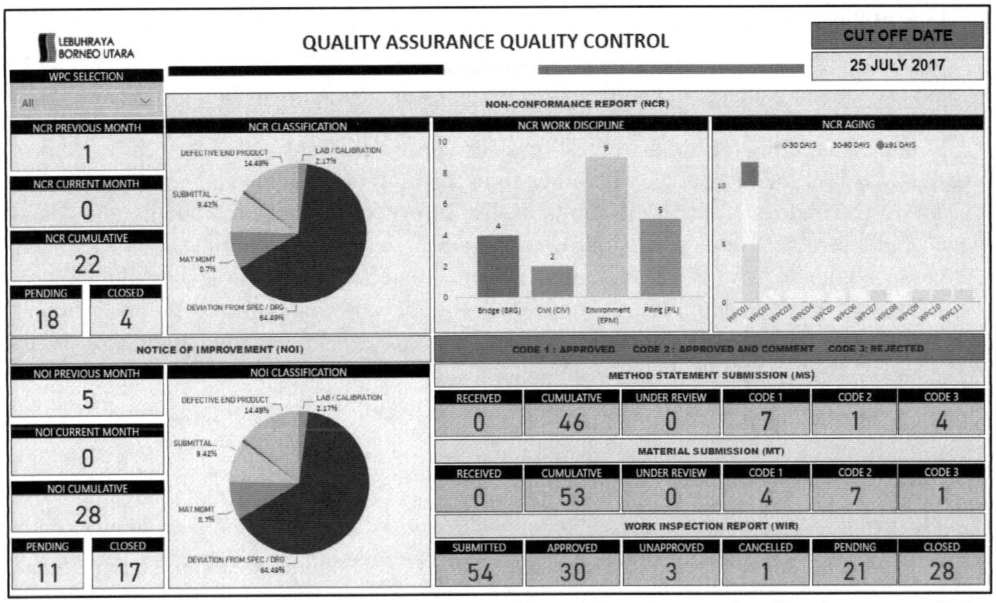

Figure 9. An example of a Project Operational Dashboard interface on Quality Assurance & Quality Control.

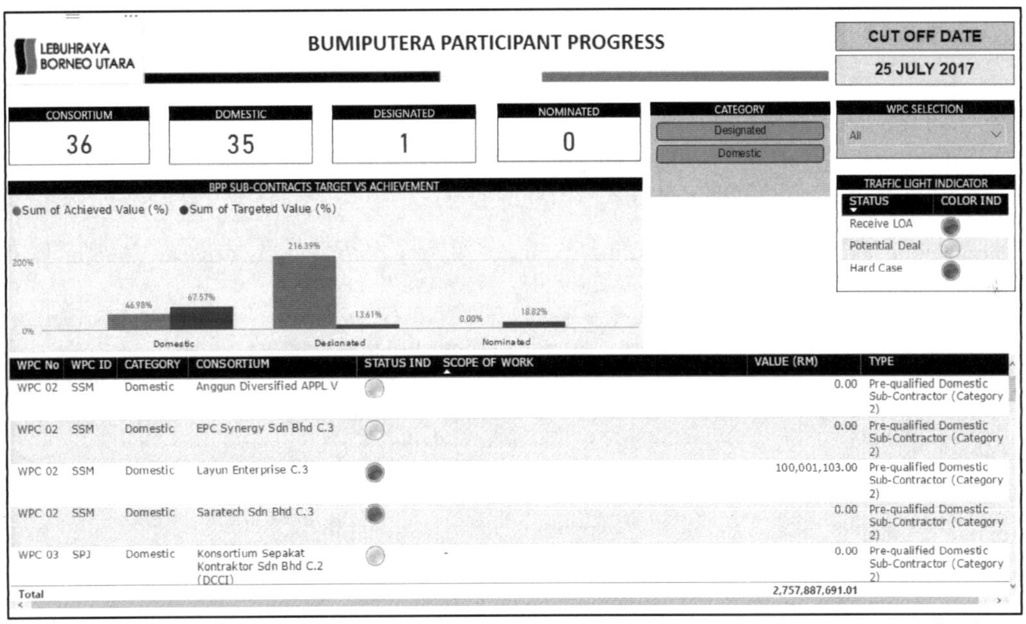

Figure 10: Example of Project Operational Dashboard Interface on Bumiputera participation.

5. Conclusion

The Common Data Environment (CDE) offers the opportunity to effectively and efficiently coordinate and collaborate various project management systems throughout the project's entire lifecycle under a single platform. This capability is important to manage, monitor, maintain and control, especially in a large infrastructure project spread out in different locations, with a myriad of processes involved from design to construction management, document management and cost control, until asset management.

In the case of the development of the Pan Borneo Highway Sarawak, CDE's key offerings include providing enterprise-level engineering content management for geographically distributed project teams, construction and contracts management, business intelligence, insight for asset-tagging and seamless integration with engineering and other desktop applications.

A CDE implementation for infrastructure projects provides measurable Returns on Investment (ROI), such as saving time in finding, validating and accessing project information. The CDE reduces project risks with a single source of truth for all project information, improves quality and consistency of all project deliverables, avoids late delivery penalties, reduces last minute rework, maximizes workforce utilization, shares work across multiple offices, and helps meet project deadlines.

In summary, a CDE is an important foundation for the coordination and collaboration of rich information and data for large scale infrastructure projects, such as the implementation of the Pan Borneo Highway Sarawak project until the operations and maintenance of the highway asset.

6. References

[1] The British Standards Institution (BSI). *Specification for information management for the capital/delivery phase of construction projects using building information modelling.* PAS 1192-2:2013 Incorporating Corrigendum No. 1, 25 (2013).

[2] Corporate Website of Bentley Systems. *The Klang Valley Mass Rapid Transit - Sungai Buloh – Serdang - Putrajaya Line.* https://www.bentley.com/en/project-profiles/2017/mrt-corporation_klan-valley-line

[3] McPartland. R 2016 *What is the Common Data Environment (CDE)?* Corporate Website of NBS Available from https://www.thenbs.com/knowledge/what-is-the-common-data-environment-cde

[4] Bentley G 2018 *Going digital to advance infrastructure delivery: The open information project.* Corporate Website of Global Infrastructure Initiative Available from https://www.globalinfrastructureinitiative.com/article/going-digital-advance-infrastructure-delivery-open-information-project

[5] Knowledge Base of CIDB Corporate Website. *BIM Guide 1 – Awareness.* Available from https://www.mybimcentre.com.my/download/bim-guide-book-1/

[6] Knowledge Base of CIDB Corporate Website. *BIM Guide 2 – Readiness.* Available from https://www.mybimcentre.com.my/download/bim-guide-book-2/

[7] Knowledge Base of CIDB Corporate Website. *BIM Guide 3 – Adoption.* Available from https://www.mybimcentre.com.my/download/bim-guide-book-3/

IOP Conf. Series: Materials Science and Engineering **512** (2019) 012002 doi:10.1088/1757-899X/512/1/012002

High performance thermoplastic and cold applied plastic road markings: how long do they last?

M H Harun[1], S Rosdi[2] and M Rosmani[1]

[1]Road Research Laboratory, Centre of Excellence for Engineering and Technology (CREaTE),Public Works Department Malaysia, Kuala Lumpur, Malaysia
[2]Engineering Department, Roadcare (M) Sdn. Bhd.,Kuala Lumpur, Malaysia

Abstract. The function of road markings is to provide guidance and warnings to motorists. These markings should be clearly visible at all times. It is empirical that the materials used should be durable especially on roads with heavy traffic to ensure that the safety and comfort of motorists are not compromised at all times, and to avoid the need to repaint regularly which is cumbersome and requires a significant amount of funds. This paper highlights some good and bad practices in laying the road markings which result in satisfactory and poor performances respectively. Field trials have been carried out using thermoplastic and cold applied plastic produced and installed by various applicators. Various factors which could possibly affect the brightness and durability of the road markings such as the quality of the inter-mix and drop-on reflective beads, different techniques in incorporating the drop-on reflective beads, variation in the laying temperature and thickness of the road markings, and the conditions of the road surface during laying, are discussed. Test results have shown that cold applied plastic provides higher retro-reflectivity than thermoplastic after four years which indicates that the former type could hold the reflective beads more firmly.

1. Introduction

Thermoplastic road markings will appear bright and glowing after installation. The ability to reflect light from vehicles' headlamps at night could help motorists to see clearly the road markings, and thus feel safe and comfortable. However, when it rains, conventional glass beads could not reflect light effectively when submerged in thin film of water on the road surface, resulting in the road markings appear dull and difficult to see. Consequently, all-weather thermoplastic (AWT) or high performance thermoplastic road markings have been introduced and included in the new specification of Public Works Department (PWD) Standard Specification for Road Works JKR/SPJ/2012 Section 6: Road Furniture, Sub-Section 6.3: Road Markings (2012) [1], hereinafter referred to as SPJ. Malaysia is the first country in Southeast Asia that uses the AWT road markings, its inception in October 2010. Befitting the tag 'all-weather', the road markings could be clearly visible in all weather conditions, especially at night and when it rains. Fish (1996) defines all-weather road markings as markings that are visible at night under dry conditions and also under rainy conditions up to 0.25 inches per hour of rainfall [2].

Figure 1. All-weather thermoplastic road markings.

However, there are dissatisfactions that the AWT road markings do not last long. This would cause discontent among the road authorities as the price of AWT road markings is three times more expensive than the ordinary thermoplastic road lines. So what could have gone wrong?

2. Road markings

AWT (or high performance thermoplastic) road markings comprise of high performance thermoplastic powder, inter-mix reflective beads and drop-on beads. In general, the performance of the AWT road markings depends on three parameters, namely materials, applicator machines and workers.

2.1.Materials

2.1.1. Material specifications. As specified in clause 6.3.2 of the SPJ, thermoplastic materials (ie. the white or yellow powder) shall comply with BS EN 1871 (2000) [3]. High performance thermoplastic powder which complies with BS EN 1871 and certified by recognised agencies should always be used; materials of inferior quality will not hold the drop-on beads firmly even though the beads are properly embedded during application. If the white thermoplastic powder change its colour to greyish or yellowish if overheated, that is a clear indication of inferior quality thermoplastic material that should be immediately rejected from sites. Substantial different in cost (~ 35%) between the superior (RM2,700/ton) and inferior (RM2,000/ton) product may encourage the contractor to use the cheaper product if enforcement is lacking. Materials without certification should not be accepted.

2.1.2. Reflective beads. The thermoplastic material should contain inter-mix reflective beads, and these beads should comply with BS EN 1424 (1998) [4]. In addition, during application on to the road surface, drop-on reflective beads are dropped (usually by gravity) shortly after the heated thermoplastic material is spread, and these beads should comply with BS EN 1423 (1998) [5]. Both BS EN 1424 and 1423 refer to glass beads, with a minimum value of refractive index (ability to refract light in water) of 1.5.

Other types of reflective material such as micro-crystalline ceramic beads* have higher refractive index of 2.4. The ability of the ceramic beads to reflect light towards the source of the light (ie. retro-reflection. In the context of road lines, it is the ability of the beads to reflect light towards vehicles' driver) is claimed to be similar to diamonds. For normal glass beads, the refractive index is relative low (usually about 1.5), but there are high performance glass beads with refractive index of 1.9. Beads of inferior quality will not be able to reflect light towards the vehicles' drivers, or reflect the light towards many other directions (diffuse reflection). During the day and when it does not rain, normal glass beads are sufficient, but at night or when it gets dark, especially when it rains, the superiority of the ceramic beads will be more pronounced. Micro-crystalline ceramic beads are claimed to be seven times brighter than glass beads.

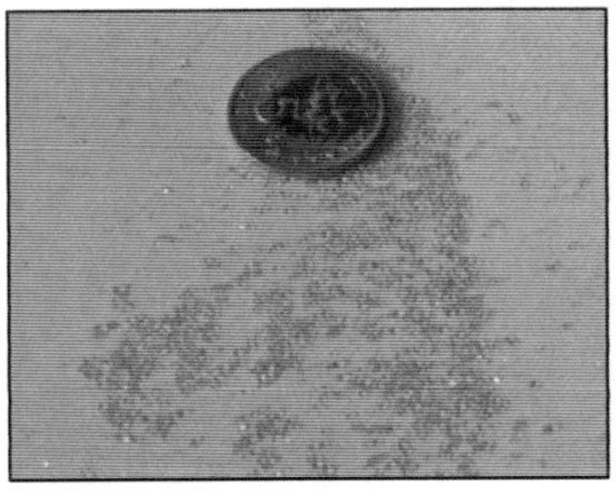

Figure 2. Glass beads.

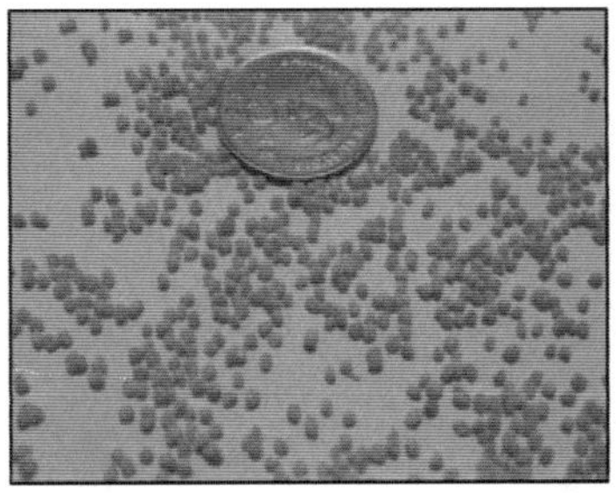

Figure 3. Micro-crystalline ceramic beads.

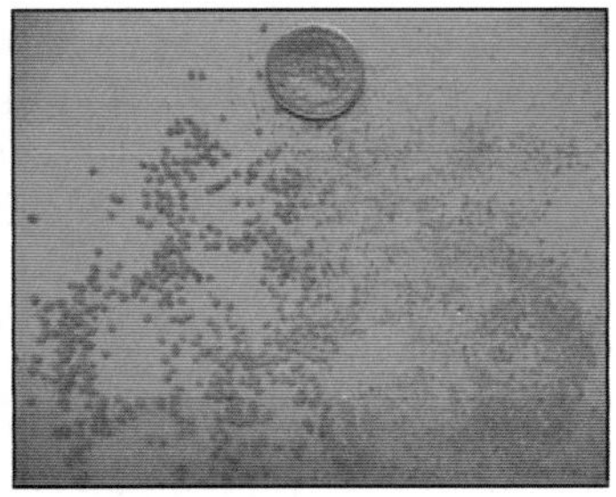

Figure 4. Comparison of size between glass.

The quantity of inter-mix beads is not specified in the SPJ. It is recommended that it should be not less than 10% by weight of the thermoplastic powder with at least 3% ceramic beads. For drop-on beads, the quantity is specified to be not less than 400 g/m^2. However, the proportion of each type of beads is not specified. It is recommended that the drop-on beads should contain at least 30% of micro-crystalline ceramic beads (refractive index > 2.4) with the remaining composition consists of 70% high performance glass beads (refractive index > 1.9).

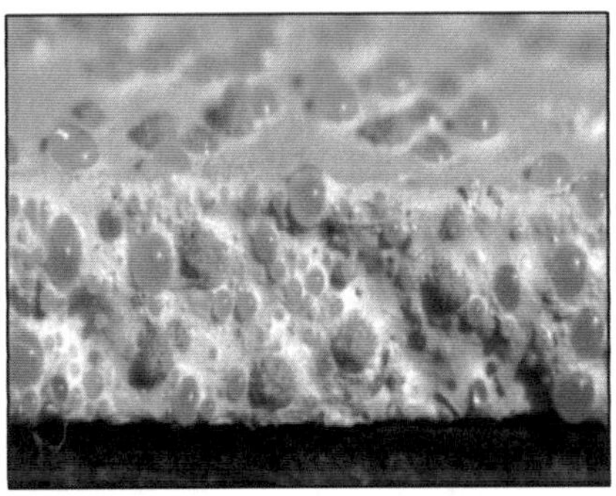

Figure 5. Inter-mix micro-crystalline ceramic beads and glass beads inside road markings.

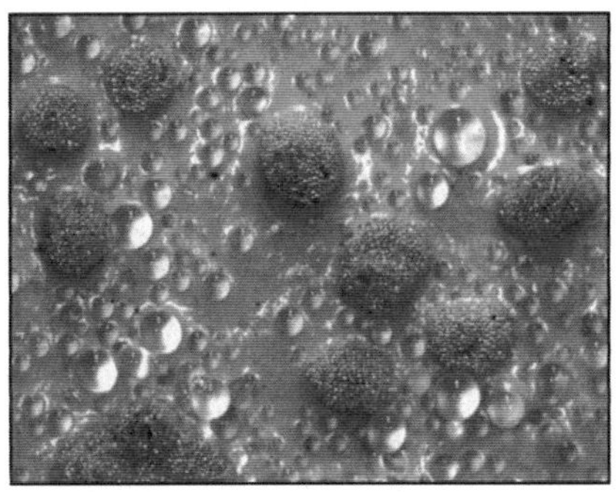

Figure. 6. Drop-on micro-crystalline ceramic beads and glass beads on road markings.

2.1.3.Performance specifications. SPJ clause 6.3.2 stipulates that the performance of the thermoplastic road line material shall comply with BS EN 1436 (2007) [6]. Among other things, the BS EN 1436 specifies; (1) Luminance factor β in dry condition not less than 0.60; (2) Luminance coefficient under diffuse illumination Qd not less than 160 mcd/m^2/lux.(Note: If the luminance factor β can be achieved, Qd does not need to be evaluated, and vice versa); (3) Coefficient of retro-reflected luminance R_L in dry condition (ASTM-E1710) [7] not less than 300 mcd/m^2/lux; (4) Coefficient of retro-reflected luminance R_L in wet condition (ASTM, E2177-11) [8]not less than 75 mcd/m^2/lux. (Note: Qd measures the brightness of the road markings in day light illumination and street light, while R_L measures retro-reflection when lighted by vehicles' headlamp.)

The SPJ specifies that the R_L shall be not less than 300 mcd/m^2/lux in dry condition, and not less than 75 mcd/m^2/lux in wet condition, within 7 days after the road markings are laid. However, after acquiring more knowledge on good and bad practices in laying road lines, the Superintendent Officer is encouraged to request the contractor to carry out the testing on the 6th or 7th day to allow any loose drop-on beads, if there is any, to detach prior to testing. Consequently, the requirement on the R_L will

IOP Conf. Series: Materials Science and Engineering **512** (2019) 012002 doi:10.1088/1757-899X/512/1/012002

be reviewed and changed accordingly. It has been suggested that the requirement for the test to be carried out 'within 7 days' be changed to 'within 10 – 20 days' while the minimum value of R_L remain unchanged.

In the opinion of the European Union Road Federation, ERF (2013) [9], a good road markings are those with performance level under dry conditions not less than 150 mcd/m^2/lux, whereas for wet conditions, the minimum performance level should be 35 mcd/m^2/lux. Apart from this, Fish (1996) [2] stated that most research has come to a conclusion that a minimum acceptable level of retro-reflectivity is around 100 to 120 mcd/m^2/lux.

Asdrubali et al (2013) observed that while the only performance parameter considered was R_L, at urban level the reflection in daylight condition Qd is equally (or more) important [10].

Figure 7. Retro-reflectometer.

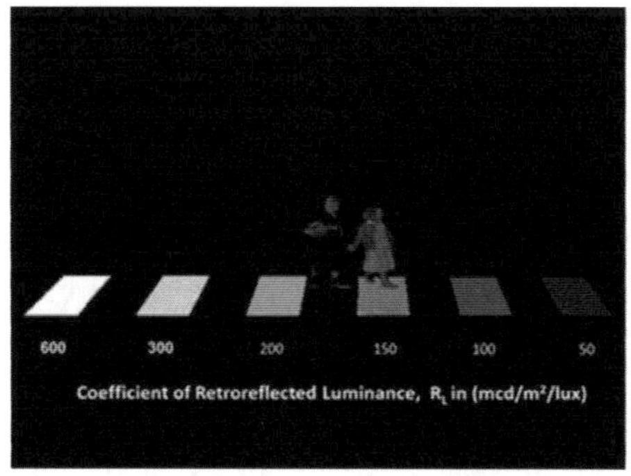

Figure 8. A range of coefficient of retro-reflected luminance.

2.2. Applicator machines

2.2.1. Drop facility for beads. If a mixture of glass beads and micro-crystalline ceramic beads is used as a drop-on reflective element, the applicator machine should be equipped with a double-drop facility with two units of dispenser. The larger ceramic beads (typically 1.1 - 1.3 mm) should be dropped first, followed by the smaller glass beads (usually 0.7 - 0.8 mm). If the smaller glass beads fall first, it would impede the larger ceramic beads from filling up the space in between the smaller beads, and this can lead to the larger beads get easily detached from the road markings by the passing traffic. Similarly, the larger ceramic beads would be loosely embedded into the road markings if both types of bead drop simultaneously from a single-drop facility with one unit of dispenser.

Figure 9. Applicator machine with double-drop facility.

Figure 10. Applicator machine with single-drop facility.

2.2.2. Height of spreader shoe. The height of the spreader shoe above the road surface should not be too high. It is recommended that the height should be 40 – 50 mm. If it is higher, there is a possibility that up to 10% of the drop-on beads will be blown away before they reach the road markings by the wind or passing traffics. It should not be too close either because the drop-on beads will not satisfactorily embed into the road markings. The target is to get the beads embedded into the road markings layer by about 50 – 60%.

A study elsewhere has shown that high retro-reflectivity could be achieved if the degree of sinking of glass beads is between 55 and 60% of their diameter. A sinking degree less than 50% weakens their resistance to tyres, while a sinking degree higher than 60% limits the retro-reflection properties [9].

Figure 11. Embedment of drop-on beads

2.2.3. Speed of application. The applicator machine should not be pushed too fast as some of the relatively light glass beads or ceramic beads will be blown away by the wind.

2.3 Workers

2.3.1. Laying thickness. The optimum thickness of the overlaid thermoplastic powder is typically 2.0 - 2.5 mm. If thinner, the drop-on beads would not stick firmly to the road markings. If thicker, most of the drop-on beads would sink into the road markings.

Figure 12. The optimum thickness is 2.0 – 2.5 mm.

IOP Conf. Series: Materials Science and Engineering **512** (2019) 012002 doi:10.1088/1757-899X/512/1/012002

2.3.2.Road surface condition. Prior to the application of the road markings, the surface of the road should be clean and dry. Surface which is dirty, sandy and wet would cause the road markings to easily peel off.
After rain, it is advisable to wait for at least 12 hours for the moisture on the road surface to dry. The formation of small holes(pinholes) on the road markings indicates that the road surface has not adequately dried when the road markings are installed. It is also recommended that road markings are installed not less than 72 hours after paving works to let the bitumen sufficiently harden.

2.3.3.Overheating. If the thermoplastic material is inadvertently overheated above the recommended mixing temperature, the viscosity during application, and thus the embedment of the drop-on beads, will be affected even though the material is allowed to cool down.

2.3.4 Mixing and laying temperatures. The mixing temperature of the thermoplastic material should be as recommended by the supplier. Usually the temperature is 200 - 220 °C. However, if the road markings are installed in the evening when the temperature is a bit cooler, or in cooler highlands, the mixing temperature should be increased up to 230 - 240 °C. Laying temperature (the temperature of the thermoplastic material as it flows out from the spreader shoe) should be similar to the mixing temperature. If applied at cooler temperatures, the drop-on beads would not stick firmly while if the temperature is higher, the drop-on beads would sink. While the R_L could be achieved when the test is carried out within seven days, the loosely attached drop-on beads would eventually detach from the road markings by passing traffics. In contrast, the R_L could not be achieved when a lot of drop-on beads sink into the road markings if the laying temperature is too hot. The applicator team should be equipped with hand-held contactless thermometers to rapidly check the temperatures.

Figure 13. The mixing temperature should be asrecommended by the material's supplier.

Cold applied plastic road markings can be laid at ambient temperatures. The material is in liquid state at relatively low tempature and consists of a variety of additives. It is fast-curing and thus can be laid more speedily - will harden and can be opened to traffic after 20 minutes. The cold applied plastic road marking is said to be extremely durable - it lasts longer than the thermoplastic.

3. Degradation of road markings
The rate of degradation of the road markings varies with traffic volume. However, how fast the retro-reflectivity decreases over a time period is still unknown. Consequently, the SPJ presently only specify the initial minimum coefficient of retro-reflected luminance R_L of 300 and 75 mcd/m^2/lux, in dry and wet conditions respectively, which are required to be achieved within seven days after laying the road markings.

3.1.Test sites

Subsequently, an actual field installation has been carried out to study the rate of degradation of the road markings with the AADT (Annual Average Daily Traffic) as the main parameter. Two states have been selected for this study namely Selangor and Melaka.

3.1.1.Selangor. Four sites have been selected with different composition of AADT with a general assumption of the classification as low, medium and high AADT. The classification considers the low, medium and high AADT volume road to experience about 1000, 4000 and 8000 AADT per lane respectively. The Level of Service (LOS), which refers to the extend of traffic congestion, is considered for the next governing parameter. The selected roads experience LOS A (no congestion – free flow with low traffic volumes and high speeds) for two sites, LOS B (no congestion – reasonably free flow, but speed depends on traffic flow) and LOS E (severe congestion – very unstable flow, possible short stoppages) for one site each.

3.1.2.Melaka. Three sites have been selected based on the composition of traffic ie. high, medium and low AADT. High AADT location is on Site1 (LOS F) Route FT 5 Masjid Tanah to Sg. Udang, medium AADT location on Site 2 (LOS E) Route FT 5Ramuan China Besar- Masjid Tanah and low AADT location on Site 3 (LOS C) Jalan Lubok Cina-Masjid Tanah. At each site, there are seven types of road marking test bars that have been installed which comprise of four different types of high performance thermoplastic and three varieties of cold applied plastic.

At each of the selected sites, the road marking test bars have been installed transversely on the pavement surface comprising of three bars of size 300 mm across a single lane width to accelerate the degradation, and one bar installed longitudinally on the edge line as the control bar. Timm and Priest (2005) [11] has shown that the wheel hit count on the wheel track area is on an average of seven times more than the wheel hit count on the edge line. Based on this, it was assumed that R_L on the wheel paths degrades seven times faster than the edge line at the trial sites. For example, R_L on the wheel paths obtained on Day 30 is equivalent to R_L obtained on Day 210 on the edge line. Nevertheless, the degradation of R_L by accelerated months is presumptuous.

Figure 14. Site 1, Route FT55, Ampang Pechah. Traffic volume 1,514 (757/lane), LOS A (Selangor).

Figure15. Site 2, Route B54, Kuala Kubu Bharu.Traffic volume 9,415 (4,708/lane), LOS A (Selangor).

Figure 16. Site 3, Route FT54, Assam Jawa.Traffic volume 21,278 (7,093/lane), LOS B (Selangor).

Figure 17. Site 4, Route FT5, Kuala Selangor.Traffic volume 35,845 (8,871/lane), LOS E (Selangor).

Figure 18. Site 1, Route FT5, Masjid Tanah. Trafficvolume 21,167 (10,584/lane), LOS F (Melaka).

Figure 19. Site 2, Route FT5, Ramuan China Besar Traffic volume 16,916 (8,458/lane), LOS E (Melaka).

Figure 20. Site 3, Route FT5, Jalan Lubok Cina. Traffic volume 11,202 (5,601/lane), LOS C (Melaka).

IOP Conf. Series: Materials Science and Engineering **512** (2019) 012002 doi:10.1088/1757-899X/512/1/012002

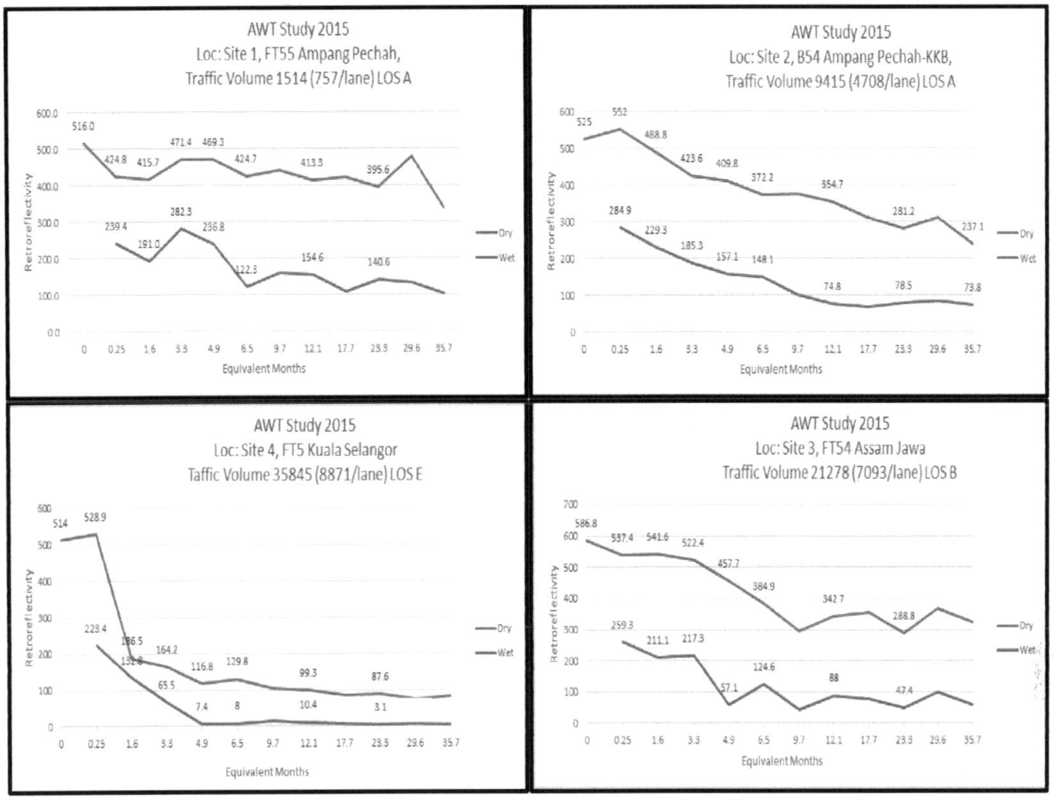

Figure 21. Three-year accelerated test results obtained from wheel paths (trial sites in Selangor).

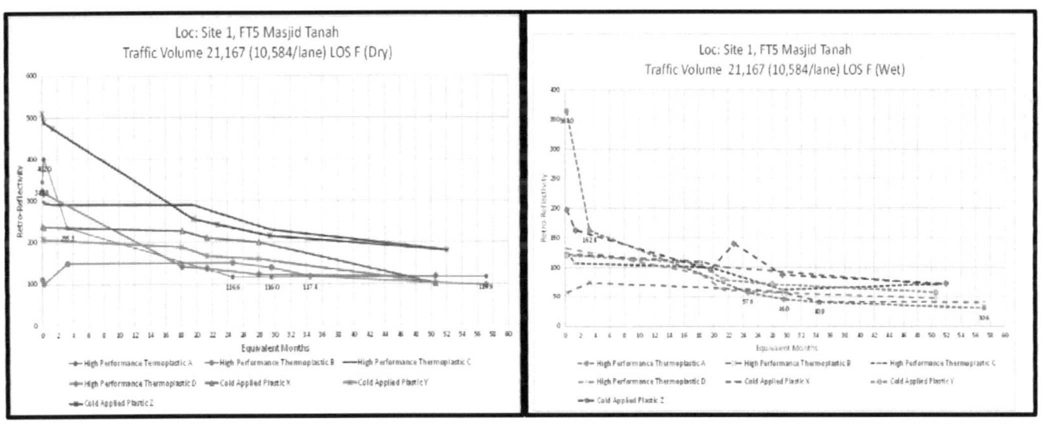

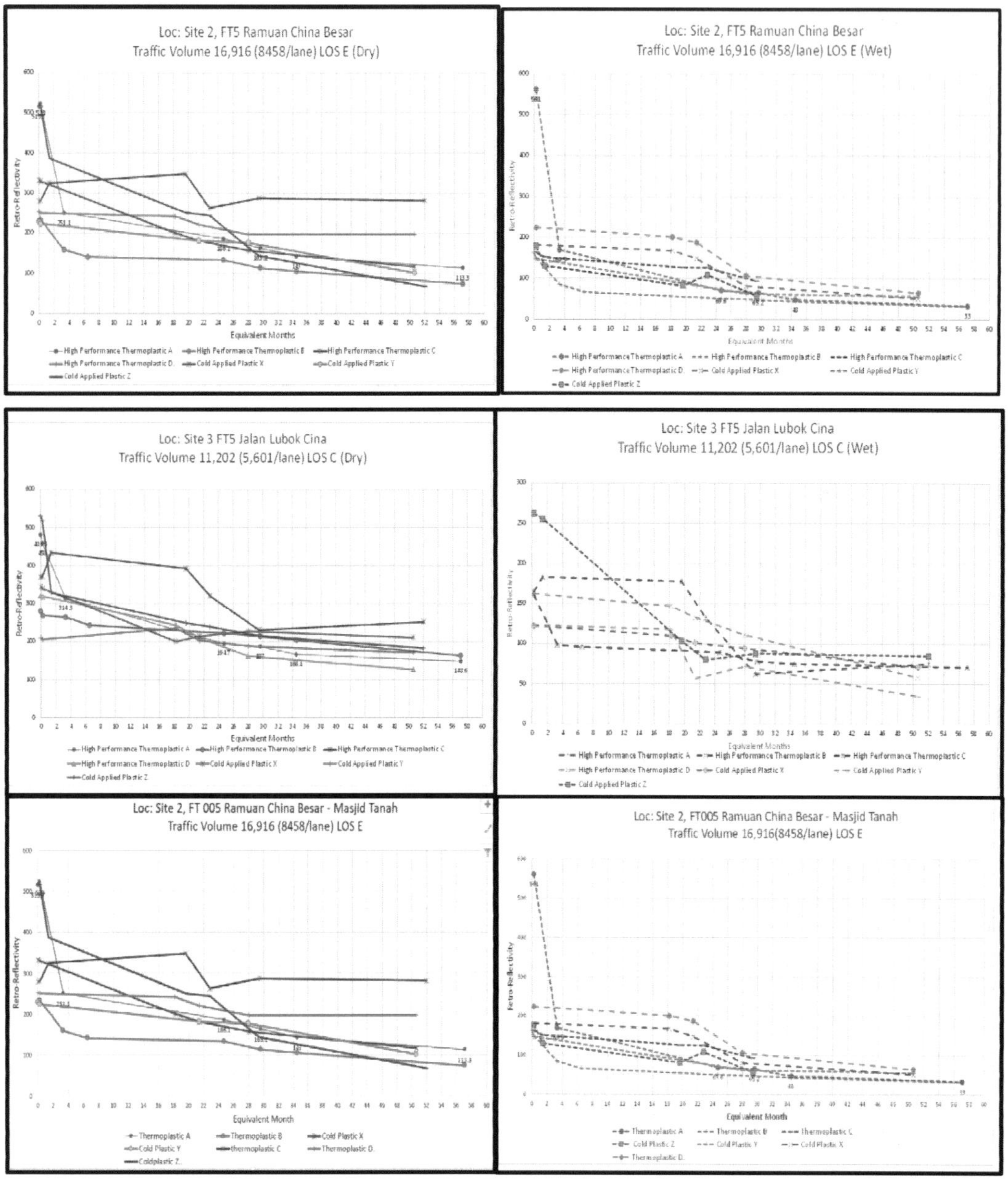

Figure 22. Four-year accelerated test results obtained from wheel paths (trial sites in Melaka).

3.3.Discussion

In general, it was observed that the transverse bars degrade drastically from Day 1 to Day 56 (equivalent month is 0.25 to 12.1). The degradation tends to stabilize after that with about 10-15% percent variation, and diminishing consistently up to Day 125 (equivalent month is 29.6). Although R_L dry dwindle, it still exceeds 150 mcd/m^2/lux after equivalent three years at trial sites with LOS A, meeting the minimum acceptable R_L as suggested by European Union Road Federation (ERF, 2013). The R_L dry and wet values show a consistent correlation, with steep degradation during early life and stabilize at later stage. The reduction of R_L is significantly affected by LOS. The trial site with LOS E shows the most drastic degradation during the first month. This could be possibly due to traffic behaviour with extreme change of vehicle speed over a short length. However, the coefficient of R_L

exceeds 100 mcd/m^2/lux after equivalent three years, meeting the minimum acceptable R_L [2]. It is worth to note that the fluctuation of R_L observed throughout the monitoring period could be attributed to dirt and debris.

Figure 21 shows that high performance reflective beads maintain a consistent trend of deterioration over time in dry and wet conditions, without the wet performance being compromised. LOS (Level of Service) highly affects the rate of deterioration and retained R_L values (referring to the high-AADT Sites 3 and 4). The high-AADT Site 4 with LOS E records passing values of R_L dry 142 mcd/m^2/lux and$R_L$wet 35 mcd/m^2/lux after 30 months. Other trial sites with lower AADT and LOS record retro-reflectivity exceeding the recommended minimum retained $R_L$values in dry and wet conditions after 36 months.

Figure 22shows that R_L dry and wet at Site 3 (low AADTLOS C) are 127.0mcd/m^2/lux and 54.3 mcd/m^2/lux respectively. These values were achieved after a period of four years. While for Site 1 with the highest AADT (LOS F), R_L dry and wet are 101.0 mcd/m^2/lux and 30.6 mcd/m^2/lux respectively after four years.

The best performance of cold applied plastic achieved R_L dry and wet of180.9 mcd/m^2/lux and 84.8 mcd/m^2/lux respectively after four years. The best performance of high performance thermoplastic (at trial sites in Melaka) achieved R_L dry and wet of 281.8 mcd/m^2/lux and 64.2 mcd/m^2/lux respectively after four years. This give an indication that the high performance thermoplastic is slightly better than the cold applied plastic. Nonetheless, it is worth to note that both types of road marking were able to last at least for four years as both R_L dry and wet still exceeded 100 mcd/m^2/lux and 35 mcd/m^2/lux respectively after that period.

However, comparison of road marking having reflective beads with similar characteristics such as thermoplastic A and cold applied plastic Z indicates that the cold applied plastic records higher R_L after four years. This shows that the cold applied plastic has a stronger bonding with the beads than the thermoplastic.

4.Conclusions

For countries like Malaysia, being located in the region with tropical climate and experiencing frequent heavy rainfall, it is imperative to have all-weather thermoplastic road markings for the safety and comfort of the road users. Some good practices which are based on past experience have been highlighted and these are recommended to be adhered to in order to consistently produce high quality road markings.

Based on trial sites in Selangor, the high performance thermoplastic records passing value of R_L dry and wet after 30 months at the high-AADT site. Based on more recent trial sites in Melaka, both the high performance thermoplastic and cold applied plastic record passing values R_L dry and wet after four years. However, the cold applied plastic provides higher R_L which indicate that it could hold the reflective beads more firmly than the thermoplastic.

References

[1] Public Works Department 2012 *Standard specification for road works JKR/SPJ/2012* (Kuala Lumpur: PWD)
[2] Fish J K 1996 Evaluation of all-weather pavement markings: Report on two years of progress Semisequicentennial? *Transportation Conference Proceedings*
[3] BS 2000 *Road marking materials – Physical properties* (British Standards Institution)
[4] BS 1998 *Road marking materials – Premix glass beads* (British Standards Institution)
[5] BS 1998 *Road marking materials – Drop on materials - glass beads, antiskid aggregates and mixtures of the two* (British Standards Institution)
[6] BS 2007 *Road marking materials – Road marking performance for road users* (British Standards Institution)

[7] ASTM International. *Standard test method for measurement of retro-reflective pavement marking materials with CEN-prescribed geometry using a portable retro-reflectometer* E1710 – 11

[8] ASTM International *Standard test method for measuring the coefficient of retro-reflected luminance of pavement markings in a standard condition of wetness.* E2177 – 11

[9] European Union Road Federation *Marking the way towards a safer future. An ERF position paper on how road markings can make our road safer*

[10] Asdrubali F, Cinzia B, Elisa M, Francesco D F and Samuele S 2013 Assessment of the performance of road markings in urban areas: The outcomes of the CIVITAS RENAISSANCE Project *The Open Transportation Journal*

[11] Timm, D H and Priest A L 2005 *Wheel wander at NCAT test track* (Alabama: Auburn University)

IOP Conf. Series: Materials Science and Engineering **512** (2019) 012003 doi:10.1088/1757-899X/512/1/012003

The effect of aggregate and compaction method on the physical properties of hot mix asphalt

Mutahar Abdul Salam Al-ammari, Fauzan Mohd Jakarni, Ratnasamy Muniandy and Salihudin Hassim

Department of Civil Engineering, Faculty of Engineering, Universiti Putra Malaysia, 43400 UPM Serdang, Selangor, Malaysia

E-mail: motaher.alamari@gmail.com

Abstract. Developing countries such as Malaysia have increased the traffic density in commercial vehicles and overloading trucks which cause tire pressure on the roadway and result into rutting and fatigue cracking in roadways. Aggregate plays an important role in influencing the performance of the flexible pavement. This study aimed to determine the effect of aggregates in terms of types and gradation together in compaction of methods on the resulting prop of laboratory performance to compare the various in term of aggregate types, gradations and compaction methods based on the physical properties and resilient modulus. As a comparative study, the study attempted to compare two types of aggregate (granite and limestone) with the two gradations (upper and lower) for each kind of the mentioned aggregates using Marshall and Gyratory Compaction Methods for asphalt mixtures. Two phases were carried out in this research: determining the optimum asphalt content and comparing the mixtures in different type of aggregate, gradation and compaction. The results of the study on the performance of the mixtures indicated that the granite upper mixtures achieved the best results of Marshall stability, fatigue cracking resist and rutting resist for compacted samples by gyratory compactor.

1. Introduction

Hot mix asphalt (HMA) basically consists of an asphalt cement binder and mineral aggregates. The binder works as a cohesive agent that binds the aggregate particles into the adherent mass. The mineral aggregates represent a stone structure that provides strength and hardness to the hot mix asphalt. The behavior of hot mix asphalt depends on the properties of the individual components and how they react to each other in the system [1]. The temperature mostly plays a significant role in the hot mix asphalt, particularly in binder asphalt and aggregate. The asphalt binder and aggregate are heated together to obtain a fluidity of asphalt binder to cover the aggregate and to dry the aggregate, respectively. A construction project has a type of mixture based on the site conditions. There are many methods used in designing hot mix asphalt such as Marshall, Hveem, and Superpave method [2].

Aggregate type and gradation play an important role in reducing intergranular void spaces. Such reduction reflects the amount of bituminous cement used as a lubricant for mineral aggregates. Moreover, the increase of the voids in the mineral aggregates, and the air voids in the mixture make it porous and water permeability [3].

Gradation is probably the most significant property of aggregates. It affects almost all the important characteristics of an HMA, including strength, stability, durability, permeability, operability, fatigue resistance, friction resistance, moisture resistance and damage [4]. It is considered a key factor in the resistance of mixture to permanent deformation. The most important concept is that gradation provides the greatest structural strength (resistance to rutting) for any given type and quality of aggregates [5]. The way aggregates are assessed influences all stability characteristics of mixtures. In addition, it was pointed that aggregate grading controls the void structure, and therefore, aggregate properties in HMA pavements oblige such aggregate mixture to drop within a specified range of gradation values. The result of a previous related study indicated that the value of maximum stability was obtained with the dense mixture was higher than the open-grain mixture [6]. The study aimed to evaluate two types of aggregates such as limestone and basalt on three varieties of aggregate gradations in hot mix asphalt (HMA). The results revealed that the mixture that was exposed to long periods of wetting and /or freezing/thawing cycles in field preferred to use basalt in preparing HMA, especially for roads, it was expected to be exposed to heavy axle loads in arid regions [7]. The purpose of this latter study was to compare the aggregates such as limestone, sandstone and granite in hot mix asphalt. The results have shown that the reduction of VMA in dens hot mix asphalt leads to increase both longitudinal and alligator cracking which influence the performance of the mixtures. Moreover, the lower VMA led to a decrease in the effective asphalt content which in turns reduced the binder film thickness. Asphalt mixtures of relatively low film thickness are mostly considered to have a lower resistance to cracking. Thereby, low VMA in asphalt mixtures is used with lower traffic volumes [8].

The compacting characteristic of hot mix asphalt (HMA) blends with different nominal maximum aggregate size (NMAS) and different gradation at different compaction temperatures. The results showed that the stability and volumetric properties of Marshall were significantly influenced by the compaction temperature, which increased the stability of Marshall presented due to the increase in the compaction temperature. In addition, mixtures with different types of gradations but with the same NMAS require different compaction energy to reach the predicted density [9]. The influence of compaction procedures using Superpave gyratory compactor (SGC) and Marshall compactor on volumetric and mechanical properties of asphalt mixture indicated that it is wrong to set up a number of gyrations with SGC to compact the specimens with a density that is similar to the obtained one through Marshall compactor because the number of cycles varies depending on the mixture type. In addition, it is not certain to use the air- void content to compact specimens with similar characteristics in-place mixtures. As observed, the mechanical properties are affected by the temperature more than the compact ability with both procedures. On the other hand, the different SGC in density of specimens between 80 C° and 160 C° was about 1-2 % while modulus reduction was between 25 % - 30 %. A higher variation in density reduction was appreciated by the Marshall compactor, which is estimated between 3% - 5 % when passing from 80 C° – 160 C° and a modulus loss 30% - 35% [10]. The air void content and the mechanical properties of compacted mixtures were evaluated using both Marshall and Superpave mixture design methods. Furthermore, the Superpave gyratory compactor observed better reproduction of compaction that occurs in the field than the Marshall compactor. However, the specimens that compacted using the Marshall compactor and Superpave compactor resulted into similar values of resilient modulus. Moreover, the optimum asphalt contents in Superpave and Marshall proved to be more technically reliable with better fatigue behavior, without compromising permanent deformation compaction [11]. Previous research also compared between traditional Marshall Design method and Superpave system design methods in the wearing course mixes in flexible pavements by evaluating the volumetric, mechanical properties and moisture susceptibility. The outcomes pointed out the estimated asphalt content for the Superpave mix design was lower than that of the Marshall Mix Design method. Therefore, the Superpave mix design was described to be more economical. Furthermore, modified asphalt played a role in improving the asphalt mixture against rutting by using a trial field section [12].

HMA mixture was also investigated with the fine and filler particles sizes of limestone aggregates with granite in coarse sizes through laboratory tests. The HMA was submitted by Marshall and Superpave methods. The results presented better mechanical behavior and gain of mechanical resistance mainly to permanent deformation [13]. Aggregate properties and filler play a significant role in asphalt mixtures behavior. In this research, three types of aggregate (limestone, gravel and basalt) with three types of mineral filler (limestone powder, cement and hydrate limestone powder) were examined. The results illustrated that the use of basalt provided the highest values of Marshall stability, stiffness modulus and indirect tensile strength of the mixture. Yet, for each flow, density and air voids were the lowest values. Likewise, the use of 4% or less of cement content as a mineral filler showed better results than those of other types of mineral fillers [14].

This paper reports a study into comparing mixtures in various type of aggregate, gradation and compaction in term of physical properties and mechanical properties for evaluating the performance of asphalt mixtures capability on resistance to permanent deformation and fatigue cracking.

2. Experimental

The main objective of this research is to compare the granite mixtures with limestone mixtures in their own gradations (upper, and lower) by two laboratory compaction techniques. It also attempts to identify the condition in types of mixture, gradations, and laboratory compaction methods that achieves the best resistance to rutting and fatigue cracking. The laboratory work was divided into two phases. In the first phase, the optimum asphalt content for each mixture was determined (Figure 1). In the second phase, the various compaction methods, different types of aggregate mixtures and gradations were compared in term of the physical properties and mechanical properties (Figure 2).

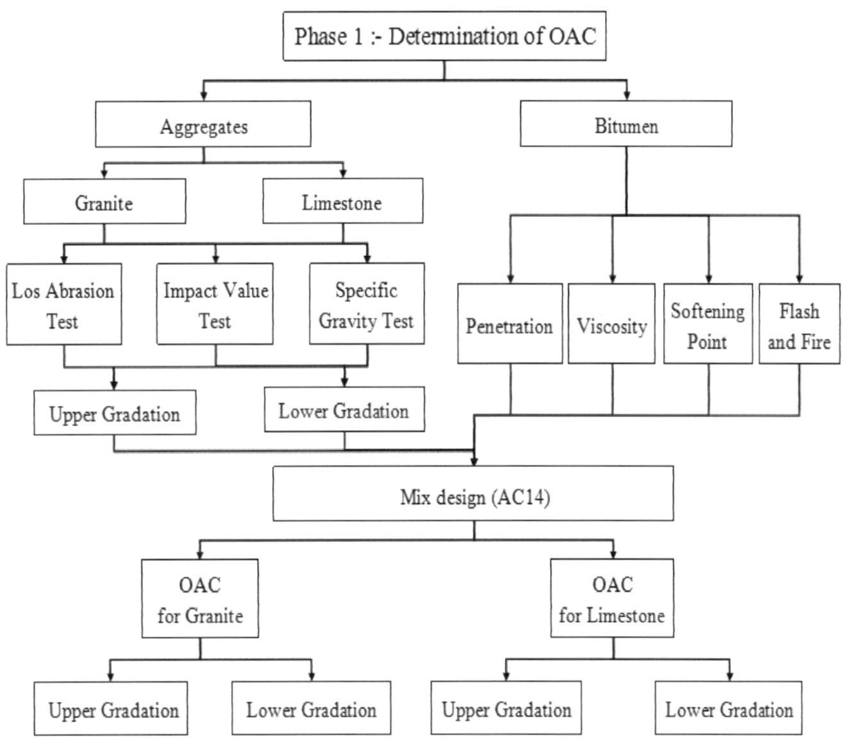

Fig. 1: The flow chart for first phase

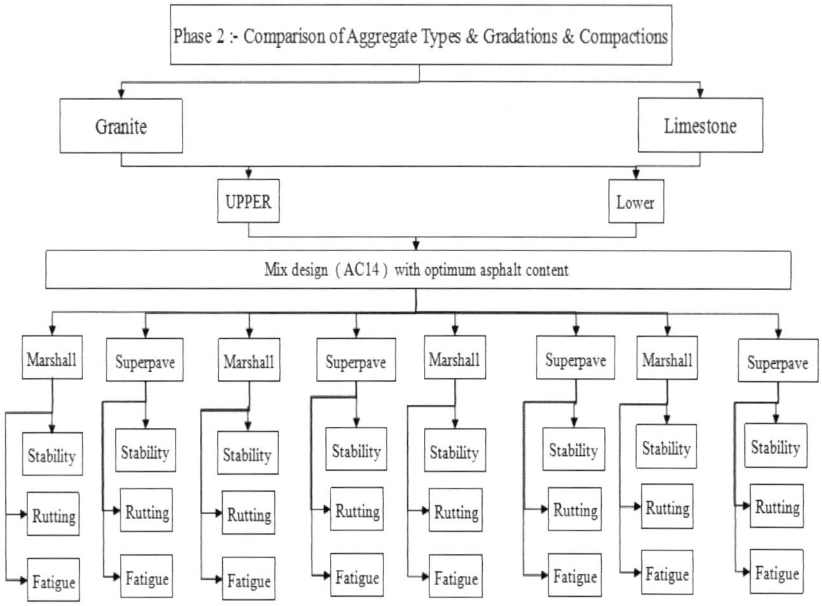

Fig. 2: The flow chart for the second phase

2.1 Materials

The materials used in this study were aggregates, bituminous binders, and fillers. All materials were prepared in accordance with the Standard Specification for Roadworks (JKR, 2008).

2.1.1. Aggregates. The aggregates used in this study are granite and limestone. Some tests such as Los Angeles Abrasion, aggregate impact value and specific gravity were also performed in the study. The aggregates were separated by a sieve analysis into different sizes to obtain the required proportion of the aggregates in each sieve accordance to ASTM C136. Moreover, blending of the aggregates was conducted to achieve the desired gradation which is within the range of ACW14 in accordance to specifications (JKR, 2008).

2.1.2. Bituminous binder. The bituminous binder for asphalt mixtures was 60-70 according to the expected climatic conditions in Malaysia. Some tests were prepared for bituminous binder, including penetration test, viscosity test, softening point test and flash and fire test. The procedures for these tests were carried out according to ASTM specifications.

2.2 Mix design

In this study, the Marshall and Superpave mixed designs were utilized for hot mix asphalt properties.

2.2.1. Marshall mix design. The mixtures in the first phase were accomplished using the Marshall mix design for determination of the optimum asphalt content. In this regard, 60 samples prepared were subjected to 75 blows that are considered for the mix design for heavy traffic.

2.2.2. Suprpave mix design. The Superpave mix design was used in this phase. A total of specimens which were prepared in this procedure were 84 samples to compare between the Marshall compactor and the Gyratory compactor using different types of aggregate in mixtures such as granite and limestone in the two gradations (upper and lower). The parameters which were considered to control the compaction effort of the Superpave machine were 600 KPa of vertical pressure and the angle of the gyration was set up to 1.25. In addition, the gyrations were applied for a rate of 30 revolutions per minute. Thereby, the number of gyrations was simulated for heavy traffic and the settings of the gyratory compactor were adjusted to stop when achieving 4 % air voids.

3. Results and discussion

The results of the laboratory tests were conducted in order to compare the physical properties of mixtures such as bulk density and air voids as well as the mechanical properties of mixtures such as resilient modulus, stability and flow, permanent deformation and fatigue cracking.

3.1. Aggregates tests

The aggregates properties play an important role in affecting the performance of hot mix asphalt (HMA) such as fatigue cracking and rutting which are caused by the poor selection of misuse of aggregates. In this research, some aggregate tests such as Los Abrasion Test, Aggregate Impact Value, Absorption and Specific Gravity conducted are reported in Figure 3.

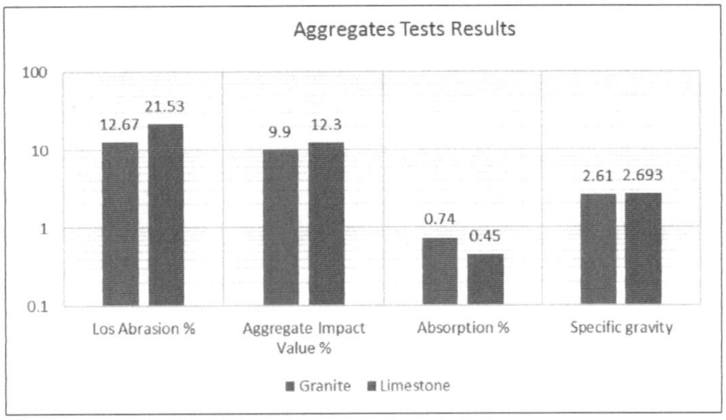

Fig. 3: Aggregates test results for granite and limestone

3.2. Bitumen tests

There are a number of tests for evaluating the properties of bituminous materials. These tests presented in Table 1 were generally conducted in order to estimate the properties of bituminous materials.

Table 1: Bitumen 60/70 Test Results

Penetration (mm)	Softening point (C°)	Flash Point (C°)	Fire Point (C°)	Viscosity (cP) at 135 C°	Viscosity (cP) at 165 C°
64.83	48.25	230	260	300.8	1531.82

3.3. The first phase

This approach was used to find out the optimum asphalt content (OAC) of each type of mixture as shown in Figure 1. The prepared specimens were analyzed to determine OAC based on the density, 4% air voids, stability and Resilient Modulus [15]. The results of the optimum asphalt content were achieved for the procedure as shown in Table 2 according to the UPM method.

Table 2: The Optimum Asphalt Content (OAC)

Type of aggregate	OAC (%) for Density	OAC (%) for Stability	OAC (%) for VTM	OAC (%) for Resilient Modulus	Average Optimum Asphalt Content
Granite upper gradation	5.3	4.7	4.5	4.5	4.8
Granite lower gradation	6	5	6.8*	4.5	5.2
Limestone upper gradation	5.1	4.3	NA	4.3	4.6
Limestone lower gradation	4.8	4.3	5.5	4	4.7

* Value not considered in calculation. NA: Not applicable

Based on the optimum asphalt content, it was observed that the granite mixtures in upper and lower gradations scored taken a higher percentage of asphalt content comparing with limestone mixtures in upper and lower gradations. In addition, the percent asphalt content of the lower gradation mixtures was higher than that of the upper gradation mixtures.

3.3.1. Marshall tests analysis for OAC. The Marshall properties were examined based on the optimum asphalt content and Marshall test analysis that were obtained in first phase using the Marshall procedure. The results are presented in Table 3.

Table 3: The Marshall Test Result

Type of Mixture	OAC (%)	Density (Kg/cm^3)	Stability (KN)	Flow (mm)	VTM (%)	VFA (%)	Resilient Modulus (MPa)
JKR/SPJ/2008 Specification	4 - 6	-	> 8 KN	2 - 4	3 - 5	70 - 80	> 2000
Granite upper gradation	4.8	2.401	24.23	2	3.05	75.5	6200
Granite lower gradation	5.2	2.298	12.49	6.9	5.92	64.1	3100
Limestone upper gradation	4.6	2.469	16.1	3.5	NA	98.6	3800
Limestone lower gradation	4.7	2.442	12.53	3.8	5.9	56.7	3845

NA: Not applicable

The above results of stability and stiffness showed that the values are in accordance with the range stated in the JKR specifications. As for the flow, the obtained results showed that all values are in the range except for granite lower gradation mixture. In term of VTM and VFA, the granite upper gradation mixture values were also in the range.

3.4. Second phase

The aim of this approach was to compare the granite mixtures with limestone mixtures in terms of physical properties such as density and air voids and mechanical properties such as resilient modulus, stability and flow, rutting and fatigue cracking of mixtures based on the results obtained from the preliminary approach. These factors showed a relatively good comparison between granite mixtures and limestone mixtures including the Marshall Compaction Method and the Gyratory Compaction Method.

3.4.1. Bulk density and air voids. The most important factor that affects the long-term durability of the hot mix asphalt (HMA) is the mix density, especially when it is accomplished with compaction. Furthermore, the density of the HMA material increased when the air void content of the mixture decreased, which implies that they are inversely proportional to each other. Therefore, the air voids of HMA in order to properly design should be within the range (3% - 5%) [16]. The results obtained from our comparison based on density and air voids are shown in Table 4.

Table 4: Marshall versus Superpave in Density & Air Voids

Type of aggregate	Density (kg/m^3)		Air voids (%)	
	Marshall	Superpave	Marshall	Superpave
Granite upper gradation	2.381	2.368	2.3	2.83
Granite lower gradation	2.344	2.374	3.34	2.1
Limestone upper gradation	2.482	2.439	0.92	2.63
Limestone lower gradation	2.484	2.478	0.72	0.96

From Table 4, it is evident that the high-density value resulted into low air voids. In addition, the air voids were less than the range (3 - 5 %) except for the granite lower gradation mixture compacted by the Marshall Compactor Method. Furthermore, all of the mixtures appeared susceptible to permanent deformation, thus excluding the granite lower gradation mixture used to the Marshall Compactor. As seen, the limestone mixtures resulted into lower values of air voids compared with granite mixtures. Therefore, the limestone mixtures need improvement to resist the rutting.

3.4.2. Height & compaction. Compaction is one of the main factors significantly affecting the asphalt mixtures. In this research, the results were observed based on the height of samples in different compaction methods such as the Marshall compactor and the gyratory compactor for each mixture as shown in Table 5.

Table 5: Marshall versus Superpave in Height and Compaction

Type of aggregate	Average Height (mm)		Compaction (%)	
	Marshall (actual)	Superpave (actual)	Marshall (actual)	Superpave (actual)
Granite upper gradation	65.7	68.3	97.7	97.17
Granite lower gradation	67.4	68.9	96.66	97.9
Limestone upper gradation	62.5	66.7	99.08	97.37
Limestone lower gradation	62.6	66.7	99.28	99.04

The above results for the height of samples indicate that the samples compacted using the Marshall compactor yielded less height comparing with the specimens compacted utilizing gyratory compactor. On the other hand, the granite mixtures were higher than the limestone mixtures. These results are in line with results of some previous studies, which indicate that the granite is stiffer than the limestone. In addition, a decrease in the height of the samples was observed, and it could be attributed to increasing the compaction of specimens except for granite lower gradation mixture. Based on the

IOP Conf. Series: Materials Science and Engineering **512** (2019) 012003 doi:10.1088/1757-899X/512/1/012003

results of compaction, the mixtures containing limestone of aggregate obtained a higher value than mixtures containing granite of aggregate.

3.4.3. Marshall stability & flow. The Marshall stability and the flow test represent a real experimental and they are not substantially correlated with the performance. However, they aid in evaluating the properties of the performance of the HMA mixtures. The differences between the stability and flow test in granite mixtures with limestone mixtures in various compaction methods are highlighted in Table 6.

Table 6: Marshall versus Superpave in Stability & Flow

Type of Mixture	Stability (KN)		Flow (mm)	
	Marshall	Superpave	Marshall	Superpave
Granite Upper Gradation	27.48	22.87	3.56	3.9
Granite Lower Gradation	16.47	17.54	3.59	4.24
Limestone Upper Gradation	16.12	15.84	4.19	4.13
Limestone Lower Gradation	13.23	11.55	5.36	4.16

In terms of the assessed performance, the aggregate upper gradation mixture gave the highest value of stability over other mixtures when it was subjected to compact using the Marshall compactor and the gyratory compactor. As far as the flow is concerned, all the limestone mixtures provided values above the range of specification using the Marshall compactor and the gyratory compactor whereas the granite mixture values were within the range of specification except for the granite lower gradation mixture using the gyratory compactor. The range of flow is normally between 2 and 4 mm.

3.4.4. Resilient modulus results. The resilient modulus test was performed using the Universal Asphalt Testing Machine, MATTA, according to the ASTM test method D4123 (ASTM, 1995). Each sample was tested at 25 C° after 4 hour-conditioning. The peak load was applied 1200 N along the vertical diameter of the sample. The test sequence containing a 5 count of conditioning pulses was followed by 5 loading pulses. The pulse period and pulse width were 3000 ms. The rise time was measured at the points of 10 % and 90 % of the peak force and force pulse. The results of resilient modulus are illustrated in Figure 4.

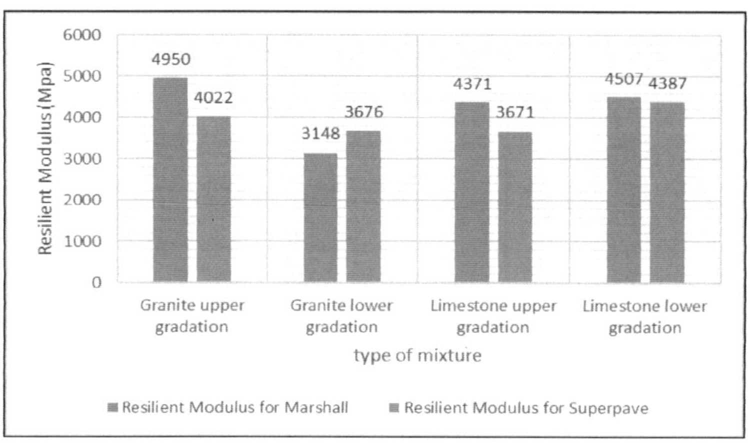

Fig. 4: Marshall versus Suprpave in Resilient Modulus

Figure 4 provides a summary of comparison in Resilient Modulus for whole mixtures in this research. The results showed that the mixtures conducted by the Marshall compactor obtained higher values of Resilient Modulus than those mixtures which were performed utilizing the gyratory compactor except for the granite lower gradation mixture as well the granite upper gradation mixtures as both demonstrated a higher Resilient Modulus than the limestone upper gradation mixtures. In contrast, the granite lower gradation mixtures have a lower value of Resilient Modulus than the limestone lower gradation mixtures. The results indicated specimens compacted at a low density normally have lower Resilient Modulus than those compacted at a higher density. Moreover, the magnitude of difference was a function of some parameters, including the particle shape and fines content.

3.4.5. Permanent deformation results. The permanent deformation of the pavement occurs by the passage of heavy vehicles as each wheel passes the pavement deflects downwards and then bounces, thus leaving an irrecoverable downwards deformation. Such deformation accumulates and after millions of passes and the pavement requires rehabilitation. The test was conducted using the Asphalt Universal Testing Machine (MATTA). The parameters relied on a haversine waveform with 200 kPa stress levels. Thereby, test temperatures of 40 ° C were selected. The load was applied for 0.5 s followed by a rest period of 0.1 s. The sample was terminated after 3600 load cycles or until the accumulated deformation reached 10%. The outcomes are shown in Figure 5.

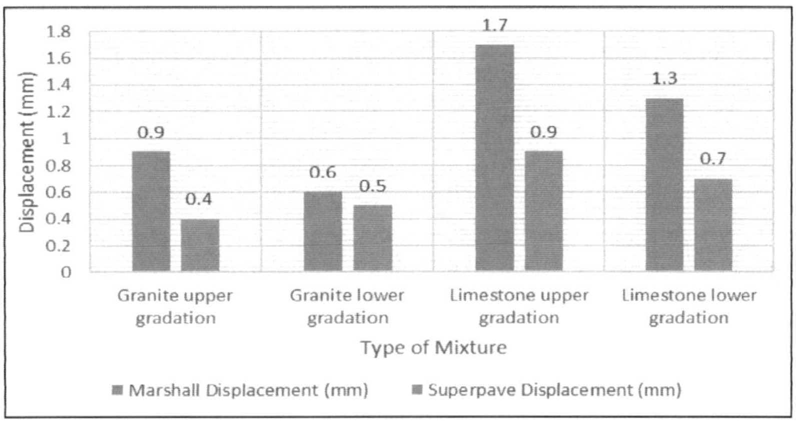

Fig. 5: The Permanent Deformation Test Results

As it can be seen from Figure 5, the samples were compacted using the gyratory compactor achieved a lower displacement than specimens compacted utilizing the Marshall compactor. Otherwise, the limestone mixtures demonstrated a higher displacement than granite mixtures. In brief, these results indicated that the stiffness and resistance to permanent deformation of asphalt mixtures strongly depend on the mixture composition and the degree of compaction. In addition, the mixtures containing a lot of coarse aggregates have shown more resistance to permanent deformation except for the granite mixtures which were compacted by the gyratory compactor.

3.4.6. Fatigue cracking. The fatigue cracking test was accomplished utilizing the Universal Testing Machine (MATTA). The tests were set out under the following conditions as100 N seating force, 0.12s loading time,1.5s rest time, 1000 N cyclic loading and the conditioned temperature 20°C c was selected. Furthermore, the tests were terminated when they reached a maximum cycle count of 3600 (1 hour) or a maximum axial displacement of 9 mm. The method for determining the fatigue characteristics of mixtures was used in this study by employing the Indirect Tensile Fatigue. The specimens were divided into two types based on their thickness. The first type was performed using approximately 40 ±5 mm of the middle of the sample and the second type was conducted utilizing the entire sample. The results of tests are presented in Table 7.

Table 7: Indirect Tensile Fatigue Results

Type of Mixture	Thickness (mm)	Displacement (mm)	Thickness (mm)	Displacement (mm)
Granite upper gradation Marshall	37	0.67	66	0.23
Granite upper gradation Superpave	35	0.53	68	0.66
Granite lower gradation Marshall	35	1.43	68	0.33
Granite lower gradation Superpave	37	0.93	68	0.69
Limestone upper gradation Marshall	35	1.1	63	0.82
Limestone upper gradation Superpave	36	0.95	66	1.3
Limestone lower gradation Marshall	37	1.19	63	0.81
Limestone lower gradation Superpave	37	1.32	66	1.05

Based on the results of Table 7, the displacement of fatigue cracking for full samples compacted by the Marshall compactor was significantly lower than the displacement of full samples compacted using the gyratory compactor. Furthermore, compaction seemed to play a significant role in determining the indirect tensile strength in the asphalt mixtures as proved by previous studies. Thereby, high compaction yielded a lower displacement of fatigue cracking. On the other hand, the type of aggregate in mixtures is one of the parameters that cause immediate fatigue cracking. Therefore, the weak aggregate such as limestone indicated a higher displacement than granite. However, as noticed from the results in Table 7, when the thickness of approximately 40 ± 5 mm in the middle of the sample was considered to investigate the indirect tensile fatigue, the results indicated that using the Marshall compactor resulted into a higher displacement than the gyratory compactor except for limestone lower gradation mixtures.

3.4.7. Summary of results. Several tests such as density, air voids, Marshall stability, flow, resilient modulus, permanent deformation and fatigue cracking were performed in the second phase of this study. These tests were performed to compare compaction methods: the Marshall compactor and the gyratory compactor in four mixtures: granite upper gradation mixtures, granite lower gradation mixtures, limestone upper gradation mixtures and limestone lower gradation mixtures. The results of the study obtained from these various tests are summarized in Figure 6.

Type	OAC %	Density(Kg/m^3)		Air Voids (%)		Stability (KN)		Flow (mm)		Resilinet Modulus (Mpa)		Rutting (mm)		Fatigue Cracking (mm)	
		Marshall	Super pave	Marshall	Super pave	Marshall	Super pave	Marshall	Super pave	Marshall	Super pave	Marshall	Super pave	Marshall	Super pave
Granite upper gradation	4.8	2.381	2.368	2.3	2.83	27.48	22.87	3.56	3.9	4950	4022	0.9	0.4	0.23	0.66
Granite lower gradation	5.2	2.344	2.374	3.34	2.1	16.47	17.54	3.59	4.24	3148	3676	0.6	0.5	0.33	0.69
Limestone upper gradation	4.6	2.482	2.439	0.92	2.63	16.12	15.84	4.19	4.13	4371	3671	1.7	0.9	0.82	1.3
Limestone lower gradation	4.7	2.484	2.478	0.72	0.96	13.23	11.55	5.36	4.16	4507	4387	1.3	0.7	0.81	1.05

Fig. 6: Summary of the Second Phase Results

4. Conclusions

The most important findings of this study are summarized based on the research objective that can aid in future as follows:

The results of comparison of the physical properties in the mixtures show the highest values of the Marshall stability achieved with the granite upper gradation mixtures as well as the flow of the granite upper mixtures and the granite lower mixtures compacted by the Marshall compactor are within the range (2-4) mm. On the other hand, the limestone lower gradation mixtures resulted into the lowest values of air voids. Otherwise, the granite lower mixtures gave the lowest values of resilient modulus compared to other mixtures.

Based on our comparison of the compaction methods, the mixtures compacted by the Marshall compactor achieved the highest values of Marshall stability except for the granite lower gradation mixtures and the flow which are not within of the range (2 – 4) mm and the granite upper gradation mixtures and granite lower gradation mixtures compacted by the Marshall compactor. All mixtures compacted by the gyratory compactor gave a higher percentage of air voids and resilient modulus values than the mixtures which were subjected to the Marshall compactor except for the granite lower mixtures.

The results of permanent deformation and fatigue cracking indicated that the granite upper gradation mixtures compacted by the gyratory compactor and the granite lower mixtures compacted by Marshall compactor fulfilled the best result of rutting resist. In addition, the granite upper gradation mixtures provided the best resistance of fatigue cracking with both compaction methods: the Marshall compactor and the gyratory compactor.

5. References

[1] West R, Willis C and Marasteanu M O 2013 Improved mix design, evaluation and materials management practices for hot mix asphalt with high reclaimed asphalt pavement content *Transportation Research Board* **752**

[2] West R, Watson C, Turner E and Casola J R 2010 *Mixing and compaction temperatures of asphalt binders in hot-mix asphalt* (Washington: Transportation Research Board)

[3] Abdullah W, Obaidat S and Abu-Sa'da N M 1998 Influence of aggregate type and gradation on voids of asphalt concrete pavements *Journal of Materials in Civil Engineering* **10**(2) 76-85

[4] Brown E, Kandhal P, Roberts S, Kim F, Lee Y and Kennedy T 2009 *Hot mix asphalt materials, mixture design, and construction* (Lanham: NAPA Research and Education Foundation)

[5] Golalipour A, Jamshidi E, Niazi Y, Afsharikia Z and Khadem M 2012 Effect of aggregate gradation on rutting of asphalt pavements *Procedia-Social and Behavioral Sciences* **53** 440-9

[6] Kalaitzaki E, Kollaros G and Athanasopoulou A 2015 Influence of aggregate gradation on HMA mixes stability *Romanian Journal of Transport Infrastructure* **4**(2) 13-22

[7] Abo-Qudais S and Al-Shweily H 2007 Effect of aggregate properties on asphalt mixtures stripping and creep behavior *Construction and Building Materials* **21**(9) 1886-98

[8] Al-Shamsi K, Hassan H and Mohammed L N 2017 Effect of low VMA in hot mix asphalt on load-related cracking resistance *Construction and Building Materials* **149** 386-94

[9] Gao Y, Huang X and Yu W 2014 The compaction characteristics of hot mixed asphalt mixtures *Journal of Wuhan University of Technology-Mater. Sci. Ed.* **29**(5) 956-9

[10] Pérez-Jiménez F, Martínez A, Miró R, Hernández-Barrera D and Araya-Zamorano L 2014 Effect of compaction temperature and procedure on the design of asphalt mixtures using Marshall and gyratory compactors *Construction and Building Materials* **65** 264-9

[11] Leandro R, Vasconcelos K and Bernucci L 2017 Evaluation of the laboratory compaction method on the air voids and the mechanical behavior of hot mix asphalt *Construction and Building Materials* **156** 424-34

[12] Jasim 2017 Comparison between Marshall and Superpave mixtures design *Al-Qadisiyah Journal for Engineering Sciences* **5**(4) 394-406

[13] De Assis S, de Queiroz B, Araújo C, Nunes K, de Melo R and Lucena L 2017 Evaluation of limestone crushed dust aggregates in hot mix asphalt *Construction and Building Materials* **148** 659-65

[14] Rashwan N 2015 Evaluation of aggregate and mineral filler types on characteristics of hot mix asphalt *International Journal of Civil Engineering and Technology*

[15] Muniandy R and Sohadi R 2001 *Highway Materials: A Guide Book for Beginners* (Malaysia: Universiti Putra Malaysia Press)

[16] Scherocman J 1996 Compacting hot mix asphalt pavements Part 2 *Road & Bridges* **34**(9)

Evaluating properties of wood ash modified asphalt mixtures

YC Bi and FM Jakarni

Department of Civil Engineering, Faculty of Engineering, Universiti Putra Malaysia, 43400 UPM Serdang, Selangor, Malaysia

E-mail : sinavamder@gmail.com, fauzan.mj@upm.edu.my

Abstract. Nowadays, sustainable development has gained more attention because humanity will face the problem of energy shortage in the near future. Biomass' consumption has increased year by year as it is a renewable material. However, there will be a lot of ash residue after combustion biomass, which can cause pollution problems. This research investigates the potential use of wood ash as filler replacement in the asphalt mixture because wood ash has similar physical and chemical properties to traditional fillers. Wood ash used in this study consisted of 100% passing the 0.075 mm sieve and 25%, 50%, 75% and 100% conventional filler (by weight) were replaced by the wood ash. Marshall Stability & Flow, Resilient Modulus, Rutting performance and Fatigue performance tests were employed to determine the influence of wood ash as filler to the hot mix asphalt (HMA) mixture. The results showed that present wood ash in the asphalt mixture could improve fatigue performance, reduce permanent deformation, and increase the resilient modulus of the asphalt mixture. Therefore, wood ash can be used as the filler substitute material for asphalt mixture.

1. Introduction

With the development of the economy and the improvement of people's living standards, passenger car has become popular choice of transportation among ordinary families. According to the Statistics portal [1], the number of passenger cars increased from 679 million in 2005 to 947 million in 2015. In addition, with the rapid development of the logistics industry, more and more large trucks are designed to carry goods to meet people's demand for various materials. Some countries have even raised legal axle loads such as South Africa, Egypt and India. It is a fact that both increase of the number of vehicles and increase of the axle load will increase pavement distress and thus reduce the pavement life. Heavier axle load truck will cause more damage to the pavement as compared to the larger number of passenger cars since the damage to the pavement is not linear with axle load increased but exponential.

In general, there are two ways to improve the performance of the pavement. First is to find a suitable mix design, and the second way is to utilize the proper material. There are three materials in asphalt mixtures namely aggregate, bitumen and filler. The choice of aggregate is largely limited by location. Modified bitumen has been populated nowadays. Filler as one of the materials also played a major role in the asphalt mixture [2]. There are two mechanisms to describe the role of filler in asphalt mixes. Firstly, fillers provide additional contact points between larger aggregates. Secondly, fillers increase the stability of the mix by increasing the viscosity of asphalt binders [3]. During the mixing, filler will blend with bitumen to form asphalt mastic, which plays a major role in controlling the mechanical properties of their mixtures [4]. The type of filler affects the performance of asphalt mixture. Mistry and Roy [5] carried out research on the effect of fly ash as filler in the hot asphalt mix, and the results shown that fly ash as the filler replacement in hot mix asphalt (HMA) can give lesser deformation and better strength compared with hydrated lime. Modarres and Rahmanzadeh [6] examined the effect of coal waste power as filler in hot mix asphalt (HMA), and the obtained results showing that the indirect tensile strength, Marshall Stability, and resilient modulus of hot mix asphalt (HMA) mixes were increased by using coal waste power as filler compare with limestone powder.

As a renewable resource, biomass is now being paid more attention by human beings. In 2017, about 5% of the total primary energy was provided by biomass fuels in the United States [7]. Because coal resources will be fully used in the near future, biomass is being used by more and more power plants as substitutes for coal. However, combustion biomass fuels will produce large amounts of ash residues. These ashes are usually in the forms of fine particles that can be easily blown away by the wind, thus cause air pollution and human health problems. Usually, these ash will be landfilled but it will wasted land and cause groundwater pollution because the ash contains some heavy metals that will enter the groundwater system with rainfall. Some researchers have attempted to use these biomass ashes as fillers in the asphalt mixture. Sargın at al. [8] investigated the effect rice husk ash as filler in hot mix asphalt (HMA) concrete. It found that the best Marshall Stability results were obtained when the filler was composed of 50% limestone and 50% rice husk ash. Al-Hdabi [9] researched on the properties of asphalt concrete mixture with rice husk ash as filler by conducting an Indirect Tensile Strength, Index of Retained Strength and Marshall Stability tests. It concluded that by replacing conventional mineral fillers with rice husk ash, the Marshall stability of hot asphalt mixtures is significantly improved. Tahami et al. [10] investigated the potential of using date seed ash and rice husk ash as filler in hot mix asphalt (HMA). The results shown that the asphalt mixtures with date seed ash and rice husk ash fillers have significantly improved stability and stiffness modulus compared to the control mixtures. However, for wood ash as the major components of biomass ash, only a few researchers tried to study the possibility of it as a partial replacement for pavement material, especially in Malaysia. This study was designed to evaluate the potential of wood ash as filler substitute material for the production of asphalt mixtures. Ranges of percentages of 25%, 50%, 75% and 100% of conventional filler were replaced by the wood ash to produce asphalt mixture. Then, those specimens were subjected Marshall Stability & Flow test, asphalt permanent deformation test, Indirect Tensile Fatigue test and Resilient Modulus test and the results were compared to control samples.

2. Materials and experiment design

2.1. Aggregates

The aggregate used in this study is sharp-edged and crushed granite aggregates which are obtained from quarry in Kajang, Malaysia. The grading of aggregate is following the JKR/SPJ/2008-S4 requirement of AC 10 [11] as shown in table 1. The average specific gravity of aggregate is 2.57.

Table 1. Gradation of Aggregate

Sieve Size	Specification		Passing
(mm)	Upper Limit	Lower Limit	(%)
14	100	100	100
10	100	90	95
5	72	58	65
3.35	64	48	56
1.18	40	22	31
0.425	26	12	14
0.15	14	6	10
0.075	8	4	6
Pan	0	0	0

2.2. Asphalt binder
The asphalt used for this study is a penetration grade of binder 60/70 (PEN 60/70) obtained from the same quarry in Kajang. The specific gravity of asphalt is 1.03.

2.3. Wood ash
The wood ash obtained from a pottery company located in Kuala Selangor. The wood ash was dried in an oven at a temperature of 110°C before use. It was then sieved through a 425µm sieve in order to remove the debris and extraneous materials. The retained wood ash then milled and sieve through a 75µm sieve. The properties of the wood ash as shown in Table 2.

Table 2. Properties of Wood Ash

Chemical Composition%	
CaO	29.53
K_2O	9.64
SiO_2	28.11
MgO	5.14
SO_3	0.97
Al_2O_3	5.14
Fe_2O_3	2.91
P_2O_5	2.48
Na_20	0.32
Specific Gravity (g/cm^3)	2.52
Median Particle Size, d_{50} (*um*)	22.486

2.4. Sample preparation
The Marshall Mix design method was employed to prepare asphalt mixture samples with 75 blows on each side of the cylindrical samples. The specified procedure in ASTM D6926 [12] for heating, mixing and compaction the asphalt cement-aggregate mixture was followed. The Optimum Asphalt Content (OAC) was found to be 5.7% for the control mixture and all hot mix asphalt (HMA) samples were prepared using this OAC to maintain consistency throughout the study. For each different percentage (0%, 25%, 50%, 75%, and 100%) of wood ash as filler, three asphalt mixture samples need to be prepared for each test. However, for Marshall Stability & Flow and Resilient Modulus tests, they can utilize the same sample because resilient modulus test is non-destructive test and the sample after resilient modulus test can still be used in Marshall Stability & Flow test.

3. Experimental procedure
3.1. Marshall stability &flow test
The purpose of the Marshall Stability& flow test is to determine the Marshall stability and flow values of asphalt mixture. The maximum allowable aggregate size in this experiment is 25mm and it is conducted based on ASTM D1559. Samples were immersed in a 60°C water bath for 30 minutes before testing. A load with a constant strain rate of 50.88 mm/min was applied to the sample until maximum load reached. The Marshall Stability and flow values read from the display of Marshall Apparatus. The corrected Marshall Stability values obtained by multiplying the Marshall stability with Height correlation ratio.

3.2. Resilient modulus test
Resilient Modulus Test is to measure of material stiffness. In pavement design, stiffness of asphalt mixture are tested by using Material Testing Apparatus (MATTA) in the resilient modulus test. And,

this test is carried out according the ASTM D4123 [13]. Samples were placed in a MATTA machine for at least 2 hours at 25°C before testing. The peak load force used in this test is 1200 N.

3.3. Rutting performance test
The rutting performance of asphalt mixture was determined by MATTA machine in asphalt permanent deformation test. The samples were subjected 200 kPa force for 3600 cycles at 40°C. The rest period and load repeat time was 0.1s and 0.5s, respectively.

3.4. Fatigue performance test
In this study, Indirect Tensile Fatigue test was adopted to test the fatigue performance of asphalt mixtures. The test machine is the same as the rutting performance test. The cyclic loading force was 1000 N for 3600 loading cycle at 20°C. And, load repeat time was 500ms with 100ms rest time.

4. Results and discussion

4.1. Marshall stability &flow test results
The Marshall Stability for all of the asphalt mixture is shown in Figure 1. From the figure, the present of wood ash will reduce Marshall Stability, and by increasing the wood ash content, the stability was decreased. That's because wood ash has lower ductility compared to traditional fillers. However, all of the stability values are still significant larger than JKR/SPJ/2008-S4 requirement for AC10 which is larger than 8 KN. The mixture produced with 25% wood ash and 100% wood ash had 6.8% and 15.1% lower stability than control group.

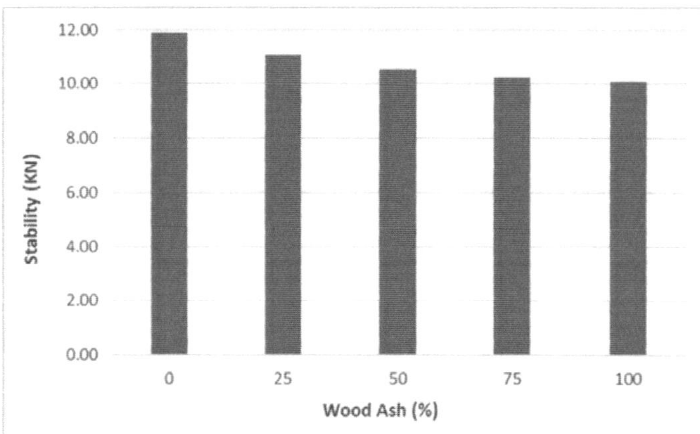

Figure 1. Marshall Stability of all asphalt mixtures.

Table 3 shows results of few parameters for different wood ash content at OAC and the JKR requirement. As shown in Table 3, all of the flow values are fulfil the requirement, and flow value is decreased with the increasing of wood ash content from 25% to 100%. It means that wood ash has lower flexibility and addition wood ash in the asphalt mixture could reduce the flexibility of the mixture. Besides, the stiffness value is increased with the increasing of wood ash content from 25% to 100%. The stiffness of control and 100% wood ash mixture were found to be 3514 N/mm and 3820 N/mm. The stiffness of the asphalt mixture can predict permanent deformation and rutting. The higher the stiffness value, the stiffer the asphalt mixture and the better resistance to permanent deformation [14, 15].

Table 3. Results and Requirements.

Parameter	Specification (JKR 2008)	Wood Ash Content				
		0%	25%	50%	75%	100%
Stability (N)	> 8000	13670	12730	12390	12340	12300
Flow (mm)	2-4	3.89	3.99	3.97	3.25	3.22
Stiffness (N/mm)	>2000	3514	3190	3121	3797	3820

4.2. Resilient modulus test results

Figure 2 shows the results of the resilient modulus of all asphalt mixtures. The figure shows that using wood ash could improve the resilient modulus of asphalt mixtures, and by increasing the wood ash content, the resilient modulus was increased. This is because wood ash has a lower specific gravity than conventional fillers and it can fill more gaps/voids in the asphalt mixture than conventional fillers. The mixture produced with 25% wood ash and 100% wood ash had 17.5% and 40% higher resilient modulus than control group. This means that the stiffness of the asphalt mixture was improved due to the presence of wood ash.

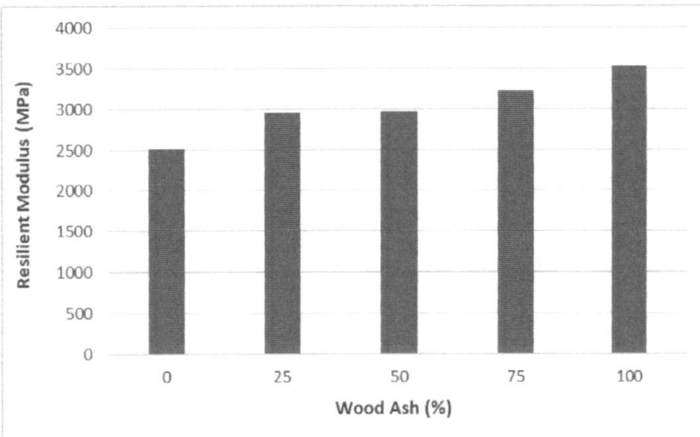

Figure 2. Resilient Modulus of all asphalt mixtures.

4.3. Rutting performance test

The final strain of rutting performance for all of the asphalt mixture is shown in Figure 3. It shows that control group have more permanent deformation compared to the Wood Ash modified asphalt mixtures. As the wood ash content increases, permanent deformation will decrease. The reason for this phenomenon is that the wood ash has low thermal sensitivity. The more wood ash content in the mixture, the less the effect of temperature on the asphalt mixture and the good permanent deformation resistant.

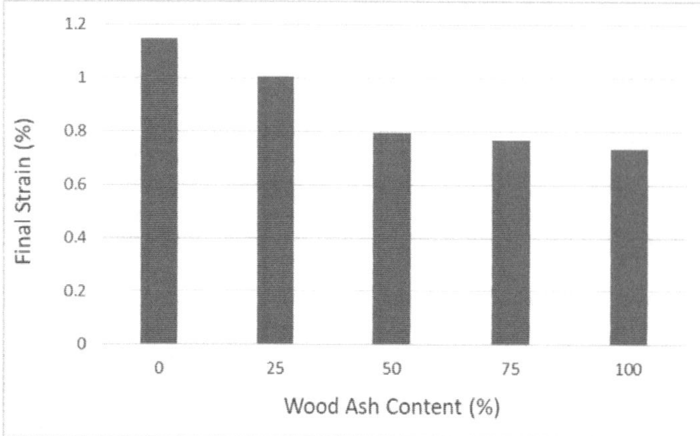

Figure 3. Final strain of rutting for all asphalt mixtures.

4.4. Fatigue performance test

Figure 4 present the final strain of fatigue for all asphalt mixtures. The figure shows that addition wood ash in asphalt mixture could reduce the final strain of samples. This means that fatigue performance can be improved by adding wood ash to the asphalt mixture. And as the wood ash content increases, better fatigue performance will be in the end. Which is because the addition of wood ash in the asphalt mixture can reduce the gaps/voids in the asphalt mixture and improve the adhesion between the asphalt binder and the aggregates [16].

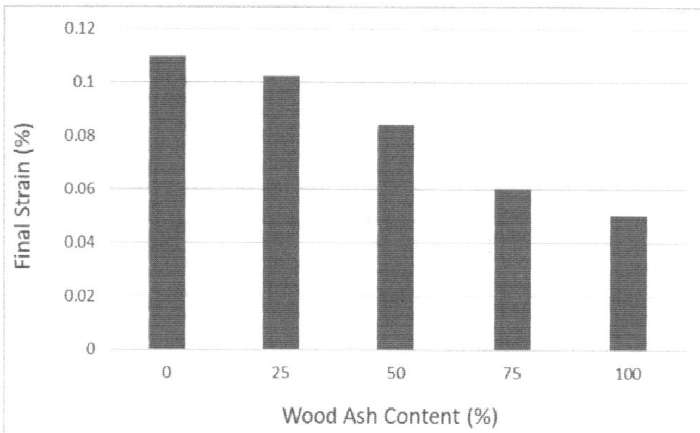

Figure 4. Final strain of fatigue for all asphalt mixtures.

5. Conclusion and recommendation

From the results, the present of wood ash can significantly improve the resilient modulus of the asphalt mixture, and by increasing the wood ash content, the resilient modulus was increased. But, the stability was decreased with increased of wood ash content. However, those stability values still fulfilled the JKR requirement. As the wood ash content increases, permanent deformation was decrease and fatigue performance was improve. Therefore, wood ash as a sustainable material and are suitable as a replacement material for conventional fillers for the production of hot mix asphalt (HMA) by using AC 10 gradation and PEN 60/70 asphalt binder. Wood ash used as a filler replacement material not only can improve properties of the asphalt mixture but also reduce environmental pollution. Further study is needed to evaluate the effect of wood ash as filler on asphalt mixture by using different gradation and binder types. In addition, wood ash may also be used as an asphalt modifier to improve the quality of the asphalt mixture.

6. References

[1] Statistic 2108 *Number of passenger cars and commercial vehicles in use worldwide from 2006 to 2015* Statista Available at https://www.statista.com/statistics/281134/number-of-vehicles-in-use-worldwide/

[2] *Wasilewska* M, Małaszkiewicz D and Ignatiuk N 2017 Evaluation of different mineral filler aggregates for asphalt mixtures *IOP Conference Series: Materials Science and Engineering* **245** 022042

[3] Wang H, Al-Qadi I, Faheem A, Bahia H, Yang S and Reinke G 2011 Effect of mineral filler characteristics on asphalt mastic and mixture rutting potential *Transportation Research Record: Journal of the Transportation Research Board* **2208** 33-39

[4] Diab A and Enieb M 2108 Investigating influence of mineral filler at asphalt mixture and mastic scales *International Journal of Pavement Research and Technology* **11** 213-24.

[5] Mistry R and Roy T K 2016 Effect of using fly ash as alternative filler in hot mix asphalt *Perspectives in Science* **8** 307-09

[6] *Modarres* A and Rahmanzadeh M 2014 Application of coal waste powder as filler in hot mix asphalt *Construction and Building Materials* **66** 476-83

[7] *Biomass* - Energy Explained 2018 Your guide to understanding energy - energy information administration Available at https://www.eia.gov/energyexplained/?page=biomass_home

[8] *Sargın Ş*, Saltan M, Morova N, Serin S and Terzi S 2013 Evaluation of rice husk ash as filler in hot mix asphalt concrete *Construction and Building Materials* **48** 390-97

[9] Al-Hdabi A 2016 Laboratory investigation on the properties of asphalt concrete mixture with rice husk ash as filler *Construction and Building Materials* **126** 544-51

[10] Tahami S A, Arabani M and Mirhosseini A F 2018 Usage of two biomass ashes as filler in hot mix asphalt *Construction and Building Materials* **170** 547-56

[11] Public Work Department 2008 *Standard Specification for Road Work (JKR/SPJ/2008)* (Kuala Lumpur: JKR)

[12] ASTM D6926-16 2016 *Standard Practice for Preparation of Asphalt Mixture Specimens Using Marshall Apparatus* (West Conshohocken: ASTM International)

[13] ASTM D4123-82 1995 *Standard Test Method for Indirect Tension Test for Resilient Modulus of Bituminous Mixtures (Withdrawn 2003)* (West Conshohocken: ASTM International)

[14] Bostancioğlu M and OruçŞ 2015 Effect of activated carbon and furan resin on asphalt mixture performance *Road Materials and Pavement Design* **17** 512-25

[15] Yilmaz M, Kök B V and Kulog˘lu 2011 Effects of using asphaltite as filler on mechanical properties of hot mix asphalt *Constr. Build. Mater* **25** 4279-86

[16] ArabaniM and Tahami S 2017 Assessment of mechanical properties of rice husk ash modified asphalt mixture *Construction and Building Materials* **149** 350-58

Acknowledgments

Authors would like to thank the Ministry of Higher Education Malaysia and Universiti Putra Malaysia for the award of Fundamental Research Grant Scheme (FRGS) FRGS/1/2017/TK06/UPM/02/1 (Project Code: 03-01-17-1894FR) as well as all the people whom are directly and indirectly involved in this study.

IOP Conf. Series: Materials Science and Engineering **512** (2019) 012005 doi:10.1088/1757-899X/512/1/012005

The important aspects of subgrade stabilization for road construction

N N Nik Daud[1], F N A Jalil[1], S Celik[2] and Z N K Albayrak[2]

[1]Department of Civil Engineering, Faculty of Engineering, Universiti Putra Malaysia, 43400 UPM Serdang, Selangor, Malaysia
[2]Department of Civil Engineering, Engineering Faculty, Ataturk University, Erzurum, Turkey

E-mail : niknor@upm.edu.my

Abstract. The subgrade stabilization for road construction especially in Malaysia is most concerns nowadays. Due to the rising development in urban area, there is a high prospect of rural area also to be developed and the important aspects of subgrade construction is aware. Many studies are being conducted which are focused on the soil stabilization techniques of problematic soils present at specific location for a road construction. In Malaysia, there are several types of soil which can be considered as problematic such as marine clay, peat soil etc. Several stabilization techniques including physical, chemical and mechanical approaches have been implemented and some of it provides a good solution either in a short term or long term application. This paper will highlight and discuss the important aspects of subgrade stabilization by reviewing previous studies such as i) the subgrade issues, ii) factors that influence the stabilization technique, iii) application of additives to stabilize the subgrade and iv) the economic valuation of additive usage.

1. Introduction

The performance of the completed road is not only depends on the pavement structural design, the subgrade support conditions and subbase layers also play a main role in it. As the foundation of pavement's upper layers, the subgrade and subbase layers help in mitigating the detrimental effects of climate and the static dynamic stresses generated by traffic. Therefore, building a stable subgrade and a properly drained subbase is vital for constructing an effective and long lasting pavement system. The subgrade stiffness and strength determination can be categorized under specific technical of geotechnical issues in pavement design.

Pavement construction on a natural subgrade is the classic case for pavement design. The subsurface profile (including depth to bedrock and groundwater table) are determined directly from the subsurface exploration program, and subgrade properties needed for the design can be taken from tests on the natural foundation soil in its in-situ condition and in its compacted state, if the upper foundation layer is to be processed and recompacted or removed and replaced during construction.

However, the alignment for most highway projects does not always follow the site topography, and consequently a variety of cuts and fills will be required. The geotechnical design of the pavement will involve additional special considerations in cut and fill areas. Attention must also be given to transition

IOP Conf. Series: Materials Science and Engineering **512** (2019) 012005 doi:10.1088/1757-899X/512/1/012005

zones such example between a cut and an at-grade section because of the potential for non-uniform pavement support and subsurface water flow.

The main additional concern for cut sections is drainage, as the surrounding site will be sloping toward the pavement structure and the groundwater table will generally be closer to the bottom of the pavement section in cuts. Stabilization of moisture-sensitive natural foundation soils may also be required. Stability of the cut slopes adjacent to the pavement will also be an important design issue, but one that is typically treated separately from the pavement design itself.

The embankments for fill sections are constructed from well compacted material and in many cases, this results in a subgrade that is of higher quality that the natural foundation soil. Drainage and groundwater issues will in general be less critical for pavements on embankments, although erosion of side slopes from pavement runoff may be a problem, along with long-term infiltration of water. The principal additional concerns for pavements in fill sections will be the stability of the embankment slopes and settlements, either due to compression of the embankment itself or due to consolidation of soft foundation soils beneath the embankment.

The unbound soil layers in a pavement provide a substantial part of the overall structural capacity of the system, especially for flexible pavements (often more than 50 percent). As shown in Figure 1, the stresses induced in a pavement system by traffic loads are highest in the upper layers and diminish with depth.

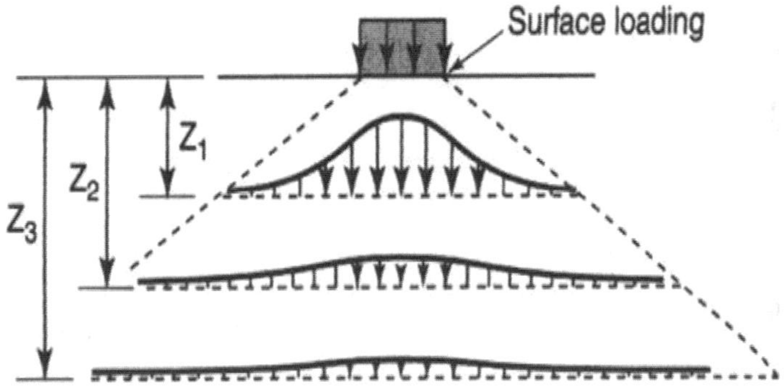

Figure 1. Attenuation of load-induced stresses with depth

Consequently, higher quality - and generally more expensive - materials are used in the more highly stressed upper layers of all pavement systems, and lower quality and less expensive materials are used for the deeper layers of the pavement (Figure 2). This optimization of material usage minimizes construction costs and maximizes the ability to use locally available materials. However, this approach also requires greater attention to the lower quality layers in the design (*i.e.*, the subgrade) in order to reduce life-cycle pavement costs. Good long-term performance of lower layers means that upper layers can be maintained (rehabilitated) while avoiding the more costly total reconstruction typically associated with foundation failures.

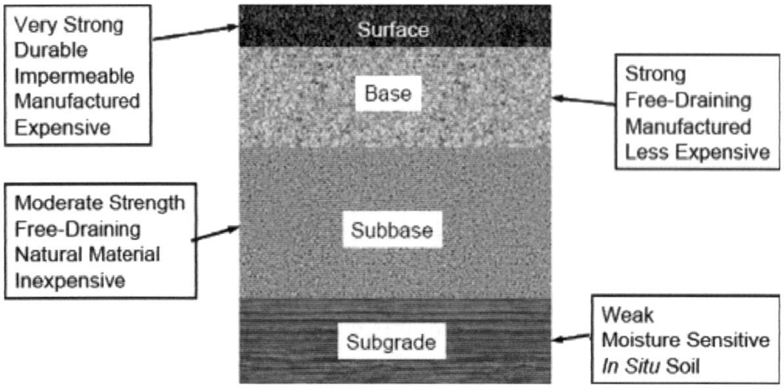

Figure 2. Variation of material quality with depth in a pavement system with ideal drainage characteristics

As in the case for all geotechnical structures, pavements will be strongly influenced by moisture and other environmental factors. Water migrates into the pavement structure through combinations of surface infiltration (*e.g.* through cracks in the surface layer), edge inflows (*e.g.* from inadequately drained side ditches or inadequate shoulders), and from the underlying groundwater table (*e.g.*, via capillary potential in fine-grained foundation soils). In cold environments, the moisture may undergo seasonal freeze/thaw cycles. Moisture within the pavement system nearly always has detrimental effects on pavement performance. It reduces the strength and stiffness of the unbound pavement materials, promotes contamination of coarse granular material due to fines migration, and can cause swelling (*e.g.* frost heave and/or soil expansion) and subsequent consolidation. Moisture can also introduce substantial spatial variability in the pavement properties and performance, which can be manifested either as local distresses, like potholes, or more globally as excessive roughness. The design of the geotechnical aspects of pavements must consequently focus on the selection of moisture-insensitive free-draining base and subbase materials, stabilization of moisture-sensitive subgrade soils, and adequate drainage of any water that does infiltrate into the pavement system.

1.1. Subgrade issues
Subgrade issues can be discussed in varies way by focusing some certain aspects and specific concerns as below topics.

1.1.1. Site conditions. The soil properties are the most concern part when deal with the original soil to construct a road or highway. Their characteristics either can be highly shrink or swell can contribute to the pavement design. Special investigation should be undertaken to identify the problem soils. Black Cotton soils (BC soils) has been a great challenge to the highway engineers in India. A study has been conducted by Ref. [1] and the outcome shows that the Black Cotton soil is very hard when dry, but loses its strength completely when in wet condition. It is also has been observed that on drying, the black cotton soil develops cracks of varying depth. As a result of wetting and drying process, vertical movements lead to failure of pavement, in the form of settlement, heavy depression, cracking and unevenness.

There is also some issues related to the minerals present in the original subgrade material. Ref. [2] had studied about the clay minerals namely smectite presents in the subgrade and contributed to the very bad engineering properties such as strength. It has been observed that once the pavement layer applied, significant damaged occurred. This has been agreed by Ref. [3] who studied the problematic rocks such as shale, weathered limestone and very hard granite which may create problems for design and construction of roads and bridges. In Southern Africa, Ref. [4] had highlighted on how to deals with road subgrade problems which are includes the type of subgrade material (*i.e.* expansive clays, collapsible soils, dispersive/ erodible soils, saline soils, karst area and soft clays) during the project planning stages. Studied shown that it is necessary to identify the problems based on available geological and soil type mapping in order to avoid major cost implications related to later remedial measures. In Turkey, the forest road is a great challenge since it is located on natural ground so that directly affected the physical structures of subgrade. This is due to the exposure to excessive load during wood transportation by heavy trucks [5]. In China, the soft soil subgrade problems have led their researcher to provide a solution by applying a reinforcement technology such as stack overpressure, plastic drainage board combined with stack preloading, cement mixer and others [6]. In Malaysia, peat soil is one of the abundant soft soil material vastly found in east coastal area of Peninsular Malaysia and a study on modifying the peat properties has been carried out to make sure it is ready for a quick construct of road [7, 8].

1.1.2. Nature catastrophic. In nature catastrophic point of view; earthquake damage, storm and flood damage and extreme weather damage are most major concerns for engineer to lay out a conceptual design especially in area which always facing this kind of issues. Lack of attention to proper flood protection and drainage systems in earlier designs contributed in many pavement failures during or after major storm events [3]. In Jammu and Kashmir, Ref. [9] had concluded that the climatic condition and topography are the major issues encountered during the construction and maintenance of roads.

1.1.3. Construction methods issue. The subgrade is actually one of the biggest factors in pavement performance. In order to get the best possible compaction of the subgrade is one of the main requirement to fulfil during the construction to maximize the longevity of the road and minimize the required pavement thickness. However, due to the variation on soils used for subgrade and the challenges involved in measuring the characteristics of these soils, the quality of subgrade compaction is sometimes less than optimal. The key is to ensure that the field compaction values come as close as possible to the laboratory values so the optimal compaction can be achieved.
Drainage problem is the main issue especially for the road that passing through hills since there are no 'catch water drains' to divert and intercept the water from the hill slope [9].

1.2. Subgrade stabilization using additives

1.2.1. Additives selection and usage. The selection of additives actually depends on the soil type also the availability of the additive as the nearest sources from the site. The additives chose must be capable of being implemented under the conditions and with the same equipment as the natural materials they replace, most frequently in various environments.

The interactions in between soil and stabilizer or additives vary with soil type and so does the extent of improvement in soil properties as a result of stabilization. Due to that matter, the efficacy of using an additives must be evaluated prior to the treatment. The bearing capacity values are the most important data for assessing the original subgrade layer. Most of the problematic subgrade type of soil is clay soil, known as a type of soil having poor technical properties. There are several studies have been conducted by adding waste materials such as waste glass [10], fine steel slag [11], cement kiln [12], coir waste [13, 14], silica fume and nano silica [15] and fly ash [16]. In general, most of additives mixed with problematic soil can be assessed in terms of their strength properties as a function both the percentage of additive in the mixture and the time of curing. However, due to the lack of some chemicals to bind the waste materials, some additional binding materials were used such as lime and cement [15, 16]. Due to the demand of a quick treatment for peat soil, studies have been conducted to mix the polymer namely as melamine urea formaldehyde (MUF) resin. The resin had shown a great effect towards the changes of peat soil properties especially in strength behaviour [8, 17]. Figure 3 shown a flow chart of prior to deciding stabilization techniques is classification of selected soil type.

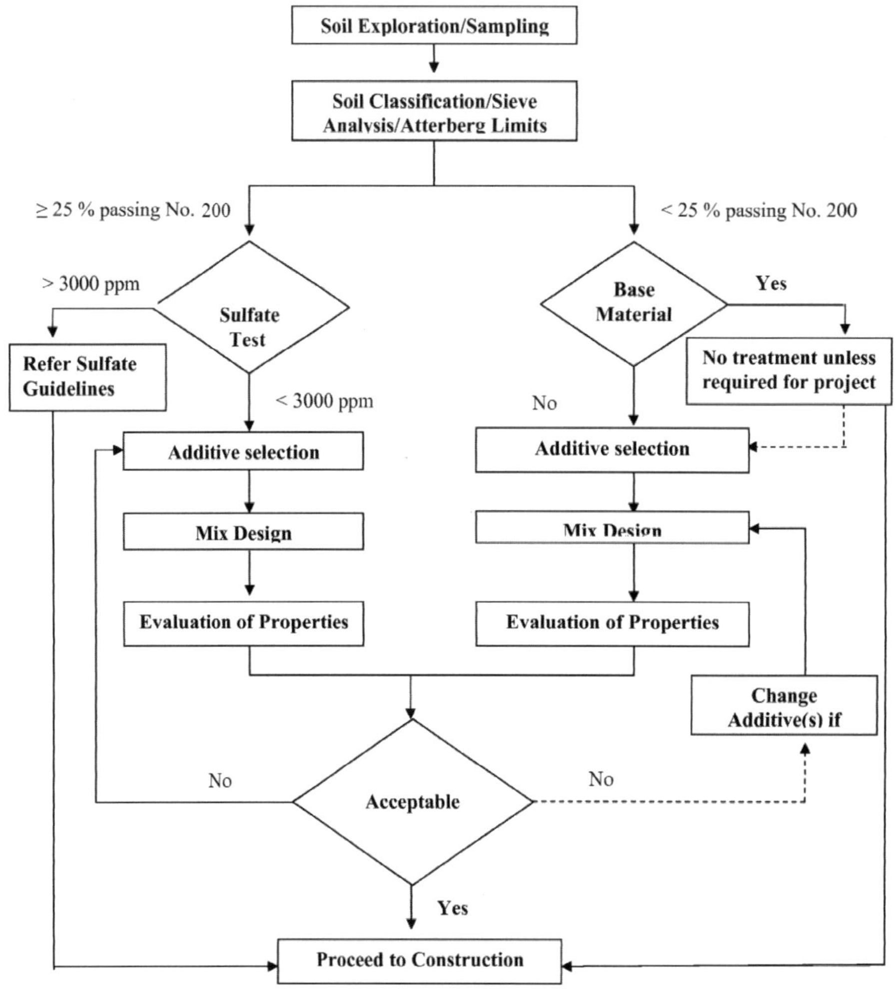

Figure 3. The decision making factor prior to deciding stabilization techniques is classification of selected soil type [18]

According to the flow chart, the main factor is focussed on the native soil classification, either fall in subgrade category or a base category material. The size distribution test should be carried out which is based on the fraction passing No. 200 sieve. If 25% or more of the soil mass passes the No. 200 sieve, then the soil is classified as a soil otherwise, it is classified as a base for stabilization purposes. The second factor to be considered is the presence of and the concentration of sulfate salts and organic materials in the soil especially when using traditional soil stabilizers.

Figure 4 shown the flow chart of decision making for selecting stabilizers to be used in subgrade soil based on the plasticity index (PI) values. The PI values were categorized as less than 15, in between 15 to 35 and more than 35. Soils with a high PI tend to be predominantly clay, while those with a lower PI tend to be predominantly silt. Soils with high plasticity index are highly compressible. Plasticity index is also a measure of cohesiveness with high value of PI indicating high degree of cohesion. Experience shows that soils with high PI are much less desirable for subgrade or base course than those having less indexes. From the chart, it can be summarized that the most appropriate additive for all category of PI is fly ash Class C and F.

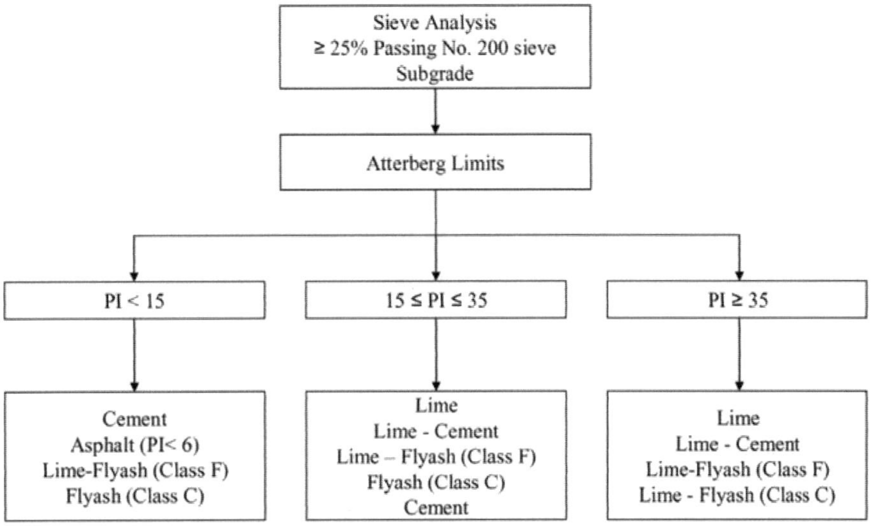

Figure 4. The decision flow chart for selecting stabilizers for use in subgrade soils [18]

1.2.2 Additives application – case study of State Highway 32, Port Washington, Wisconsin [19]

In this subtopic, a step by step procedures were presented in flow chart (Figure 5) and will be discussed, emphasized on how the application and evaluation of additives mixture samples in road/ highway construction.

Laboratory Testing: For this case study, five soil samples were taken from the site and brought back to the laboratory for compaction and California Bearing Ratio (CBR) testing. The soils classify as sandy lean clay, clayey sand or clayey gravel with sand according to the ASTM D 2487. At the same time, the Class C fly ash sample was tested for their chemical compound.

Mix Design: Then the mix design was prepared by applying 10% fly ash mixture due to a great performance of the mixture. The selected design consisting of a 0.3 m thick stabilized layer with a fly ash content of 10% and water content of 12 – 14% (1% wet of optimum water content – standard Proctor effort).

Construction Methods: The construction stage begin by spreading the fly ash onto the subgrade in a 0.1 m thick layer using a lay-down truck that was specially designed for fly ash application with minimal dust generation [20]. A special equipment – road reclaimer was used to mix the fly ash into the subgrade to a depth of 0.3 m. It is equipped with the water boom in front of the mixing tines that is used to add water to the mixture as needed to achieve the desired water content (12 – 14%). Then the mixture was compacted using 4 – 6 passes of a self-propelled tamping foot compactor. Followed by the surface smoothed with a motor grader and compacted again using a vibratory compactor with a steel drum. The compaction process was continued until the dry unit weight of the mixture exceeded 18.8 kN/m^3 (minimum dry unit weight tested in the laboratory). It is estimated that the compaction was completed within 1 to 2 hours of mixing.

The construction of the overlying pavement layers began 1 day after the subgrade was stabilized. The overlying layers consisted of (bottom to top) 0.15 m of crushed aggregate, 0.05 m of recycled asphalt pavement (RAP) and 0.23 m of Portland cement concrete.

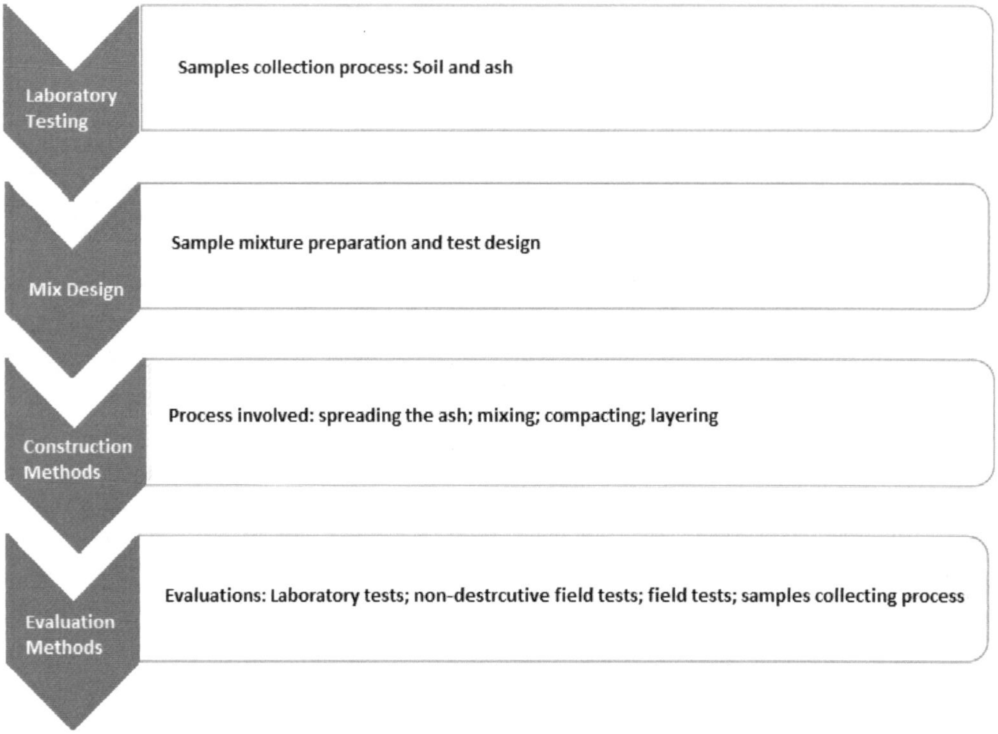

Figure 5. The flow chart of road construction works and testing involved

Evaluation Methods: The evaluation tests started in the laboratory condition when the samples were testing to determine several parameters such as bearing ratio and capacity of the mixture using soil and fly ash collected in the field. Non-destructive field tests were conducted to determine the stiffness and modulus of the subgrade. The field tests were conducted directly on the subgrade using a soil stiffness gauge (SSG) and on the completed pavement structure using a falling weight deflectometer (FWD).

In order to collect samples on the in situ mixture for laboratory testing, two methods were used i) grab sample of the mixture immediately after mixing (similar state in lab condition which is 2 hours after mixing process) and ii) coring the stabilized subgrade after 7 days with a coring tool used to collect samples of concrete.

1.3. Economic valuation in additive usage

High demand on road and highway construction of all over the world especially in a very challenging area when the presents of problematic soils categorised as a weaker subgrade. These type of original subgrade essentially requires thicker layers whereas stronger subgrade goes well with thinner pavement layers.

Ref. [21] had studied on the economic point of view for soil stabilization in road making industry in India. The pavement thickness and saving of thickness due to stabilizers usage also had been explained. It is found that the cost of lime and lime and fly ash including haulage per kilometre of road distance reduce compare to without stabilization design. The road thickness also shown a reduction/ saving when the additives applied to the subgrade. Table 1 tabulated the comparison of cost values and savings of thickness for a lime and lime mixed fly ash additives usage to stabilize subgrade soil.

Table 1. Comparison of cost analysis for lime and lime mixed fly ash of soil stabilisation

Subjects	Lime 5%	Lime 5% and Fly Ash 35%
Quantity	53.5 tonnes	125.5 tonnes
Cost of lime per tonne	RS 3000	RS 1500
Cost of lime stabilization per km	RS 160, 500	RS 188, 280
Approx. of lime-soil blending cost	RS 180, 000	RS 200,000
Saving in road thickness	240 mm	217 mm
Saving in material	720 m^3	651 m^3
Cost of compacted	RS 290/m^3 – RS 540/m^3	RS 290/m^3 – RS 540/m^3

2. Conclusions

This paper had presents and discussed the important aspects of subgrade stabilization focusing on the application additives such as chemical, waste materials, fly ash and others by reviewing previous studies. These include i) the subgrade issues, ii) factors that influence the stabilization technique, iii) application of additives to stabilize the subgrade and iv) the economic valuation of additive usage. In general, the application of additives to stabilize subgrade have a good impact as an alternative to reduce cost and at the same time to make sure the sustainability of stabilized subgrade.

3. References

[1] Seehra S S 2008 *Practical Problems of Highway Construction in Black Cotton Soil Areas and In-Place Remedial Measures: A Case Study* Retrieved from: https://www.nbmcw.com/tech-articles/roads-and-pavements/307-practical-problems-of-highway-construction-in-black-cotton-soil-area.html

[2] Marino G G 2018 *Soil Provides Poor Road Construction Support (Engineering Update, Marino Engineering Associates, INC)* Issue 7 Retrieved form: http://www.meacorporation.com/wp-content/uploads/project-bulletin-07.pdf

[3] Khandker A A, Anagnostou J T and Deo P 2013 Geotechnical challenges in highway engineering in twenty first century: lessons from the past experiences and new technologies *7th International Conference on Case Histories in Geotechnical Engineering* Chicago, 29 April – 4 May 2013

[4] Paige-Green P 2008 Dealing with road subgrade problems in Southern Africa *The 12th International Conference Of International Association For Computer Methods And Advances In Geomechanics (IACMAG)* India, 1-6 October 2008 pp 4345 - 53

[5] Eroglu H, Acar H and Eker M 2003 An evaluation of the forest road superstructure in Turkey *XII World Forestry Congress 2003* Quebec City, Canada

[6] Cheng L and Zhai X 2015 Highway soft soil subgrade treatment research *International Journal of Engineering Research And Applications* **5** (6) 7-10

[7] Nik Daud N N, Daud M N M and Muhammed A S 2018 Rice husk ash (RHA) as a partial cement replacement in modifying peat soil properties *International Conference on Engineering and Technology (Intcet 2017)* Putrajaya, Malaysia

[8] Daud M N M and Nik Daud N N 2017 Effect of wet and dry condition of muf polymers on strength properties of treated peat soil *Proceedings of The 1st Global Civil Engineering Conference*

[9] Rashid M, Bhat S H and Bahsir I A 2017 Road construction, maintenance challenges and their solutions in kashmir *Irrigation & Drainage Systems Engineering* **6**(1) 1-5

[10] Nebojsa D, Zoran B and Prolovic V 2012 Waste glass as additive to clayey material in subgrade and embankment of road pavement *Facta Universitatis Series: Architecture and Civil Engineering* **10**(2) 215-22

[11] Aldeeky H and Al Hattamteh O 2017 Experimental study on the utilization of fine steel slag on stabilizing high plastic subgrade soil *Hindawi: Advances in Civil Engineering* 1-11

[12] Mosa A M, Taher A H and Al-Jaberi L A 2017 Improvement of poor subgrade soils using cement kiln dust *Case Studies in Construction Materials* **7** 138 – 43

[13] Peter L, Jayasree P K, Balan K, and Raj A 2016 Laboratory investigation in the improvement of subgrade characteristics of expansive soil stabilised with coir waste *Transportation Research Procedia* **17** 2016 558 – 66

[14] Nik Daud N N, Jalil F N A, Sadiq M A and Azmi J A 2016 Influence of agricultural wastes on shear strength properties of soil *MATEC Web of Conference* **47** 1-7

[15] Mostafa A E A, Ouf M S, and Elgendy M F 2016 Stabilization of subgrade pavement layer using silica fume and nano silica *International Journal of Scientific & Engineering Research* **7**(3) 573 – 81

[16] Athanasopoulou A and Kollaros G 2011 Use of additives to improve the engineering properties of swelling soils in Thrace, Northern Greece *WIT Transactions on Engineering Sciences, Material Characterisation V* **72** 327- 38

[17] Nik Daud N N and Daud M N M 2015 Characterisation of peat soil treated with polymer resin by unconfined compressive strength test *Journal of Advanced Civil Engineering Practice and Research* **1**(1) 11 – 17

[18] Little D N and Nair S 2012 *Recommended Practice for Stabilization of Subgrade Soils and Base Materials* (The National Academies Press)

[19] Trzebiatowski B D, Edil T B and Benson C H 2004 *Case Study of Subgrade Stabilization using Fly Ash: State Highway 32, Port Washington, Wisconsin* (Baltimore: Recycled Materials in Geotechnics, American Society of Civil Engineers) pp 123 – 36

[20] Bin-Shafique S, Benson C and Edil T 2002 *Leaching of Heavy Metals from Fly Ash Stabilized Soils used in Highway Pavements (Geo Engineering Report* 02-14) (Wisconsin: University of Wisconsin-Madison)

[21] Sood S and Bansal S 2015 Evaluation of construction materials for soil stabilization in road making industry – a techno economic study *International Journal of Engineering Research & Technology* **4**(5) 404 – 10

Acknowledgement

The authors would like to thank the Department of Civil Engineering of Universiti Putra Malaysia, Malaysia and Department of Civil Engineering of Ataturk University, Turkey. Also to all the people whom are directly and indirectly involved in this study.

IOP Conf. Series: Materials Science and Engineering **512** (2019) 012006 doi:10.1088/1757-899X/512/1/012006

The usage of cement for soil stabilisation in construction of low volume roads in Malaysia

Roziawati Razali and Mohamad Shukor Che Malek

Road Branch, Public Works Department Malaysia, Kuala Lumpur, Malaysia

E-mail: roziawati.jkr@1govuc.gov.my & mshukor.jkr@1govuc.gov.my

Abstract. There are about 25% of the total road networks in Malaysia is an unpaved road which is contained gravel or earth road [1]. These are commonly referred to as the low volume roads or unpaved rural road or agriculture roads which required a very minimum standard of design and construction due the low traffic impact and function of the road as an accessibility tool to the local communities. Difficulties in transporting of the materials to the construction site due to the distance from the quarry or construction material source do occur together with the requirement of suitable machinery. Soil stabilisation was identified as an alternative method to facilitate the design and construction requirement in achieving the implementation of the project where respect to cost, time and quality. Currently, soil stabilisation method was not popular in Malaysian construction industry players especially among road designer because the conventional method of pavement design always been the most priorities. Perhaps this may due to the absence of the specification purposely for soil stabilisation works. This paper will present the Malaysian experiences of implementing cement as soil stabiliser in road construction project for rural and low volume roads. The study had been conducted in the road construction project of Jalan Pos Sinderut, Kuala Lipis, Pahang. The scope of the study covers all the stages that involved in the implementation of soil stabilisation using cement this project including design, procurement, and construction.

1. Introduction

There are about the total of 216,837 kilometers of roads length in Malaysia including federal and state roads. From this total length of road, there is 52,801 kilometer of unpaved roads that consist of earth and gravel roads [1]. Majority of the roads are falling into the low volume roads category and usually surrounded by rural development as it promotes access to economic and social services, generating agricultural income and productive employment. Long-term performance of pavement structure is significantly impacted by the stability of the underlying soil usually refer to in-situ subgrade that often not achieve the required level of performance under traffic loading and environmental demands [2]. Soil stabilisation can be used to treat the several inches of soil or aggregate surface or other properties of the in-place soil that not achieves the required level of anticipated traffic. Soil stabilisation method also helps to reduce the construction cost by reducing the hauling and transportation of import materials on site due to recycling of the existing materials. The construction of soil stabilisation also required the minimum type of construction machine so that the problem of accessing the machine to the construction site especially at the remote area will be ease.

Basically, soil stabilisation is the alteration of any inherent property of a soil to improve its engineering performance. It is the treatment of natural soil to improve the engineering strength and properties [3]. The most commonly used stabiliser in road construction has been traditionally cement, lime or asphalt products but due to the improvement and innovation of the product by the industry, the non-traditional stabiliser product was developed such as chloride, resins, sulfonated oil, synthetic polymer and various biological enzymes[2],[4].

Currently, the used of soil stabilization in Malaysia is still at the initial stage and not so common compared to the conventional construction method such as using crusher run and others typical road construction materials. The application of soil stabilisation in road construction was still limited to the certain purpose maybe due to the deficiency of the contractor's competency, uncertain about the effectiveness of the technique for the large scale of project and also lack of the suitable machinery for the purpose of soil stabilisation works. Application of soil stabilisation method in Malaysia usually for construction of estate or plantation roads and rural road but not in major roads. The roads were designed and constructed based on the low traffic volume but sometime it will do subjected to heavy load. Due to the benefits of having stabilisation technique is able to increase the existing subgrade strength with economical cost, the government of malaysia through public works departments together with ministry of rural and regional development has decide to use the method in the road construction for the rural area. The construction of the road using soil stabilisation technique was studied and monitored to identify the suitability of the method, the effectiveness of application and the performance compare to conventional method. The study also managed to observe and record all the activities included during planning, designing, procurement, and construction.

1.1. Objective of the study

The objectives of this study are as follows:

i. To identify the effectiveness of using soil stabilisation method in road construction for low volume and rural road.
ii. To observe the preferable of machinery to be used in soil stabilisation works in the rural area.
iii. To identify the performance of soil stabilisation compared to conventional technique.

2. Background of the project

The Project of Menaiktaraf Jalan di Pos Sinderut, Dun Jelai, Cameron Highland, Pahang was one of the projects in the 10th Malaysian Plan under the provisions of the Ministry of Rural and Regional Development. The Public Works Department has been appointed as an implementor for the project. The scope of the project has comprised the upgrading of 35 kilometers of an existing road according to JKR R1 Standard. An existing road condition was earth roads without proper drainage and its result to the severe damage on road surface condition and has poor driving quality. This existing route is used by indigenous people as a medium of communication which is only suitable for pedestrians, motorcycles or four-wheel drive vehicles. Figure 1 below showed the existing road conditions at the project site before the construction.

Figure 1. The condition of an existing road before construction.

The project site is located in Kuala Lipis District, Pahang Darul Makmur as shown in Figure 2. Through the result of in situ California Bearing Ratio (CBR) that had been conducted on the existing subgrade before start the construction works, it was identified 12 locations of soil stabilisation as showed in Table 1 and Figure 3.

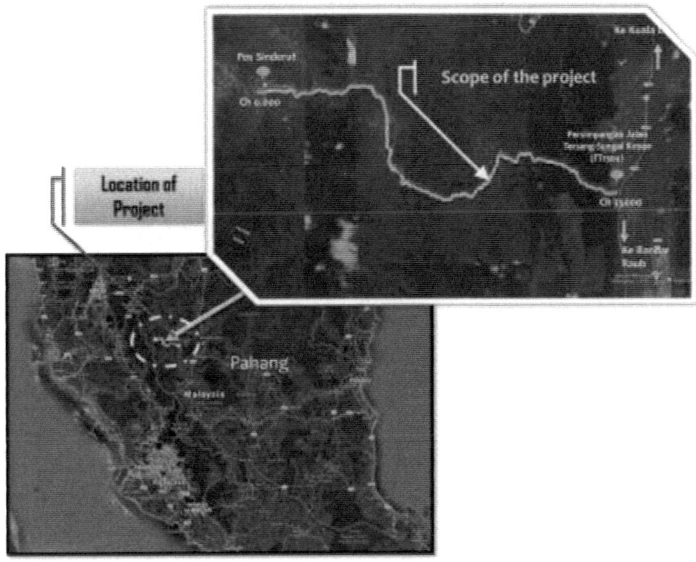

Figure 2. Location of the project.

The selected sections of soil stabilisation are represented as the lowest subgrade value of CBR strength. Site investigation including mackintosh probes, hand auger was conducted to determine the existing soil profiles and its bearing capacity meanwhile soil classification test have been carried out in the laboratory in order to obtain the soil properties. Generally, the existing ground soil consists of yellowish brown sandy clay with little gravel. The average of in-situ California Bearing Ratio (CBR) test base on BS1377 for the soil stabilised sections are ranges of 8 – 20% when tested using Dynamic Cone Penetrometer (DCP).

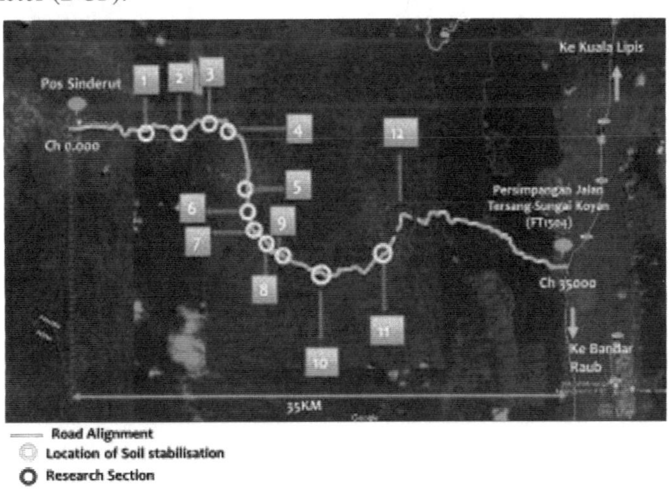

Figure 3. Location of soil stabilisation and research section.

Table 1. Location of soil stabilisation section.

Section	Start CH	Finish CH	Length (m)
1	5875	6285	410
2	6650	7850	1200
3	8650	8950	300
4	9651	9951	300
5	11950	12350	400
6	13170	13904	734
7	14150	14950	800
8	15350	15550	200
9	16744	17044	300
10	19419	20127	708
11	21724	23074	1350
12	23816 (research section)	25416	1600

The research section (Section 12) was divided into four (4) segment which is each segment was stabilised with a different percentage of the stabiliser. Initially, there are four (4) types of soil stabiliser was selected to be incorporated with an existing soil material on site for this study. The soil stabiliser namely portland cement, bitumen emulsion, hydrated lime and cement/fly ash blended were initially selected as the soil stabiliser to be used in this project. From the test report and analysis, it was found that the used of bitumen emulsion and lime was not suitable with the existing soil and it does not shows any significant improvement on the strength of the soil.

The mix design report also showed that the used of four percent (4.0%) of cement as soil stabiliser was achieved the minimum requirement of the strength of the stabilised layer. With that, 4% cement was applied on the selected of 11 sections except for research section. For study purposes, the research section at Section 12 was divided into four segments which is each segment are applying 3%, 4%, 5% and 6% of cement as showed in Figure 4 to identify how the soil stabilised perform in actual field when the content of the cement content was insufficient or more than the optimum content.

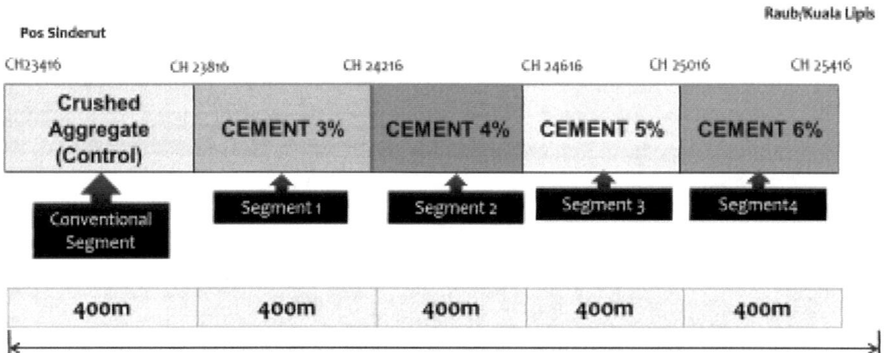

Figure 4. Segmentation of soil stabilization research section.

3. Structural design

The pavement structure was designed according to Arahan Teknik Jalan 5/85 (Pindaan 2013) and Design Guide for Alternative Pavement Structure Low Volume Roads [6],[7]. The pavement was designed base on the requirement of local traffic necessity which is in the range of low traffic volume. There are two types of structural designs which are conventional design using crusher run as road base material and stabilised an existing subgrade soil to became road base as showed in Figure 5 and 6 below. Both surface structures were paved using double-surface dressing layer.

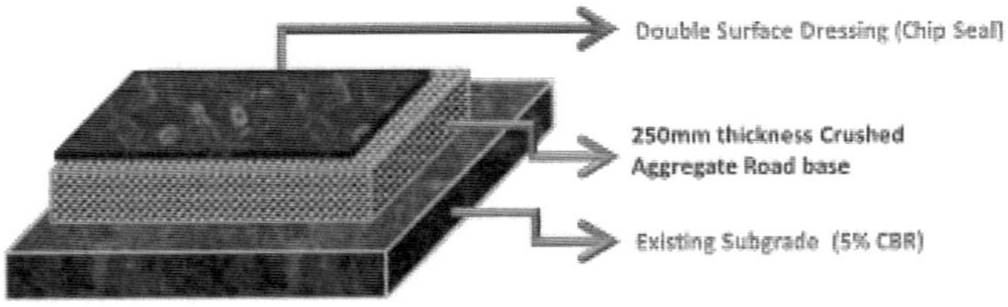

Figure 5. Conventional structure design.

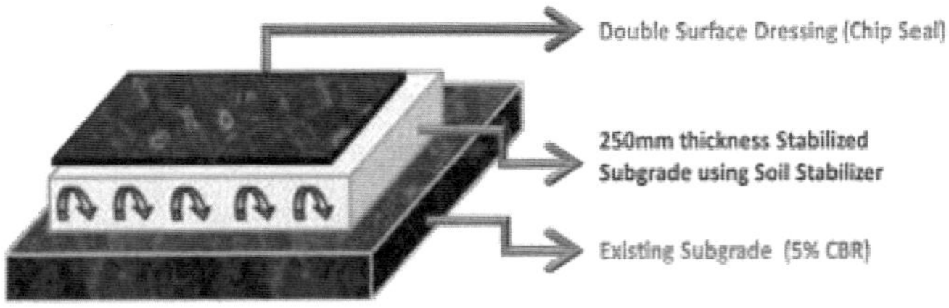

Figure 6. Soil stabilisation structure design.

4. Construction

The construction of soil stabilisation project using cement as the stabilising agent for this project was successfully completed by early 2016. The construction of soil stabilisation section started from site preparation to ensure that the soil surface is free from any foreign material such as grass, sticks, and rubbish. The prepared stretch should be compacted and free from boulders rocks or any other hard material to the designed depth and accessible by the stabilisation machinery. The stabilisation work was including spreading of the binding agent, mixing, compacting, shaping, testing and curing. Supply of cement in bulk or bags will be transferred into the purposed built spreader. After stabilisation areas have been prepared, the required quantity of cement was spread over the area. To achieve uniform spreading in line with the mixing requirement, the cement was spread in areas using a mechanical spreader. During the mixing process, the stabiliser machine was used to mix the existing soil materials with cement and water to the design thickness. The purpose of this mixing pass, however, is to incorporate moisture so that the required chemical reaction can take place so that even strength gain is achieved. The optimum moisture content was achieved by a connected water truck with the stabiliser machine.

After the mixing process, the motor graded was used to grade and form the desired cross fall as well as achieving finished surface levels. Finally, the vibratory compactor was used to compact the stabilised layer to achieve the desired density. The stabilised layer need to be cured before it was paved with double surface dressing.

The application of surface dressing started with spraying the polymer modified emulsion on the stabilised layer surfacing and then the single size of stone chipping precoated with adhesion agent was spreading through the road surface in two layers. The construction process was showed in Figure 7 below.

Figure 7. Soil stabilisation process.

5. Testing

5.1. Soil parameter

Laboratory and field testing was carried out before and after construction. Laboratory testing consisted of material testing to identify the characteristic of the soil and mix design to identify the suitable type stabiliser, the content of stabiliser agent and the optimum moisture content. The parameter of the soil materials was showed in the Table 2 below.

Table 2. Result of soil testing.

Soil parameter	Average
California Bearing Ratio (CBR), %	16
Plastic Limit	39
Plastic Index	33
Optimum Moisture Content (OMC), %	20
Maximum Dry Density	1.672
Passing 0.075mm sieve size	64%
Soil type	Yellowish brown sandy clay with little gravel

5.2. Mix Design

Initially, it was suggested to apply several types of conventional soil stabilizer namely cement, lime, and bitumen emulsion. Referring to the guide of selection of the stabilising agent for various plastic index value and content of fine. Figure 8 it showed that the soil material was usually suitable to be stabilized using lime.

IOP Conf. Series: Materials Science and Engineering **512** (2019) 012006 doi:10.1088/1757-899X/512/1/012006

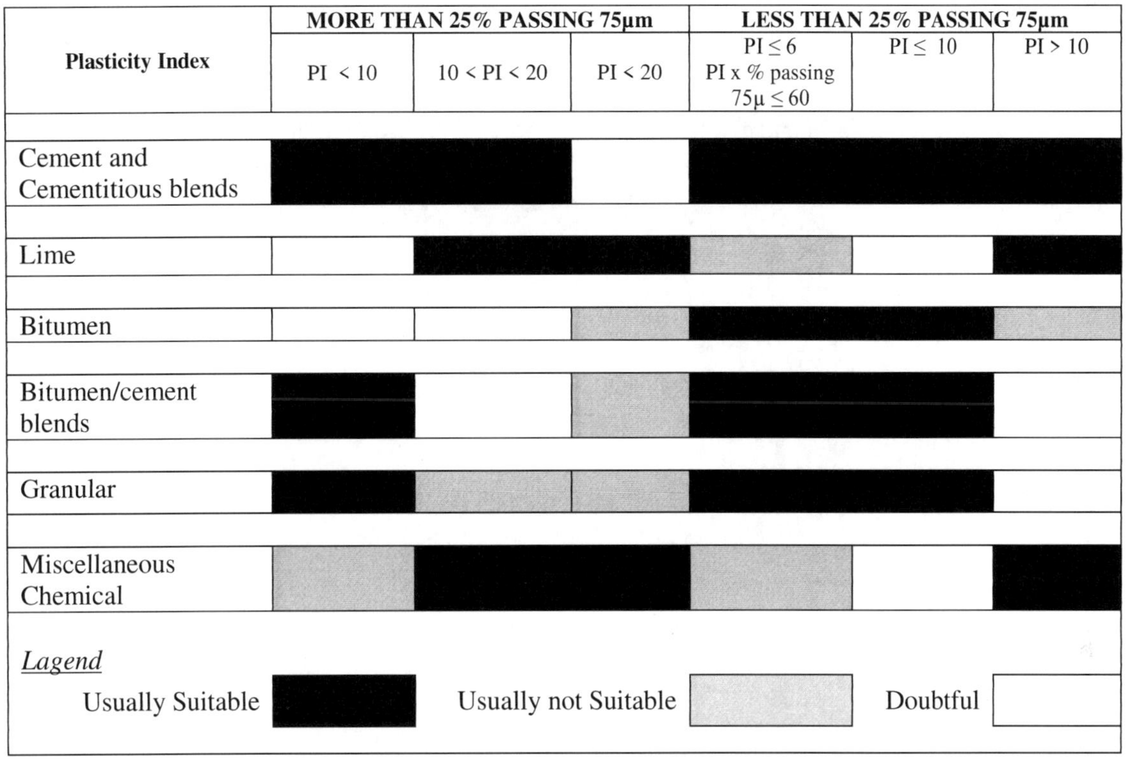

Figure 8. A guide to the selection of the stabilising agent for various PI values and content of fines [5].

However, the result from the initial mix design test showed that the existing soil material on site was compatible stabilised using cement compared to lime and bitumen emulsion. The second stage mix design process was carried out using Ordinary Portland Cement to identify the optimum cement content, optimum moisture content and maximum dry density of the soil stabilised mixture.

The result of the mix design was showed in Figure 9 and 10 below. The mix design samples were tested for Unconfined Compressive Strength (UCS) and California Bearing Ratio test in the laboratory. Sample 1, Sample 2 and Sample 3 were represented the different locations of the sample taken from the construction site. An existing material from the site was stabilised with 4%, 6% and 8% by mass of Ordinary Portland Cement. The Unconfined Compressive Strength (UCS) was carried out after cube molds cured for 7 days according to B.S. 1881. California Bearing Ratio (CBR) according to ASTM D 1883-07 test procedure also was carried out to identify the optimum stabiliser content in the mix.

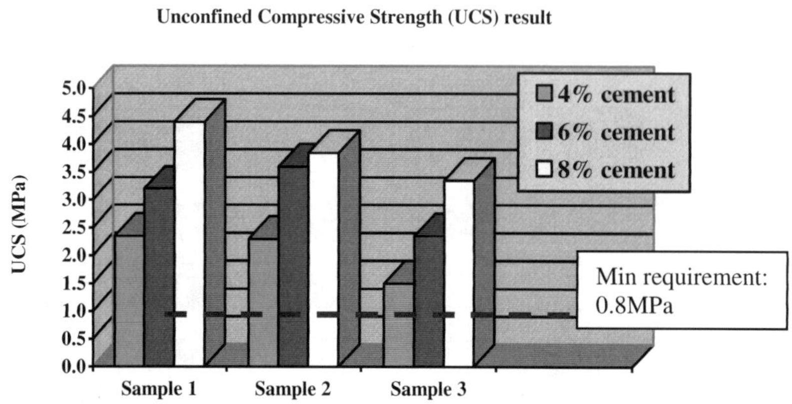

Figure 9. Unconfined Compressive Strength (UCS) result.

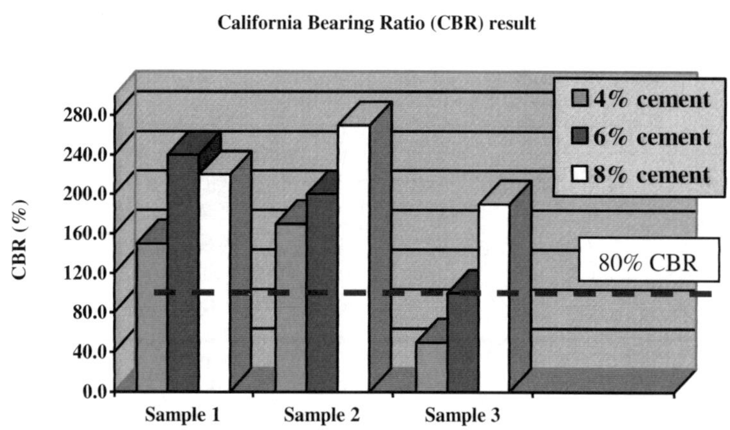

Figure 10. California Bearing Ratio test result (CBR).

From the report, it showed that most of the samples were complied with the minimum requirement of UCS and CBR values as stated in the specification which is respectively 0.8MPa and 80%. All CBR result comply with the specification except for sample 3 at 4% cement that suspected due to handling error since the property description and proctor results are consistent with other samples. It was concluded that all the test for the mix design was achieved the minimum requirement of soil stabilised strength at 4% cement content. Therefore, 4% of cement content by dry mass was applied for soil stabilisation project at the project site.

6. Result & analysis

Field testing was carried out to identify the performance of the stabilised pavement structure before and after construction. The performance of soil stabilised pavement structure was monitored before and after construction by conducting the field testing apparatus using Clegg Hammer and Lightweight Deflectometer.

The result in Figure 11 and 12 obviously showed that the used of cement increase the CBR value and the modulus of the existing subgrade material. The value of subgrade CBR was increased up to 400% after one month stabilised while the E-Modulus was increasing up to 200% after one month.

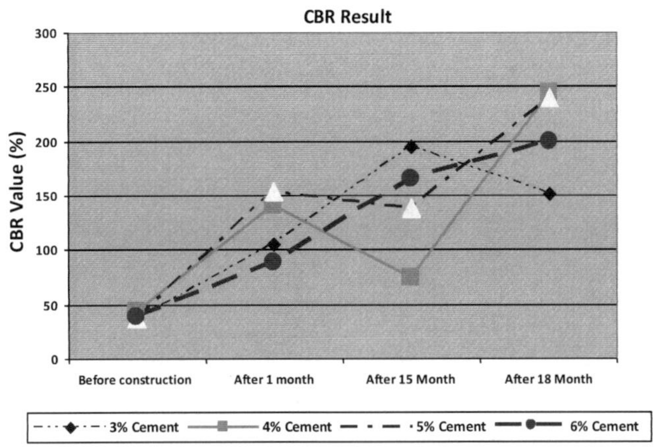

Figure 11. California Bearing Ratio test result (CBR).

From the result showed in Figure 11, the value of Subgrade was increased for all section after 15 months stabilised except for the section of 4% and 5% cement. This is maybe the effect of the slope failure that occurred during this due to high water table along with this section but the value of CBR was still above the value of CBR after 1-month construction.

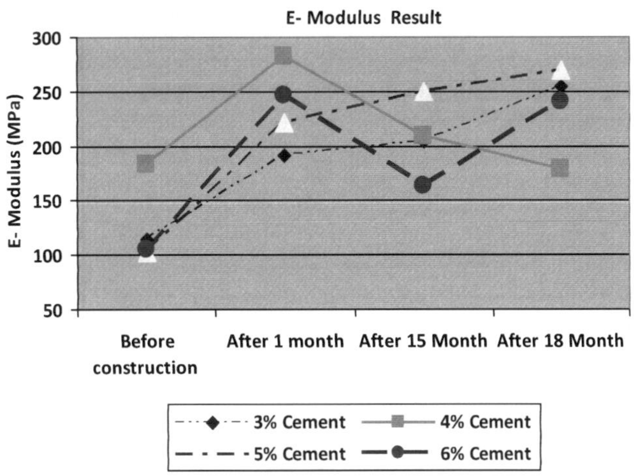

Figure 12. E-Modulus Result.

Field testing at Control Section was carried out after 12-month construction and it was showed that the Modulus value was 162MPa when tested using Lightweight Deflectometer. This result showed the increasing of Modulus value which is about 120% compared to the Modulus value of subgrade before construction.

From the surface condition survey that had been carried out randomly at site, it was showed that the most of area of the stabilised section still in good condition compared to the conventional section. Figure 13 showed the condition of the site at the stabilised section and conventional section after 18-month construction. The pavement was overlying with double surface dressing as the final surfacing throughout the length. Figure 13 showed that the surface materials of both sections peeled off and exposed the lower layer. This is may be due to the poor construction quality and quality control during construction of surface dressing itself. Therefore, the road surface which is consisted of the unbound layer was exposed to the water penetration became breakable to the high surface loading.

Conventional Section **Stabilised Section**

Figure 13. The condition at the site after 18 months.

7. Conclusion

Based on the findings, result, analysis, observation of the implementation of soil stabilisation technique in this project, it can be concluded as followed:

i. To ensure that a uniform and the homogenous stabilised mix was achieved, it very important that the suitable machine became necessary but the condition of the remote construction site area may cause the difficulty on bringing the machine inside. It was suggested to allow the use of any simple machine or purpose-built machine as long as it can provide the homogenous stabilised mixture.
ii. The mix design of for soil stabilisation should be always to be carried out at every project due to the difference of soil properties required different types and content of stabiliser agent.
iii. Soil stabilisation using cement increase the existing subgrade after construction complete and performed well compared to the conventional method.

iv. It was suggested to have a particular Malaysian design manual and specification in soil stabilisation to assist the engineer and construction industry to have a better design and quality control regarding soil stabilisation.
v. The implementation of soil stabilisation technique should be extended to the higher hierarchy standard of the road such as federal or state road.

8. References

[1] Cawangan Senggara Fasiliti Jalan 2016 *Buku statistik jalan edisi 2016* (Kuala Lumpur: Jabatan Kerja Raya).

[2] Little D N and Nair S 2009 *Recommended practice for stabilization of subgrade soils and base materials - Final task report for NCHRP project 20-07* (Texas: National Cooperative Highway Research Program, Transport Research Board of the National Academies Texas Transport Institute, Texas A&M University College Station).

[3] Garber N J and Hoel L A 2001 *Traffic and highway engineering third edition* (United State of America: University of Virginia).

[4] Maureen A K 2009 *Stabilization selection guide for aggregate and native surface low volume roads* (U.S Department of Agriculture: National Technology & Development Program)

[5] REAM 2010 *Manual on cold in place recycling - Workshop 1* (Kuala Lumpur: Road Engineering Association of Malaysia)

[6] JKR 2013 *Manual for the structural design of flexible pavement ATJ 5/85 (Pindaan 2013)* (Kuala Lumpur: Jabatan Kerja Raya Malaysia)

[7] Cawangan Kejuruteraan Jalan dan Geoteknik 2012 *Design guide for alternative pavement structures low volume roads*. (Kuala Lumpur: Jabatan Kerja Raya).

IOP Conf. Series: Materials Science and Engineering **512** (2019) 012007 doi:10.1088/1757-899X/512/1/012007

Use of geophysics for site investigations and earthworks assessments

M Azrief Azahar, N Farhan Zakiran Mahadi, Qusanssori Noor Rusli, N Narendranathan and E C Lee

Infra Tech Geo Solutions (M) Sdn Bhd, Petaling Jaya, Selangor, Malaysia

Email: azrief@infratechgeo.com

Abstract Site investigations for various infrastructure is traditionally done using boreholes drilled through various strata, logging of boreholes, sampling of strata and laboratory testing to assess engineering properties and infer behaviour of slopes, soft soils, embankments and foundations. Geophysics is still not used as a standard tool by most engineers due to the perception that it is not well-established science or that the results are unreliable. This paper outlines how Engineering and Geophysics can be integrated by having a well-balanced team with all the team members understand the fundamentals of each area. This is demonstrated by case histories for site investigations and earthworks quality assessment. The variety of geophysical methods available for site investigations, their applications, interpretations are presented. The paper describes how cost of investigations can be reduced and the risk of ground variations can be reduced by applying geophysics in geotechnical investigations.

1. Introduction

In typical infrastructure projects such as roads, airports, pipelines etc. the cost of underground works can be between 20% to 40% of the project cost. However, world over the main cause of cost and time disputes is the fact that the ground conditions were found to be different to what was presented in the tender using limited number of boreholes for site investigations.

The use of geophysical investigations supplemented with selective intrusive investigations at critical locations offer an economical means of obtaining an assessment of the subsoil strata. Echo sounding, seismic wave measurements, electromagnetics and microgravity are some techniques that have been used with success as they can provide indication of soft sediments and hard layer profiles. The data obtained from intrusive investigations such as boreholes and Cone Penetration tests are typically used to relate geophysical data with observed engineering properties of the ground. These are used to develop a 2-D and 3-D model of the subsoil profile to show the variable distribution of the various subsoil layers.

2. Geophysical techniques

Geophysical method applies the principle of physics, which is used in studying the earth. Some of the geophysical methods that are commonly used are resistivity, seismic, gravity, magnet and electromagnet. According to [1], geophysical technique offers the chance to overcome some problem inherent in more conventional ground investigation techniques. In most equipment testing, the geophysical methods apply a non-destructive testing which can reduce cost and time of the project [4].

There are several other specialised geophysical techniques, but all of these are not covered in this paper.

An endemic issue in the civil engineering industry is that Civil Engineers are not trained in the principles of geophysics or even possess in-depth knowledge of geology. The similar applies to geologists and geophysicists whereby their disciplines do not adequately cover the type of challenges civil and geotechnical engineers face in projects. Hence these disciplines tend to work in silos with in dequeue appreciation of how each expertise area can contribute to solving their challenges.

2.1 Seismic techniques

Seismic techniques utilise the ways in which vibrations travel through materials. The surface wave techniques measure the density of the materials below the surface, irrespective of any strength inversions. The Multi-Channel analysis of surface waves (MASW) technique as in Figure 1 measures the surface waves generated from a source, such as sledge hammer. The user measures the propagation velocities of those surface waves with an array of geophones, and then finally deduces shear-wave velocity (Vs) variations below the surveyed area that is most responsible for the analysed propagation velocity pattern of surface waves [2] as in Figure 1. Shear-wave velocity (Vs) is one of the elastic constants and closely related to Young's modulus and therefore is beneficial to geotechnical engineers [3] as in Table 1. Under most circumstances, Vs is a direct indicator of the ground strength (stiffness) and therefore is commonly used to derive load-bearing capacity [2].

Figure 1. Data acquisition at site using MASW technique

Other techniques that make use of surface waves are the CSWS (Continuous surface wave system) [5] and SASW (Spectral analysis of surface waves) [6]. These techniques both produce 1D profiles using a small array of geophones with the CSWS having the advantage of the ability to control the wave frequency through a vibrating plate. This element of control allows the operator to focus measurements on certain depths, such as depths that might correspond to a bedrock interface and gain a higher resolution of data.

IOP Conf. Series: Materials Science and Engineering **512** (2019) 012007 doi:10.1088/1757-899X/512/1/012007

Table 1. Relationship between geological material and shear wave velocity (Vs) based on SASW measurement

Classification of Geological material	Shear wave velocity, Vs (m/s)
Very Soft	84 – 107
Soft Soil	107 – 137
Moderately soft soil	137 – 183
Hard Soil	183 – 274
Very hard/stiff soil	274 – 366
Highly weathered rock	366 – 610
Slightly / moderately weathered rock	610 – 2743

2.2 Resistivity techniques

This technique also known as electrical imaging as in Figure 2 which aim to build up a picture of the electrical properties of the subsurface by passing an electrical current along many different paths and measuring the associated voltage [7]. From these measurement, the true resistivity of the subsurface can be estimated. The ground resistivity is related to various geological parameters such as mineral and fluid content, porosity and degree of water saturation. Sedimentary rocks are usually more porous and have higher water content, thus they normally have lower resistivity values compared to igneous and metamorphic rocks as in Figure 3. Unconsolidated sediments (overburden) generally have even lower resistivity values than sedimentary rocks, the resistivity value is dependent on the porosity as well as the clay content.

Figure 2. Data acquisition at site using Resistivity technique

(a)

(b)

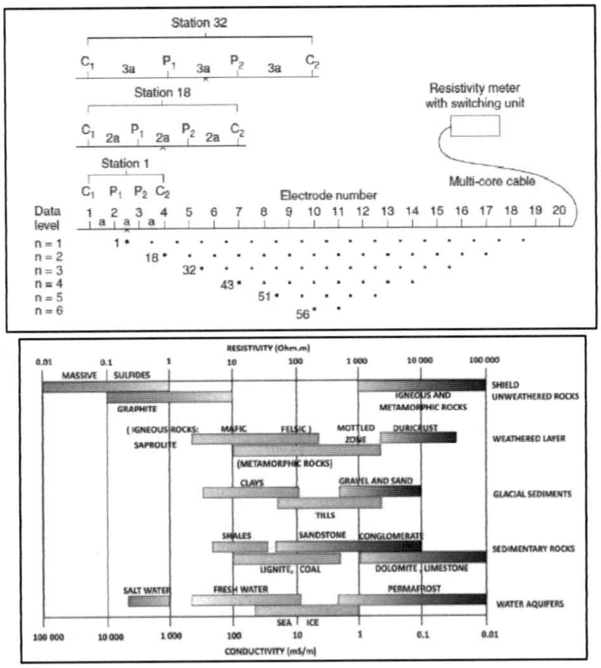

Figure 3. (a) The arrangement of the electrode for 2-D electrical survey and the sequence of measurements used to build up a pseudo section, **(b)** Resistivity and conductivity values for some of the earth materials. (after Palacky, 1987)

2.3 Gravity techniques

Gravity survey as in Figure 4 is a measurement of the gravitational potential field in a series of different locations. The objective of this survey is to relate the density differences to anomalous gravity changes [8]. The gravity anomaly depends on densities, density relates volume to mass and the variation can cause different in mass and gravity acceleration as in Figure 5. Any geological condition that results in a horizontal variation in density will cause a horizontal variation in gravity. The anomaly gravity changes show horizontal density differences of subsurface rocks or materials and could be used to determine the subsurface structure [9].

Figure 4. Data acquisition at site using gravitometer

In engineering and geotechnical applications, gravity surveying is used to locate natural cavities (such as caves, caverns) and man-made subterranean openings such as abandoned mines, tunnels). Natural cavities may be air-filled, water-filled, soft-sediment filled or partially filled.

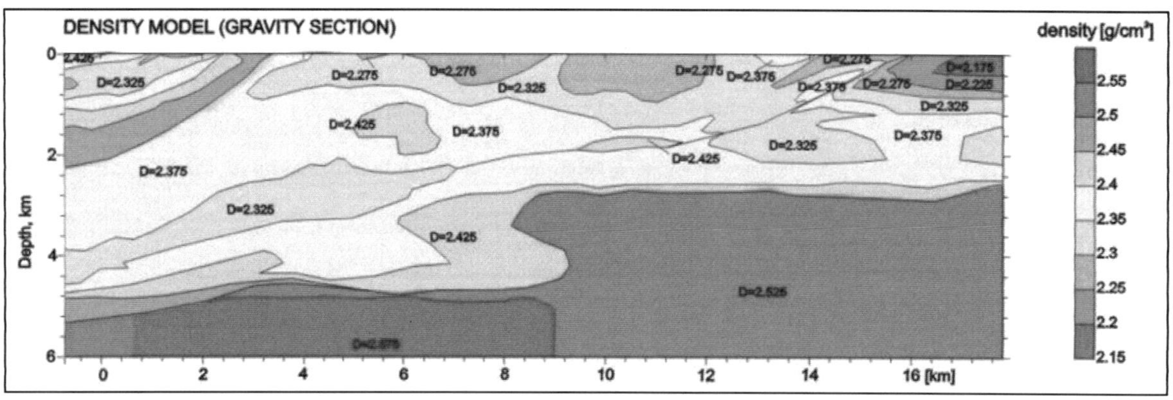

Figure 5. Shown below is an example of a study using gravity measurements for a residential development to check for cavities.

IOP Conf. Series: Materials Science and Engineering **512** (2019) 012007 doi:10.1088/1757-899X/512/1/012007

2.4 Borehole geophysics

Downhole geophysical logging is a technique used to determine the physical properties and distribution of soil and rock surrounding a borehole annulus. These measurements record naturally occurring physical phenomena or they may use an artificial physical source such as electrical, nuclear and acoustic, to excite the medium and measure the response to the excitation. Borehole geophysics can be used to obtain valuable data including information on geological conditions and in-situ physical parameters in drill holes. The amount and benefit of this information is determined by longing suite, borehole conditions, geological parameters, interpreter's experience and application of current technology as in Figure 6.

The choice of appropriate geophysical methods requires an understanding of the geological environment and the borehole conditions. Very seldom can a geological property be identified from a single geophysical log (BS7022, 1988). It usually involves a combination of methods to identify each property of interest.

	Nuclear				Electrical						Acoustic	Seismic wave		Structure	Electro-magnetic
Information required / Log measurement	Natural Gamma	Gamma Spectroscopy	Gamma Gamma	Neutron-Neutron	Focused Electrode	Focused Induction	Dipmeter	Non Focused Electrode	Single Point Resistance (SPR)	Spontaneous Potential (SP)	Acoustic Borehole Televiewer	Suspension PS	Downhole Seismic	Caliper (3-arm or 4-arm)	Ground Penetrating Radar
Bulk Density			A												
Correlation	A	A	A	A	B / D	B & P	B / D					B / D	B / D		B & P
Diameter / Hole Volume											B / D			B	
Elastic Properties			A									B / D	B / D		
Fluid Flow															
Formation Dip and Direction							B / D				B / D				
Formation Fluid Quality					B / D	B & P		B / D		B / D					
Formation Resistivity					B / D	B & P		B / D							
Fractures							B / D				B / D	B / D	B / D		B & P
Lithology	A	A	A	A	B / D	B & P		B / D		B / D		B / D	B / D		B & P
Permeable Zones						B & P		B / D		B / D				B	
Porosity			A	A	B / D							B / D	B / D		
Shale Content	A	A		A						B / D					
Uranium / Thorium / Potassium	A	A													
Water Level					B / D				B / D						

LEGEND :

- A — No restrictions on hole
- B — Open hole
- B & P — Open hole or with plastic casing
- D — Fluid filled hole

Figure 6. The applicability and limitation of relevant methods.

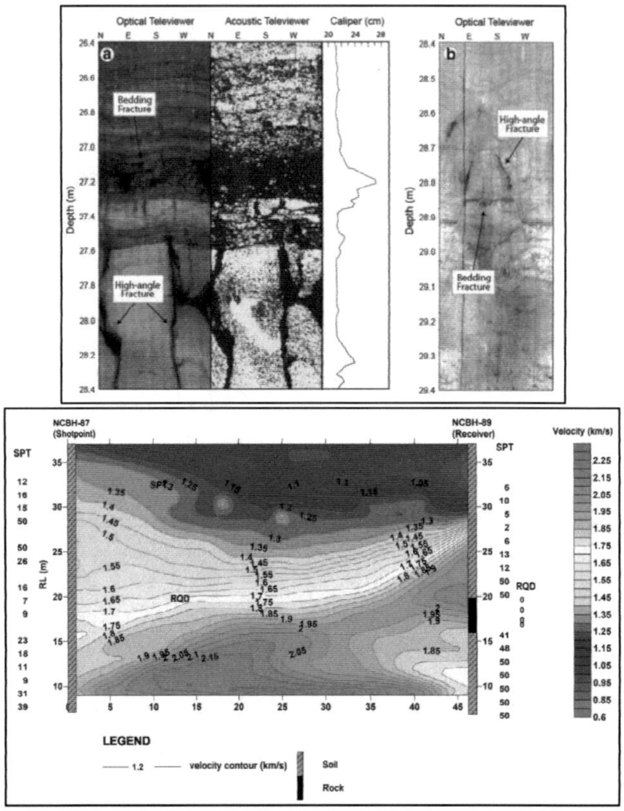

(a)

(b)

Figure 7. (a) Acoustic Televiewer was used in determine discontinuities and rock quality design [12], **(b)** Cross hole seismic was used in mapping subsurface geological information.

3. Applications in infrastructure projects

Civil and Geotechnical Engineers need to understand the basic principles associated with data acquisition, data processing and interpretation for engineering parameters. This will enable the proper selection of the appropriate method for the specific situation to be investigated. Due to the lack of this understanding there is a perception of geophysics as an unreliable approach. This can be overcome by the Civil designer, Geotechnical Engineer, Geophysicist and Geologist working to gather as a team with mutual understanding of each other's capabilities and expertise to derive best value in terms of reducing ground variation related issues on projects.

3.1 Rollinson Road, North Coogee, Australia

The area was planned for development, at the south of Fremantle. The land surface was partly been compacted by impact roller. The MASW processing scoped is to address the effect of compaction, competence of the ground and information about subsurface condition. Three lines (156m each) were proposed as in Figure 8. The data along one of the lines were recorded in the opposite direction to others.

IOP Conf. Series: Materials Science and Engineering **512** (2019) 012007 doi:10.1088/1757-899X/512/1/012007

Figure 8. Location of survey line at site

The S-wave velocity sections as in Figure 9 show the S-wave velocity is generally over 250m/s from the surface. There is hardly any area where the S-wave velocity is slower than 250m/s. Therefore, the ground is quite competent with Young's Modulus over 300MPa.

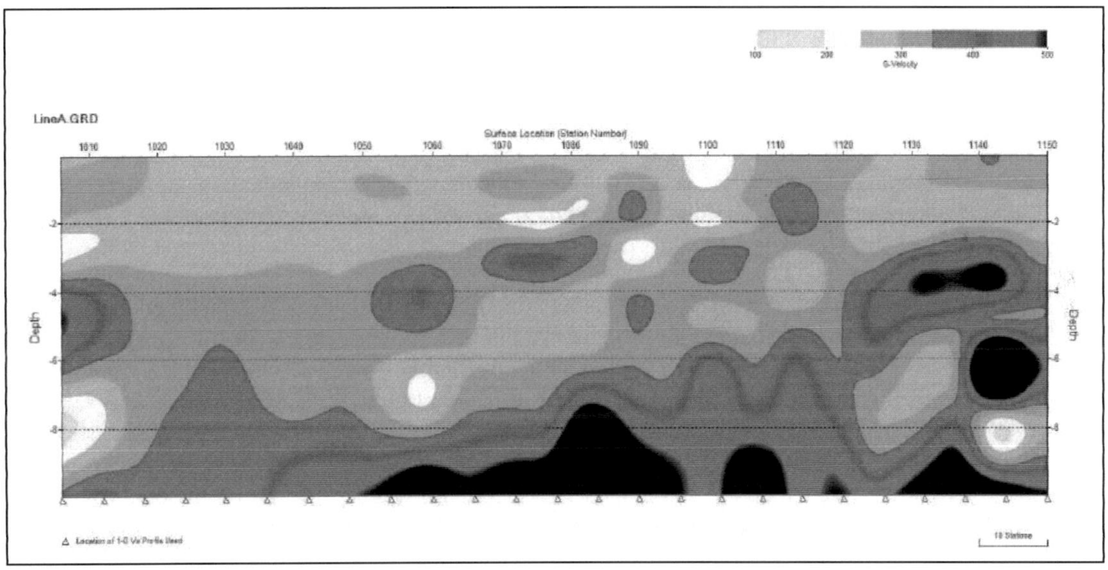

Figure 9. MASW data from site investigation

3.2 Mount Magnet Leinster road, Western Australia

This 180km road was to be constructed as a 2-lane single carriageway sealed road. There were several cuttings and valley fills. The site as in Figure 10 had weathered residual soils but there were also cemented layers that might require hard digging, ripping or even blasting. If this is not captured adequately in each cutting the chances of cost and time overruns due to encountering hard materials would be very high. Similarly, if each valley floor is not assessed adequately for loose or soft deposits then the fill embankments could settle. The site investigation would normally require at least 360 boreholes to obtain a good assessment of each cutting and valley section. The site investigation consisted of 180km of seismic refraction and 50 bore holes. The seismic velocity profiles were interpreted for hard layers, rippability and blasting and presented as colour coded long sections. This enabled the civil designers to have a more accurate calculation of common excavation, ripping excavation and excavation requiring blasting.

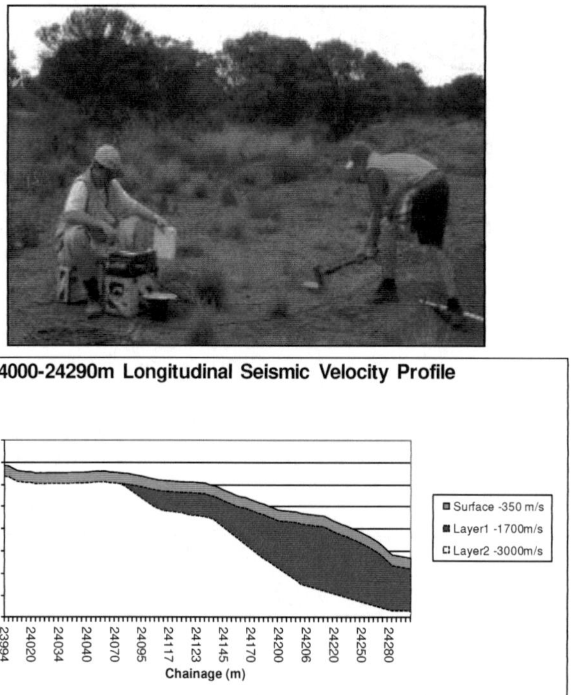

Figure 10. Data acquisition and geological profile from seismic survey

3.3 Pulau Pangkor, Perak, Malaysia.

The project consisted of the construction of a network of underground sewer pipes and a regional sewerage treatment plant along Jalan Pasir Bogak [10]. The depth to inert level of the proposed sewer pipes are designed to be as deep as 15m over the valley of Pangkor Island. The estimated length of the sewer pipe network is approximately 4.5 km. The proposed alignment traverses along existing road alignments, road shoulders, crossing existing buildings, culverts and other existing infrastructure facilities. The project objective was to recognize boulders in the subsurface in the specific area and the identification of the weathered and fresh bedrock interface in each section of the alignment. To achieve this objective, 3 methods were used that is Continuous Wave Surface System (CSWS) as in Figure 11, resistivity as in Figure 12 and Ground penetrating radar (GPR). Based on this, locations of boulders were mapped to reduce the tunnelling uncertainties.

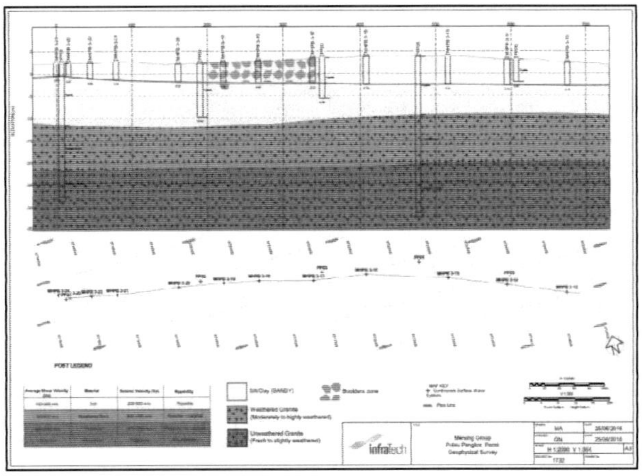

Figure 11. Final finding and location of boulder and bedrock depth.

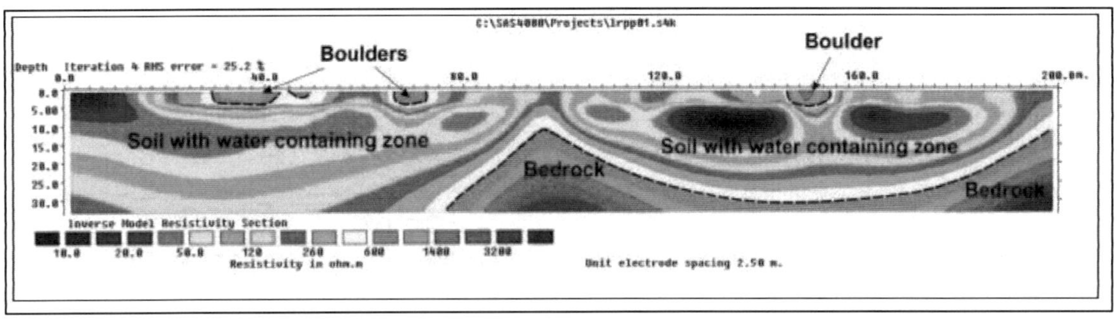

Figure 12. Resistivity survey along alignment to detect the boulder location

3.4 Jalan Chan Sow Lin, Sungai Besi, Kuala Lumpur, Malaysia.

The site is located along Jalan Chan Sow Lin, occupying a plan area of 2.1 acres. The site was planned for the development of 2 blocks of 40-storey residential towers with a podium and a 1-level basement. This Geological Formation mainly consists of limestone with possible presence of cavity and pinnacle structures. The initial plan was to drill at least one borehole at each column location. This was perceived as the only way to ensure cavities can be detected. Later MASW scanning was proposed along all column grids as shown in the scanning layout plan for 22 scan lines as in Figure 12. This was followed by drilling 5 numbers of boreholes. This integrated technique could identify features like cavity and pinnacles that could easily have been missed with traditional investigation only as in Figure 14. Overall findings showed that the area have shallow rock head between 4m – 7m. Potential cavities were also detected at some locations. This finding was correlated well with the existing boreholes.

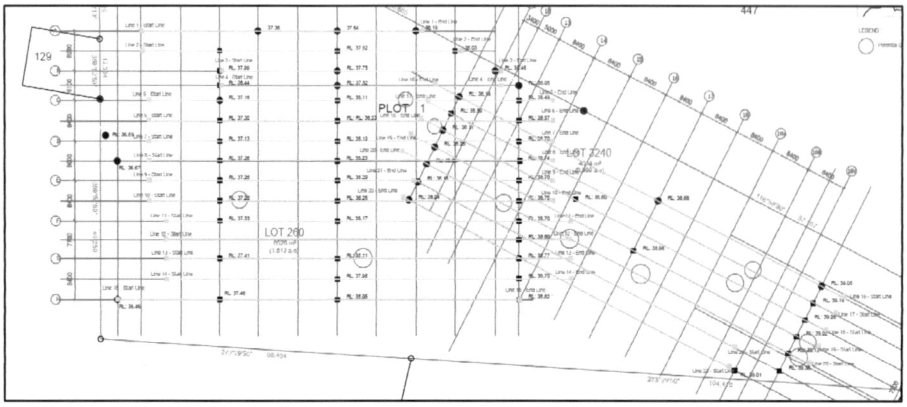

Figure 13. Scan line location plan

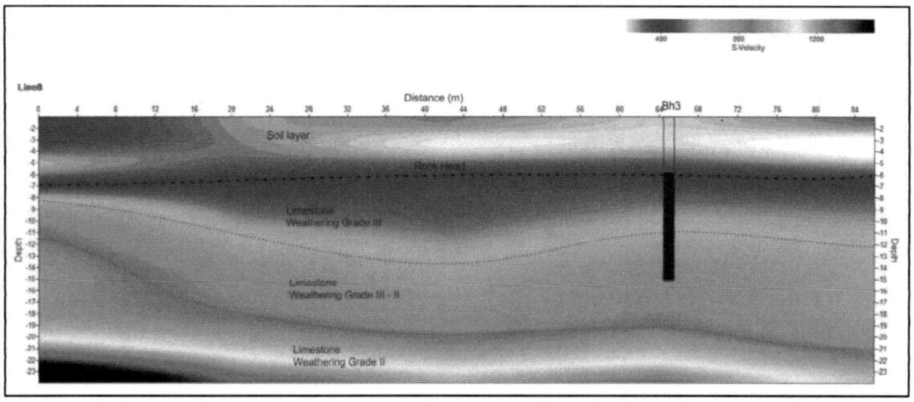

Figure 14. Correlation between MASW result & borehole

3.5 Bagan Datuk, Perak, Malaysia.

This project initially covering a plain area of approximately 900acres. Area of Bagan Datuk is covered by quaternary sediment. The geology of the area consisted generally depositions of clay and silt. It also known as soft soil area. Since the site was a former meander plain of the nearby Sg Perak deep buried channels were expected with high variation of the soft soil layer depths. The challenge faced was to capture the ground variation with many boreholes (450 bore holes based on a one bore hole per 2 acres) which would be costly and time consuming. To reduce cost and time for investigation while reducing the risk of missing out on large ground variations, an integrated approach was adopted. This integrated approach included Boreholes and Geophysical methods. The objective was to identify soft soil depth variation and geotechnical properties. 20 nos. of borehole were done with 40 nos. of resistivity survey line [11]. This approach provides adequate information to assess the depth of ground improvement required, depth of piling in various parts of the site. Figure 15 shows the typical outputs from this approach.

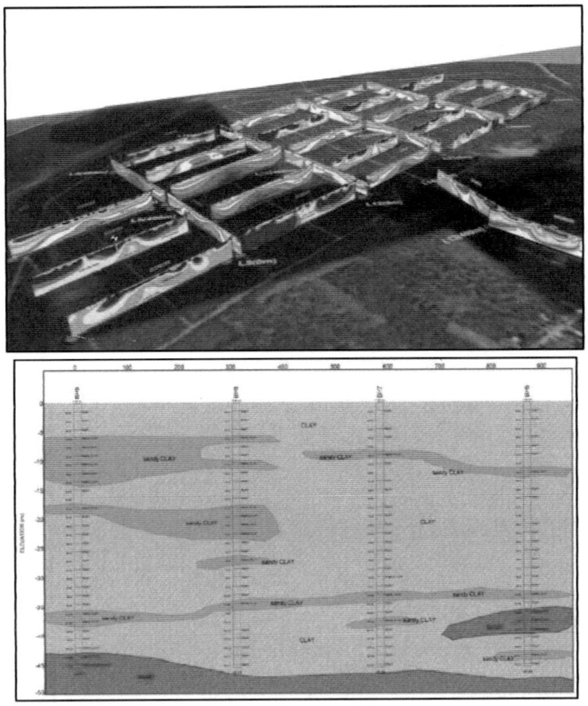

Figure 15. ER line survey (400m each) at of Bagan Datuk, Perak and subsurface cross-section profile showing the lithological variations based on correlation of borehole and geophysics investigation.

4. Advantages and limitations of geophysical methods

The advantages of this technique are fast, low cost, and covers a large area intensively. It is the best way to capture ground variation which is practically not possible by traditional bored holes. Another advantage is a good correlation between engineering properties and geophysical parameter (Resistance, Vs etc). However sometime correlations are not fully explored and established for all types of soils. Furthermore, each technique has its resolution issue and depth limitations. These limitations must be understood to enable the selection of the appropriate technique.

5. Summary

The use of geophysics in geotechnical and geo environmental site investigations requires a multi-disciplinary approach between Civil and Structural designers, Geotechnical Engineers, Geophysicists and Geologists. If carried out with an in-depth understanding of the applications and limitations of each methods to what the engineers required to assess, then effective results can be obtained. This eventually leads to reduced ground variation risk while saving time and cost. Consequently, will also increase the quality of the site investigation. Intrusive boreholes must be used in correlation with the findings. With appropriate geophysical scanning of sites, the number of boreholes reduced, and the variation risk can also be greatly reduced.

6. References

[1] Clayton C R I, Matthews M C and Simons N E 1995 *Site Investigation. Second Edition* (UK: Blackwell Science Ltd)

[2] Park C B, Miller R D, Xia J and Ivanov J 2007 Multichannel analysis of surface waves (MASW) active and passive methods *The Leading Edge* **26**(1) 60-4

[3] Xia J, Miller R D and Park C B 1999 Estimation of near surface shear-wave velocity by inversion of Rayleigh waves *Geophysics* **64**(3) 691-700

[4] Setchell S, Azahar M A, Rusli Q N, Lowe N 2016 The use of Non-Invasive Geophysical Techniques to build 3-Dimensional ground models and reduce the cost and risk of site investigation *Southeast Asian Geotechnical Conference* Kuala Lumpur

[5] Menzies B and Matthews M 1996 The continuous surface-wave system: a modern technique for site investigation *Proceedings Special Lecture: Indian Geotechnical Conference* Madras, pp 11-4

[6] Stokoe K H, Wright S, Bay J and Roesset J 1994 Characterization of geotechnical sites by SASW method *Geophysical Characterization of Sites* pp 15-25

[7] Barker R D 1997 Electrical imaging and its application in engineering investigations *Geological Society Engineering Geology Special Publication* (12)

[8] Parasnis D S 1986 *Principle of Applied Geophysics* (New York: Chapman and Hall)

[9] Samsudin A R 1990 *Geofizik Konsep dan Penggunaan* (Kuala Lumpur: Dewan Bahasa dan Pustaka)

[10] Azahar M A, Rusli Q N, Salleh M A and Mahadi N F Z 2018 The Use of Non-Invasive Geophysical Techniques in Detecting of Boulder and Bedrock at Pulau Pangkor, Perak, Malaysia *Electronic Journal of Geotechnical Engineering* **2**(23)

[11] Ariffin M H, Kayode J S, Azahar M A, Jamil H and Sabri S F A 2018 Data for in-situ Industrial Site characterization with the applications of combined subsurface and surface mapping *Elsevier Data in Brief* **18** pp 1864-8

[12] Des Roches A J, Butler K E and Pelkey S 2012 Influence of fracture anisotropy and lithological heterogeneity on wellfield response in a fluvial sandstone aquifer of the Carboniferous Moncton Subbasin, Canada *Hydrogeology Journal* **21**(3)

Comparative study of moisture damage performance test

Adaweiah Taib[1], **Fauzan Mohd Jakarni**[1], **Muhammad Fudhail Rosli**[1], **Nur Izzi Md Yusoff**[2] and **Maniruzzaman Abd Aziz**[3]

[1] Department of Civil Engineering, Faculty of Engineering, Universiti Putra Malaysia, 43400 UPM Serdang, Selangor, Malaysia
[2] Department of Civil and Structural Engineering, Faculty of Engineering and Built Environment, Universiti Kebangsaan Malaysia 43600 Bangi, Selangor, Malaysia
[3] Department of Geotechnics and Transportation, Faculty of Civil Engineering, Universiti Teknologi Malaysia, 81310 Skudai, Johor, Malaysia

E-mail: adaweiahtaib@gmail.com & fauzan.mj@upm.edu.my

Abstract. This paper presents a comprehensive literature on moisture damage performance tests of asphalt mixtures. Moisture damage recognized as a cause for pavement distresses. The most common forms of moisture damage are adhesive failure between bitumen and aggregates. Adhesive failure is due to bitumen coating the aggregates are completely displaced by moisture, and stripping becomes visible in the asphalt mixtures. However, both quantitative and qualitative test do not focus on measuring the fundamental properties of the asphalt mixtures related to the moisture damage mechanisms of adhesion and cohesion. The results of the quantitative tests measured quantitatively, minimize the subjective evaluations of the test results. The quantitative are Tunnicliff and Root Conditioning (ASTM D4867) and Modified Lottman (AASHTO T283) test provide evaluation of moisture susceptibility of the asphalt mixture on the basis of strength and/or modulus ratio before and after conditioning. Other tests, such as Hamburg Wheel Tracking Device (HWTD) (AASHTO T324-04) provide measurement of moisture damage in terms of permanent deformation of the specimens. This paper is a review of the effectiveness of the selected available moisture damage performance tests. The analysis was conducted based on the success rate for each moisture damage performance tests.

1. Introduction

Road users in Malaysia feel uncomfortable driving on highway cause by existence of pavement distresses especially after monsoon season. These pavement distresses mostly happen due to moisture damage. Moisture damage or moisture susceptibility in asphalt pavements remain a main concern in the pavement construction almost a few decades ago. The common term used for moisture damage in asphalt pavements is stripping which a primary cause is of distress in asphalt pavement. Stripping commonly occur when moisture infiltrates the pavement and weakening the bond between the aggregate and bitumen. Stripping cause the reduction of pavement shear strength and could lead to several distress such as raveling, rutting, fatigue cracking [1], [28] of the binder to the pavement surface, which could decrease pavement's skid resistance.

In [1] stated that various tests had been performed to evaluate moisture-susceptible asphalt mixture since 1930s. However, these tests are still producing wavering result in evaluating moisture susceptible in asphalt mixture. Solaimanian [27] pointed out that the tests can be classify into those evaluating the affinity between aggregate and asphalt in loose mix condition and those accustomed to evaluate the moisture sensitivity in compacted mix condition. At present accessible moisture damage performance test that had been standardized, for example, Boiling Water test (ASTM D3625) and Static Immersion test (ASTM D1664/AASHTO T182) only depend on subjective evaluation while Modified Lottman test (AASHTO T283) and Immersion Compression Test (AASHTO T165) rely on the principle of relative assessment of mechanical properties (indirect tensile strength, resilient modulus, marshal stability or compression strength) of conditioned sample and unconditioned sample. Meanwhile, other tests such as Saturation Ageing Tensile Stiffness (SATS) test [12], Environmental Conditioning System (AASHTO TP34) and Hamburg Wheel Tracking Test are also recognized to be used in assessing moisture damage performance of asphalt mixture.

This paper discussed the selected test methods for evaluating moisture susceptibility on various combinations of asphalt mixtures.

1.1 Moisture damage

1.1.1 General background of moisture damage. Moisture damage is manifested by infiltration of moisture or water through the asphalt binder-aggregate adhesion within an asphalt mixture. This circumstance may lead to a reduction of adhesion among the asphalt binder and aggregate known as stripping and may cause some of distress in pavement such as fatigue cracking and rutting. [10] described that moisture damage is one of the main cause that lead to early rehabilitation of asphalt pavement. They also stated that thermodynamic, chemical, physical, and mechanical processes are likely involved in this action. [26] describes that it is inevitable to dismiss the fact of existing of water in asphalt pavement and found that a few factors can become significant to the existing of water in pavement. Water can penetrate the pavement from the surface through cracking of pavement as explained by [9] that the water on pavement will infiltrate through small crack at the surface of pavement, connection of the air-void mechanism, from the bottom caused by increasing of ground water table, or from the sides. Besides, heating of aggregate throughout mixing process poorly conducted may cause existing of water in pavement.

Study by [25] summarized that moisture damage is a very indescribable type of distress and symbolizes an acclimatizing phenomenon when the pavement is exposed to the water or moisture. When water or moisture is in contact with bitumen and aggregates, the asphalt mixtures will lose its structural strength and stiffness. This problem will lead to pavement distress such as ravelling, stripping, fatigue cracking, surface wear and rutting, for longer term may lead to extensive damage and the serviceability of the pavement may be reduced.

As a conclusion, even though numerous research have concluded that moisture damage of asphalt mixture is tend to the adhesive failure than the cohesive failure, the basic understanding regarding to these failure are very indescribable to illustrate and researches on these two failure are limited. Thus, moisture damage mechanisms continue to become the most complicated distress and not completely understood.

1.1.2 Moisture damage mechanism.birgis [23] found that there were a few contributing factors that known to be as the failure mechanism in asphalt mixtures include pore pressure, hydraulic scour, detachment, displacement, spontaneous emulsification, and environmental factors. [8] explained that these mechanisms may act independently or coincides with each other that lead to adhesion failure in asphalt mixtures. Moreover, their studies also describe that other possible mechanism for stripping probably due to osmosis cause by the existing of salts or salt solution in the aggregate pores that forms an osmotic pressure gradient that sucks moisture through the bitumen. Two major failure types: loss of adhesion or loss of cohesion occurs from the action of these mechanisms [9].

Table 1. Factor influencing moisture damage

Factors	Determining Characteristics	Favourable Properties	References
Aggregate Properties	Surface texture, mineralogy, porosity, surface moisture, surface chemical composition and surface coating	Rough surface texture, carbonaceous aggregate, low silica content, optimum amount of porosity, surface dry aggregate, no coating	[18], [16], [17], [11]
Bitumen Characteristics	Asphalt film thickness, viscosity, physical and chemical structure	High asphalt film thickness, High viscosity, existence of phenol and nitrogen	[18], [8], [11]
Construction Method	Compaction method, drainage system, air void mechanism,	Adequate compaction, proper drainage system, low air void percentages, adapt water resistance additives on each layer of pavement	[16], [8], [11], [21]
Environmental Condition	Climates, Environmental temperature	Warm climates, mild temperature (low rate of changing in temperature), no freeze-thaw cycles	[18], [11]
Imposed traffic load	Traffic load	Low traffic	[18]

1.2 A Review on Moisture Susceptibility Test

Development of tests to evaluate moisture susceptibility of asphalt mixtures have been started since 1930s [29]. It has been conclusively been shown that since that time, a number of tests had been implemented in order to identify the proneness of asphalt mixtures to moisture damage [4], [11]. The test procedures in the present have tried to resemble the loss of strength that possibly occurs in the pavement so that the premature distress of asphalt mixtures can be recognized prior to construction. Diab and You (2013) expressed that even though continuous improvement on moisture susceptibility tests has been made in clarifying and understanding the mechanisms of moisture damage, a reliable and practical laboratory method that can simulate moisture damage in the field is still needed for agencies such as state highway agencies.

It is generally difficult to develop a laboratory test procedure that can fully simulate the field condition such as environmental condition, traffic and construction practices. Diab and You (2013) in his research state that some efforts have been made so far to develop a test procedure that would precisely determine the susceptibility of an asphalt pavement to moisture damage, however none of the moisture susceptibility test have been accepted widely due to lack of repeatability, difficulty of the process, expensive equipment, or lack of quantitative results. In [4] described that moisture susceptibility tests have a "conditioning and "evaluation" phase. The conditioning phase is a process to simulate the deterioration action on asphalt pavement in the field including environment condition, traffic load repetition, climate condition (humid and hot climate), air void level and others. For evaluation phase, the asphalt mixtures sample will then be assessed by visual evaluation (qualitative evaluation) and physical test (quantitative evaluation). In the visual evaluation, the percentages of retained asphalt coating is then determined after the conditioning process. While, physical test evaluation consist of strength or modulus and a ratio between the result from conditioned sample by the result from unconditioned sample is computed. If the ratio is less than standardized value, the sample will be clarified as moisture susceptible.

And [13] described in [27] that moisture susceptibility tests can be divided into two categories which is test on loose mixtures (qualitative test) and test on compacted mixtures (quantitative test). Following tests are national standard that currently been used widely by public agencies including AASTHO and ASTM.

- AASHTO T 165/ASTM D 1075 Effect of Water on Compressive Strength of Bituminous Mixtures
- AASHTO T 283/ASTM D 4867 Resistance of Compacted Asphalt Mixtures to Moisture-Induced Damage
- ASTM D 3625 Effect of Water on Bituminous-Coated Aggregate using Boiling Water
- ASTM D 4867 Effect of Moisture on Asphalt Concrete Paving Mixtures
- AASHTO T 324 Hamburg Wheel-Track Testing of Compacted Hot-Mix Asphalt.

1.2.1 Test on Loose Mixture (Qualitative Test). These types of tests are conducted on bitumen-coated aggregates by immersing samples into water. Some examples of these tests are boiling test, film strip, and static/dynamic immersion tests. Benefit of these tests is they are only consuming short time and less costly to conduct comparing with tests on compacted samples. However, these tests are not able of simulate pore pressure, traffic condition, and mix design properties to justify moisture susceptibility of asphalt mixture. The results are mostly qualitative and clarification of the results tends to be subjective as it is reliant on the evaluator's judgement and experience. Besides, correlation between these types of tests to field performance of asphalt pavement is still unreliable.

In addition, these tests are suitable to be used for comparison between different aggregate-bitumen mixtures or uses of different anti-stripping additive to evaluate compatibility, stripping and strength of adhesion of asphalt mixtures. Mixtures that not achieved required standard of these tests will be considered fail and have higher probability to strip and should not be used. Though, successful results not necessarily mean that the mix can be used, as the effects of other factors are not taken into consideration in these tests. Most popular test conducted on loose samples currently used such as Static Immersion test (AASHTO T182) and Boiling Water test (ASTM D3625). Table 2 will provide explanation on the only established standard test on loose mixture according to AASTHO and ASTM.

1.2.2 Test on Compacted Sample (Quantitative Test). Stated in [27] that this type of test are performed on laboratory-compacted samples or taken from field in the form of cores or slabs. Some of the tests that currently established as standard tests and widely used according to ASTM or AASHTO are immersion-compression test (ASTM D1075/AASHTO T165), Modified Lottman test (AASHTO T183) and Tunnicliff-Root test (ASTM D4867). Other tests such as Hamburg wheel tracking test (AASHTO T324), Environmental conditioning system (ECS) (AASHTO TP34), Simple performance test (SPT), Asphalt Pavement Analyser (APA), moisture induced sensitivity test (MIST) and saturated ageing tensile stiffness (SATS) are also taken into consideration but rarely used due to lack of standardization in the procedures used in term of sample preparation and complexity of the procedures. The main benefits of these tests is that it can assess the physical and mechanical properties while the traffic action and pore pressure effects can also be considered [27]. The results provided are measured quantitatively and this will reduce the higher variability of the test results due to visual evaluation. However, the weaknesses from these tests are it involved very expensive and complex testing equipment, take longer time to perform and demanding more laborious test procedures. Summarize for some of these tests that currently been widely used by highway authority is briefly explained in the following table 3.

1.3 Comparison of Previous Moisture Susceptibility Test on Various Asphalt Mixtures

From table 4 to table 6, it can be concluded that asphalt mixture consist of limestone aggregate will give more resistant to moisture damage. This can be seen from result shown in table 4 provided by [19], asphalt mixture consist of limestone aggregate produce the highest tensile strength ratio (TSR) which is 61.7% compare to asphalt mixture consist of slate aggregate which is 48.6% and granite which is 58.5% with the same type of bitumen without any anti-stripping additives after conditioning by Modified Lotmann test. While in table 5 after immersion compression test, it can be seen from data provided by [21] that asphalt mixture consist of limestone give the highest Marshall stability ratio which is 98% compare to asphalt mixture consist of granite (89.1%), sandstone (87.8%) and Harwar Quartzite (86.5%). Whereas in table 6, after boiling water test being conducted on loose asphalt mixture, the result by [7] shown that asphalt mixture consist of limestone produce the highest percentage of aggregate remain coated by bitumen which is 98.7% and 98.4% compare to other asphalt mixture which consist of quartzite (59.7%), granite (84.2%) and andesite (13.5%).

Furthermore, addition of anti-stripping additive will increase asphalt mixture resistant to moisture damage exponentially. It can be seen from the research carried by [5] in table 4 that the TSR value recorded by asphalt mixture with addition of anti-stripping additives such as limestone dust is very high which is 96% compare to asphalt mixture without anti-stripping additives which is recorded to be 48% and 60% with constant type of bitumenand aggregate. This circumstance also can be seen in research by [19] that also shown high TSR ratio ranging from 72% to 95.2% in asphalt mixtures with addition of anti-stripping additives compare to without any additives which is in range of 48% to 61.7%. While in other test such as immersion compression test, addition of antistripping additive also give the same result. With respect to the table 5, it was found that with the same type of aggregate and bitumen, the TSR value for asphalt mixtures in addition of anti-stripping additives is higher than without any addition of anti-stripping additives [21]. By referring to table 6, result from boiling water test also shown the same situation. Percentage of aggregate remain coated with bitumen is higher in asphalt mixtures with the addition of anti-stripping additives compare to the asphalt mixture without any additives with the fixed aggregate and bitumen.

Table 2. Test methods on loose asphalt mixture

Test name	Measured parameter	Approach of the test	Description of test procedure
Static immersion (AASHTO T1&2)	Percentage of aggregate remain coated after static immersion in water	Focusing on adhesion bond failure	This test required a sample of asphalt mixture been immersed in a jar filled with 600 mL of distilled water after been cured for 2 hours at 60oc and cooled to room temperature. The jar is then capped left settled in a 25oc water bath for 16 to 18 hours. The degree of stripping is visually evaluated while the mixture still in the jar.
Boiling water (ASTM D3625)	Percentage of aggregate remain coated after boiling in water	Focusing on adhesion bond failure	The test involves placing loose sample of asphalt mixtures into boiling water and being stirred using glass rod. After 10 minutes, the mixture is left to cool while the stripped bitumen is detached away. Then, the mixture is removed from the water and being dried in room condition

In general, types of bitumen also have a significant impact in moisture resistant to moisture damage. Uses of modified bitumen will increase the moisture resistant towards moisture damage. This view is supported by [3] who writes that increasing of binder grade will then lead to reducing of retained stiffness of bitumen. This circumstance will result in reduction of resistant towards moisture damage on asphalt mixture.

Table 3. Test methods on compacted asphalt mixture

Test name	Measured parameter	Description of test procedure
Immersion compression test (ASTM D1075/AASHTO T165)	The ratio of average strength of conditioned specimens over controlled specimens is being used as a parameter to measure loss of strength cause by moisture damage	Core specimens are compacted with a double plunger at 3,000 psi for two minutes to achieved air void content at 6%. The procedure involve six specimens which been divided equally into two group known as control group and conditioned group. The control group is been left dry while the specimens in conditioned group is being immersed in a water bath at 120oF (49oC) for four days or at 140oF (60oC) for one day. Then, the compressive strength of the specimens from both groups is being measured at 77oF (25oC) at a loading rate of 0.05 in./min per inch of height. So, for Marshall mix design specimen with 4 inches diameter, loading rate of 0.2 in./min will be used
Lottman test	Ratio of test values conditioned specimen to control group specimen (tensile strength ratio,TSR) including freeze and thaw cycle	Nine compacted Marshall specimens of 100mm in diameter and 63.5mm in height which are being divided into 3 groups equally. • Group 1: Control group, dry • Group 2: Vacuum saturated at 660mmHg with water for 30-minutes. • Group 3: Vacuum saturation followed by freeze cycle at - 18°C for 15 hours and then subjected to a thaw at 60°C for 24 hours. After the conditioning procedure, the Resilient Modulus (MR) and/or Indirect Tensile Strength Test (ITS) are conducted on each specimen. The loading rate for testing at 55o F (13oC) is 0.065 in./min while 0.150 in./min are being used for testing at 73oF (23oC)
Tunnicliff – Root Test (ASTM D4867)	Ratio of test values conditioned specimen to control group specimen (tensile strength ratio,TSR) without freeze and thaw cycle	Improvising from Lottman test •Load rate increases to 2 in/min from 0.065 in/min •Test temperature increases from 55oF (12.8oC) to 77oF (25oC) •Presaturation of 55% -80% compared to an infinite level in Lottman test •Removing freeze cycle condition
Modified Lottman Test (AASHTO T283)	Ratio of test values conditioned specimen to control group specimen (tensile strength ratio,TSR) with/without freeze and thaw cycle	The procedure combines features of both the Lottman and Tunnicliff and Root procedures. The Lottman procedure attempts to achieve a 100 percent saturation level in it specimens, while the Tunnicliff and Root procedure attempts to control the level of saturation between 55 and 80 percent. Concern that oversaturation induces damage in specimens that is not associated with moisture damage but rather with the oversaturation of the specimen, meant that for the Modified Lottman procedure the degree of saturation was decreased to between 60 to 80 percent. As the saturation level achieved by partial vacuum is primarily responsive to the magnitude of the vacuum and relatively independent of the length of time, this reduced saturation was achieved by reducing the partial vacuum from 600 mm Hg to 508 mm Hg.

IOP Conf. Series: Materials Science and Engineering **512** (2019) 012008 doi:10.1088/1757-899X/512/1/012008

Table 4. Modified Lottman test

Mixture Design	Aggregate	Bitumen	Anti-stripping additives	Strength/Criteria Ratio			References
				Min. Req. (%)	Test result (%)		
					Water Immersion @60°C for 24 hours	Freeze - thaw@-18°C for 16 hours	
M	Granite	Pen 60/70	Quarry dust	70	82.0	70.9	[6]
M	Granite	Pen 60/70	Ordinary Portland Cement	70	83.3	74.6	[6]
M	Granite	Pen 60/70	Polymer Modifier	70	86.2	76.7	[6]
M	L(MLDG)	Pen 60/70	Calcium Hydroxide	80	-	68	[5]
M	L(MLDG)	Pen 60/70	Limestone dust	80	-	96	[5]
M	L(MLDG)	Pen 60/70	No additives	80	-	48	[5]
M	L(MLDG)	Pen 80/100	Calcium Hydroxide	80	-	60	[5]
M	L(MLDG)	Pen 80/100	Limestone dust	80	-	96	[5]
M	L(MLDG)	Pen 80/100	No additives	80	-	40	[5]
SP	Granite	PG 64	Not available	80	82.2	-	[2]
SP	Granite	PG 70	Not available	80	94.7	-	[2]
M	Granite	PG 64	Not available	80	99.8	-	[2]
M	Granite	PG 70	Not available	80	97.3	-	[2]
SP	Slate	PG 64-22	No additive	80	48.6	-	[19]
SP	Slate	PG 64-22	Hydrated lime	80	80.8	-	[19]
SP	Slate	PG 64-22	Amine	80	95.2	-	[19]
SP	Slate	PG 64-22	Phosphate Ester	80	83.5	-	[19]
SP	Limestone	PG 64-22	No additive	80	61.7	-	[19]
SP	Limestone	PG 64-22	Hydrated lime	80	80.9	-	[19]
SP	Limestone	PG 64-22	Amine	80	81.2	-	[19]
SP	Limestone	PG 64-22	Phosphate Ester	80	72.0	-	[19]
SP	Granite	PG 64-22	No additive	80	58.5	-	[19]
SP	Granite	PG 64-22	Hydrated lime	80	85.7	-	[19]
SP	Granite	PG 64-22	Amine	80	81.2	-	[19]
SP	Granite	PG 64-22	Phosphate Ester	80	79.0	-	[19]
SP	L+G (less angular)	PG 64-22	No additive	80	-	69	[20]
SP	L+G (less angular)	PG 64-22	Hydrated lime	80	-	77	[20]
SP	L+G (more crushed)	PG 70-28	No additives	80	-	79	[20]
SP	L+G (more crushed)	PG 70-28	Hydrated lime	80	-	85	[20]
SP	L+G (more crushed)	PG 70-28	Fly ash	80	-	91	[20]

Note: L(MLDG)=Limestone (Mid limits of dense graded) L+G = Limestone+Gravel

Table 5. Immersion compression test

Mixture Design	Aggregate	Bitumen	Anti-stripping additives	Strength/Stability Ratio		References
				Min. Req. (%)	Test result (%)	
M	Granite	VG 30	No additives	70	89.1	[21]
M	Granite	VG 30	Hydrated lime	70	96.8	[21]
M	Limestone	VG 30	No additives	70	98	[21]
M	Sandstone	VG 30	No additives	70	87.8	[21]
M	Sandstone	VG 30	Hydrated lime	70	97	[21]
M	Harwar Quartzite	VG 30	No additives	70	86.5	[21]
M	Harwar Quartzite	VG 30	Hydrated lime	70	94.5	[21]
SP	Crushed stone	PG 64-16	Class C Fly Ash	70	95	[9]
SP	Crushed stone	PG 64-16	Class F Fly Ash	70	112	[9]
SP	Crushed stone	PG 64-16	Cement Kiln Dust	70	95	[9]
SP	Crushed stone	PG 64-16	Hydrated lime	70	93	[9]
SP	Crushed stone	PG 64-16	HP Plus (amine chemical)	70	97	[9]
SP	Crushed stone	PG 64-16	No additives	70	97	[9]

Notes: M = Marshall; SP = Superpave; PG = Performance grade; VG = Viscocity grading

Table 6. Boiling water test

Mixture Design	Aggregate	Bitumen	Anti-stripping additives	Strength/Criteria Ratio		References
				Min. Req. (%)	Test result (%)	
SP	Limestone+Gravel (less angular)	PG 64-22	No additive	90	85.0	[20]
SP	Limestone+Gravel (less angular)	PG 64-22	Hydrated lime	90	94.0	[20]
SP	Limestone+Gravel (less angular)	PG 64-22	Fly ash	90	95.0	[20]
SP	Limestone+Gravel (more crushed)	PG 70-28	No additives	90	98.0	[20]
SP	Limestone+Gravel (more crushed)	PG 70-28	Hydrated lime	90	99.0	[20]
SP	Limestone+Gravel (more crushed)	PG 70-28	Fly ash	90	99.0	[20]
M	Granite	VG 30	Hydrated lime	95	>95	[21]
M	Sandstone	VG 30	Hydrated lime	95	>95	[21]
M	Limestone	VG 30	Hydrated lime	95	>95	[21]
M	Delhi Quartzite	VG 30	Hydrated lime	95	>95	[21]
M	Harwar Quartzite	VG 30	Hydrated lime	95	>95	[21]
SP	Quartzite	Pen 60/70	No additives	95	59.7	[7]
SP	Quartzite	Pen 60/70	Hydrated lime	95	96.5	[7]
SP	Quartzite	Pen 60/70	Zycosoil	95	98.6	[7]
SP	Granite	Pen 60/70	No additives	95	84.2	[7]

Table 6. Boiling water test (continued)

SP	Granite	Pen 60/70	Hydrated lime	95	98.6	[7]
SP	Granite	Pen 60/70	Zycosoil	95	98.6	[7]
SP	Andesite	Pen 60/70	No additives	95	13.5	[7]
SP	Andesite	Pen 60/70	Hydrated lime	95	98.6	[7]
SP	Andesite	Pen 60/70	Zycosoil	95	99.2	[7]
SP	Limestone	Pen 60/70	No additives	95	98.4	[7]
SP	Slag limestone	Pen 60/70	No additives	95	98.7	[7]
M	Lime stone (Mid limits of dense graded)	Pen 60/70	Calcium Hydroxide	95	96	[5]
M	Lime stone (Mid limits of dense graded)	Pen 60/70	Limestone dust	95	96	[5]
M	Lime stone (Mid limits of dense graded)	Pen 60/70	No additives	95	89	[5]
M	Lime stone (Mid limits of dense graded)	Pen 80/100	Calcium Hydroxide	95	95	[5]
M	Lime stone (Mid limits of dense graded)	Pen 80/100	Limestone dust	95	95	[5]
M	Lime stone (Mid limits of dense graded)	Pen 80/100	No additives	95	90	[5]

Notes: M = Marshall; SP= Superpave; PG= Performance grade; Pen = Penetration; VG = Viscocity grading

2. Conclusions

This study indicates that asphalt mixtures consist of limestone aggregate, modified bitumen and addition of anti-stripping additives will provide more resistant towards moisture damage. This result supported by [22], [30] and [6] in their previous research. Hydrated lime tends to be the most popular among others anti-stripping additives because it had been proven effective in increasing resistance of asphalt mixture towards moisture. While polymer modified bitumen are most popular to be used in asphalt mix design because it can sustain moisture damage more than commonly used asphalt binder. Generally, Modified Lottman test, Immersion Compression test and Boiling Water test can be expected to be reliable test on evaluating the moisture susceptibility of asphalt mixture.

3. References

[1] Alam M, Vemuri N, Tandon V, Nazarian S and Picornell, M 1998 *A Test Method For Identifying Moisture Susceptible Asphalt Concrete Mixes* (Texas: The University Of Texas At El Paso)

[2] Ahmad J, Abdul Rahman M Y, Hainin M R and Hossain M 2011 Comparative evaluation of hot-mix asphalt design methods *Int. J. Pavement Engineering* **13** 89-97

[3] Airey G, Collop A, Zoorob S and Elliott R 2008 The influence of aggregate, filler and bitumen on asphalt mixture moisture damage *Construction And Building Materials* **22** 2015-24

[4] Al-Swailmi S H 1992 *Development Of A Test Procedure For Water Sensitivity Of Asphalt Concrete Mixtures*

[5] Abo-Qudais S 2007 The effects of damage evaluation techniques on the prediction of environmental damage in asphalt mixtures *Building And Environment* **42** 288-96

[6] Aman M Y, Shahadan Z and Noh M Z M 2014 A Comparative study of anti-stripping additives in porous asphalt mixtures *Jurnal Teknologi* **70**

[7] Amelian S, Abtahi S M and Hejazi S M 2014 Moisture susceptibility evaluation of asphalt mixes based on image analysis *Construction And Building Materials* **63** 294-302

[8] Birgisson B, Roque R, Tia M and Masad E A 2005 *Development And Evaluation Of Test Methods To Evaluate Water Damage And Effectiveness Of Antistripping Agents* (Oregon State of University)

[9] Boyes A J 2011 *Reducing Moisture Damage In Asphalt Mixes Using Recycled Waste Additives* (San Luis Obispo : California Polytechnic State University)
[10] Caro S 2009 Coupled micromechanical model of moisture-induced damage in asphalt mixtures *J. Materials In Civil Engineering* **22** 380-88
[11] Choi Y 2007 *Case Study And Test Method Review On Moisture Damage*
[12] Collop A 2004 Development of the saturation ageing tensile stiffness (SATS) test *Proc. The Ice-Transport* **157** pp 163-71
[13] Copeland A R 2007 *Influence Of Moisture On Bond Strength Of Asphalt-Aggregate Systems* (Doctoral dissertation, Vanderbilt University)
[14] Copeland A R, Youtcheff J and Shenoy A 2007 Moisture sensitivity of modified asphalt binders : factors influencing bond strength *Trans. Research Record: J. The Trans. Research Board* 18-28
[15] Diab A and You Z 2013 Development of a realistic conditioning and evaluation system to study moisture damage of asphalt materials *Airfield And Highway Pavement 2013@ Sustainable And Efficient Pavements Asce* pp 1008-17
[16] Emery J and Seddik H 1997 *Moisture Damage Of Asphalt Pavements And Antistripping Additives: Background Document*
[17] Hanz A, Bahia H, Kanitpong K and Wen H 2007 *Test Method To Determine Aggregate/Asphalt Adhesion Properties And Potential Moisture Damage*
[18] Hicks R G 1991 Moisture Damage In Asphalt Concrete *Trans. Research Board*
[19] Khosla N P, Birdsall B G and Kawaguchi S 2000 Evaluation of moisture susceptibility of asphalt mixtures: conventional and new methods. *Trans. Research Record: J. Trans. Research Board* **1728** 43-51
[20] Kim Y R, Pinto I and Park S W 2012 Experimental evaluation of anti-stripping additives in bituminous mixtures through multiple scale laboratory test results *Construction And Building Materials* **29** 386-93
[21] Kumar P and Anand P 2012 Laboratory study on moisture susceptibility of dense graded mixes *J. Trans. Engineering* **138** 105-13
[22] Lavin P 1999 *Comparison Of Liquid Antistrip Additives And Hydrated Lime Using Aashto T-283* Construction Chemicals Technical Director Arr-Maz Products
[23] Little D N and Jones D 2003 Chemical And Mechanical Processes Of Moisture Damage In Hot-Mix Asphalt Pavements *Trans. Research Board National Seminar* San Diego, California pp 37-70
[24] Lu Q 2005 *Investigation Of Conditions For Moisture Damage In Asphalt Concrete And Appropriate Laboratory Test Methods* (California : University Of California Transportation Center)
[25] Miller J S and Bellinger W Y 2003 *Distress Identification Manual For The Long-Term Pavement Performance Program*
[26] Santucci L 2002 *Moisture Sensitivity of Asphalt Pavements* Tech Topics (Ucberkley's Institute Of Transportation Studies)
[27] Solaimanian M, Harvey J, Tahmoressi M and Tandon V 2003 Test Methods To Predict Moisture Sensitivity Of Hot-Mix Asphalt Pavements *Transportation Research Board National Seminar* San Diego, California pp 77-110
[28] Terrel R L and Al-Swailmi S 1994 *Water Sensitivity of Asphalt-Aggregate Mixes: Test Selection*
[29] Terrel R L and Shute J W 1989 *Summary Report On Water Sensitivity*
[30] Uddin W 2003 Viscoelastic characterization of polymer-modified asphalt binders of pavement applications *Applied Rheology* **13** 191-99

Acknowledgments
Authors would like to thank the Ministry of Higher Education Malaysia and Universiti Putra Malaysia for the award of Fundamental Research Grant Scheme (FRGS) FRGS/1/2017/TK06/UPM/02/1 (Project Code: 03-01-17-1894FR) as well as all the people whom are directly and indirectly involved in this study.

IOP Conf. Series: Materials Science and Engineering **512** (2019) 012009 doi:10.1088/1757-899X/512/1/012009

Challenges in proof load test of bridges – Malaysian experience

KMS Ku Mahamud[1], CC Lim[2], SK Ng[1], MN Shahruddin[1] and MQ Jamil[1]
[1]Evenfit Consult Sdn. Bhd., Kuala Lumpur, Malaysia
[2]Jabatan Kerja Raya (JKR), Kuala Lumpur, Malaysia

E-mail: kusani1960@gmail.com

Abstract. It is a known fact that theoretical calculation for assessing the capacity of an existing bridge is grossly underestimating its actual capacity. The reasons are, among others, the safety factors adopted and assumptions made in modelling. The recommended alternative approach to assess the actual capacity of the bridge is by performing a proof load test. Proof load test is a useful tool where a bridge is incrementally loaded to a final targeted load. This approach has been used to check the bridges in Sarawak for their compliance with Weight Restriction Order (WRO) 2003. In these tests, bridges with Evaluation Load Ratings (ELR), which fall short of Public Work Department Malaysia (JKR) requirement related to WRO 2003 and hence recommended for replacement, were proof load tested with the intention to preserve these bridges. This paper presents the experience in conducting the proof load test on a few bridges in Malaysia. It describes the methodology for the proof load tests and the challenges faced during the tests. The proof load tests showed that the tested bridges can carry the full JKR standard of Medium Term Assessment Load (MTAL) loading, which is well within the legal load limits allowed under WRO 2003.

1. Introduction

Bridge testing is a common bridge management tool used to support major bridge management decisions, whether to retain or replace a bridge. A bridge test can also be used to establish an initial signature on the performance of a newly constructed bridge. Supplemental load test, which is a type of bridge testing, is sometimes utilised to validate and calibrate a computer model to improve the estimation of a bridge load carrying capacity [1]. Another type of bridge testing known as proof load test can be used to verify the safe load carrying capacity of existing bridges subjected to a new load requirement [2, 3].

JKR has relied on the analytical strength evaluation as one of the three criteria for bridge replacement [3]. It is thus common that bridge owners, frustrated by failure of their bridges in the theoretical strength evaluation, resorted to testing their bridges. However, analytical strength evaluation most often than not, under estimate the actual safe load carrying capacity of the bridge; as exemplified by a load test carried out by JKR in Year 1991 [4]. In this test, a bridge, commissioned for replacement was load tested to failure. The failure load was found to be much higher than the theoretical capacity.

2. Proof Load Test

Almost all the load test carried out in Malaysia involved proof loading test. The authors had participated in the proof load tests while in JKR and in the private practice. The experience of the authors in proof load test was summarised and compiled in Road Engineering Association of Malaysia's (REAM) Publication "A Manual on Bridge Asset Management" [5]. A load test is a complex operation, which requires an extensive planning, analyses, specialised equipment, test vehicles, concrete blocks and lifting crane, and large manpower utilisation. Figure 1 below presents the workflow for a proof load test.

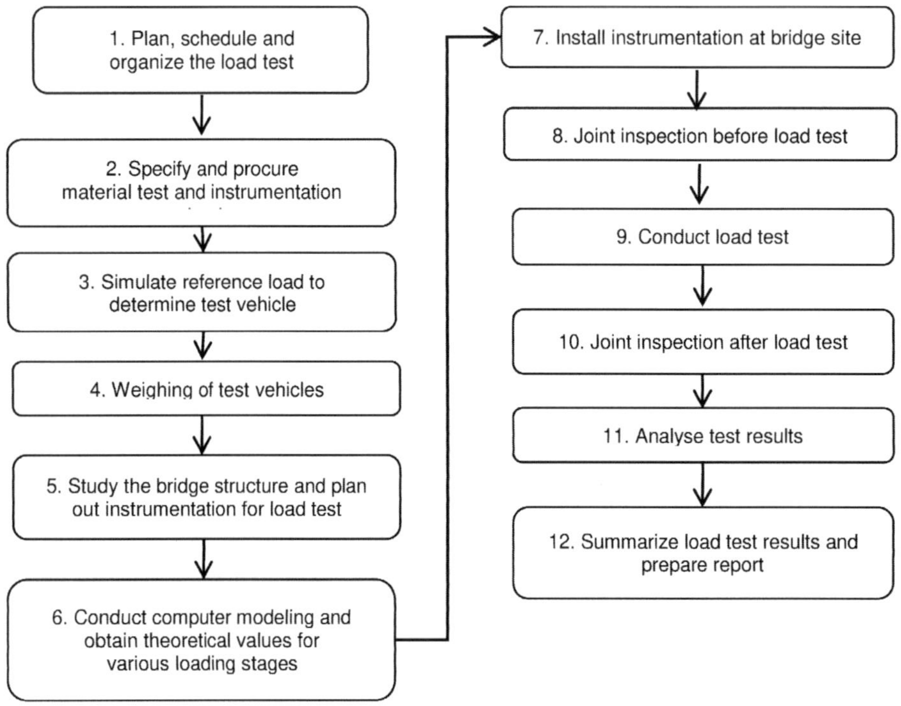

Figure 1. Workflow of Proof Load Test [5]

The workflow in conducting a proof load test can be categorised as:

i. Before the test (step 1 – 8),
ii. During the test (step 9),
iii. After the test (step 10 – 12).

Before the test, planning, scheduling and organizing of the load test are necessary to ensure that the load test would be conducted smoothly and in a safe manner. During the test, data acquired from the instrumentations was used to monitor the bridge responses under load to prevent the bridge from undergoing any damage.

After the test, the results were analysed and conclusion made. Besides verifying the safe load carrying capacity of the bridge by the proof load, analysis of the test result may reveal the behaviour of the bridge much as composite action, lateral load distribution characteristic.

3. Challenges

The fundamental concept of a proof load test is rather simple; it comprises the loading of the bridge in question incrementally to the intended proof load. However, as the well-known bridge testing engineer, Baidar Bahkt had stated in his paper "Bridge testing: A surprise every time", testing of bridges very often would encounter problems and challenges [6].

IOP Conf. Series: Materials Science and Engineering **512** (2019) 012009 doi:10.1088/1757-899X/512/1/012009

3.1. Pre-Test Phase

3.1.1. Compliance to the local authorities' requirements. Compliance to the local authorities' requirements is one of the main challenges in conducting a bridge testing. The main concerns to the local authorities are the safety of road users and public complaints resulted from the traffic obstruction during the load test. Therefore, the local authorities impose stringent guidelines to be followed in order to ensure the safety of the road users. Traffic management plan (TMP) that meets all the requirements along with schedule of work at site need to be prepared for the approval of the local authorities. Usually there would be several revisions to the proposed TMP in the process of getting the approval from the local authorities. The challenge is in having to obtain the approval from the local authority within the scheduled time frame without too many rounds of submission. In certain cases, the local authority may even require inputs from a professional traffic consultant to come out with an acceptable TMP.

The local authorities would usually allow for a full road/bridge closure at daytime if alternative road is available for traffic diversion. Temporary road diversion and traffic management plan (TMP) need to be designed and endorsed by a qualified traffic management officer (TMO). The TMP proposal would then be reviewed and commented by the road safety audit prior to the approval by local authority.

In case there is no alternative road for traffic diversion, the local authorities would only allow the road/bridge to be intermittently closed during the load test. In this situation, the requirements imposed are even more stringent. The load test is allowed to be carried out only at night and very often, the test needed to be stopped before 6.00 am when the traffic begins to build up. This is to minimize disruption to traffic. In addition to that, the maximum duration allowed for the closure in most cases is not more than 30 minutes.

3.1.2. Availability of suitable test vehicles. JKR has established a standard procedure of proof load test involving a standard test vehicle comprising a low-loader trailer with two tandems at the back (see Figure 2) [7]. The use of the standard vehicle rather than having the concrete blocks loaded directly on the bridge deck has a few advantages, such as:

i. Loading and unloading of concrete block onto the test vehicle could be carried out away from the bridge.
ii. Removal of the test load from the bridge could be done quickly with the test vehicle.
iii. The axles impose on the bridge as point loads would be more deterministic than that due to a distributed load from the concrete blocks, which may have uncertain load distribution due to the development of arching effect within the concrete blocks.

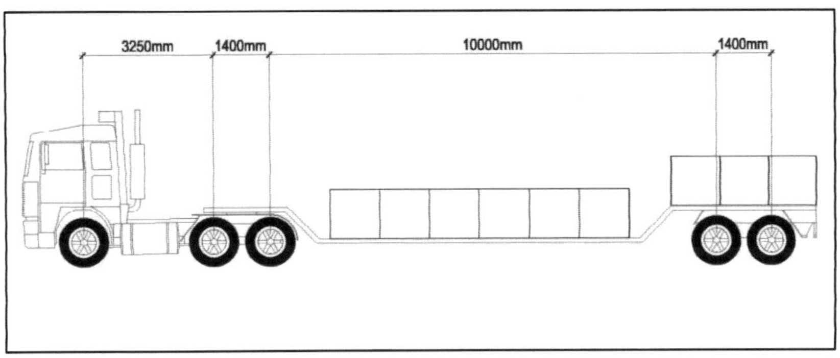

Figure 2. Configuration JKR Standard Test Vehicle

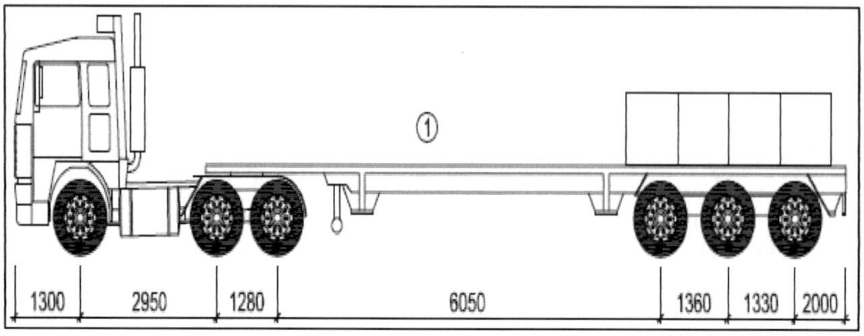

Figure 3. Configuration of test vehicles (high-loader) which are easily available in the market

Low-loader trailer has better capacity in carrying the concrete blocks as compared to the other types of trailer. However, it is hardly available in some part of the country. In some cases, to get as many as 4 to 6 numbers of identical configurations and dimensions of low-loader trailers for a load test is a big challenge. Considering this difficulty, high bed trailers were chosen as an alternative, as they are easily available in the market. However, the high bed trailer which tends to have the payload concentrated on the rear axles would often have the difficulty moving under heavy load. The use of standard test vehicles with known axle load under each loading level would avoid the need to weigh the test vehicle prior to the test.

Prior to carrying out the load test at site, the test vehicle has to be weighed to get the load from each axle (See Figure 3). The weighing of test vehicle is usually done at the nearest Jabatan Pengangkutan Jalan (JPJ) weighbridge station (see Figure 4). The axle loads are then used in the bridge modelling to obtain the theoretical responses of the bridge when subjected to load from the test vehicle. However, the test vehicles supplied at site for the load test sometimes are of different dimensions from the vehicle weighed prior to the load test. As a result, the targeted load effect might not be as accurate as desired. In this case, if the dimensions of the test vehicle is significantly different from the vehicle that had been weighed, the test vehicle shall have to be reweighed and the bridge model redone.

Figure 4. Weighing of test vehicle in Stesen Penguatkuasa JPJ

IOP Conf. Series: Materials Science and Engineering **512** (2019) 012009 doi:10.1088/1757-899X/512/1/012009

3.1.3. Superstructure over river or high in elevation. Superstructure that crosses over the river or is high in elevation often poses many challenges in a proof load test. They are:-

i. Site measurement for dimension verification,
ii. Collection of samples for material test,
iii. Installation of instruments,
iv. Joint inspection.

When as-built drawing of the bridge is not available, a comprehensive site measurement would be necessary. Other than that, sampling and testing of bridge material are also required to assess the in-situ material properties of the existing bridge elements. All these are needed for the structural modelling and analysis. In addition, joint inspection to check if there is new damage to the bridge is usually carried out at arm's length distance.

Therefore, in order to carry out the aforementioned activities, construction of a cantilevered working platform is required at locations of interest as a means of access (see Figure 5). The installation of cantilevered working platforms would often require a contractor specialized in scaffolding installation.

Figure 5. Working platform at a superstructure that crosses over river and high in elevation

Also, a linear Variable Displacement Transducer (LVDT) is not suitable to be used at a bridge that crosses over river or high in elevation. This is because the LVDT requires an independent and firm support so that it can measure the movement of the bridge component accurately. A survey instrument such as total station with prisms would be often used as an alternative. However, this would compromise the precision of deflection reading to certain degree.

3.1.4. Safe procedure for loading and unloading. During the planning stage, a safe procedure for loading and unloading has to be established. The challenge is to ensure that the loading and unloading activities would not cause damage to the bridge at any stage of loading and unloading activities. To minimise the risk of a bridge being loaded beyond its capacity, the proof load is often applied incrementally in stages with smaller increments towards the later stages.

One should also bear in mind that any mistake in unloading sequence of one span could result in a higher bending effect to the adjacent span of the bridge. For example, to obtain the maximum hogging moment at pier, the test vehicle would be loaded incrementally on the left and the right side of the pier (see Figure 6).

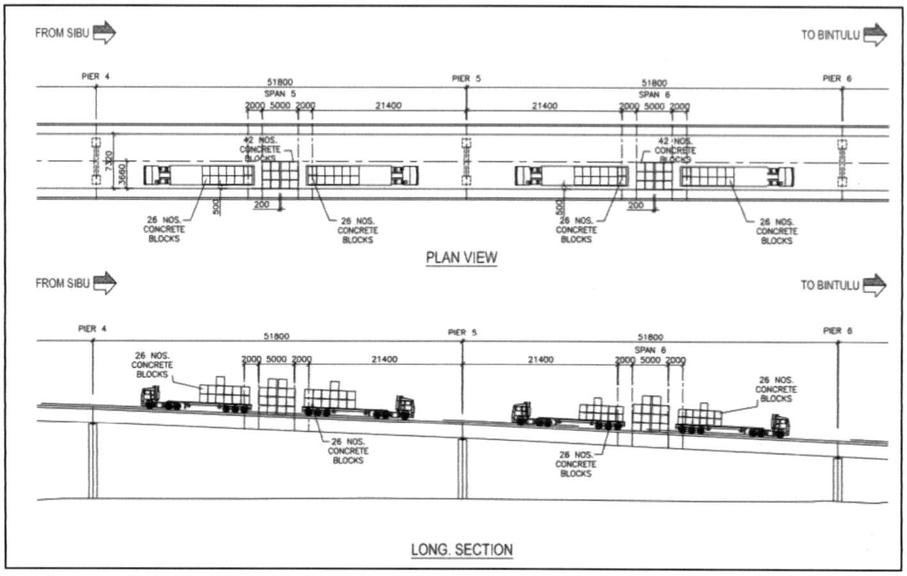

Figure 6. Load vehicle for hogging moment at Pier 5

During unloading, removal of test load in Span 5 could result in an increase in bending effect in Span 6. Thus, to avoid from this situation, it is advised that the test load in both spans to be removed concurrently. To establish this load sequences, the movement and sequence of load crossing over the bridge span to the positions of test loads are analysed and studied. The safe loading and unloading sequences are identified and any related procedures are formed. The time taken to complete the whole cycle of loading and unloading are analysed to ensure that the proposed load stages in loading and unloading procedure could be done within the test period.

3.2. During the Test Phase.

During the test, common problems include slow progress of works, machineries breakdown, actual measurement conflicting with theoretical values and sign of failure occurs.

3.2.1. Load test within a short time frame. Usually, the time available for the load test is short, especially when the load test is only allowed to be conducted at night and need to be stopped in the morning. In this situation, efficient management of the resources is crucial to ensure that the test could be carried out smoothly and successfully within the short time frame. Time analysis is carried out in order to get a best work sequences that suite to that short time frame. The work and time schedule is then established to guide the tester during the load test. This schedule is used as a baseline to monitor the progress of works during the test.

Sometimes, inexperienced worker could cause slower work progress. Other than that, high volume of traffic during the load test could also delay the progress of works during the test. It resulted in longer time required to ease the traffic after every 30 minutes of closure.

3.2.2. Machinery problems. Machinery problems related to the test vehicles and crane can happen during the test. It is thus important to prepare a backup plan during the planning stage, e.g. having an extra test vehicle and crane.

3.2.3. Measurement deviated from the theory. The use of instruments in bridge load testing is to monitor the behaviour of critical bridge members during the test. The measured readings are closely monitored and compared to the theoretical values to ensure that the bridge behaves in a linear elastic way when subjected to the test load. Precautionary action would be taken when the measured readings significantly exceeded the theoretical values or the bridge started to show non-linear behaviour. This is to ensure that the bridge is not be overloaded or deformed excessively during the test. However, sometimes the readings are not as expected, for example:

i. readings that are out of order of the magnitude,
ii. strain readings that shows the location is in compression when theoretically it was expected to be in tension and vice versa.

The instrument could also be damaged due to impact from boat, barge, vehicles moving under the bridge. If the bridge being tested is located in rural area, there is possibility for wild animals fiddling with the instrument thus damaging it.

3.2.4. Accident. One should anticipate for accident that could happen during the load test. Therefore, an accident response plan should be prepared prior to the load test, which shall include, but not limited to the following:

i. The action to be taken to minimize the risk of accident,
ii. Action to be taken when accident occur to minimize the damage.

3.3. Post-Test Phase. If a bridge has been successfully tested to the final targeted load, and did not show any signs of failure, i.e. behaving in linear-elastic way to the load increments and no damage on the bridge after the load test; this is actually an indicator that the bridge has the capacity of at least the targeted test load.

Deviation of some of the test results from the theories is quite a common occurrence in proof load test. It is because the model developed did not be behave exactly as the actual bridge. However, as long as the measured values did not deviate significantly from the theoretical values; and behaved linearly with the incremental load stages, the test results can be considered as acceptable.

24-hour readings of deflections and strains are taken one day before and after the load test, without loads, to study the behaviour of strains and deflections due to temperature changes. The measured strains and deflection should be corrected to compensate the variations affected by temperature changes.

4. Conclusions

Proof load test involves the positioning of the load (known as "proof load") to a bridge, to verify that the bridge is capable of sustaining it without causing severe damage or permanent deformation to the bridge. Though this appears simple, the actual challenges come from the planning, inspection and execution of the test, not only technical but also logistics aspects, which requires extensive planning, analysis and specialized equipment.

In this paper, the challenges encountered by the authors while conducting a proof load test are presented. The intention is to share the authors' experience in conducting proof load test, as a reference for engineers who may perform such tests in the future.

5. References

[1] National Steering Committee for the Load Testing of Bridges 1998 *Guidelines for the supplementary load testing of bridges* (Great Britain: Institution of Civil Engineers)
[2] Casas J R and Gomez J D 2013 Load Rating of Highway Bridges by Proof-loading *KSCE Journal of Civil Engineering* **17**(3) 556-67
[3] Tharmabala T, Toutant D, Ku Mahamud KMS and Salleh R 1995 *Strength Evaluation and Load Testing of Bridges in Malaysia*
[4] Jabatan Kerja Raya *Terms of Reference for Bridge Assessment Procedure* (Kuala Lumpur: JKR)
[5] Khoo C L, Ng S K and Ismail A 1992 *Full Scale Load Testing to Failure of RC Rigid Frame Type Bridge*
[6] Road Engineering Association of Malaysia (REAM) *Manual on Bridge Assets Management* (Kuala Lumpur: REAM)
[7] Bakht B and Jaeger L G 1988 Bridge Testing - A Surprise Every Time *Structural Research Paper SRR-88-04*
[8] Ku Mahamud K M S 2000 *Bridge Load Testing – The Malaysian Experience*

IOP Conf. Series: Materials Science and Engineering **512** (2019) 012010 doi:10.1088/1757-899X/512/1/012010

Carbon footprints calculator of highway pavement rehabilitation: The quantification of carbon emissions per unit activity

Che Muhammad Fatihi Hafifi Che Wahid[1], Eeydzah Aminudin[1], Muhd Zaimi Abd Majid[1], Mohd Rosli Hainin[1], Mohd Khairul Idham Mohd Satar[1], Muhammad Naqiuddin Mohd Warid[1], Rozana Zakaria[1], Muhammad Rozaid Ramli[1], Rosli Mohamad Zin[1], Zainura Zainon Noor[1], Ain Naadia Mazlan[1], Omar Sedeeq Yousif[1], Monh Neardey[1], Azman Mohd Jais[2], Rohaya Abdullah[2], Yazlin Salfiza Mohd Yazid[1] and Nadzrol Fadzilah Ahmad[2]

[1] School of Civil Engineering, Faculty of Engineering, Universiti Teknologi Malaysia, 81310 Johor, Malaysia
[2] Lembaga Lebuhraya Malaysia, Wisma Lebuhraya, KM-6, Jalan Serdang - Kajang, 43000, Malaysia

E-mail: eeydzah@utm.my

Abstract. Carbon dioxide that emitted throughout the construction lifespan is crucial from development of design, construction, up to demolition stage. Focusing in the rehabilitation of highway pavement, the activities that normally involved are from milling, roller up to dumping the wastes asphalts. The aim of this paper is to estimate the carbon footprint emission in unit of tonnes for carbon dioxide equivalents (t CO_2-eq/km-lane). The case study involved based on five selected projects for north–south bound expressway of Malaysia. The results show that highest carbon footprint were from the Mill and Pave 250 mm (MP 250) with 117.94 t CO_2-eq/km-lane. This is due to its depth of pavement with 140 mm dense bituminous macadam, 60 mm asphalt concrete base course and 50 mm asphalt concrete wearing course as compared as other Mill and Pave. The result followed by MP 200, MP 110 and MP 50 at 95.19, 54.10, and 27.37 t CO_2-eq/km-lane respectively. The critical activities involved during the construction shows the asphalt concrete production as the highest generation of emission with 53.78 t CO_2-eq/km-lane, followed by lorry transportation of asphalt waste, lorry transportation of asphalt concrete, fuel consumption from machineries, neomed coat, and water consumption with the value of 28.58, 24.50, 10.83, 0.23, and 0.01 t CO_2-eq/km-lane respectively. The results from this study considered as references guidelines for carbon footprint calculator to calculate emission as compared to other expressway in Malaysia.

1. Introduction

As Malaysia transforms into a high-income generation, there is a need for national development strategy in line with current megatrends of the world focusing in climate change issues. One of main agenda of Malaysia's involvement was pledging in cutting down the national carbon emission intensity to 45 per cent by 2030 based on 2005 emission levels. As since June 2017, 150 countries have ratified the Paris Climate Agreement regarding this issue. Align with pledging, the country initiatives were extended to the development of The National Green Technology Policy of greenhouse gases emission in order to ensure sustainable development approach can significantly progress besides providing the major improvement, especially in construction sector.

Different methods to reduce the emissions of pollutants in various countries proposed such as used eco tax and carbon tax in promoting the activities involved in line with low environmental impacts. These emission reduction methods require accurate measurement [1]. Therefore, this study focuses on the development of Carbon Footprint Calculator tool for highway construction in Malaysia, which will create tools that allow every sector (contractor, concessionaire and authority) to calculate, monitor and receive feedback on their personal emissions. In addition, these carbon footprint calculators are not focuses on the amount of emissions generated but also providing detail activities related to construction such as list of machinery and quantity of material needed in each activity.

Carbon footprint technology is still evolving, especially with regard to the accuracy of estimations and transparency of calculations methods, as a means of benchmarking and comparison. Hence, there is a need to explore, in more detail, the consequences of personal choices before setting a normative concept of responsible behaviour toward emissions reductions [2].

The purpose of this research is to estimate the carbon footprint emission in the unit of tonnes of carbon dioxide equivalents (t CO_2-eq/km-lane) based on the activities involved during the construction of rehabilitation of pavement. Additionally, a selection on few first highways will be evaluated for the Carbon Footprint Assessment tool and will be used as a baseline study for highway organization in planning the setting target for carbon emission reduction for upcoming years. Lastly, with this tool it will provide an outlook on the trends of the GHG emissions that can be utilized to come out with a better strategy to manage green highway in Malaysia.

2. Research methodology

This study attempts to propose a Life Cycle Assessment (LCA) method to estimate CO_2-eq emission. The data was being collected which record the detail consumptions of materials and energy. In the study scope, the sources of emission divided into three stages: raw material production, material transportation and onsite construction. The total emissions are the sum of each stage:

$$Q = Q_1 + Q_2 + Q_3 \quad (1)$$

Where, Q is the total amount of CO_2 emission from highway construction; Q_1, Q_2, and Q_3 is the amount of CO_2 emission from raw material production, material transportation and onsite construction, respectively. The unit of Q, Q_1, Q_2, and Q_3 is t CO2-eq/km-lane.

Table 1. Sources of carbon footprint throughout the entire life cycle of highway maintenance.

Onsite construction	Embodies carbon	
	Material transportation	Raw material production
Milling machine	Lorry	Asphalt concrete
Paver	Truck	DBM40
Tandem Roller	Van	ACBC28
Tyre Roller		ACWC20
Bobcat		Tack coat
Lorry		Prime coat
Gen Set		
Compressor		
Water tanker lorry		
Diamond cutter		

The sources of emission produce during construction activities normally is the usage of diesel engines in on and off-road construction machinery [3]. The possible sources of carbon footprint throughout the entire life cycle of highway maintenance show in Table 1 below. Embodied emissions also should consider in term life cycle assessment. The embodied emissions is the GHG emissions which the sum of all the energy required to produce any goods or services, considered as if that energy was incorporated or 'embodied' in the product itself.

3. Data collection

This study focuses on the machineries and materials used in rehabilitation phases that cover input and output in terms of quantities of materials and fossil fuel consumption for machineries. The pavement rehabilitation type in this study is Mill and Pave 250mm (MP 250), Mill and Pave 200mm (MP 200), Mill and Pave 110mm (MP 110), and Mill and Pave 50mm (MP 50) as show in Table 2. Table 3 show the average route length, quantities of material asphalt such as dense bitumen macadam (DBM40), asphaltic concrete binder course (ACBC28), and asphaltic concrete wearing course (ACWC20), neomed coat and quantity of water in difference type of pavement rehabilitation in one day's construction were obtained from the respective project officers via interviews. Table 4 summarizes the quantity of fuel consumption from machineries for one day construction. The types of lorry used at the site and the fuel used of the lorry are shown in Table 5.

Table 2. Type of pavement rehabilitation in mill and pave work.

Type	Description
MP 250	Removal of the asphalt layer and base material to a depth of 250mm and replacement with 140mm DBM40, 60mm ACBC28 and 50mm ACWC20
MP 200	Removal of the asphalt layer and base material to a depth of 200mm and replacement with 90mm DBM40, 60mm ACBC28 and 50mm ACWC20
MP 110	Removal of the asphalt layer to a depth of 110mm and replacement with 60mm ACBC28 and 50mm ACWC20
MP 50	Removal of the asphalt layer to a depth of 50mm and replacement with 50mm ACWC20

Table 3. The data obtained from the respective site engineer via interviews[a].

Type	Average length (meter/day)	Asphalt (tonne/day)	Neomed coat (kg/day)	Water (litre/day)
MP 50	330	155	600	3000
MP 110	260	258	200	3000
MP 200	176	313	200	3000
MP 250	140	310	200	3000

[a] The width of one lane is 4.1 meter.

Table 4. The quantity of fuel consumption from machineries for one day construction.

Machineries	Average fuel usage (litre/day)					
	Sub con A	Sub con B	Sub con C	Sub con D	Sub con E	Average
Milling machine	100	200	100	200	200	160
Paver	30	50	50	60	60	50
Tandem Roller	10	50	30	20	20	26
Tyre Roller	10	50	30	20	20	26
Bobcat	10	50	20	30	30	28
Lorry (sprayer)	80	80	50	50	50	62
Generator set	5	10	5	5	5	6
Lorry (compressor)	80	80	50	50	50	62
Compressor	10	10	10	10	10	10
Lorry (water tanker)	80	80	80	50	50	68
Diamond cutter	3	3	3	3	3	3
Van	30	33	25	30	30	30

Table 5. The types and the fuel used of the lorry for transport the material[b].

Transportation	Average distance per one trip (km/trip)	Average fuel usage for one trip (litre/trip)
Lorry transport the asphalt from plant *tyre 10 : 21-26 ton/trip *tyre 12 : 27-36 ton/trip	60	100
Lorry transport the asphalt waste *tyre 10 : 9.5 m3/trip *tyre 12 : 16 m3/trip	60	100

[b] The fuel used in the processes was brought by lorry trucks from the supply source, which is 60 km away. This detail is regarded as the fuel transportation inventory value.

4. Data analysis

For normal average consumption of energy per unit time and the equipment running time for a particular unit process can be obtained by multiplying the average energy consumption per unit of time by the running time of equipment, the fuel consumption by the equipment can be obtained [4]. Then, the total consumption can be determined by summing the consumption of all equipment used in the same unit process.

The normal emission factors estimates for water is 0.376 kg CO_2/m^3 [5], asphalt concrete 0.0238 kg CO_2/kg [6] [7] [8], emulsified bitumen neomed 0.16 kg CO_2/kg [9] and diesel 3.164 kg CO_2/kg [9] [10] [11]. It can be calculate by multiplying the activity data with emission factor. Normally the CO_2 gas emitted from machinery is converted into CO_2-eq by multiplying the weight of the CO_2 being measured by its respective Global Warming Potential (GWP) [4]. The GWP value for CO_2 is quantified as one. As for the machineries involved during construction, the amount of fuel were identify based on consumption of fuel taken into account to correspondent the amount of CO_2 emissions. The total emission and energy factors are presented based on calculated per kilometre. Figure 1 shows the calculation on calculate the energy consumption of machineries.

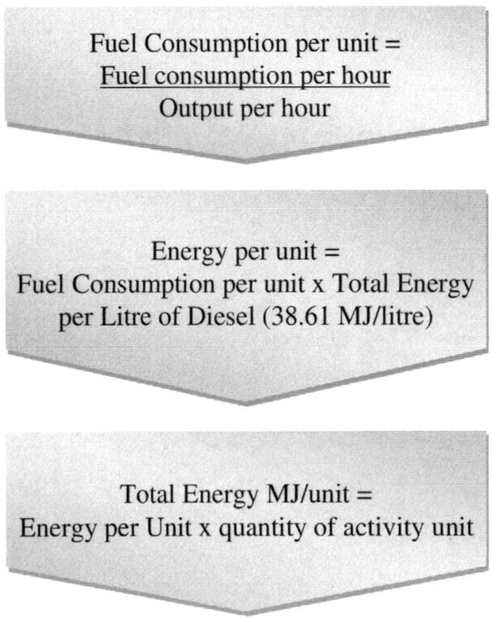

Figure 1. Justify the caption. Calculation of energy consumption.

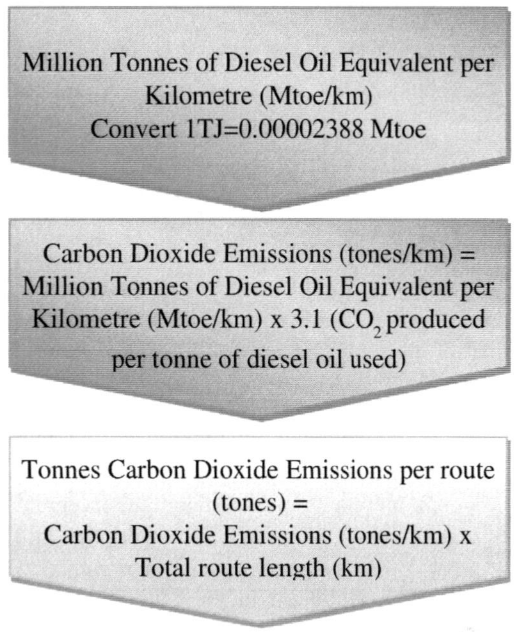

Figure 2. Calculation of carbon dioxide from diesel used.

Diesel fuel has a Lower Calorific Value (LCV) of 35.1 MJ/litre where the energy contained within the fuel. It has a Precombustion Value (10%) of 3.51 MJ/litre when it takes a total of 10% of the diesel energy to get the diesel from the earth to usable form. Therefore the 'Total Energy per Litre of Diesel is 38.61 MJ/litre' where it includes a precombustion value of 10% [12]. The amount of 38.61 MJ/litre will be an assumption if all the machines used the diesel as the fuel. Moreover, figure 2 shows how the calculation will be used to calculate the CO_2 emission of machineries. From the total Energy, 'Total Energy per Kilometre (TJ/km)' the value will be convert into the 'Million Tonnes of Diesel Oil Equivalent per Kilometre' by the conversion factor turning TJ into Mtoe (1 TJ = 0.00002388 Mtoe). The next will be identify as 'Carbon Dioxide Emissions (tones/km)'. This can be measure by multiplying the 'Million Tonnes of Diesel Oil Equivalent per Kilometre (Mtoe/km)' by 3.1. Standard of Environment Impact Assessment shows that there are 3.1 tonnes of carbon dioxide has been produced by per tonne of diesel oil used [13]. Then, the amount of carbon emission per route can be obtained with this calculation.

5. Result and Discussion

The carbon footprints of machineries in road pavement rehabilitation originates from different types of machineries used on the different layers, such as milling work, the DBM layer, the ACBC layer, and the ACWC layer. The produced carbon footprint is associated with embodies carbon and transportation values. Figure 3(b) shows the total carbon footprint emitted from pavement rehabilitation. The highest carbon footprint in pavement rehabilitation work is from Mill and Pave 250mm (MP 250) a total of 117.94 t CO_2-eq/km-lane. Because due to its highest depth of pavement with 140 mm dense bituminous macadam, 60 mm asphalt concrete base course and 50 mm asphalt concrete wearing course, followed by MP 200, MP 110 and MP 50 at 95.19, 54.10, 27.37 t CO_2-eq/km-lane, respectively. In the difference activity, asphalt pavement material produce the highest carbon footprint 53.78 t CO_2-eq/km-lane, followed by lorry transportation of asphalt waste from site construction to site disposal, lorry transportation of asphalt concrete from plant to site construction, fuel from machineries, neomed coat, and water consumption emit 28.58, 24.50, 10.83, 0.23, and 0.01 t CO_2-eq/km-lane respectively. The total result for the carbon footprint emission through all activity is shown in figure 3(a). Figure 3(b) show the water consumption and neomed coat has the lowest carbon footprint emission, because its use is lowest among all materials.

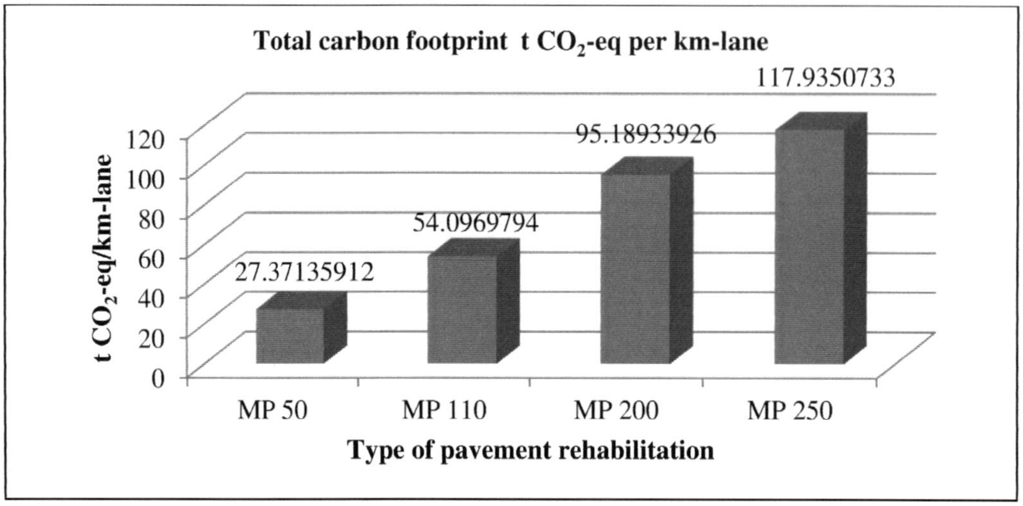

(a)

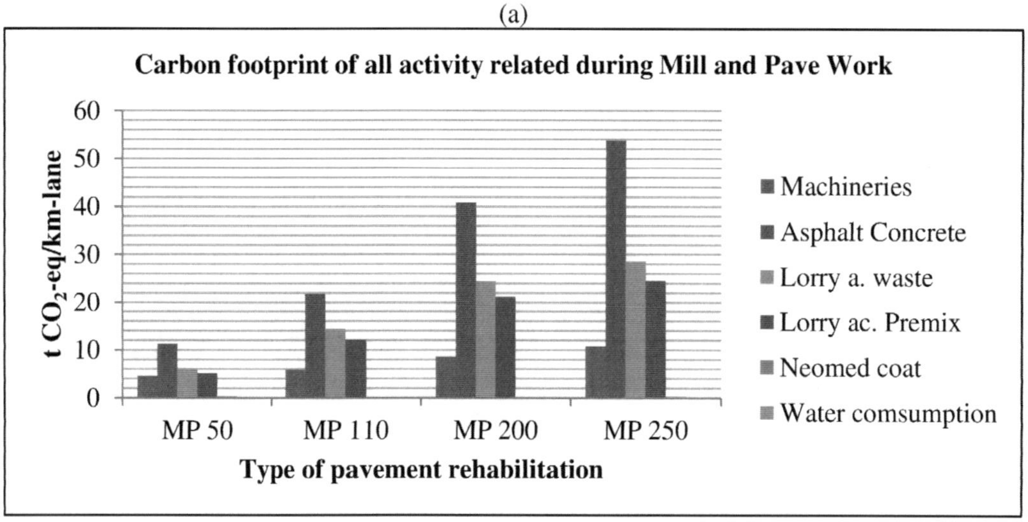

(b)

Figure 3. Carbon footprint emission of the pavement rehabilitation: (a) Total activity, and (b) Separated

6. Conclusions

Overall, the major sources of CO_2 emission generated for Malaysian highway came from material production, followed by onsite construction, and transportation stage. The estimated emission factor of MP 250, MP 200, MP 100, and MP 50 is 117.94 t CO_2-eq/km-lane, 95.19 t CO_2-eq/km-lane, 54.10 t CO_2-eq/km-lane, and 27.37 t CO_2-eq/km-lane, respectively. This indicates that a proper invention on the technology for future is needed in producing the highway is necessary especially in the mill and pave work in order to cut down the emission produces from it sources.

7. References

[1] Cheng J C 2011 A Web Service Framework for Measuring and Monitoring Environmental and Carbon Footprint in Construction Supply Chains *The Twelfth East Asia-Pacific Conference on Structural Engineering and Construction* 14 pp 141–47

[2] de Aguiar T R S, Fearfull A and Sanagustín Fons M V 2016 Calculating the carbon footprint: Implications for governing emissions and gender relations *Account Forum* **40** (2) 63–77

[3] Heidari B and Marr L C 2015 Real-time emissions from construction equipment compared with model predictions *Journal of the Air & Waste Management Association* **65**(2) 115-25

[4] Noland, R. B., & Hanson, C. S. (2015). Life-cycle greenhouse gas emissions associated with a highway reconstruction: A New Jersey case study *Journal of Cleaner Production* **107** 731-40

[5] Shimizu Y, Dejima S and Toyosada K 2012 The CO2 emission factor of water in Japan *Water* **4** (4) 759–69

[6] Thives L P and Ghisi E 2017 Asphalt mixtures emission and energy consumption: A review *Renew. Sustain. Energy Rev.* **72** 473–84

[7] Yu B 2013 Environmental Implications of Pavements: A Life Cycle View (Doctoral Dissertation, University of South Florida)

[8] Mukherjee A 2011 *Carbon Footprint for HMA and PCC Pavements* (Michigan: Michigan Technological University)

[9] Huang Y, Ning Y, Zhang T and Wu J 2016 Measuring carbon emissions of pavement construction in China *Sustainability* **8**(8) 1–13

[10] Sarbring 2014 *A Carbon Footprint Assessment on Construction and Maintenance Operations for the Port of Gothenburg* (Master thesis, Chalmers University of Technology)

[11] Kitakyushu 2016 *Feasibility Study on Joint Crediting Mechanism Project For Realization for a Low-Carbon Society in Asia Establishment of Base for Low-carbon Project Expansion in Iskandar* (Johor Bahru: Iskandar Malaysia)

[12] Stripple H 2001 *Life Cycle Assessment of Road: A Pilot Study for Inventory Analysis Second* (Stockholm: IVL Swedish Environmental Research Institute)

[13] Rahman N B A 2013 *Practical Approach for Reduction of Fuel Consumption on Earthwork Highway Machineries* (Doctoral dissertation, Universiti Teknologi Malaysia)

Acknowledgments

The authors would like to express their deep gratitude to the Malaysian Highway Authority (LLM) and Universiti Teknologi Malaysia for supporting financial grant, Cost Centre No: R.J130000.7322.4B322, R.J130000.7322.4B326, R.J130000.7322.4B332, R.J130000.7322.4B334, & R.J130000.7322.4B331. MyGHI research team members and Department of Construction Management, UTM together with opportunity for the research.

IOP Conf. Series: Materials Science and Engineering **512** (2019) 012011 doi:10.1088/1757-899X/512/1/012011

Organosilane technology for sustainable road infrastructure

Mikhil Ranka

Zydex Industries Pvt Limited, Vadodara, Gujarat, India

E-mail: mikhilranka@zydexindustries.com

Abstract. New Organo-Silane nanotechnology, apart from other green benefits, enables dropping of mixing and compaction temperatures of Hot Mixes. Although, the concept of Warm Mix Asphalt (WMA) has been around for a while, the available WMA technologies only talk of drop in mixing temperature and not the compaction temperature, which the new Organo-Silane technology does. It is therefore prudent to call this technology, a 'Warm Compaction' technology. In addition, this 'Warm Compaction' technology improves strength, fatigue resistance and water resistance, leading to potential extension of pavement life, contributing to sustainability. This paper presents the summary of laboratory studies made at the IMT *(Instituto Mexicano del Transporte)*, Mexico, and Braunschweig University, Germany to assess stiffness modulus, fatigue-resistance and water-resistance of this 'Warm Compaction' technology at different mixing and compaction temperatures for optimization.

1. Green road

We have taken this earth on lease from our future generations. It is our duty to return it to them in liveable condition if not better. Striving for sustainability through prudent use of limiting natural resources and restraint in emissions is the call of the day.

As it is said, 'Nations don't build roads. Roads build a nation'. It is very important to develop road infrastructure at a brisk pace, if we want to attain / sustain high economic growth. At the same time, it is equally important to ensure that such infrastructure development is sustainable. Prudent use of limiting natural resources therefore, becomes our responsibility now, more than ever before.

There may be many ways of bringing in sustainability in building roads. Here are some of the ways in which Zydex Nanotechnology contributes to sustainability in road construction:

- Reducing the usage of limiting resources like bitumen, aggregates etc.
- Extending the life cycles of the roads, so as to defer the demand for such resources
- Reducing fuel consumption and low temperature operation leading to reduction in emission

This paper discusses the 'Low Temperature Operation' aspect of the Zydex Nanotechnology in particular.

2. Conventional warm mix asphalt and its limitations

The concept of Warm Mix Asphalt (WMA) has been around since the 90's. It was developed in Europe in response to EEC countries signing the 1997 Kyoto Treaty to reduce greenhouse gases.

A warm asphalt mix process (WAM) has been developed in Europe and was reported at the First International Conference of Asphalt Pavements in Sydney, 2000. A more complete report was given by [2] at the Eurobitume congress in 2000. [1] describes an innovative warm mixture 3 process that was tested in the laboratory and evaluated in large-scale field trials (in Norway, the UK and the Netherlands) with particular reference to the production and laying of dense graded wearing courses [2]. Their work resulted in the development of WARM Foam, Warm Asphalt Mix with foamed bitumen [3]. At the Eurobitume congress in 2004, [4] introduced the use of a synthetic zeolite additive to produce warm mix asphalt. The zeolite creates a foaming effect that results in a higher workability of the mix [4]. Warm mixes have received some attention in Europe and Australia since around 2000. The pavement industry in North America started to give warm mixes some interest a few years later and in June 2005 the National Center for Asphalt Technology (NCAT) published two reports about the use of Sasobit, a synthetic wax, and Aspha-min, a synthetic zeolite, in warm mix asphalt [5,6].

Lower mixing temperature in case of WMA, is certainly a huge advantage, but the technology does have certain challenges as follows.

- Lower temperatures used for WMA can result in incomplete drying of the aggregates and the resulting trapped water in the coated aggregates may cause moisture damage.
- Finding the right balance between lowering the production temperatures, applying anti-stripping agents and achieving a sufficiently moisture resistant asphalt mixture might be a challenge when using WMA

This is probably the reason behind WMA technology not catching up all over the world at the pace initially anticipated.

3. Organosilane 'warm compaction' technology

The available WMA technologies only talk of drop in mixing temperature and not the compaction temperature. The new Organosilane technology now allows reduction in both, mixing as well as compaction temperature.

In addition the new Organosilane technology also improves coating efficiency, leading to faster and complete coating of even the fines. This is because the Organosilane chemistry reduces the surface tension of bitumen for faster and better wetting.

The workability of the Organosilane mixes is also observed to be better and the compaction easier.The low temperature mixing, low temperature compaction and the other benefits mentioned above are verified by laboratory studies.

4. Case study - transfer center for the road sector (tsw), Braunschweig university, Germany

Goal of the examinations carried out here was to document the influence of the binding agent ZycoTherm® on the mixing process and to document the compression properties of asphalt.

For this purpose, two types of asphalt that are common in Germany, an asphalt binding agent AC 16 B S with 50/70 and an asphalt concrete for asphalt surface layers AC 11 D S with 25/55-55 A, were mixed with and without ZycoTherm® and compressed at different temperatures. An overview of the variants produced is specified in table 5-1.

During the mixing processes, the power consumption of the laboratory mixer was recorded and the time that the degrees of coating of 50 %, 75 %, 90 % and 100 % were reached was noted. The times for reaching the degree of coating of 50 % are specified in table 5-1. Therefore, the mixed material is coated quicker when ZycoTherm® is added. On average, 20 % less time is required. With regard to the mixing performance, for AC 16 B S a tendency was recognised that less performance was required when adding ZycoTherm®.

The compression resistance does not show any difference for both asphalt types when adding ZycoTherm®, as documented in table 5-1. The raw density of the asphalt mixture produced can be considered as equal for both types of asphalt. The densities by volume of the roller compressed asphalt test plates only indicate small differences. The void content is specified as outcoming result in table 5-1.

Table 1. Compositions of the type of asphalt applied

Asphalt		AC 16 B S	AC 11 D S
Bitumen		50/70	25/55-55 A
Bindemittelgehalt	M.-%	4,3	6,0
Anteil ZycoTherm®, bezogen auf den Bindemittelgehalt	M.-%	0,10	0,15
Gesteine		Gabbro	Gabbro
> 16,0 mm	M.-%	2,8	-
11,2 - 16,0 mm	M.-%	28,5	0,5
8,0 - 11,2 mm	M.-%	12,1	20,2
5,6 - 8,0 mm	M.-%	12,8	10,8
2,0 - 5,6 mm	M.-%	15,2	23,3
0,063 - 2,0 mm	M.-%	22,0	37,6
< 0,063 mm	M.-%	6,6	7,6

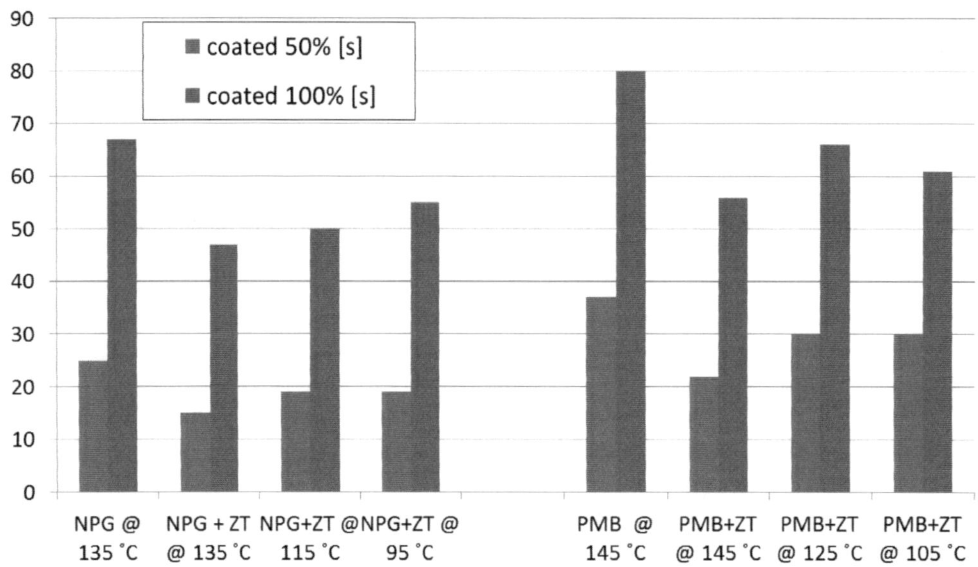

Figure 1. Coating efficiency of ZycoTherm at different mixing temperatures

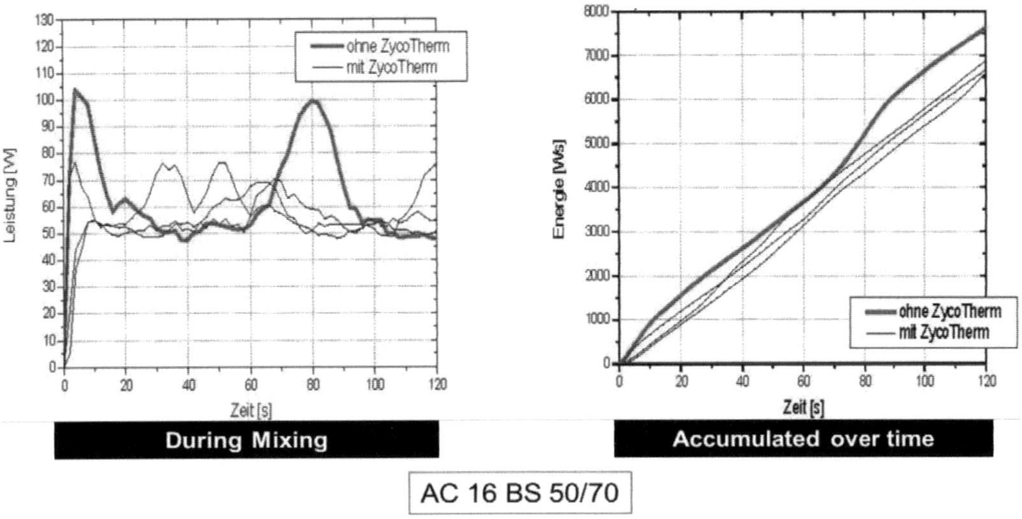

Figure 2. Effect of ZycoTherm on power / energy consumption while mixing

Table 2. Effect of ZycoTherm on Compaction Resistance and Voids

Mix Type	Binder Type	C / ZT	Compaction Temp ^{0}C	50 % Coating Sec	Compaction Resistance 21Nm	Void Content Vol %
AC 16 BS	50/70	C	135	25	41.6	7.0
		ZT	135	15	43.5	7.1
			115	19	42.8	7.2
			95	19	41.3	6.9
AC 11 DS	25/55-55 A	C	145	37	37.3	4.3
		ZT	145	22	34.2	4.6
			125	30	36.5	5.0
			105	30	36.5	5.3

Innovationsgesellschaft Technische Universität Braunschweig mbH iTUBS

IOP Conf. Series: Materials Science and Engineering **512** (2019) 012011 doi:10.1088/1757-899X/512/1/012011

Table 3. Overview for producing the asphalt variants

Variant	Mix type	Binder type	Addition of ZycoTherm	Compac-tion Temp.	50% Coating achieved	Compaction resistance	Void content
-	-	-	-	^{0}C	s	21 Nm	Vol.-%
1a	AC16 B S	50/70	No	135	25 s	41.6	7.0
1b			Yes	135	15 s	43.5	7.1
1c			Yes	115	19 s	42.8	7.2
1d			Yes	95	19 s	41.3	6.9
2a	AC11 D S	25/55-55 A	No	145	37 s	37.3	4.3
2b			Yes	145	22 s	34.2	4.6
2c			Yes	125	30 s	36.5	5.0
2d			Yes	105	30 s	36.5	5.3

For AC 16 B S, no tendency could be specified with regard to the void content. The void contents are within a narrow range so that this can be considered as equal.

The void contents for AC 11 D S increase with reducing compression temperature. This increase however, is not within the testing accuracy so that here, only one tendency can be specified, there is no statistical security. Deviations within these ranges can still be considered as close to practice.

As a summary, it can be stated that when adding ZycoTherm® you can reduce the time for coating the stones and with a significantly reduced compression temperature, you can achieve an almost consistent void content.

5. Case study – instituto mexicano del transporte (imt)

Parameters for testing are as following;

- Asphalt binder: AC 20 (PEMEX, Mexico) with ZycoTherm: 0.075%; reference OBC 5.5% and with ZycoTherm OBC 5.8%; Aggregates: La Canada Quarry
- Sample size: 50mm H X 63 mm W X 380 mm L
- Air voids: 6%
- Test temperature: 20 ^{0}C
- Deformation Frequency: 10 Hz for four levels - 300, 400, 500 & 600 με (represents pavement deformation under normal transit condition)
- Asphalt mix fatigue life failure criteria: Number of cycles in which the beam stiffness modulus decreases 50% compared to its initial value

For all levels of deformation (300, 400, 500 and 600 μ.), samples with optimum content of Zycotherm (.075%) presented extended life for fatigue, so the incorporation of Zycotherm significantly favors the asphalt mix. The amount of load repetitions recorded for the Zycotherm samples were increased by 37 to 180% depending on the level of deformation. It was also observed that Zycotherm addition to the asphalt mix modifies the slope of the fatigue law, which means that Zycotherm samples are less susceptible to deformation levels, compared with the reference samples.

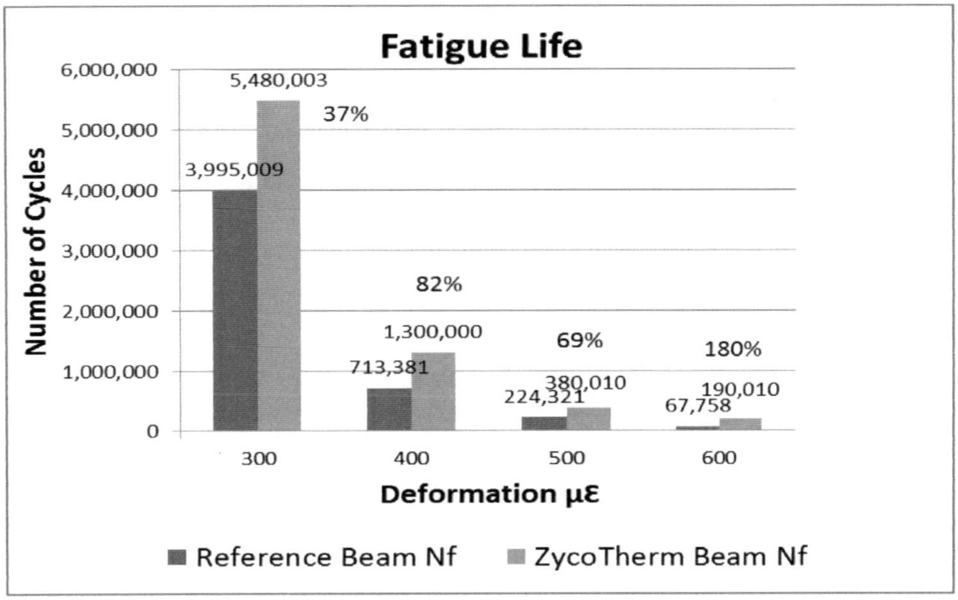

Figure 3. Higher Number of Cycles means Higher Fatigue Life

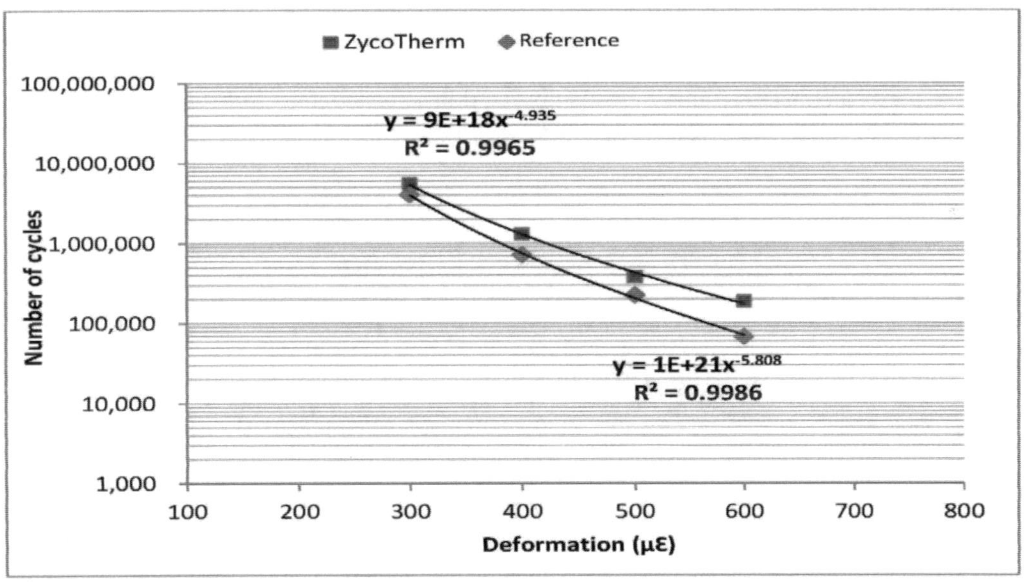

Figure 4. Wohler curve (fatigue law)

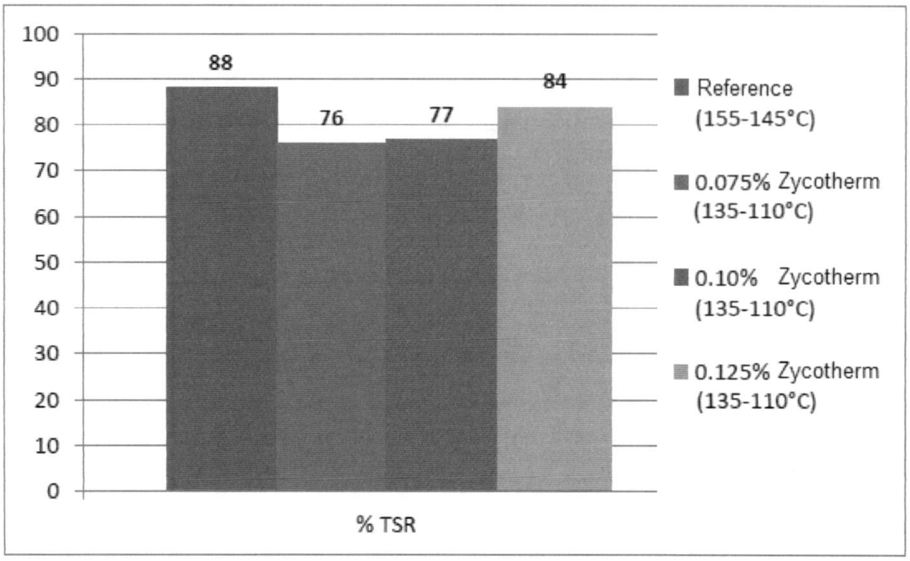

Figure 5. Effect of ZycoTherm on TSR

It was observed that with the addition of Zycotherm in the highest dosage (0,125%) the TSR specification was achieved (≥80%), even when mixing and compaction temperatures were reduced to 135° and 110°. Asphalt mixes with 0,075% and 0,1% presented slightly inferior to 80% TSR values. The conditioned specimens with additive presented 2 to 3 saturation cycles to achieve saturation levels between (70-80%). For all missing dosages, it can be estimated that TSR results will be acceptable for temperature reduction. It was observed that for mixing 135°C and compaction 110°C temperatures, TSR increased when the Zycotherm dosage was increased as well.

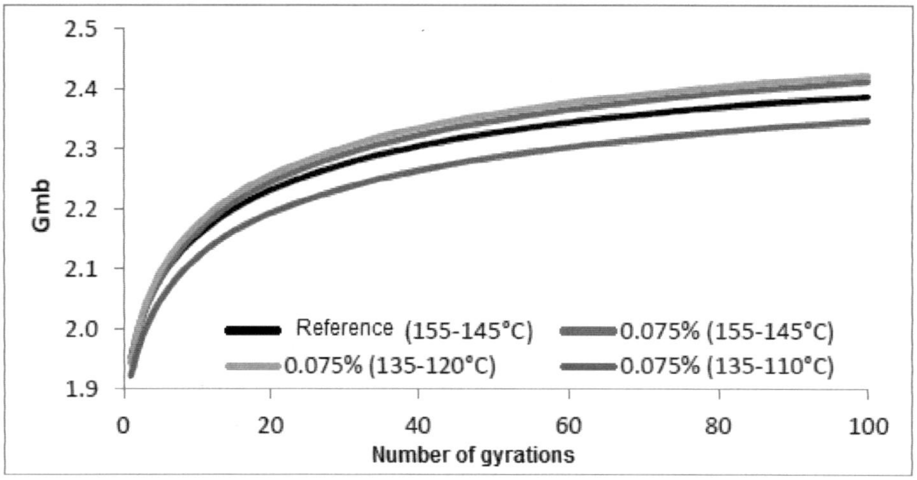

Figure 6. Effect of ZycoTherm (@0.075%) on compaction density

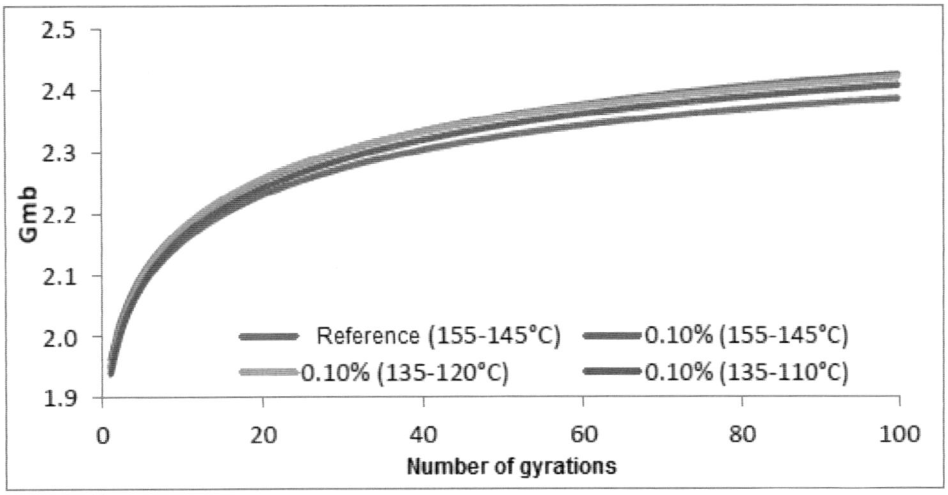

Figure 7. Effect of ZycoTherm (@0.1%) on compaction density

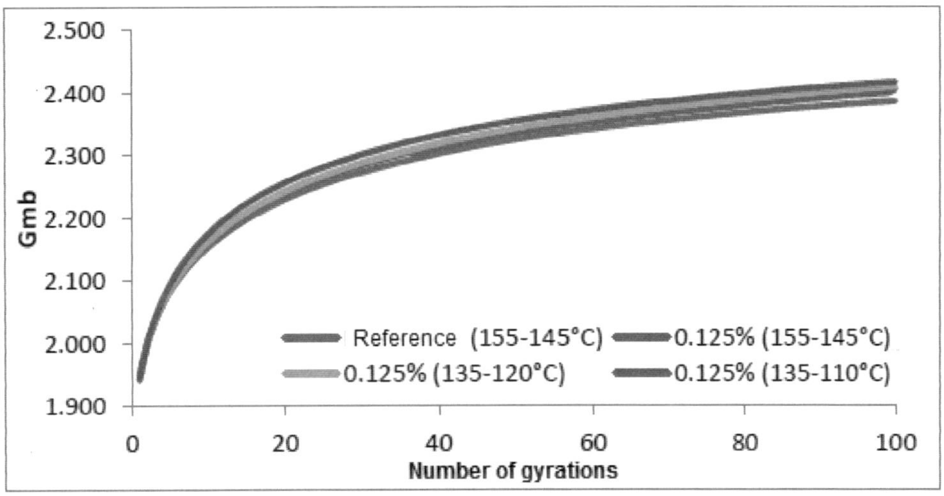

Figure 8. Effect of ZycoTherm (@0.125%) on compaction density

When Zycotherm is used a greater compaction can be achieved for the same and lower temperatures than the reference mix. If mixing and compaction temperatures are reduced benefits in time can be obtained; more time for hauling and waiting before compaction is possible with the Zycotherm addition. Fuel savings during the mixing process and less carbon footprint benefits can be obtained as well.

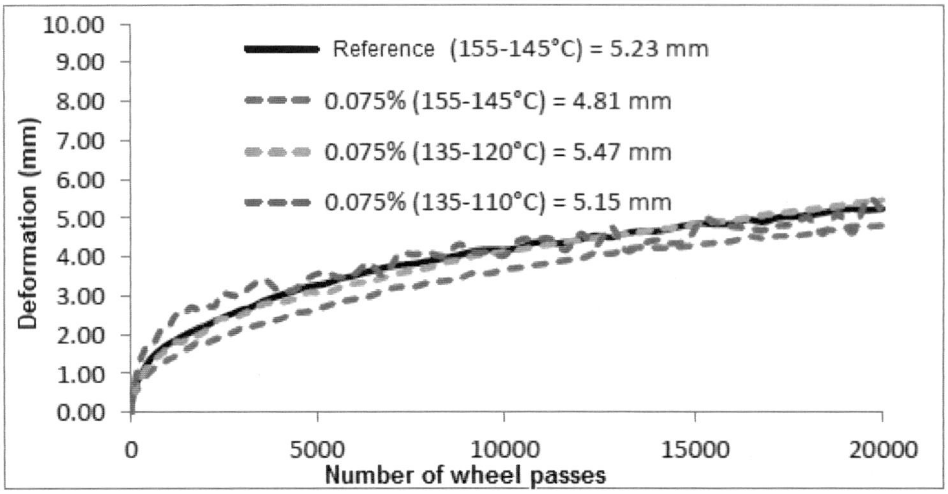

Figure 9. Wheel Tracking performance for ZT @ 0.075%

When the 0,075% additive dosage is incorporated into the asphalt mix, it is observed that deformation is maintained with acceptable tolerance, for similar temperatures as the reference mix. When mixing and compaction temperatures are reduced, deformation is still within high transit levels, according to AMAAC specifications. There is not affectation in rutting behavior when lowering mixing and compaction temperatures.

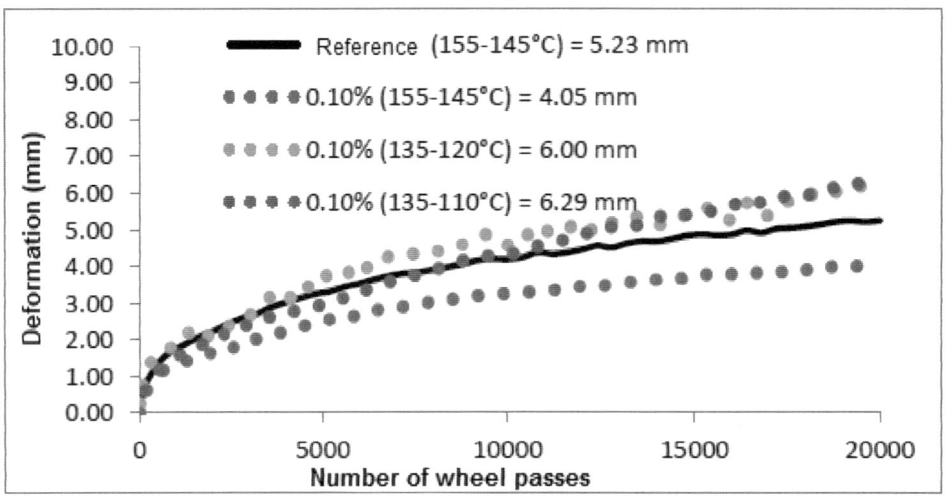

Figure 10. Wheel Tracking performance at ZycoTherm @ 0.1 %

The asphalt mix with Zycotherm present a superior behavior when produced and compacted at the same temperatures as the reference mix. When adding 0,1% of Zycotherm to the mix the mixing temperature can be lowered in 20°C and compaction temperature to 35°C, deformation is slightly affected, but it is still whitin the high transit levels and this results may be between the test dispersion. If mixing and compaction temperatures are reduced benefits in time can be obtained; more time for hauling and waiting before compaction is possible with the Zycotherm addition.

IOP Conf. Series: Materials Science and Engineering **512** (2019) 012011 doi:10.1088/1757-899X/512/1/012011

6. Conclusion
In summary, addition of ZycoTherm, the Organosilane additive to bitumen

1. Allows the mixing temperature to be lower
2. Gives equivalent or better compaction at lower compaction temperatures.
3. Improves coating efficiency, leading to faster and complete coating.
4. Reduces mixing effort and saves energy.
5. Results in acceptable levels of TSR at lower mixing temperatures.
6. Gives excellent rut resistance at lower mixing and compaction temperatures.

Overall, the Organosilane technology, in addition to lower mixing temperatures, allows lower compaction temperature and gives all the benefits stated above.

7. References

[1] Australian Asphalt Pavement Association (AAPA) *Advisory note 17 – Warm Mix Asphalt – a-state-of-the-art review* Available from http://www.aapa.asn.au/docs/no17.pdf

[2] Koenders B G, Stoker D A, Bowen C, de Groot P, Larsen O, Hardy D and Wilms K P 2000 Innovative processes in asphalt production and application to obtain lower operating temperatures *2nd Eurasphalt & Eurobitume Congress* (Barcelona, Spain)

[3] Webpage of Shell Global Solutions Available from *http://www.shell.com. Accessed November 2005.*

[4] Barthel W, Marchand J P and Von Devivere M 2005 *Warm asphalt mixes by adding asynthetic zeolite* Available from Eurovia. www.asphamin.com.

[5] Hurley G C and Prowell B D 2005 *Evaluation of Aspha-min® zeolite for use in warm mix asphalt. NCAT Report 05-04* (USA: National Center for Asphalt Technology, Auburn University)

[6] Hurley G C and Prowell B D 2005 *Evaluation of Sasobit® for use in warm mix asphalt. NCAT Report 05-06* (USA: National Center for Asphalt Technology, Auburn University)

IOP Conf. Series: Materials Science and Engineering **512** (2019) 012012 doi:10.1088/1757-899X/512/1/012012

Assessing the impact of grenade explosion towards Phnom Penh's road network reliability

T H Law[1], C P Ng[2] and Hussain Hamid[1]

[1] Department of Civil Engineering, Faculty of Engineering, Universiti Putra Malaysia, Serdang,Selangor, Malaysia
[2] Department of Civil Engineering, Faculty of Engineering, Universiti Pertahanan Nasional Malaysia, Kuala Lumpur, Malaysia

E-mail: lawteik@upm.edu.my

Abstract. Under normal circumstances, the motorist optimal route choice is based on the perception of shortest route between the successive destinations. However, during disaster, some links and nodes are not accessible and consequently motorists reroute their commutes based on limited road network performance information. Therefore, the entire road network reliability cannot be optimized. Therefore, an efficient traffic diversion plan is required to improve the road network reliability. We used the Technique for order of Preference by Similarity to Ideal Solution (TOPSIS) to rank all nodes in the study area based on their values of betweenness and closeness. The study aims to assess the impact of grenade explosion towards Phnom Penh's roadnetwork reliability in Chamkar Mon district and a traffic diversion plan then wasdeveloped to keep the traffic movement unaffected. The results indicated that there are 4 nodes their TOPSIS rank index value (RIV) increased greater than 25%, while, on the other hand, there are 9 nodes with TOPSIS RIV value decreased greater than 25%. The results also showed that 6 nodes with TOPSIS RIV persists comparatively high before and after the incident. These information was used as the basis of the design of the traffic diversion plan.

1. Introduction

The impact of natural and manmade disasters on road network reliability has gained attention of researchers especially in recent years [1-2]. Road network reliability is concerned with the probability a road network provides a specific standard of performance [3]. It can be divided into three types: connectivity (or terminal), capacity and travel time reliability. Connectivity reliability refers to the probability that the road network nodes stay connected in the deteriorated road network [4], while the capacity reliability concerns the probability that the road network can accommodate a given level of traffic demand at a stated service level [5]. Travel time reliability is defined as the probability that travel time between two given nodes of a road network will not greater than a specified time period [6]. This indicates the higher the variability of travel time, the lower will be the travel time reliability. Under normal circumstances, the motorist optimal route choice is based on the perception of shortest route between the successive destinations. However, during disaster or emergency situations, some links and nodes are not accessible or unsustainably used, and consequently motorists reroute their commutes based on limited road network performance information. Therefore, the entire road network performance cannot be optimized. In view of this, an efficient traffic diversion plan is required to improve the road network reliability.

Identifying influential nodes in complex networks has drawn significant attention in the field of network analysis [7-9]. Betweenness centrality and closeness centrality are two common topological measurements used to identify influential nodes. Betweenness centrality is a measure of number of shortest paths that pass through a particular node. Closeness centrality is a measure of the sum of geodesic distance of a node to all other reachable node in the network. Therefore, the more central a node is, the closer it is to all other nodes in the network. In road transportation, a high betweenness and closeness of a node implies more of the shortest road paths via the given node, indicating a larger selective probability of this node when effectively circulating traffic in the road network. However, the values of betweenness and closeness for a node may change significantly as a result of disasters or emergencies. Together, these sources of information will be used to design traffic diversion plan during disasters and emergencies. The fundamental concept for developing this plan is to divert the traffic from more congested nodes to those operating within or under capacity.

An apparent grenade explosion in Phnom Penh's Chamkar Mon district, near the Tuol Sleng prison museum, on 6 September 2016 has caused injuries to at least four people and rattled the city. The Tuol Sleng prison musuem is located within the city center of Phnom Penh, and therefore, this incident was greatly affect traffic in the area (See in Figure 1). The study area is surrounded by four main roads: the Preah Sihanouk, Samdech Monireth, Mao Tse Toung and Preah Monivong. We used the Technique for order of Preference by Similarity to Ideal Solution (TOPSIS) to rank all nodes based on their values of betweenness and closeness. This study aims to assess the impact of grenade explosion towards Phnom Penh's road network reliability using TOPSIS technique and a traffic diversion plan then was developed.

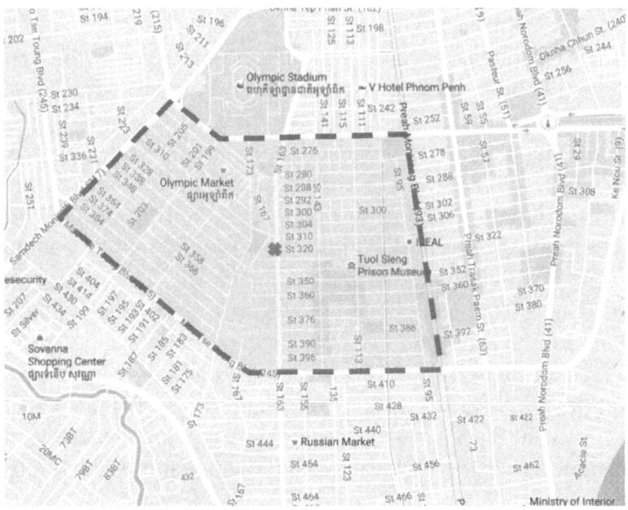

Figure 1. Study location

2. Literature Review

Viewed from a more general perspective, reliability refers to the probability that a system, a device or component able to deliver expected service under the operating conditions for the intended time period [10]. A road network can be regarded as reliable when it can function in good condition and able to handle different levels of disruption for a specific time [11]. Reliability is widely used to assess a road network's performance under disaster conditions [12]. The application of network reliability under disaster or emergency conditions is subjected to pre-disaster and post-disaster [13]. Disaster phases can be recognized based on the demand nature. These include: the evacuation phase at the inception and before the disaster, aid distribution after the disaster and post-disaster. During evacuation, a reliable network is required to transport people to safe areas. In the aid distribution phase, a reliable network allows the timely delivery of humanitarian supplies to the disaster-affected areas. A reliable network is essential during the post-disaster recovery to allow individuals resume their daily activities. This study is related to evacuation reliability.

Connectivity reliability is concerned with the probability that network nodes remains connected when links in the network break down. Konak and Smith pointed out that this reliability measure can be further divided into three groups on the basis of the number of nodes covered: two-terminal reliability (the probability of the present of a path between origin and destination nodes), all-terminal reliability (the probability that each node of the network is linked with all other nodes) and k-terminal probability (the probability that each two nodes in a subset k are linked). Iida and Wakabayashi defined connectivity as each link is represented with a binary state value of operating or failed; that is they are either in operation or not, or more broadly, offer an acceptable service or not. Due to its binary character, connectivity reliability does not consider changes in link capacity and, therefore it appears to be more suitable for acute events, such as earthquake. Moreover, Nicholson indicated that this reliability analysis is suitable is appropriate for sparse and uncongested road network, such as interstate highway system.

Capacity reliability is a measure to assess the performance of a deterioratable road network, when link capacity is depended on random variations as a result of user routing choices [5]. It can be defined as the probability that the network can successfully accommodate a certain traffic demand at an acceptable service level [1]. It is noted that the estimation of the relationships between the road network elements which are sensitive to stochastic variation and the targeted performance, especially studies included larger road networks [17]. The capacity reliability concept was employed to assess road network reliability as a result of disaster [18]. Previous studies indicated that the concept of capacity reliability and travel time reliability could be utilized as a valuable network tool to plan and evaluate traffic management program in urban areas [19].

A large volume of literature has been published on travel time reliability. Such reliability is concerned with the probability that a trip between a given origin destination (OD) pair can be accomplished less than a certain threshold time period given daily stochastic variation of travel demand [20]. This implies that an increase in time variance the lower would lead to a decrease in travel time reliability [21]. Carrion and Levinson indicated that road travel time can be separated into two components as free flow time and additional time. The first component is concerned with the time taken to complete a trip under free flow condition, while the second component regards an increase of travel time as a result of variations in the traffic conditions. They pointed out these variations can be categorized as predictable (e.g. peak-hour congestion) or unpredictable (e.g. traffic accidents). The predictable variations can be predicted by travellers and therefore necessary adjustments can be taken to offset the extra costs caused by congestion. The unpredictable variations are caused by the uncertainty of travel time. Wong and Sussman explained that this uncertainty can be divided into three types as follows: (1) variation between seasons and days of the week; (2) variation by changes in travel conditions due to weather and road incidents; and (3) variation regarded to motorist's perception.

3. Methodology

In this section, we present the methodology used to design a traffic diversion plan during disasters or emergencies. We first present the network centrality measures (betweenness centrality and closeness centrality) used to identify influential nodes in the study road network, and then present the application of TOPSIS to evaluate the node importance. Data used to compute the network centrality measures were obtained from Google Map service.

3.1 Network Centrality Measures

Network centrality measures are indicators to measure a node's importance in a network. Two commonly used network centrality measures are closeness centrality and betweenness centrality [24].

The betweenness centrality is a measure of the proportion of the shortest paths that pass through a particular node. It can be written as:

$$BC(v) = \sum_{i \neq j \in V} \frac{\rho_{ij}(v)}{\rho_{ij}} \quad (1)$$

where ρ_{ij} is the number of shortest paths between any two nodes i and j, while $\rho_{ij}(v)$ denotes the number of shortest paths between nodes i and j that pass through node v. In regards to road traffic, a high betweenness centrality of a node indicating a large influence on the transfer of traffic through the road network, under the assumption that traffic primarily flows over the shortest paths between them. This measure can be normalized in the interval between zero and one by dividing it by $(n-1)(n-2)/2$, where n represents the number of nodes in the network.

Unlike betweenness centrality which measures the proportion of the shortest paths that pass through a node in a network, closeness centrality refers to the inverse of the average geodesic distance between a node and other nodes in the network [25]. The expression is as follows:

$$C_C = \frac{n-1}{\sum_{i \neq j \in V} d_{ij}} \quad (2)$$

where d_{ij} is the minimal path length from node i to j. In general, a node with a high degree of closeness centrality has shorter geodesic distance to other nodes in the network and thus it is more accessible. This centrality measure is a more comprehensive tool to rank important node as it considers the entire geographical structure of the network.

3.2 Techniques for Order Preference by Similarity to Identical Solution

Betweenness and closeness centrality measures are widely used to evaluate node importance in a network but with limitations. One of the limitations of closeness centrality indicated by Wei et al. is that it cannot be used for disconnected networks. This is because two nodes in different components of the same network do not have a finite distance between them. Although the betweenness centrality can be applied to disconnected networks, it does not assign any value to nodes do not lie on the shortest path between any two other nodes [27]. This implies that if only one centrality measure is employed, then the rankings of influential nodes may be different by employing different centrality measure. In view of this the present study used both closeness and betweenness centrality to evaluate node importance. This was achieved by using the technique for order preference by similarity to identical solution (TOPSIS) developed by Hwang and Yoon.

The TOPSIS aims to provide guide to the decision-maker in the process of making the selection between alternatives. It fully uses the information of attribute, offers a cardinal ranking of alternatives, and attribute preferences need not be independent [29]. According to this technique, the rankings of alternatives is on the basis of the shortest Euclidean distance from the positive ideal solution (PIS) and the farthest from the negative ideal solution (NIS) (also known as anti-ideal solution). The PIS refers to a solution that maximizes the benefit criteria/attributes and also minimize the cost criteria/attributes, whereas the NIS is the one which minimizes benefit criteria/attributes and maximizes the cost criteria/attributes [30]. The final ranking is obtained by considering simultaneously the distances to both the PIS and the NIS, and a preference order is ranked according to their relative closeness [31].

The TOPSIS procedure is described in the following steps. The first step is to determine the decision matrix of *n* criteria/attributes and *m* alternatives. The second step is to normalize the decision matrix, while the third step is to weight the normalized decision matrix. We suppose that closeness and betweenness centrality have the same importance under evaluation. This implies that the weights for the two centralities are equal to 0.5. In step four, we determine the positive ideal and negative ideal solutions using the decision matrix. Next, we calculate the Euclidean distances from the positive ideal and negative ideal solutions, respectively. In the final step, the relative closeness of each alternative to the ideal solution is determined. Rank the alternatives in regard to the relative closeness to the ideal

IOP Conf. Series: Materials Science and Engineering **512** (2019) 012012 doi:10.1088/1757-899X/512/1/012012

solution in descending order. The alternative with higher value is regarded to be more important as they are closer to the positive ideal solution.

4. Results and Discussions

In the next section, we discuss results on the change in the centrality measures, while the design of traffic diversion plan is discussed in Section 4.2.

4.1 The Centrality Analysis Results

The closeness and betweenness centralities were computed for 175 nodes in the study road network, and then the TOPSIS rank index were computed based on both closeness and betweenness centralities. As shown in Table 1, there are 4 nodes their TOPSIS rank index value (RIV) increased by more than 25%. Therefore, it is expected a significant increase in traffic via these nodes after the grenade explosion. However, it is important to note that even though there is a significant increase in TOPSIS RIV for Nodes 75519 and 75520, their TOPSIS RIV after the incident is relatively small. This incidated that the incident's impact on these two nodes' performance was marginal. Conversely, there are 9 nodes their TOPSIS RIV decreased by more than 25% (Table 2) and thus it is anticipated a drastic decrease in traffic via these nodes after the incident. However, it is interesting to note that there are 6 nodes their TOPSIS RIV remained relatively high before annd after the incident (Table 3).

Table 1. Nodes' TOPSIS RIV increased by more than 25%

	Before grenade explosion			**After grenade explosion**			
Node	Betweenness	Closeness	TOPSIS	Betweenness	Closeness	TOPSIS	%
75464	0.128	0.096	0.589	0.199	0.097	0.829	41%
75474	0.094	0.093	0.447	0.134	0.094	0.568	27%
75519	0.017	0.077	0.141	0.039	0.077	0.195	39%
75520	0.017	0.076	0.138	0.041	0.076	0.194	41%

Table 2. Nodes' TOPSIS RIV decreased by more than 25%

	Before grenade explosion			**After grenade explosion**			
Node	Betweenness	Closeness	TOPSIS	Betweenness	Closeness	TOPSIS	%
75473	0.144	0.089	0.651	0.076	0.085	0.337	-48%
75475	0.077	0.086	0.369	0.043	0.083	0.218	-41%
75476	0.138	0.085	0.623	0.109	0.082	0.462	-26%
75592	0.052	0.061	0.231	0.041	0.061	0.169	-26%
75611	0.016	0.067	0.106	0.009	0.065	0.078	-26%
75612	0.025	0.069	0.141	0.036	0.068	0.104	-26%
75622	0.043	0.058	0.191	0.031	0.058	0.132	-31%
75626	0.043	0.056	0.187	0.031	0.056	0.128	-31%
75627	0.033	0.053	0.143	0.021	0.054	0.088	-39%

Table 3. Nodes' TOPSIS RIV remains relatively high before and after the incident

	Before grenade explosion			**After grenade explosion**		
Node	Betweenness	Closeness	TOPSIS	Betweenness	Closeness	TOPSIS
75469	0.182	0.098	0.819	0.201	0.098	0.836
75470	0.222	0.099	1	0.241	0.101	1
75471	0.198	0.095	0.891	0.189	0.096	0.789
75477	0.151	0.091	0.685	0.157	0.092	0.659
75489	0.181	0.098	0.817	0.185	0.098	0.776
75528	0.166	0.089	0.747	0.172	0.091	0.717

4.2 Design of Traffic Diversion Plan

Figure 2 shows the distribution of more influential nodes in the study area after the incident. The green cross indicates nodes' TOPSIS RIV decreased by more than 25%, while the red cross represents nodes' TOPSIS RIV increased by more than 25% after the grenade explosion. The purple triangle sign represents nodes' TOPSIS RIV remained relatively high before and after the incident.

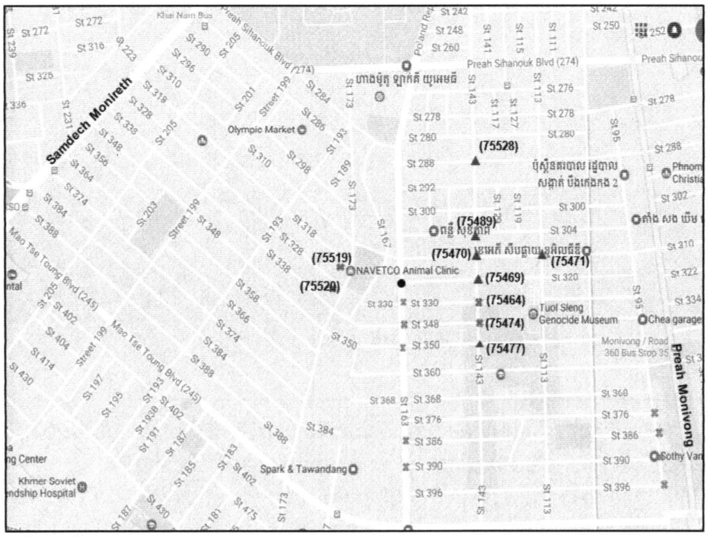

Legend

- : decreased by more than 25%
- : increase by more than 25%
- : Grenade explosion location
- : Nodes' with relatively high TOPSIS index value before and after the incident

Figure 2. The distribution of more influential nodes

The incident location, which is represented by the black circle in Figure 2, was not accessible to traffic after the incident. Therefore, the traffic originally planned to pass through this node must reroute to the next shortest route. According to the traffic diversion concept, traffic moving within the study road network shoud be diverted to the nodes with significant decrease in TOPSIS RIV, but not to the nodes with significant increase in TOPSIS RIV as well as nodes with relatively high TOPSIS RIV before and after the incident. This would improve traffic circulation during the incident. Figure 3 illustrates the recommended diversion routes designated in the study road network. The blue line shows the diversion route from the North to the South and vice versa. The green line indicates the diversion route to the West, towards the Tuol Sleng Prison musuem, while the blue line indicates the diversion route to the East, towards the Samdech Monireth road.

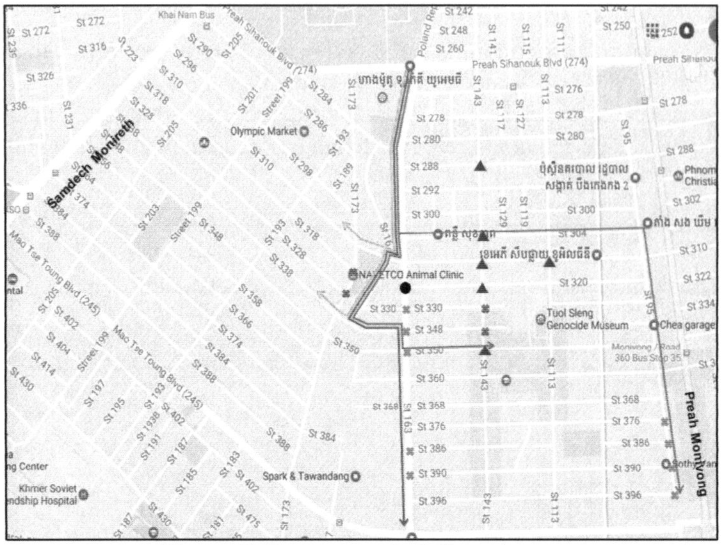

Figure 3. Recommended diversion routes

5. Conclusion and Recommendations

This study assessed the impact of grenade explosion towards Phnom Penh's road network reliability and a traffic diversion plan was then designed based on the assessment results. Results indicated that there were 4 nodes their TOPSIS RIV increased greater than 25%, while, on the other hand, there were 9 nodes with TOPSIS RIV value decreased greater than 25% after the incident. The results also showed that 6 nodes with TOPSIS RIV persistently high before and after the incident. These information was used as the basis of the design of the traffic diversion plan to keep the traffic movement unaffected after the incident.

A limitation of this study is that the TOPSIS model omitted distance between the node and other traffic attraction places, such as shopping areas, schools and tourist attraction places (such as The Tuol Sleng genocide musuem). It is anticipated that more traffic will be attracted to nodes closer to these traffic attraction places even the nodes are not placed on the shortest path. Besides, node and link capacity should also be included in the TOPSIS model for two reasons. Second, an increase in node or link capacity would increase the traffic flow. Consequently, this would attract more traffic to higher capacity nodes and links. Third, an increase in TOPSIS RIV may have different impact on the performance of node/link with different capacity. This may have implication on the design of traffic diversion plan. Therefore, further research should be conducted to investigate into the effect of the above mentioned factors on identifying influential nodes and hence improving the design of traffic diversion plan.

Road network are vulnerable to manmade and natural disasters that can cause a massive destruction to the infrastruture. Consequently, this poses a threat to vehicle travel on a degraded road network and prevent access for the emergency services, aids workers and relief supplies. Therefore, it is crucial to understand road network reliability, know the most influential nodes and the consequences to network disruption in order to improve road network design and develop traffic diversion plans during the disaster. This is necessary to achieve a sustainable road network.

6. References

[1] Chen A, Kasikitwiwat P and Yang C 2013 Alternate capacity reliability measures for transportation networks *Journal of Advanced Transport* **41**(1) 79-104

[2] Edrissi A, Nourinejad M and Roorda M J 2015 Transportation network reliability in emergency response *Transportation Research Part E* **80** 56-73

[3] Michael A P and Susilawati S 2012 Remoteness and accessibility in the vulnerability analysis of regional road networks *Transportation Research Part A: Policy and Practice* **46** 761-71
[4] Bell MGH 2000 A game theory approach to measuring the performance reliability of transport networks *Transportation Research Part B* **34**(6) 533-46
[5] Chen A, Yang H, Lo H K and Tang W H 2002 Capacity reliability of a road network: an assessment methodology and numerical results *Transportation Research Part B* **36**(3) 225-52
[6] Berdica K 2002 An introduction to road vulnerability: what has been done, and should be done *Transport Policy* **9** 117-27
[7] Zeng A and Zhang C J 2013 Ranking spreaders by decomposing complex networks *Physics Letters A* **377**(14) 1031-35
[8] Liu J G, Ren Z M and Guo Q 2013 Ranking the spreading influence in complex networks *Physica A* **392** 415-19
[9] Bae J and Kim S 2014 Identifying and ranking influential spreaders in complex networks by neighborhood coreness *Physica A* **395** 549-59
[10] Husdal J 2004 Reliability and vulnerability versus costs and benefits *Proceedings of the second international symposium on transportation network reliability* Strasbourg
[11] Wakabayashi H and Iida Y 1992 Upper and lower bounds of terminal reliability in road networks: an efficient method with Boolean algebra *Journal of Natural Disaster Science* **14**(1) 29-44
[12] Fang S and Wakabayashi H 2011 Travel time reliability for recovery activity immediately after disaster *Procedia Social and Behavioral Sciences* **20** 621-29
[13] Chen L and Miller-Hooks E 2012 Resilience: an indicator of recovery capability in intermodal freight transport *Transportation Science* **46**(1) 109-23
[14] Konak A and Smith A E 2006 Network reliability optimization, *Handbook of Optimization in Telecommunications* (Springer: New York)
[15] Iida Y and Wakabayashi H 1989 An approximation method of terminal reliability of road network using partial minimal path and cut set *Proceedings of the Fifth WCTR Yokohama* pp 367-80
[16] Nicholson A 2003 Transport network reliability measurement and analysis *Transportes* **XI** 49-62
[17] Dauda ROS2010 Investment in education and economic growth in Nigeria: An empirical evidence *International Research Journal of Finance and Economics* **55** 158-69
[18] Sumalee A and Kurauchi F 2006 Network capacity reliability analysis considering traffic regulation after a major disaster *Networks Spatial Economics* **5** 205-19
[19] Taylor M P 1999 Dense network traffic models, travel time reliability and traffic management II: application to network reliability *Journal of Advanced Transportation* **33** 235-51
[20] Clark S and Watling D P 2005 Modelling network travel time reliability under stochastic demand *Transportation Research B* **39(2)** 119-40
[21] Nocholson A J and Dalziell E 2003 Risk evaluation and management: A road network reliability study *Proceedings of the 1st International Symposium on Transportation Network Reliability Oxford* pp 49-59
[22] Carrion C and Levinson D 2012 Value of travel time reliability: A review of current evidence *Transportation Research Part A* **46** 720-41
[23] Wong H K and Sussman J M 1973 Dynamic travel time estimation on highway networks *Transportation Research* **74(4)** 353-70
[24] Freeman L C 1977 A set of measures of centrality based on betweenness *Sociometry* **40** 35-41
[25] Freeman L C 1978 Centrality in social networks: conceptual clarification *Social Networks* **1** 215-39
[26] Wei D, Deng X, Zhang X, Deng Y and Mahadevan S 2013 Identifying influential nodes in weighted networks based on evidence theory *Physica A: Statistical Mechanics and Its Applications* **392**(10) 2564-75

[27] Opsahl T, Agneessens F and Skvoretz J 2010 Node centrality in weighted networks: generalizing degree and shortest paths *Social Network* **32** 245-51
[28] Hwang C L and Yoon K 1981 *Multiple attribute decision making: Methods and applications* (New York : Springer-Verlag)
[29] Yoon K P and Hwang C L 1995 *Multiple attribute decision making: An introduction* (Thousand Oaks CA: Sage Publications)
[30] Wang Y M and Elhag T M S 2006 Fuzzy TOPSIS method based on alpha level sets with an application to bridge risk assessment *Expert Systems with Applications* **31** 309-19
[31] Wu D S 2006 A TOPSIS data mining demonstration and application to credit scoring International *Journal of Data Warehousing and Mining* **3** 1-10

Acknowledgment
This research was funded by Ministry of Education Malaysia (Grant no. IPPTN/KPT/CLMV/2016/R12(3)).

IOP Conf. Series: Materials Science and Engineering **512** (2019) 012013 doi:10.1088/1757-899X/512/1/012013

The benefits of National Intelligent Transportation Management Centre (NITMC) establishment in Malaysia

Nurul Saffrena Musa, Noor Mazlan Muhammad Noor and Jamilah Mohd Marjan

Highway Planning Unit, Ministry of Works, Kuala Lumpur, Malaysia

E-mail: saffrena@kkr.gov.my

Abstract. Malaysia has been embracing Intelligent Transportation System (ITS) technology and has started the effort by establishing a few traffic management centres (TMC) which are operated by different government agencies for transportation data collection. For examples, Kuala Lumpur City Hall (DBKL) has been operating Integrated Transport Information System (ITIS) where traffic data is collected in Kuala Lumpur and Klang Valley area., In addition, Malaysian Highway Authority (MHA) has TMCs at all tolled highways to collect and manage traffic data and Land Public Transport Land Public Transport Agency (APAD) is in charge for public transport management including bus and light rail train. Due to the operation and maintenance of these centres are maintained by multiple agencies separately, transportation data from them are not coordinated and integrated. Therefore, the in-silo working culture causes difficulty in data collection and analysis for historical data record, real-time information and future prediction. In conclusion, National Intelligent Transport Management Centre (NITMC) is important as the centre that will coordinate, integrate, collect, analyse and disseminate these data from multiple agencies from government and private agencies. The data analysed in NITMC is expected to benefit all relevant stakeholders including the public by providing up-to-date and reliable data to them.

1. Introduction

Malaysia has a nationwide road network includes Peninsular, Sabah, Sarawak and Federal Territory of Labuan. In general, road in Malaysia is categorized into two categories which are Federal and State Road. Both of these roads are developed and maintained by Federal and State Government respectively. On the other hand, the number of vehicle has gradually increased over past few years. This increment affect the need to increase the road capacity too. Currently, Level of Service (LOS) is used as a basic indicator to show the condition of the traffic flow at specific road. LOS A until LOS C is very much preferable as the traffic can move without long delay. Nevertheless, once the traffic flow reach LOS D to LOS F, the traffic is halted and highly congested. Table 1 below shows the further detailed description of each LOS [1].

One of the ways to mitigate congestion problem is centralization of all Intelligent Transport System (ITS) centres nationwide. Currently, in Malaysia, there are a few ITS which are operated and maintained by several agencies. For example, Kuala Lumpur City Hall (DBKL) has established Integrated Transport Information System (ITIS) where traffic data in Kuala Lumpur and Klang Valley is channelled for monitoring and supervising purposes. Meanwhile, Malaysian Highway Authority (MHA) which is under Malaysian Ministry of Works (MOW) has traffic management centre (TMC) at all tolled highways to collect and manage traffic data and Land Public Transport Agency (APAD) which is under Ministry of Transportation (MOT) is in charge for public transport management including bus and light rail train.

Table 1: Type of LOS and Description [1]

Type of LOS	Description
A	Free flow with individual users virtually unaffected by the presence of other vehicles in the traffic stream. This is a condition of free flow with low volume and high speed of vehicle travel on the highways
B	Stable traffic flow with a high degree of freedom to select speed and operating condition but with some influence from the other users.
C	Restricted flow that remains stable but with significant interaction with others in the traffic stream. The general level of comfort and convenience decline noticeably at this level. Speed and manoeuvrability are closely controlled by the higher volume. Most of the drivers' freedom to select their own speed, change lane or pass are restricted.
D	High density flow in which speed and freedom to manoeuvre are severely restricted meanwhile comfort and convenience have decline. This level of service represents unstable flow with operating speed being maintained, though considerably affected by changes in operating condition.
E	Unstable flow at or near capacity levels with poor level of comfort and convenience. This level represents operation at lower operating speed with volume at or near the capacity of the highways. Flow is unstable and stoppage may occur for a momentary duration.
F	Forced traffic flow in which the amount of traffic approaching a point exceeds the amount that can be served. LOS F is characterized by poor time travel with low comfort, convenience and increase accident exposure. This condition describes a forced flow operation at low speed where volume has reached the capacity level. Speed is reduced substantially and stoppage may occur for short or long periods of time because of the downstream traffic condition.

2. Government Agencies and Their Roles in Transportation

Generally, at an urban area such as Kuala Lumpur, traffic congestion is a common problem especially during peak hour. Most roads in Kuala Lumpur are surrounded by buildings and houses and this condition restricts the capability for the road expansion. Based on traffic count data obtained from Road Traffic Volume Malaysia 2016 (RTVM 2016) [2], Kuala Lumpur has recorded LOS F at several major road as tabulated in Table 2 [2]. The data was collected at the road connecting between Kuala Lumpur to adjacent state. According to new vehicle registration record by Malaysian Road Transportation Department (JPJ) as shown in Figure 1, there is an increase in amount of new car and motorcycle registered from year 1986 to 2016 with total of motor vehicle from 4,439,949 units to 27,163,120 units respectively [2]. This shows amount of new vehicle grows by seven times in urban areas in 30 years. However, the infrastructure capacity does not increase in parallel with the traffic growth due to space limitation and financial constraint.

Table 2: Traffic Count Data in Kuala Lumpur

Census Station Number	Road Location	Level of Service (LOS)
WR101	Kuala Lumpur – Kuala Selangor (Jalan Kepong)	E
WR102	Kuala Lumpur – Ipoh	F
WR103	Kuala Lumpur – Ipoh	F
WR107	New Klang Valley Expressway(NKVE) – Middle Ring Road 2 (MRR2)	F

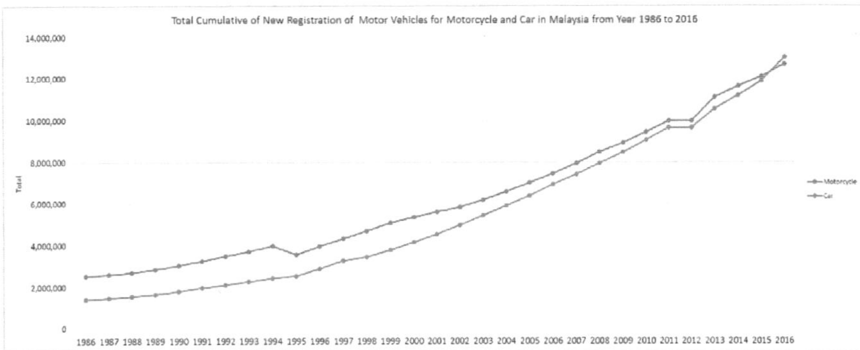

Figure 1. Total Cumulative of New Registration of Motor Vehicles (Motorcyle and Car) from Year 1986 to 2016 [2]

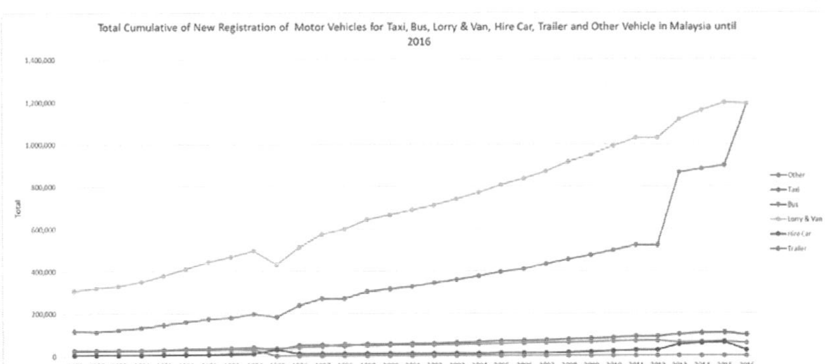

Figure 2. Total Cumulative of New Registration of Motor Vehicles from Year 1986 to 2016 (Taxi, Bus, Lorry and Van) [2]

APAD is in charge for public transportation including bus, train, light rail and also taxi services. APAD plays a major role in providing the public with comfortable and practical public transportation infrastructure with intra-city and inter-city connectivity. The commuter can reach destination without the need to drive private vehicle via public transportation and thus reduce the number of private vehicles on the road. Therefore, one of the ways to encourage people to use the infrastructure is to provide them with reliable journey planner. The journey planner can provide real-time and accurate routes option, bus and train schedule and also estimated fare. Mobile application is much more

IOP Conf. Series: Materials Science and Engineering **512** (2019) 012013 doi:10.1088/1757-899X/512/1/012013

preferred as it is accessible by most commuters. With these information, the commuters will likely to opt for public transportation in their everyday routine.

Besides that, the first-and-last-mile of public transportation and connectivity between different public transportation modes are important to encourage the public to use them in daily basis. The distance between the stations and residential areas must be close enough for walking distance. If the location between both areas are far, sufficient parking space and hailing services need to be provided for commuters. Public transportation also contributes to greener environment since it will reduce private vehicle on the road.

On the other hand, DBKL has established a centre named ITIS. ITIS is responsible to monitor the traffic condition on Kuala Lumpur and certain Klang Valley area. The total length of the road under DBKL jurisdiction is approximately 1,539.46 km [3]. ITIS has installed an amount of 250 surveillance cameras for traffic monitoring and key junctions [4]. The information obtained by ITIS is beneficial for other government agencies such as Malaysian Public Works Department (JKR) for highway planning and APAD for bus and taxi route planning.

On the other hand, all tolled highways in Malaysia is operated by highway concessionaires under MHA. The total of tolled highway length is 1,988.21km in Peninsular Malaysia [3]. Since year of 2017, all tolled highways implement electronic toll collection (ETC) and eliminate cash payment as to reduce bottleneck traffic at toll plazas. These toll payments are collected and recorded by authorized highway concessionaires. The implementation of ETC in all tolled highways have improve the traffic flow especially highways within the urban city. Besides the successful of ETC implementation, MHA also install surveillance cameras and Variable Message Signs (VMS) to share the real-time information to the road user if there is any construction, accident or slow traffic ahead. VMS will show traffic information and advice about alternate route available. These valuable information stored by MHA has not been utilized up to its maximum potential. For example, the recorded data in previous years can be used to predict traffic growth in a few years ahead.

All of these agencies play a major role in managing and controlling transportation in Malaysia. Integration of transportation data between these agencies and NITMC is seen as one step forward towards transportation improvement urban cities and create a better lifestyle for the public in general.

3. The Data Management in NITMC

3.1 Data Sources

One of the main objective of NITMC's establishment is to coordinate multiple agencies, collect, analyse and disseminate transportation data among them. The set of data can be utilized to generate a new set of data which is useful for the government and public. In order to succeed, communication between existing control centres under these agencies under the same platform which is NITMC. Besides that, NITMC need to provide a common platform that is ready to integrate, compile and analyse the collected data. Table 3 below is the type of data needed for NITMC from several parties.

On the other side, crowdsourcing is another source of transportation data that can be channelled to NITMC. Crowdsourcing means the source of information is obtained the public through internet. The public can provide data to NITMC by reporting the information such as accident, vehicle stopped at roadside and potholes. Smartphone is one of the efficient and fast way for crowdsourcing since smartphones are equipped with Global Navigation Satellite System (GNSS) and mobile internet. GNSS and mobile internet are needed to enable reports to be sent with location and time of incidents. In addition, pictures on site can also be attached with the report. One of the famous mobile application is Waze. Waze recently added a new feature which the public can send report on potholes. Besides that, Waze can also be used to report on traffic condition, accidents and police on duty ahead of the road. NITMC can collect and utilize data from Waze.

Table 3: Data from government agencies

AGENCY	TYPE OF DATA
Ministry of Works	Road length, traffic count, infrastructure inventory
Ministry of Transportation & Land Public Transport Agency (APAD)	Private and commercial vehicle registration, road transport enforcement and Public transportation data
Malaysian Highway Authority	Traffic count, electronic tolling collection, infrastructure inventory
Local authority	Infrastructure inventory, surveillance camera, enforcement

In addition, industrial player can also play their part to contribute for NITMC success by providing data of fleet tracking to back-end system in NITMC. The data is important for NITMC for data collection and analysis in order to manage the route and schedule of fleet for congestion and accidents control especially during peak hour.
All of these data sources for NITMC can be stored at data warehouse and open limited access to private sector and public.

3.2 Communication Tools Between Vehicles, Roadside Infrastructure and NITMC
Internet of Thing (IOT) devices such as Radio Frequency Identification (RFID) can be a communication tool for data sharing between vehicles, roadside infrastructure and NITMC. For example, RFID tag is unique to each vehicle in terms of its serial number. For example, if one owner has two cars, then there is two RFID with different serial number for both cars.

RFID tag serial number will connect with RFID reader at the road and respond to back-end system to view stored data such as vehicle ownership, vehicle's plate number, backend payment system, road tax and also summon record.

Currently, MHA is working on the implementation of Multi-Lane-Free-Flow (MLFF) by using RFID. The implementation of MLFF will be conducted in several phases. Back in earlier days, all toll plazas only accept cash as a toll payment and yet to implement electronic tolling collection (ETC). ETC is a use of prepaid magnetic card called Touch N' Go (TNG) and as an option, there is an infrared device called Smart Tag (STAG) which hold the prepaid magnetic card that can help the user to pass through the toll plaza without need to stop and touch the card reader. Lanes for TNG and STAG user are separated at toll plaza to ensure smooth traffic flow at STAG lane. Since March 1997, the first ETC was implemented at Jalan Pahang Toll Plaza and a year later, ETC implementation was extended to toll plazas along North-South Expressway (PLUS) route. In year 2017, all tolled-highways replace cash payment method with ETC [6].

In addition, MHA continues to move forward in ETC improvement with MLFF implementation at all tolled-highways. The first phase towards MLFF is to create RFID – TNG lane with barrier and STAG – TNG lane with barrier. Initially, RFID tags was distributed to exempted vehicles for introduction and testing purposes. These exempted vehicles can pass through both lane. Exempted vehicles consists of emergency vehicles,

Later, the next phase is designation of RFID special lane with barrier. This special RFID lane is for vehicles installed with RFID tags only. After the RFID is ready to launch technically and legally, RFID tags will be distributed to all types of vehicle and become compulsory for each vehicle to have

RFID tag. The RFID tag will be distributed in stages due to large amount of vehicles in Malaysia. Then, the third phase is open Single-Lane-Free-Flow (SLFF) without barrier if the legal framework is ready. Legal framework is important as to ensure all highway users pay for the toll or driving any vehicle without RFID tags. Lastly, by early year 2020, MLFF is expected to fully be operated at all tolled-highways. Toll plazas will be replaced with RFID reader and camera speed at gantry.

On the other hand, RFID tags function and benefits are not limited to tolled-highways only. Normal roads can also utilize the advantages of RFID. Roadside infrastructure will be equipped with RFID reader for traffic count purposes, speeding meter sensors, pedestrian crossing sensors and so on. In addition, these abilities are beneficial for journey planning as they can determine current traffic flow condition and spread the information to on-the-road vehicles as to avoid congested area. Therefore, communication between Vehicle-to-Infrastructure (V2I), Vehicle-to-Vehicle (V2V) and Vehicle-to-Centre (V2C) which is NITMC are crucial to fully utilized of the data.

4. The Outputs of NITMC

The NITMC's outputs are as below:

a) The real-time traffic data can be retrieved on the spot. The real-time data can be used for journey planning. If the real-time data collected is sufficient, journey planner can be created for public use. Journey planner may designed to give the user a single mode or multimodal transportation options.

b) Control the accuracy of traffic information spread to the public. Road users can send report to NITMC. If there are multiple similar reports received, the information sent may be identified as true and information verification can be done through CCTV.

c) Increase traffic management efficiency. Currently, traffic is monitored through CCTV and is directly connected to TMC at tolled highways and ITIS for Klang Valley area. The video feed from the CCTV is important as to determine the traffic condition and identify any occurrences happened on the road that could be the root cause of the traffic congestion at that time. Next, VMS play an important role to inform the road user about accidents or any events that affect the road so the road user can opt for other route to reach their destination.

d) The NITMC establishment can help to strengthen the enforcement related to road and transport. For example, RFID tag can store data related to vehicle ownership, road tax and link to bank accounts for toll payment. The stored data can help the enforcers to track any law violators through detection from RFID reader installed at predetermine locations. NITMC will collect data from RFID tag detector and alert the relevant authority for their further action.

e) Historical transportation data can be stored and utilized for data forecasting. RFID tag can also be used for traffic count and the data can be classified in accordance to the vehicle classes. The traffic count data can be collected all year round and stored for forecasting traffic trend.

f) Nationwide common platform for data-sharing among government agencies, private sector and also public. All parties are encouraged to share data so the data collected by NITMC has wider range of data sources.

5. Conclusion

In a nutshell, NITMC is considered as the best alternative to improve traffic and transport management in Malaysia. This will eliminate working in-silo culture in government agencies and narrow the gap among government, private and public community. In the end, all parties from government, private and also public will gain benefits from collaboration and cooperate in making NITMC a successful data centre.

6. References

[1] Cawangan Jalan JKR Malaysia 2017 *Arahan Teknik (Jalan) 8/86* 27 (Kuala Lumpur: JKR)

[2] Ministry of Works 2016 *Road Traffic Volume Malaysia (RTVM)* (Kuala Lumpur: MoW) pp 254-59

[3] Cawangan Senggara Fasiliti Jalan JKR Malaysia 2017 *Statistik Jalan Edisi 2017* (Kuala Lumpur: JKR) pp 11

[4] Omar M 2006 *Managing the Kuala Lumpur Road Network with the Integrated Transport Information System* Available from https://www.piarc.org/ressources/documents/actes-seminaires06/c14-malaisie06/8623,TS31-Omar.pdf

[5] Ministry of Works 2016 *Malaysian Intelligent Transport System Blueprint (2018-2023)* 23-25 (Kuala Lumpur: MoW)

[6] Ismail M 2009 *Memastikan Keselesaan Perjalanan Satu Malaysia* Available from http://www.kkr.gov.my/files/scan0122.pdf

[7] Hashim A B 2017 *Smart Highways* Lembaga Lebuhraya Malaysia

[8] PIARC *Malaysian Roads* Available from https://www.piarc.org/ressources/documents/1216,road-network-in-malaysia-v2.pdf

[9] *Smart Mobility* Available from http://www.vt.bgu.tum.de/fileadmin/w00bnf/www/VKA/2014_15/150212_Smart_Mobility_v5_TUM.pdf

[10] *Map Of Malaysia* Available from https://1carent.com/map-of-malaysia

Implementation of Malaysian National Axle Load Policies through Weight Restriction Orders 1989, 2003 & 2017

MN Shahruddin[1], CC Lim[2], SK Ng[1], KMS Ku Mahamud[1] and MH Uzir[1]
[1]Evenfit Consult Sdn. Bhd., Kuala Lumpur, Malaysia
[2]Jabatan Kerja Raya (JKR), Kuala Lumpur, Malaysia

E-mail: nasrullah.evenfit@gmail.com

Abstract. Axle Load Study in 1987 had led to the formulation of Weight Restriction Order (WRO) 1989 for truck legal axle weights based on Short Term Axle Load (STAL) policy, which considers HA loads with no checking done on HB load. Medium Term Axle Load (MTAL) policy was later implemented in 2003 by the introduction of WRO 2003. MTAL are loadings derived from HA load or 45 units HB loading guided along the centreline. A new assessment load model to evaluate existing bridge carrying capacity is proposed in the new axle load study conducted in 2016. The findings were accordingly gazetted in 2017 with the issuance of WRO 2017. Oblivious of these studies, many Malaysian engineers today are not informed of the rationales or basis of the practices, especially those related to bridge assessment. The purpose of this paper is to revisit these studies and discuss pertinent decisions made with regard to the national axle load policies. Based on the comparison between the previous and new assessment procedure, bridges that were rated as capable of safely carrying the load equivalent to MTAL are capable of safely carrying the vehicles that are in compliance with WRO 2017.

1. Introduction

Axle Load Study in 1985 – 1987 (Study '87) among others aims at increasing the legal commercial vehicle dimensions and weight limits in Motor Vehicle Regulations to be compatible with the vehicles available in the world market [1]. Motor Vehicle Regulations of Malaysia prior to Study '87 were deemed outdated as compared to other regulations worldwide and there was doubt in the existing bridge capacity to carry the load from vehicles available in the world market.

In Study '87, a new bridge loading standard known as Long Term Axle Load (LTAL) and a Special Vehicle (SV) model were proposed. This was subsequently published as a bridge design standard by JKR [2, 3]. As for the new Motor Vehicle Regulations, the increase in the axle and gross vehicle weight limits should comply with the bridge loading standards. However, to immediately permit vehicles equivalent to LTAL loading as the basis for the new Motor Vehicle Regulations would require many existing bridges to be strengthened or replaced [1, 4]. Therefore, a more practical step-by-step approach towards the long-term regulations, which involves the introduction of Short Term Axle Load (STAL) and Medium Term Axle Load (MTAL) policy, was proposed.

Transportation companies have been complaining the inconveniences caused by the different axle load limits along Federal Roads due to the enactment of WRO 2003. Therefore, a new Axle Load Study was conducted in 2016 (Study '16) to formulate a coordinated programme for the road transport sector in relation to the proposed plan to increase the commercial vehicular configurations and legal weight limits. The study also resulted in the formulation of a new procedure for the assessment of existing bridges. The vehicle weight limits in WRO were also accordingly revised following the recommendation given in the Study '16 through the issuance of WRO 2017.

2. Weight Restriction Order (WRO)

In Malaysia, all vehicle design and construction are required to comply with WRO, which consists of:

Schedule 1 – Outline the allowable axle loads, in relation to the axle configuration of vehicles and list of Federal Roads given in Schedule 2. Vehicles that comply with the axle configuration and axle load limits in this schedule are regarded as normal vehicle

Schedule 2 – Categorizes Federal Roads into several lists, i.e. List I, List II, List III, etc. Maximum load limits stated in Schedule 1 are different for each lists

Schedule 3 – List of gazette number for WRO

Schedule 4 – Provides the information on abnormal vehicles that are allowed to travel on Federal Roads stated in Schedule 5. These vehicles are not allowed to travel uncontrollably and shall be guided by pilot vehicles

Schedule 5 – Provides the information on Federal Roads permitted for abnormal vehicles

The enactment of WRO will ensure that the load effects from the worst combination of vehicles allowed to travel on the bridge are within the expected bridge capacity.

3. The Study '87

3.1. General

The Study '87 on bridges involves inspection of 966 bridges in Peninsular Malaysia and studies of bridge design practices in Malaysia [1]. It was reported that approximately 72 % of these bridges were incapable of safely carrying the worst likely combination of the vehicles available in the world market. From Study '87, three axle load policies termed as STAL, MTAL and LTAL were proposed. LTAL policy requires all new bridges to be designed to LTAL and SV loading [1, 2]. Whereas, existing bridges were to be controlled through two interim axle load policies termed as STAL and MTAL [1].

STAL policy was to be implemented immediately while at the same time, bridge upgrading works were to be carried out. MTAL policy was to be introduced once the bridge stocks are capable of safely carrying the load equivalent to MTAL. STAL and MTAL correspond to WRO 1989 and WRO 2003, respectively. As such, STAL and MTAL were the assessment loading standards for measuring the load carrying capacity of existing bridges that simulate the traffic loading regulated by their respective WRO [4].

3.1.1. STAL Policy. STAL policy was implemented through the issuance of WRO 1989 [4, 5]. In this policy, it was expected that majority of the existing bridges in Malaysia were capable of safely carrying the load equivalent to STAL. Generally, bridges located within the Federal Roads specified in List VI of Second Schedule in WRO 1989 have the capacity of at least for STAL. Table 1 summarizes the proposed load limits in Study '87 and gazetted load limits for roads specified in List VI of Second Schedule in WRO 1989. Accordingly, 74 bridges rated as substandard to the STAL in Study '87 were recommended to be replaced [1].

Table 1. Vehicle weight limits for Federal Roads in List VI of Second Schedule, WRO 1989 [4, 5]

Type of vehicle	Number of axle	Axle configuration	Load limit (tonne)	
			Study '87	WRO 1989
Rigid	2 axle		17.5	14 – 16
	3 axle	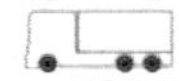	21	18 – 21
	4 axle		28	20 – 25
Articulated	3 axle		27	22 – 26
	4 axle		32	22 – 32
	5 axle		36	22 – 36
	5 axle	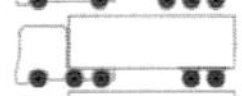	36	22 – 34
	6 axle		39	30 – 38

3.1.2. MTAL Policy. MTAL policy was implemented through the issuance of WRO 2003 [4, 6] when majority of the bridges in Malaysia were deemed to be capable of safely carrying the load equivalent to MTAL. The enactment of WRO 2003 increased the legal load limits to be equivalent to MTAL for selected Federal Roads in Peninsular Malaysia, which were listed in List I of the Second Schedule in WRO 2003. Federal Roads with bridges that were inadequate for MTAL but have the capacity of at least for STAL were categorized in List II of the Second Schedule in WRO 2003.

Since Sabah, Sarawak, and Labuan were not included in Study '87, the axle weight limits for Federal Roads in these states were restricted to List II of the Second Schedule. Table 2 summarizes the proposed load limits in Study '87 and gazetted load limits for roads specified in List I of Second Schedule in WRO 2003. Additionally, load limit and vehicle arrangement for 7 axle articulated vehicle were introduced in WRO 2003, which was not proposed in Study '87.

Table 2. Vehicle weight limits for Federal Roads in List I of Second Schedule, WRO 2003 [4, 6]

Type of vehicle	Number of axle	Axle configuration	Load limit (tonne)	
			Study '87	WRO 2003
Rigid	2 axle	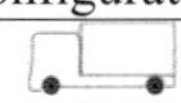	19	16 – 18
	3 axle	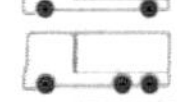	26	20 – 26
	4 axle	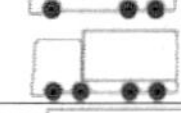	33	25 – 33
Articulated	3 axle	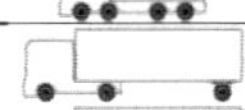	30	26 – 30
	4 axle		38	27 – 37
	5 axle		40	39
	5 axle	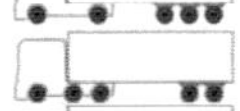	40	28 – 40
	6 axle		44	34 – 44
	7 axle	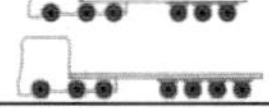	–	53

While vehicular loads were controlled by Road Transport Department (JPJ) vide WRO, existing bridges had to be assessed of their load carrying capacity. The load carrying capacities for majority of the bridges in Malaysia are expected to be of least STAL and MTAL whilst STAL policy and MTAL policy, respectively, were in effect [1].

3.2. Bridge Assessment

3.2.1. The Procedures. Bridge Replacement Policy in JKR Terms of Reference (TOR) for Bridge Assessment Procedure [7] imply that the decision to replace, widen or strengthen a bridge can only be executed if any one of the following criteria was not satisfied:-

i. The condition of the existing bridge based on visual inspection

ii. The hydraulic capacity of an existing bridge based on site investigation and discharge capacity
iii. the existing bridge capacity based on theoretical calculation

The theoretical strength evaluation was to be in accordance with the procedure proposed in a study by JKR in 1995 [8]. Evaluation Load Rating, ELR, which represents the ratio of the available live load capacity of a member to the effect of reference live load can be obtained using Equation 1 [7, 8].

$$ELR = \frac{R^* - D^*}{L^*} \tag{1}$$

where:
R^* = factored resistance
D^* = factored dead load effect
L^* = factored reference live load effect

The TOR specifies that the reference live load shall be of LTAL [7]. Hence, ELR_{LTAL} represents the ratio of the available live load capacity to the effect of LTAL as shown in Equation 2. Based on Equation 2, it is logical to infer that if a bridge has ELR_{LTAL} of less than 1, it is theoretically not capable of safely carrying the load equivalent to LTAL.

$$ELR_{LTAL} = \frac{R^* - D^*}{L^*_{LTAL}} \tag{2}$$

The implementation of STAL policy through WRO 1989 requires most of the bridges in Malaysia to be capable of safely carrying the load equivalent to STAL [1, 9]. This means that ELR_{STAL} of the existing bridges need to be more than or equal to 1. However, as per the TOR, the rating was calculated based on the effect of LTAL and it was considered that a bridge with ELR_{LTAL} of 0.8 or higher is capable of safely carrying at least the load equivalent to STAL, i.e. the worst possible combination of vehicles allowed in WRO 1989.

MTAL policy was later implemented through WRO 2003, which increases the limits on legal axle load and gross vehicle weight. This requires most of the bridges in Malaysia to be capable of safely carrying the load equivalent to MTAL [1, 9]. The relationship between ELR_{MTAL} and ELR_{LTAL} is represented by a β ratio, which varies with the loaded length as shown Figure 1. Therefore, if a bridge has ELR_{LTAL} of at least β, the bridge is deemed capable of safely carrying at least the load equivalent to MTAL, i.e. the worst possible combination of vehicles allowed in WRO 2003. The procedure in developing the β ratio is explained in a paper published in 2013 [9].

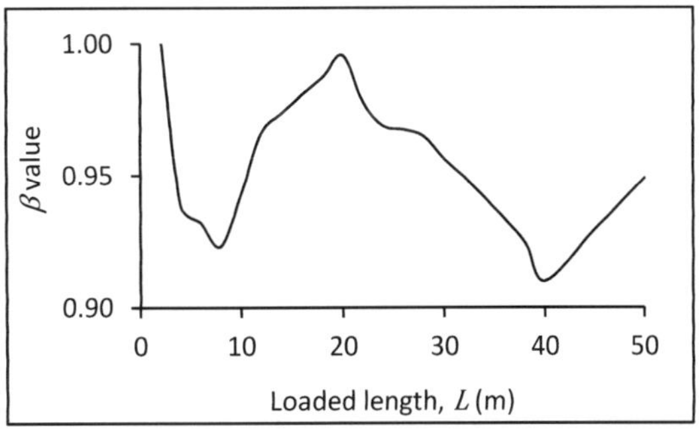

Figure 1. β ratio of ELR_{MTAL}/ELR_{LTAL} [9]

3.2.2. Assessment Load. STAL, MTAL and LTAL each consists of a uniformly distributed load, ω and knife edge load, KEL which represents the effects of all normal vehicles that are permitted by the WRO to travel uncontrolled on the road. The KEL was to be taken as 100 kN per lane [3, 4] whereas Figure 2 shows the curve for load, ω according to STAL, MTAL and LTAL along with the curves for load, ω in BS 153 and BD 21/84 for comparison.

The derivation of load, ω for STAL was governed by the strength of the nation's bridges built before 1972 that were designed to BS 153 or its equivalent with no checking done on HB load [4]. As shown in Figure 2, the load curve for STAL was technically a BS 153 load curve that has been modified at loaded length of 4 – 9 m.

Composite line of bending moment effect from HA load of BS 153 and equivalent HB load was used to derive the vehicle dimension and weight limits. Then, the shear force effect from these vehicles were determined to derive equivalent UDL, which was then proposed to be used as the load, ω in the MTAL [4] shown in Figure 2.

The derivation of load, ω for LTAL was based on the loading curve in BD 21/84 where modification has been made to suit a fixed notional lane width of 2.5 m. The equation to derive the loading curve given in BD 21/84 was multiplied by the lane width factor, F_w for 2.5 m lane width to obtain the equation for load, ω in LTAL [2, 3].

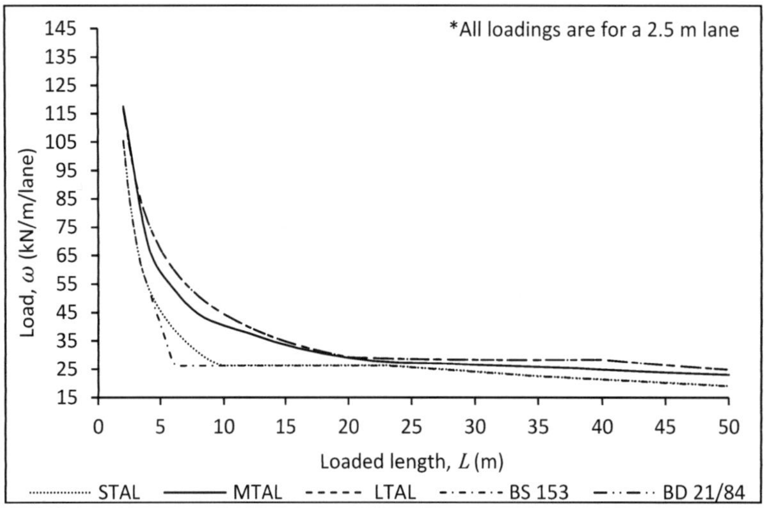

Figure 2. Load, ω according to STAL [4], MTAL [4], LTAL [3], BS 153 [10] and BD 21/84 [11]

The notional lane width for STAL, MTAL and LTAL was to be fixed at 2.5 m as opposed to the varying notional lane width usually used in other standards [4]. This was to avoid from the reduction in the total live load with the increase of carriageway width. The anomaly in total load due to varying notional lane width is illustrated in Figure 3.

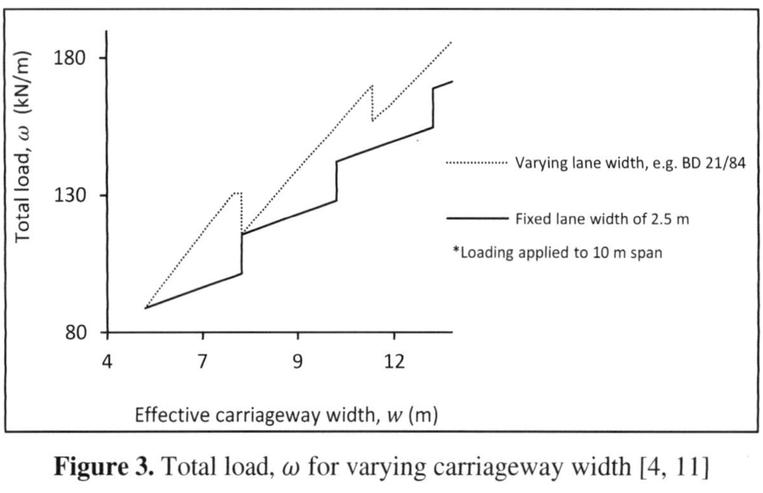

Figure 3. Total load, ω for varying carriageway width [4, 11]

4. The Study '16

4.1. General

From Study '16, a new assessment load was developed for the assessment of existing bridges. One of the reasons was the previous assessment load that was developed based on the Study '87, might have unnecessary safe-sided assumptions [12]. The Study '16 involves the review of weight-in-motion (WIM) data and the limits given in WRO 2003. WIM data was used as it can provide a more realistic representation of the types, distributions and overloading situations of vehicles in Malaysia. The weight limits given in WRO 2003 also had been proposed to be revised accordingly [13]. The proposed weight limits were then applied in WRO 2017. Table 3 summarizes the proposed load limits in Study '16 and gazetted load limits for roads specified in List I of Second Schedule in WRO 2017.

Table 3. Vehicle weight limits for Federal Roads in List I of Second Schedule, WRO 2017 [13, 14]

Type of vehicle	Number of axle	Axle configuration	Load limit (tonne)	
			Study '16	WRO 2017
Rigid	2 axle	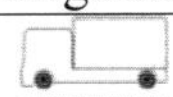	19	16 – 19
	3 axle	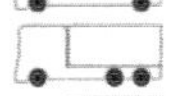	27	20 – 27
	4 axle	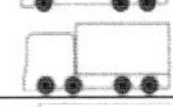	–	25 – 33
Articulated	3 axle	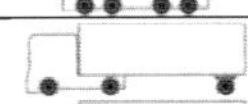	31	26 – 31
	4 axle		39	27 – 39
	5 axle		41.5	40
	5 axle	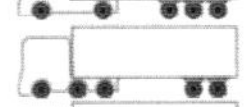	45	28 – 45
	6 axle	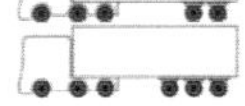	50	34 – 50
	7 axle	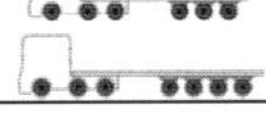	–	53

4.2. Bridge Assessment

4.2.1. The Procedure. In Study '16, the new assessment procedure for evaluating a bridge capacity is given in Technical Note No. 15 (TN15) [15]. Essentially, the strength of a bridge is indicated by a C rating that can be obtained using Equation 3 [15], which represents the ratio of the maximum theoretical live load capacity of a member to the assessment live load. If a bridge has C value of less than 1, then it should be classed as 'Immediate Risk' and treated as the highest priority for remedial management.

$$C = \frac{\omega_R^* - \omega_D}{\omega_L^*} \tag{3}$$

where:
ω_R^* = factored resistance in terms of load intensity
ω_D = unfactored dead load intensity
ω_L^* = factored reference live load intensity

4.2.2. Assessment Load. The new assessment load to be used in evaluating the load carrying capacity of a bridge was given in Technical Note No. 1 (TN1) [12] of the Study '16. The notional lane width to be used in bridge assessment generally follows Eurocode [12, 16] as shown in

4.2.3. ***Table 4***. It is reasonable to infer that the notional lane width is always equal to 3 m except for the carriageway width between 5.4 – 6 m. The load model was derived for two types of routes, namely Federal Route and Port & Cement Route. However, it was proposed that the load model for Federal

Route to be used for the assessment of all bridges [12]. The formula for the load, ω and KEL in the new assessment load is summarized in

Table 4. *Notional lane width for assessment in Study '16 [12]*

Carriageway width, w (m)	Number of lanes	Lane width, w_l (m)	Remaining width (m)
$w < 5.4$	$n = 1$	3	$w - 3$
$5.4 \leq w < 6$	$n = 2$	2.7 to 3	0 m
$6 \leq w$	$n = \text{Integer}\left(\frac{w}{3}\right)$	3	$w - 3n$

Table 5. New assessment load in Study '16 [12]

Routes	ω (kN/m/lane)	KEL (kN)
Federal Routes	$40\left(\frac{1}{L}\right)^{0.2}$	180
Port and Cement Routes	$42\left(\frac{1}{L}\right)^{0.2}$	190

Study '16 also recommended the use of equivalent uniformly distributed load, $EUDL$ for comparison of different load models, i.e. STAL, MTAL and LTAL to the new assessment load. $EUDL$ for an assessment load model can be calculated using Equation 4.

$$EUDL = \omega + \frac{2*KEL}{L} \quad (4)$$

5. Comparisons of the Procedures and Load Models

The difference between the previous and new procedure is summarized in Table 6. C is essentially similar to the old rating, ELR. The equation used to obtain the rating in both procedures indicates that the judgement is to be made based on the ratio of available live load capacity to the reference live load. However, the available live load capacity in previous procedure was obtained by subtracting the factored resistance with factored dead load, whereas in the new assessment procedure, it is to be subtracted with unfactored dead load. This means that there was no safety factor applied on the dead load in calculating C, thus giving a higher value of available live load capacity, and therefore, a higher C rating.

Table 6. Difference between the previous and new procedure in bridge assessment [4, 7, 15]

Procedures	Study '87 / Previous procedure	Study '16 / New procedure
Rating	$ELR = \frac{R^* - D^*}{L^*}$ (Refer 3.2.1)	$C = \frac{\omega_R^* - \omega_D}{\omega_L^*}$ (Refer 4.2.1)
Notional lane width	Fixed at 2.5 m	3 m except at 5.4 – 6 m carriageway width where notional lane is between 2.7 – 3 m
ω and KEL	$EUDL$s as per Study '16 recommendations are as shown in Figure 4	
Application of load	Full ω and KEL applied on two notional lanes and 0.6 times ω and KEL to all other notional lanes. Remaining areas not covered with notional lane to be applied with minimum load of 5 kN/m². Refer Figure 5 and Figure 6	

Figure 4 shows the *EUDL* for STAL, MTAL, LTAL and the new assessment load in a single notional lane. The new assessment load is generally equivalent to MTAL with minor difference. However, the notional lane of 2.5 m in MTAL and 3 m in the new assessment load meant that the load intensity per area in MTAL is actually greater than the new assessment load.

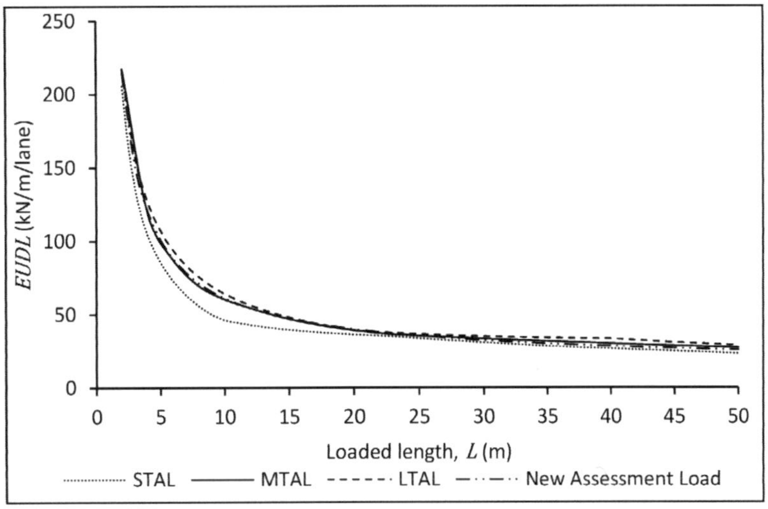

Figure 4: *EUDL* on a single notional lane [4, 15]

The applications of load, based on the previous and new procedure are as shown in Figure 5 and Figure 6, respectively. Using the load arrangement shown in these figures, the total load acting on the bridge was calculated and shown in Figure 7. Figure 7 shows that the total load from previous load model, i.e. MTAL and LTAL, is greater than the new load model; signifying a higher reference live load used in previous procedure. This could be attributed to the safe-sided assumptions as stated in Study '16 [12].

Therefore, if a bridge has been previously assessed to the previous procedure and rated as capable of safely carrying the load equivalent to MTAL or LTAL, then it is capable of safely carrying the vehicles that are in compliance with WRO 2017. This is due to the fact that the limits in WRO 2017 was derived from the new load model.

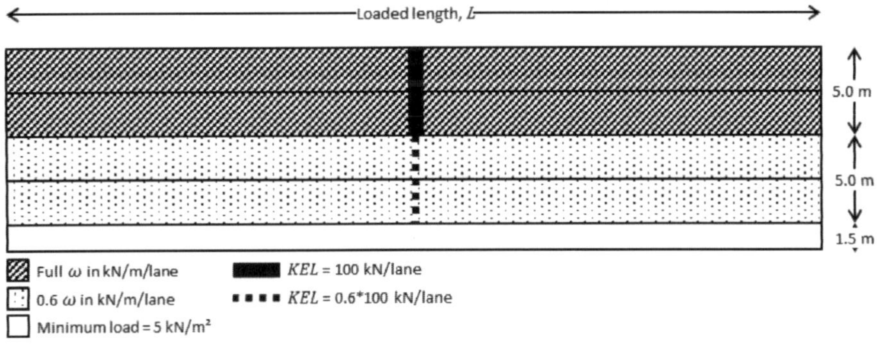

Figure 5. Load distribution plan on a bridge with 11.5 m carriageway (Study '87) [4]

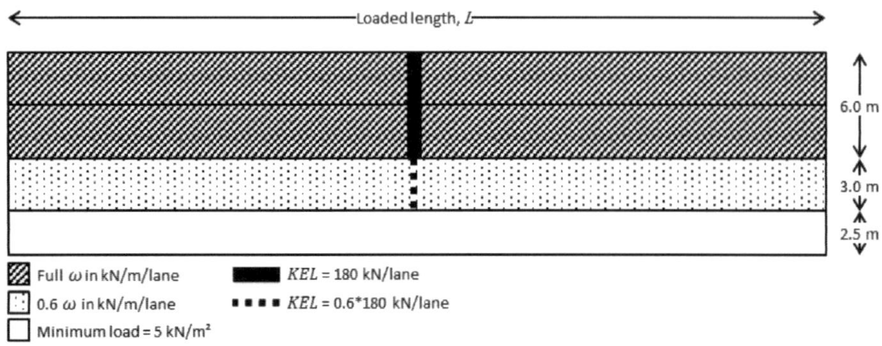

Figure 6. Load distribution plan on a bridge with 11.5 m carriageway (Study '16) [15]

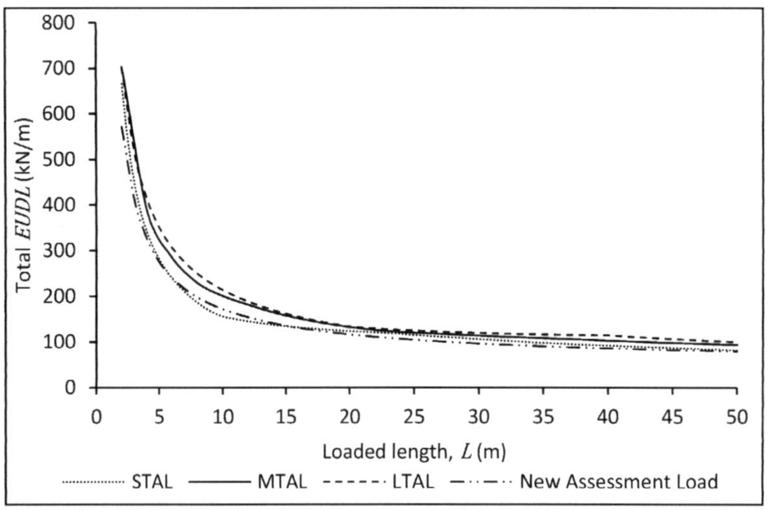

Figure 7. Total *EUDL* on the bridge [4, 15]

6. Conclusions

This paper gives general idea on how the STAL, MTAL and LTAL policies originated from, and its relation to the WRO. STAL policy and MTAL policy are load policies made to ensure that the vehicles permitted in its corresponding WROs are within the safe carrying capacity of the existing bridges. LTAL policy is the futuristic policy that specifies new bridges to be designed to LTAL loading. JKR has adopted LTAL loading in the design of new bridges until around 1997. Study '16 has come up with a new procedure for bridge assessment and a higher load limit has been allowed in WRO 2017 as a result from the study. Based on the comparison between the previous and new assessment procedure, bridges that were rated as capable of safely carrying the load equivalent to MTAL or LTAL will also be capable of safely carrying the vehicles that are in compliance with WRO 2017.

7. References

[1] Rendel Palmer and Tritton Ltd. 1989 *Final Report for Axle Load Study*

[2] Ng S K, Mohd Hisham M Y and Lim C C 1990 Standard JKR Specification for Bridge Loading *Road Engineering Association of Asia and Australasia (REAAA) Conference* Kuala Lumpur, March 1990

[3] Public Works Department of Malaysia 1989 *DJ 1/89. JKR Specification for Bridge Live Loads* (Kuala Lumpur: Public Works Department of Malaysia)

[4] Rendel Palmer and Tritton Ltd. 1989 *Final Report for Axle Load Study: Technical Note 20*

[5] *Weight Restrictions (Federal Roads) Order 1989 - P.U. (A) 478* (Malaysia)
[6] *Weight Restrictions (Federal Roads) (Amendment) Order 2003 - P.U. (A) 275* (Malaysia)
[7] Jabatan Kerja Raya *Terms of Reference for Bridge Assessment Procedure* (Kuala Lumpur: JKR)
[8] Dessau International Ltd., Ranhill Bersekutu Sdn. Bhd. and Jabatan Kerja Raya 1995 *Determination of the Structural Capacity of Existing Bridges in Peninsular Malaysia*
[9] Lim C C, Ng S K and Jasmani Z 2013 Developing an Assessment Criterion for Medium-Term Axle Load Bridge Capacity in Malaysia *REAAA International Conference* Kuala Lumpur, 28 March 2013
[10] BS 1972 *BS 153: Part 3A - 1972 Steel Girder Bridges* (Online: British Standard)
[11] Department of Transport 1993 *BD 21/84 The Assessment of Highway Bridges and Structures*
[12] Jurutera Perunding Zaaba Sdn. Bhd. 2016 *Axle Load Study (Bridges) on Federal and Major State Routes in Peninsular Malaysia: Technical Note No. 1: Bridge Assessment Loading*
[13] Jurutera Perunding Zaaba Sdn. Bhd. 2016 *Axle Load Study (Bridges) on Federal and Major State Routes in Peninsular Malaysia: Technical Note No. 18: Recommendation of Maximum Axle Loading and Vehicle Dimension for Design of Future Roads and Highways*
[14] *Weight Restrictions (Federal Roads) (Amendment) Order 2017 - P.U. (A) 226* (Malaysia)
[15] Jurutera Perunding Zaaba Sdn. Bhd. 2016 *Axle Load Study (Bridges) on Federal and Major State Routes in Peninsular Malaysia: Technical Note No. 15: Bridge Capacity Evaluation Procedure*
[16] BS 2003 *BS EN 1991-2: Actions on structures - Part 2: Traffic loads on bridges* (Online: British Standard)

IOP Conf. Series: Materials Science and Engineering **512** (2019) 012015 doi:10.1088/1757-899X/512/1/012015

The motorcycle to car ownership ratio: inflation and it's indirect effects

May Yen Chu[1] ,Teik Hua Law[1,2], Hussain Hamid[1,2], Siong Hook Law[3] and Jin Chai Lee[4]

[1] Road Safety Research Centre, Faculty of Engineering, Universiti Putra Malaysia, Serdang, Malaysia

[2] Department of Civil Engineering, Faculty of Engineering, Universiti Putra Malaysia, 43400 UPM Serdang, Selangor, Malaysia

[3] Faculty of Economy, Universiti Putra Malaysia, 43400 UPM Serdang, Selangor, Malaysia

[4] Faculty of Engineering, Technology & Built Environment, UCSI University, 56000 Cheras, Kuala Lumpur, Malaysia.

E-mail: mayyenchu006@gmail.com

Abstract. In the initial phase of economic development, motorcycle ownership rise with the growing demand in transportation. However, as income rose and perhaps due to convenience, safety and prestige, motorists opt to purchase car than motorcycle. Under high inflation, motorists tend to choose cheaper mode of transportation, like motorcycles and mopeds, which in turn leads to higher ownership ratio of motorcycle to passenger car (MTPC). On the other hand, economic studies indicate that the correlation between inflation and economic development is nonlinear; it is negative for high levels of inflation, but positive for low levels of inflation. Consequently, this may lead to a rise in MTPC ownership ratio at some levels of income and a decline at others. This study focused on understanding how inflation affects the MTPC ownership ratio indirectly and the factors underlie this relationship. The data used in this analysis contained a sample of 76 countries at various levels of economic development growth over the 51-year period between 1963 and 2013 using panel data analysis. The indirect effect of the inflation on the MTPC ownership ratio varied in accordance with increases in the inflation. Policy implications of the study were discussed in the conclusion part of the study.

1. Introduction

The existing empirical evidence indicated that there is a Kuznets curve between motorcycle ownership and income growth. Motorcycle ownership rose with per capita income at a lower level of GDP, but decreased with per capita income at a higher level of GDP [1-4]. The literature explained motorcycle ownership increased with the transportation demand in the beginning of the economic development. However, as income rose motorists opt to purchase car than motorcycle. This may due to convenience, safety and prestige. The inverted U-shaped relationship is not mainly explained by per capita income, but also by the inflation rate [4]. Under high inflation, motorists tend to choose cheaper mode of transportation (i.e. motorcycles and mopeds), which leads to higher ownership ratio of motorcycle to passenger car (MTPC) [4]. Economic studies indicate that the correlation between inflation and economic development is nonlinear; it is positive for low levels of inflation but negative for high levels of inflation [5-8]. Consequently, this may lead to a rise in MTPC ownership ratio at some levels of income and a decline at others.

This paper discussed the indirect impact of inflation on the MTPC ownership ratio, transmitted via the impact of inflation on per capita income.

2. Review on economic growth and inflation

Previous studies showed that inflation has detrimental impact on economic development in medium and long term [9-11]. Nevertheless, several literatures suggested that the relationship between economic growth and inflation in long term is nonlinear. The correlation between these two variables is positive or nonexistent below certain threshold levels, but hampers the economy when inflation is above that threshold level [5-7, 12].

The first author who classifies this non-linear link was Fischer (1993). He concluded that inflation boost economic growth when it is below a threshold value, but has a negative impact if it is above. Various literatures summarized that the inflation-growth impacts are less apparent at low rates of the inflation but significant negative at high inflation rate [12-13]. The results are depending on the inflation threshold level and it changes according to the group of countries being assessed (exp. developing, developed, or a mixture of both). The inflation threshold is lower in developed than in developing countries [13].

Economic literature found that inflation hinders economic growth once surpasses inflation threshold; however it promote growth before the threshold level. As previous studies found that MTPC ratio has inverted-U shape relationship with economic growth. Therefore, an indirect impact from inflation affecting economic growth is existing hence shaping MTPC ratio. This assumption is presented in the following section.

3. Methodology

The indirect effect of inflation on MTPC ownership ratio can be reflected on the effect of inflation on per capita income and the subsequent effect of per capita income on MTPC ownership ratio. The indirect effect can be expressed as,

$$\frac{dMTPC}{dCPI} = \frac{\delta MTPC}{\delta GDP}\frac{\delta GDP}{\delta CPI} \tag{1}$$

where MTPC is the MTPC ownership ratio, CPI represents consumer price index, and GDP is a measure of real GDP per capita.

This study utilized the panel data regression methods to estimate the equaton 1 equation model. Due to the high accessibility and availability of cross-country time series data sets, panel regression modeling has becoming more favorable method. Panel data are more competent due to its better estimations and more informative. Linearity between variables is being reduced [15]. First, linear panel model is used to estimate the impact of CPI on per capita income (refer to equation 2). Then, the effect of per capita income and CPI on MTPC ownership ratio is estimated using panel model with the AR(1) disturbance term (refer to equation 3). In the panel regression analysis, the Hausman test is used to assess the suitability of a fixed or random effects model under the null hypothesis that there is no correlation between country-specific intercept and the independent variables in the model. When the null hypothesis is rejected, fixed effect estimator is chosen. The two equations are expressed as,

$$\ln GDP_{it} = \varphi_i + \alpha_1 CPI_{it} + \alpha_2 Y_{it} + \varepsilon_{it} \tag{2}$$

$$\ln MTPC_{it} = \omega_i + \beta_1 CPI_{it} + \beta_2 \ln GDP_{it} + \beta_3 (\ln GDP_{it})^2 + \beta_4 X_{it} + \beta_5 T + \zeta_{it} \tag{3}$$

where T denotes a time trend, ω_i and φ_i represent country fixed effects; ζ_{it} and ε_{it} refer to the error term; The sub-index i refer to a country and the sub-index t refers to time; Equation 2 consists of a few control variables, Y, specifically total of imports and exports of goods and services, physical capital stock per worker, and human capital index. X, in Equation 3 specifies other control variables, which include road density and urban to rural population ratio.

4. Data

This study included 1900 annual observation of an unbalanced panel dataset. It is consisting of 77 countries from the year1963-2013 (51 years); some variables were missing in the sample for some countries and years. The countries list included in this study is presented in Table 1.

The depended variable for equation 2 was the per capita Real Gross Domestic Product (GDP) (US$ 2011 constant prices: Chain series) was used as a proxy for per capita income. There will be a positive relationship between openness of trade and economic growth [17]. According to the Solow growth model, the capital stock has positive impact on economy growth [17-18]. There were two different types of capital stock, human and physical capitals respectively were included in equation 2. Human capital refers to the productive skills, education level, and work-related capabilities in the employment force while the physical capital stock refers to the essential infrastructure (i.e. water supply, buildings, roads, tools and technology). Physical capital stock per worker (CPW) was calculated by dividing physical capital stock by the number of workers. It was calculated as 'rkna'divided by 'emp', where 'rkna' was the capita stock at constant 2011 national prices and 'emp' was the number of persons engaged in millions [18]. The consumer price index (CPI), a measurement for inflation was used as a proxy for power to purchase vehicles [19].

MTPC ownerships ratio, the dependant variable in equation 3, was obtained by dividing the total of motorcycles and mopeds by the total of passenger cars (for a specific country and the year). The other two control variables for equation 3 the total road length per thousand populations and were the urban to rural population ratio. The urban to rural population ratio was used to investigate the impact of population concentration on MTPC ownership ratio. The total road length per thousand population was to explain the travel patterns and vehicle ownership [20]. Table 2 shows the data sources and definitions used in this analysis.

Table 1. The list of the countries and the available years of data.

No.	Country	Observation	Period From-To
1	Australia	30	1963-2004
2	Austria	45	1963-2010
3	Bahrain	26	1987-2013
4	Bangladesh	9	1990-2003
5	Belgium	48	1964-2011
6	Benin	5	1992-1996
7	Botswana	14	1981-2005
8	Brazil	3	2000-2004
9	Brunei	17	1991-2011
10	Bulgaria	20	1986-2011
11	Cameroon	13	1968-2008
12	Canada	30	1969-2009
13	China	19	1990-2013
14	Colombia	20	1983-2013
15	Costa Rica	25	1984-2013
16	Cote d'Ivoire	10	1966-2007
17	Cyprus	32	1975-2013
18	Czech Republic	21	1993-2013
19	Denmark	34	1977-2013
20	Ecuador	20	1984-2013
21	Egypt	19	1982-2010
22	Estonia	18	1996-2013
23	Finland	50	1963-2013
24	France	40	1963-2010
25	Germany	18	1991-2013
26	Greece	40	1971-2013
27	Hong Kong	20	1981-2013
28	Hungary	22	1991-2013
29	Iceland	51	1963-2013
30	India	41	1965-2012
31	Indonesia	34	1966-2013
32	Ireland	42	1965-2013
33	Israel	27	1987-2013
34	Italy	38	1963-2005
35	Japan	49	1963-2011
36	Jordan	11	2003-2013
37	Kenya	33	1966-2013
38	Korea, Rep.	38	1971-2013
39	Laos	12	1990-2002
40	Latvia	17	1995-2013
41	Lithuania	19	1995-2013
42	Luxembourg	27	1975-2009
43	Malawi	3	1980-1982
44	Malaysia	48	1963-2012
45	Malta	10	1998-2009
46	Mauritius	31	1980-2013
47	Mexico	15	1990-2013
48	Mongolia	10	1994-2003
49	Morocco	34	1969-2011
50	Nepal	3	2010-2012
51	Netherlands	43	1963-2012
52	New Zealand	38	1971-2013
53	Nigeria	11	1973-1996
54	Norway	49	1963-2013
55	Pakistan	38	1967-2012
56	Panama	16	1977-2009
57	Peru	11	2000-2013
58	Philippines	26	1981-2011
59	Poland	24	1990-2013
60	Portugal	17	1963-2003
61	Romania	19	1990-2012
62	Slovakia	18	1995-2013
63	Slovenia	22	1992-2013
64	South Africa	23	1967-2001
65	Spain	45	1967-2013
66	Sri Lanka	21	1969-2010
67	Swaziland	14	1987-2003
68	Sweden	37	1973-2013
69	Switzerland	41	1963-2013
70	Syria	18	1963-2010
71	Thailand	33	1967-2011
72	Togo	14	1980-2007
73	Tunisia	16	1983-2013
74	United Kingdom	39	1975-2013
75	United States	43	1970-2013
76	Vietnam	12	2000-2011

Table 2. Data sources for variables in equation 2 and 3.

Data source	Definition	Variable
International Road Federation	Motorcycle to passenger car ownership ratio	MTPC
	Per capita Real GDP (In mil. 2011US$)	GDP
Penn World Table version 9.0	Physical capital stock per worker	CPW
	Human capital index	HC
	Consumer price index (2010 = 100)	CPI
World Development Indicator	Ratio of urban population to rural population	URBRUR
	Road density (km per 1000) population)	ROAD
	Sum of import and export goods and services (% of GDP)	TRADE

5. Results and Discussions

5.1. Model of inflation and per capita income

Table 3 explained the statistic for variables in equation 2 while Table 4 showed three models were used to describe the effect of inflation on per capita income. In order to assess the quadratic relationship between inflation and the per capita income, a quadratic term was added to the natural logarithm of CPI. Model A (in Table 4) model the linear and quadratic terms of inflation to investigate the link between per capita income and inflation. Trade, CPW and HC were included to assess the robustness of the estimation of inflation (Models B and C in Table 4).

Table 3. Description statistics for all variables used in equation 2

Variable	Obs	Average	Std. Dev.	Minimum	Maximum
GDP	1900	18671.11	13961.01	412.797	84288.27
CPI	1900	56.47824	33.07998	1.004284	131.8465
HC	1900	2.586621	0.65639	1.039415	3.726472
TRADE	1900	78.01422	49.33161	7.529721	455.4151
CPW	1900	174479.5	122320.9	5.548396	564.9913

The estimated results in Models A and B (in Table 4) revealed that the coefficient for inflation was significantly positive whereas the coefficient on its squared term was estimated to be negative. This shows that the per capita income increased with inflation at a lower level and decreased with inflation at a higher level. The turning point happened at CPI value of 76.2 (at 65 percentile of CPI). 'Trade', 'CPW' and 'HC' are estimated to be significantly positive, in line with expectations. We used Modal A to estimate the pure effect of inflation on MTPC ownership ratio.

The Hausman test showed that a fixed-effect estimator was more suitable in Models A and B, whereas a random-effect estimator was more appropriate in Model C. The inverted-U relationship disappeared once the data were controlled for the human capital index. Interestingly, the CPI coefficient in Model C was negative, and the CPI squared was positive (U-shaped). However, the slope of the per population income and CPI relationship changed from negative to positive at lower CPI levels(turning point 20.5). This showed that the CPI had a positive impact on per population income throughout most of the sample range.

Table 4. The effect of inflation on per population income.

Variables	Model A	Model B	Model C
ln(CPI)	3.537 *	3.518 *	-0.145*
$(\ln(CPI))^2$	-0.408 *	-0.468 *	0.0240*
TRADE		0.005*	0.0008*
CPW		0.006*	0.001*
HC		0.	1.099***
Constant	2.320 *	1.725*	6.375*
No. of observation	1824	1824	1900
No. of groups	76	76	76
R squared (within samples)	0.8134	0.8572	0.8272
Hausman test (chi-square)	8721.54*	13251.51*	N.A.
Random- or Fixed-Effect	FE	FE	RE
Turning point	76.2	43.0	20.5

Remarks:
*Significant at 1%
RE-Random-effect, FE- Fixed-effect.

5.2. Model of consumer price index and the MTPC ownership ratio

Wherever Table 5 explained the statistics for all variables used in equation 3. The estimated results in Table 6 confirmed a significant inverted U-shaped relationship between the MTPC ownership ratio and the per capita income. This result is consistent with previous study which reported that MTPC ownership ratio inclined with per capita income at a lower level and declined with income at a higher level [4]. The threshold level for the MTPC ownership ratio was estimated in the range of US$2800 to US$3770.

Table 5. Descriptive statistic for variables included in Equation 3

Variable	Obs.	Average	Std. Dev.	Minimum	Maximum
MTPC	1900	1.042	5.443	0.00115	77.0123
GDP	1900	18623.48	13884.32	412.79	84288.27
CPI	1900	56.366	32.967	1.004	131.847
URBRUR	1900	178.19	3992.71	0.135	100054
ROAD	1900	390520.8	1031661	1150	6550896

The coefficient for the inflation in Table 6 is significantly positive. Inflation raises the price of goods and services, consequently decreasing buying power [20]. As a result, during high inflation stage, travelers choose to obtain lower running cost and price vehicles (i.e. motorcycle and mopeds). Therefore, this will cause a rise in the MTPC ownership ratio.

Model B (in Table 6) showed that the total road length per thousand population is positive on the MTPC ownership ratio. This shows that the MTPC ownership ratio rises with total road length per thousand population. And the road network growth started in urban areas with lower number of lanes and roads. A higher road density is associated with increasing local accessibility and motorcycle ownership. It is anticipated to increase motorcycle more rapidly compare to car ownership at this growth stage

Table 6. Estimates of MTPC

Variables	Model A	Model B
ln(GDP)	2.0137***	1.8997***
$(\ln(GDP))^2$	-0.1345***	-0.1274***
ln(CPI)	0.1278***	0.1392***
ln(ROAD)		0.0162
ln(URBRUR)		-0.0831***
Constant	-9.1502***	-8.8551***
No. of observation	1900	1900
No. of groups	76	76
R squared (within samples)	0.0822	0.0831
Random or Fixed Effects	RE	RE

Remarks:
*Significant at 1%
RE-Random-effect, FE- Fixed-effect.

In Model C, the coefficient for urbanization was significantly negative. This shows that the MTPC declined with the level of urbanization of a country. As urbanization increases, private cars are expected to be preferable to motorcycles, as they are suited to long-distance trip. The estimated results for Model A were used in estimating the pure impact of CPI on the MTPC ratio.

5.3. Estimation of the indirect effects of consumer price index

The indirect effect is calculated from the product of δMTPC/δGDP and δGDP/δCPI (as indicated in equation 1). The results in Model A (Table 4) show that the turning point for the CPI is at 76.2, while the estimated turning point for the per capita GDP is reached at US$1,779.

The indirect impact of inflation on MTPC can be divided into four different phases (refer to Figure 1) varies according to the CPI level. A rise in CPI levels is associated with increases in per capita GDP (when the CPI value is below 6.4, phase I), and this would lead to a rise in the MTPC ownership ratio. The indirect impact is estimated to be positive in the first phase, implying that an increase in the CPI increases the MTPC. During this stage (at a low income level and low economic activity), motorcycle ownership was more common than car ownership because of the higher cost of cars, as indicated by [4].

However, when the CPI value is between 6.4 and 76.2, increases in per capita GDP (as a result of increases in CPI) would lead to a decrease in the MTPC ownership ratio (phase II). This is because per capita GDP has already passed the turning point of US$1,779. The indirect impact changed from positive to negative in the second phase (when CPI growth exceeded 6.4). This shows that due to safety and prestige, more commuters readily switched from motorcycles to private cars at higher per population income levels

As the CPI continued to rise (exceeded the threshold of 76.2, turning point), increases in CPI would hinder the growth of economic and hence increased the MTPC ownership ratio (phase III). The indirect impact changed from negative to positive during this third phase. The reduction in per population income (due to inflation rate rises) arose with a reduction in purchasing power. Therefore, motorists tended to choose vehicles (i.e. mopeds and motorcycles) with lower purchase, maintenance, and running costs in periods of high inflation. This resulted increased in motorcycle ownership at a higher inflation level.

Per capita GDP decreases continuously as the CPI value exceeded 604.19 and this leads to a reduction in the MTPC ownership ratio (phase IV). The indirect impact was estimated to be negative at this phase. Due to resource scarcity, motorcycle ownership will likely to decline and private car ownership can be expected to be low at this stage. The above discussion can be illustrated in Table 7.

Table 7. The impact of CPI on MTPC (Elasticities)

lnCPI	CPI	Estimated ln(per population GDP)	$\frac{\partial MTPC}{\partial GDP}$	$\frac{\partial GDP}{\partial CPI}$	$\frac{\partial MTPC}{\partial GDP} x \frac{\partial GDP}{\partial CPI}$	Phase
1.6198 (5%)	5.00	6.9794 ($1,074)	0.1357	2.2151	0.3006	I
3.2593 (25%)	26.05	9.5139 ($13,547)	-0.5463	0.8768	-0.4790	II
4.0987	60.26	9.9623 ($21,211)	-0.6669	0.1915	-0.1278	II
4.6052 (90%)	100.48	9.9546 ($21,049)	-0.6648	-0.2219	0.1476	III
6.9176*	1012.32	7.2587 ($1,420)	0.0606	-2.1097	-0.1278	IV

Remarks:
* (High CPI, not in data range)

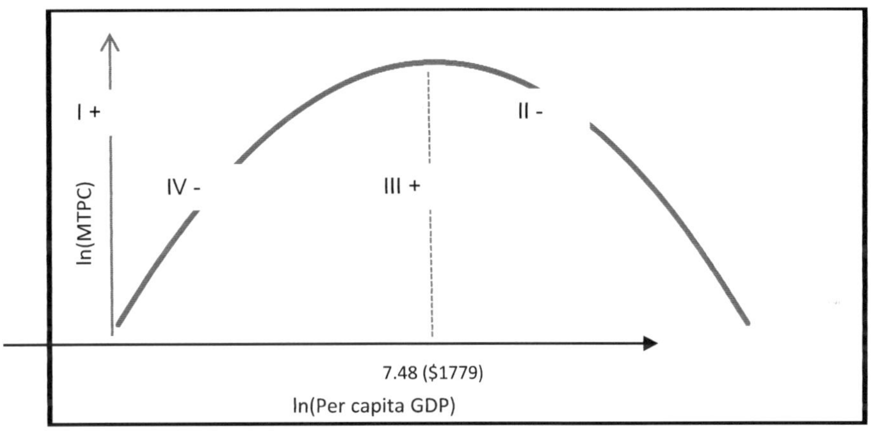

Figure 1. The relationship between the MTPC and the per population GDP.

6. Conclusion

This paper assessed the indirect impact of the inflation on the MTPC ownership ratio via the effect of the inflation on per population income and the resultant effect on MTPC ownership ratio. The result indicates that MTPC ownership ratio varies accordance to the rise of CPI at different level. The key finding of this study was that inflation has a non linear effect on MTPC ownership ratio, positive at earlier phase and turn to negative once per capita income is higher. It turned back to positive once reached higher inflation and further negative elasticity which is at resource scarcity phase.

As regards to policy implications, lower level of inflation is associated with a high number of MTPC ownership ratio (phase I). The increased of large motorcycle population and high mobility of motorcyclists indicates that economic growth at lower income levels would lead to a rise in motorcycle deaths [21]. A similar problem also arises at higher inflation rates (phase III). However, the increase of motorcycle ownership at this stage is due to a decline in per capita income. To combat the excessive motorcycle safety problems, some injury prevention and risk control interventions such as segregate motorcycles via the provision of motorcycle lanes, encourage modal shift from motorcycle to public transportation, raising the rigorous of motorcycle helmet laws and enforcement intensity should be implemented. Nonetheless, for countries that passing through phase II, it is anticipated that passenger car ownership rises rapidly. To adapt the rise of passenger cars they should redirect their country resources to improve and upgrade their road transport infrastructure (i.e. extension of road networks, building more highways and upgrading public transport).

Several limitations of this study should be noted. First, the quadratic function used to describe the link between the MTPC and per capita income as well as between the CPI and per capita income. Once obtained the threshold point, the decline rate is similar with the previous incline rate. This seems unlikely to happen because the downward and upward of (CPI-income and MTPC ratio-income) shape are impacted by several aspects. Thus, for future study other curve fitting methods (i.e. spline function) is suggested. Second, this study assumed only two main private transport modes which are motorcycles and passenger cars. In reality, other transport modes (i.e. bicycling, walking and using public transport) could be chosen to fulfill travel needs. Thus, a possible extension of this study is to assess the modal shift from motorcycles to the above mentioned transport modes and to identify the factors that promote this modal shift.

7. References

[1] Senbil M, Zhang J and Fujiwara A 2007 Motorization in Asia – 14 countries and three metropolitan areas *Int. Assoc. Traffic Saf. Sci. Res.* **31**(1) 46–58

[2] Pongthanaisawan J and Sorapipatana C 2010 Relationship between level of economic development and car ownerships and their impacts on fuel consumption and greenhouse gas emission in Thailand *Renew. Sustain. Energy Rev.* **14**(9) 2966–75

[3] Tuan V A 2011 Dynamic interactions between private car and motorcycle ownership in Asia: a cross-country analysis *J. Eastern Asia Soc. Transport. Stud.* **9** 541–56

[4] Law T H, Hamid H and Goh C N 2015 The motorcycle to private car ownership ratio and economic growth: a cross-country analysis *J. Transport Geography* **46** 122–28

[5] Fischer S 1993 The role of macroeconomic factors in growth *Journal of Monetary Economics* **32** 45–66

[6] Sarel M 1996 Nonlinear effects of inflation on economic growth *IMF Staff Papers* **43** 199–215

[7] Ghosh A and Phillips S 1998 Warning: inflation may be harmful to your growth *IMF Staff Papers* **45** 672–710

[8] Gillman M and Kejak M 2005 Contrasting models of the effect of inflation on growth *Journal of Economic Surveys* **19** (1) 113–36

[9] Bruno M and Easterly W 1998 Inflation crises and long-run growth *Journal of Monetary Economics* **41** (1) 3-26

[10] Barro R J 2013 Inflation and economic growth *Annals of Economic and Finance* **14** (1) 85-109

[11] Bittencourt M 2012 Inflation and economic growth in Latin America: Some panel time-series evidence *Economic Modelling* **29** (2) 333-40

[12] Khan M S and Senhadji A S 2001 Threshold effects in the relationship between inflation and growth *IMF Staff paper* 48 pp 1-21

[13] Villavicencio A L and Mignon V 2011 On the impact of inflation on output growth: Does the level ofinflation matter? *Journal of Macroeconomics* **33** (3) 455-64

[14] Hsiao C 2003 *Analysis of Panel Data, second ed.* (Cambridge, UK: Cambridge University Press)

[15] Solow R 1956 A contribution to the theory of economic growth *Quarterly Journal of Economics* **70** 65–94

[16] Doucouiagos C and Ulubasoglu M A 2006 Economic freedom and economic growth: Does specification make a difference? *European Journal of Political Economy* **22** 60–81

[17] Caselli F 2004 Accounting for cross-country income differences *NBER Working Paper No. 10828*

[18] Bento A M, Cropper M L, Mobarak A M and Vinha K 2005 The effects of urban spatial structure on travel demand in the United States *Rev. Econ. Stat.* **87** (3) 466–78
[19] Chauhan G S 2017 Corporate financing and deleveraging of firms in India *IIMB Management Review* **29** 170-87
[20] Manzoor M M, Siddiqui A A, Sattar A, Fahim M and Rasheed S 2011 Impact of inflation on household consumption – a case of Pakistan *Int. Res. J. Finance Econ.* **68** 161–67
[21] Law T H, Noland R B and Evans A W 2009 Factors associated with the relationship between motorcycle deaths and economic growth *Accid. Anal. Prevent.* **41** (2) 234–40

Acknowledgments
This study was sponsored by Universiti Putra Malaysia Research University Grant (Grant No. GP/2017/9564000) and UCSI University Pioneer Scientist Incentive Fund (Project code: Proj-In-FETBE-030).

IOP Conf. Series: Materials Science and Engineering **512** (2019) 012016 doi:10.1088/1757-899X/512/1/012016

A study of self-similarity in vehicular arrival pattern

Hooi Ling Khoo

Department of Civil Engineering, Faculty of Engineering and Science, Universiti Tunku Abdul Rahman, Bandar Sungai Long Campus, 43000, Kajang, Selangor, Malaysia.

Abstract. Traditionally, the vehicle arrival pattern is modeled based on Poisson assumptions in the road traffic studies. However, recent research has shown that the vehicle arrival patterns exhibit characteristics of self-similarity. This finding will affect delay calculation, queue lengths and other variables that are used in traffic engineering which were based on Poisson assumptions. In this paper, empirical studies are carried out to investigate the existence of self-similarity characteristics in vehicle arrival pattern for traffic flow on highways and isolated signalized intersections. Traffic flow that is recorded from site is processed and analysed. A software is adopted to analyse the possible existence of self-similarity by computing the Hurst parameter from the data set. Hypothesis testing approach is then adopted to evaluate whether the findings are statistically significant. The results confirmed that the vehicle arrival pattern for arterial road exhibit self-similar pattern, and its corresponding headway is heavy-tailed. This finding is important especially in traffic signal design on arterial road.

1. Introduction

Vehicle arrival pattern is one of the fundamental studies in traffic engineering. Most of the traffic applications developed later is based on the knowledge of the arrival pattern, for example the vehicle generation process in traffic simulation model [1], the analysis of the capacity and performance of an unsignalized intersection [2], the determination of the level of service on highways [3] and the optimization of traffic signal setting. This shows that a well understanding of the vehicle arrival pattern is an important issue in the traffic flow theory.

The vehicle arrival pattern is defined as a sequence of vehicles arrive on a highway passing through a specific spot during a specific time period of a day, measured at successive times, spaced at a uniform time interval. In short, vehicle arrival is the number of vehicles counted/observed continuously at a certain fixed point on the highway, during some specific time interval (say every minute) over a time period (say 24 hours). From the statistical analysis point of view, the vehicle arrival pattern can be regarded as a time series or a stochastic process. The identification of the vehicle arrival pattern can expose the time headway distribution of the vehicles, in which the time headway is referred to as the separation between two consecutive vehicles in time.

Poisson process has already been used to model the vehicle arrival pattern for decades. In addition to the road traffic, it has also been adopted to describe the arrival patterns of data packet. Nevertheless, started from a seminar paper by [4], the data network researchers have discovered that most data packet arrival patterns are self-similar process as opposed to the Poisson process [5]; a time series or a stochastic process is called the self-similar if its statistical properties remain invariant under time (or space) rescaling and it often implies the heavy tailed distributions. Leland et al. [4] demonstrated that Ethernet LAN traffic shows the self-similar behavior, while Paxson and Floyd [6] found that Poisson process fails to model the WAN arrivals. They found that the Poisson process model underestimates the burstiness of TCP traffic over a wide range of time scales seriously. Following this, the VBR traffic [7], Local Area Network (LAN) traffic [8] and World Wide Web (www) traffic [9] are found to show the self-similarity characteristics in the data packet arrival patterns. This shows that the conventional assumption of Poisson process is incorrect since it fails to model the data packet arrival patterns accurately.

In road traffic, few researches investigate the self-similarity characteristics of the vehicle arrival patterns. Nagatani [10] is the only work on the traffic self-similarity study. Using the simulated data in a laboratory, he found that a single vehicle passing through a sequence of traffic light with varying cycle time shows the self-similarity behavior. This study shows that there is a possibility of existence of self-similarity in road traffic. Meng and Khoo [11] investigated the arrival on highways and found that the highway traffic exhibit self-similarity pattern. Inspired by these new findings, we examine whether the arterial road vehicle arrival pattern can be described as a self-similar process too.

This objective of the study is to investigate the vehicle arrival pattern on arterial roads in Malaysia. Vehicle arrival pattern is collected from a 4-leg junction in Setapak area, Kuala Lumpur through video camera. Hypothesis testing is adopted to study the significant of the existence of self-similarity pattern. The self-similarity is evaluated by estimating its Hurst parameter using Selfis software. Results indicated that the vehicle arrival pattern at arterial road is self-similar distributed. This means that the conventional way of optimizing traffic signal timing might need to be revised since the arrival pattern is self-similar.

2. Self-similar Process

Self-similarity, in general, describes a phenomenon where a certain property of an object is preserved with respect to scaling in space or time.

Let $X = \{X_t : t = 1, 2, \dots,\}$ be a discrete time covariance stationary stochastic process or a stationary time series with mean μ and variance σ^2. In terms of a highway vehicle arrival pattern, X_t can be interpreted as the number of vehicles passing through a specific spot on a highway during time period t on a day. Given the stochastic process X, we can define the m-aggregated stochastic process $X^{(m)} = \{X_i^{(m)}, i = 1, 2, \dots,\}$,where m is a positive integer, by averaging the original stochastic process X over non-overlapping blocks of size m, namely,

$$X_i^{(m)} = \frac{1}{m} \sum_{t=(i-1)m+1}^{im} X_t,\ i = 1, 2, \cdots, \tag{1}$$

We say X is self-similar with Hurst parameter H $(0.5 < H < 1)$, if for all positive integer m, $X^{(m)}$ has the same distribution as X rescaled by factor m^{H-1}, i.e.,

$$X_t =_d m^{1-H} \sum_{t=(i-1)m+1}^{im} X_t, \text{ for all } m \geq 1 \tag{2}$$

where symbol $=_d$ stands for equality in distribution. Eqn. (2) means that any aggregated stochastic process is distributionally self-similar in the sense that the distribution of the aggregated stochastic process is the same (expect for the change in scale) as that of the original one. This means that X will perceive its pattern with respect to different aggregate scale. In the layman words, X is like a fern leave or a Koch curve in which the shape of the leave or curve remains the same if it is view in different zoom-in scale. For short, X is called the self-similar process throughout the rest of this paper.

If X is a self-similar process, according to eqns. (1)-(2), the mean and variance of each random variable in the m-aggregated stochastic process $X^{(m)}$ can be calculated by [see, page 54 of Beran (1994)]

$$E\left(X_i^{(m)}\right) = \mu,\ i = 1, 2, \cdots \tag{3}$$

$$\mathrm{var}\left(X_i^{(m)}\right) = \frac{\sigma^2}{m^{2-2H}}, \; i = 1, 2, \cdots \tag{4}$$

More interestingly, for all $m \geq 1$, the autocorrelation functions for stochastic processes X and $X^{(m)}$ are identical with the following explicit expression [see, pages 51-52 of Beran (1994)]:

$$r^{(m)}(k) = r(k) = \frac{1}{2}\left[(k+1)^{2H} - 2k^{2H} + (k-1)^{2H}\right], \forall k \geq 1 \tag{5}$$

where the autocorrelation functions $r^{(m)}(k)$ and $r(k)$ are mathematically defined by

$$r^{(m)}(k) = \frac{E\left[\left(X_{i+k}^{(m)} - \mu\right)\left(X_i^{(m)} - \mu\right)\right]}{\left(\sigma^2 / m^{2-2H}\right)}, \; k \geq 1 \tag{6}$$

$$r(k) = \frac{E\left[\left(X_{i+k} - \mu\right)\left(X_i - \mu\right)\right]}{\sigma^2}, \; k \geq 1 \tag{7}$$

Recall that the variance of sample mean $\overline{z}$ of a random variable z satisfies $\mathrm{var}(\overline{z}) = \sigma_z^2 / m$, where m is the sample size and σ_z^2 is the variance of random variable z, provided that samples are drawn independently. Therefore, eqn. (4) clearly indicates the statistically dependence among random variables $X_t, t = 1, 2, \cdots$, in the self-similar process X because the Hurst parameter $1/2 < H < 1$. However, the autocorrelation functions shown in eqn. (5) reflect the long-range dependence of the self-similar process, that is,

$$\lim_{k \to \infty} r(k) = 0 \tag{8}$$

$$\sum_{k=1}^{\infty} r(k) = \infty \tag{9}$$

According to eqns. (8) and (9), it can be seen that the autocorrelation function decays to zero as k intends to infinity so slowly that it is not summable. The autocorrelation function $r(k)$ in terms of magnitude is equivalent to function $H(2H-1)k^{2H-2}$ for the large argument k, that is,

$$\lim_{k \to \infty} \frac{r(k)}{H(2H-1)k^{2H-2}} = 1 \tag{10}$$

Let parameter $\beta = 2 - 2H$, we then have $0 < \beta < 1$ as Hurst parameter $1/2 < H < 1$. Eqn. (10) sometimes is rewritten as

$$r(k) \sim H(2H-1)k^{-\beta}, k \to \infty \tag{11}$$

where symbol ~ stands for the equivalence in the sense of eqn. (10).

Eqn. (10) implies that the autocorrelation function of a self-similar process follows a power-law. The power decay is slower than exponential decay, and since $0 < \beta < 1$, the sum of the autocorrelation values of a self-similar process approaches infinity. Note that the Poisson process has infinite variance, stationary and independent increments therefore it is asymptotically self-similar with Hurst parameter equals exactly to 0.5. Hence, it does not possess the long-range dependence according to eqn. (5). If Hurst parameter of a stochastic process is close to 0.5, the process resembles a Poisson behavior [8] otherwise, the closer the Hurst parameter to one, the greater the degree of self-similarity. Perfect self-similar process has the Hurst parameter value equivalent to 1. Karagiannis and Faloutsos [12] further pointed out that the Hurst parameter cannot be calculated in a definitive way. Nevertheless, it can be estimated from real observations.

3. Methodology

The methodology involved in this research is divided into three categories, i.e. data collection, data processing, and data analysis. Data collection is carried out by taking the video images at the intersection for 24 hours for 2 weeks. Then, the data is processed to obtain the vehicle arrival volume for the specified interval for analysis. Selfis software is used to estimate the Hurst parameter to quantify whether the vehicle arrival pattern exhibits self-similar pattern or not. Hypothesis testing is then adopted to ensure that the analysis was statistically significant. The details of the methodology adopted in this research is presented below.

3.1 Data Collection and Processing

Two isolated junctions are chosen for data collection, namely (1) intersection between Jalan Tumbuhan and Jalan Genting Klang, and (2) intersection between Jalan Taman Ibu Kota and Jalan Langkawi. Both intersections are located in Setapak, Kuala Lumpur as shown in Figure 1. These intersections are ideal for this study because the traffic flow of these intersections are not low (i.e traffic flow of 300 veh/hr/lane). In this paper, the lanes in which the vehicles are moving straight across the inter sections are considered. For the first intersection, there is a total of 10 lanes from 4 approaches, and for the second intersection, there is a total of 8 lanes. Therefore, in total, the data are being collected from 18 lanes for both intersections. The data collection was carried out from 13 August 2015 to 26 August 2015.

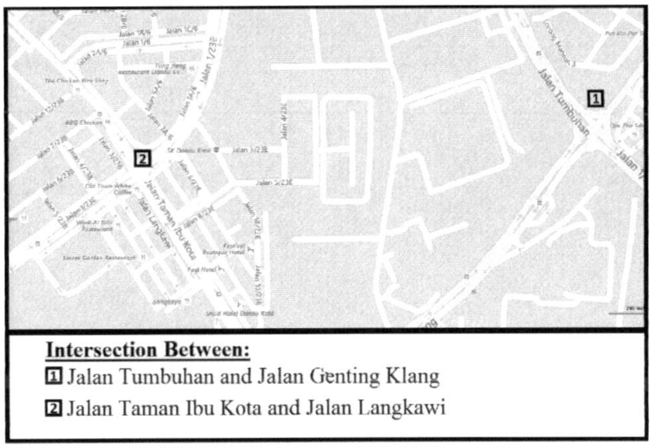

Figure 1. Location of Intersections

Video cameras are set up on 4 arms of the intersections and the traffic flow is recorded continuously for 24 hours for two-week period. This ensures that the data obtained is continuous. The video cameras are set up on the light poles around 3 meters from the ground. This ensures the video cameras to capture incoming vehicles from approximately 500 m distance from the light poles. The statistical characteristics for both intersections are tabulated in Table 1. The first intersection is labeled as S1 and the second intersection is labeled as S2. Since there are four approaches for each intersection, each approach is named as C1, C2, C3 and C4 respectively. From Table 1, it can be observed that the maximum vehicle flow rate for the first intersection is approximately 350 vehicle/hour, whereas the second intersection has a lower rate of 270 vehicle/hour.

For video recordings that are affected by factors such as road constructions, stationary vehicles by the roadsides and blurry video recordings, these video recordings are not used due to its inaccurateness in measuring the traffic flow. Due to this limitation, the timeframe for the second intersection was limited from 7:00:00 AM to 7:00:00 PM, while for the first intersection, 24-hour data are investigated. However, this study ensures that a total data set for 10 days was used. Therefore, there are 18 lanes that are being investigated and the total datasets used are 180.

For video collected from the sites, a reference line is added. As mentioned earlier, the reference line is important because it is the point whereby the timestamp of the vehicle is recorded. When the front wheels of the vehicle passed through the reference line, its timestamp is recorded. The timestamps for each vehicle is tabulated and analyzed to obtain its vehicle arrival patterns and time headways.

Table 1. Statistic Summary for Traffic Data Collected

Intersection	Lane	Investigation Period	Average Flow Rate (veh/hr/lane)		Total Number of Vehicles Per Investigation Period (veh)
			Min	Max	
S1C1	1	24 Hours	37	356	5467
	2	24 Hours	81	425	7321
	3	24 Hours	27	364	5356
S1C2	1	24 Hours	22	311	4117
	2	24 Hours	31	398	5531
S1C3	1	24 Hours	17	270	3804
	2	24 Hours	42	381	5805
	3	24 Hours	18	348	4868
S1C4	1	24 Hours	17	356	4328
	2	24 Hours	21	355	4481
S2C1	1	7AM - 7PM	157	272	2489
	2	7AM - 7PM	200	355	3381
S2C2	1	7AM - 7PM	94	186	1697
	2	7AM - 7PM	146	310	2675
S2C3	1	7AM - 7PM	105	206	1785
	2	7AM - 7PM	163	302	2749
S2C4	1	7AM - 7PM	113	224	1960
	2	7AM - 7PM	175	324	3000

3.2 Data Analysis

3.2.1 Poisson Distribution Test. The first part of the data analysis is focused on investigating whether the vehicle arrival pattern exhibits Poisson distribution pattern. The hypothesis testing for Poisson distribution is as followed:

H_0 : The vehicle arrival pattern follows Poisson distribution

H_1 : The vehicle arrival pattern does not follow Poisson distribution

11

4. Result and Discussion

Figure 2 shows the traffic volume plot vs time for different time scales, i.e. ranging from 1 minute, 2 minutes, 5 minutes, 10 minutes, 30 minutes to 60 minutes interval. It could be observed that the volume pattern is self-repeating for different time frame. This fulfils the definition of self-similarity which warrants the need to further investigate the potential existence of self-similarity in the vehicle arrival pattern.

The results of the hypothesis testing carried out to test the validity of self-similarity existence is shown in Table 2. It shows the average Hurst value for each lane of the intersections. It can be seen that for the first intersection, the average Hurst value estimated by Whittle estimator is more than 0.95, while the average Hurst value estimated by Whittle estimator is more than 0.75 for the second intersection. This shows that the first intersection exhibits a higher degreeness of self-similarity as compared to the second intersection. The results obtained are statistically significant at p=0.01. It is also worth mentioning that since the Hurst parameter is more than 0.5, the hypothesis testing of Poisson distribution of vehicle arrival pattern is rejected.

Table 2. Hurst value estimated

Intersection	Hurst value		
	Lane 1	Lane 2	Lane 3
Junction 1, Approach 1	0.9754	0.9728	0.9759
Junction 1, Approach 2	0.9592	0.96	-
Junction 1, Approach 3	0.9627	0.9622	0.9728
Junction 1, Approach 4	0.9851	0.9833	-
Junction 2, Approach 1	0.7293	0.7355	-
Junction 2, Approach 2	0.7982	0.7946	-
Junction 2, Approach 3	0.7982	0.7946	-
Junction 2, Approach 4	0.7533	0.7829	-

This study is important because vehicle arrival pattern is one of the assumptions in traffic signal design. The conventional assumption is that vehicle arrival follows Poisson distribution. However, this study shows that the vehicle arrival pattern on urban/arterial street exhibits self-similarity. As such, the traffic signal design using Webster Method might need a revision based on this finding.

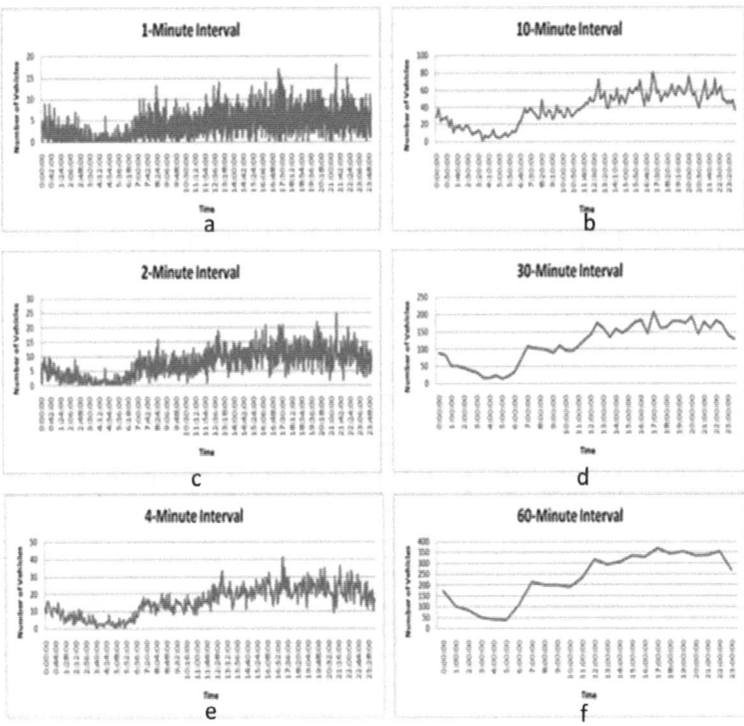

Figure 2. Traffic Volume Pattern At Different Time Frame

5. Conclusion

This study investigates the existence of self-similarity in vehicle arrival pattern. An empirical study was carried out in which vehicle arrival pattern at the isolated intersection was collected and analysed with Selfis software. The hypothesis testing on the Hurst value obtained shows that the vehicle arrival pattern exhibits self-similarity pattern, with statistical significant level of p=0.001. This finding is important as the existing traffic signal design is based on the assumption of Poisson distribution. A revisit of the traffic signal design based on self-similarity arrival pattern is needed.

6. References

[1] Lieberman E and Rath A K 2007 *Traffic Flow Theory: A State-of-Art Report* Available from www.tfhrc.gov/its/tft/tft.htm
[2] Sullivan D P and Troutbeck R J 1994 The use of Cowan's M3 headway distribution for modeling urban traffic flow *Traffic Engineering and Control* **35** 445-50
[3] HCM 2000 *Highway Capacity Manual* (Washington: Transportation Research Board)
[4] Leland W E, Taqqu M S, Willinger W and Wilson D V 1994 On the self-similar nature of ethernet traffic (extended version) *IEEE-ACM Transactions on Networking* **2** 1-15
[5] Park K and Willinger W 2000 *Self-similar Network Traffic and Performance Evaluation* (England: John Wiley &Sons)
[6] Paxson V and Floyd S 1995 Wide-area traffic: the failure of poisson modeling *IEEE/ACM Transactions on Networking* **3** 226-44
[7] Kyu B K and Elwalid A 1996 The importance of long-range dependence of VBR video traffic in ATM traffic engineering: myths and realities *Proceedings of the ACM Sigcomm'96* pp 3-14
[8] Willinger W, Taqqu M S, Sherman R and Wilson D V 1997 Self-similarity through high variability: statistical analysis of ethernet lan traffic at the source level *IEEE/ACM Transactions on Networking* **5** 71-86
[9] Crovella M and Bestavors A 1996 Self-similarity in world wide web traffic: evidence and possible causes *International Conference on Measurement and Modeling of Computer System (ACM SIGMETRICS)* Philadelphia
[10] Nagatani T 2005 Self-similar behaviour of a single vehicle through periodic traffic lights *Physical A* **347** 673-82
[11] Meng Q and Khoo H 2009 Self-similar characteristics of vehicle arrival pattern on highways J. Transp. Eng. **135**(11) 864–72
[12] Karagiannis T and M Faloutsos 2002 *SELFIS: A Tool for Self-similarity and Long-range Dependence Analysis* The First Workshop on Fractals and Self-Similarity in Data Mining: Issues and Approaches (in KDD), Edmonton, Canada

Acknowledgement
This research is supported by Ministry of Education, Fundamental Research Grant Scheme, 2016.

IOP Conf. Series: Materials Science and Engineering **512** (2019) 012017 doi:10.1088/1757-899X/512/1/012017

Road surface condition monitoring system utilizing motorcycle (ROCOM) – System development, validation and field test

Muhammad MarizwanAbdul Manan, Muhammad Ruhaizat Abd Ghani

Road Safety Engineering and Environment Research Centre, Malaysian Institute of Road Safety Research (MIROS), Kajang, Selangor, Malaysia

E-mail: marizwan@miros.gov.my

Abstract. Motorcycles are the riskiest mode of travel in Malaysia, however motorcycles are also very sensitive to the road surface condition. Thus, by taking advantage of this, a system of applications that analyses motorcycle motion and mapped out risky road sections was developed, i.e. ROCOM. The system consists of three major components, i.e. acceleration data logger smart phone app, risk mapping web-based app and a visual tracking app utilizing the phone. ROCOM is able to detect adverse acceleration (> 2g) on the road surface similar to the High Accuracy GPS Data Logging for Vehicle Testing (VBOX). Risk mapping validation along a section of the motorcycle lane along Federal Route 2 shows that ROCOM have the similar risk-mapping pattern, but superior tracking capability that the VBOX. The pilot and field test results showed that ROCOM works best when mounting the smartphone on the motorcycle handle bar or basket, and with successful detection rate of 62% of various road anomalies. In the future, once the system is fully developed, ROCOM intends to give road authorities cheaper and faster alternative to inspect road surface condition and also through crowdsourcing, motorcyclists can be warned of risky road condition ahead of their path.

1. Introduction

According to the World Health Organization (WHO), almost to a quarter (24.1%) of the world's road traffic deaths occur among riders of powered-two-wheelers (PTW) [1], or commonly identified in Asian region as motorcycle [2]. Of these motorcycle fatalities, the South-East Asia region (i.e. mostly low- to middle-income countries) has the highest rate of 49.9%, compared to only up to 10.9% motorcyclist fatalities in high-income countries in the European region. Malaysia is a good example to study motorcycle safety, as it typifies the countries with safety problems for motorcyclists and in Malaysia, the risk of a motorcycle rider of being fatally injured about 3 times higher than the risk of a car driver, 6 times higher than pedestrian and nearly 50 times higher than of a bus passenger [3].

One of the contributing factors to PTW crashes is the condition of the road surface. Road surface anomalies, such as potholes, speed bumps, railroad crossings, joints, can cause problems for passing vehicles and can affect road users safety [4]. Road defects such as rutting, water ponding, potholes, corrugation, etc. affect the safety of motorcycles [5-8]. In Malaysia, motorcycle single-vehicle fatal crashes are almost four times more likely than those motorcycle involve in multi vehicle fatal crashes to occur on road with rutting and corrugation [2].

To ensure that roads are free from defects, road maintenance activities have to be done systematically and cost-effectively by the road authorities. Municipalities around the world spend millions of dollars to maintain and repair their roadways [9], thus road quality assessment plays a key role in optimizing infrastructure management and road maintenance operations [4]. At the same time, informing drivers on road real conditions, in terms of the presence of bumps, potholes, or other anomalies, has a great importance in order to make the transportation system more safe, efficient and comfortable. In addition, informing drivers of hazardous road conditions [4] especially at night or when lighting is poor would be a useful for motorcyclists.

In the last few years, mobile sensors and computing systems have been developed to collect data on road quality with the final aim of giving real-time information to road users [4]. The monitoring of road conditions can be done by means of several sensors embedded in the mobile devices and nowadays, most of these smartphones/tablets use a three-axis accelerometer to collect acceleration data due to their motion on road anomalies and a GPS receiver to obtain location information of the road segment in question [4,10]. In this way, each event can be located allowing to create a database of road anomalies by means of a central server to which data can be collected by road users, e.g. such as motorcyclists [11].

Motorcycles are the riskiest mode of travel but on the flip side, it is very sensitive to the road surface condition. Thus, by taking advantage of this characteristic, we aim to develop a system of software applications that analyses motorcycle motion and mapped out risky road sections for motorcyclists.

2. Literature review

Applications for road surface defect detection are not something new. These intelligent transportation system (ITS) technologies have been well documented by Vittorio et al. [10], Vittorio et al. [4], Eriksson et al. [9] and Alessandroni et al. [12]. This ITS tech are beneficial to drivers navigation system when informing of hazardous road conditions especially at night or when lighting is poor [12]. However, these currents developed ITS devices for road defect detection; services and applications are usually designed for four-wheeled vehicles [13] and rarely for motorcycles.

Earlier studies, Vittorio et al. [10], Vittorio et al. [4] and Eriksson et al. [9], used accelerometer sensors on the phone purely to detect extreme values (x, y and z axis g values) during vehicle movement over road anomalies and they compared those to a normal driving conditions along a well-maintained road. The outcome of this technology would help road authorities to map the information of road defects efficiently as proven successfully by Alessandroni et al. [12]. This tech however is design for cars, while motorcycles are still at risk without this similar ITS tech.

As explained by Vittorio et al. [4], the accelerometer is a sensor that can detect linear accelerations along 1, 2 or 3 axes, by measuring the inertial forces, which can be represented with a 3- dimensional Cartesian frame with fixed x'y'z', embedded in the mobile device. The accelerometer provides acceleration values in three directions: longitudinal (a_x), lateral (a_y) and vertical (a_z). When the vehicle is in a stationary condition, the only acceleration registered is along the z axis, and it is equal to gravity:

$$a_x = 0\ m/s^2\ ;\ a_y = 0\ m/s^2\ ;\ a_z = 9.81\ m/s^2 = 1g \qquad (1)$$

Previous studies such as in Vittorio, et al. [4], Kitani [13] and González, et al. [14], relied on the orientation of the accelerometer sensors so that the x-axis identified the longitudinal direction, the y-axis the transversal one and the z axis the perpendicular direction to the xy plane.

3. Method

This research project requires the team to develop a system of application, which comprise of a mobile, a web-based application and a stand-alone software. After the development of ROCOM, data validation and field test was carried out in order to test and evaluate the system. Once the system has been developed, the system undergoes data validation and field test.

3.1 System development

Conceptually, a mobile app (ROCOM Data Logger) in the smartphone records the acceleration of the motorcycle's motion and also video of the ride when placed correctly on the motorcycle. The acceleration data is transferred via Internet into a web-based application (ROCOM Risk Mapping) for risk mapping, while the video data is transferred into a stand-alone software (ROCOM Visual Tracking) for visual tracking analysis (see Figure 1). The development of video tracking capability requires us to integrate the smart phone's camera function with the GPS and built-in map software. This capability was suited as a stand along software components of the system.

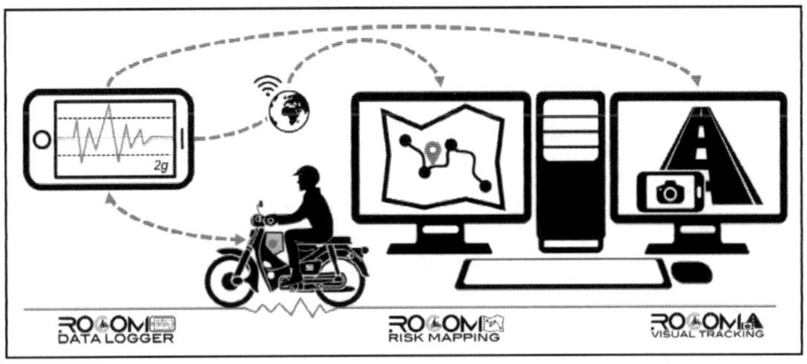

Figure 1. Schematic diagram of ROCOM

3.2 Data validation

The data validation was carried out utilizing the High Accuracy GPS Data Logging for Vehicle Testing (VBOX) as the baseline equipment. The location of the field test was along a section of the exclusive motorcycle lane along the Federal Highway (FR2). The data validation is based on two aspect, i.e. risk mapping validation and acceleration data validation.

3.3 Pilot and field test

Before the field-test being conducted, a pilot tests were conducted on 15th November 2016 (9.30 am), in the city of Kajang in the Kajang district. The objective of the pilot test was to test the capability of ROCOM to detect the road defects depending on the way of mounting the device on the motorcycle. The way of mounting of the phone on the motorcycle were as follow:

1. On the handle bars
2. In the motorcycle basket
3. In the rider's backpack

This test requires us to control the road environment (based on the selected sites) and also the rider. Thus, the pilot test employed two selected riders with good riding fitness and experience and also two well-maintained motorcycles. During the tests, they were required pass along the test route, which consisted of various road defects objects demanding abrupt braking and weaving.

The field tests were conducted on the 25th November 2016 (9.00 am) at Jalan Seksyen 2/15, Kajang Utama, Kajang, Selangor (see Figure 2). The objective of the tests was to test the capability of ROCOM to obtain the acceleration values for each object of interest. Fifteen qualified riders passed through a 400 m test route containing eight obstacles with their motorcycles equipped with a mobile smartphone. To obtain variations in the data collected the type of riders and motorcycles were varied. The riders were not familiar with the test route. This was in order to maintain the Naturalistic riding behaviour – the riders to perform the test rides with their usual riding behaviour without anticipating the obstacles ahead. The tests required us to control the road environment the best possible by fixing the road environment and road obstacles. The eight obstacles along the test route were as follows: 1) weaving area, 2) road hump number 1, 3) uneven surface number 1, 4) abrupt braking area, 5) road hump number 2, 6) uneven surface number 2, 7) road hump number 3, and 8) pothole (see Figure 2). The riders were told to pass through obstacle 1, 2 and 3 without stopping until obstacle 4, where they had to perform an abrupt breaking and then proceed to the rest of the obstacles until reaching the end of the test route.

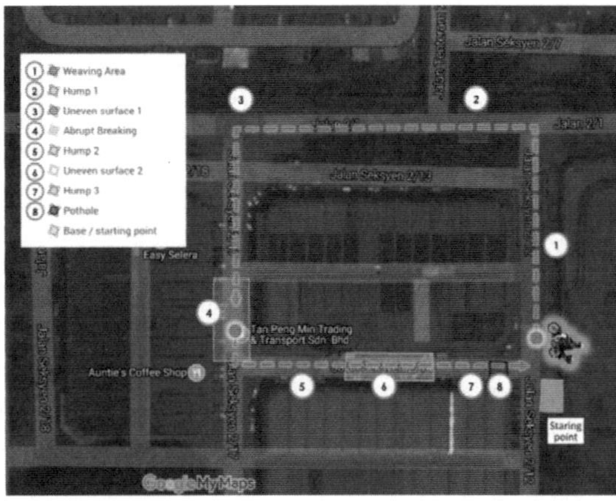

Figure 2. Field test route of ROCOM

Figure3. Photos at field test – obstacle 1 and 4

4. Result

The subsection below shows the results of the developed applications, field test and data validation results.

4.1 ROCOM Data Logger

The mobile application software development was done using Apple's IOS system software. The ROCOM Data logger app developed and designed in such way that that the rider would not need to have the smart phone to be in exact orientation of x, y and z, i.e. in order to detect high or extreme acceleration or g-force with respect to the road defects or anomalies. Instead, in other words, the motorcyclists would just mount the phone on the motorcycle (e.g. on the handle bars or place securely in the basket) in any position.

Figure 4. Screen capture of ROCOM Data Logger app

In principle, the app would detect and record a vertical acceleration when passing a bump, or an undulating road section, or a horizontal acceleration when abruptly weaving past a pothole or object. For the high or extreme acceleration, we have set the ROCOM minimum threshold of 2g, which is equivalent to 19.62 m/s2. This threshold is based on previous findings such as in *eCall* and *Wreck Watch* as seen in [15, 16].

Our system was designed in such a way that the mobile app would automatically collect the data once it is activated, and at the end of the ride, the data will be saved into a text file, ready to be transferred via email, Bluetooth, cloud, etc.

4.2 ROCOM Risk Mapping

ROCOM Risk Mapping application is a web-based system that was developed as a tool to analyze the data (i.e. acceleration forces) from ROCOM Data Logger and display the location geographically of high threshold acceleration or road anomalies on a dedicated map.

The web base it consists of three interfaces: Overview, Map and Data display (see Figure 5). The Overview interface displays some details about the rider's travelled distance, his or her average speed, total data collected and number of accelerations exceeded the susceptible limit of 2.0g. The Map interface reveals the risk mapping or the route that the rider has travelled (indicated by the green line) with the location where the occurrence of acceleration exceeded the susceptible limit of 2.0g. Lastly, the Data display section reveals a more detail information in terms of the acceleration figures and graph, and each GPS coordinate with respect to the sequential timestamp on when and where the data was logged by ROCOM.

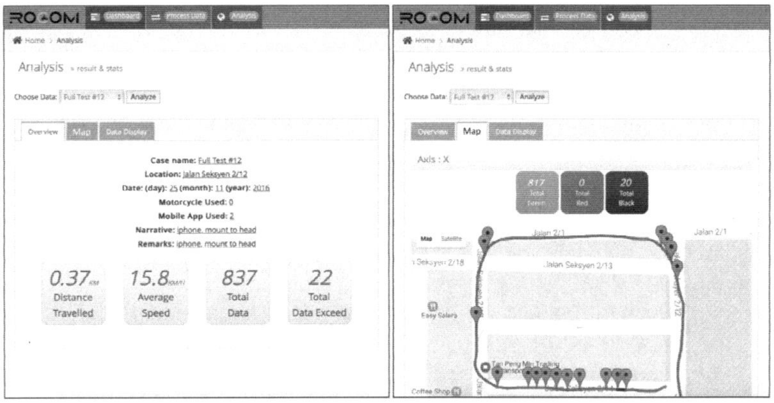

Figure 5. Screen capture of ROCOM Risk Mapping app

4.3 ROCOM Visual Tracking

Video tracking is one of the value-added functions the ROCOM system user can observe their route and verify the risky location and to view the road surface condition with respect to the high acceleration obtain from ROCOM Data Logger and ROCOM Risk Mapping app.The video, maps, and line chart panel run according to video timestamp and raw data point (see Figure 6).

Figure 6. Screen capture of ROCOM Visual Tracking

4.4 Data validation result

Data validation was carried out utilizing the VBOX as the baseline. The location of the field test was along a section of the exclusive motorcycle lane along the Federal Highway (FR2), from the Jalan Templer – Persiaran Narayanan interchange to the LDP interchange. The total distance travel was 6.11km with the average speed of travel along the exclusive motorcycle lane of 22.7km/h. During the validation, ROCOM and the VBOX was synched and run together for three times along the same road section, same motorcycle and same mounting procedure.

4.4.1 Acceleration data validation

Validating each and every point from ROCOM and VBOX is a strenuous task. Moreover, not all of the points for each trial run are the same due to the fact that different rider goes on a different path despite the constant type of motorcycle. Thus, we have chosen one location that has similar coordinate for both devices in order to validate the acceleration readings.

Table 1 shows the acceleration data comparison between ROCOM and VBOX. The single data on the same location has shown that all of the x-axis acceleration value for both ROCOM and VBOX exceeded the 19.60m/s2 threshold for each trial. The acceleration differences are between 3.0m/s2 to 9.7m/s2 compared to the VBOX data. This shows that ROCOM is able to detect adverse acceleration or vibration on the road surface similar to VBOX.

Table 1. Acceleration data validation for ROCOM and VBOX

Trial	1		2		3	
Latitude Longitude	3.0851 101.6236		3.0851 101.6233		3.0851 101.6237	
Device	#1 VBOX	#1 iPhone	#2 VBOX	#2 iPhone	#3 VBOX	#3 iPhone
Acc. on X (m/s^2)	*20.1 (2.05)*	*29.8 (3.04)*	*27.6 (2.81)*	*22.2 (2.26)*	*21.7 (2.21)*	*24.7 (2.52)*
Acc. on Y (m/s^2)	11.9 (1.21)	-4.4 (-0.45)	*-20.7 (-2.11)*	-10.1 (-1.03)	0.8 (0.08)	-7.9 (-0.81)
Acc. on Z (m/s^2)	11.9 (1.21)	13.3 (1.36)	*-20.7 (-2.11)*	-0.1 (-0.01)	0.8 (0.08)	5.7 (0.58)

( ): The value in the brackets are the value of the equivalent g value
Acc. : Acceleration
The red italic number indicate that the value of the acceleration (m/s2) exceed the 19.62 m/s2 threshold

4.4.2 Risk mapping validation

Figure 7shows the risk mapping comparison between ROCOM and VBOX for each trial. From the figure, we can see that both ROCOM and VBOX have the similar risk mapping pattern, i.e. most of the red markers are located similarly between trials. However, the VBOX route mapping (the green line) is inferior to the ROCOM, by which the VBOX route does not coincide with the map compared to ROCOM (see Figure 7). This may be due to the low accuracy of the standalone VBOX GPS system where else the iPhone has several complementary technologies on board that work in conjunction with the GPS chip to master the phone's location, i.e. accelerometer and gyroscope, Wi-Fi Tracking, compass, Barometer and M10 Motion Coprocessor.

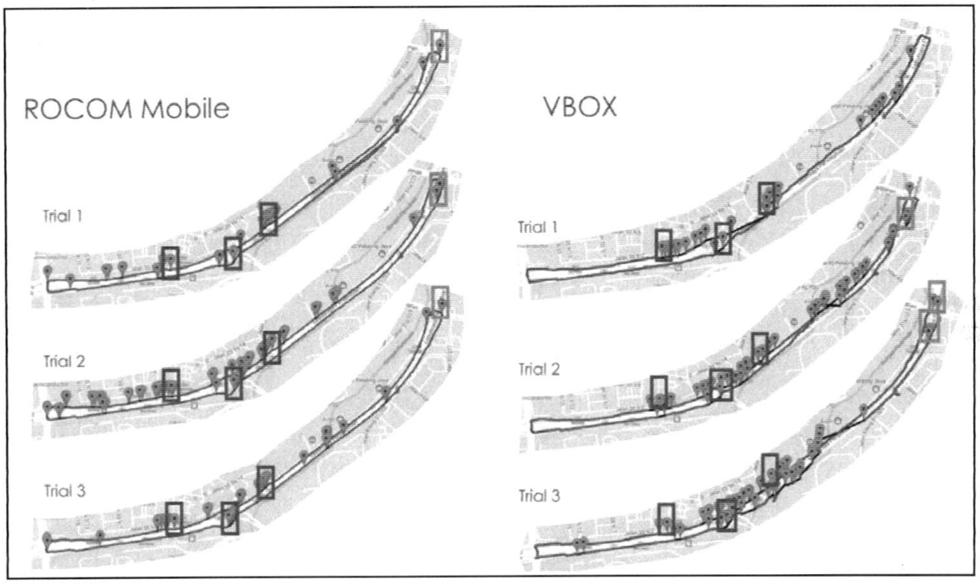

Figure 7. ROCOM Mobile and VBOX risk mapping comparison for each trial

4.5 Pilot and field test results

The pilot test results showed that mounting the smartphone on the motorcycle handle bar and fasting it on the motorcycle basket with the ROCOM activated had the highest success rate of detecting the obstacles compared to placing the smartphone in the backpack.

To evaluate the detection capabilities of ROCOM, data of each individual trial was recorded with a frequency of 0.01 Hz and for each obstacle its acceleration readings were observed and match (see Figure 8). Those exceeding acceleration readings exceeding 2g were counted and compared with other rider.

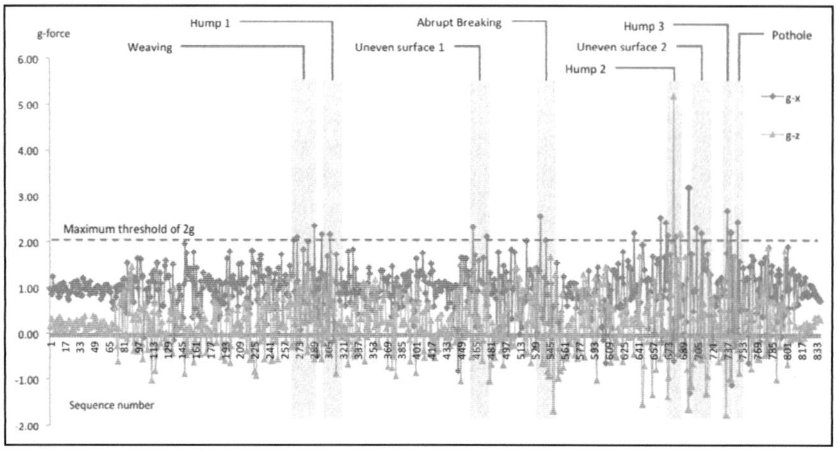

Figure 8. The acceleration graph matched with the location of the obstacles

Results from the field test indicated that ROCOM is able to detect various obstacles with the average of successful detection rate of 61.9% while the highest successful detection rate (80%) occurred when riders passed through uneven surface and road humps.

5. Discussion

The aim of the work described in this paper was to develop mobile- and web-based application software that analyses motorcycle motion data gathered from a mobile app and mapped out the hazardous road sections for motorcyclists. We have developed a system that consists of three major components, i.e. ROCOM Data Logger app, which utilizes a smart phone to collect acceleration data, ROCOM Risk Mapping app, which is a web-based application, and ROCOM Visual Tracking, which is a stand along software. Data validation results shows that ROCOM is able to detect adverse acceleration (> 2g) or vibration on the road surface similar to the High Accuracy GPS Data Logging for Vehicle Testing (VBOX). Risk mapping validation along a section of the motorcycle lane along Federal Route 2 shows that not only ROCOM have the similar risk-mapping pattern, but its route tracking capability on the map is far superior that the VBOX. The pilot and field test results showed that ROCOM works best when mounting the smartphone on the motorcycle handle bar or basket, and it can detect various road anomalies with successful detection rate of 62%, with high detection rate when passing through uneven road surfaces.

The results from our test experiment show that ROCOM has similar capabilities as software developed by [17], [10] and [9]. For example, [17] and [4] has developed a dedicated sensing unit for the system and have evaluated the availability of a smartphone as the sensing device with the system, which resulted that the sensors on the mobile phone were enough to sense the motions as similar to the dedicated sensing unit. On the other hand, many literatures have suggested that filtering equation be embedded in the software algorithm such as in [10] and [11]. However, ROCOM adopts a simpler approach by detecting any road defects or obstacles only when the threshold of 2g is reached.

Further tests and calibrations of ROCOM are needed in terms of finding the ideal condition (e.g. appropriate speed) for accurate detection. Although we have found that mounting on the handlebars is suitable for a better detection rate, the mounting equipment must be placed so that it is close to the handed in order to mitigate the effects of vibrations. Thus, we recommend that ROCOM should have custom made mounting equipment.

6. Conclusion

The results from our data validation and field test show that ROCOM has similar capabilities as software developed by [17], [10] and [9], that could record road surface condition and map out the risky location.

The study also finds that ROCOM Visual Tracking app was able to proof the accuracy of road surface detection from raw data taken by ROCOM Data Logger. While the ROCOM Risk Mapping application shows that not only ROCOM have the similar risk-mapping pattern with the High Accuracy GPS Data Logging for Vehicle Testing (VBOX), but its route tracking capability on the map is far superior that the VBOX.

Once ROCOM is fully develop, it intends to give road authorities a cheaper and a quicker alternative to inspect road surface in the future. Road authorities can form an inspection team consist of motorcyclists mounted with ROCOM and inspect their road on a regular or schedule basis without breaking their yearly budget. Risk mapping in ROCOM can give an accurate risky location for motorcycle and allow road authorities prioritized their road maintenance. Most important of all, ROCOM can be upscaled for mass usage and crowdsourcing among motorcyclists so that they can be warned of potholes, cracks or corrugation ahead of their destination.

7. References

[1] WHO 2013 Global Status Report on Road Safety 2013: Supporting a decade of action (Geneva: World Health Organization)

[2] Abdul Manan M M, Várhelyi A, Çelik A K and Hashim H H 2017 Road characteristics and environment factors associated with motorcycle fatal crashes in Malaysia *IATSS Research* 1-14

[3] Abdul Manan M M and Várhelyi A 2012 Motorcycle fatalities in Malaysia *IATSS Research* **36** (1) 30-9

[4] Vittorio A, Caruso M V, Danieli G, Festa D C, Giofrè V P, Iuele T and Vaiana R 2012 A mobile application for road surface quality control: Uniqualroad *Procedia - Social and Behavioral Sciences* **54** 1135-44.

[5] Elliott M A, Baughan C J, Broughton J, Chinn B, Grayson G B, Knowles J, Smith L R and Simpson H 2003 Motorcycle safety: A scoping study *TRL report* **581** (UK: TRL)

[6] Haque M M, Chin H C and Huang H 2009. Modeling fault among motorcyclists involved in crashes *Accident Analysis & Prevention* **41** (2) 327-35

[7] Haworth N 2012 Powered two wheelers in a changing world, challenges and opportunities *Accident Analysis & Prevention* **44** (1) 12-18

[8] Forkenbrock G, Hoover R L, Gerdus E, Buskirk T R V and Heitz M 2014 *Blind spot monitoring in light vehicles - system performance* Report No. DOT HS **812 045** (Washington: National Highway Traffic Safety Administration)

[9] Eriksson J, Girod L, Hull B, Newton R, Madden S and Balakrishnan H 2008 The pothole patrol: Using a mobile sensor network for road surface monitoring *Proceedings of the MobiSys '08* Breckenridge, Colorado, USA

[10] Vittorio A, Rosolino, V, Teresa I, Vittoria C M, Vincenzo P G and Francesco DM 2014 Automated sensing system for monitoring of road surface quality by mobile devices *Procedia - Social and Behavioral Sciences* **111** 242-51

[11] Lee U and Gerla M 2010 A survey of urban vehicular sensing platforms *Computer Networks* **54** (4) 527-44

[12] Alessandroni G, Klopfenstein L C, Delpriori S, Dromedari M, Luchetti G, Paolini B D, Seraghiti A, Lattanzi E, Freschi V, Carini A and Bogliolo A 2014 Smartroadsense: Collaborative road surface condition monitoring *Proceedings of the UBICOMM 2014: The Eighth International Conference on Mobile Ubiquitous Computing, Systems, Services and Technologies*

[13] Kitani T 2013 A prototype of a vehicle-motion sensing system with a smartphone and dedicated sensing devices *Proceedings of the Bicycle and Motorcycle Dynamics 2013: Symposium on the Dynamics and Control of Single Track Vehicles* Narashino, Japan

[14] González A, O'brien E J, Li Y Y and Cashell K 2008 The use of vehicle acceleration measurements to estimate road roughness *Vehicle System Dynamics* **46** (6) 483-99

[15] Bliss T and Breen J 2009 Country guidelines for the conduct of road safety management capacity reviews and the specification of lead agency reforms, investment strategies and safe system projects *The World Bank Global Road Safety Facility* Washington DC
[16] Linins A and Berjoza D 2010 Experimental investigation in acceleration of vehicles *Proceedings of the Engineering for Rural Development*, Jelgava
[17] Kitani T, Miyazawa A, Kamimura T, Shiomi A and Watanabe T 2012 A motion sensing system to grasp a motorcycle's behavior with sensors as mounted on a smartphone *Proceedings of the International Workshop on Informatics* (IWIN2012) pp 115-20

Acknowledgements
The authors would like to express their appreciation to the Director of Road Safety Engineering and Environment Research Centre, Dr Siti Zaharah Ishak for giving permission to conduct this project and extending their support to produce this report. The authors would also like to express special thanks to the team members and research assistants Siti Nurhidayah Binti Musa, Azzemi Bin Aziz, Mohd Saifullah Bin Mohd Sohi, Zulfadhli Bin Sharifudin, Norhazlina Binti Razali, and Muhammad Ridzuan Bin Basah for their help and contribution in completing the project.

IOP Conf. Series: Materials Science and Engineering **512** (2019) 012018 doi:10.1088/1757-899X/512/1/012018

The readiness of JKR Sarawak to adopt ISO 39001 road traffic safety (RTS) management systems

Imelda Jelani

Civil Engineer, Road Branch, Public Work Department Sarawak, Malaysia

E-mail: imeldaj@sarawak.gov.my

Abstract. The recently launched ISO 39001 is a management system for the safety of road traffic. It specifies requirements to enable an organization that interacts with the road traffic system, to reduce death and serious injuries related to road traffic crashes which it can influence. This paper looked at the readiness of JKR Sarawak to adopt ISO 39001 Road Traffic Safety (RTS) Management Systems. This paper highlights the current (RTS) practices and identified those which fulfilled the ISO 39001 requirements. This paper also investigated the challenges in achieving certification based on the six steps of a basic plan-do-check-act model for continuous improvement. The finding shows that the organization do fulfill parts of the requirement, however, it is not ready to adopt the best practice in road traffic safety management due to challenges in certification. Measurable RTS objectives are proposed to overcome the identified challenges.

1. Introduction

Malaysia's economic growth accelerated to 5.8% in 2017, the country's highest annual growth rate since 2014 [1]. Rapid economic growth is usually connected to a rapid expansion of road transportation [2]. Sarawak in particular, is currently expanding its existing road network to connect the urban and rural areas.While the people enjoy the connectivity, it also leads to road crashes, injuries, and fatalities. According to the statistics published by Kementerian Pengangkutan Malaysia, in 2007, Malaysia recorded 6282 deaths compared with 3913 deaths in 2017. For Sarawak, there are 316 deaths in 2007, the highest deaths recorded 443 in 2012, and 236 in 2017 [3]. Although the figures show a decrease in the number of deaths, it is relatively high and calls for an effective measure.

The United Nations (2011) concur that having an effective and efficient road safety management system is crucial to support the implementation of measures that improve legislation, roads, vehicles, road-user behavior, and post-crash response [4]. The US Transport Research Board (2011) agrees that management is the direction of resources to attain defined objectives [5]. This means that having an effective management system would ensure that all department or personnel will collaborate to achieve a common set of goals.

2. Background

Malaysia adopted the Road Traffic Safety (RTS) management system ISO 39001[6] and later named it as MS ISO 39001:2013, to address the nation's road-safety related issues. It is fully aligned with the

Safe System approach to road safety, which is embedded in the UN Global Plan for Road Safety, and the UN sustainable Development Goals . ISO 39001 is the template for best practice road traffic safety management in an organizational context [7] and has been adopted by countries like Sweden and Japan. The implementation in both countries are reported to be effective and successful in increasing the awareness and interest on the part of top management in addressing road safety.

Jabatan Kerja Raya (JKR) Sarawak is currently adopting MS ISO 9001 which is a quality management system. Unlike ISO 39001[8], it is a more general quality system that can be applied to every level of an organization. ISO 39001 is fully aligned with ISO 9001 Quality Management System, the difference is the former provides direction on key road traffic safety issues and a strong focus on achieving better results for the interim and long-term. With success stories from countries like Sweden and Japan, JKR Sarawak should adopt ISO 39001 as a Road Traffic Safety Management System to address traffic safety issues during the planning, construction and operation of the road.

This paper will determine the readiness of JKR Sarawak to adopt the ISO 39001 to address the state's road-safety related issues. MIROS [9] claims that organization that have implemented the ISO 9001 can easily integrate ISO 39001 into their system. In the context of this study, the focus is on the Intervention stage: Road Network, planning, design, construction and maintenance of road.

3. Related Literature

In 2008, the Organization for Economic Cooperation and Development (OECD), and the International Traffic Forum published "Towards Zero: Ambitious Road Safety Targets and the Safe System Approach" [10]. This report noted the importance of RTS management systems and referenced a codified management system that was developed, and subsequently published in 2009, by the World Bank Global Road Safety Facility (WBGRSF). The WBGRSF publication drew on a comprehensive review of successful jurisdictional RTS management practice to develop an RTS management system framework as in figure 1. It is neutral to organizational structures and cultures, and presents three inter-related elements, namely institutional management functions, interventions and results.

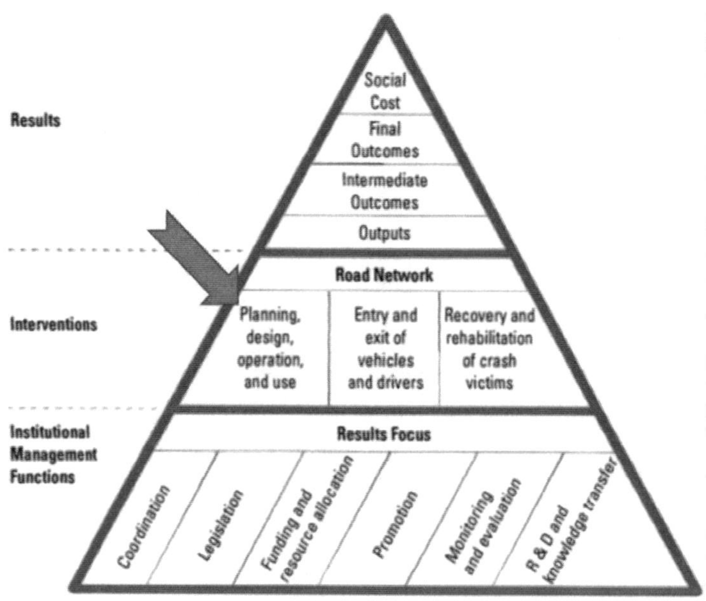

Figure 1. World Bank Global Road Safety Facility Road Safety Management System Framework

This paper aims to answer the following research questions:

1. What are the current JKR Sarawak Road Traffic Safety Management practices?
2. What are the challenges faced by JKR Sarawak in adopting ISO 39001?
3. How ready is JKR Sarawak in adopting ISO 39001 as its Road Traffic Safety Management practices?

4. Methodology

This is a qualitative case study and data is gathered through document analysis. The rationale for document analysis lies in its role in methodological and data triangulation, the immense value of documents in case study research, and its usefulness as a stand-alone method for specialized terms of qualitative research [11]. Quality manuals for ISO 39001, and manuals prepared by JKR Sarawak based on MS ISO 9001 are potential source for empirical data for this study.

5. Findings

5.1 Current RTS Practices

This section will identify the current RTS practices in compliance with MS ISO 9001. The next section will analyse the differences and similarities with ISO 39001 practices.

JKR Sarawak organization is structured based on the conventional lines with a headquarters set-up divided into specialist technical and non-technical branches, three (3) regional offices and twelve (12) divisional organizations.

JKR Sarawak Headquarters is responsible for the overall administration and policy making. Road Branch is an engineering authority for Sarawak on matter relating to road. It undertakes the task of ensuring sustainable development, planning, design and setting standards for better and safer road network in Sarawak. Road Safety and Traffic Engineering Section is specifically designated to oversee road safety matters.

JKR Regional Offices undertakes the task for all projects implementation. Road and Civil Engineer Asset Branch maintain and rehabilitate the roads to the required level of service. This office represents JKR Sarawak to participate in the Majlis Keselamatan Jalan Raya (MKJR). MKJR is a council that brought together relevant parties concerning road safety, namely, the Police, Road Transport Department, Health Department, etc to discuss road safety issues and come up with short term and long term solutions.

The current RTS practices are categorized into four stages: planning stage; design stage; construction stage and maintenance stage. The practices will be acknowledged accordingly.

5.1.1 Planning Stage. In the planning stage, project proposal is received from the client (state or federal agency) and the JKR divisional offices. These proposals can be a new project or an upgrading project. New project is concerned with mobility and safe movement of goods and people. Upgrading project usually addresses the increasing number of traffics and improving the safety features of existing roads.

Planning officer will assess and identify safety problems that might arise with the concept of the proposed project, such as, road network implications, alignment and other traffic engineering features and make necessary improvement to the proposal. The proposal would be submitted to relevant ministry for approval.

5.1.2 Design Stage. The design team is to discuss with design committee on the scope of work, design requirements, road standards and geometric criteria to be adopted and construction cost estimate. The design team can be in-house design team or appointed design consultant.

All detailed design works are to be carried out in accordance with the standard JKR Terms of Reference for Detailed Ground Survey, Engineering Design and Documentation for Sarawak Roads (CETRDES, Rev.11/96), relevant Arahan Teknik Jalan (ATJ), other technical standards and best engineering practices (collectively referred as guidelines). Drawings and report are submitted in stages: conceptual, preliminary and final.

The detail of traffic engineering features such as the cross section elements, carriageway layout, intersection layout, horizontal and vertical alignment and traffic control options are established based on the ATJ published by JKR Malaysia.

A Road Safety Audit (RSA) is also to be carried out on the design. The RSA Auditor can be an independent certified RSA Auditor, Road Designer from within the department or by a member of the project's design team. Arising from the findings and recommendations of the Road Safety Audit Report, amendments shall be made to the design where necessary, if any.

5.1.3 Construction Stage. The JKR Regional Office will be responsible for all projects implementation and field operations that include site supervision and coordination. During the construction, traffic management is the biggest concern so as not to cause much inconveniences to road users, thus, a contractor is required to submit the traffic management plan to be checked by certified road safety auditor. The plan is prepared by certified traffic management officer. JKR Regional Office will endorsed the plan prior to the implementation on-site.

5.1.4 Road Maintenance. JKR divisional office is responsible to monitor and ensure the implementation of maintenance work being carried out accordingly as required in the agreed performance standards.

Inspection is carried out on the existing roads to identify safety problems which develop due to normal 'wear and tear' from traffic operation. Routine and periodic maintenance will be carried out by the road maintenance concessionaire. This is to improve safety, and, conversely, to remove all known safety defects.

5.2 Practices that Conform with ISO 39001

Document analysis is applied to identify the current RTS practices that conform with ISO 39001. The analysis is based on the six steps plan-do-check-act (PDCA) in figure 2.

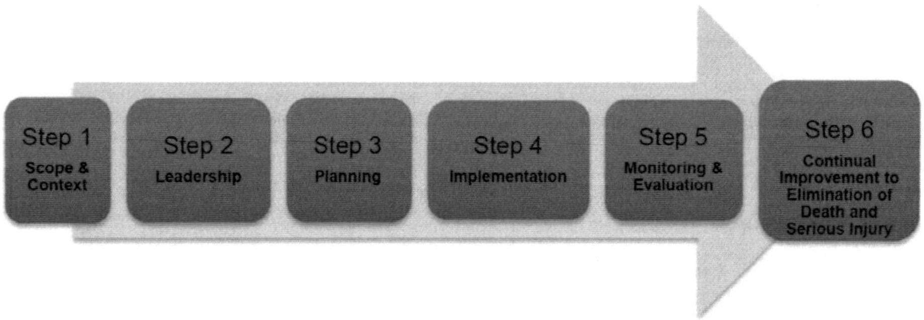

Figure 2. Six Steps Plan-Do-Check-Act. (Adapted from [6])

Table 1. Comparison of ISO 39001 RTS Management System and Current RTS Practices

Step	ISO 39001 RTS Management Systems	Current RTS Practices
Step 1: Scope and Context	Identify the impact the organization can have on road traffic safety (RTS), map that impact across interested parties, and determine the organizational scope of an RTS management system (refer clause 4 of ISO 39001).	JKR Sarawak's Road Branch is the engineering authority ensuring sustainable development, planning, design and setting standards for better and safer road networks in Sarawak.
Step 2: Leadership	Establish leadership commitment by adopting a long-term vision to eliminate death and serious injury and providing resources to establish, implement, maintain and continually improve the RTS management system towards these ends. Establish, document and communicate RTS policy, and assign organizational responsibilities (refer clause 5 of ISO 39001).	Road Safety and Traffic Engineering Section is specifically designated to oversee road safety matters. The RSA Auditor can be an independent certified RSA Auditor, Road Designer from within the department or by a member of the project's design team.
Step 3: Planning	Determine risks and opportunities through assessment of current performance and identify the RTS performance factors that are relevant to the agency. Set objectives and targets for each performance factor and develop action plans (refer clause 6 of ISO 39001).	Planning officer will assess and identify safety problems that might arise with the concept of the proposed project, such as, road network implications, alignment and other traffic engineering features and make necessary improvement to the proposal. The detail of traffic engineering features such as the cross section elements, carriageway layout, intersection layout, horizontal and vertical alignment and traffic control options are established based on the guidelines.
Step 4: Implementation	Implement and operate the RTS management system, and ensure that sufficient capacity is provided for the objectives and targets to be met (refer clauses 7 and 8 of ISO 39001).	Road Branch is an engineering authority for Sarawak on matter relating to road. JKR Regional Office will be responsible for all projects implementation and field operations that include site supervision and coordination. JKR divisional office is responsible to monitor and ensure the implementation of maintenance work being carried out accordingly as required in the agreed performance standards. Each office is responsible in maintenance of records and reports within their scope of work.
Step 5: Monitoring and Evaluation	Monitor and evaluate RTS performance, conduct internal audits and periodic reviews of the RTS management system to identify opportunities for continual improvement (refer clause 9 of ISO 39001).	Not practiced.
Step 6: Continual Improvement to Elimination of Death and Injury	Improve the RTS management system on a continual basis following review of RTS performance, and of the RTS management system itself (refer clause 10 of ISO 39001).	Not practiced.

5.3 Challenges to Certification Based on 6 Steps PDCA

The six steps PDCA are: (1) Scope and context; (2) Leadership; (3) Planning; (4) Implementation; (5) Monitoring and Evaluation; (6) Continual Improvement to Elimination of Death and Injury. It is a structured guideline for ISO 39001 certification.

Based on the analysis tabulated in Table 1, it is deduced that some JKR Sarawak RTS practices do conform with ISO 39001 RTS Management Systems. Step 1 is practiced as the scope and context is specified. Steps 2 to 4 show that the current practices produced the desired outcome of the relevant departments. However, there are some discrepancies found. In Step 2, a 'long term vision to eliminate death and serious injury' is absent. The absence of this vision resulted in the inability to 'establish, document and communicate Road Traffic Safety policy'. These two criteria are important step for documentation and their absence affect the next step (Stage 3 - Planning). JKR Sarawak is unable to effectively 'determine risks and opportunities through assessment' as there are no official documentation available for reference. This leads to the challenges in setting objectives and target for RTS practices. The implementation stage (stage 4), shows that the relevant offices are able to implement and operate the current management system, with documentation.

However, there are no practices recorded in Steps 5 and 6. Step 5 is monitoring and evaluation, a very crucial step in identifying opportunities for continual development. There is no evidence that suggest JKR Sarawak identifies a measurable outcome for monitoring purposes, thus, evaluation cannot be executed. The absence of practices in Step 5 leads to the non-existence of practices in Step 6. Any form of improvements is not interpreted because there is no benchmark available for review purposes. This further explained the challenges faced by JKR Sarawak in obtaining the certification for ISO 39001.

6.0 Conclusion

In conclusion, although ISO 39001 is fully aligned with ISO 9001 Quality Management System, JKR Sarawak is not ready to adopt it in the near future. JKR Sarawak must acknowledge the crucial role of setting a long term vision with measurable objectives. These objectives may differ from one department to another, but the outcome is similar, which is to eliminate death and injury. Having measurable RTS objectives will help to evaluate the RTS performance and effectiveness of the system. The suggested measurable RTS objectives are as follow: 1) reduction in the number of complaint in regards to safety issues, 2) high compliance to the safety requirements in road project, 3) an increase of road users' satisfaction, and 4) a decline in the number of accidents caused by road conditions for a defined time frame. Subsequently, RTS management system itself can be improved.

7. References

[1] World Bank 2017 *Global Road Safety Facility (GRSF) annual report 2017* (Washington: World Bank Group)

[2] Luoma J and Sivak M 2012 *Road Safety Management in Brazil, Russia, India and China* (Michigan: The University of Michigan Sustainable Worldwide Transportation)

[3] Kementerian Pengangkutan Malaysia 2017 *Buku Statistik Keselamatan Jalan Raya* (Kuala Lumpur)

[4] United Nations 2011 *Improving global road safety* (Note by the Secretary-General) 66th session of the General Assembly, 30 September 2011

[5] Transportation Research Board 2015 *Annual Report* (Washington: TRB)

[6] Small M and Breen J 2017 *Startup Guide to ISO 39001. Road Traffic Safety Management Systems* (Switzerland: International Organization for Standardization)

[7] Bliss T and Breen J 2009 *Implementing the Recommendations of the World Report on Road Traffic Injury. Prevention Country Guidelines for the Conduct of Road Safety Management Capacity Reviews and the Specification of Lead Agency Reforms, Investment Strategies and Safe System Projects* (Washington: World Bank Global Road Safety Facility)

[8] ISO 2012 *Road Traffic Safety (RTS) Management Systems - Requirements with guidance for use* (Online: International Standards Organization)

[9] MIROS 2013 *ISO 39001:2012 Implementation in Malaysia Project Progress* (Kuala Lumpur: MIROS)

[10] Economic Co-operation and Development 2008 *Towards zero: Ambitious road safety targets and the safe system Approach* (France: OECD/ITF)

[11] Bowen GA 2009 Document analysis as a qualitative research method *Qualitative Research Journal* **9** (2) 27-40

Red light running motorcyclists at signalized intersection in Malaysia: An empirical study

Muhammad Marizwan Abdul Manan, Syed Tajul Malik Syed Tajul Arif, Ho Jen Sim, Norfaizah Mohamad Khaidir, Hawa Mohamed Jamil & Muhammad Ruhaizat Abd Ghani

Road Engineering and Environment Research Centre, Malaysian Institute of Road Safety Research, Kajang, Malaysia

E-mail: marizwan@miros.gov.my

Abstract. This study investigates factors associated with red light running motorcycle (RLR-MC) behavior during their approach and crossing of signalized 3-leg and 4-leg intersections along majors in Malaysia. Twenty-seven (27) intersections with Pre-timed traffic light (PTTL) and Actuated traffic light (ATL) were selected and observed during peak and off peak hour period. In general, the average rate of RLR-MC was 3.61%, with the highest rate of RLR-MC recorded was 22%, while the lowest rate was 0.6%. Observations have shown three-ways of RLR-MC approached the signalized intersection, i.e. (1) approaching the SI with Weaving/Lane Splitting, (2) approaching the SI from center of the lane, and (3) approaching the SI from the left side/the shoulder. There are three ways RLR-MC crossing the signalized intersection, i.e. (1) by Illegal maneuver (i.e. Illegal U-turn, contra-flow, prohibited left-turn), (2) by stopping first at or before the stop line, and (3) without stopping before the stop line. 'Crossing the SI without stopping at the stop line' is the majority with 51.15% from total observation. It is recommended that SI with high RLR-MC be equipped with adequate traffic island, replace SI signal type to actuated traffic signal and installing traffic light pole instead of gantry.

1. Introduction

According to the World Health Organization, close to a quarter (24.1%) of the world's road traffic deaths occur among motorcyclists or powered-two-wheelers (PTW) [1]. Of these motorcycle fatalities, the South-East Asia region (i.e. mostly low- to middle-income countries) has the highest rate with 49.9%, compared to only up to 10.9% motorcyclist fatalities in high-income countries in the European region. Sixty five percent of the world's motorcycles are in Asia, whereas Europe and North America account for only 16% [2]. The 4 countries with the highest numbers of motorcycles per 1,000 of population are Malaysia, Thailand, Cambodia and Japan [2, 3]. In developing and low to middle-income countries, such as those in the Asian region, motorcycles are used and exposed frequently as they are relatively affordable to buy and run [1, 2]. Hence, the high number of motorcycles on Asian roads is reflected in their high proportion of fatality crashes.

Malaysia may be the best example to study motorcycle crashes, as it typifies the countries with safety problems for motorcyclists and its data is close to the average, i.e. 47% of registered vehicles are motorcycles and 59% of the victims of reported accident fatalities are motorcyclists [1, 4]. In 2013, Malaysia ranked number 5 in the world among countries with a high percentage of motorcyclist fatalities, i.e. more than 50% of the total road fatalities are associated with motorcycles [1]. In addition to this, Malaysian motorcycles represent about 20% to 35% of motorized vehicle fleet, which are over-represented in road traffic crashes (as similar to Singapore [5] and Indonesia [6]); accounting for about 47% of total road traffic crashes and 59% of road fatalities [7]. On average, more than 70 motorcyclists are killed in road traffic crashes every week, and more than ten motorcycle riders are killed every day [1, 7].

1.1 Motorcycle Safety at Intersection in Malaysia

Although the great majority of motorcycle fatal crashes in Malaysia are reported on straight road sections (66%), motorcycle fatal crashes reported on major signalized intersections (i.e. 3-leg and 4-leg signalized intersection) is the next most common (18.5%) [7]. Table 1 shows the percentage of motorcycle fatal crashes at intersections based on the total reported crash cases from 2010 – 2011. Based on it, all type of signalized intersections have a higher motorcycle fatal crash rate (23.9%: 4-leg, 19.8%: 3-leg) compare to the un-signalized intersections ($p<0.01$), despite both un-signalized 4-leg and 3-leg intersection have a higher number of reported crash cases. Moreover, if we narrow down these motorcycle fatal crash rates based on various road jurisdiction classifications (Table 2) and various area type (Table 3), **signalized intersection**s are shown to have a statistically significantly higher fatality risk (i.e. fatality rate) than un-signalized intersection ($p<0.05$).

Table 1. The percentage of motorcycle fatal crashes at intersections

Intersection type	Signalized	Un-signalized
4-leg	23.9% * (n = 522)	17.4% * (n = 1,404)
3-leg	19.8% * (n = 334)	15.9% * (n = 5,326)

n: Total number of reported crash cases involving motorcycle since 2010 – 2011.
* Statistically significant difference according to Chi-sq. test ($p < 0.05$)
Source: PDRM (2011), analyzed by MIROS

Table 2. The percentage of motorcycle fatal crashes at intersections based on various category of road jurisdiction

Intersection at various road jurisdiction	Signalized	Un-signalized
Federal	20.0% * (n = 436)	15.8% * (n = 2,247)
State	26.8% * (n = 164)	15.2% * (n = 1,866)
Municipal	23.7% * (n = 249)	17.0% * (n = 2,004)
Others	19.4% * (n = 36)	16.4% * (n = 671)

n: Total number of reported crash cases involving motorcycle since 2010 – 2011. Others: Private roads or local roads,
* Statistically significant difference according to Chi-sq. test ($p < 0.05$)
Source: PDRM (2011), analyzed by MIROS

Table 3. The percentage of motorcycle fatal crashes at intersections based on various area type

Intersection within the vicinity of various area type	Signalized	Un-signalized
Residential	25.0% * (n = 164)	16.5% * (n = 1,664)
Commercial	22.0% * (n = 205)	18.1% * (n = 930)
Industrial	33.3% * (n = 33)	30.4% * (n = 191)
School	17.4% * (n = 23)	12.1% * (n = 248)
No development	20.5% * (n = 448)	14.9% * (n = 3,819)

n: Total number of reported crash cases involving motorcycle since 2010 – 2011.
No development signifies an area with very few populations, e.g. plantation, forest, etc.
* Statistically significant difference according to Chi-sq. test ($p < 0.05$)
Source: PDRM (2011), analysed by MIROS

2. Literature Review

One of the factors that influence fatal crashes among motorcyclists may be due to the red light running (RLR). Local studies such as [8] and [9] have found out that, red light running among Malaysian motorcyclists MCRLR is high compared to other road users and the reasons for this problem are varied and complex. A cross-sectional study was conducted in Selangor, Malaysia to identify road traffic-light violations and the findings shows that motorcycle recorded higher violation of traffic lights with 13.5% motorcycles compared to four-wheeled vehicles with 11.1%, and the odds ratio analysis shows that the violation increased by 1.24 times for motorcycles compared with four-wheeled vehicles [8]. Factors influencing MCRLR in Malaysia are found to be more on during the weekends [8], on intersection without enforcements (e.g. Red light running cameras, police, etc.) [8, 10], longer cycle time (>160s), four-phase signal timing, four-leg signalized intersections and high motorcycle volume [9].

The known behaviors associated with motorcycle risky behavior observed at intersection studies such as in Abdul Manan and Várhelyi [11], Haque, et al. [12] , Haque, et al. [13], Hawa, et al. [10], Clarke, et al. [14], de Rome, et al. [15] and Radin Umar [16], are: weaving between motorist, running red lights, speeding, splitting while speeding, not stopping at the stop line, not wearing helmet, wearing dark clothing and not using headlights. However, among all the risky behavior discussed previously, red-light running among road users stood out as the most extreme or highly risky behavior, which leads to a crash. This is because road crashes occur when there are conflicting vehicle movements [17-19]. Many research in developed countries, e.g. U.S., UK, European countries, etc., has put more focus on discovering the driver (including motorcyclists) characteristics and road traffic environment factors associated with red light runners behavior. The driver's characteristics which are associated with red light runners behavior are found to be: young drivers/riders [20, 21], male [22, 23], less likely to wear safety belts [23], poor driving records [24, 25], under the influence of alcohol [26, 27] and drive smaller or older vehicle [28, 29]. Red light runners are also more likely to be younger than 30- years old, have a record of moving violations, are driving without a valid license and/or have consumed alcohol [19, 28]. As for road traffic environment factors, Porter and England [23] and Martinez and Porter [22] concluded that higher traffic volumes at intersection can be associated with increase in red light violations among car drivers. Furthermore, an intersections with higher traffic volumes, drivers would have more chances to be in a following position in the traffic flow at the onset of yellow, thus the red light running rate and crash risk would increase [30]. Moreover, speeding car drivers were more likely to run red lights compared to non-speeding drivers and 90% of them were mostly located between the distances of 61.3m to 128.0m from the intersection [30].

Many factors, such as rider characteristics influencing or associated with red light running among motorcyclists are yet to be uncovered, especially in Malaysia. Research on traffic and environment factors that are influencing red light runners among car passenger are widely being reported but limited on motorcycle riders. As for Malaysia, research on motorcycle rider behavior and characteristics associated with red light running is still limited, and much focus was put on finding the associated factors from the traffic and road environment perspective, as seen in Law, et al. [9], Kulanthayan, et al. [8], Che Puan and Ismail [31] and Hawa, et al. [10].

2.1 Aims of current study

The current study investigates motorcycle rider behavior and characteristics factors associated with motorcyclist red light running behavior at signalized intersection and suggest countermeasures in order to curb the risky behavior. Given this focus, it was of particular interest to identify motorcyclists' risky behavior prior and during their red light running behavior and factors associated with red light running motorcycle.

3. Methodology

This study was conducted in early 2016 and ends in mid 2017, with 3 main phases: (1) Site selection criteria and identification, (2) Site verification and data collection, (3) Statistical Analysis. This study only focused on red light running motorcycle (RLR – MC) on selected signalized intersections on 3 and 4-legged intersections along major roads in Malaysia.

3.1 Site selection criteria and identification

There are two types of traffic light signal time in Malaysia based on Kulanthayan, et al. [8] observation. The most common is the pre-timed traffic light (PTTL) with the signal timing cycle length usually falling between 45 and 120 seconds. The timing for each signal is determined based on traffic volume and traffic patterns in each particular area. The second type is a traffic light with a sensor, i.e. actuated (ATL). This system maximizes the efficiency at a traffic junction by allocating green time for each approach to a traffic junction according to traffic demand. This means that if the sensor detects that the demand of a particular approach is higher, it will redistribute the green time accordingly to optimize the usage of the traffic junction. Another type of traffic light has a countdown timer (CTTL). It is a two-digit time indicator, placed on the pole above the traffic signal. Its purpose is to reduce congestion at traffic junctions, help motorists to have a better understanding of the traffic flow and helps motorists to be aware of the remaining time left on the green phase. This study however, do not take into account the CTTL type due to is rarity and uncommon along main roads in Malaysia.

This study considered only two types of traffic light signal time, i.e. Pre-timed traffic light (PTTL) and Actuated traffic light (ATL), on each 3-leg and 4-leg signalized intersection. As for the location, we prioritized the signalized intersections that connect a major road with a minor road, for example a Primary road with a Secondary road, or Collector with a Local roads, etc. The location of these signalized intersections were also narrowed down into nearby an industrial, a residential or commercial area. This is based on the fact that the fatality rate involving motorcycles at signalized intersection are higher compared to their respective category. Additional selection criteria for each signalized intersection (SI) are as follow:

1. The SI is on a straight geometry road section and on a flat terrain. This is to ensure the site has no sight distance limitation and the crossing road users are not affected or influence by it.
2. The major road on the SI has a minimum of 10% of motorcycles from the traffic volume. This is to ensure that the site has sufficient number of observation of motorcycles.
3. The SI shall have a vantage point for the researches to conduct the observations and to collect speed data obscurely from other road users.
4. The SI has some history of traffic crash involving motorcycle or the SI is observed to have RLR behaviour among motorcyclists passing through. This criterion is to justify the site's current risky condition.

This study is a nationwide study, thus to reflect and represent Malaysia, the SI was sampled from major towns in Malaysia respective to the region, i.e. Kuala Lumpur or Shah Alam (Central region), Penang (Nothern region), Johor Bahru (Southern region), and Kuantan (Eastern region).

In order to effectively and efficiently perform the data collection, each city, e.g. Kuala Lumpur and Shah Alam, was screened (via Google maps and Google Street view) in terms of SI availability and SI geometry type, while being cross checked with the crash history from the police data base. The plan was to select at least four SI with each type of SI signal time (PTTL and ATL) and each for 3-leg and 4-leg intersection on every major town in their respected region. For example, the central region has three major cities (Kuala Lumpur, Petaling Jaya, Shah Alam), thus the screened number of intersection would be 48 (3 cities x 2 SI signal time type x 2 of SI type x 4 SI) for each region. Thus the total number of SI planned for screening was 192 (4 region x 48 SI).

3.2 Site verification and data collection

After the rigorous screening process, we had to verify each of the selected sites in order to conform to the site selection criteria by conducting a preliminary observation on each sites. Once the site fulfils the selection criteria, it undergoes the data collection process.

3.2.1 *Method of data collection.* To capture the observation of each motorcyclists passing from Scenario 1 and up through Scenario 2, two synchronized camera position on a strategic location in the vicinity of the SI. These cameras was mounted on a customized retractable (i.e. telescoping pole), which can lifted as high as 4 to **7** meters depending on the site condition and restriction. The reason to have a 7 meters observation view point is to capture all activity within the intersection, which also covers all intersection legs. The time of recording was conducted during peak (7.00am – 8.00am, 5.00pm – 6.00pm) and off peak hour (10.00am – 11.00am, 4.00pm – 5.00pm) with duration of 1-hour each.

3.3 Types of variables for acquisition. Once all the sites have been identified, each signalized intersection's characteristics, infrastructure measurements, signal timings, traffic volume and motorcycle characteristics were captured and categorized accordingly (see Table 4).

Table 4. Variables on signalized intersection characteristics

No.	Factors	Variables (continuous / categorical)
1.	SI type	3-leg, 4-leg
2.	SI signal type	Actuated, Pre-timed
3.	Red / Green time category [a]	Actuated, 40s-80s, 80s–120s, 120s–160s, 160s–200s
4.	Amber time category	Actuated, 1s, 2s, 3s
5.	Major / Minor road signal infrastructure	Signal on Pole, Signal on Gantry
6.	Total num. of lanes on Major / Minor road	2 lanes, 4 lanes, 5 lanes, 6 lanes
7.	Speed limit on Major / Minor road	50km/h, 60km/h, 70km/h, 90km/h
8.	Number of traffic island	0, 1, 2, 3, 4, 5
12.	Major road crossing length [ab]	0m – 10m, 10m – 15m, 15m – 20m, 20m – 25m
15.	% of MC passing on Major / Minor leg	< 15%, 15% - 25%, 25% - 50%, > 50%
16.	Num. of vehicles on Major leg 1 & 2	- Count -
17.	Motorcycle type based on c.c.	<250cc, >250cc
18.	Motorcycle number of occupancy [a]	1, 2, > 2
19.	Location of RLR-MC	From Major road, From Minor road
20.	Destination of RLR-MC	To Major road, To Minor road
21.	Gender	Male, Female
22.	Helmet wearing	With helmet, no helmet

Notes:
a – indicates that the factors were measures on site and later categorized into few categories.
b – Crossing distance is measured from the stop line to stop line for each leg.
Traffic volume: Number of vehicles passing the intersection during 1hour of observation
Num. of vehicles on Major leg 1 and leg 2: Count of vehicles on each leg during each RLR-MC occurrence

For every red-light running motorcycle, all of its characteristics, movement and motorcyclists' riding behavior was identified and recorded (see the variables in Table 4). As part of the data collection and analysis process, the research team developed state-of-the-art prototype software called MECHRED (Motorcycle Characteristics and Motorcyclists' Red light running Behavior Data Management Software), dedicated for data collection and management of motorcycle characteristics and motorcyclists' red light running behavior at signalized intersection. MECHRED consists of two video interfaces and a data logger interface (see Figure 1). For every passing motorcycle in the video, the observer goes through each factor (see Table 4) and chooses the suitable option for the motorcycle and motorcyclist. The observers (MECHRED users) are the main author, co-authors and six highly trained research assistants who conducted the data collection. The MECHRED's user enters the observed information, e.g. motorcycle type, motorcyclists' gender, motorcycle movement, etc., from the video interface.

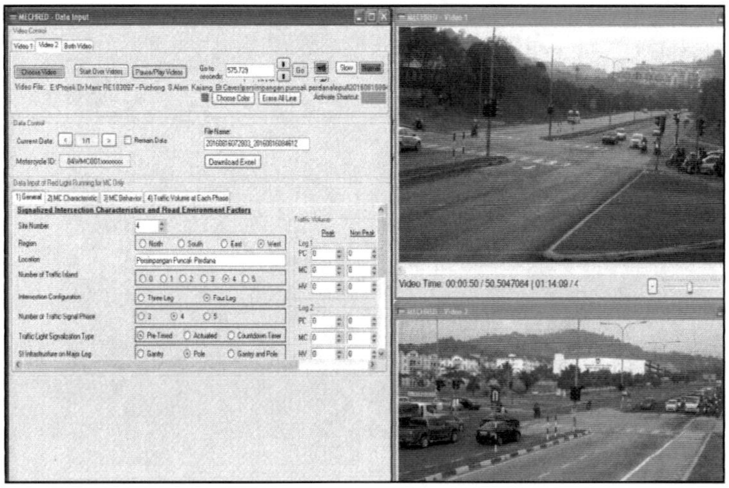

Figure 1. Screenshot of the MECHRED software

3.4 Statistical Analysis

The main objective of this research intends to find the factors associated with red light running motorcycle (RLR-MC) at signalized intersections. The analysis for this study catered multivariate categorical data which looked into the associated SI factor variables in **Scenario 1**, i.e. red light running motorcyclist behavior when approaching the SI and the associated SI factor variables in **Scenario 2**, i.e. red light running motorcyclist when crossing the SI (see Figure 1). In analyzing these variables, a Multiple correspondence analysis (MCA) was used in order to clustered or find a pattern relating RLR-MC behavior to the SI characteristics, infrastructure, dimension, signal timing, traffic volume and motorcycle characteristics.

Multiple correspondence analysis (MCA) are methods for analyzing observations on categorical variables. MCA is usually viewed as an extension of simple correspondence analysis (CA) to more than two variables. CA analyses a two-way contingency table; MCA and JCA analyze a multi-way table. Multiple Correspondence Analysis (MCA) is also a part of a family of multidimensional descriptive methods (e.g., clustering, factor analysis, and principal component analysis) revealing patterning in complex datasets when we dispose more qualitative variables (ordinal, or nominal) in [32]. As mention in Kalayci and Basaran [33], MCA is used to represent datasets as "clouds" of points in a multidimensional Euclidean space; this means that it is distinctive in describing the patterns geometrically by locating each category of analysis as a point in a low-dimensional space. The results are interpreted on the basis of the relative positions of the categories and their distribution along the dimensions; as categories become more similar in distribution, the closer (distance between points) they are represented in space [33]. MCA can be a particularly powerful one as it "uncovers" groupings of categories in the dimensional spaces, providing key insights on relationships between categories, without needing to meet assumptions requirements such as those required in other techniques widely used to analyse categorical data (e.g., Chi-square analysis, Fischer's exact test, -statistics, and ratio test) [33].

According to [34] and [33], for the proximity between categories we need to distinguish two cases. First, the proximity between levels of different nominal variables means that these levels tend to appear together in the observations. Second, the proximity between levels means that the groups of observations associated with these two levels are themselves similar.

4. Results

This section discusses the results and findings of the study. This section is divided into four subsections: Site selection, Analysis on Scenario 1, Analysis on Scenario 2 and Overall Analysis.

Figure 2. Photos of RLR-MC at selected sites

Table 5. Selected location of signalized intersection

Region		State	N	Location	SI type 4L	SI type 3L	Crash history
i	Southern	Johor Bharu	1	Persimpangan Jalan Padi Mahsuri (Bandar Baru Uda)	√		6
			2	Persimpangan Jalan Pendidikan (Taman Universiti		√	81
			3	Persimpangan Kampung Ungku Mohsin (Jalan	√		10
			4	Persimpangan Kampung Melayu (Jalan Tampoi)		√	0
ii	Northern	Perlis	5	Persimpangan Jln Tuanku Syed Sirahjuddin		√	0
			6	Persimpangan Jln Tuanku Syed Alwi	√		5
			7	Persimpangan Jln Tuanku Syed Putra		√	0
			8	Persimpangan Masjid Batu 2	√		2
		Kedah	9	Persimpangan Jln Kubang Rotan (Jalan Alor Star –	√		0
			10	Persimpangan Jln Tun Razak (Alor Star)		√	18
		Penang	11	Persimpangan Jalan Sri Tanjung Pinang		√	6
			12	Persimpangan Jalan Sungai Dua		√	9
			13	Persimpangan Jalan Jelutong	√		167
iii	Eastern	Kuala Terengganu	14	Persimpangan Medan Selera Batu Burok (FR3)	√		2
			15	Persimpangan Jalan Batu Burok (FR3) - Jalan Tok	√		0
			16	Persimpangan Kamarudding (FR3) - Jalan Pusara		√	0
		Kuantan	17	Persimpangan Jalan Beserah - Jalan Air Putih (FR2)	√		0
iv	Central	Selangor	18	Persimpangan Tesco Bukit Puchong		√	0
			19	Persimpangan Puncak Perdana	√		0
			20	Persimpangan Jln UPM Sri Serdang		√	68
			21	Persimpangan Jln Wan Siew Sg Chua	√		51
			22	Persimpangan Padang Jawa	√		47
			23	Persimpangan Persiaran Kewajipan, Subang Jaya		√	350
			24	Persimpangan Persiaran Sukan S.Alam		√	60
			25	Persimpangan Uitm (Pintu 2) Shah Alam	√		0
			26	Persimpangan Jln Batu Caves	√		102
			27	Persimpangan Sunway Suria	√		0

Crash history = Total number of crash from 2012 – 2014, N: Site number
SI: Signalized intersection, 4L: four leg signalized intersection, 3L: three leg signalized intersection

IOP Conf. Series: Materials Science and Engineering **512** (2019) 012019 doi:10.1088/1757-899X/512/1/012019

4.1 Site selection

The total number of SI planned for screening was 192 (4 region x 48 SI). However, the number of SI after the screening was decreased down to 50 due to site selected criteria, e.g. most of them do not have a significant reported crash history and unsuitable road geometry. Results from the preliminary observation, we have narrowed down to **27** sites, as seen in

Table 5. Forty-eight per cent (40%) from the screened 50 sites was rejected due to insufficient number of red-light running motorcycle and unsuitable vantage point for observation. Figure 2 shows photos of RLR-MC occurrences at selected sites.

4.2 Analysis of red light runner among motorcycle

As mention previously, data collection was being conducted during peak and off-peak hours, thus, the number of data sets for this study increased up to 54. Based on Table 6, the average rate of RLR-MC is 3.61%, by which the highest rate of RLR-MC recorded was 22% at Site 24, while the lowest rate was 0.6% at Site 8 during off peak hours. We also record RLR among other vehicle, which is 1.2%. Moreover, average rate of number of red light running motorcycles during peak hours is higher (4.66%) than the off peak (3.6%).

Table 6. Number of RLR MC at the selected sites based on peak and non-peak hour

Site	Traffic vol. - Major leg		Num. of RLR-MC		Site	Traffic vol - Major leg		Num. of RLR-MC	
	Off peak	Peak	Off peak	Peak		Off peak	Peak	Off peak	Peak
1	1,055	1,249	20 (1.90%)	18 (1.44%)	15	2,771	2,844	45 (1.62%)	37 (1.30%)
2	1,583	2,063	49 (3.10%)	94 (4.56%)	16	2,195	2,950	107 (4.87%)	145 (4.92%)
3	2,226	2,635	111 (4.99%)	161 (6.11%)	17	2,940	3,047	109 (3.71%)	136 (4.46%)
4	2,751	3,006	27 (0.98%)	38 (1.26%)	18	3,818	4,058	55 (1.44%)	139 (3.43%)
5	2,250	2,010	33 (1.47%)	46 (2.29%)	19	2,194	3,528	106 (4.83%)	370 (10.49%)
6	963	1,292	26 (2.70%)	6 (0.46%)	20	2,644	3,327	92 (3.48%)	79 (2.37%)
7	1,624	1,979	16 (0.99%)	36 (1.82%)	21	1,358	1,326	37 (2.72%)	27 (2.04%)
8	840	1,188	5 (0.60%)	21 (1.77%)	22	878	1,159	30 (3.42%)	22 (1.90%)
9	1,894	2,209	106 (5.60%)	154 (6.97%)	23	2,449	3,084	27 (1.10%)	116 (3.76%)
10	1,434	1,930	33 (2.30%)	44 (2.28%)	24	2,007	3,008	443 (22.07%)	677 (22.51%)
11	2,964	3,116	54 (1.82%)	50 (1.60%)	25	2,571	2,630	38 (1.48%)	62 (2.36%)
12	1,918	2,490	79 (4.12%)	94 (3.78%)	26	1,489	2,141	53 (3.56%)	55 (2.57%)
13	2,549	2,163	227 (8.91%)	276 (12.76%)	27	3,449	3,404	112 (3.25%)	188 (5.52%)
14	2,110	2,840	15 (0.71%)	21 (0.74%)	Ave	2,108	2,469	76 (3.61%)	115 (4.66%)

Note: (%):
The rate of red light running motorcycle from the total volume count in percentage.
Ave. : Average, Traffic volume: Vehicle count for one hour period

Figure 3 shows the distribution plot of RLR-MC based on the major road traffic volume. Based on the regression line, the slope for the rate of RLR-MC during the peak hour is steeper than the off-peak hour. In other words, the rate of RLR-MC occurrence increases as the traffic volume along the major road increases, especially during peak hour.

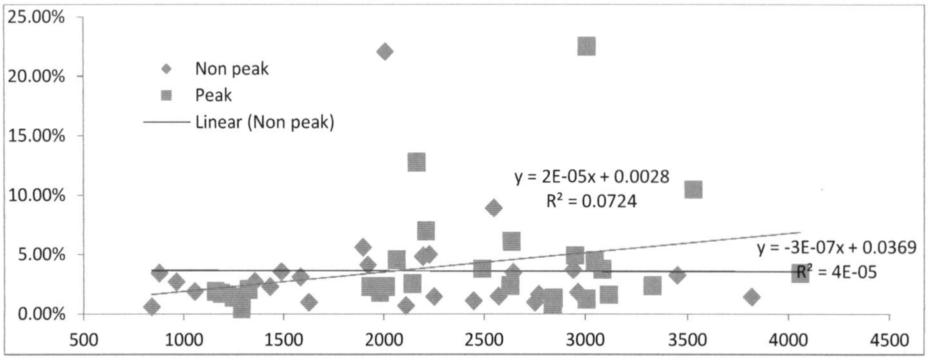

Figure 3. Distribution of RLR-MC rate based on the major road traffic volume

Table 7 shows that 3-leg SI has the highest RLR-MC rate (60.8%) compared to 4-leg SI, especially during peak hour period. However, non-peak hour period has the highest RLR-MC rate (66.9%) compared to peak hour period, especially at 4-leg. This analysis however, is not statistically significantly different ($p>0.05$), or in other term, RLR-MC on 3-leg SI occurrences are not much difference on 4-leg SI, regardless the period.

Table 7. Number of RLR-MC based on period category and type of SI

Period category	Four Leg	Three Leg	Grand Total
Non-peak	1,375 (*40.1%*), (66.9%)	680 (*39.2%*), (33.1%)	2,055
Peak	2,057 (*59.9%*), (66.1%)	1,056 (*60.8%*), (33.9%)	3,113
Grand Total	3,432	1,736	5,168

(%): Percentage in italics indicates the rate of RLR-MC based on type of SI
(%): Percentage in bold indicates the rate of RLR-MC based on period category,
Pearson $chi^2(1) = 0.3650$ $Pr = 0.546$

4.3 Observation of RLR-MC approach and crossing

Our observation on all of the selected sites has shown that there is a distinct or typical type of RLR-MC movement in both scenarios. In scenario 1, there are three type of movement performed by RLR-MC: Approaching the SI with Weaving / Lane Splitting, Approaching the SI from the center of the lane, and Approaching the SI from the left or shoulder (see Figure 4, note that the figure also apply for 3-leg SI). RLR-MC approaching the SI with Weaving / lane splitting is a rare behavior which require motorcyclists to move in between vehicles that are waiting at the SI with fast pace and often weave between vehicles when the spaces in between are limited. RLR-MC approaching the SI from the center of the lane is typical behavior seen during our observation. This behavior is seen when a RLR-MC move consistently in the middle of the lane or carriageway (either near the lane line or near the median) until the motorcyclist reaches the stop line to run the red light. RLR-MC approaching the SI from the left side or on the shoulder is another typical behavior in Scenario 1. This behavior occurs when a RLR-MC move consistently on the left side of the carriageway (near the edge line) or on the shoulder until the motorcyclist reaches the stop line to run the red light.

As for scenario 2, there are also three typical types of RLR-MC movement: Crossing the SI by Illegal maneuver (i.e. Illegal U-turn, contra-flow, prohibited left-turn), Crossing the SI by stopping first at or before the stop line, and Crossing the SI without stopping before the stop line (see Figure 4, note that the figure also apply for 3-leg SI). RLR-MC crossing the SI by illegal maneuver is a movement occurred when the RLR-MC make an illegal U-turn during the red-phase or moves in the opposite direction of the traffic (i.e. contra flow) into either into the major or minor road. RLR-MC crossing the SI by stopping first at or before the stop line is one of the most common RLR behavior among motorcyclists. Motorcyclists in this scenario are seen to stop before or at the stop line before running the red light. Motorcyclists are also seen more observant of the opposite traffic (by turning their heads couple of times) before crossing it during the red phase. More often than not these RLR-MC are considered as running over the red light because they are already in the red phase of the SI. Another type of crossing detected during our observation, is the second most common RLR behavior among motorcyclists. This behavior shows that RLR-MC did not stop at the stop line or sometime slows down before they run the red light. More often than not, this kind of behavior is seen as trying to beat the red light rather than running over the red light, by which they are in the transition stage of traffic light phase from green to red or amber to red.

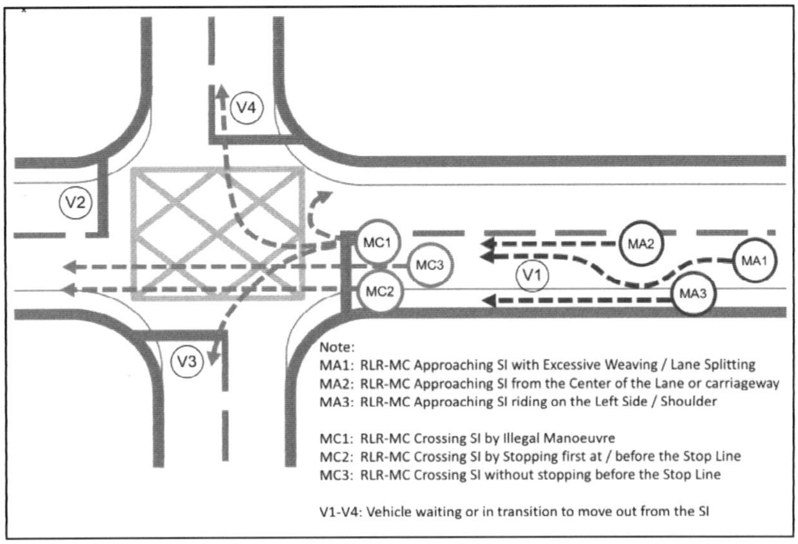

Figure 4. The behaviour of RLR-MC when approaching and crossing the SI

4.3.1*Association of Scenario 1 and 2 and signal timing*. In order to see the association of Scenario 1 and 2, we refer to Table 8 and Figure 5. As seen in Table 8, there is a statistically significantly different between the behaviours in Scenario 1 and 2 ($p<0.05$). In order to look at the association between these two scenarios, the MCA plot in Figure 8 was used. Figure 8 shows that RLR-MC who moves along the centre of the lane are likely to stop at the stop line before run the red light at a 4 leg signalized intersection. Meanwhile, those motorcyclists whom ride on the shoulder are like to run the red light without stopping at the stop line, i.e. beat the red light. These scenarios are like to occur at 3-leg signalized intersection.

Table 8. Number of RLR-MC based on Scenario 1 and Scenario 2

RLR-MC approaching the SI (Scenario 1)	RLR-MC crossing the SI (Scenario 2)			
	Illegal maneuver	Stopping first at stop line	Without stopping at stop line	Grand Total
Weaving / lane splitting	7	20	62	89 (1.72%)
From the center of the lane	23	1,629	806	2,458 (47.57%)
From the Left Side or Shoulder	71	774	1775	2,620 (50.71%)
Grand Total	101 (1.95%)	2,423 (46.89%)	2,643 (51.15%)	5,167

Pearson chi2(4) = 722.0462 Pr = 0.000
Note that the Table 14 was not used as a reference in interpreting the MCA plot but rather to tabulate in order to calculate the statistical difference via Person Chi2

From the cluster of variable (see Figure 5: left side) which sits closely together in the plot, those RLR-MC who approach the SI by riding on the center lane and run the red light after stopping at stop line, are mostly found on 4 leg SI which are associated with green time of 20s – 40s and 60s – 80s, red time of 80s – 120s and amber time of 1s and 2s. RLR-MC behavior such as not stopping at the stop line, preforming lane splitting / weaving and riding on the shoulder are not associated with any of the signal timings, however, those RLR-MC who approach the SI via riding on the left or shoulder are closely associated with traffic signal green time of 10s – 20s. Where else, Illegal maneuver are associated with amber time of 3s and red time of about 40s to 80s. Interestingly enough, the all of the actuated signal time phase are plotted away from the behavior variables which may indicate that RLR-MC behavior are not associated or rarely occurs on SI with actuated signals.

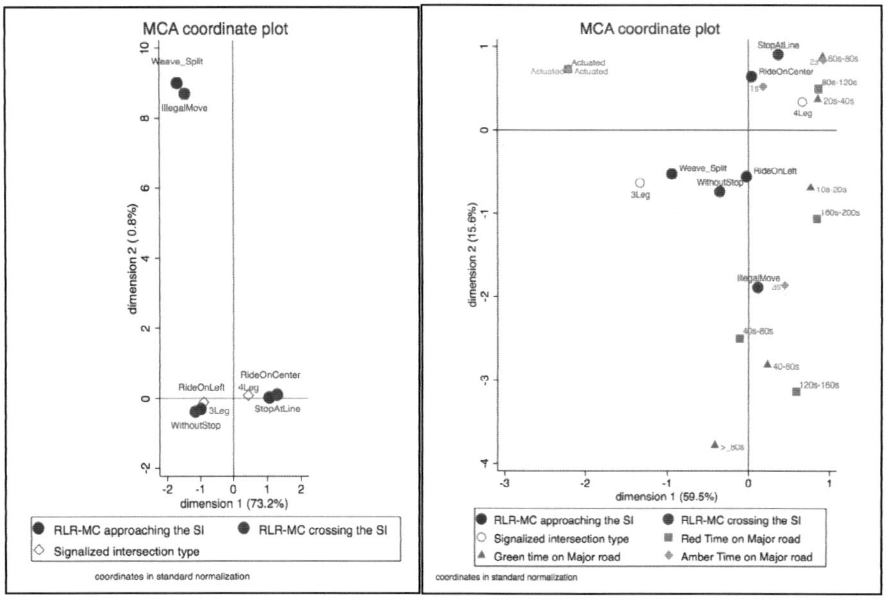

Figure 5. MCA plot for RLR-MC crossing and approaching the SI and association with signal timing

4.3.2*Analysis of RLR-MC based on motorcycle characteristics and movement.* From the MCA plot in Figure 6, it is clear that male RLR-MC performs most of the behaviours (approaching by riding on the centre or the left side, RLR before stopping or without stopping) and with or without a pillion rider. However, male RLR-MC are more incline to display these behaviour on 4-leg SI, where their most frequent RLR movement are when they are moving from a major to minor road and also minor to minor road. As for the female RLR-MC, they can be considered as an outlier in this analysis but the plot show that they are closer to behaviours such as approaching the SI on the left or shoulder or running the red light without stopping. On 3-leg SI, we can see that most of the RLR movement occurs when the RLR-MC crosses from a major road to a major road (i.e. straight movement), and they are occasionally associated with behaviour such as approaching the SI by weaving or lane splitting or crossing it via illegal movement.

Based on the bottom left quadrant of the graph, we can see that RLR-MC that are approaching the SI by riding on the left or shoulder and crossing the SI without stopping at the stop line are associated with: 3-leg SI with a total of 4-lanes on the major roads with 70km/h speed limit (see Figure 6). RLR-MC that are approaching the SI by riding on the center lane and stopping at the stop line before running the red light are closely associated with: 4-leg SI with 4-lanes minor roads with median and along major road with 60km/h speed limit. In addition to this, RLR-MC that performs illegal maneuver are also associated with these variables, but can be found more on SI with minor road with no shoulder. As for those RLR-MC that perform weaving or lane splitting, there are no clear association with any of the infrastructure variable as it is plotted away from it.

Variables such as: road with 50km/h speed limit, SI with 1, 2, 3 and 4 traffic island, 6-lanes major roads, 3-lanes and 5-lanes minor roads, SI with pole traffic signal, and major road with shoulder, are plotted away from these behavior variables, which may indicate that RLR-MC behaviors are not associated or rarely occurs on SI with these infrastructures.

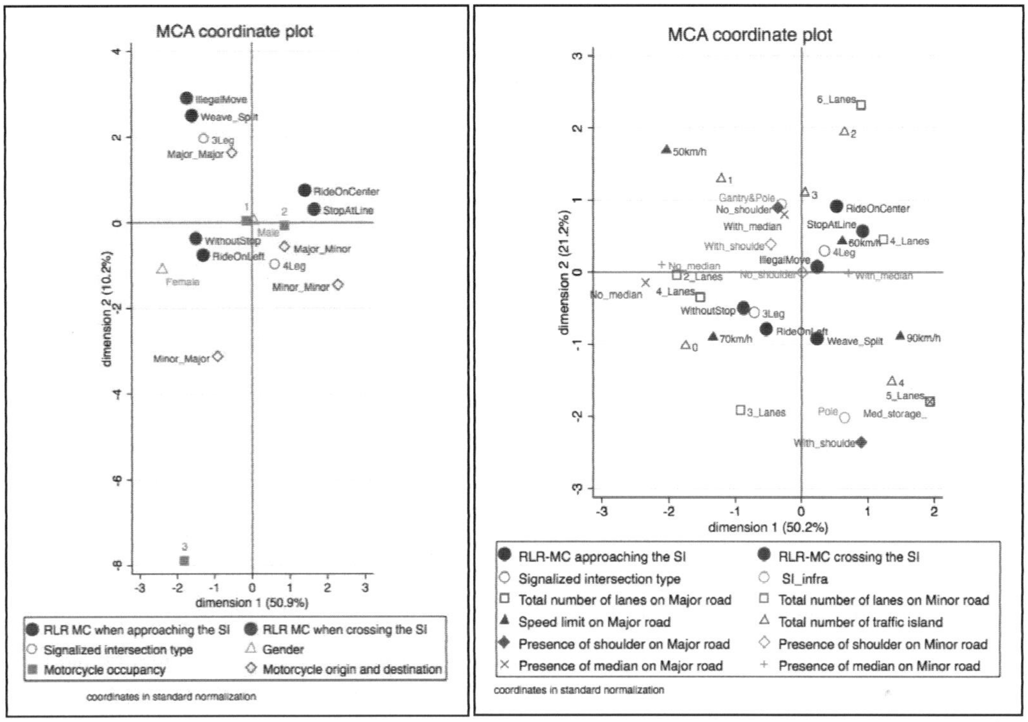

Figure 6. MCA plot for RLR-MC behaviour, motorcycle and infrastructure characteristics

4.3.3*Analysis of RLR-MC based on number of vehicles waiting*. Figure 7 indicates that general, as the number of vehicles waiting increases either on the major or minor road, the number of RLR-MC occurred at 3-leg SI decreases significantly. When comparing the number of vehicles waiting on each road (leg), Figure 7 shows that 90.9% of RLR-MC occurred at 3-leg SI, when zero to five vehicles are waiting on the minor road, while 54.5% of RLR-MC occurred when zero to five vehicles are waiting on the major road instead. However, when there are five to ten vehicles waiting on the minor road, RLR-MC rate decrease down to 7.7% while 22.6% RLR-MC occurred when five to ten vehicles waiting on the major road.

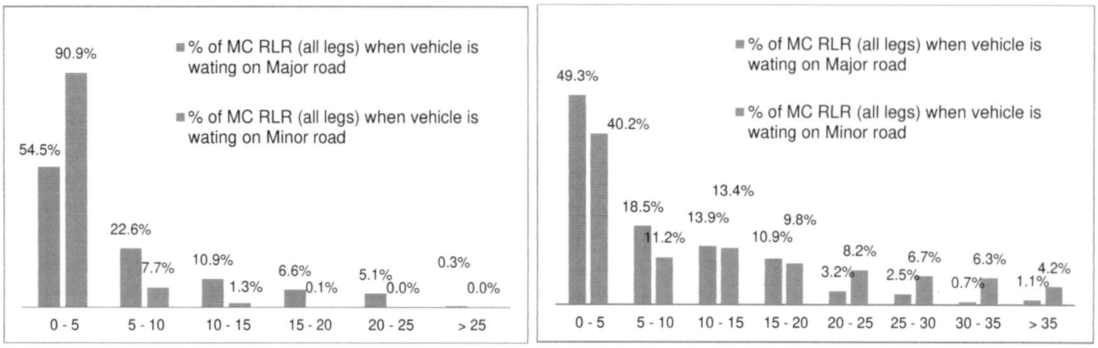

Figure 7. Rate of RLR-MC at 3-leg (left) and 4-leg (right) SI based on the number of vehicle waiting

When comparing between the two types of SI, the 4-leg SI has a lower rate of RLR-MC when there are zero to five vehicles waiting on both major and minor road (see Figure 7). In contrast to 3-leg SI, the rate of RLR-MC is higher (49.3%) when there are zero to five vehicles waiting on major road compared to those same number waiting on the minor road (40.2%). In addition, as the number of vehicle increases on the minor road, 4-leg SI has a higher rate of RLR-MC occurrence compared to 3-leg SI, but has a lower rate of RLR-MC when there is an increase of vehicle waiting on the major road.

5. Discussion

This study investigates the factors that are associated with red light running motorcycle (RLR-MC) behavior during their approach and crossing the signalized 3-leg and 4-leg intersections along majors in Malaysia and develops countermeasures in order to curb these risky behavior. This study was conducted in early 2016 and ends in mid 2017, where only **27** intersections with Pre-timed traffic light (PTTL) and Actuated traffic light (ATL) were selected based on a strict selection criteria and observed during peak and off peak hour period. As part of the data collection and analysis process, the research team developed state-of-the-art prototype software called MECHRED, dedicated for data collection and management. In analyzing these variables, a Multiple Correspondence Analysis (MCA) was used in order to clustered or find a pattern relating RLR-MC behavior to the SI characteristics, infrastructure, dimension, signal timing, traffic volume and motorcycle characteristics.

In general, the average rate of RLR-MC was 3.61%, by which the highest rate of RLR-MC recorded was 22%, while the lowest rate was 0.6% from the total traffic volume. This rate is quite small as compared to the done by Kulanthayan, et al. [8], however our rate remains higher as compared to other vehicle (RLR by other vehicle is 1.2%), which also corresponds to the study done by Kulanthayan, et al. [8]. In addition, this study show that RLR-MC on 3-leg SI occurrences are not much difference on 4-leg SI, regardless of the peak and off peak period. Furthermore, the rate of RLR-MC occurrence increases as the traffic volume along the major road increases, especially during peak hour.

Our observations have shown that there are three-ways RLR-MC approached the signalized intersection, i.e. (1) approaching the SI with Weaving or Lane Splitting, (2) approaching the SI from the center of the lane, and (3) approaching the SI from the left side or on the shoulder. Among these approaching behaviors, 'approaching the SI with Weaving or Lane Splitting' is the least common one with only 1.72% from the total observation.

Among all the RLR-MC approaching behaviors, the weaving or lane splitting is a considered a form of an aggressive riding behavior (see Shinar [35]).

RLR-MC crossing the SI without stopping at or before the stop line is the most dangerous among all behavior. We would classify this RLR-MC crossing without stopping as a form of risk taking behavior, because the RLR-MC would not stop and most of the time beat the red light after they slow down while have a 'quick look' of the traffic (i.e. turn their head quickly to scan the traffic). Moreover, most of the time, they are seen risking their lives by beating the red light with high speed in order to surpass the incoming crossing vehicles.

Our analysis has shown that motorcyclists who run the red light without stopping are associated with 3-leg SI with a total of 4-lanes on the major roads with 70km/h speed limit. This risky behavior however, is also not associated with any of the signal timings or any of the dimension variables, which indicate that this behavior occurrence may be highly dependent by the motorcyclist's personality or other factors previously mention. After viewing the analysis on this particular behavior, we can see that RLR-MC who cross the SI without stopping at the stop line requires a SI which has a high-speed limit (70km/h) and also location that has RLR-MC behavior as a norm.

We also detected another risk-taking behavior is the RLR-MC crossing via illegal maneuver. This is a rare behavior, which only records up to 1.95% from the total observation, but its movement stood out as risk-taking behavior because this RLR-MC behavior is unpredictable to other road user at the signalized intersection. Our observation has revealed that RLR-MC performs sudden illegal U-turn during the red-phase and sometimes moves in the opposite direction of the traffic (i.e. contra flow) into either into the major or minor road. Based on the analysis, we can generalize that RLR-MC crossing via illegal maneuver are mainly perform from those crossing from a major road into a major road and when the SI has a long and predictable amber time and red time and combined the presence of high motorcycle volume along the major road.

Our analysis has shown that there is an association between the approaching and crossing behaviors. It seems that, at a 4 leg signalized intersection, RLR-MC who moves along the center of the lane are likely to stop at the stop line before run the red light. Meanwhile at 3-leg signalized intersection, those motorcyclists whom ride on the shoulder are like to run the red light without stopping at the stop line. This shows that by riding on the shoulder or center of the lane, a motorcyclist may have to opportunity to be away from the main traffic and maintaining a distance further away from the crossing vehicles when they run the red light. The only difference is that, a shoulder on a 3-leg SI provides the RLR-MC more space to cross without stopping.

The number of vehicles waiting on the major or minor road influences the red light occurrence among motorcyclists. Our analysis shows that as the number of vehicles waiting increases either on the major or minor road, the number of RLR-MC occurrence decreases significantly. Between the 2 types of SI, we can generalize the rate of RLR-MC would significantly drop when there are 5 – 10 vehicles waiting on either the major or minor road. This shows that the RLR-MC are cautious on the presence of vehicle regardless of the road type, and would stop for the red light as the number of vehicle waiting increases.

The limitation of this study was more towards the site selection process, where we had to undergo painstaking process of visiting each screened location in order to verify its conductivity of data and location suitability. This process has taken a long time to accomplished and has consumed much of our resources. We hope that in the future that all information pertaining the intersection design, crash information and traffic volume in Malaysia to be collected and stored in a database so that it can be useful for future research such as this.

6. Conclusion and Recommendations

This study intends to investigate factors that are associated with motorcyclist red light running behavior at signalized intersection and develops countermeasures in order to curb the risky behavior. Despite the low rate of RLR-MC reported in this study, their aggressive approach and risk-taking crossing should not be taken easy. Thus, we are recommending some measures in order to curb the behaviour:

- Construct traffic island (i.e. protective left turn channelizing island) in order to restrict illegal movement and also discourage RLR-MC crossing the SI without stopping.
- Replace the SI signal type from pretimed to actuated signal time phase. This is to make the SI signal phase less predictable thus discourage RLR-MC.
- Building traffic light pole instead of gantry is encouraged on location where there is a high volume of motorcycle. This is to restrict their view of the traffic light ahead and discourage them from approaching the SI aggressively (via. Weaving or lane splitting)

7. References

[1] WHO 2013 *Global status report on road safety 2013: supporting a decade of action* (Geneva: World Health Organization)

[2] Haworth N 2012 Powered two wheelers in a changing world, challenges and opportunities *Accident Analysis & Prevention* **44** (1) 12-18

[3] Senbil M, Zhang J and Fujiwara A 2007 Motorization in Asia - 14 Countries and Three Metropolitan Area *IATSS Research* **31**(1) 46 - 58

[4] Abdul Manan M M, Jonsson M T and Várhelyi A 2013 Development of a safety performance function for motorcycle accident fatalities on Malaysian primary roads *Safety Science* **60** 13-20

[5] Haque M M, Chin H C and Debnath A K 2012 An investigation on multi-vehicle motorcycle crashes using log-linear models *Safety Science* **50**(2) 352-62

[6] ESCAP 2009 *Road Safety, Economic and Social Commission for Asia and the Pacific* (New York: ESCAP)

[7] Abdul Manan M M and Várhelyi A 2012 Motorcycle fatalities in Malaysia *IATSS Research* **36** 30-39.

[8] Kulanthayan S, Phang W K and Hayati K S 2007 Traffic Light Violations among motorists in Malaysia *IATSS Research* **31** (2)

[9] Law T H, Xu L and Radin Umar R S 2003 Factors influencing red light runners among motorcyclist in Malaysia *Journal of the Eastern Asia Society for Transportation Studies* **5**

[10] Hawa M J, Akmalia S, and Sharifah Allyana S M R 2014 *The Effectiveness of Automated Enforcement System in Reducing Red Light Running Violations in Malaysia - Pilot Locations* (Kuala Lumpur: MIROS)

[11] Abdul Manan M M and Várhelyi A 2015 Exploration of motorcyclists' behavior at access points of a Malaysian primary road – A qualitative observation study *Safety Science* **74** 172-83

[12] Haque M M, Chin H C and Lim B C 2010 Effects of Impulsive Sensation Seeking, Aggression and Risk-Taking Behaviors on the Vulnerability of Motorcyclists *Asian Transport Studies* **1** (2) 165 - 80

[13] Haque M M, Chin H C and Huang H 2009 Modeling fault among motorcyclists involved in crashes *Accident Analysis & Prevention* **41**(2) 327-35

[14] Clarke D D, Ward P and Truman W 2005 Voluntary risk taking and skill deficits in young driver accidents in the UK *Accident Analysis & Prevention* **37** (3) 523-29

[15] de Rome L, Ivers R, Fitzharris M, Du W, Haworth N, Heritier S and Richardson D 2011 Motorcycle protective clothing: Protection from injury or just the weather? *Accident Analysis & Prevention* **43** (6) 1893-900

[16] Radin Umar RS 2005 The Value of Daytime Running Headlights Initiavites on Motorcycles Crashes in Malaysia *Transport and Communication Bulletin for Asia and the Pacific* **74** 17-31.

[17] ITE 1999 *Traffic Engineering Handbook: Fifth Edition* (Washington: Institute of Transportation Engineers)

[18] JKR 2002 *Road Safety Audit: Guidelines for the Safety Audit of Roads and Road Projects in Malaysia* (Kuala Lumpur: Public Works Department Malaysia (JKR))

[19] Elvik R, Høye A, Vaa T and Sørensen M 2009 *The Handbook of Road Safety Measures 2nd Edition* (Bingley, United Kingdom: Emerald Group Publishing Limited)

[20] Lawson SD 1991 *Red-light running: accidents and surveillance cameras* (U.K: The Automotive Association Foundation for Road Safety Research)

[21] Retting RA, Ulmer RG, and Williams AF 1999 Prevalence and characteristics of red light running crashes in the United States *Accident Analysis & Prevention* **31** 687-94

[22] Martinez KL and Porter BE 2006 Characterizing red light runners following implementation of a photo enforcement program *Accident Analysis & Prevention* **38** (5) 862-70

[23] Porter BE and England KJ 2000 Predicting red-light running behavior: a traffic safety study in three urban settings *Journal of Safety Research* **31** (1) 1-8
[24] Green FK 2003 *Red light running* (Vermont South: ARRB Transport Research)
[25] Retting RA, Williams AF, Farmer CM and Feldman AF 1999 Evaluation of red light camera enforcement in Oxnard, California *Accident Analysis & Prevention* **31** 169-74
[26] Bonneson J and Zimmerman K 2004 *Red-light-running handbook: An Engineer's guide to reducing red-light-related crashes* (Texas: Texas Department of Transportation)
[27] Romano E, Tippetts S and Voas R 2005 Fatal red light crashes: the role of race and ethnicity *Accident Analysis & Prevention* **37** (3) 453-60
[28] ITE 2003 *Making Intersection Safer: A toolbox of engineering countermeasures to reduce red-light running* (Washington, USA: Institute of Transportation Engineers)
[29] Retting RA and Williams AF 1996 Characteristics of Red Light Violators: Results of a Field Investigation *Journal of Safety Research* **27** (1) 9-15
[30] Elmitiny N, Yan NX, Radwan E, Russo C, and Nashar D 2010 Classification analysis of driver's stop/go decision and red-light running violation *Accident Analysis & Prevention* **42** 101-11
[31] Che Puan O and Ismail CR 2010 Dilemma zone conflicts at isolated intersections controlled with fixed–time and vehicle actuated traffic signal systems *International Journal of Civil & Environmental Engineering* **10** (3)
[32] Greenacre M and Blasius J 1994 *Correspondence Analysis in the Social Sciences* (Elsevier)
[33] Kalayci N and Basaran MA 2014 A Combined Approach Using Multiple Correspondence Analysis and Log-linear Models for Student Perception in Quality in Higher Education, *Procedia Economics and Finance* **17** 55-62
[34] Aitchison J and Greenacre M 2002 Biplots of compositional data *Applied Statistics* **51** 375-92
[35] Shinar D 1998 Aggressive driving: the contribution of the drivers and the situation *Transportation Research Part F: Traffic Psychology and Behaviour* **1** 137-60

Music and al-Quran sounds effects on driver behaviour due to road safety purposes

J Md Diah[1,3], E A Azmi[1,2], N A Yahya[3] and A R Zulkifli[2]

[1] Head, Malaysia Institute of Transport (MiTRANS), UiTM Shah Alam, Malaysia
[2] Lecturer, Faculty of Civil Engineering, Universiti Teknologi MARA, Shah Alam, Malaysia
[3] Senior Lecturer, Faculty of Civil Engineering, Universiti Teknologi MARA Shah Alam, Malaysia

E-mail: jezan@salam.uitm.edu.my

Abstract. The road safety issues had cover many issues such as speeding, mobile phone, motorcyclist and other issues that may related with safety on road. The road users had differently behaviour while driving their vehicles on the road and may have distractions that can cause an accident, such as using hand phone, hearing the music, eat, drink and etc. This paper describes one approach to explore the driver behaviour based on listening music and Al Quran recitation. In traffic engineering, the approach of evaluating driver behaviour is based on traffic parameters that are normally considered and measured such as by vehicle type, speed, gap and others. The method use in this paper is using vehicle with V-Box and Race Logic technique for data collection. The data analysis are focused on speed trend and patterns indicate the music and Al Quran recitation that effects on driver behaviour. Five types of the categorization during driver drive the vehicles (e.g. car) were presented and discussed. The finding shows that this study enables the relationship between music and Al Quran recitation due to speed parameter and it show how significance that driver behaviour reaction (e.g. aggressiveness) on them especially due to road safety purposes.

1. Introduction

In traffic engineering, the approach of evaluating driver behaviour is based on traffic parameters that are normally considered and measured are vehicle type, speed, gap and others. To relate the relationship between traffic engineering and music can be as a complicated task. [1] showed that during driving, most drivers like to hear music while they drive. [4] had researched on the effect of music types on driver behaviour based on happy, sad, neutral music and no-music cases. Through this study, results showed that happy music distracted drivers the most as their mean speed unexpectedly decreased and their lateral control deteriorated. Sad music influenced drivers in a different way as they drove slowly and kept their vehicle in its lane. While [2] had made study on driver behaviour only based on No Song – no music, Balada – slow music, Etnik & Creative – traditional music and Pop Rock – heavy music. Other researcher, [3] is based on brain signal analysis using different type of music in order to improve certain functions of human body physiologically and psychologically.

While [5] had develop the heart rate detection of individual autonomic response on music and Quran recitation. This paper generally describes an approach of evaluating driver behaviour based on music and Al Quran recitation effects. The objective of this study is to develop the effect of music and Al Quran recitation that effect on driver's driving behaviour based on speed trends. Five types of music categorization (No Song – no music, Balada – slow music, Etnik & Creative – traditional music, Pop Rock – heavy music and Al Quran recitation) taken while drivers were driving their vehicles (e.g. car) are presented and discussed. The method use in this paper is using vehicle with V-Box and Race Logic technique for data collection. This scope of this study enables the relationship of without music, with music, Al Quran recitation and speed parameters that indicate the significance of driver behaviour regarding with the one of the factor of road safety purposes.

2. Methodology

Figure 1 shows the process and sequence of data collection technique employed in this study. A suitable site was identified that suits the criteria specified in the study. The site selection is based on daytime under normal traffic conditions where the road type was four-lane divider. For this study, only one driver had been selected, a male, aged between 20 to 25 years old who uses a car as the vehicle. The driver needs to do four times with different sound categories, which are driving with No Song, Balada, Ethnic & Creative, Pop Rock and Al Quran recitation categories. Five (5) types of music categorization taken while driving the vehicle (e.g. car) was tested along three (3) km of the same site. The equipment that were used in the study were V-Box with its accessories, computer for data recording and retrieval, and Race Logic software. This paper had a limitation of test procedure in determining the result and analysis of music effect on driver driving behaviour based on speed trends.

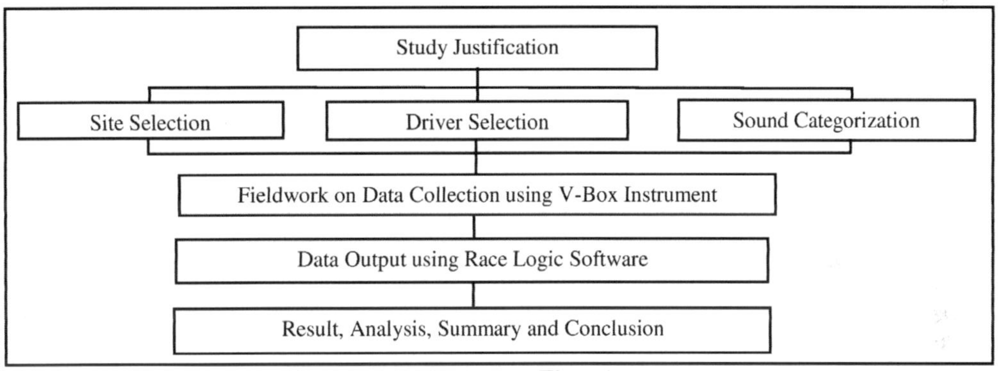

Figure 1. Methodology Flowchart

2.1 Fieldwork on Data Collection using V-Box Instrument

The V-Box components are V-Box 'ANT' socket, central controller (cables included), battery charger and V-Box GPS. Figure 2 shows the set-up and calibration of V-Box during fieldwork can be refer to [6].

Fundamental to any traffic engineering study is to ensure accurate and adequate traffic information (data) being collected and analyzed according to the study objectives. In this study, the data collected was running speed by using V-Box instrument. Normally, this instrument is put at the back seat. For the first time, it was used to test whether the result appears while driving the car. After all were tested to satisfaction, the driver needed to do four times with different music categories which are driving with No Song, Balada, Ethnic & Creative, Pop Rock categories and Al Quran recitation. It should be noted that, before the process in identifying the data collection, the instrument is the most important to be prepared before use. The instrument can be used by checking it the day before field data collection is needed (e.g. whether the battery is full charge if the instrument using power supply).

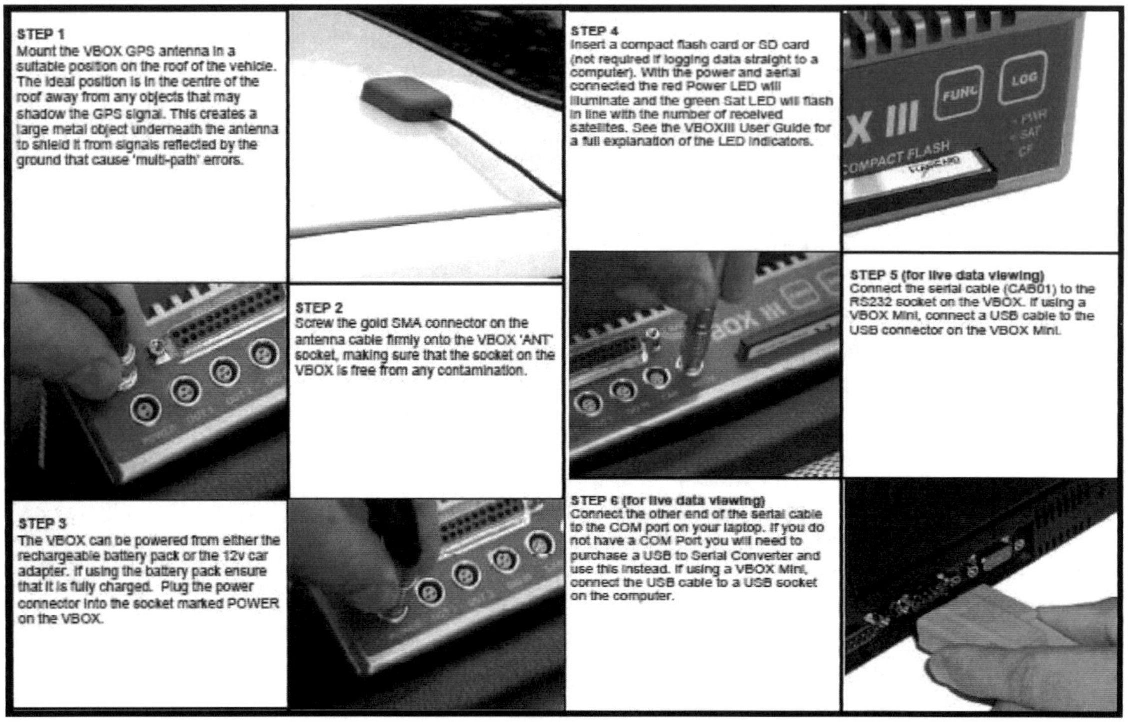

Figure 2. The set-up of V-Box Instrument

2.2 Data Output using Race Logic Software

Vehicle movement (running speed data) was stored directly in the laptop/computer. Data stored in the laptop/computer was retrieved using race logic software (see Figure 4). The data will then be analysed in the laboratory/office (see Figure 5). The default format which comes from Race Logic is in *.VBO format. This format can be converted into *.XLS format for excel user friendly to analyse the result due to speed and etc. From Figure 5, there are four outputs: No. 1 shows the parameters profiles (e.g. speed of vehicle one distance travelled). No. 2 shows the exact location where the parameters were highlighted. No. 3 shows which data that the user can select for generate the results. No. 4 run and generates the output results based on data selected in No.3. In this study, the speed profiles were categorized into 4 types: No Song Category, Balada Category, Etnik & Creative Category, Pop Rock and Al Quran recitation Category. Figure 3 shows an example of output results from the used of V-Box Instrument (e.g. No Song category).

Report Generator - JezanInterface (modified)

Speed(km/h)	Speed(km/h)	Time(s)	Dist(m)	Graph run	Description
36.05	36.05	254.39	2995.00		Distance Line
35.00	35.00	254.90	3000.00		Distance Line
35.08	35.08	255.42	3005.00		Distance Line
35.17	35.17	255.94	3010.00		Distance Line
33.34	33.34	256.48	3015.00		Distance Line
30.86	30.86	257.04	3020.00		Distance Line
28.67	28.67	257.64	3025.00		Distance Line
26.55	26.55	258.31	3030.00		Distance Line

Figure 3. Output Results (e.g. No Song category)

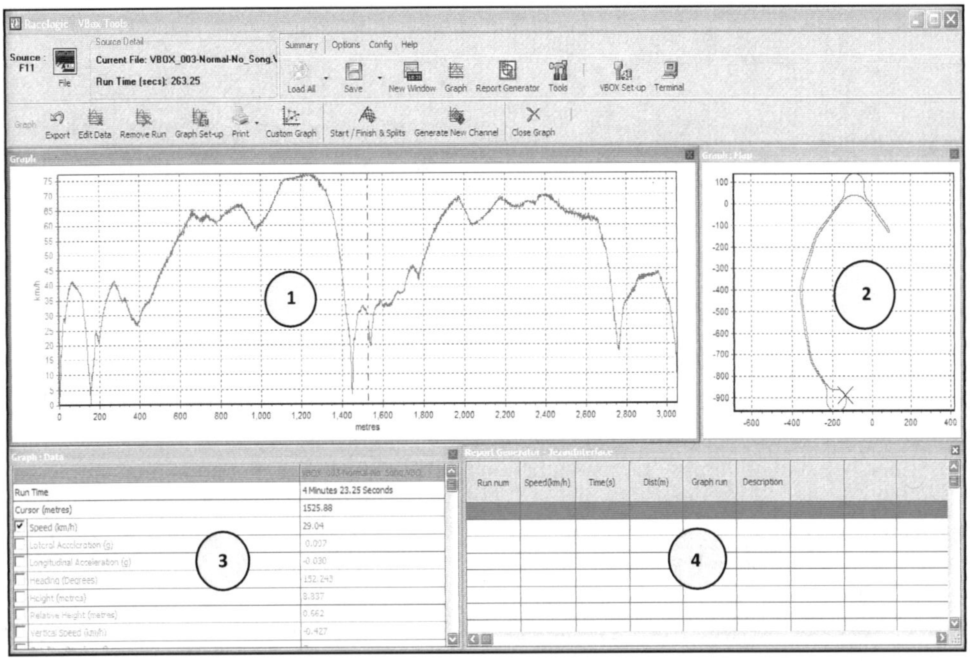

Figure 4. The traffic data result from Race Logic program

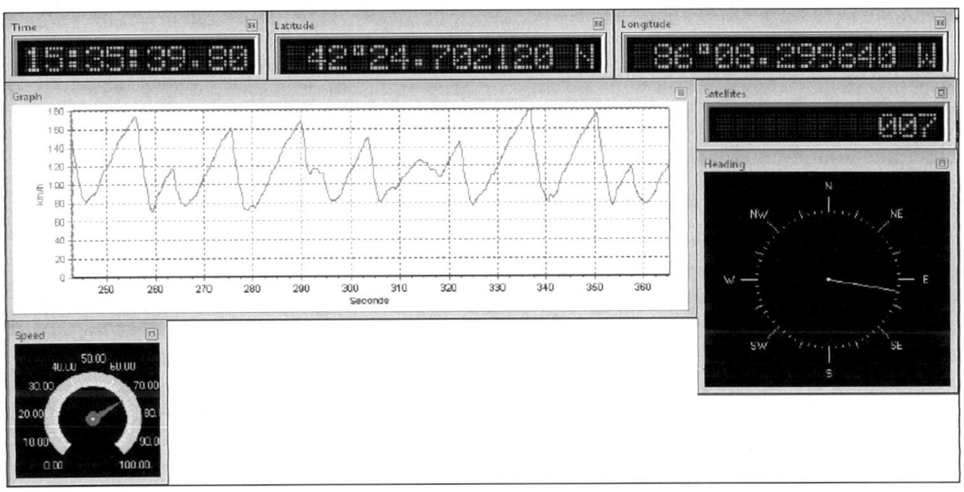

Figure 5. The traffic data retrieve from Race Logic program

3. Results and discussions
The following section show the relationship between running speeds related to music.

3.1 Relationship of Music and Driver Behaviour based on Speed Parameter
The traffic speed data was retrieved using Race Logic program, the format in (*.VBO) format. The traffic speeds were detected by output of V-Box instrument. Table 1 shows the statistics of skewness and kurtosis for all parameters and the values are small which means that the data is approximately normal. Summary of the Descriptive Statistics performed using MINITAB Software from the Master Database is as shown in Table 1. From Table 1, the values for skewness and kurtosis are relatively small for all variables which confirmed that the data is approximately normal.

Table 1. Summary of the overall descriptive statistics

Variable (Category)	Speed Sample	Mean	Median	Standard Deviation	Minimum	Maximum	Skewness	Kurtosis
No Song	579	52.9	60.5	16.259	8.0	76.9	-0.52	-0.89
Balada	579	49.9	53.7	14.483	1.7	71.5	-0.62	-0.48
Etnik & Creative	579	52.0	56.6	14.503	6.2	71.9	-0.64	-0.45
Pop Rock	579	58.0	62.3	16.470	4.6	81.2	-0.54	-0.60
Al Quran Sound	579	27.6	28.9	15.129	2.5	54.1	-0.16	-0.66

Figure 6a – 6e shows the histogram and cumulative frequency distribution curve of observed vehicles speed results based on different types of music in Race Logic program. It shows that a similar pattern on driver behaviours while driving and hearing different types of music with different speed magnitudes. Comparing the mean speed obtained from the cumulative frequency distribution curve, it was observed that the categories of mean speed in No Song = 52.9 km/hr, Balada = 49.9 km/hr, Etnik and Creative = 52.0 km/hr, Pop Rock = 58.0 km/hr and Al Quran recitation = 27.6 km/hr. Thus, it can be deduced that the driver may behave aggressive in Pop Rock categories, while more relaxed or comfortable driving can be found in Al Quran recitation categories.

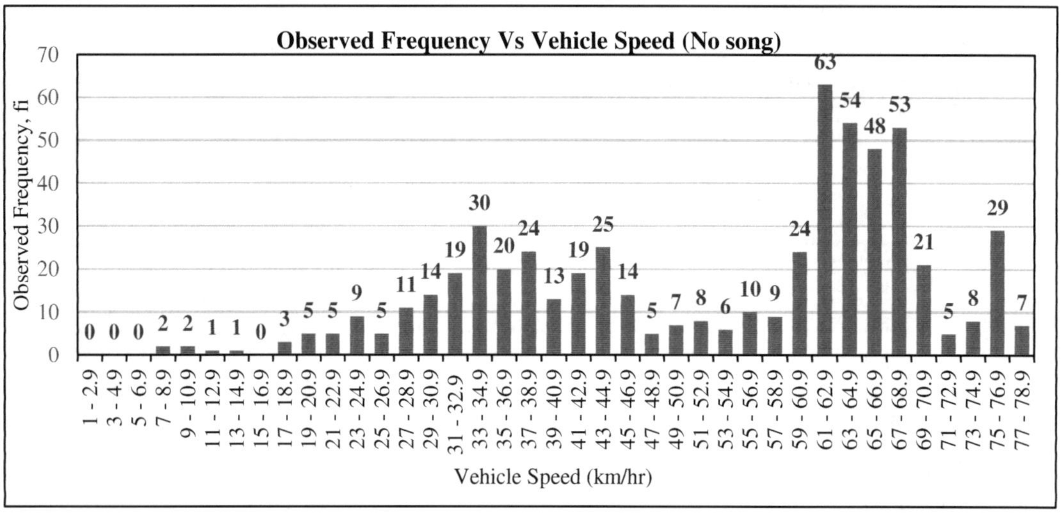

Figure 6a. The histogram of traffic speed data for No Song category

IOP Conf. Series: Materials Science and Engineering **512** (2019) 012020 doi:10.1088/1757-899X/512/1/012020

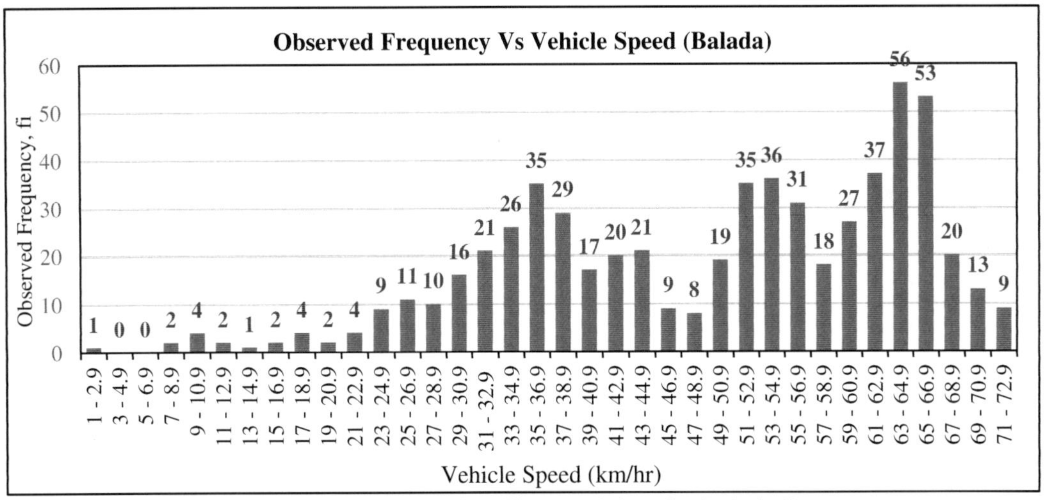

Figure 6b. The histogram of traffic speed data for Balada Song category

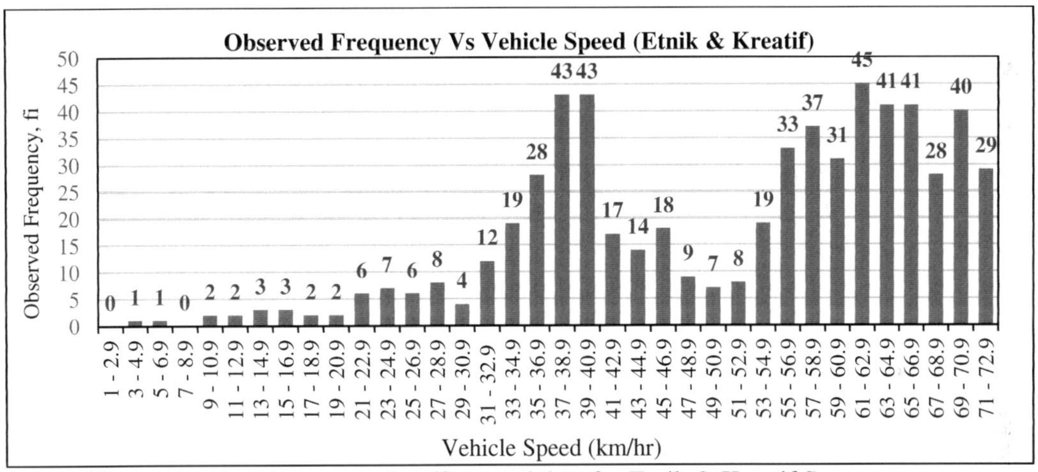

Figure 6c. The histogram of traffic speed data for Etnik & Kreatif Song category

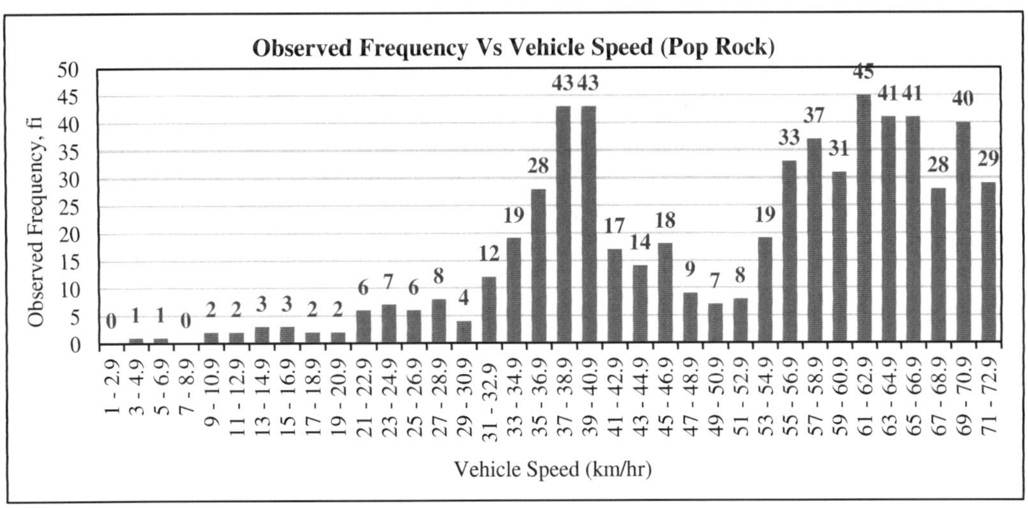

Figure 6d. The histogram of traffic speed data for Pop Rock Song category

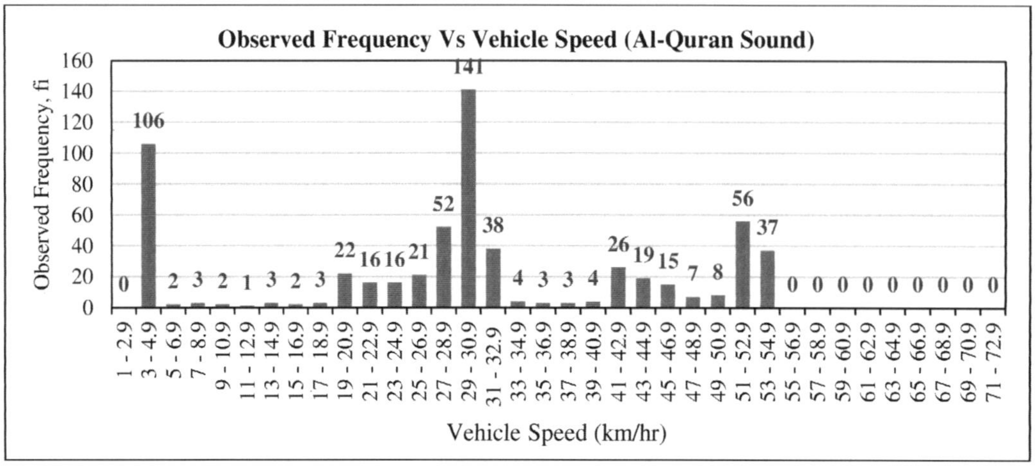

Figure 6e. The histogram of traffic speed data for Al-Quran Sound category

The results are summarized in Figure 7. The results show that the relationship of music and driver behaviour can be predicted through speed parameters. Based on road safety issues, this study can give good impact to driver behaviour based on their aggressiveness on road in normal traffic condition.

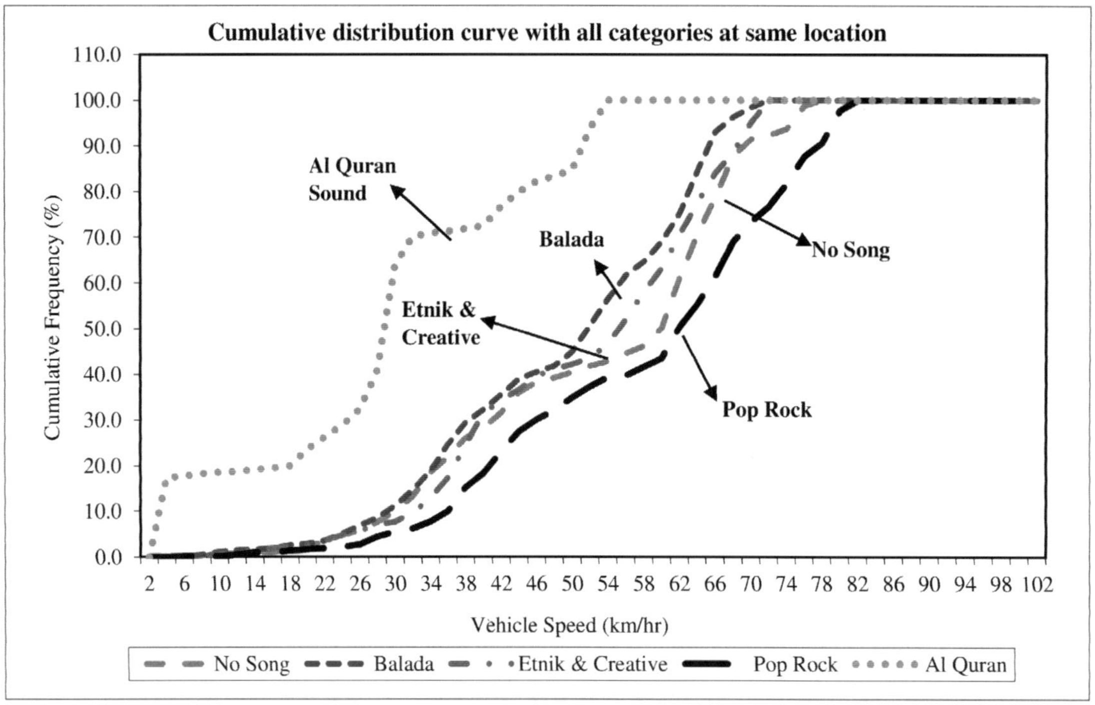

Figure 7. Cumulative distribution curve with all categories at same location

4. Conclusions

This paper described the technique of analysing and evaluating the effect of music on driver behaviour using V-Box instrument and Race Logic software based on speed data for traffic movement along a 3 km road length taken as case study. The study focused on the patterns and variations of vehicle running speed when a driver was subjected to different types of music (i.e. magnitude of the music). The instrumentation used was able to relate the vehicle speed which the driver had full control and the music played/listened during driving. Similar speed trends shown for all the five (5) categories, with respect to location but of different magnitude, with pop rock showing higher values possibly indicating that the driver acted and reacted to being affected by the music.

There is a lot of potential for future research in this area related to driver alertness and road safety such as different gender and age type, road category and last but not least long-distance travel. In this study, the approach of evaluating driver behaviour was based on vehicle running speed data. Sample of data were presented and discussed where the observed speed patterns and cumulative frequency of vehicles speed was noted. However, the main advantages of this methodology of data analysis using v-Box can be upgraded in additional instrument such as medical instrument (blood pressure devices or etc.). Hence, it can give more relationship of music affects in measuring and identifying the driver behaviour alertness. Next research will be in developing a model of driver behaviour with such parameters that affected the characteristic of the driver. Finally, the finding shows that this study enables the relationship between music and Al Quran recitation due to speed parameter and it show how significance that driver behaviour reaction (e.g. aggressiveness) on them especially due to road safety purposes in normal traffic condition.

5. References

[1] Dibben N and Williamson V J 2007 An exploratory survey of in-vehicle music listening. *Psychology of Music* 1–19
[2] Md Diah J, Abdul Rahman M Y and Adnan M A 2009 The Study of Music Effects on Driver Behaviour *Conference Proceeding of Faculty Music* UiTM
[3] Mohd Nasir S A and Wan Mahmud W M H 2015 Brain signal analysis using different types of music *Int. J. of Integrated Engineering* **7** (3) 31-36
[4] Pêcher C, Lemercier C and Cellier J M 2009 Emotions drive attention: Effects on driver's behaviour *Safety Science* **47** 1254–59
[5] Rezal M, Mengko T L R and Jofizal J 2009 The development of heart rate variability analysis software for detection of individual autonomic response on music and Quran recitation *WC 2009 IFMBE Proceedings 25*
[6] Racelogic Limited 2009 *V-Box Tools Software Manual* (United Kingdom: Racelogic Limited)

Acknowledgement

The authors gratefully acknowledge the financial supports by Universiti Teknologi MARA (UiTM) under Exploration Research Grant Scheme (ERGS) (600-IRMI/ERGS (36/2012)). Authors would also like to thank especially to Mr. Mohd Fahkery Hassan who directly involved in this research.

Characteristics of heavy goods vehicles (HGV) accidents in Malaysia

R Hamidun, A P Wah Hoong, A Roslan, A Shabadin and H Jamil

Malaysian Institute of Road Safety Research, 43000 Kajang, Selangor, Malaysia

E-mail: rizati@miros.gov.my

Abstract. An increasing trend of freight road traffic in Malaysia resulted in more than a thousand fatal accidents involving heavy goods vehicles (HGV) every year. Of these fatal accidents involving HGV, more than 80% resulted in deaths for other road users. This situation creating a significant need to better understand the accident characteristics of this vehicle group. Utilizing five years of historical accidents data from the year 2011 to 2015, this study attempt to examine the factors associated with HGV accident resulted in fatalities. Results from the binary logistic regression indicated that the smaller vehicles have significantly higher odds of resulting fatal outcome. Comparing other vehicles with the bus, fatality odds ratio for the car, motorcycle and cycle are 1.929, 2.423 and 3.626 respectively. The likelihood of being fatal in an accident involving HGV is significantly higher on the straight road as compared to the T/Y junction or others (interchange & roundabout). Accidents involving HGV on expressway were also found to have higher odds compared to other road hierarchies. The result of this study can aid safety planners to plan safety management for HGV transport.

1. Introduction

Heavy goods vehicles (HGVs) are referred to large vehicles that used to transport goods by road. In Malaysia, the HGV can be classified into three main class, including trailer (articulated lorries), rigid lorry (2 or more axles lorry with over than 2.5 ton's permissible gross weight) and small lorry (2 axles small lorry or pick-up with less than 2.5 ton's permissible gross weight). The Malaysia fleet of HGV is composed of almost million units (data in 2016) cumulating a total of 200 kilometres of travelled distance per day as well as an Average Annual VKT (AAKT) of approximately 70,000 km [1]. Travelled distance by HGV is expecting to be grow in line with the growth of the ecommerce market in Malaysia.

The evolution of HGV accidents in Malaysia has shown a continuous up-and-down trend. Figure 1 below show the five years' trend of road accidents involving HGV in Malaysia. Despite of a small numbers of HGV in traffic volume, accident involving HGV results in more than 1000 deaths each year in Malaysia. Out of that numbers, involvement of HGV in accident resulting more than 80% of second vehicle fatalities. This shows that HGV accident has a substantial impact on the safety of other road users.

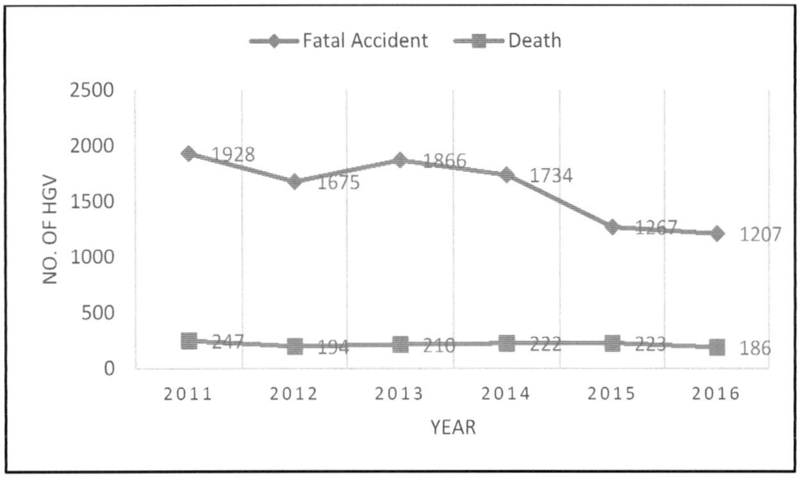

Figure 1. Number of fatal road accidents and death involving HGV [5]

In Europe, accident involving HGV resulted 4,500 deaths or about 18% of road accident fatalities in year 2013 [2]. Road accident fatalities involved HGV is also become a concern in Sweden since it contribute half of car occupant deaths [3]. In related to fatality risk, study conducted by Buddhavarapu, Banerjee, & Prozzi (2013) proved that the probability of fatal accident is higher at location with a higher percentage of truck traffic.

The large dimensions of these HGV, and their big masses contribute enormously to the severity of their impact with other vehicles involved in the road accidents [2]. Since HGV often travel at lower speeds than other vehicles, the differential of speed may potentially lead to rear-end accidents [6].

Study focus on the risk of motorcyclist by turning lorries conducted by [7] concluded that lorry drivers perceived that they drive more cautiously when a motorcyclist was riding nearby the lorry. However, in terms of the location of blind spots, most drivers perceived that the rear is the blind spot area of a lorry without knowing that the front and both sides of a lorry are also blind spots. Improve design of rear view mirror in HGV could reduce the blind spot which in turns leads to reduce the possibility of the road accidents [8].

Due to large dimension, HGV have operating limitations such as large blind spots, long stopping distances, and limited manoeuvrability that make it essential for other vehicles to put extra focus on safety. HGV involvement in fatal accidents creating a significant need to better understand the impact of this vehicle group to other road user. Thus, this study attempt to examine the characteristic HGV accidents and its associated factors resulted in fatalities to other road users.

2. Method

Road accident data obtained from Royale Malaysia Police (RMP) extracted from MROADS (MIROS Road Accidents Database System) were used to examine the characteristics of HGV accidents in Malaysia. The five years' accident data related to HGV for the years 2011-2015 were analysed quantitatively using SPSS. HGV accident data for the five years is then filtered to select a subsample of 7026 cases involving HGV versus other vehicles, which classified into four main categories as shown in Figure 2. These four categories of vehicles namely bus, car/van, motorcycle and bicycle. Accident cases involving lorry versus lorry and single lorry accidents were also excluded for analysis.

The effect of the variables on the likelihood of fatality outcome were explored using the binary logistics model. Dichotomous level of injury severity (fatal vs non-fatal) for accidents involving HGV in Malaysia is used as the dependent variable in the model analysis. The odds ratio (OR) were used as the model outcome to indicate the probability that the categorical variable to a predictive factor. The greater value of OR (more than 1) indicate that the variable is related with higher odds of the outcome, and vice versa.

Variables with large amount of missing data such as type of traffic system, type of light condition, type of location and type of collision were dropped for model analysis. The final model considered 7026 accident cases involving HGV without any missing values for all independent variables considered for analysis.

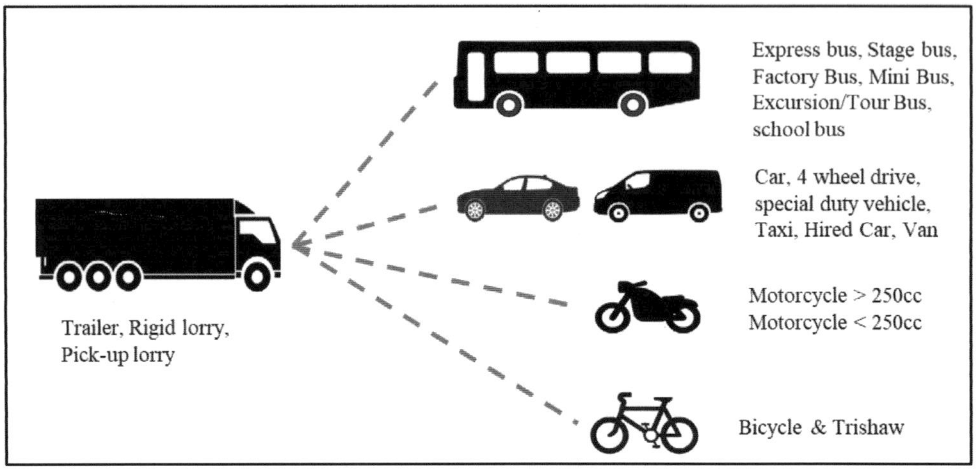

Figure 2. HGV Accident and other vehicles

3. Results and Discussion

3.1. Accident involving HGV with other vehicles

Out of 7026 accident analysed, there were 3787 fatal and 3239 non-fatal cases, which include serious and slight injury of the driver/s or passenger/s of the vehicles. Table 1 shows that majority of HGV accidents involved motorcyclists, which contribute for 2311 motorcycle fatalities. Involvement of HGV also causing about 1372 fatalities in car accidents, 56 fatalities in cycle accidents and 48 fatalities in bus accidents. Motorcycle fatalities in accidents involving HGV is the highest in Malaysia which constitute of 61% fatalities. As comparison to other countries such Europe, involvement of heavy vehicles resulted 47% of car occupants killed in road accidents that involved HGV and buses [2].

Table 1. HGV accident involving other vehicles by injury severity

Vehicle types	Fatal cases (n =3787)		Non-fatal cases (n=3239)	
	n	%	n	%
Bus	48	1.3	68	2.1
Car	1372	36.2	1275	39.4
Motor	2311	61.0	1864	57.5
Cycle	56	1.5	32	1.0
Total	3787	100.0	3239	100.0

The distribution of accident for three different types of HGV (trailer, rigid, small) involving other types of vehicles is summaries in Table 2. Based on data obtained, the most frequent type of HGV collide with other vehicles is the rigid lorry, which having permissible gross weight of over than 2.5 ton's. From 5058 of the rigid lorry accident, 3047 cases involved motorcycle, 1852 cases involved car/van, 77 cases involved bus and the rest 62 cases involved bicycle.

Table 2. Type of lorry and other vehicles cross tabulation

HGV types	Other vehicles				Total
	Bus	Car	Motor	Cycle	
Trailer	36	640	776	18	1470
Rigid	77	1852	3047	62	5038
Small	3	155	352	8	518
	116	2647	4175	88	7026

The HGV accidents by type of collision with other vehicles were summarised in Table 3. Cross tabulation of collision type by vehicle types show that the high percentage of HGV accident experienced rear end collision with buses and motorcyclists. Head on collision are the most common type of collision for HGV accident involving car/van. While angular and side swipe collision contribute the majority for HGV accident involving motorcycles.

Table 3. Collision type and other vehicles cross tabulation

	Other vehicles (%)				Total
Collision type	Bus	Car/Van	Motorcycle	Bicycle	
Head on	0.2	13.2	8.4	0.1	21.9
Rear end	0.9	11.6	16.1	0.2	28.8
Right angle	0.0	2.0	3.7	0.1	5.9
Angular	0.2	7.7	17.2	0.4	25.6
Side swipe	0.1	2.1	10.4	0.4	13.0
Others	0.0	0.3	0.5	0.0	0.8
Hitting	0.1	0.7	0.7	0.0	1.5
Out of control	0.0	1.1	1.4	0.0	2.5
Total	1.7	38.8	58.3	1.2	100

Table 4. Road type and other vehicles cross tabulation

Road Type	Other vehicles				Total
	Bus	Car/Van	Motorcycle	Bicycle	
Expressway	56	464	430	0	950
Federal	43	1325	1600	29	2997
State	13	622	1296	31	1962
Municipal	3	151	571	19	744
Other	1	85	278	9	373
Total	116	2647	4175	88	7026

For a period of 2011 to 2015, most of the HGV accident involving bus occurred at expressway with 56 cases (refer Table 4). For the HGV accident involving car, majority of cases occurred at federal road. A large amount of HGV accident involving motorcycles occurred at federal and state road. While for the HGV accident involving cyclist, about 31 cases occurred at state road and none at expressway. Overall, federal road had been recorded the highest number of HGV accident involving other vehicles.

The distribution of HGV accident involving other types of vehicles as shown in Table 5 allows identification of the types of vehicles most likely to be involved in the HGV accidents at different set of road geometry. Analysis indicate that 65.5% of the HGV accident occurred at straight road where majority of the accident involving motorcyclist and car. At bending road, HGV accident involving car contribute the highest percentage (8.2%) as compared to other types of vehicles. Apart from this, HGV accident involving motorcycles are frequently occurred at T/Y junction (9.9%).

Table 5. Road geometry and other vehicles cross tabulation

Road geometry	Other vehicles (%)				Total
	Bus	Car/Van	Motorcycle	Bicycle	
Straight	1.4	25.7	37.6	0.8	65.5
Bend	0.2	8.2	7.3	0.1	15.8
Cross junction	0.0	1.3	4.1	0.1	5.6
T/Y junction	0.0	2.4	9.9	0.2	12.6
Other	0.0	0.2	0.4	0.0	0.6
Total	1.7	37.7	59.4	1.3	100

Looking to the type of road and geometry, most of the HGV accident occurred on a straight section of the federal road (27.3%), followed by state road (16.8%) and expressway (12.5%). For other type of road geometry (bend and junctions), HGV accident is most likely to occurred on the federal road, see Table 6.

Table 6. Road geometry and other vehicles cross tabulation

Road geometry	Road type (%)					Total
	Expressway	Federal	State	Municipal	Other	
Straight	12.5	27.3	16.8	6.1	2.8	65.5
Bend	0.9	7.6	5.2	1.1	1.0	15.8
Cross junction	0.0	2.3	1.5	1.2	0.5	5.6
T/Y junction	0.2	5.3	4.3	1.9	0.9	12.6
Other	0.0	0.2	0.1	0.2	0.1	0.6
Total	13.5	42.7	27.9	10.6	5.3	100

3.2. Model result

Logistic regression analysis was used to quantify the fatality risk factors associated with HGV accident with other vehicles. The effect of variables that influence the likelihood of fatal accidents involving HGV is represented by the odds ratio. The fatality odds ratio from model analysis is summarised in Table 7.

The likelihood of being fatal in an accident involving HGV is significantly higher on the straight road as compared to the T/Y Junction or others (interchange & roundabout). In related to the road geometry, it was also found that bend section was a statically significant factor of HGV accident fatality at 10% when comparing to straight section (OR=1.134).

Table 7. Model summary

Attributes	Odds ratio	(95% CI)	Sig.
Road geometry			
Straight			
Bend	Reference		
Cross Junction	1.134	0.991-1.299	0.068**
T/Y Junction	0.908	0.735-1.123	0.374
Other	0.617	0.531-0.716	0.000*
	0.529	0.281-0.995	0.048*
Road types			
Expressway	Reference		
Federal	0.494	0.422-0.579	0.000*
State	0.486	0.411-0.575	0.000*
Municipal	0.800	0.649-0.987	0.038*
Other	0.468	0.364-0.602	0.000*
Vehicle types			
Bus	Reference		
Car	1.929	1.310-2.841	0.001*
Motor	2.423	1.646-3.566	0.001*
Cycle	3.626	2.024-6.498	0.000*

*Note: * statically significant at 5% level*
*** statically significant at 10% level*

Accident involving HGV on the expressway is more likely to result in fatalities compared to other road types including federal road, state road, municipal and others. This could be due to the higher speed operated by vehicles when travelling on the expressway, as compared to other types of road. HGV with heavy mass when travelling with higher speed would create big momentum. When collide with other vehicle, an energy from this big momentum transform into big impact that resulted in higher fatality odds.

The likelihood of fatality for other vehicles except for bus is significantly higher when involve in HGV accidents. Car had odds of fatality that were 1.929 (p-value < 0.05) times higher than bus. Motorcycle and cycles had more odds of fatality, OR = 2.423 (p-value < 0.05) and OR = 3.626 (p-value < 0.05) compared to bus respectively. This indicates that the risk of fatality is higher for the smaller vehicles involved in the HGV accidents. Sharing traffic system with HGV could be dangerous for other smaller vehicles due to greater collision impact [3]. In addition, involvement of heavy vehicles would increase 5.6 times fatality odds to pedestrians [9].

4. Conclusion

This study focus on the analysis of the HGV accident in Malaysia. Accident involving HGV would have resulted more than 80% fatalities to other vehicles. Of those vehicles involved in fatal cases, motorcycle is the most common vehicle become a victim in HGV accident. Majority of HGV accident involving motorcycles experience angular and rear end collision.

Based on the 5 years' accident data, association between selected variables and the odds of fatality for HGV accidents were analysed using logistics regression. Results from model analysis indicate that accident involving HGV is significantly higher on the expressway. As the highest hierarchy of road, vehicles traveling on the expressway are operating with higher speed. Heavier vehicles like HGV traveling in higher speed would having greater momentum to cause greater injury to other vehicles occupant. In addition, a collision of HGV with smaller vehicles would resulted in higher fatality odds.

5. References

[1] Malaysia Institute of Road Safety Research 2017 *Vehicle Kilometre Travelled (VKT) Study For Goods Vehicle: Pilot Study At PUSPAKOM* (Kuala Lumpur: MIROS)

[2] Evgenikos P, Yannis G, Folla K, Bauer R, MacHata K and Brandstaetter C 2016 Characteristics and Causes of Heavy Goods Vehicles and Buses Accidents in Europe *Transp. Res. Procedia* **14** 2158–67

[3] Björnstig U, Björnstig J and Eriksson A 2008 Passenger car collision fatalities - with special emphasis on collisions with heavy vehicles *Accid. Anal. Prev.* **40** 158–66

[4] Buddhavarapu P, Banerjee A and Prozzi JA 2013 Influence of pavement condition on horizontal curve safety *Accid. Anal. Prev.* **52** 9–18

[5] Polis Diraja Malaysia 2016 *Royal Malaysia Police Annual Report, 2016* (Kuala Lumpur, Malaysia: PDRM)

[6] Mohd Radzi MA 2016 *Rear-end Markers (REMs) Issues involving Trucks and Lorries in Malaysia* (Kuala Lumpur: MIROS)

[7] Musa M 2017 *Awareness and Prevalence of Vehicle Blind Spot Issues among Lorry Drivers Awareness and Prevalence of Vehicle Blind Spot Issues among Lorry Drivers* (Kuala Lumpur: MIROS)

[8] Pitchipoo P, Vincent D S, Rajini N and Rajakarunakaran S 2014 COPRAS decision model to optimize blind spot in heavy vehicles: A comparative perspective *Procedia Eng.* **97** 1049–59

[9] Hamidun R, Roslan A and Sarani R 2017 Exploring factors for pedestrian fatalities at junctions in Malaysia *Pertanika J. Soc. Sci. Humanity* **25** 1833–40

IOP Conf. Series: Materials Science and Engineering **512** (2019) 012022 doi:10.1088/1757-899X/512/1/012022

Motorcycle lane: how to judge if that is necessary

W H Alvin Poi[1], A Shabadin[1], H Jamil[1], A Roslan[1] and R Hamidun[2]

[1] Research Officer, Malaysian Institute of Road Safety Research, Kajang, Malaysia
[2] Director, Malaysian Institute of Road Safety Research, Kajang, Malaysia

E-mail: alvinpoi@miros.gov.my

Abstract. Malaysia has suffered great social and economic losses due to traffic injuries motorcyclists. Segregation of motorcyclists from other motorized traffic is one approach carried out in Malaysia. The non-exclusive motorcycle lane or NEMCL, is one facility provided along the major trunk roads to serve that purpose. However there are not many studies on the effectiveness of the facility thus far, especially in terms of providing justification to build the facility. This study attempted to examine the effectivess of NEMCL by estimating the numbers of killed and seriously injured (KSI) that could be saved with the provision of the facility. Crash statistics and several key road attributes were analysed for road segments with and without NEMCL using statistical regression. Negative binomial model developed in this study revealed that NEMCL is effective in reducing the crash risk of motorcyclists. Area type, proportion of motorcycles in the traffic and the presence of NEMCL were found to significantly influence the crash rates. The model was used to develop criteria to determine the cost effectiveness of providing NEMCL for existing roads and new road projects, based on the benefits and the costs invested to build and maintain the facility over a period of time.

1. Introduction

Disorderliness on the road due to limited space for motorcycling is one of major contributing factors to the rising numbers of casualties among motorcyclists over the years. Crash statistics from 2005 to 2015 released by the Royal Malaysian Police revealed that motorcyclists (including the pillion riders) and pedestrians account for 62% on average of the total road deaths - more than half of deaths occurred on rural roads [1]. The most common types of collision that cause deaths among motorcyclists are head-on, intersection and run-off. Youths aged 16 to 30 constitute half of the total number of deaths among motorcyclists. Many of those who died are young working adults, and perhaps the only breadwinner for the family. Their contribution to the nation socio-economic progress is cut short prematurely, costing the country in lost earnings. A model developed by the International Road Assessment Programme (iRAP) estimates that the cost of fatal and serious road traffic crashes could be as large as 5% of Gross Domestic Product (GDP) for a middle income country like Malaysia [2].

Due to the greater vulnerability of this group of road users in sustaining severe injuries in the event of a crash, the motorcycle lane was built along certain stretches on the federal roads and expressways in Malaysia. Compared to an exclusive lane where motorcyclists are segregated by physical measures, a non-exclusive lane uses only painted markers or solid line for segregation [3]. The first motorcycle lane (exclusive type) in the world was built along the Federal Highway Route 2 in Malaysia in the early seventies under the World Bank Project [4], after which more and more such lanes were constructed. A study showed that such facility on average reduced motorcycle crashes by 39% and fatality by 600% [5]

Non-exclusive motorcycle lane which is without physical separator (designed as close as possible to the available design guideline) were found to have reduced motorcycle crash risk by as much as 83% [6]. Up to date, there are approximately 135km of exclusive lanes and 110km of non-exclusive lanes along the expressways and major federal roads in the country [7].

However a major challenge faced by the authorities is how to provide – in an affordable and feasible manner – a road facility that is forgiving for motorcyclists and yet does not disrupt the safe use of other road users. Crash occurrences have been reported on motorcycle lane, indicating that there is room for improvement in relation to the design and maintenance of the facilities. The cost to build and maintain the facility is no small amount and sound justifications are required before huge allocation is invested by the Government. The guideline published by the Road Engineering Association of Malaysia (REAM) in 2007 has been regularly updated to include current best practices in design, as well as research findings on the effectiveness of motorcycle lanes.

The return of investment for any construction project is extremely crucial for a developing country like Malaysia. The provision of non-exclusive motorcycle lane on new major roads should be considered if the benefits exceed or equals that of the costs incurred as far as safety outcome is concerned. The benefits can be measured in terms of the number of lives saved which is converted into values of money, termed the Value of Statistical Life (VOSL). The ratio of VOSL to the total costs of providing the motorcycle lane can be used to judge if the project should be considered. The estimation of the benefits of the motorcycle lane is thus important but still lacking in Malaysia.

This paper aims to provide some evidences on the effectiveness of providing non-exclusive motorcycle lane using statistical regression approach as well as to suggest the use of criteria for deciding the provision of the facility. Hence the objectives of this study are to develop a motorcycle crash prediction model and to develop criteria for determining the return of investment of providing non-exclusive motorcycle lane (hereafter referred to as NEMCL). These criteria can supplement the use of the existing REAM guideline to justify if the safety outcome justifies the amount of investment required for a specific project.

2. Effectiveness of motorcycle lane

As mentioned above, two types of facilities are currently provided in Malaysia to segregate motorcyclists from other motorized traffic, namely the exclusive lane and the NEMCL. The exclusive lane can be found on both expressways and non-expressways while the non-exclusive lane is only built on non-expressways. Warrants for providing these facilities are included in Guidelines for Motorcycle Facilities published by REAM in 2011, criteria of which include crash history and traffic volume. Other considerations include the number of access points and junctions, as well as type of carriageway [3].

Despite these warrants, various designs of the lane are found across the country, prompting investigation into the effectiveness of each design. The effectiveness of the lane in reducing crash and injury risk among motorcyclists greatly depends on the usage, in addition to its characteristics. Besides, factor such as prevalence of misuse by other vehicles could also contribute to the measure of effectiveness of the facility, especially the non-exclusive type. Unfortunately research works in these areas is lacking in the country. A study by revealed that benefits of providing the exclusive lane far outweigh the costs of building and maintaining it, by about 3 to 5 times [4, 8]. This study compared the annual savings made from preventing crashes with the estimated costs of initial implementation and annual maintenance costs. The benefit-to-cost ratio (BCR) values, which were more than one clearly show that provision of the exclusive lane is a highly cost effective countermeasure in addressing motorcycle safety issues.

On the other hand, to the best knowledge of the authors, no published work which examines the effects of NEMCL on crash and injury risk has been found in the literature. As such, no conclusion can be made thus far on the potential of the facility in reducing motorcycle related crashes. In fact, crash records published by the Royal Malaysia Police for 2006 and 2007 showed that there was an increase in motorcycle fatalities on roads with the facility [6, 9]. Many factors should be considered in determining the effectiveness of such facility, particularly the usage rates among motorcyclists. It is bias to expect a reduction in motorcycle related crashes on road sections with such facility where the compliance rate is low. Certainly the crash risk of motorcyclists would not decrease if they still remain in close contact with other vehicles on the main lanes although they are given a chance to distance themselves. The effect of junctions on the usage rates of paved shoulder by motorcyclists was studied in an earlier work where continuous paved shoulder at junctions can reduce conflicts between motorcycles and other vehicles [10]. Besides increasing the usage rates, the treatment was also found to be able to reduce motorcycles' average and operating speeds in the vicinity of junctions. Though it can act as means for segregation, the law, except for emergency situations, does not permit the use of paved shoulder by motorcycles.

Many studies on motorcycle safety worldwide were carried out to identify contributing factors and several developed crash prediction models using crash data in Malaysia. These models revealed that the design of the road environment as well as the traffic speed contributed significantly to the crash rates [11, 12, 13, 14, 15]. However none of these studies covered the effect of NEMCL, which may probably due to the limited crash data occurring on road sections with the facility. Moreover the pure effectiveness of NEMCL in influencing the crash rates is difficult to be assessed due to other factors that exist within the road and traffic system. Many of these factors especially related to the riding behaviour are complicated to be studied and as such very often unobserved or purportedly omitted from the analysis. Nevertheless such regression techniques are still useful to provide indications as to how a road attribute influence the crash outcomes, as far as motorcycling safety is concerned.

One advantage of using regression modelling in road safety studies is the ability of deriving crash modification factors (CMFs). A CMF for a countermeasure indicates the proportion of the expected number of crashes after the implementation of the countermeasure at a specific site [16]. Depending upon the study from which the CMF is estimated, it can be developed and applied for all crash types and locations, as well as for specific site characteristics, crash types and severities. It is of course more beneficial if CMFs are developed for each site condition, each crash type or each crash severity as more accurate estimation can be performed in a cost-benefit analysis. CMF had been developed and used in road safety engineering for quite a number of countermeasures. All these CMFs were developed based on studies on the effectiveness of each countermeasure to reduce crashes. A CMF can be in the form of a numerical value, equation, graph, or a combination of these.

3. Data and methods

3.1. Road segmentation and crash data

In order to achieve the specific objectives of this study, sections along major federal roads with NEMCL were identified based on information obtained from the Public Works Department (PWD). Together with the availability of survey data collected from the iRAP Malaysia Pilot Study, the following road sections were selected for this study;

1. Federal Route 1 from Gurun to Sungai Petani.
2. Federal Route 5 from Pekan Nenas to Batu Pahat.

The location of motorcycle related crashes (fatal and serious injury) obtained from the Royal Malaysia Police for the selected road sections from 2009 to 2015 were compiled and plotted on a map. Using the survey data from the pilot study, which recorded attributes for every 100 m, the observed routes were further segmented based on characteristics shown in table 1. The number of crashes were then recorded for each road segment.

3.2. Regression modelling

It has become a standard practice to use statistical models to describe and predict the occurrence of road crashes based on traffic volumes and road geometrics. Traffic engineers have been using such approach to decide on the appropriate countermeasures to be implemented. However, majority of these models are developed by conventional linear regression techniques which contain inherent errors that limit their application and accuracy [17]. Due to the limitations, Poisson and Negative Binomial regressions were considered in this study.

An exponential function is a common formulation, since this function ensures that the expected number of crashes is a positive number. Thus, crash frequency may be written as a multiplicative function for which the value of the exponents can be estimated directly by measuring the explanatory variables on the logarithmic scale as in Equation 1 where β is the coefficient estimation and Xi are the independent variables, e.g. road traffic properties such as traffic volume, percentage of motorcycles or area type, etc.

$$E(Y) = exp\ [\Sigma\ \beta_i\ ln(X_i)] \tag{1}$$

Table 1. Road attributes used in segmentation.

Factors	Description	Source of data
Traffic volume	Average of 16 hours manual count conducted twice annually	Highway Planning Division, Ministry of Works
Traffic composition	The percentage of motorcycle in traffic stream	Highway Planning Division, Ministry of Works
Area type	Urban or rural	iRAP Malaysia Pilot Study survey data
Carriageway type	Single or dual	iRAP Malaysia Pilot Study survey data
Number of lane	One, two or more than two lanes in each travel direction	iRAP Malaysia Pilot Study survey data
Access point density	The number of access point per kilometre	iRAP Malaysia Pilot Study survey data
Presence of NEMCL	Yes or no	iRAP Malaysia Pilot Study survey data

3.3. Development of motorcycle lane provision criteria

The crash prediction model developed based on the crash history of motorcycle crashes recorded along the study sites was used to generate criteria for the provision of motorcycle lane. These criteria were developed on the principle of ensuring positive return of investment, whereby the savings in terms of the number of lives saved equals to or exceed the total cost of constructing and maintaining the facility over a period of time. The benefits of lives saved used were estimated in the form of money, derived from the value of statistical life (VOSL) of an average Malaysian. In road safety, the VOSL is widely used to estimate the burden of socio-economic losses to a country based on the gross domestic product per capita of the people in that particular country [18]. On average, the VOSL (a fatality) is estimated by the following equation:

$$VOSL = 70\ x\ GDP/capita \tag{2}$$

Thus, the savings in the cost of crashes is calculated by multiplying the VOSL with the expected number of fatalities saved after a treatment is implemented. In this study, the difference between the expected number of fatalities on a road section with and without motorcycle lane is taken as the number of fatalities saved due to the provision of motorcycle lane (conservatively assuming that a single KSI crash will involve a single fatality).

The difference between the expected number of fatality, D was calculated using the crash prediction model for each level of motorcycle percent as follow:

$$D = E(Y_0) - E(Y_1) \tag{3}$$

Where $E(Y_0)$ is the expected number of KSI crashes at sections without motorcycle lane and $E(Y_1)$ is the expected number of KSI crashes at section with motorcycle lane.

The total value of D as calculated using the VOSL will determine the associated total cost of constructing and maintaining the motorcycle lane for different benefit-to-cost ratios (BCRs). A BCR value of one means that the total values of lives saved equals that of the total cost of providing the motorcycle lane, which ought to be the minimum return of investment to be considered. The criteria when the benefits are expected to be two and three times the costs (BCR = 2 and BCR = 3) were also generated.

4. Results and discussions

4.1. Effectiveness of NEMCL

A total of 88 road segments were generated with a total length of 77km ranged between 0.1km and 8.0km. The breakdown in table 2 shows that the percentage distribution of all controlling road attributes was not fairly equal. This is because of the difficulty in identifying road segments with similar characteristics in the vicinity of NEMCL. However the distribution is sufficient enough for statistical modelling purposes except for the number of lane where segments with more than one lane constitute only less than 20%. As such this variable was not considered in the regression analysis. Table 3 shows the descriptive statistics of these variables.

Table 2. Breakdown of road attributes as percentage of number of road segments.

No.	Attributes	Breakdown	
1.	Traffic volume (AADT)	10,000 – 20,000	32%
		20,000 – 30,000	25%
		30,000 – 40,000	25%
		40,000 – 50,000	18%
2.	Motorcycle percentage (%)	Less than 20%	52%
		More than 20%	48%
3.	Area type	Rural	33%
		Urban	67%
4.	Number of lane (each direction)	One	83%
		More than one	17%
5.	Access point density (no. per km)	Less than five	50%
		More than five	50%
6.	Presence of NEMCL	Yes	60%
		No	40%

The variance of the crashes per km variable is far greater than its mean, and therefore Poisson model would not be the best choice. The negative binomial model instead was used to develop the crash prediction model. An analysis of correlations among the variables was performed to determine if high collinearity occurs among them. If collinearity occurs between two variables, one of them can totally predict the other and vice versa. In other words, a regression model with highly correlated variables (considered to be when the Pearson correlation value is more than 0.5) can only predict how well the entire set of independent variables predicts the dependent variable, but may not provide the true effect of each individual correlated variable.

The correlation analysis showed that the traffic volume and motorcycle percentage were highly correlated. Due to the importance of traffic volume in any crash prediction model, it was included in the analysis initially but was found to be not significant. Table 4 shows the regression coefficients for the best fit model with three predictor variables. An interaction term between motorcycle percent and presence of NEMCL was introduced to examine the effects of motorcycle percentage on crash outcome with and without NEMCL. Note that the dispersion parameter was found to be significantly different from zero which confirms that the negative binomial distribution offered better prediction than the Poisson model.

Table 3. Descriptive statistics of variables included in regression analysis.

Continuous variables	Minimum	Maximum	Mean	Std. Dev.
Number of total KSI crashes/km	0	8	2.0	2.3
Traffic volume (AADT) (*aadt*)	11133	43306	27097	10489
Motorcycle percentage (%) (*mcpct*)	12	28	20	6
Access point density (no./km) (*apd*)	0	35	7	7
Categorical variables	Description		Percentage	
Area type (*area*)	Rural		33	
	Urban		67	
Presence of NEMCL (*mclane*)	Yes		60	
	No		40	

Table 4. Regression coefficients for the best-fit model.

Variables	Estimated coefficient[a]	z-statistics	Confidence Interval	
Constant	0.220 (0.547)	0.40	-0.852	1.292
Area (*area*)	0.736 (0.435)[c]	1.69	-0.117	1.589
Presence of NEMCL (*mclane*)	-1.142 (0.452)[b]	-2.52	-2.028	-0.255
mcpct.mclane	0.029 (0.134)[b]	2.12	0.002	0.056

[a] standard errors are in parentheses
[b] significant at 95% confidence level
[c] significant at 90% confidence level

The regression results revealed that area type, percentage of motorcycle and presence of NEMCL were significant in influencing the crash rates among motorcyclists. As expected, urban area was found to have higher crash rates due to the higher traffic volume (including higher proportion of motorcycles) and higher access point density, in line with an earlier study conducted on Malaysian primary roads [15]. Despite having no detailed crash information on the location of those motorcycle crashes (whether occurred on NEMCL or out of NEMCL), these results suggest that the presence of NEMCL had proven effective in reducing motorcycle crash rates (note the negative coefficient of variable *mclane*). This is strongly believed to be due to the fact that the usage of the facility is high among motorcyclists [19] which decreases the likelihood of being hit by other motorists.

However the results indicated that the effectiveness of NEMCL is higher when the percentage of motorcycle is low. When the percentage of motorcycle increases, the effectiveness of NEMCL slightly reduces (note the positive coefficient of the interaction term variable of *mcpct.mclane*), which may probably due to more motorcyclists not using the facility as the occupancy increases. As such the likelihood of motorcycle crashes become higher whenever its percentage increases despite the presence of NEMCL especially in urban areas. In other words, the provision of NEMCL alone is not enough and should not be treated as the only intervention to tackle motorcycle crashes in urban areas.

For a better representation of the regression results, the model can be written as:

$$\text{Crashes/km} = e^{0.736(area) + 0.029(mcpct.mclane) - 1.142(mclane)} \tag{4}$$

Where:

area = 1 when the area type is rural; 2 when the area type is urban
mclane = 1 when NEMCL is not present; 2 when NEMCL is present

4.2. Development of motorcycle lane provision criteria

The model developed in this study was used to develop justification criteria for provision of NEMCL. As the costs for building and maintaining the facility vary by locations, these criteria were developed with the aim of providing an alternative consideration to build the based on the current costs. The estimated costs to build an extra lane (both directions) including the necessary road marking and traffic signs were RM2 million per km (in urban areas where adequate shoulder width is usually limited), and RM400,000 (both directions) for separator marking and road studs (in rural areas where adequate shoulder width is converted to NEMCL). The minimum criteria therefore would be taken as when the costs equal the safety benefits in terms VOSL. The VOSL for an average Malaysian was determined to be RM2,953,930, based on the GDP/capita of RM42,199 for Malaysia in 2017 [20].

In developing the criteria, the number of fatality saved was taken as the difference in the number of KSI crashes between with and without NEMCL at every level of motorcycle percentage. Based on the develop model above, the maximum fatality saved is 1/km (with a value of RM2,953,930) for urban area and 0.5/km (RM1,476,965) for rural area. Therefore, the maximum BCR that could be obtained is about 1.5 for a 1km section of urban area and 3.7 for a 1km section of rural area. These BCR values however will subject to change depending on the current costs of providing the facility, the percentage of motorcycles as well as the GDP/capita.

Figure 1 and Figure 2 show the criteria developed for urban and rural areas for a range of motorcycle percentage. The range used was limited by the samples of the 88 road segments included in this study. BCR values of 1 to 3 were used to reflect the practical range achievable based on the model developed. The lowest criteria that should be considered was when BCR = 1 as the benefits obtained will be equal to the cost incurred. For example, in urban area and when the motorcycle percentage is 20%, the cost to construct and maintain the NEMCL per km over a 7-year period must not exceed RM3,155,000. The return will be twofold if the cost can be kept at RM1,577,000 at the same level of motorcycle percentage.

Based on these criteria, the limits in the cost for providing and maintaining NEMCL per km is less rigid for urban area. This is because of the higher crash rates observed in urban areas, and thus the estimated number of fatality saved was also higher. In other words, the value of benefits are higher if NEMCL is provided in urban area. For example, if RM1,000,000 is allocated to provide NEMCL, the return of investment will be RM3,000,000 for urban areas compared to only RM1,700,000 for rural area with 20% motorcycle.

These criteria can be used as a tool to justify allocation and put up a business case for the government to spend on NEMCL, which supplements the existing warrants stipulated in the REAM guideline. These criteria provide scientific evidence on the safety benefits that can be achieved with available level of funding for various levels of motorcycle composition. It can be useful for new road projects or even for upgrading of existing roads where information on the costs and traffic composition are readily available.

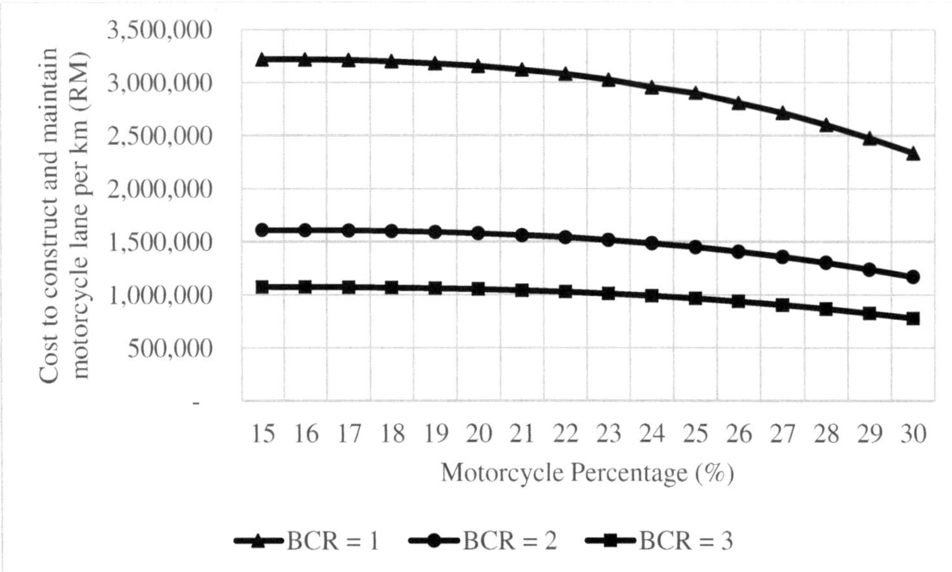

Figure 1. NEMCL provision criteria for urban area.

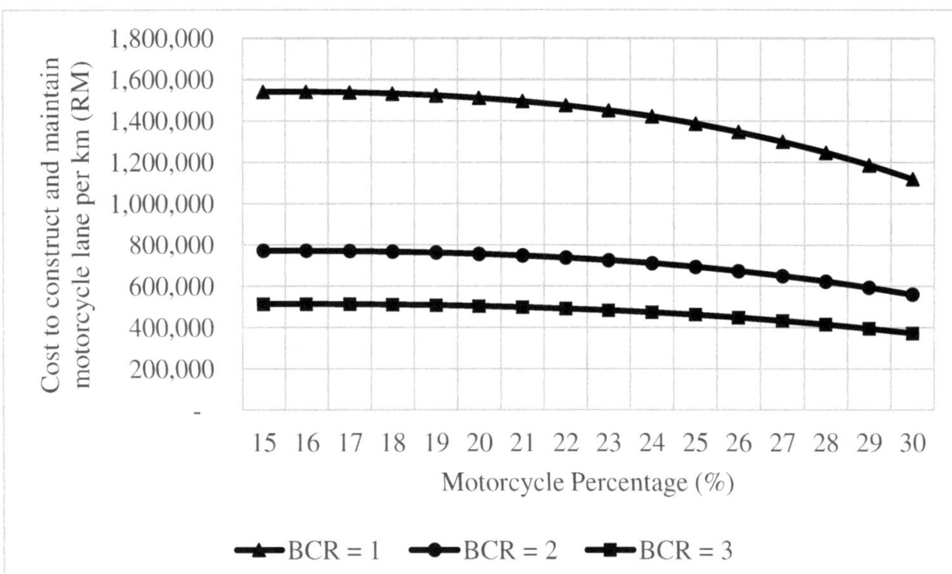

Figure 2. NEMCL provision criteria for rural area.

5. Conclusions

This study aims to examine the safety effectiveness of non-exclusive motorcycle lane as well as to generate criteria for provision of the facility. A total of 88 road segments were identified based on several road attributes that include the presence of NEMCL along several sections on FT001 and FT005. These segments were identified in the vicinity of NEMCL in various length. Motorcycle crash data from 2009 to 2015 were compiled and mapped along these road segments to produce a dataset for statistical regression analysis. The negative binomial distribution was found to be able to fit the crash data with several road attributes. The type of area, the percentage of motorcycle and the presence of NEMCL were found to have significant effects on the crash rates.

The regression results revealed that area type, percentage of motorcycle and presence of NEMCL were significant in influencing the crash rates among motorcyclists. Regardless of the presence of NEMCL, the model showed that crash rates was significantly higher in urban areas due to high traffic exposure. The presence of NEMCL was found to be effective in reducing motorcycle crash rates by increasing the separation between motorcyclists and other motorized traffic. The high utilization rate of the facility as revealed by an earlier study confirmed the effectiveness of NEMCL. However this study did not account for the effect of the design of the facility which influences the utilization rate.

In order to study the return of investment for providing NEMCL, this study developed criteria based on the motorcycle percentage and the overall construction cost for urban and rural areas. The limiting criteria were based on three BCR values (BCR = 1, BCR = 2 and BCR = 3). These critera can be used to judge if provision of NEMCL is necessary for new road projects or existing roads based on the estimated cost required and the expected motorcycle composition in the traffic stream. The lowest return of investment is when the safety benefits equals that of the total cost to construct and maintain the facility. As such in any given project requirements, the provision of NEMCL can be decided based on the BCR values obtained through these criteria. Also, the BCR values can be used to identify priority among various development projects, especially when available budget is limited.

The use of only 88 road segments may not be sufficient to portray the effectiveness of NEMCL in reducing crash rates. It is recommended that more road segments are used, with good representation of the various design of NEMCL. Good quality of crash data will definitely improve the model accuracy. But as far as this study is concerned, the available evidence gathered so far by the police (crash data), Highway Planning Division (traffic data) and MIROS (road attributes) revealed that NEMCL is recommended especially in urban areas.

6. References

[1] Royal Malaysian Police 2013, 2014, 2015, 2016, *Statistical Report Road Accident* (Kuala Lumpur: PDRM)

[2] McInerney R 2014 *The global cost of road crashes: Fact Sheet* Available from http://www.irap.org/en/about-irap-3/research-and-technical-papers.

[3] Road Engineering Association of Malaysia 2011 *Guidelines for motorcycle facilities* (Kuala Lumpur: REAM)

[4] Radin Sohadi R U and Barton E V 1997 Preliminary cost-benefit analysis of the exclusive motorcycle lane in Malaysia *REAAA Journal*

[5] Radin Sohadi R U 2006 Motorcycle safety programmes in Malaysia: how effective are they? *Int. J. of Injury Control and Safety Promotion* **13**(2) 71-79

[6] Poi A W H 2016 *Evaluation of non-exclusive motorcycle lane in Malaysia* (Kuala Lumpur, MIROS)

[7] Public Works Department of Malaysia 2013 *Lorong motosikal. Taklimat bagi membincangkan cadangan penyelesaian kepada isu lorong motosikal di jalan raya dan lebuh raya.* Presentation conducted at Ministry of Works, Malaysia

[8] Radin Sohadi R U, Mackay M and Hills B 2000 Multivariate analysis of motorcycle accidents and the effects of exclusive motorcycle lanes in Malaysia *J. Crash Prevention and Injury Control* **2**(1) 11–17

[9] Abdul Manan M M and Varhelyi A 2012 Motorcycle fatalities in Malaysia *IATSS Research* **36** 30–39

[10] Ahmad Saman M S and Radin Sohadi R U 2005 The effectiveness of a continuous paved shoulder to reduce motorcycle conflicts at junctions *Road Engineering Association of Asia and Australasia Journal* **12**(2) 41–49
[11] Abdul Manan M M, Varhelyi A, Celik A K and Hashim H H 2017 Road characteristics and environment factors associated with motorcycle fatal crashes in Malaysia *IATSS Research*
[12] Harnen S, Wong S V, Radin Sohadi R U and Wan Hashim W I 2003 Motorcycle crash prediction model for non-signalized intersections *IATSS Research* **27**(2) 58-65
[13] Ibrahim M K A, Hamid H, Law T H and Wong S V 2018 Evaluating the effect of lane width and roadside configurations on speed, lateral position and likelihood of comfortable overtaking in exclusive motorcycle lane *Accident Analysis and Prevention* **111** 63-70
[14] Abdul Manan M M 2014 Motorcycles entering from access points and merging with traffic on primary roads in Malaysia: Behavioural and road environment influence on the occurrence of traffic conflicts *Accident Analysis and Prevention* **70** 301-13
[15] Abdul Manan M M, Jonsson T and Varhelyi A 2013 Development of safety performance function for motorcycle accident fatalities on Malaysian primary roads *Safety Science* **60** 13-20
[16] Gross F, Persaud B and Lyon C 2010 *A guide to developing quality crash modification factors* (Washington: United States Department of Transportation)
[17] Bowman B L, Vecellio R L and Miao J 1995 Vehicle and pedestrian accident models for median locations *J. Trans. Engineering* **121**(6)
[18] McMahon K and Dahdah S 2007 *The true cost of road crashes: valuing life and the cost of a serious injury* (Hampshire: International Road Assessment Programme (iRAP))
[19] Ishak S Z, Poi A W H, Mohamad Khaidir N and Abdul Ghani M R 2017 An evaluation on the effectiveness of motorcycle lanes in Malaysia *Proc. 15th REAAA Conference* Bali, Indonesia, 22-24 March 2017
[20] International Monetary Fund, *World Economic Outlook Report 2018*, Available from www.imf.org/external/pubs/ft/weo/2018.

Acknowledgments
The authors would like to thank the Road Safety Engineering and Environment Research Centre of Malaysian Institute of Road Safety Research and to all the people whom are directly and indirectly involved in this study.

IOP Conf. Series: Materials Science and Engineering **512** (2019) 012023 doi:10.1088/1757-899X/512/1/012023

Road safety audit – what we have learnt?

Abu Mansor SN, Ahmad Saman MS, Tengku Razman TMS and Masnel H

Road Department, Public Works Department Malaysia, Kuala Lumpur, Malaysia

E-mail: nurilam.jkr@1govuc.gov.my

Abstract. The traffic accident problem in Malaysia is one of the most serious problem facing the nation. In the effort to reduce traffic accidents, 'Re-Active' initiatives, such as the 'Accident Blackspot' Program, 'Pro-Active' measures such as Road Safety Auditing and improved road design practices is implemented. In early 1997, Road Safety Audit (RSA) was started in JKR and a manual was produced as the primary guide for Road Safety Auditing by JKR and its Consultants engaged in this work. Since then, the implementation of RSA has revealed issues on poor RSA management, common repetitive mistakes by designers and the absence of warrant application. This paper discusses these issues and relates them to the study carried out. This study attempts to compare local RSA practices with other countries in finding the solutions to the above problems. It has been shown that having a policy, clear roles & responsibility and proper RSA management is vital in ensuring road safety is adequately catered for. Furthermore, frequent updating of design guidelines and building up of competency of road designers can substantially reduce the issue on quality of road design.

1. Introduction

Road Safety Audit (RSA) is an engineering technique aimed at identifying potential safety problems during planning and design of projects. RSA can identify potential safety hazards before they become accident prone locations [2]. The RSA definitions may differ from one country to another.

Road safety audit is a formal procedure for assessing accident potential and safety performance in the provision of new road schemes, and schemes for the improvements and maintenance of existing roads – United Kingdom (UK) [4].

A RSA is a formal safety performance examination of an existing or future road or intersection by an independent audit team. It qualitatively estimates and reports on potential road safety issues and identifies opportunities for improvements in safety for all road users - United State of America (USA) [4].

The definition of Road Safety Audit (RSA) in Malaysia is different from that of other countries. Road Safety Audit for a road project may be defined as the formal examination of its planning, design and construction stage and also of the characteristics and operation of an existing road, by independent and qualified examiners, to identify any potentially unsafe feature or operational arrangement that may adversely affect the safety of any road user - Malaysia [4].

The traffic accident problem in Malaysia is one of the most serious problems facing the nation. The implementation of RSA faces many challenges, amongst which are poor RSA management, common repetitive mistakes by designers and the inconsistent application and implementation of RSA. This paper attempts to compare local RSA practices with other countries in finding viable solutions to the above problem. The countries which have been selected are due to their advanced practices in implementation of Road Safety Audit (RSA).

2. Literature review

2.1 Road Safety Audit Practices in Other Countries

2.1.1 United Kingdom. The concept of road safety audits originated in the United Kingdom during the 1980s. In 1987, the United Kingdom (UK) Department of Transport formulated strategies directed towards achieving a one-third reduction in the number of annual highway casualties by the year 2000. In 1988, the UK passed a legislation requiring all road authorities in mainland Britain to take necessary steps to reduce crashes on new roads. This requirement led to the development of two key publications: A Road Safety Code of Good Practice (Local Authorities Association, 1989) and Guidelines for the Safety Audit of Highways (Institution of Highways and Transportation, 1990, revised 1996) [2].

In 1991, the UK Department of Transport made road safety audits mandatory for all national trunk roads and freeways. It currently remains the responsibility of the individual highway organizations to determine what to audit and when, as a function of their highway programs, design procedures, and type of project [2].

2.1.2 Australia. In Australia, the National Association of Road Transport and Traffic Authorities is known as AUSTROADS. In 1994, AUSTROADS released a publication entitled, *Road Safety Audit*. This publication establishes a broad set of guidelines for a national road safety audit program. It includes widely adopted checklists, developed through close interaction with Transit New Zealand, which are used to ensure all areas of safety are considered when conducting a road safety audit. Individual states are incorporating road safety audits at different rates throughout Australia [2].

The state of Victoria's road agency, Victoria Roads Corporation (VicRoads), considers the road safety audit to be an integral component of the quality management process. Road safety audits are carried out from project conception to construction completion on all projects costing in excess of A$5 million (CDN $4.8 million). Furthermore, VicRoads randomly audits 20 percent of other construction projects at one or more stages and 10 percent of maintenance work [2].

The Roads and Traffic Authority (RTA) is responsible for road safety in New South Wales. RTA published a road safety audit manual as part of the New South Wales quality management approach in 1991. Twenty percent of existing roadways within all regions are to be audited to "identify deficiencies in existing roads and identify priorities for action" (Roads and Traffic Authority, 1991). Furthermore, twenty construction projects, varying in project size and stages, are to be audited every year within each region [2].

2.1.3 New Zealand. Transit New Zealand (TNZ) is the national road agency responsible for the maintenance and improvements to the New Zealand highway network. In 1989, TNZ created an Authority whose main objective is the provision of an integrated and safe highway network. After reviewing the practices and procedures of road safety audits developed by the UK and Australia, TNZ published a document entitled, Safety Audit Policy and Procedures [2].

This publication states that all projects costing more than NZ$5 million (CDN$4.2 million) would be audited from project conception to construction completion. TNZ mandated that road safety audits would be conducted on a 20 percent sample of state highway projects, however, there are no guidelines for the identification of projects to be included in the sample [2].

2.2 Road Safety Auditor

The Road Safety Auditor must be independent, so that the design is viewed with 'fresh eyes'. Nonetheless, good communication between the parties must be established and maintained if the audit is to be done effectively and without wasted time and effort. Further, the sensitivity associated with having design work 'judged' should be recognised. Auditors need to be objective in their assessments, yet sensitive to the fact that no one likes criticism. Designers and clients need to consider audit recommendations objectively and gain from the experience [6].

2.2.1 Qualifications. Road safety audits should be conducted by an individual or team with adequate knowledge of road safety engineering principles and practices, experienced in accident investigation and recommending preventive measures, traffic engineering and road design. Consequently, members with experience in enforcement, maintenance, and human factors can be added to the team on a project by project basis and at different audit stages. Human factor expertise may, in selected areas, contribute to a road safety audit by providing an understanding of the interactive nature of user behavior with the road environment [2].

2.2.2 Experience. It is imperative that the audit team has substantial collective experience in the key areas noted in the previous section. While audit checklists serve to identify critical items/areas to be considered, they should only be considered memory aides for individuals with a wealth of experience and not an exhaustive listing of issues [2].

2.3 Problems and Issues

2.3.1 Inadequate Information About the Scheme. Sometimes the Designer does not provide sufficient information to enable the Auditors to properly understand what is to be built. This can largely be avoided by requiring Designers to provide full section drawings (vertical and horizontal alignment) and large-scale drawings of all junctions, bridges, etc., as well as an explanation of what standards have been used, and how safety issues have been addressed. Drawings of standard cross-sections and standard junction layouts are unacceptable on their own. Designers should also be encouraged to show the location of traffic signs, safety barrier and other roadside furniture [9].

Lack of appropriate information to carry out Road Safety Audit is a common issue within the questionnaire responses, highlighting the need for a clear Road Safety Audit Brief and well defined commissioning process [1].

2.3.2 Lack of Time for Auditing. The scheme preparation programed must allow sufficient time for audits to be done at the appropriate stages. The time between design completion and issue of invitations to tender is particularly critical. This is frequently too short a time to enable a proper audit and follow-up to be done [9].

There are a number of issues on details for a Road Safety Auditor dealing with some development schemes. Some of these are related to the quality of the design. As there is often pressure to complete these audits within a short period of time, the Road Safety Audit Team Leader is then faced with a choice – whether to reject the commission due to insufficient information or to work on the information provided and point out the deficiencies [1].

2.3.3 Misunderstanding of Roles and Responsibility

2.3.3.1 Road Safety Audits Is Being Used as A Design Checks and Other Informal Reviews. Deficiencies should only be identified by the audit team. It is not within an audit's mandate for a redesign or recommendation to be made to mitigate a deficiency. This responsibility will rest with the project owners or their design staff. Auditors may suggest plausible options, but it is not their responsibility to make specific recommendations nor to promote a particular solution. The primary task should be for auditors to 'describe the problem and what it entails' and let the project owners/designers make the decision. Informal reviews should be a part of the normal design process, separate from the scope of services which RSA provides [2].

2.3.3.2 Road Safety Audits Is Viewed as A Check Of Compliance Design Standards. Compliance to standard alone does not guarantee a road to be safe because the standard gives minimum values which do not allow rooms for errors among drivers. Furthermore, the standard itself may out dated. [2].

3. JKR's experiences

In early 1997, Road Safety Audit (RSA) was started in JKR and a manual was produced as the primary guide for Road Safety Auditing by JKR and its Consultants engaged in this work.

Since then, the implementation of RSA faces many challenges due to poor RSA management, such as late appointment of RSA Auditor and subsequently leading to inappropriate timing for auditing in each stage of audit. Furthermore, the appointment of the Auditor is by the consultant responsible for the design and the consultant may hesitate to expedite the appointment because the audit will be another checking process on the result of their works. From the consultant's perspective, more checking will only result in more amendments to be done and will delay the completion of the design works. The time between design completion and tender stage is particularly critical. This is frequently too short to enable a proper audit and follow up to be done.

Moreover, there is no specific warrant to identify project criteria and type for choosing the relevant Road Safety stages involved in the project. In most cases, project managers normally assume that the requirement of all stages of RSA is mandatory for all project. As a result, over dependence on the services of Auditors by the designers occurs and this sometimes causes confusion on the roles and responsibilities between designers and Auditors.

3.1 Result of Study

A desk study was carried out on the road safety deficiencies highlighted by Road Safety Auditors involving 32 projects comprising of RSA stages 1, 2 and 3 for each project. The sampling involves equal numbers of internal and external audit reports. Furthermore, the desk study involves 13 nos. (41%) upgrading of roads and 19 nos. (59%) new roads as per figure 1 below.

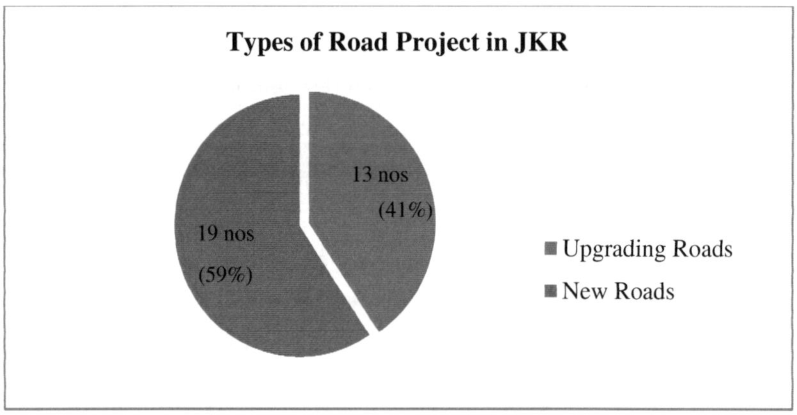

Figure 1. New and upgrading road projects

Based on the observation carried out, there were 29 numbers of road safety deficiencies highlighted by road safety auditors. Road network effect, provision for road users with special needs and general geometric standards were found to be the most frequent deficiencies found in RSA Stage 1, which is 20 nos. (69%), 5 nos. (17%) and 4 nos. (14%) respectively as per figure 2.

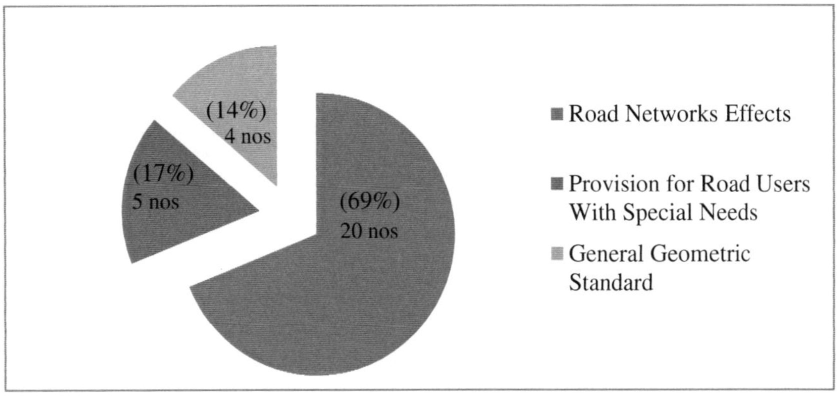

Figure 2. Deficiencies audit item in stage 1

A closer observation on road network effect issues above found that inappropriate placement of project terminals constitutes 9 nos. (45%) of the deficiencies found in the road network effect. This is clearly shown in the figure 3.

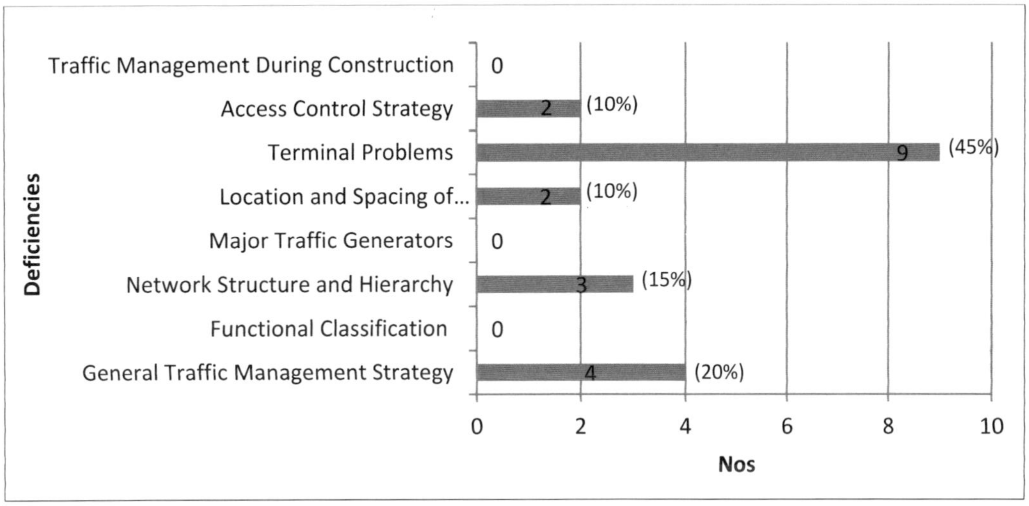

Figure 3. Deficiencies found in the road network effect

As for RSA Stage 2, a total of 58 numbers deficiencies were highlighted in the sample reports. It was observed that issues involving horizontal and vertical alignment constitute the highest number of safety deficiencies, i.e. 37 nos. (64%). This is followed by cross-section 15 nos. (26%), interchanges and intersections 4 nos. (7%), and design criteria 2 nos. (3%) as shown in figure 4.

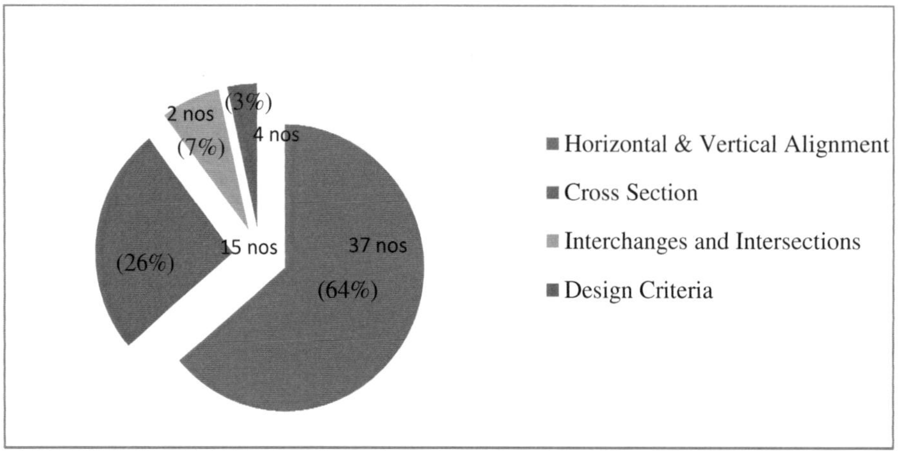

Figure 4. Deficiencies audit item in stage 2

Based on detailed observation it was found issues involving sub-standard horizontal curves and vertical gradients constitute 28 nos. (76%) of the deficiencies found in the horizontal and vertical alignment category. This is clearly shown in figure 5.

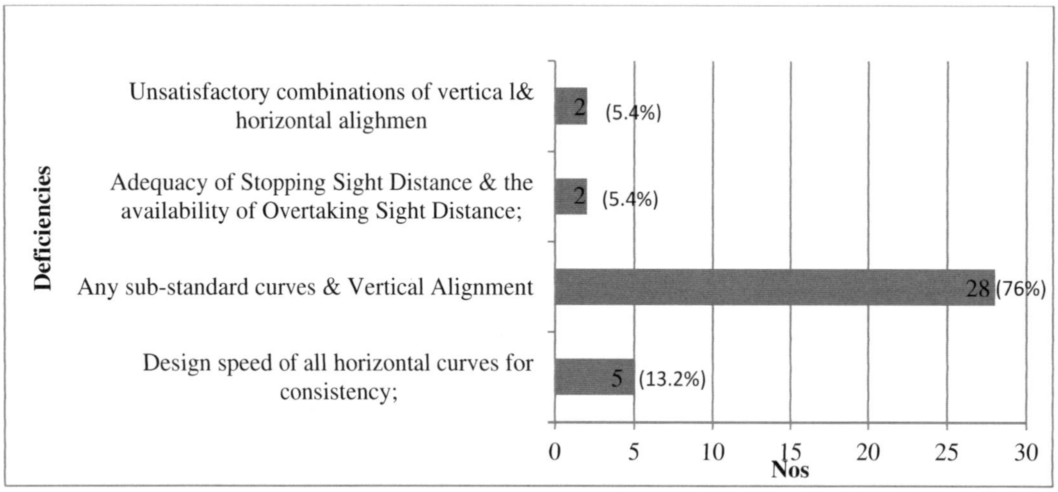

Figure 5. Deficiencies Found in the Horizontal & Vertical Alignment

The study has also indicated that 92 number of safety deficiencies were found in Road Safety Audit Report Stage 3 involving various issues. It was found that issues involving traffic signing and road marking are 30 nos. (33%) and road side safety provision 20 nos. (22%) contributed to 50 nos. (55%) of the total deficiencies as shown in figure 6.

IOP Conf. Series: Materials Science and Engineering **512** (2019) 012023 doi:10.1088/1757-899X/512/1/012023

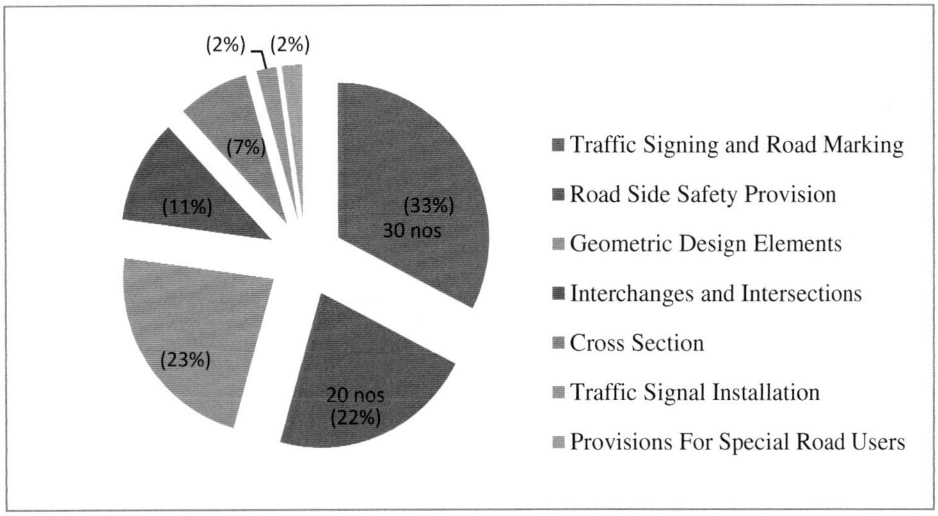

Figure 6. Deficiencies Audit Item in Stage 3

Figure 7 shows that issues involving inadequate numbers of signage and inappropriate usage of road markings 15 nos. (50%) were the main concerns highlighted in auditing of signage and road markings. This is followed by inadequate detailing of barriers in auditing of road side safety provisions 13 nos. (65%) as figure 8.

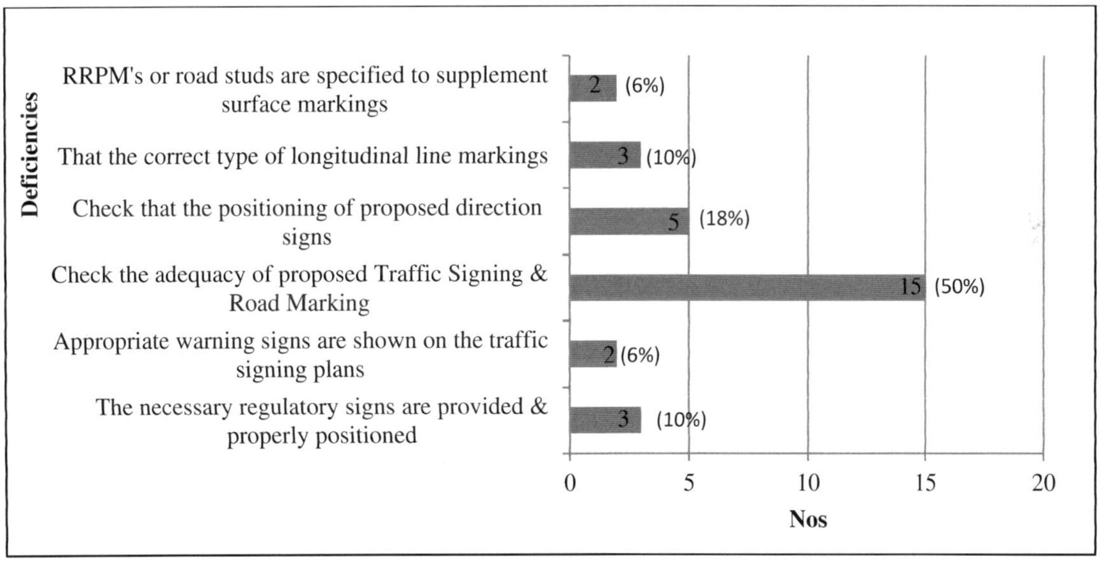

Figure 7. Deficiencies Found in Traffic Signing & Road Marking

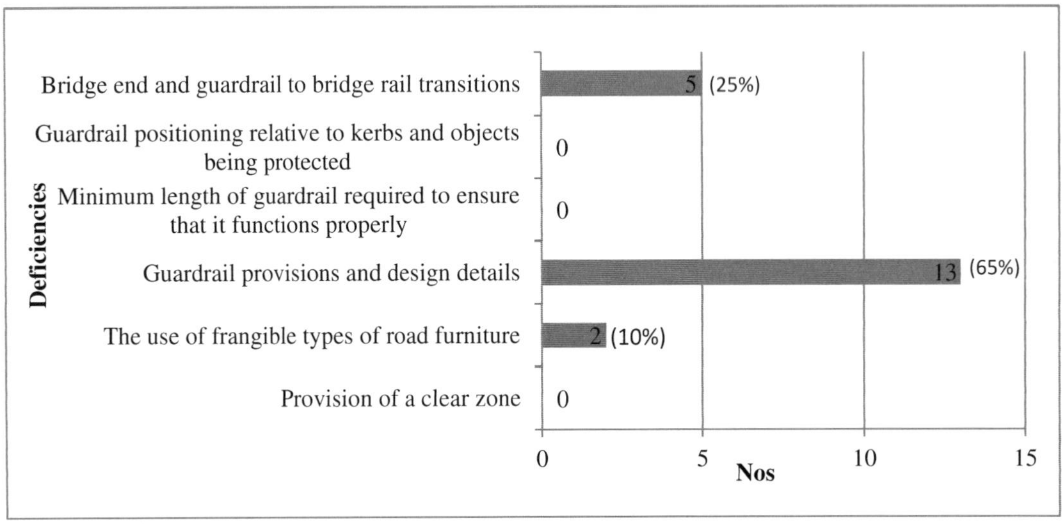

Figure 8. Deficiencies Found in Roadside Safety Provision

Most of the deficiencies that were found in RSA 1,2 and 3 involved 54% in-house design and the remaining 46% those designed by Consultants. Meanwhile, 57% of the deficiencies involved new road projects.

3.2 Discussion and Recommendation

Result of the study above shows that most of the RSA Stage 1 issues involve terminals of road project. This type of safety deficiency is typical of both roads designed by in-house designers and those involving new road projects. This is possibly due to the current practice of limiting the design of the road works within the specified project limit only. This is further aggravated by inadequate transition at the interfacing of new and existing road.

The issue of substandard road curve is a common road safety deficiency especially in RSA Stage 2, as shown in figure 9. The RSA Guidelines (1997) states that a substandard road curve is a curve with a speed difference of more than 10-15km/hr from its approaches [3]. It normally occurs on the first curve after a long straight section. In addition, sometimes the horizontal curve length is also inadequate for the vehicle to maneuver the curve safely.

Meanwhile in RSA Stage 3, road safety deficiencies commonly involve issues of traffic signing and road marking and road side safety provision, as per figure 10, figure 11 and figure 12 respectively.

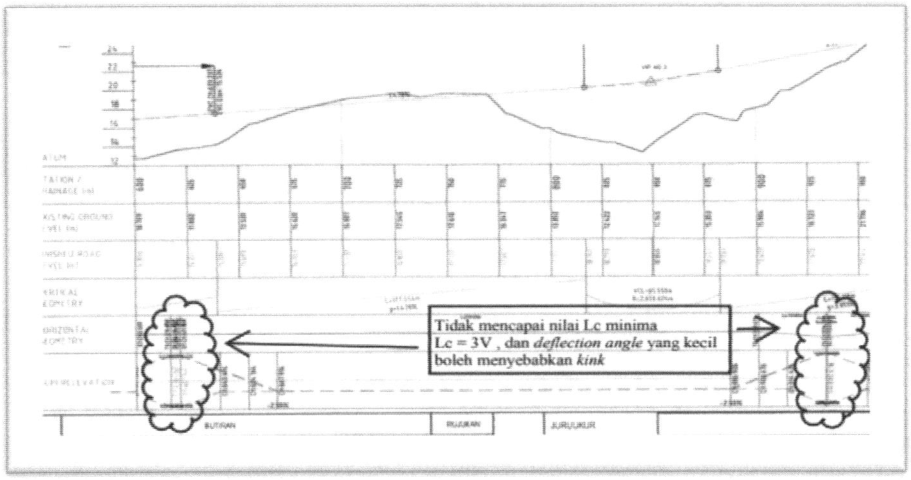

Figure 9. Inadequate Length Curve (Lc)

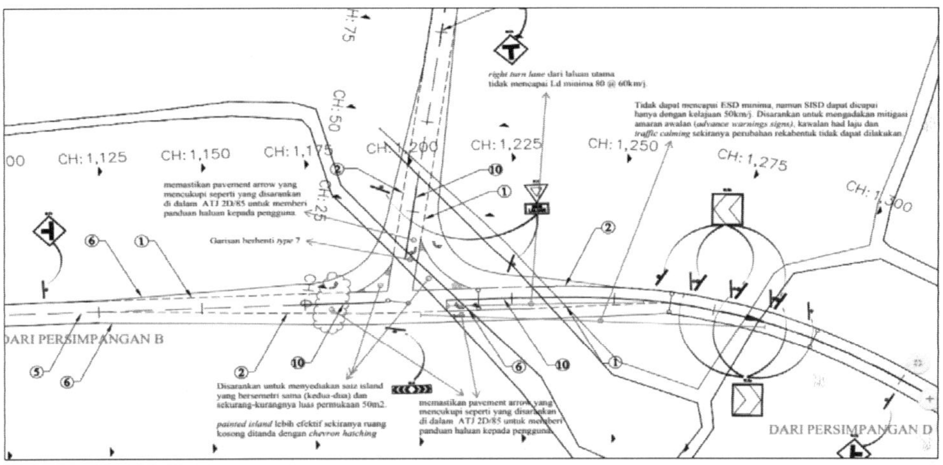

Figure 10. Inadequate Type and Traffic Signing Location

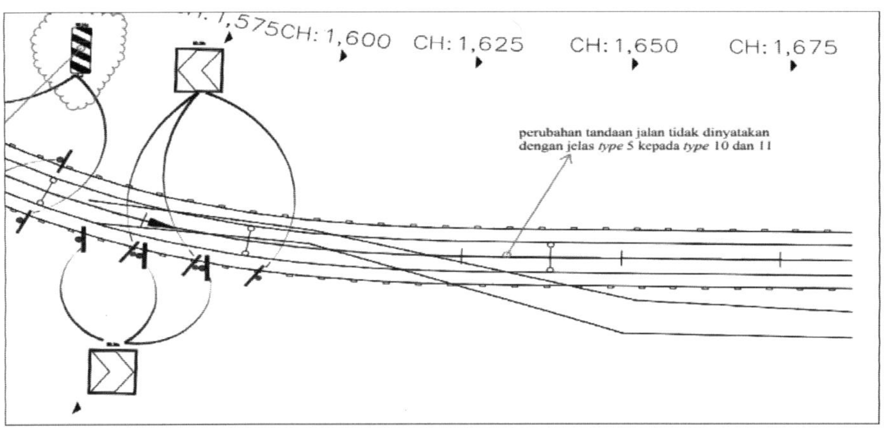

Figure 11. Inadequate Type of Road Marking

IOP Conf. Series: Materials Science and Engineering **512** (2019) 012023 doi:10.1088/1757-899X/512/1/012023

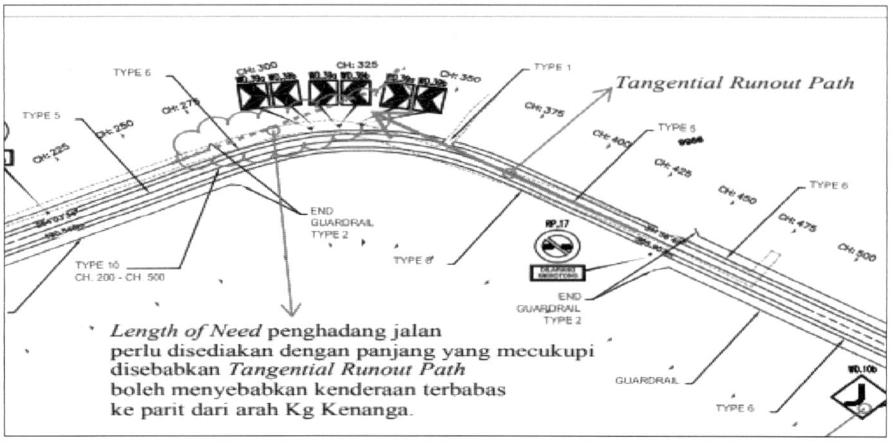

Figure 12. Barrier Requirements at Curve Area (*Length of Need - LON*)

The issues on safety deficiencies which are repeatedly reported in RSA reports need to be looked into to minimize further occurrences. Hence, JKR has embarked on a policy to review the design guidelines every 5 years to ensure it is being updated to the current best practices. Furthermore, JKR has established its own training facility, known as Center of Excellence in Engineering and Technology for JKR's staff (CREaTE) and started to introduce training modules in road design. This is part of the effort to increase the competency in road design among JKR engineers and its technical staffs.

Currently JKR has its own design verification processes, however there is a need to improvise so that issues involving design discrepancy and shortcomings are addressed early before the road safety audit processes. This is crucial because the roles and responsibility between designer and road safety auditor is often unclear. This problem also happens in other countries, where the road safety auditor is also considered as a road designer [2]. It is important to note that road safety audits are not informal design checks or inspections and do not replace the duties of a road designer. Informal reviews should be a part of the normal design process separate from the services a RSA provides.

Obviously, the implementation of RSA in Malaysia has revealed issues on poor RSA management which also occurs in other countries [9]. The practice of late appointment of RSA Auditor, leads to late execution of the audits or inappropriate timing of auditing. Recognizing this as an issue, JKR has introduced RSA as part of additional services of the Consultants. This means the appointment is carried out early during the planning phase and carried through during construction to ensure continuity. In addition, to facilitate tracking the RSA processes throughout the project duration, RSA activity is also included as a part of milestone in the project's work program.

Besides that, the existence of an objective criteria/warrant helps to optimise the usage of resources and also avoid over dependence on road safety auditor. In other countries, criteria/warrant are used to identify those schemes or projects that should be audited to optimize the resources involved [2] and when to audit [5]. Subsequently, the absence of warrant in the existing procedure may result in poor RSA management. Therefore, the development of warrants is a necessity to improve the overall execution of RSA.

JKR has also carried out many other initiatives to improve the implementation of RSA. Among them are RSA training, an accreditation system for the selection of qualified auditors, production of Guidelines for The Safety Audit of Roads and Road Project in Malaysia and guideline on reporting. However, there is a need to improvise further especially in establishing a hierarchy ranking of safety deficiencies to enable quicker decision making, improved documentation system to facilitate traceability and retrieval of RSA decisions.

4. Conclusion

This paper has discussed the implementation of Road Safety Audit in Malaysia in comparison with other countries. Some examples of road safety deficiencies are commonly highlighted in RSA reports. Among them are; inappropriate placement of project terminals, issues involving horizontal and vertical alignment and also issues involving inadequate numbers of signages and inappropriate usage of road markings. This issue is certainly unique in Malaysia compared to other developed countries. Hence, the RSA processes should be tailored to the need of the country. To solve these issues, we need to improve the competency of our designers, improvise our design verification processes, review design guidelines and standards every 5 years to ensure it is being updated to the current best practices. In addition, poor RSA management, warrants and recommendations were also discussed.

5. References

[1] David T 2008 *Road Safety Audit* (London: The Institution of Highways & Transportation)

[2] Eric H and Frank W 1999 *Road Safety Audit Guidelines* (New Brunswck: University of New Brunswick)

[3] PWD 1997 *Guideline for The Road Safety Audit of Roads and Road Project In Malaysia* (Kuala Lumpur: PWD)

[4] Ahmeda I, Puana O C and Ismaila C R 2013 A comparative review of road safety audit guidelines of selected countries *Jurnal Teknologi (Sciences & Engineering)* **65**(3) 67–74

[5] Mainroads Western Australia 2015 *Policy and Guidelines for Road Safety Audit* Available from https://www.mainroads.wa.gov.au/Documents/Policy%20and%20Guidelines%20for%20Road%20Safety%20Audit.RCN-D15%5E23159750.PDF

[6] Robert M and Judd E 2002 *Road Safety Audit, Austroads 2nd Edition* (Sydney: Austroads Incorporated)

[7] Van der Kooi R 1999 *Road Safety Audit, Tools, Procedures and Experiences: A Literature Review and Recommendations* (Leidschendam: SWOV)

[8] Tomaz T and Marko R 2012 Typical Deficiencies in Traffic Safety and Irregularities of Slovien Roads *SIIV-5th International Congress – Sustainability of Road Infrastructure*

[9] The Republic of Uganda Ministry of Works, Housing and Communications 2004 *Road Safety Audit Manual* (Berkshire: TRL Limited)

The study on the level of knowledge of the approving authority in the practice of traffic impact assessment (TIA) report submission in Malaysia: case study in Johor

L Wei May[1,3], R Rahman[1,3], M F Hassan[2] and M Rohani[3]

[1] Department of Infrastructure and Geomatic Engineering, Faculty of Civil and Environmental Engineering, Universiti Tun Hussein Onn Malaysia, 86400 Parit Raja, Batu Pahat, Johor, Malaysia
[2] Pejabat Tanah Kluang, 86000, Kluang, Johor
[3] SDRC (Smart Driving Research Center), Faculty of Civil and Environmental Engineering, Universiti Tun Hussein Onn Malaysia, 86400 Parit Raja, Batu Pahat, Johor, Malaysia

E-mail: limweimay@gmail.com

Abstract. Road Engineering Association of Malaysia (REAM) gazetted traffic impact Assessment (TIA) Guidelines in 2011. TIA is a systematic and scientific process used as early identification to analyse the traffic impacts generated by a development and control with necessary mitigation measures. This study aims to identify level of knowledge of the Approving Authority in the practice of TIA report submission by questionnaire distribution and the data collected is analysed using Statistical Package for the Social Sciences software. This research study focuses on the Approving Authority because of their involvement in the evaluation of TIA report and approval of planning proposal. Lacking of knowledge in TIA understanding will cause ineffectiveness in the TIA implementation. Hence, the scope for this research is the local authorities and the Public Work Department in the state of Johor, Malaysia. This will contribute to the significant of the understanding of the Approving Authority in the practice of TIA and to determine any improvements required to effectively implementing TIA in all stages. This study shows obvious differences between the level of knowledge state PWD (95%) and the local authority (51%) that hinders the intention to have the local authority to act as the Approving Authority in TIA.

1. Introduction

Traffic Impact Assessment is a systematic and scientific process used as the early identification to foresee traffic impact generated from a development and to mitigate the impact by providing practical solutions. A development will invariably bring about unwanted impact to the traffic of the developing location and the surrounding area. At some point, the cumulative development will generate additional vehicular trips that will adversely affect the traffic condition of the prevailing road network. In order to address this matter, Traffic Impact Assessment is a common platform for the authority to ensure the development of the town is in controlled orderly manner. Potential issues, such as, congestion, can cause hazard to the safety of users on the road. Faheem 2012 stated that traffic congestion causes delayed in travel times, air pollution and accidents that are more cost-consuming [1].

Lack of knowledge has been a concerning issue in the practice of TIA, the success of the implementation of TIA process greatly depends on the expert or technicians in the government and consultants having sufficient knowledge on TIA. Lack of potential staffs in the local offices, limited experts in private sector and only a few famous companies have traffic engineers experienced in TIA will leads to cases where the report is approved despite the obvious overestimated of the traffic analysis. This also hinders the intention to distribute the responsibility to handle TIA to local agencies to control land developments. Therefore, government must train their people to develop enough knowledge to accomplish the target in efficiently implementing TIA. It is important that assessments are not viewed merely as a hurdle for development approval. With no incentive to modify the existing TIA, this may result in the lack of accountability for the development due to assumptions and decisions inherent in the preparation of TIA. Therefore, to accomplish an efficient implementation of TIA, the conduct of TIA must have a well-developed human resource in the related organisation such as the approving authority in ensuring the integrity of the practice of TIA. The lack of knowledge of the involving parties will be a hinder to produce meaningful TIA [2]-[5].

1.1. Traffic Impact Assessment in Malaysia

Malaysia has applied Traffic Impact Assessment (TIA) since early 90', with the first draft TIA Guidelines proposed in 2005. In 2011, Road Engineering Association of Malaysia (REAM) gazetted the latest guidelines. TIA report is an element in the Development Proposal Report in compliance with section 21A of the town and Country Planning ACT 1976 (ACT172). The guidelines ensure the implementation of TIA is uniform throughout the country with the standardised preparation of TIA reports. The report will provide the necessary technical links with the Environmental Impact Assessment and the Road Safety Audit. Figure 1 illustrates the TIA process for the application of a development planning permission in Malaysia. From the figure, the approving authority will proceed with planning approval after evaluation and acceptance of the TIA report. In the guidelines, the approving authority defined as the local authority and the road authority who had the authority to act according to the law. Besides, those eligible to do the professional TIA report must be accredited TIA Assessor registered with the Board of Engineers, Malaysia and meets the criteria for accreditation as listed in the guidelines. The main aim of implementing TIA in Malaysia is to have a professional evaluation on the adverse impact on the proposed development and its surrounding the traffic and transportation network. Then, execution of the appropriate mitigation measures to relief the forecasted impact to restore the level of service (LOS) of the road condition to its baseline. The baseline of the road condition should be a minimum of LOS D [6]-[8].

Transportation is an important daily mobilisation, with the unceasing development, traffic accidents and road congestion have become more significant and pronounce issues. Malaysia as a developing country is no exceptionally for this issues. Johor occupied the second largest populated state in Peninsular but also having the highest number of death caused by road accident. The best is to intercept these traffic safety problems at the initial stage of planning and this can be achieved through well-coordinated traffic impact management. Given Malaysia have well-established Guidelines for TIA since 2011 as a planning tools to control the traffic impacts from developments, however the number of road accident are still high. With Federal Route 50, FT050, connecting Batu Pahat-Kluang-Mersing recorded the highest number of accidents in the state of Johor [9]. Thus, Johor is selected as the case study with the district of Batu Pahat including the town of Yong Peng and the district of Kulai as the study radius to conduct this research. This research aims to identify the level of knowledge of the Approving Authority in the practice of TIA report submission by using questionnaire method and to analyse the collected questionnaire data using Statistical Package for the Social Sciences software.

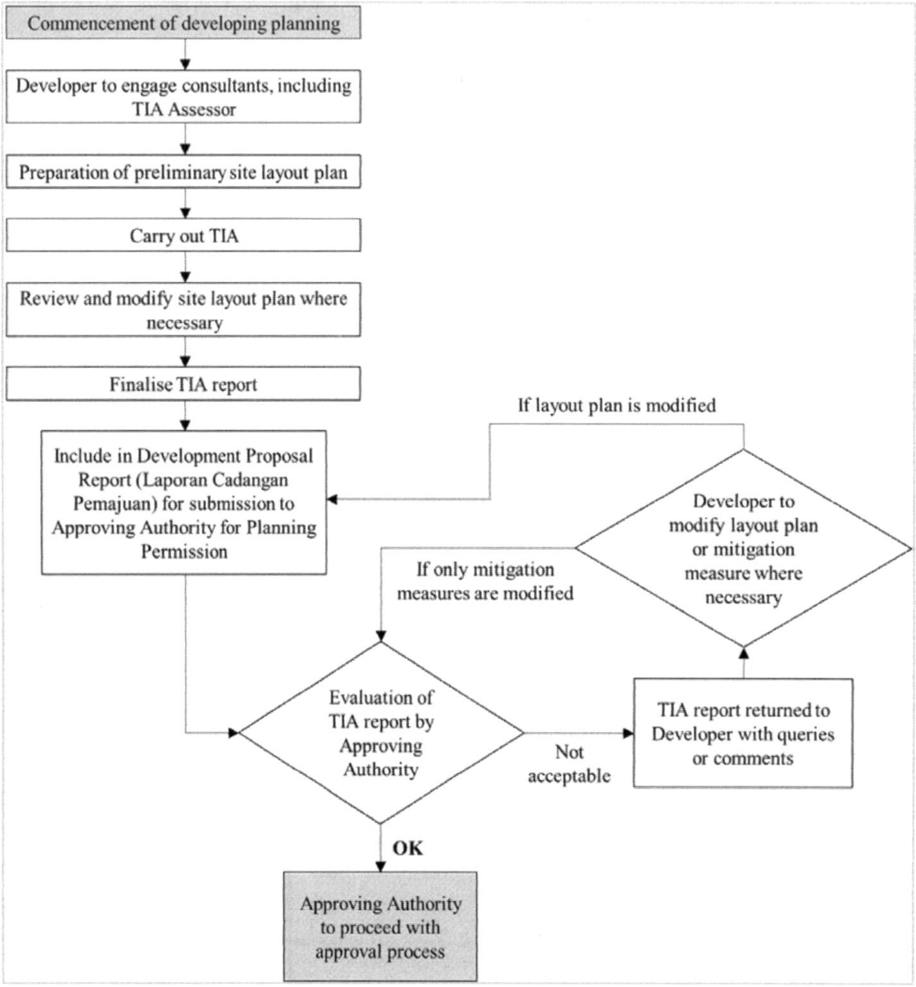

Figure 1. TIA in planning permission application process. Source: [6]

2. Materials and Methods

Johor as the study scope for this research, Table 1 depicts the respective location where the questionnaire is distributed. Department such as the road department and technical team that involved in the practice of TIA is chosen as the respondents, a total 44 respondents is collected. Pilot study is conducted to ensure the data collected is able to achieved the aim of this study, modifications are done during the actual research by changing the format of testing the respondents' level of knowledge using true and false statements. The analysis is done using Statistical Package for the Social Sciences software to relate to the demographic characteristics of the respondents.

Table 1. Scope of Study

Department	Name
Local authorities	Majlis Perbandaran Kulai (MPK)
	Majlis Perbandaran Batu Pahat (MPBP)
	Majlis Daerah Yong Peng (MDYP)
District Public Work Department	JKR(D) Kulai
State Public Work Department	JKR (S) Johor (Road Department)

3. Results and Analysis

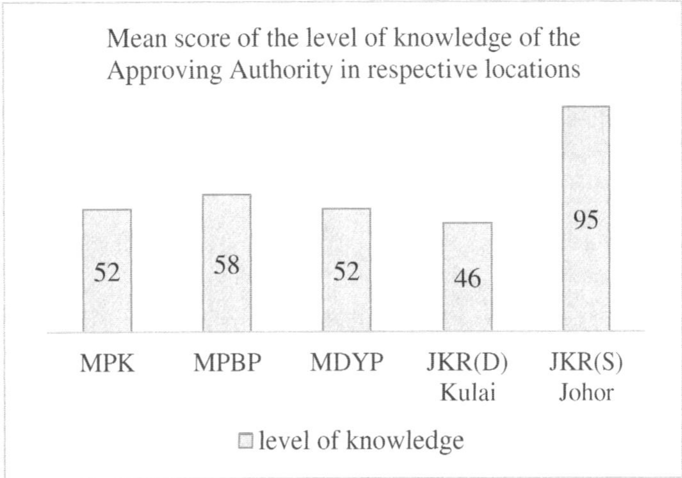

Figure 2. Level of knowledge

Table 2. Variables tested against level of knowledge

Level of knowledge in the practice of TIA report submission against	R-square regression
Age	0.014
Education Level	0.015
Working experience in current designation or department	0.151
Working experience on profession regarding TIA	0.105

R-square regression closer to 1 indicates that there are relationship between the variables, the values for R-square obtained from the analysis are closed to 0 This shows the level of knowledge is regardless of education level, working experience and age as presented in Table 2. In Figure 2, the state PWD achieved outstanding level of knowledge in the practice of TIA report submission of 95%. While, the district PWD, district council and municipal council score 51% in average. The obvious differences between the level of knowledge is an obstruction to the intentions of implementing TIA locally.

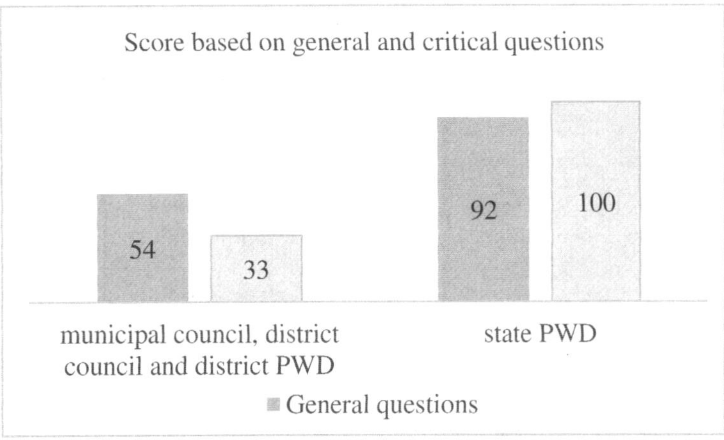

Figure 3. Level of knowledge based on question categories

The common question that was incorrect by the state PWD is a general question regarding the requirement of TIA presented in Figure 3. The only question that was incorrect is regarding the requirement of TIA. In the Malaysia TIA guidelines stated that not all development requires TIA especially in rural areas, only required when the development meets the trigger level as in Table 3. However all the respondents from state PWD answered true that all development proposals need TIA to control the negative traffic impacts. These is also one of the issue discussed in [10], TIA study should include all study region, rural and urban because as development progress, the traffic impacts from the rural will be more pronounce and severe.

Table 3. Criteria and corresponding trigger levels to warrant TIA. Source: [6]

No	Criteria	Trigger Levels
1	Peak Hour Trip Gerenation (Commuter peak)	150 added vehicles per hour (2-Way)
2	Off-Peak Hour Trip Gerenation (Generator peak occurs at the off-peak period)	200 added vehicles per hour (2-Way)
3	Size of residential development	200 dwelling units
4	Size of Commercial development	45.500 sq.ft. (gross floor area)

Note: Trip Generation Rates shall be based on the Malaysian Trip Generation Manual published by the Highway Planning Unit of the Ministry of Works, Malaysia.

Whereas for the other departments, the results are relatively low (33%), especially on the 7 critical questions regarding the role of Approving Authority, the role of TIA Assessor, the criteria of warranting TIA and the baseline condition after mitigation measure. This questions are the basic principle and must know for the practice of TIA. Other 13 questions are the general questions which consist of the purpose of TIA and others related questions after the TIA report is in progress.

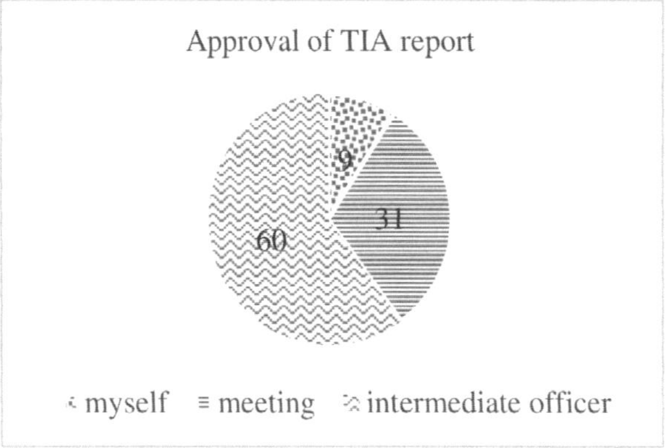

Figure 4. Approval of TIA report

The state PWD agrees that TIA report approval should be done in a meeting. However, most approval are given by their intermediate officer, 60% as shown in Figure 4. The Malaysia TIA guidelines stated the local authority or the road authority is given the power to act according to the law regarding TIA. This statement is agreed as 91.4% of the respondents agreed that the current legal jurisdiction has provided them the authority. This will affects the implementation of TIA as many is unclear about the role in TIA. This will also arises the question of the doubt whether the report is evaluated sufficiently with one's knowledge as majority of them scored average level of knowledge.

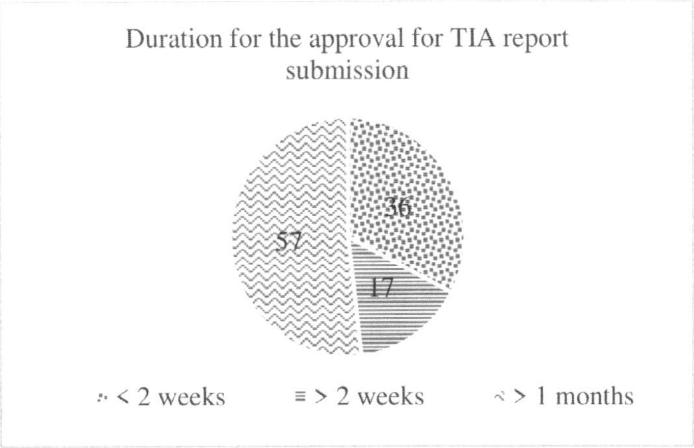

Figure 5. Duration for the approval for TIA report submission

The purpose of having Guidelines so process for approval within 14 days. Figure 5 shows 26% of the reports are approved within 2 weeks. Delay could be due to the problems by the consultant or the developer (48.3%), lack of in-house capacity to evaluate the reports (25.7%) and postpone of regular meeting (14.3%) presented in Table 4.

Table 4. Problems in TIA report submission

Type of problem	Percent
Problem by the consultant or developer	48.3
Lack of in-house capacity to evaluate the reports	25.7
Postpone of regular meetings	14.3

About 85.7% recommended monitoring programme after TIA report submission is approved, this will help continue monitor the traffic impacts generated by the development and provide necessary data for future development. 85.7% suggested to have compulsory courses or programmes on TIA organised by road provider such as Public Work Department conducted by the Road Engineering Association of Malaysia or Institution of Engineers, held constantly to train the personnel involved in TIA. 91.4% agree to have competency test for the Approving Authority. By having capable and accredited TIA Approving Authority, the implementation of TIA could be more effective as the Approving Authority are able to practice TIA independently.

Due to the lack of information in the Malaysia TIA Guidelines, in the *Garis Panduan untuk Memproses Permohonan Pembangunan Tepi Jalan Persekutuan* clearly stated that reports are to be processed within 14 days which include the TIA reports [11]. However in the Malaysia TIA guideline the duration for the approval of TIA report is not stated and there are no referencing in the guideline to direct users to the related details regarding this matter. This is agreeable to the improvement of more detailed information to evaluate the TIA report submission and having checklist in the TIA Guidelines (91.4% respectively), Table 5. This will assist the Approving Authority in the completion of TIA report submission approval within 14 days.

Table 5. Improvements

No	Agreement to the following statement	%
1	Is the current legal jurisdiction provides the authority the power to compel the submission of TIA reports when being challenged the relevant statutes?	91.4
2	Compulsory training courses or programmes on traffic impact assessment should be held constantly to train personnel involved in this department.	85.7
3	A board should be created with all the important members especially with the present of TIA consultant or assessor to evaluate TIA reports.	91.4
4	Competency test should be held for those who wish to be the Authority that approve the TIA report submission.	85.7
5	The current guidelines should provide more detailed information on the method to evaluate the TIA report submission.	91.4
6	Adding checklist to the current TIA guidelines to standardise the evaluation process.	91.4
7	Monitoring programme to continue monitor the traffic impact after the TIA report submission is approved.	85.7

TIA current process was lack of reviewing by the authorities, meanwhile the guidelines should be clearly defined. Sustainable transportation system for a town can be achieved by a coordinated effort of the stakeholders, namely the local authority (the municipality), project developers, practicing traffic consultants and the public at large. Main problem is the authorities have poor coordinating control in the provision of a sustainable transportation system since there is a lack of reviewing by authorities. Lack of review and monitoring is a key issue for the practical application of TIA due to the lack of accountability of the TIA report to the accuracy of the report. Thus, provision of timely and quality report will help to reduce administration and rework. The implementation of a comprehensive transport system from the planning stage is vital because when the problem aggregated in future, the developers are not around and not liable to the proposed infrastructure. Besides, compared to developed countries, the guidelines are more detailed with standardised forms and checklist of elements to be included in the report, some with template to present the data. Design years for some countries depends on the scale of the development categorised into high and medium impact development [12]-[17].

4. Conclusions

TIA is important to ensure the safety of the end users on road with proper planning in the initial stage of design. The proper conduct of TIA will prevent rework in later stage. This study found that regardless of education level, working experience and age, the obvious different between the level of knowledge of the road authority, state PWD (95%) and the local authority (51%). The lack of knowledge in the local authority hinders the intention to have the local authority to act as the approving authority in handling TIA to control traffic impacts in the local development. This study also indicated for the government to give the local authority the responsibility to carry out TIA independently is not effective. In return, causes delay in the TIA report submission. Delay are due to the problems by the consultant or the developer, lack of trained staffs to evaluate the reports, postpone of regular meeting and lack of information in the guidelines for TIA.

This resolved to the need to impose accreditation for approving authority, only qualified approving authority with sufficient qualifications in terms of knowledge and experience. This including having them to participate in intensive training to develop adequate knowledge in TIA Courses organised by any road provider such as Public Work Department conducted by the Road Engineering Association of Malaysia or Institution of Engineers to ensure the quality and relevant of the training. This allow the authorities to accumulate sufficient experience in TIA to be in track with the trend of TIA and to be consistent in the TIA process. This study also recommended approval must be done in meeting that involved the TIA Assessors, more detailed information in the guidelines, checklist and monitoring programme to ensure the implementation of TIA is properly conduct and not merely for the planning approval.

5. Recommendations

While doing this research study, there are few improvements to achieve better result outcomes such as follows:

1. The questionnaire can be in multiple-choice test to reduce the error of getting the correct answer without knowing, the chance of being correct by guessing is reduced to 1/4 as compared to true and false test where the chance are 1/5 of being correct by guessing and not knowing the correct answer [18].
2. The scope of the respondents rather than Approving Authority, can be expanded to Developers and TIA Assessors as they also play vital roles in the practice of TIA. Their opinion and recommendations contribute greatly in the implementation of TIA.
3. Besides the study on the role of the parties involved in TIA, the TIA Guidelines and TIA case study or approved TIA report submission can be further research to provide the overall understanding on the lacking in the TIA practice.
4. The categories of the level of knowledge also should been identify by giving grading or grouping. In general, a lot of improvements can be done involving all parties that involve in this matter. JKR as one of the road providers play the biggest contribution to make sure this TIA guidelines been fulfilled.

6. References

[1] Faheem M I 2012 A Frame work for traffic impact analysis *Int. J. of Engineering, Economics and Management* **1**(1) 5

[2] Yayat K D, Kombaitan B and Purboyo H H 2016 Traffic impact assessment practice in Indonesia *Proc. Social and Behavioral Sciences* **227** 75-80

[3] Limapornwanitch K, Montalbo Jr C M, Hokao K and Fukuda A 2005 The implementation of traffic impact assessment in Southeast Asian cities: case studies of Thailand and the Philippines *J. of the Eastern Asia Society for Transportation Studies* **6** 4208-23

[4] Van Rensburg J F and Van As S C 2004 Issues with traffic impact assessments *Proc. of the 23rd Southern African Transport Conf. (SATC 2004)* **12** 15

[5] Azra Nuzhat 2014 Implementation of traffic impact assessment in developing countries: case study of Bangladesh *Proc. of the Intl. Conf. on Future Trends In Civil and Structural Engineering*

[6] Road Engineering Association of Malaysia 2011 *Guidelines for Traffic Impact Assessment* REAM-GL 10/2011. 1st ed. (Malaysia: REAM)

[7] Adnan Z, Riza Atiq A, Amiruddin I and Azlina I 2008 Interactive Traffic Impact Assessment study for Kuantan *23rd ARRB Conference – Research Partnering with Practitioners, Adelaide Australia*

[8] Laws of Malaysia 2006 *Town and Country Planning Act, 1976 (Act 172). The Commissioner of Law Revision Malaysia* (Malaysia: Percetakan Nasional Malaysia Bhd)

[9] Ministry of Tansport 2015 *Transport Statistic Malaysia 2015*

[10] Kazaura WG and Burra MM 2017 Land use change and traffic impact analysis in planned urban areas in Tanzania: The case of Dar es Salaam City *Current Urban Studies* **5** 1-19
[11] Jabatan Kerja Raya 2011 *Garis Panduan untuk Memproses Permohonan Pembangunan Tepi Jalan Persekutuan*, ATJ 3/2011. 3rd ed. (Malaysia: JKR)
[12] Botha JL 2005 Trends in the traffic impact assessment process. *Proc. of the 24th Southern African Transport Con. (SATC 2005)* **11** 13
[13] Cooley K, De Gruyter C and Delbosc A 2016 A best practice evaluation of traffic impact assessment guidelines in Australia and New Zealand *38th Australasian Transport Research Forum (ATRF)*
[14] Transport Canberra and City Services. 2016 *Guidelines for Traffic Impact Assessment version 4.0.* Available from https://www.tccs.act.gov.au/__data/assets/pdf_file/0009/991989/Transport-Impact-Assessment-Guidelines.pdf
[15] Volusia Transportation Planning Organization 2009 *Transportation Impact Analysis (TIA) Guidelines*
[16] Kirstine C, Chris D G and Alexa D 2016 A best practice evaluation of traffic impact assessment guidelines in Australia and New Zealand *Australasian Transport Research Forum 2016 Proc.*
[17] Clark I D and Faid M 2007 Guidelines for undertaking Transport Assessments in New Zealand and Australia *Australian Inst. of Traffic Planning and Management (AITPM) National Con.*
[18] Hunt D P 2003 The concept of knowledge and how to measure it *J. of Intellectual Capital* **4** 100-13

Acknowledgments

The authors are grateful to the member of the district council, municipal council, district Public Works Department and Johor state Public Works Department for their valuable time in contribution and involvement in the collection of data. This study is supported by the Research Management Center (RMC) research grant of Universiti Tun Hussien Onn Malaysia.

IOP Conf. Series: Materials Science and Engineering **512** (2019) 012025 doi:10.1088/1757-899X/512/1/012025

Use of continuous speed profiles to investigate motorcyclists' speed choice along exclusive motorcycle lane

Mohd Khairul Alhapiz Ibrahim[1] , Hussain Hamid[2], Teik Hua Law[2] and Shaw Voon Wong[3]

[1]Road User Behavioural Change Research Centre, Malaysian Institute of Road Safety Research, 43000, Kajang, Selangor, Malaysia
[2]Department of Civil Engineering, Faculty of Engineering, Universiti Putra Malaysia, 43400 UPM Serdang, Selangor, Malaysia
[3]Department of Mechanical Engineering, Faculty of Engineering, Universiti Putra Malaysia, 43400 UPM Serdang, Selangor, Malaysia

E-mail: mkhairul@miros.gov.my

Abstract. The occurrence of speed-related crashes on the exclusive motorcycle lanes in Malaysia is a matter of concern. This study sought to address this concern through the analyses of speed data collected using GPS-based instrumented motorcycles. Consenting motorcycle riders (N = 29) were asked to ride the instrumented motorcycles along a 20 km stretch of an exclusive motorcycle lane along Federal Highway 2 (F02) in the city of Shah Alam. Analyses of the continuous speed profiles revealed significant inputs regarding the exclusive motorcycle lane. The geometric design of the facility is currently allowing motorcyclists to travel with high overall traveling speeds, with frequent and potentially unexpected speed changes. The acceleration profiles on tangents revealed that the motorcyclists were able to reach a cruising speed in only 50.3% of the acceleration trips, whilst other trips were characterized by sudden and rapid changes in speeds. Further, deceleration profiles on tangents indicated that the participants rode onto the transition elements with high approach speeds. The findings of this study highlighted the risk of a crash due to high traveling speeds and large speed reduction, especially during frequent transitions between tangential sections and horizontal curves.

1. Introduction

Exclusive motorcycle lanes have long been introduced in Malaysia to segregate motorcycles from the main traffic and eliminate the risk of collision with the four wheelers. In a study conducted more than ten years ago, the initial construction of exclusive motorcycle lanes was found to be very cost effective with proven success in reducing the numbers of motorcycle crashes and fatalities [1]. However, there is no latest literature that looks into the latest impact of the exclusive motorcycle lanes on crash reduction among motorcyclists. In light of the worrying statistics on motorcycle casualties on Malaysian roads, their segregation from the main traffic should be a priority. Thus, the utilisation of the lane among the motorcyclists should be promoted.

However, many previous studies highlighted one major concern that could affect the safety of the motorcyclists who use the facility; the absence of motorcycle-specific design standards [2–5]. Without the motorcycle-specific design standards, the current design criteria and safety level could not be thoroughly assessed. On the other hand, the occurrence of crashes along the facility has been reported previously. For instance, [6] reported that side-swipe collisions and run-off-road single motorcycle crashes were the most common type of crashes recorded on the exclusive motorcycle lane along the Federal Route 2 in Shah Alam. [5] discovered that 64.6% of the reported crashes along the exclusive motorcycle lane were run-off-road crashes and crashes involving fixed roadside objects, which could point to the loss of control or unsuitable speed for the prevailing road conditions.

This study set out to investigate the impact of geometric design on the operating speeds along an exclusive motorcycle lane. One way to assess the safety level of a facility is by looking into variability in operating speeds [7,8]. A geometric design that is not in accordance to drivers' expectation is manifested through a large speed disparity along different segments characterized by sudden speed changes. This lack of consistency in design is postulated to increase the crash rate along a facility [9]. Research on motorcycle operating speeds on the exclusive lane has been mostly restricted to analyses of spot speed data [10]. The experimental work presented in this study provides one of the first investigations into how the operating speeds were affected during tangent to curve transitions. The results could be very beneficial for speed management and engineering treatment of affected sections of the lane.

2. Methodology

Traditionally, the speed distribution of a traffic stream on a particular roadway is measured at a specific location through spot speed studies. The present study aims to improve the depth of spot speed data by using continuous speed profiles, collected along continuous points.

2.1 Instrumentation of the test motorcycle

GPS-based instrumented motorcycles were used to record the operating speeds and riding behaviours of the participants. The main motivation of using the GPS-based instrumented motorcycles was to acquire a database of real-time naturalistic riding data comprising second-by-second speed profiles, thus enabling the analysis of speed disparity along different segments of the lane. The instrumented motorcycles were modified from the original Honda Wave 100 cc models. The motorcycle data acquisition unit (DAQ) was approximately 111.8 mm (W) x 91.4 mm (D) x 25.4 mm (H) in size, weighed around 135 g and powered by 12 V direct voltage. Two weatherproof miniature cameras were used to capture real-time video images of the forward and rear view. A Global Positioning System (GPS) device (update rate: 5 Hz) was used to record participants' operating speeds. All data were synchronized and stored in a 6 GB memory card. Figure 1 shows the overview of the motorcycle instrumentation. Detailed descriptions of the instrumented motorcycle can be found in [11].

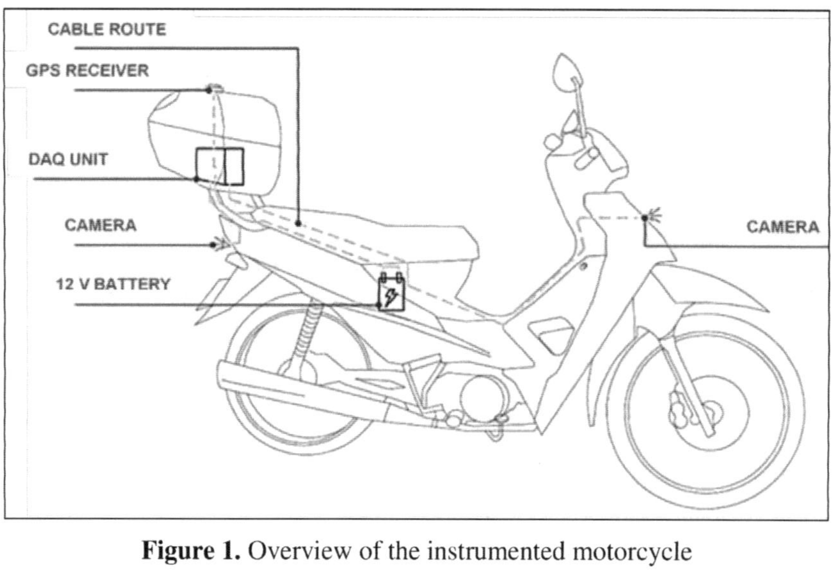

Figure 1. Overview of the instrumented motorcycle

2.2 Location of data collection
Data collection was conducted on a 20 km stretch of an exclusive motorcycle lane along the Federal Highway Route 2, which is one of the longest in Malaysia and recorded the highest motorcycle volume [5]. The average lane width along the selected route was 3.05 m (*SD* = 0.16 m) with the average riding time of 20 minutes. The final dataset consists of speed profile data belonging to 29 consenting participants with an average age of 25.6 years old (*SD* = 5.7), and average riding experience of 7.8 years (*SD* = 5.5). Data collection was conducted in a low volume traffic with good weather and clear visibility to accommodate participants' free-flow speeds. Participants rode the instrumented motorcycle starting from a petrol station near Batu Tiga toll plaza towards Klang and made a U-turn at an intersection immediately after Sungai Rasau toll plaza, towards the end of data collection point.

2.3 Construction of a database of riding speeds
The continuous speed data and recorded video footage recorded by the instrumented motorcycle were manually examined using a data management software. Any changes in participants' riding speed were tracked and recorded into a spreadsheet to create a database of a second-by-second speed profile for each participant. During a transition between two or more elements, the final element was considered as such only when the speed stopped increasing or decreasing at the approach point of that element. If the change in speed ended in the middle of a curve, for example, the next element connecting the curve (i.e., a tangent) was recorded as the final element. The corresponding lane geometry data consisting of lane width, length of tangent and radius of selected curves were also added to the database to increase the depth of data analysis. The width of lanes were physically measured on sites, whilst the curve radius and the length of tangent were estimated using Google Earth software. Figure 2 displays the operating speed profile and the instances of speed changes of a participant, on a certain section of the data collection route.

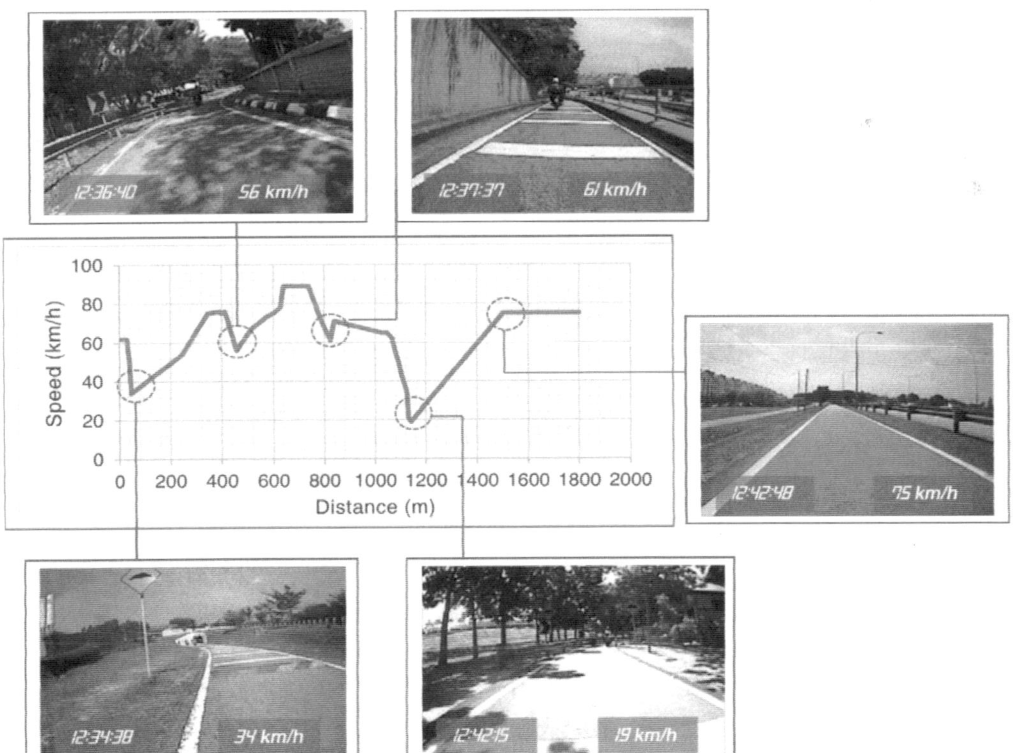

Figure 2. A sample of continuous speed profile and the instances of speed changes

3. Results

Continuous speed profiles involving 1839 speed data were generated for further analysis of acceleration and deceleration behaviours along tangential sections and during transitions between different elements.

3.1 Acceleration profiles on tangents

The speed database was filtered for acceleration events that ended on tangential sections to examine the tangent acceleration profiles. The possibility of participants reaching the cruising speeds on tangents was also examined. Table 1 lists the average acceleration rate, time and distance by groups of final speeds. Out of the total 340 recorded acceleration trips, participants were able to reach the final speeds of a minimum 70 km/h in 82.9% of the trips, indicating a high overall speed distribution on tangents. A cruising speed was defined as the speed that was maintained for a considerable distance by the participant beyond the point of reaching the maximum speed. In other words, the cruising speed is the final desired speeds of an acceleration trip. Only 50.3% of all acceleration trips were the trips where participants reached their cruising speeds on tangents. Other trips were characterized by sudden changes in speeds, whereby the participants had to reduce their speeds rapidly.

Table 1. Average acceleration rate, time and distance by groups of final speed

Final speed (km/h)	50-60	60-70	70-80	80-90	90-100
Number of trips	4	54	134	120	28
Average acceleration rates (m/s^2)	0.48	0.35	0.35	0.36	0.32
Average acceleration time (s)	14.5	15.7	17.8	22.6	25.7
Average acceleration distance (m)	120.2	112.4	138.9	195.8	275.6
Average initial speed (km/h)	32.8	49.9	54.8	59.3	64.2
Average speed increase (km)	24.3	16.7	21.4	25.6	28.9
85th percentile of final speed (km/h)			88.0		

3.2 Deceleration profiles from tangents to transition elements

With high overall speed distribution on tangents, it was found that majority of the deceleration events from tangents to transition elements were initiated from the initial speeds of more than 70 km/h. Table 2 lists the average deceleration rate, time and distance by groups of initial speeds. The average speed reduction across all groups of initial speeds was low and more than 85% of participants managed to ride at 73 km/h after deceleration. These findings could indicate that the majority of participants rode onto the transition elements such as a horizontal curve following the tangents with dangerously high approach speeds.

Table 2. Average deceleration rate, time and distance by groups of initial speed

Initial speed (km/h)	30-50	50-70	70-90	90-110
Number of trips	65	184	432	37
Average deceleration rates (m/s^2)	1.57	0.71	0.75	0.95
Average deceleration time (s)	2.8	6.2	7.3	7.0
Average deceleration distance (m)	12.0	26.8	43.7	35.7
Average final speed (km/h)	23.9	51.6	61.5	76.8
Average speed decrease (km)	14.2	12.3	17.7	16.1
85th percentile of final speed (km/h)		73.0		

Figure 3 displays the speed cumulative frequency distribution plots for the final speeds in acceleration and deceleration profiles on tangents.

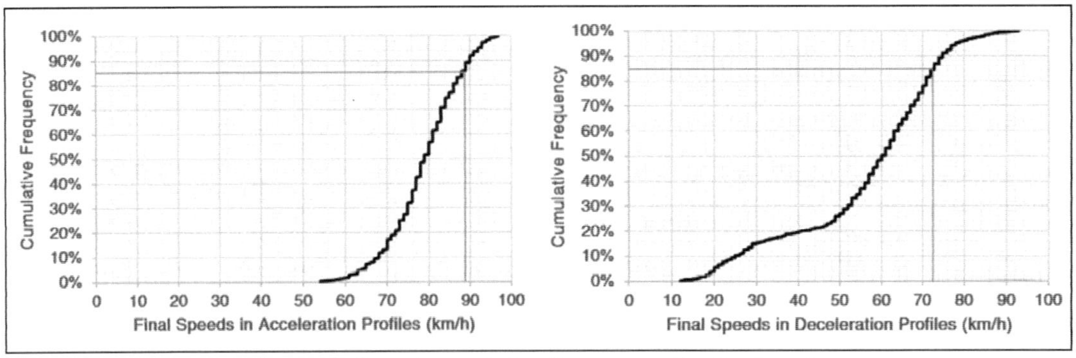

Figure 3. Speed cumulative frequency distribution plots

3.3 Speed profiles during tangent to horizontal curve transitions

The speed change data were normalized with the number of available curves along the route and it was found that the majority of speed changes occurred when participants travelled from tangents to left horizontal curves. There were eight horizontal left curves (LC) along the data collection route. The majority of participants were found to reach their lowest speeds during the transition at the points of curvature (PC) and started to increase their speeds towards the middle of the curve. Table 3 lists the descriptive statistics for the eight horizontal left curves investigated in this study.

Table 3. Descriptive statistics for left horizontal curves ($N = 8$)

Parameter	Mean	*SD*	Median	Minimum	Maximum
Curve radius (m)	64.5	105.9	202.0	18.0	332.0
Average lane width (m)	3.06	0.06	3.08	2.97	3.15
Length of preceding tangent (m)	806.8	429.8	650.0	384.0	1635.0
Initial speed on preceding tangent (km/h)	77.3	8.1	77.0	50.0	97.0
Speed at the beginning of curves (km/h)	63.1	16.5	66.5	20.0	93.0
Speed changes (km/h)	-14.2	15.1	-11.0	-59.0	14.0
Rate of speed changes (m/s^2)	-0.65	0.59	-0.56	-4.17	0.35
Distance travelled during speed changes (m)	33.7	41.5	19.2	0.28	226.9

The speed changes during tangent-to-horizontal curve transitions were found to be overwhelmingly in the form of speed reduction although some data points indicated that participants accelerated through the curves. Table 4 lists the descriptive statistics of speed changes data during tangent-to-horizontal curve transitions.

Table 4. Descriptive statistics of speed changes for individual horizontal left curves

Site ID	Curve Radius (m)	*N* of Events	Speed Changes (km/h)			
			Mean	*SD*	Minimum	Maximum
LC1	97.0	29	-11.6	4.2	-19	-2
LC2	180.0	29	-13.5	7.0	-32	-3
LC3	18.0	29	-48.6	6.4	-59	-26
LC4	298.0	27	-1.0	5.2	-11	14
LC5	281.0	27	-2.7	5.0	-13	7
LC6	256.0	29	-10.8	6.3	-23	3
LC7	202.0	29	-15.2	5.7	-29	-2
LC8	332.0	29	-8.5	6.5	-32	0

Box plots were drawn to examine the distributional characteristics of the speed changes across each individual horizontal curves. Outliers were removed from the dataset prior to the construction of the box plots.

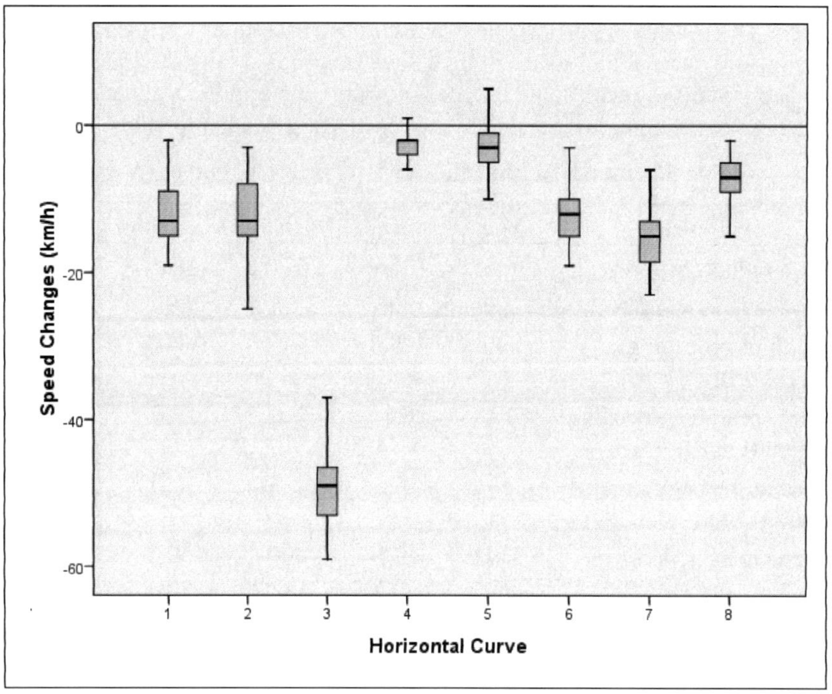

Figure 4. Speed cumulative frequency distribution plots

Based on the box plots, most speeds reduction were in the range of 0-20 km/h (mean = 14.2 km/h) except for LC3 (mean = -48.6 km/h). In addition, the box plots were comparatively tall (except for LC4), which suggests a large variance in the magnitude of speed changes on individual curves. The large speed disparities could also suggest high inconsistencies in speed changes across participants. Compared to other curves, the biggest reduction in speeds was observed at LC3. Among the horizontal curves, the radius of LC3 was the lowest (i.e., the sharpest curve), which could explain the large mean speed reduction. Site inspection indicates that LC3 is located immediately before a petrol station (see Figure 5).

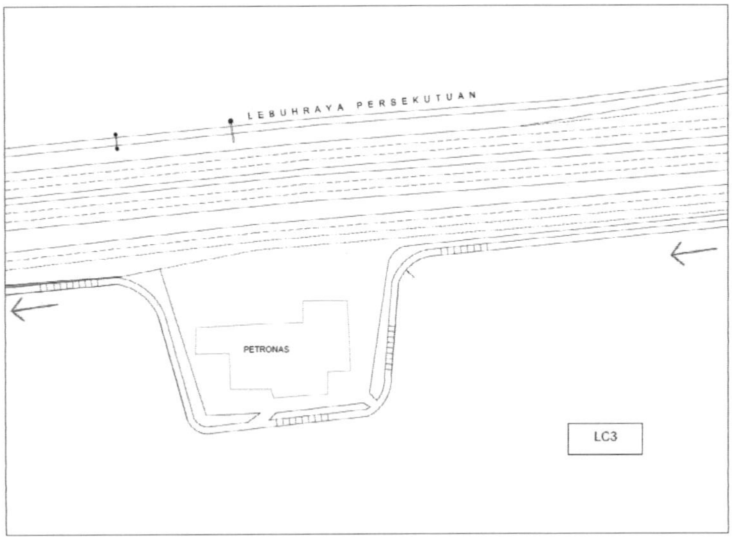

Figure 5. Location of LC3 on the lane

The researchers postulated that the initial speeds on tangents and the geometric design parameters of the lane could have direct or confounding effects on the magnitude of speed reduction. Thus, an Analysis of Covariance (ANCOVA) test was conducted to examine their relationships, with initial speeds as the covariate. Results of ANCOVA suggested that the initial speeds were significantly related to the variations in speeds during the transition, $F(1, 218) = 41.306$, $p < 0.001$. Further, when the effects of the initial speed were controlled, the effects of horizontal curves were found to be significant in predicting the speed changes, suggesting that the change in speed can also be explained by the geometry of individual curves, $F(7, 218) = 225.195$, $p < 0.001$. The homogeneity of variances in speed changes across each curve was confirmed using Levene's test, $F(7,219) = 1.120$, $p = 0.351$.

4. Discussion and Recommendations

As mentioned in the literature review, the proven role of the exclusive motorcycle lane in reducing the number of motorcycle crashes should not be left eroded by the increased risk of a crash along the facility. This study set out with the aim of investigating motorcyclists' speed choice on the lane, focusing on the impact of the lane geometry. The most obvious finding to emerge from the analysis is that motorcycle riders travelled with high overall speeds along tangents and on the approach point during a transition to other elements. Their speed profiles indicate that the change in operating speed occurs frequently and within short travel distance. On top of that, almost half of the acceleration trips were interrupted by sudden and rapid changes in speeds. The speed disparity during tangent-to-horizontal curve transitions was found to be affected significantly by the initial speeds on tangents and the lane geometry. Some of the issues emerging from these findings relate specifically to the risk of a speed-related crash and the need for speed management on the lane. Based on the findings of this study, an introduction of curve-warning signs, accompanied by advisory speed limit signs is strongly recommended at locations where motorcyclists may not be able to judge the safe speed. In addition, traverse bars may be placed before the approach of the transition points along the lane.

5. References

[1] Radin Umar R S 2006 Motorcycle safety programmes in Malaysia: How effective are they? Int. *J. Inj. Contr. Saf. Promot.* **13** 71–9

[2] Hussain H, Radin Umar R S and Ahmad Farhan M S 2011 Establishing speed–flow–density relationships for exclusive motorcycle lanes Transp. Plan. Technol. **34** 245–57

[3] Hussain H, Radin Umar R S, Ahmad Farhan M S and Dadang M M 2005 Key components of a motorcycle-traffic system – a study along the motorcycle path in Malaysia *IATSS Res.* **29** 50–6

[4] Law T H and Radin Umar R S 2005 Determination of comfortable safe width in an exclusive motorcycle lane *East. Asia Soc. Transp. Stud.* **6** 3372–85

[5] Tung S H, Wong S V, Law T H and Radin Umar R S 2008 Crashes with roadside objects along motorcycle lanes in Malaysia *Int. J. Crashworthiness* **13** 205–10

[6] Radin Umar R S 1996 *Accident diagnostic system with special reference to motorcycle accidents in Malaysia* (Doctoral dissertation, University of Birmingham)

[7] Camacho-Torregrosa F J, Pérez-Zuriaga A M, Campoy-Ungría J M and García-García A 2013 New geometric design consistency model based on operating speed profiles for road safety evaluation *Accid. Anal. Prev.* **61** 33–42

[8] Fitzpatrick K, Elefteriadou L, Harwood D W, Collins J M, McFadden J, Anderson I B, Krammes R A, Irizarry N, Parma K D, Bauer K M and Passetti K 2000 *Speed Prediction for Two-Lane Rural Highways* (McLean, VA)

[9] Lamm R, Wolhuter K, Beck A and Ruscher T 2001 Introduction of a new approach to geometric design and road safety *Proc. of 20th South African Transport Conference*

[10] Faezi S F, Hussain H and Davoodi S R 2011 Predicting Speed Model of Horizontal Curves on Exclusive Motorcycle Lane in Malaysia *Aust. J. Basic Appl. Sci.* **5** 590–8

[11] Ibrahim M K A, Hamid H, Law T H and Wong S V 2018 Evaluating the effect of lane width and roadside configurations on speed, lateral position and likelihood of comfortable overtaking in exclusive motorcycle lane *Accid. Anal. Prev.* **111** 63–70

Mix design and application of porous asphalt pavement using Japanese technology

H Nakanishi [1], M O Hamzah[2], M RMohd Hasan[2], P Karthigeyan[3] and O SHaur[3]

[1]TAIYU Kensetsu Co., Ltd. 5-14-2 Kanayama, Naka-ku, Nagoya, Japan
[2]School of Civil Engineering, UniversitiSains Malaysia, 14300 NibongTebal, Penang, Malaysia
[3]Kamunting Premix Asphalt Plant Sdn Bhd, Lot 4188, Jalan Perusahaan 3, Kamunting Industrial Estate, 34600 Kamunting, Perak, Malaysia

E-mail: nakanishi@taiyu.co.jp

Abstract. Early applications of porous asphalt in Malaysia took place in the 1990's on the JKR roads. However, the performance of the material has not been encouraging, primarily due to ravelling and clogging. Initial porous asphalt experience in Japan was imported from Europe in 1987 and encountered similar problems. However, these problems have been successfully resolved by using a high viscosity modified asphalt developed in Japan. Now, porous asphalt has been popular in Japan. This paper presents the design concepts of the technology. The aggregate design incorporates a gap between 2.36mm and 4.75mm to increase the percentage of connected air voids; hence improved permeability and better resistance to clogging. Bonding between aggregate particles is significantly enhanced using TPS modifier that results in approximately binder grade comparable to PG82. The modified binder is characterized by its high mechanical properties, good anti-stripping and anti-aging properties. The results of the tri-axial compression test are introduced to explain the durability of the porous asphalt. In view of the superior laboratory mixture properties, a field trial was conducted on the North-SouthExpressway. The monitoring results after sixmonths in service demonstrate excellent condition. This paper also deliberates on the successful applications of Japanese technology in Asia.

1.Introduction

For several years, the Universiti Sains Malaysia and TAIYU Kensetsu Company Limited have worked together to adapt Japanese porous asphalt pavement technology to improve the performance of porous asphalt pavements in Malaysia. The research venture culminated into a field trial on kilometre 188.6 to 189 on the North-South Expressway. Earlier porous asphalt application in Malaysia in the 1990's was biased to European porous asphalt technologies. However, there are many complaints on its lackof durability and short service life primary due to clogging and ravelling. In the Japanese context, porous asphalt technology was imported from Europe in 1987, but various improvements in modified asphalt and mix proportion of aggregates have been successfully performed to fit climate and traffic conditions in Japan. The improved modified asphalt and porous mixture have been successfully implemented and tried in China [1], Korea [2], Taiwan [3, 4], and others after year 2000.

Laboratory tests at the USM Highway Engineering laboratory affirmed the need for high viscosity modified asphalt and mix design technologies to ensure durability and sustainability of porous asphalt in Malaysia. The research findings have been presented to international [5] and domestic [6] conferences.

This paper presents the fundamentals and characteristics of Japanese porous asphalt pavement technology. The paper also dwells on the porous asphalt mixture shear strength via tri-axial compression test and the mechanism of strength generation and sustainability of the pavement [7], which have been published by the main author and his team. The monitoring test results from a field trial carried out on the South-North expressway based on Japanese technology will be deliberated.

2.Porous asphalt mixture

Porous asphalt mix design implicates selection of an appropriate aggregate grading and blending it with a target binder content. The stability of its aggregate gradation relies on maximising the stone-to-stone contact of the coarse aggregate matrix. The asphalt mortar comprising of asphalt, fine aggregates, and mineral filler; binds the coarse aggregates together. Therefore, asphalt binder, aggregates, and mineral filler are essential components that must be considered.

2.1. Properties of TPS modified asphalt

The Japanese porous asphalt emphasises on the asphalt binder performance. Accordingly, the asphalt binder used must exhibit higher strength or higher viscosity even when subjected to high temperatures. The Japanese standard on asphalt properties for porous mixtures was not initially influenced by the Strategic Highway Research Program (SHRP) requirements. However, the modified binder used that meets the Japanese standard is equivalent to PG82.

The modifier used to strengthen conventional binder penetration grade 60/70 properties is TPS. TPS is a bitumen modifier that can be manually added into thepugmill of the asphalt batch mix plantand let to homogeneously mix with the porous asphalt mixture within one minute. The TPS modified asphalt meets the Japanese standard, and it exhibits better performance in terms of softening point, viscosity at 60 °C, besides other excellent properties. TPS is being employed widely in Asian countries. A detailed explanation on the properties of TPS modified asphalt was presentedat an international conference held in Singapore [5] based on extensive laboratory tests conducted at USM.

2.2. Influence of asphalt property to the sustainability of porous pavement

The durability and sustainability of porous asphalt pavement based on Japanese technology is evident from field trials monitoring results conducted on a motorway in Bangkok, Thailand in December 2015. The site was paved with Thai and Japanese porous asphalt technologies using identical asphalt mixing plant, mix proportion, site location, paving period, paving company and paving machinery but different binders. The Thai porous asphalt used PG 76 binder prepared in Thailand, while the Japanese porous asphalt used TPS modified asphalt equivalent to PG 82.

Plate 1. Ravelling & Pothole on PG76 **Plate 2.** Soundness on PG82

As shown in Plates 1 and 2, after two years in service, the differences in performance is obvious. A lot of ravelling and pothole have occurred on the track area where PG76 was used, while the porous asphalt pavement using TPS modified asphalt remains in good condition. This takes place despite the TPS porous pavement being placed on the slowest lane where traffic loading and time of loading were much more severe. The positive results indicate the positive role of the TPS modified binder since all construction variables and weather conditions are identical.

Figure 1 shows the difference in shearing strength of PG 76 and PG 82 binders. The PG76 results represent typical values since no test data on DSR of TPS modified asphalt used at Bangkok and Thai PG 76 is available. The data shown in Figure 1 is the DSR test result carried out for PG 82 at USM. Regarding PG 76 produced in Thailand, the lower limit values for PG 76 is used, which was supposed to be from the DSR data of PG82. If the temperature of the pavement is assumed to be 65 °C, the elastic component of the complex shear modulus (G*/sin δ) at 65°C of PG 76 is approximately 2.8 kPa, while that of PG 82 is estimated to be 7.5 kPa. The difference in DSR is remarkable, and this explains the differences in the performance of the porous pavement.

The G*/sin δ derived from the DSR test result can be relatedwith therutting property of the mixture. However, according to the research carried out through tri-axial compression test, there is a close relationship between the shearing strength and the Cantabro loss that could be used to evaluate the ravelling property. Thoughit will be explained in Section 3, the results of the field trial in Bangkok corroborate this idea.

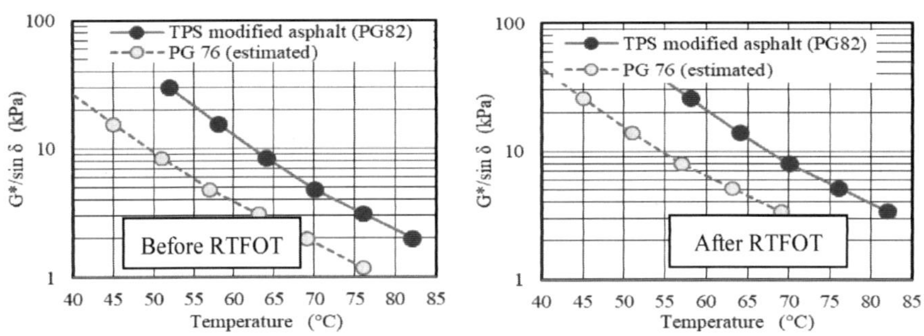

Figure 1. Comparison of DSR between PG 76 and PG 82

2.3. Mix proportion and gradation of porous asphalt mixture

Ensuring proper air voids is essential for a durable porous asphalt mixture. More importantly, these air voids must be interconnected to sustain the drainage function. The ranges of synthetic gradation of the porous asphalt mixture in Asian countries are almost the same.

Aggregate gradation envelope for porous asphalts is typically expressedin terms ofan upper and lower limits. It is not right to assume that any gradation that lies within this envelope would result in a durable porous mixture. To the layman, these gradations are categorized as open-graded. However, more correctly, it should be categorized as not open-continuous but rather an Open-Gap gradation.

This is because of the need to incorporate a gap between 2.36 mm and 4.75 mm to increase the connected air voids percentage and to enhance the mechanical strength of the mixture. By gap gradation means reducing the amount of medium-sized aggregate from 2.36 mm to 4.75 mm. This is because the existence of the medium-sized aggregate sometimes disturbs the formation of the stone-to-stone contact and the source of mix strength and stability; hence decreasing the connected air voids. From experience, the difference between the passing weight percentage at 4.75mm and 2.36mm should preferably be less than 6%.

2.4. Properties of porous asphalt mixture

The porous asphalt mixtures have been designed to conform to specifications used in each country namely; Malaysia, Vietnam, and Thailand. Different aggregate types have been used according to the country as shown in Table 1. However, in all countries, between 4 to 5 % calcium carbonate was used as a mineral filler. Asphalt binder used for these mixtures was TPS modified asphalt, which consists of 88% pen grade 60/70 base bitumen and 12% TPS modifier. In addition, anti-stripping agent, Tough Fix, was used for granite aggregate in Malaysia in weight percentage of 0.3 % to binder content.

The mix design follows several distinct steps beginning with determining the proportion of aggregates to attain 20 % air voids, followed by the determination of the design asphalt content based on the asphalt flow down test. The detail of this procedure has been presented elsewhere [5]. Table 1 shows, the synthetic gradation, asphalt content, air voids and connected air voids, and mechanical properties used according to country.

Table 1. Properties of porous asphalt mixture (in Asian countries)

Item		Unit	Malaysia (In-house @USM)	Vietnam (In-house test)	Thailand (QC on Dec. 2015)	Standards In Japan
Types of aggregate used	Coarse Aggregate		Granite	Basalt	Limestone	
	Fine aggregate		Crushed granite dust	River sand	limestone dust	
Synthetic Gradation	19mm	%	100 (20)[a]	100	100	100
	13.2mm	%	91.6 (14)[a]	92.6	100	90-100
	9.5mm	%	61.5 (10)[a]	47.1	64.2	
	4.75mm	%	18.9 (5)[a]	17.3	15.6	11-35
	2.36mm	%	16.4	14.5	15.0	10-20
	0.075mm	%	5.7	5.0	5.1	3-7
Asphalt Content (TPS content)		%	4.7 (12%)	4.9 (12%)	4.5 (12%)	4 ~ 6
Max. Theoretical Density		g/cm³	2.457	2.614	2.552	-
Density		g/cm³	1.959	2.082	2.043	-
Air Void		%	20.3	20.4	19.9	Around 20
Connected Air Void		%	18.5	18.5	16.9	(Min. 13)[b]
Marshall Stability		kN	8.89	4.17	5.16	Min. 3.43
Flow Value		1/10mm	34.9	22.0	18.0	-
Permeability Coefficient		cm/sec	0.495	0.383	-	Min. 0.01-
Cantabro Loss at 20 °C		%	9.1	12.1	7.6	Max. 20
Dynamic Stability		pass/mm	21,094	10,500	7,000	Min. 3,000
ITSR by DASM[c]			0.84			

Note: [a] Figure in parentheses shows the size of sieve used in Malaysia.
[b] Recommended value.
[c] DASM stands for Dynamic Asphalt Stripping Machine developed at USM.

As shown in Table 1, proper open-gap gradations have been designed for all mixtures. It can be seen that the percentage of medium-sized aggregate from 2.36mm to 4.75mm is very small (less than 3%). The connected air voids exceed much more than the 13% recommendation value. These porous asphalt mixtures also meet all other specified values.

3. Shearing strength of porous asphalt mixture

From our experiences, deteriorations of the porous asphalt mixture occur easily when subjected to high temperature. In practice, the mechanical strengths of porous asphalt mixture, especially shear strength, reduces as temperature increases despite pavements being exposed to no changes in traffic loading. This strength generation mechanism can be explained from the experimental research results on shearing strength by tri-axial compression test [7].

In this section, the results and knowledge derived from the tri-axial compression test are drawn. This research was published in the Japanese Pavement Industry magazine. Nevertheless, the results and knowledge are again presented in this paper since it did not make freely accessible to the public internationally.

3.1. Tri-axial compression test for porous asphalt mixture

This test has been carried out at Research Laboratory of TAIYU Kensetsu Co., Ltd. The photo of the equipment setup, test parameters and typical results are shownin Plate 3, Table 2 and Figure 2, respectively.

Plate 3. Test situation of Tri-axial test

Table 2. Test conditions of Tri-axial test

Item	Test conditions
Size of specimen	φ5×10cm
Casting method	By compaction
Density	Standarddensity ±1%
Test temperature	60 °C
Curing temperature	60 °C, 30 minutes
Loading rate	2mm/min
Side pressure (kgf/cm^2)	0,0.5, 1.0,1.5, 2.0

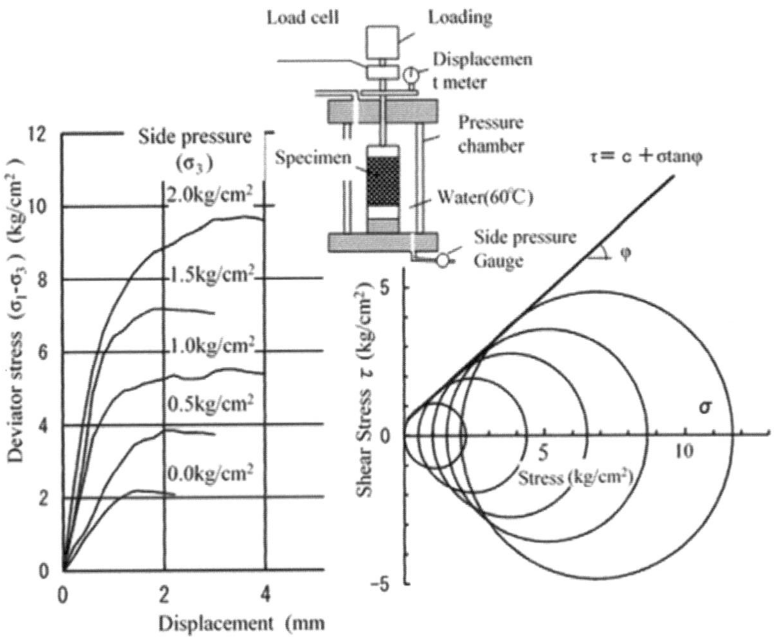

Figure 2. An example of the tri-axial compression test results

Four porous asphalt mixtures prepared with different mix proportions and air voids at approximately 16 %, 22 %, 26 %, and 30 %; were tested. The binder viscosities at 60°C used ranged from 58,500 poise (5,850Pa·s) to 60,000,000 poise (6,000,000 Pa·s). The tri-axial tests were carried out in the temperature-controlled chamber at 60°C using the tri-axial compression machine used for testing soils. Specimens with the same mix proportion were subjected to five levels of side stress (σ_3). A linear relationship was found between the side stress and deviator stress, σ_1-σ_3, and the results are similar to those obtained from the soil test. The internal friction angle (φ) was obtained from the gradient of the straight line, and cohesion (C) is obtained from the intercept of the straight line.

In the study, the unit for stress is kg/cm^2. Tri-axial compression tests have been carried out for all mixtures in which the combinations of the mix proportion of aggregates and the viscosity at 60 °C of asphalt used are respectively different, cohesion "C" and internal friction angle "φ" for each mixture have been experimentally obtained. All tests were carried out at 60 °C.

3.2. Shearing strength of porous asphalt mixture

The shear strength "τ" is expressed in terms of cohesion "C" internal friction angle "φ" and vertical stress "σ" in the form of $\tau = \sigma \tan(\varphi) + C$. Porous asphalt mixture with higher shear strength will be more stable and can better resist external forces. Table 3 shows the test results of the tri-axial compression test and other mechanical properties on each mixture condition. Figure 3 and Figure 4 show the cohesion and internal friction angle, respectively. The results imply that the deterioration of the porous asphalt mixture follows the "Coulomb's Failure Criterion" too as in soil mechanics. Therefore, porous asphalt mixture with higher cohesion and higher internal friction angle should be more sustainable and durable.

Table 3. Test results of tri-axial compression test (at 60°C, σ_1=6.4 kg/cm^2)

Mix proportion	Air void (%)	Viscosity at 60°C (poise)	Test results of Tri-axial compression test: Cohesion (C:kgf/cm^2)	Internal friction angle (φ: degree)	Shearing strength (τ: kgf/cm^2)	Dynamic stability (pass/mm)	Cantabroloss (%)
A	16.9	58,500	0.74	40.0	6.10	3736	9.4
	16.6	1,170,000	0.78	40.7	6.28	5907	7.8
	15.8	9,230,000	0.85	41.8	6.58	7000	4.3
	15.2	60,000,000	0.91	43.0	6.87	10800	3.1
B	22.0	58,500	0.43	40.9	5.97	2167	14
	21.7	1,170,000	0.57	41.8	6.28	4875	9.4
	21.6	9,230,000	0.64	42.0	6.41	5250	6.6
	22.2	60,000,000	0.65	42.8	6.58	7875	6.1
C	26.9	58,500	0.34	40.8	5.86	1768	26.2
	26.4	1,170,000	0.39	40.9	5.93	3627	19.3
	25.9	9,230,000	0.41	41.2	6.01	4717	11.4
	25.9	60,000,000	0.47	41.9	6.21	5727	8.7
D	29.7	9,230,000	0.27	39.4	5.53	2245	33.2
	29.8	60,000,000	0.29	40.2	5.69	3150	27.6

It is essential to knowthe factors influencing cohesion "C".A linear relationship can be identified between the cohesion and change in viscosity at 60°C when plotted on a semi-logarithmic graph, and it can be demonstrated by a multivariate formula using viscosity at 60°C and airvoids. The formula is shown in Equation (1).

$$C = 1.459 - 0.0413Vo + 1.465 \times 10^{-9}Vi \quad (\gamma: 0.979) \quad \cdots\cdots \quad (1)$$

where: C: Cohesion (kgf/cm^2), Vo: Air voids (%), Vi: Viscosity at 60°C (poise).

Equation (1) implies that cohesion increases as air voids decrease and increase in viscosity at 60 °C.

On the other hand, as shown in Figure 4, there is no linear relationship between the internal friction angle and air voids. However, the internal friction angle seems to attain a maximum value when at air voids ranging from 20% to 23%.

Figure 5 shows the relationship between asphalt viscosity at 60 °C, air voids percentage, and shear strength at 60 °C.

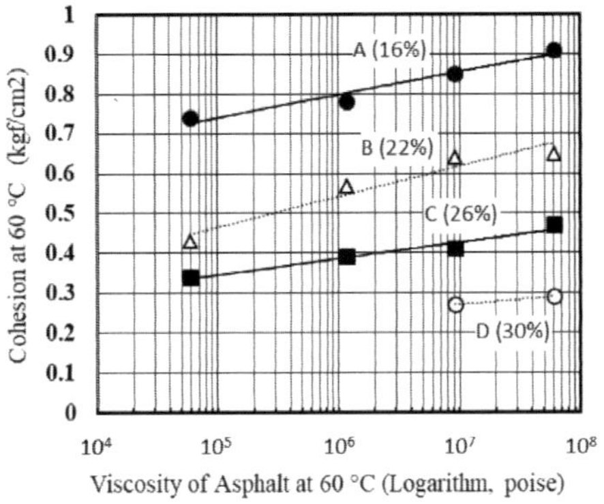

Figure3. Cohesion

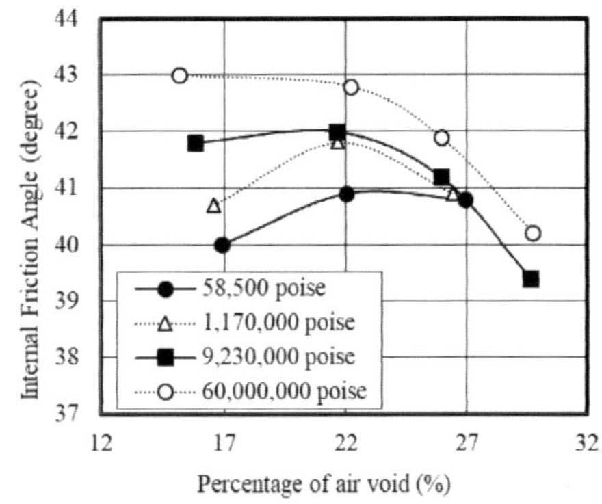

Figure 4. Internal friction angle

In this assessment, vertical stress by tire contact pressure is approximately 6.4kg/cm^2 which is equivalent to 8tons wheel load,was adopted. The shear strength is obtained from the cohesion and internal friction angle in regards with the "Coulomb's Failure Criterion" thatis considered to betrue even for the porous asphalt mixture. The shearing strengths for each combination of mixtureswere then calculated and shown in Table 3. It indicates that if the shearing strength of the porous asphalt mixture is greater than the shearing stress generatedon the pavement surface by the wheel loading, deteriorations of the porous pavement, like ravelling, rutting, and potholes, would not be arising.

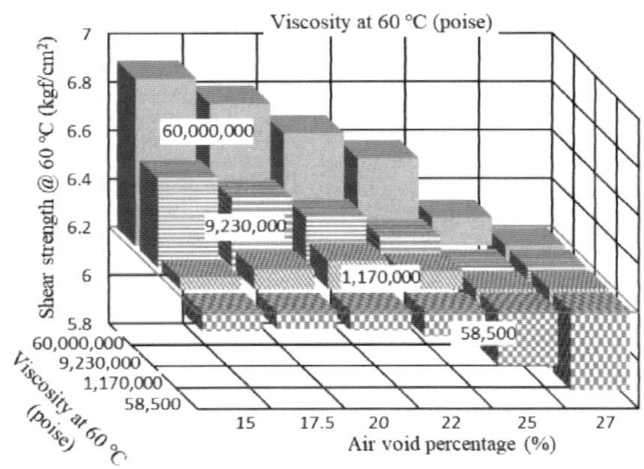

Figure 5. The relationship between asphalt viscosity at 60°C, air void percentage and shear strength

The relationships between shearing strength and other mechanical properties of the porous asphalt mixture, like Dynamic stability and Cantabro loss, are shown in Figure 6 and Figure 7.

The above data demonstrate that porous asphalt mixture which has a higher shear strength will exhibit greater mechanical property. The shear strength was derivedbased on the cohesion and angle of internal friction. In addition, higher cohesion and angle of internal friction can be obtained by using highly modified asphalt binder and proper mix proportions.

The results indicate that the shear strength of the porous asphalt increases along with the increase in G*/sinδ. Figure 6 and Figure 7 show that both rutting and ravelling resistance is dramatically improved along with the increase in the shearing strength.

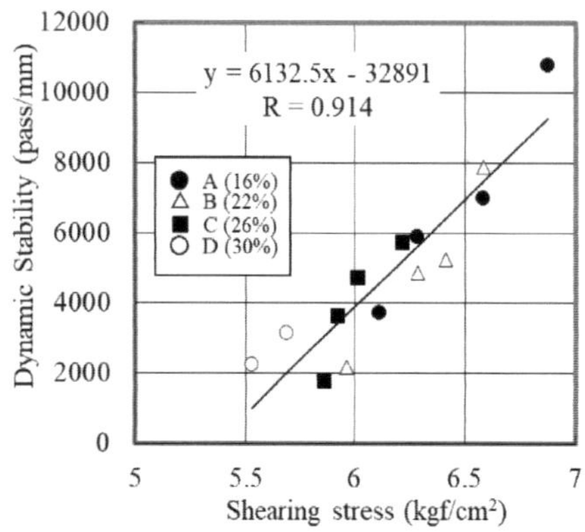

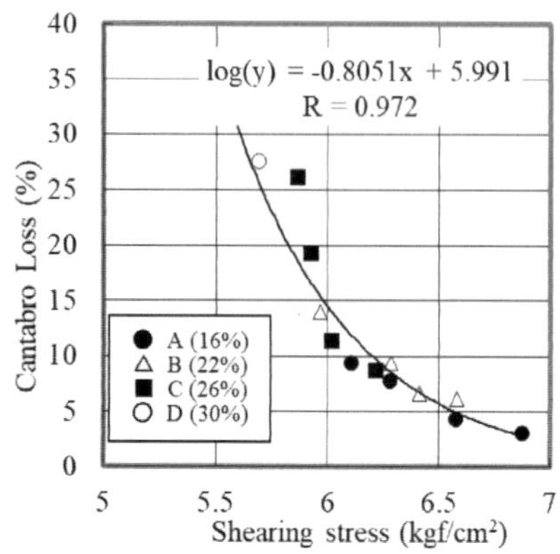

Figure 6. Shearing strength vs Dynamic stability **Figure7.** Shearing strength vs Cantabro

4. Field trial and monitoring results of porous asphalt mixture

4.1. Porous asphalt mixture used for field trial

Porous asphalt pavement based on Japanese technology has been paved on km 188.6 to km 189 on the southbound of North-South Expressway. The 400m stretch was allocated for the trial section in conjunction with the required maintenance by PLUS. Table 4 shows the porous asphalt used for 1st field trial and 2nd field trial. Japanese anti-stripping agent has been employed for these field trials, too.

Table 4. Mix proportion and synthetic gradation of porous asphalt mixture for field trial

		1stField trial	2ndfield trial	PLUS standard
Aggregate Mix proportion(%)	10-14 mm	68.8	74.0	
	5-10mm	17.2	10.5	
	0-5 mm	10.0	11.3	
	Mineral Filler	4.0	4.2	
Synthetic gradation(%)	20mm	100	100	100
	14	96.5	95.6	80-100
	10	56.3	50.2	
	5	18.7	17.6	11-31
	2.36	10.5	14.2	10-21
	0.425	5.6	7.7	
	0.3	5.3	7.0	
	0.15	4.9	6.0	
	0.075	4.3	4.9	2-7
Asphalt content (TPS content)		4.7 (12%)	4.7 (12%)	

The gradation for the 1st field trial does not incorporate a gap in the fine aggregate gradation. Rather, the gradation type is continuous. However, the porous asphalt mixture for the 2nd field trial has been improvedto include a gap in the aggregate gradation. The properties of the mixture are shown in Table 5. Photographs of construction of field trial are shown in Plate 3, Plate 4 and Plate 5. The paving method is similar to that used in the conventional mixture.

Table 5. Mixture properties just after paving for field trial

Item		1st field trial	2nd field trial	PLUS standard (Japanese standard)
Marshall test results	Density (g/cm^3)	1.944	1.963	-
	Air void %	22.2	21.3	20-22
	Connected air void %	18.8	17.2	≧ 13 % [a]
	Marshall stability (kN)	10.31	9.22	-(≧3.43)[b]
	Flow value (mm)	2.32	2.34	-
Cantabro loss ratio (%)		20.6	15.5	Max. 20
Permeability coefficient (cm/sec)		0.286	0.237	- (≧0.01)[b]
Dynamic stability (pass/mm)		10,138	11,307	- (≧3,000)[b]
In-situ permeability (Japan way) (ml/15sec)		1536.1	1521.4	- (≧1,000)[b]

Note: [a] Recommendation value, [b] Japanese standard

Plate 3 Lifting

Plate 4 Rolling

Plate 5 The surface of porous asphalt

4.2. Monitoring results after six months passed

Due to constraints on road occupancy, monitoring was carried out only on the fast lane, and the results are shown in Table 6. Visual observations were performed on the whole area of the field trial. The monitoring of the slow lane will be carried out roughly one year from the construction date. The monitoring situations are shown in Plates 6 to 9.

Table 6 Monitoring results after six months on the fast lane.

Test item		Monitoring results								Standard
Rutting		Rutting is not seen on the whole area of field trial								
Skid resistance	KM	188.65 RHS	188.7 LHS	188.75 RHS	188.8 LHS	188.85 RHS	188.9 LHS	188.95 RHS	Average	≧55.0 (BPN)
	Value	66.6	62.1	64.0	64.4	59.5	61.1	67.4	63.6	
Permeability at site	KM	188.65	188.7	-	188.8	-	188.9	188.95	Average	≧1000 /15 sec
	RHS	1503.8	1510.1	-	1547.7	-	1534.5	1493.8	1522.2	
	LHS	1550.4	1542.4	-	1513.9	-	1502.5	1522.8		
Visual Observation	Cracking	It is not visible in the whole area of the field trial.								
	Ravelling	It is not seenin the whole area of the field trial.								
	Stripping	Surface remains black in colour.								
	Clogging	It is not visible on travelled lane and road shoulder.								

Note: Japanese style on-site permeability tester was used. Measuring time to flow down 400cc water and calculating water volume flowing down for 15 seconds.

Plate 6 Rutting **Plate 7** Permeability **Plate 8** Full view **Plate 9** Black surface

The porous asphalt mixture used for the field trial remained in excellent condition as expected. The monitoring results after six months proved that the performance of TPS porous asphalt pavement was almost similar to the time when it was first paved six months earlier, while the top surfacings remained black in colour.

5. Conclusion

The porous asphalt pavement tends to deteriorate when subjected to high temperature. That is because the mechanical properties come down along with the increase in temperature. Therefore, the porous asphalt mixture which would be able to endure even in the hot temperature is needed in tropical countries like Malaysia. The key elements which influence the durability and sustainability of the porous asphalt pavement are the asphalt used and the mix proportion of aggregates.

The adequacy of the asphalt used for the porous asphalt can be expressed by the various mechanical properties, like Performance Grade, softening point, viscosity at 60 °C, and others. The PG 82 of performance grade asphalt binder or bitumen with the softening point more than 80 degrees C are preferable. Particularly, the ravelling problem that is a concern in Malaysia could be diminished by using asphalt with the high mechanical property. Furthermore, the selection and adoption of an anti-stripping agent should be deliberated.

The other key element for the porous asphalt pavement is to introduce a gap within the aggregate gradation between 2.36mm and 4.75mmsieve sizes. Whereby, the Open-Gap gradation provides a steady state of stone-to-stone contact and improve the mechanical property of a porous asphalt pavement. Furthermore, it improves the size of pores and the connectivity between the airvoids, which is preferable for the self-cleaning and the sustainability of its function.

6. References

[1] Nakanishi H 2002 China Study on improvement in durability of function for porous asphalt pavement *Journal of Guangxi Communication Science & Technology* **27** 7-12
[2] Nakanishi H 2002 Korea Application and experience of special asphalt mixture in Japan *Proc. of the Korea Society of Pavement Engineers Conference* pp 5-57
[3] Nakanishi H et al 2001Taiwan Study on improvement in durability of porous asphalt mixture, *Proc. The 11th Conference on Pavement Engineering* pp 634-42
[4] Nakanishi H Takei S Goto K 2001 Strength generation mechanism for porous asphalt pavement *Proc. The 11th Conference on Pavement Engineering* pp 646-57
[5] Hamzah M Nakanishi H Mohd Hasan M 2016 Proposed Japanese mix design methodology for porous asphalt using modified binder *Int. Conf. Mairepav 8*
[6] Hamzah M Nakanishi H 2016 KL Presentation to PLUS in-house seminar on Japanese porous asphalt technology - Lessons from porous asphalt field trial *Int. Conf. GEOTROPIKA* Bangkok
[7] Nakanishi H Takei S Goto K 1995 Japan Proposal on improvement in functional sustainability of porous asphalt pavement *Road construction* **571** pp 40-53

Acknowledgement
A series of research and trials were attempted for the porous asphalt pavement, and we express our heartfelt gratitude to many individuals and organisations for their understanding and support. We thank PLUS and OPUS for the understanding and appreciating the challenges that we faced and gave us a suitable site and opportunity to carry out the field trial. UEM EDGENTA Berhad through the Infrastructure sector cooperatively supported our activities to develop new porous asphalt pavement in Malaysia. We highly appreciated the cooperation and support for the evaluation tests at the site.

Chemilink in-situ rehabilitations for various pavements in Malaysia

Y F Tan[1], D Q Wu[2], C H Lim[3] and EArasu[4]

[1] Project Coordinator, Chemilink Technologies Group, Singapore
[2] CEO, Chemilink Technologies Group, Singapore
[3] Senior Business Manager, Chemilink Technologies Group, Singapore
[4] Project Engineer, Chemilink Technologies Group, Singapore

E-mail: yihfun_tan@chemilink.com

Abstract. Due to increased traffic volume and conventional pavement designs with poorer subgrades under tropical environment and conditions, most roads are relatively over-loading and the repair or maintenance works to existing pavements become a focus .A super-fast and super-strong technology is required to strengthen the existing pavements in short period with lesser disturbances to both public traffic and operations. Chemical in-situ rehabilitation has become the trend of quick repairs in Malaysia. Chemilink systems have more than 25 years track records in tropical region, its product has been officially approved by Federal Jabatan Kerja Raya (JKR) and Lembaga Lebuhraya Malaysia (LLM) since 2016 in JKR Material List (JMAL). This rehabilitation method can strengthen various pavement base/sub-base courses and achieve higher performances, such as California Bearing Ratio(CBR) values with more than 120% within 24 hours, it also become more popular for pavement upgrading/strengthening and maintenance in airfield and seaport too. It significantly increase overall performances and durability of various pavements with higher technical parameters, faster construction, longer working life and finally overall cost effectiveness. It can utilize the in-situ and local materials in green and sustainable manner. The application procedure include site investigation, mix-design, rehabilitation process, and immediate and long-term performances.

1. Introduction

Through decades of economic development, the volume of transport has increased dramatically from 8.8million to 28.2million in Malaysia. The design capacity using conventional way sometimes cannot fulfil the increased traffic volume/frequency and loading, as some roads originally may have only been designed for usage of P3, but may currently be under P2 or even P1 conditions because of rapid traffic development. Furthermore, this case may worsen as Malaysia is in the tropical region with high water table and rich rainfall, and there are a lot of marine clay, peat and swampy soils which will greatly affect the bearing capacity and performances of pavements.

Various soil improvement methods have been employed to build roads in these high-water table and soft ground areas, such as vertical drains/consolidation, compaction, grouting, and geo-synthetics reinforcement but with limited effectiveness and success. [1]

Frequent pavement maintenance, repair and/or upgrading for heavy traffic roads have become a tough routine work for JKR and the Town Council. In many existing airports under tropical environment in South-East Asia, a simple re-surfacing cannot meet operational requirements with heavier traffic in terms of frequency and aircraft loading, and thus effective strengthening of the base and sub-base of airfield pavements becomes crucial and necessary especially for those in swampy and soft ground areas[2], [3]. The conventional method with replacement of existing quarry materials has significant technical and operational disadvantages for various pavements under heavy operations.

A super-fast and super strong technology introduced by Chemilink is therefore required to strengthen the existing pavements in a short time period with less disturbance to both public traffic and airport/seaport operations.

Since 1995, Chemilink Technologies Group (CTG) has entered Malaysia market and written a new page of rehabilitation history in Malaysia. In the year 1995, JKR and IKRAM tested the soil stabilization product called Chemilink SS-108 in Alor Gajah, Melaka and proved with a testimonial, stated that soil which was stabilized by SS-108 can be easily strengthened from 15% CBR (California bearing ratio) to over 100% CBR within short period but without surface cracking issues that results from cement stabilization. Forthepast 20 over years, Chemilink has not only assisted JKR in solving many critical technical problems, but has also provided total solutions to Kementerian Luar Bandar dan Wilayah (KKLW), Malaysia Airport Holding Berhad (MAHB), Ministry of Agriculture (MOA) and many other public and private sector store solve different engineering issues [4], [5], [6], [7].

Out of many projects, the Senai Airport runway/taxiway widening with highly-clayey soils [8];strengthening of Port Klang container yard under soft and thick marine clay [9], Penang International Airport runway and taxiway upgrading under both higher water table and soft ground conditions [10],are some of the most difficult but successful projects in Malaysia, not to mention Singapore Changi Airport runway widening projects [11] which has featured on Discovery Channel and broadcasted globally since2008.Due to the superb overall results and excellent long-term performances for road projects of JKR in Malaysia for past many years including performance tests such as the in-situ CBR and Falling Weight Deflectometer (FWD) test conducted by the State and Federal JKRs from time to time to inspect the performance of Chemilink rehabilitated or stabilized base or sub-base courses, the Federal JKR, together with the Highway Authorityof Malaysia (LLM) has issued the first approval of Chemilink SS-108 soil stabilization product system (JMAL) in Malaysia (Figure 1). This directly and indirectly proves that this innovative system is the first proven solution for rehabilitation and stabilization of soils and crusher run base/sub-base, especially in our tropical climate and conditions. This solution has also been widely used for roads under peat and swampy areas to settle pavement problems with minimized R&R work and creating a "Floating" Semi-Rigid Platform, which is a proven and trusted solution endorsed by JKR Brunei for the past 20 years. In Perak, where the State has a lot of paddy fields, Chemilink stabilized roads can be found nearby with satisfactory performances and durability.

KEPUTUSAN MESYUARAT JAWATANKUASA BERSAMA (JKR-LLM) KELULUSAN PRODUK BAGI KERJA-KERJA JALAN BIL.2/2016

SYARIKAT	: Advanced Stabilization Technologies Resources PLT
PRODUK	: Chemilink Soil Stabilization
MODEL	: SS-108
BAHAN	: Soil Stabilizing Agent
PENGILANG	: Chemilink Technologies Group PTE LTD (Singapura)
KEPUTUSAN	: **LULUS**

ULASAN:

i. Produk telah mematuhi *Standard Specification For Road Works Section 4: Flexible Pavement* (JKR/SPJ/2008-S4 / JKR 20403 0003 07) dan *LLM/GP/T5 08 : Guideline for Malaysia Toll Expressway System - Design Standard.*

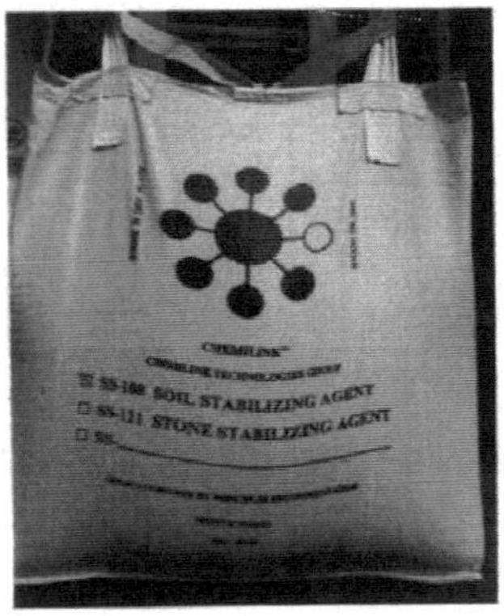

Figure 1. Chemilink Approval under JKR Material List (JMAL).

2. In-Situ rehabilitation and "floating" semi-rigid platform effects

Commonly, in road construction, the unsuitable in-situ soft soils or damaged crusher run materials need to be replaced by imported quarry (crusher run) materials as unbound base layer, but due to some limitations, it is difficult for the quarry materials to perform well and have good durability, and such materials may be much expensive in certain districts as well. Moreover, over-exploitation of natural quarry materials would surely cause negative issues to the natural environment. For urban areas and operating airports or seaports, the problem of operational breaking off is more serious than pavement repairing costs as smooth operations without breaking off is the most important and top priority for various operators. Sometimes, the repairing or maintenance working window is only a few hours and there is no way for traditional methods to repair the pavements quickly and effectively.

For the effective solution to all similar problems, the In-Situ Rehabilitation (ISR) is the best method to fulfil all the requirements including technical performances, green and environmental conditions, overall cost effectiveness within the limited construction time. With ISR, limited fresh natural quarry materials, and much less or even no waste disposals are required. By mixing appropriate percentage (few percentage only) of stabilizers or stabilizing agents with in-situ soils or crusher runs, the agent will improve the engineering properties of soils or crusher runs such as the volume stability, various strengths and moduli, stress-strain relationship, permeability and durability. After strengthening the properties, the rehabilitated or stabilized layer can function like the sub-base or base course of various pavements as the bounded platform. This is the green and environmental-friendly technology which minimizes the usage of fresh resources and dumping of existing unsuitable materials, maximizing the use of in-situ existing materials up to 100%, and significantly improving the overall performances to fit or even update real pavement capacities.

There are a lot of conventional stabilizing agents, such as cement, lime, bituminous materials, coal fly ash, slag, polymer types in liquid form and so on. In a tropical region like South East Asia, due to a lot of swampy and soft grounds, and low land areas together with reverse climate conditions such as rich rainfall as well as high water table, peaty and problematic soils, these conventional agents are

IOP Conf. Series: Materials Science and Engineering **512** (2019) 012027 doi:10.1088/1757-899X/512/1/012027

often not effective or with some fatal limitations. A stabilization system with the type of polymer modified cementitious stabilizing agent called Chemilink SS-108 sub-series has been specially developed in Singapore and successfully applied in this tropical region and other countries in Asia for infrastructural construction, especially for various pavements since the 1990s, with hundreds of engineering projects [12, [13], [14], and this proven system has also been endorsed by some national public works authorities, including the Malaysia Federal JKR and LLM as shown in Figure 1. [15], [16]

The base or sub-base rehabilitated or stabilized by Chemilink SS-108, can forma "floating" semi-rigid platform, which will function as a continuous slab with sufficient strengths, as if it were floating on the existing soft grounds. The concept of this floating semi-rigid platform is like a boat, where it floats on peat or marine clay with high water content or in high-water table area. The rehabilitated/stabilized layer is also almost impermeable to water and thus it is difficult for water from the bottom to penetrate the inside of there habilitated stabilized layer(s) and subsequently to damage the base/sub-base courses and even up to the surface layer. The platform can also minimize differential settlements as it acts as the whole bounded platform rather than individual loose materials.

3. Design criteria and construction

3.1.Rehabilitationdesign and new QC criteria

The standard design for ISR with crusher run or equivalent is as follow:

•Base (7-day): CBR □90-120%, and/or UCS □3.0-6.0 MPa (UCS: unconfined compressive strength)

•Achievable Stiffness Modulus (28-day):2,000-8,000MPa

•Chemical Binder Dosage: 1.50-3.00% (based on the dry weight of the soil or crusher run to be treated)

However, the new QC criteria has been introduced as below which evolutionally changed and strengthened the QC assurance under time contains:

•CBR≥120%within24-hour

After rehabilitation or stabilization within 24 hours, the in-situ CBR test can be carried out before laying of asphalt concrete surface.

3.2.ISR construction procedure

The installation process of ISR is very simple and easy, from spreading, in-situ mixing, to compaction completion. First of all, for the road or pavement to be rehabilitated, the existing premix has to be removed until the base layer. Then spread the stabilizing agent evenly either mechanically or manually onto the surface of the in-situ soil or crusher run to be stabilized, followed by in-situ mixing using an advanced self-running mixer called Stabilizer or Recycler. Two rounds of mixing are generally required in both dry and wet mixings. The dry mixing is mainly to ensure that the stabilizing agent is evenly mixed with in-situ soil or crusher run, and the wet mixing is to achieve OMC (optimum moisture content) status of the mixture for proper compaction, and also for further mixing homogeneity. Finally, the mixture is compacted by roller or other appropriate compactor after it is levelled with a motor grader. In-situ CBR test can be conducted within 24 hours and there is immediate laying of asphalt concrete after the in-situ CBR test.

Figure 2(a)(i). Manual Spreading **Figure 2(a)(ii).** Mechanical Spreading

IOP Conf. Series: Materials Science and Engineering **512** (2019) 012027 doi:10.1088/1757-899X/512/1/012027

Figure 2(b)(i) Dry Mixing

Figure 2(b)(ii) Wet Mixing

Figure2(c)(i) Compaction

Figure 2(c)(ii) Finished Layer of StabilizedBase

Figure 2(d)(i) In-Situ CBR Test

Figure 2(d)(ii) Laying of Asphalt

Figure 2(d)(iii) Immediate OpeningConcrete Photo to Traffic

4. Project case studies

4.1. Various pavement testing results

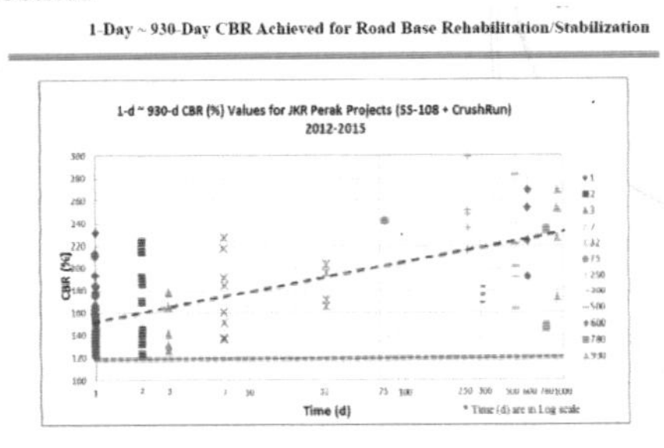

Figure 3a. Perak State CBR Testing Results

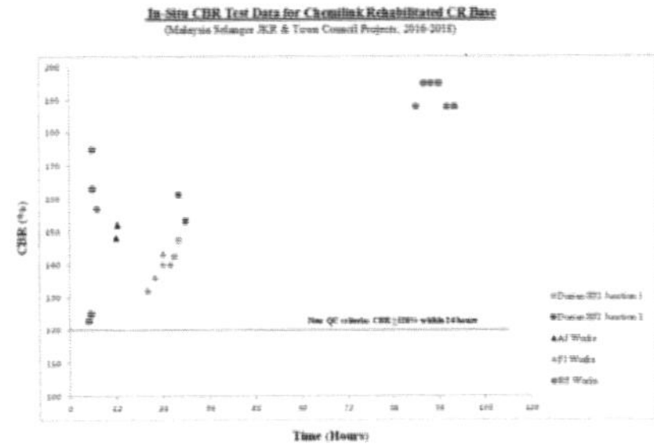

Figure 3b. Selangor State CBR Testing Results

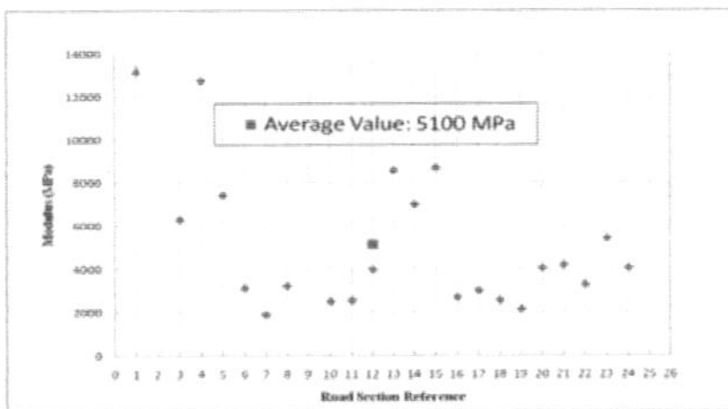

Figure4a. Average Stiffness Modulus

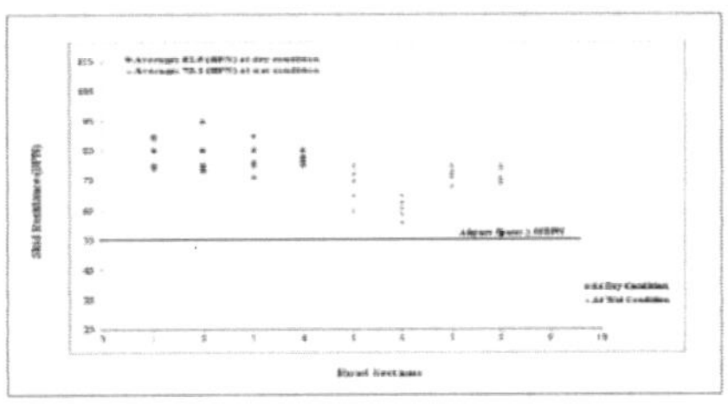

Figure 4b. Average Skid Resistance

As shown in Figure 1, JKR Perak has recorded 3 years CBR data of the rehabilitated base continuously from 2012 to 2015 to get the strength development from 1 day up to 930 days. Figure 3a and 3bshow that overall CBR on the first day is more than 120% and then, CBR increases consistently. Similarly, most of the CBR Testing Results at Selangor were taken within 24 hours, and the minimum CBR is more than 120%. Figure 4a shows that the average Stiffness Modulus of several operational projects with rehabilitated base is 5,100MPa, which is between the Modula of asphalt concrete and cement concrete and thus has properties of a semi-rigid platform. For the skid resistance of the stabilized base (Figure 4b), the average surface skid resistance data under wet and dry conditions are 73BPN and 82BPN respectively, which is much higher than the standard airfield/highway pavement requirement of 55BPN and furthermore verifies that the rehabilitated base has better binding effect with the bottom of upper asphalt concrete layer.

The bonding between asphalt concrete and road base is important, however, the type of test cannot be carried out at site. From understanding, roughness of stabilized layer is an important criteria for good bonding, as such, skid resistance test is carried out to determine the value of skid resistance and estimate the bonding of asphalt concrete with stabilized road base. No debonding of Chemilink stabilized road base and asphalt concrete layer has been found in any projects or sections.

4.2.SenaiAirport runway and taxiway widening

Senai International Airport runway (a half-strength design) and taxiways (a full-strength design) were widened in 2007-2008 after similar works were done successfully in Singapore Changi Airport

[11].The local soils were nearly 100% clayey soil (liquid limit up to 88%, plastic index up to 46%, and moisture content up to 42%, about two times of OMC). With SS-108 soil stabilization, these "unsuitable" clays were strengthened and met all the technical requirements with no defects in any form detected during and after about 5 years in full airport operations. Average UCS is 2.1 MPa, CBR is 120%, resilient modulus is 6,000 MPa and compaction degree is 98%. (Figure 5a &5b after [8]).

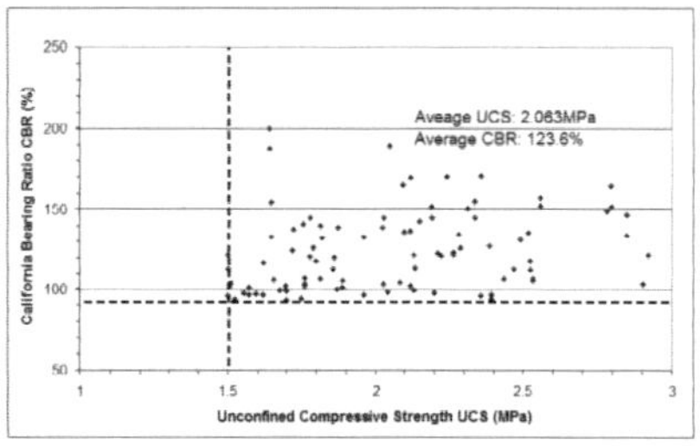

Figure5aCBR vs. UCS Result of Senai

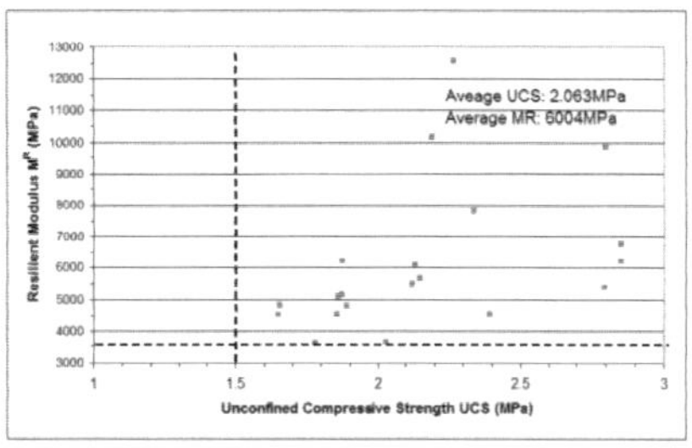

Figure 5b Resilient Modulus vs UCS Senai Runway and Taxiway Widenings

4.3. Penang Airport taxiways rehabilitation

As shown in Figure 6, two layers of the base course of total thickness of 600mm were rehabilitated or stabilized, in which the lower base (Base-2) with 300mm thickness was optional if the existing crusher run materials had reasonable CBR values of not less than 30% to 60%. At the design stage, the rehabilitated materials requirements for both Bases-1 & -2 were at minimum CBR of 180-200% and UCS of 3-5MPa at 7-day in laboratory conditions, meeting most international and local standard requirements.

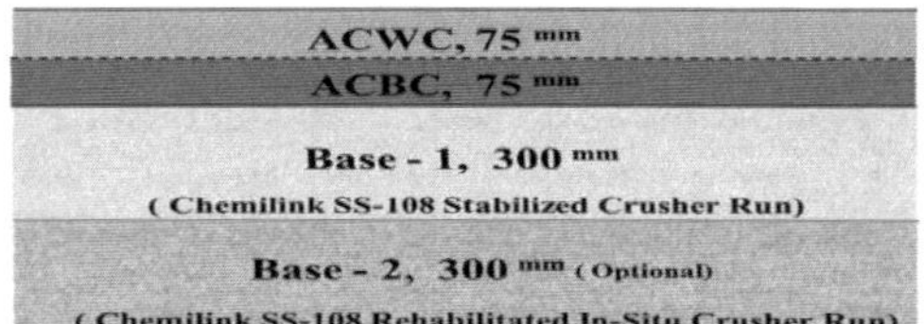

Figure 6. Design of Taxiways Strengthening by In-Situ Rehabilitation

In this Penang International Airport taxiway strengthening project, the 24-hours CBR test criteria was endorsed to shorten the duration of taxiway closing instead of a normal 7-day test [10]. This 24-hours testing can reflect the early strengths of the rehabilitated/stabilized base in meeting the minimum requirements and then proceed to another stage as soon as possible. This is also to avoid long time idling and closing of the working area.

In normal cases, if the working quality had failed and was only known 7 days later when the airport operations are in process already, the resulted damages would be irrecoverable, and the overall losses could be significant. With the 24-hour testing data as shown in Figure 7, it shows that the average in-situ CBR result is 172% within 24 hours, which is 40% higher than the new QC criteria with requirement of 120%. Figure 7 also shows that the CBR value of the base before ISR only ranges from 20% to 60%, and thus the quick strengthening effect looks very impressive.

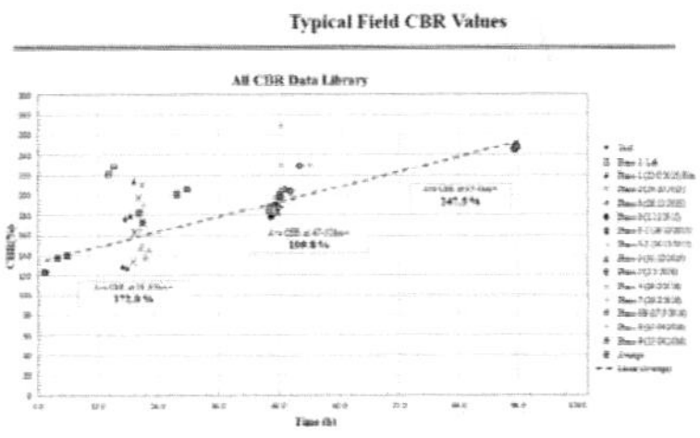

Figure7a In-Situ CBR Results

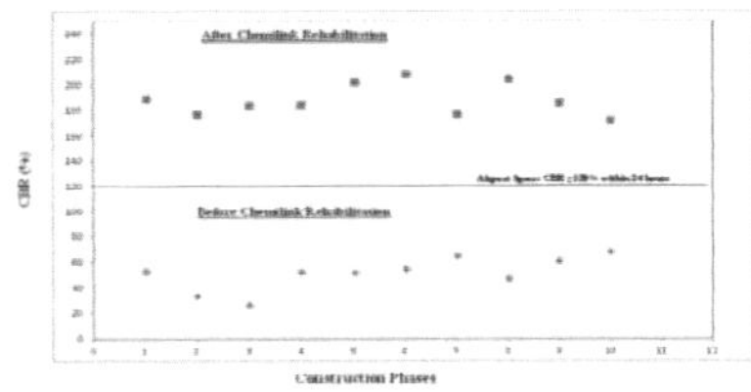

Figure 7b. Comparison of Average CBR Values before and after Strengthening at Each Construction Phase.

4.4.Strengthening of Port Klang container yards

Over the years, the deterioration and serious differential/total settlements of the compacted crusher-run sub-base and concrete surface layer due to factors such as soft marine clay foundation and seepage rendered the container stacking yard unserviceable. The in-situ crusher-run material was recycled using SS-108 in either two layers of 200 mm thickness each or one layer of 300 mm thickness. The project Phase-I was completed in 2011. The average UCS value (7-d) achieved is 2.9 MPa (Specs ≥ 2.0 MPa) and CBR is 140% (Specs ≥ 120%) (Figure 8). The overall performances are satisfactory to date [9]. The Phase II works were completed in 2013 with satisfactory results as well.

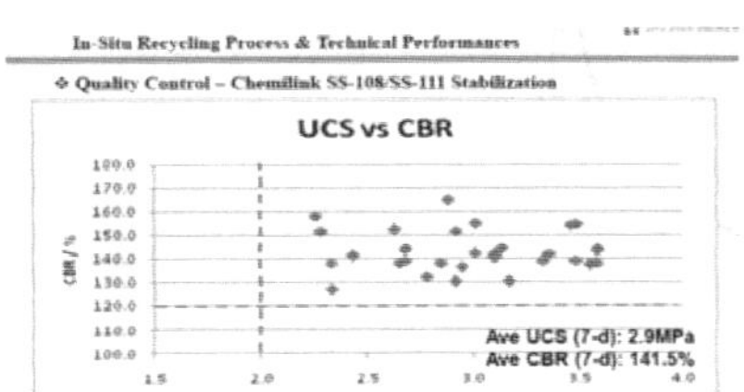

Figure 8. CBR and UCS Results of Port KlangContainer Yard.

5. Conclusions

Chemilink Stabilization is the proven technology, especially over swampy and soft ground areas, in South East Asia for the past 20 over years. The in-situ rehabilitation/stabilization technology and system have significantly promoted the re-use and recycle of the in-situ materials so as to reduce dumping of existing material and the import of fresh quarry materials.

Chemilink is the first approved rehabilitation/stabilization solution with the sub-series product in Malaysia under JMAL. Not limited to JKR and LLM, Chemilink also assisted Malaysia Airport and Seaport in solving the quick repairing difficulties with higher strengths, shorter construction duration, much superior technical performances and durability, and finally overall cost effectiveness. The rehabilitated base/sub-base courses are satisfactory, and no defects or failures have been found during the long-term operations.

6. References

[1] Mitchell J K and Katti R K 1981 Soil improvement – state-of-the-art *Proc. of the 10th Inter. Conf. On SMFE* **1** pp 261-317

[2] Wu D Q 2017a Chemical-soil stabilization for sub-grade improvement *Symposium on Airfield Infrastructure of Airports – Engineering and Future Development, SAIA 2017* Singapore Aviation Academy, 02-03 October 2017

[3] Wu D Q 2017b In-Situ Rehabilitation (ISR) method for pavement strengthening *Symposium on Airfield Infrastructure of Airports – Engineering and Future Development, SAIA 2017*, Singapore Aviation Academy, 02-03 October 2017, Singapore

[4] Wu D Q and Tan P C 2008 Chemilink Stabilization Technologies for Roads and Airfields *Seminar on Soil Subgrade Stabilization, Public Works Department (JKR) of Malaysia and Road Engineering Association of Malaysia* Kuala Lumpur, July 15 2008

[5] Lee M, Tan P C, Daud and Wu D Q 2010 Green approach to rural roads construction – stabilization of in-situ soils and construction wastes *The 7th Asia Pacific Conference on Transportation and the Environment* Indonesia, 03-05 June 2010

[6] Huang F, Romy T, Wu D Q and Shazali B 2014 A quick repair approach for damaged roads in West Malaysia *9th Malaysian Road Conference* Kuala Lumpur, 10-12 November 2014

[7] Wu D Q and Zhang Y L 2016 Pavement strengthening by in-situ rehabilitation & semi-rigid pavement methods *Proceedings of Bridging the East and West: Theories and Practices of Transportation in the Asia Pacific* China, pp 123-32

[8] Wu D Q, Shaun Kumar and Tan P C 2008 Chemical-clay stabilization for runway widening at Sultan Ismail International Airport, Malaysia *13th Singapore Symposium on Pavement Technology (SPT 2008)* Singapore, 23 May 2008

[9] Tan P C, Daud Lee M and Wu D Q 2011 Pavement rehabilitation by in-situ recycling – a case study on seaport container yards and road *The 16thSingapore Symposium on Pavement Technology* (SPT 2011) Singapore, 27 May 2011

[10] Wu D Q, Hamid M A and Aminnudin B Y 2017 In-situ rehabilitation for taxiway strengthening in operational airport *The 10th International Conference on Road and Airfield Pavement Technology* Hong Kong, 08-10 August 2017

[11] Koh M S, Lim B C and Wu D Q 2005 Chemical-soil stabilization for runway shoulders widening at Singapore Changi Airport *4th Asia Pacific Conference on Transportation and Environment (4th APTE Conference*) China, 8-10 November 2005

[12] Yong T C and Wu D Q 1999 Chemical stabilization for road construction in Brunei Darussalam *the First International Conference on Transportation for Developing Countries on Threshold of the 21st Century* Vietnam, 18-19 November 1999

[13] Liu Q, Wu D Q and Gui Z 2004 The application of non-standard stabilizers to the base course of rural roads *1st International Conference on Sustainable Waste Management* Singapore, 10-12 June 2004,

[14] Wu, D Q 2011 A Green and Effective Approach for Pavements in Tropical Region *The 24th ICTPA Annual Conference & NACGEA International Symposium on Geo-Trans* California 27-29 May 2011

[15] CPRU-Brunei 1999 *General Specification for Pavement Stabilization* (Brunei: Ministry of Development)

[16] Suhaimi H G and Wu D Q 2003 Review of Chemical Stabilization Technologies and Applications for Public Roads in Brunei Darussalam *Journal of Road Engineering Association of Asia and Australia* **10** 42-53

IOP Conf. Series: Materials Science and Engineering **512** (2019) 012028 doi:10.1088/1757-899X/512/1/012028

Towards innovative design and construction standards for lime stabilised subgrades

S Young[1], G Ismail[2] and A Chong[3]

[1]Managing Director, Stabilised Pavements of Malaysia Sdn Bhd, Shah Alam, Malaysia
[2]Director/General Manager, Stabilised Pavements of Malaysia Sdn Bhd, Shah Alam, Malaysia
[3]Technical Director/Transportation, AECOM Perunding Sdn Bhd, Malaysia

E-mail: syoung@stabilisedpavements.com.my, ghazali@stabilisedpavements.com.my, chongsi@hotmail.com

Abstract. Recycling of existing and virgin pavement materials has been carried out successfully in Malaysia since the 1980's using the process of Insitu Stabilisation. The majority of specified works have been completed using General Purpose cement as the binding agent for use in granular base and sub base materials. The use of lime is becoming more popular in Malaysia as asset owners realise the importance of creating sound platforms to build upon with pavement structures. In pavement design, the two most critical inputs that influence the overall thickness of a pavement structure are the traffic loadings and the design subgrade CBR. Subgrade stiffness is the controlling variable that can be designed for improvement and should be enhanced wherever possible. Lime is particularly suited to dramatically improve the bearing capacity, reduce the volumetric instability and decrease the moisture susceptibility of clayey, expansive and plastic materials which are commonly encountered in pavement subgrades. Due to a current lack of design methodologies and construction specifications in Malaysia incorporating lime stabilised materials, this paper will explore the influencing factors for design of lime stabilised subgrades with an emphasis on thickness design and mix design parameters.

1. Introduction

Whether building a new road or rehabilitating an existing road, subgrade stiffness is by far one of the highest risk elements to ensure the finished pavement meets design life expectations. Pavement designers recognise that there is great sensitivity in the outcome of a pavement design when the subgrade stiffness (CBR) changes. It is therefore critically important to ensure that selection of an appropriate design subgrade CBR is well understood during the modelling phase.

Unfortunately, we too often see pavement designs demonstrated by drawings that show the design subgrade CBR with a chosen value, but in reality this is rarely understood. For example, if a design subgrade CBR is shown as 5%, what thickness of subgrade is this modelled over? How does this information translate to the construction phase if subgrade treatment is required?

Another example is where many of our design manuals stipulate that based on a given subgrade CBR a minimum pavement thickness is to be considered. The most common understanding of this concept in Malaysia is that the design subgrade is modelled over a thickness of 300mm. There are numerous international papers, design guidelines and technical notes that have demonstrated through various forms of research that this is not adequate. In fact the most common approach internationally is to model at least 1m of subgrade in the determination of a Design CBR value for input into the pavement design process.

The first section of this paper will explore this in more detail with examples showing the differences in pavement design life when 300mm or 1m of subgrade material is modelled. The second phase of this paper will then explore the two common methods of improving subgrade properties, being Remove and Replace or Insitu Stabilisation. Finally a case study will be presented which explores the use of lime to treat marine clays on a project site in Pulau Penang.

2. Design of stabilised subgrade layers

The 'Manual on Pavement Design' (Arahan Teknik Jalan 5/85) was published in 1985 and was largely reliant on the use of empirical design methods that did not consider specific material properties. In 2013 JKR released a revised design guideline titled, 'Manual for the Structural Design of Flexible Pavement' (ATJ 5/85 Pindaan 2013) which promotes the use of mechanistic design modelling techniques.

Chapter 3.4 of the 2013 manual highlights the layered elastic analysis software programs that were utilised in the development of some of the manual outputs. Review of many of these software programs reveals that the modelling of subgrade layers takes into consideration the bearing capacity of up to 1m in most cases. In some international jurisdictions, up to 1.5m is considered [9] so that there is minimal risk of premature failure to the pavement structure as a result of underestimating the bearing capacity of the subgrade.

One of the most significant changes for designers was the minimum properties required for subgrade materials [5] with a minimum design CBR of 5% required for pavements with traffic categories of T1 to T3 and a design CBR of 12% required for pavements with traffic categories of T4 or T5 (>10.0E+06).

For the majority of pavement rehabilitation projects carried out in Malaysia, preliminary testing of subgrade stiffness using soaked CBR testing is generally done on the top 300mm of the subgrade. If test results show values less than CBR5 or CBR12 as the case may be, then it is often the same 300mm that is specified as requiring treatment to improve the subgrade strength to at least the design value. Here lies the inherent problem in that the remaining 700mm of untreated subgrade retains a 'non-complying' strength which has a direct impact on the ability of the overall subgrade to support the pavement structure for the designated design period.

Once the top 300mm of unsuitable soil has been treated sufficiently, according to the various layered elastic analysis programs used for design modelling, the entire 1m of subgrade must still be considered. In practice often the top 300mm is analysed for conformance against the design subgrade CBR – which is incorrect. Whilst there are many methods that have been published to describe how to model the subgrade stratum using different CBR values within the profile, JKR already have an accepted approach shown in Equation 1 from the 1985 Design Manual.

$$\text{CBR} = \left[\frac{h_1\, CBR_1^{\;1/3} + h_2\, CBR_2^{\;1/3} + \ldots + h_n\, CBR_n^{\;1/3}}{100} \right]^3 \qquad (1)$$

where:

$CBR_1, CBR_2 \ldots CBR_n$ = CBR of soil strata 1, 2…n

$h_1, h_1 \ldots h_n$ = thickness of soil strata 1, 2…n (cm)

where: $h1 + h2 + \ldots + hn$ = 100cm

The example below demonstrates the sensitivity of considering only the top 300mm instead of a 1m profile of the subgrade material using the JKR approach.

2.1. Example 1

Consider a pavement where testing of the subgrade reveals that the CBR does not meet the minimum requirement of 12% and therefore some treatment is required.

The design approach is to show a treated 'Top Subgrade' with a conforming CBR of >12% sitting on top of the underlying 'Lower Subgrade' with a CBR of >5%.

The resulting design is as shown in Figure 2.

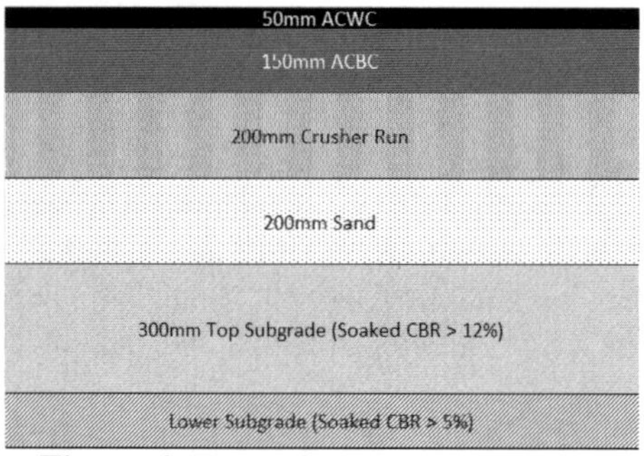

Figure 1. Example Pavement Design

Conventional thinking suggests that this design has a conforming subgrade when the top 300mm is treated (using insitu stabilisation or replacement) and results in a CBR of at least 12%.

Using the approach of considering 1m of subgrade material for modelling, we can determine the Design Subgrade Stiffness using the JKR formula as follows:

$$\text{Design Subgrade CBR} = \frac{[(0.3\text{m} \times 12^{1/3}) + (0.7\text{m} \times 5^{1/3})]^3}{1.0}$$

$$= 6.6\%$$

If we conservatively adopt the Design CBR as being 6% over the full 1m subgrade profile, we can immediately see that this is half of what is actually intended, being a minimum CBR of 12%. The impact of this discrepancy has a significant effect on the design life of the pavement.

Without considering individual layer properties within the pavement, if we apply Structural Layer Coefficients in line with Arahan Teknik Jalan 5/85 as a simplified way to demonstrate the effect on design life, we can show the Equivalent Thickness of the structural layers in the pavement configuration as:

Table 1. Pavement Equivalent Thickness

Layer Type	Thickness	SLc	Ta
ACWC	50	1.0	50
ACBC	150	1.0	150
Base	200	0.32	64
Subbase	200	0.23	46
	600		310

Using an equivalent thickness of 110mm for the base and subbase layers we can now determine the effect on design life using the Thickness Design Nomograph, also extracted from Arahan Teknik Jalan 5/85 in Figure 2.

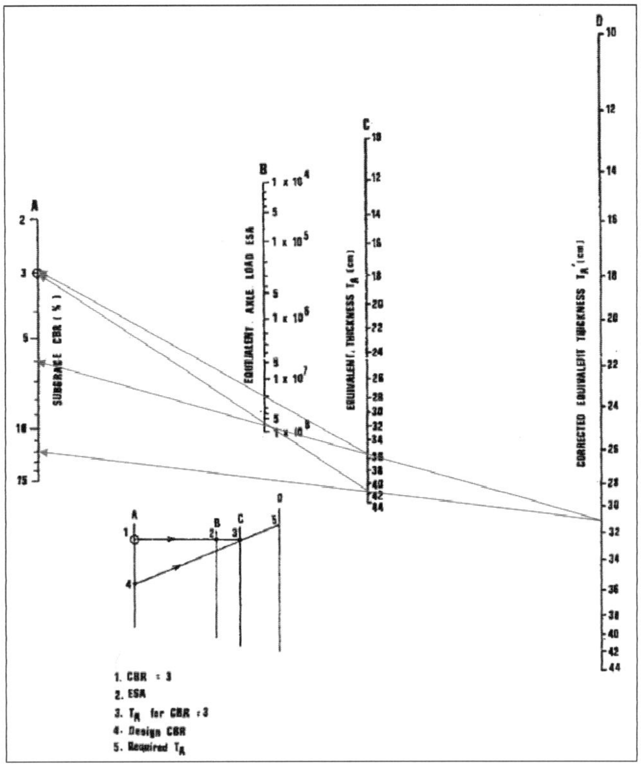

Figure 2. Thickness Design Nomograph

The resulting Design Traffic Loading for each scenario becomes:

1m Subgrade Analysis (CBR6): 2.1E+07

Top 300mm Subgrade Analysis (CBR12): 8.0E+07

This equates to a factor of 3.81 difference in design life. If this pavement was designed over a 10 year period using a subgrade CBR of 12% (derived from the top 300mm only), then it may be expected in reality that this pavement would fail in permanent subgrade deformation after 2.6 years (10/3.81) when modelling with a subgrade CBR of 6% (derived from the full 1m profile).

2.2. *Example 2*

Consider a pavement with an existing clay subgrade of CBR2 with the objective of determining the minimum CBR improvement required for the top 300mm of subgrade if it was designed to be stabilised using lime to achieve an equivalent overall Design CBR of 5% throughout 1m of the subgrade.

$$\text{Design Subgrade CBR} = 5\% = \frac{[(0.3\text{m} \times T_{300}^{\ 1/3}) + (0.7\text{m} \times 2^{1/3})]^3}{1.0}$$

Where T_{300} = Top 300mm subgrade layer CBR

Rearranging this equation,

T_{300} = 21%

This result is demonstrated in the cross section below.

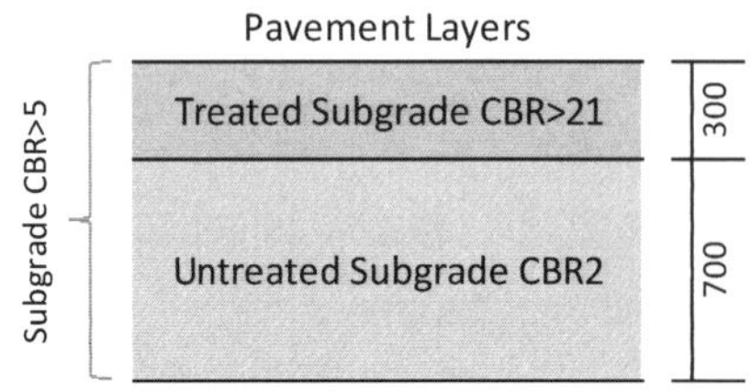

Figure 3. Design Cross Section

The importance of this outcome is to recognise that when a subgrade is designed to have a minimum CBR value (5% in the above example) and the existing material does not meet the design requirement (CBR2 in the above example), it is not just the top 300mm that needs to be improved to a CBR5 to obtain compliance.

3. Design recommendation

Whilst it is appropriate to nominate treatments to only the top 300mm of subgrade layers, it is important to understand that the subgrade beneath this treated layer will still contribute to the overall subgrade bearing capacity of the pavement layers above. If a Subgrade Design CBR of say 5% is used in a modelling program or chosen off empirical charts, then the top 300mm must be treated so as to obtain a resulting CBR far in excess of 5% to account for the underlying untreated subgrade.

Given the simplistic example above, there would be 3 options available to achieve an overall Design Subgrade CBR of 5%:

i. Stabilise the top 300mm with a minimum CBR requirement of at least 30%.
ii. Remove the top 300mm of subgrade and replace using material with a minimum CBR of at least 30%.
iii. Remove 1m of subgrade and replace using material with a minimum CBR of 5%.

3.1. Insitu stabilisation method

The process of using insitu stabilisation as an alternative to traditional remove and replace options provides greater benefits in a number of areas, such as:

- Recycling pavement materials
- Minimising use of quarried products
- Minimising disposal
- Reduced construction cost
- Less construction time
- Reduced energy use
- Less trucks on road network which means less fuel & pollution
- Reduced CO_2 emissions by up to 40% over conventional methods [11]

Further, it has been widely published [3] that the use of lime to treat expansive clay soils provides significant improvements to the material properties, including increased strength, reduced plasticity, increased workability, increased durability and reduced consolidation settlement characteristics. With the correct amount of lime addition, swell is typically reduced close to zero in most cases [3] which provides significant advantages for the use of the treated material.

The process uses specialised mixing equipment with rotating mixing drums that efficiently mix the existing material with a chemical powder binder (eg. lime or cement) which are spread at design application rates in front of the mixing machine. Figure 4 shows a cross section of the mixing process.

Figure 4. Mixing Process

The project photographs below demonstrate examples of subgrade stabilisation that is carried out all over the world, with most developed, developing and under developed nations choosing to stabilise subgrade materials.

Figure 5. Subgrade Stabilisation International Project Examples

To ensure appropriate controls are in place for the construction phase using the stabilisation method, JKR's Standard Specification for Roadworks [6] provides sound guidance in Chapter 4.10 for all aspects related to the process.

The process of stabilisation can be applied to the rehabilitation of existing pavements where unsuitable subgrades are required to be improved, as well as on greenfield sites where the same condition exists.

The three alternatives determined in section 3 above can be further analysed in terms of the benefits such as material quantities and cost. The alternative options to produce an overall Design Subgrade CBR of 5% were:

i. Stabilise the top 300mm with a minimum CBR requirement of 30%.
ii. Remove the top 300mm of subgrade and replace using material with a minimum CBR of 30%.
iii. Remove 1m of subgrade and replace using material with a minimum CBR of 5%.

An assumed volume to be treated must be made first. In this case we will assume a 5km length x 9m width of road is to be rehabilitated and the above options have been generated.

IOP Conf. Series: Materials Science and Engineering **512** (2019) 012028 doi:10.1088/1757-899X/512/1/012028

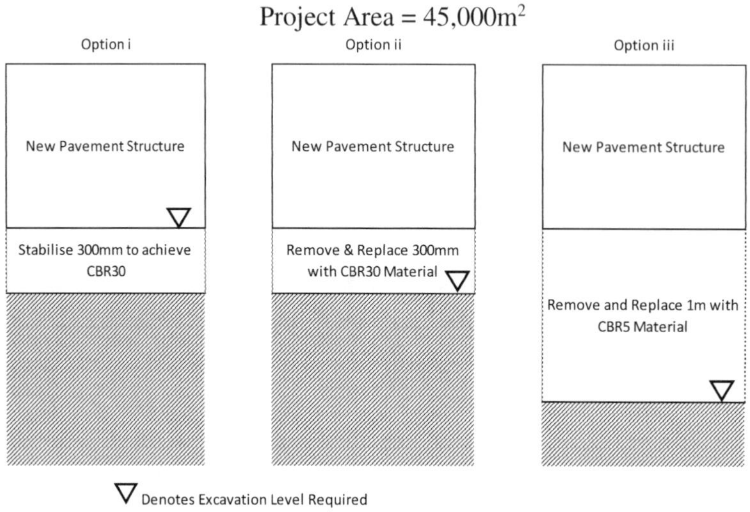

Figure 6. Equivalent_Subgrade Treatment Options

Comparisons can now be made in respect of disposal and imported material quantities as well as cost. It is assumed that the existing level is unknown and any additional excavation required to get down to the top of the subgrade or platform level is a constant among all 3 options.

Table 2. Material Comparisons of Options

	Option i	Option ii	Option iii
Depth of Excavation from Top of Subgrade	0	300	1,000
Disposal quantity (metric tonnes)	0	27,000	90,000
Import quantity (metric tonnes)	0	27,000	90,000

RM/SQM	Option i	Option ii	Option iii
Stabilisation Cost (CBR30)	20	0	0
Excavation & Disposal Cost	0	18	60
Import Cost (CBR30 quality)	0	12	0
Import Cost (CBR5 quality)	0	0	10
TOTAL	**20**	**30**	**70**

It is clear that considerable benefits exist from using the onsite subgrade material and improving the CBR property simply by adopting the stabilisation method. Designers are encouraged to consider this form of material improvement against the traditional method of remove and replace which represents considerably higher environmental degradation and higher costs to the road owner.

4. Case study - Pulau Burung, Penang

Penang Council district is largely built on marine clay which presents long term issues for the construction and maintenance of any load induced structure due to the highly reactive and expansive properties inherent with this material. In 2017 Penang Council issued instructions for the construction of multiple new land fill cells to facilitate ongoing waste collection.

Figure 7. Landfill Site (Cell 1)

For the first time, Penang Council decided to stabilise the existing marine clay to reduce the effects of the expansive influences and to reduce the upfront capital cost compared to previous design solutions which incorporated significant removal of marine clay and replacement with better quality material, covered by geotextile fabric.

Figure 8. Marine Clay

4.1. Mix design

Samples were taken from the site to analyse the material properties and then determine an appropriate mix design to improve the CBR. Host material properties are shown below.

Composition:	Silty Clay 86%
	Sand 14%
Atterberg Limits:	Plastic Limit 43%
	Liquid Limit 71%
	Plasticity Index 28%
4 Day Soaked CBR:	1%

Lime was then added to calculate the minimum quantity of lime required to ensure the chemical reaction remains permanent upon the pH reaching at least 12.4 and then plateauing [12]. This was achieved with the lime demand (or lime saturation) test which produced a minimum lime content of 4%. Soaked CBR testing was then undertaken at the 4% lime application rate to determine what the increase in CBR would be. An average value of 37% was obtained. This level of increase is not uncommon with the incorporation of lime between 3% and 6% achieving increases in strength from 2.5 to 11 times of the untreated soil [8].

The subsequent construction process utilised specialised equipment with a purpose built stabilising machine, capable of mixing the lime into the marine clay at 300mm thickness in single lift.

IOP Conf. Series: Materials Science and Engineering **512** (2019) 012028 doi:10.1088/1757-899X/512/1/012028

Figure 9. Insitu Stabilisation of Marine Clay with 4% Lime

With the forward thinking of Penang Council and their decision to utilise insitu stabilisation as a means of treating the marine clay satisfied the consultants brief as well as significantly minimising further waste materials being sent to landfill and eliminating the need to import other materials from offsite.

5. Summary

Lime stabilisation of subgrade materials in Malaysia is growing in acceptance with asset owners and designers recognising the benefits of improving the engineering properties of the untreated soil. Benefits include enhanced working platforms, reduced pavement thickness, reduced overall construction cost and construction time, reduced risk of premature failure from expansive and reactive subgrade materials and the ability to minimise the volume of imported materials.

For subgrade improvement through stabilisation technology to be successful, access to suitable design and construction standards are necessary. The current JKR construction specification [6] facilitates minimum requirements for cold-in-place recycling (CIPR), however section 4.10 does not deal directly with treatment of subgrade materials from a mix design or construction perspective.

It is recommended that JKR in consultation with REAM develop new design and construction standards suitable for successful implementation of subgrade stabilisation in Malaysia.

6. References

[1] British Standard BS 1924, Part 2 1990 *Stabilized Materials for Civil Engineering Purposes*

[2] Eades J L and Grim R E 1966 A quick test to determine lime requirements for lime stabilization *45th Annual Meeting, Transportation Research Board of the National Academy of Sciences*

[3] Jawad I T, Taha M R, Majeed Z H and Khan T A 2014 *Soil Stabilization Using Lime: Advantages, Disadvantages and Proposing a Potential Alternative*

[4] JKR, Arahan Teknik Jalan 5/85 *Manual on Pavement Design* (Kuala Lumpur: JKR)

[5] JKR, ATJ 5/85 (Pindaan 2013) *Manual For The Structural Design Of Flexible Pavement* (Kuala Lumpur: JKR)

[6] JKR 2008 *Standard Specification For Road Works, Section 4: Flexible Pavement* (Kuala Lumpur: JKR)

[7] JKR 2013 *Standard Specification For Roadworks, Section 2: Earthworks* (Kuala Lumpur: JKR)

[8] Khairul K and Kok KC 2004 Lime stabilized Malaysian cohesive soils *Jurnal Kejuruteraan Awam* **16**

[9] Queensland Department of Main Roads 2009 *Pavement Design Manual, Chapter 5, Subgrade*

[10] Vorobieff G and Murphy G 2003 a new approach to pavement design using lime stabilised subgrades *21st ARRB Conference*

[11] Young S 2014 The most sustainable quarry we have is the one we drive on every day: using stabilisation to provide greener roads *9th Malaysian Road Conference* Kuala Lumpur

IOP Conf. Series: Materials Science and Engineering **512** (2019) 012029 doi:10.1088/1757-899X/512/1/012029

Comparison of bitumen multiple-stress creep recovery test values from the viewpoint of different test parameters

Benešová Lucie and Valentin Jan

Department of Road Structures, Faculty of Civil Engineering, Czech Technical University in Prague, Czech Republic

E-mail: jan.valentin@fsv.cvut.cz

Abstract. Multiple Stress Creep and Recovery (MSCR) test is a suitable test for assessing the resistance of bituminous binders to permanent deformation. This is one of the most frequently used tests on Dynamic Shear Rheometer (DSR) in the world because it describes in a suitable way the nonlinear viscoelastic behaviour of the bituminous binder. The test is performed at higher temperatures and is particularly suitable for modified bituminous binders. The method is described in the European standard EN 16659, which is valid in the Czech Republic since October 2016. The principle of the test is based on multiple loading and recovering of a bitumen sample, which determines the percentage of elastic recovery and non-recoverable creep compliance J_{nr} of the bituminous binder. The paper deals with the assessment of the influence of the different input parameters set on the DSR device - three selected stress levels, two test temperatures and a different geometry of the measuring device. The paper also compares different number of loading cycles applied during the test. Conclusions are made with respect to selecting appropriate test temperature and to consider most suitable test geometry. In case of additional stress level further laboratory assessments are needed.

1 Introduction

The durability of an asphalt mixture is largely dependent on the performance behaviour of the used bitumen. Basic empirical tests can characterize or classify bituminous binders, but in order to more accurately describe their behaviour, functional tests are needed, for example by using Dynamic Shear Rheometer (DSR).

In recent years there has been an increasing demand on road traffic in most parts of Europe, which lead to the necessity for developing bituminous binders with improved performance behaviour if compared to conventional paving grade bitumen. This causes the progress and use of a wide range of bitumen modifiers which improve the performance of the basic bitumen and thus the asphalt mixture [1, 2, 3, and 4]. The most widely used technical standards for bituminous binders in the Czech Republic as well as in many other countries I Europe are based on the behaviour of neat binders that cannot properly characterize potentials of modified binders.

One of the fundamental failures of asphalt pavements at higher operating temperatures are permanent deformations (rutting). Since the 90s′ rutting prediction of asphalt mixtures was based on

G*/sin δ bitumen criterion. This parameter works reliably for conventional speed and moderate traffic volume, but for pavements with slowly moving loads and/or high traffic volumes it has been shown that it does not correspond sufficiently to permanent deformations [4, 5, and 6]. Values of complex shear modulus (G*) and the shift factor (δ) are measured in the linear viscoelastic domain, but permanent deformations are not a linear viscoelastic phenomenon and therefore the characteristic G*/sin δ does not sufficiently correlate with permanent deformations [7]. From the perspective of permanent deformation of bituminous binders there is a need to pay attention on the area of nonlinear behaviour of bituminous binders, which include J_{nr} parameter obtained during the MSCR test. According to the authors [8, 9] parameter J_{nr} has a significantly better correlation with rutting measurements on asphalt mixtures than conventional oscillatory frequency sweep test.

However, since bituminous binders are complex viscoelastic materials determined parameters are very sensitive to boundary conditions and pre-defined test parameters. In case of dynamic tests which can be performed on DSR these parameters are related to selected test temperature(s), used test geometry including the thickness of tested bitumen sample, level of applied stress, number of cycles as well as range of gathered data which are considered for final calculations.

2 Multiple stress creep recovery test (MSCR)

Multiple Stress Creep and Recovery test is a test of repeated loading and relaxation/recovery, using a device called Dynamic Shear Rheometer (DSR). The test is used for measuring the accumulated strain in the bitumen within the prescribed stress level. The MSCR test uses a constant stress creep of 1 second duration followed by a zero stress recovery of 9 seconds duration. This cycle is according to technical standards in Europe or the U.S. repeated 10 times for each stress level. Figure 1 presents the typical creep and recovery cycle in the test.

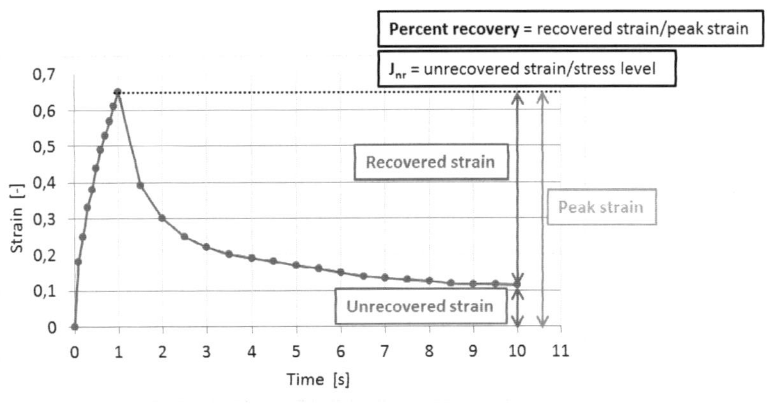

Figure 1. Typical creep and recovery cycle

Two most important parameters of bituminous binder calculated from the measured strain at individual stress levels are non-recoverable creep compliance (J_{nr}) and percent recovery (R). Jnr is a measure of the amount of residual strain left in the specimen after repeated creep and recovery, relative to the amount of stress applied. Parameter Jnr is currently perceived as the best rheological parameter for the assessment of the susceptibility of bituminous binder to resist permanent deformation in asphalt mixture. The lower the J_{nr} is, the greater proportion of the stress the bitumen is able to absorb and the less susceptible to plastic deformation it is. Nevertheless, there are no threshold values neither in the Czech technical standard nor in the European standard. Besides the mentioned characteristics, stress sensitivity parameters R_{diff} and $J_{nr,diff}$ are calculated. Golalipour [10] indicated that determine the stress level is not a simple task and no one level can be clearly defined as most appropriate of what a binder is exposed to in a real mixture. Therefore it seems most reliable to use the sensitivity parameter ($J_{nr,diff}$). The $J_{nr,diff}$ is important mainly when pavement is exposed to unexpected traffic loads or extremely high temperatures. In case the $J_{nr,diff}$ value is greater than 75 %, the bituminous binder is classified as a very stress sensitive material, which is not recommended for use in the construction of a pavement structure, [11].

MSCR is a test that has received a lot of attention and holds great promise for predicting rutting resistance of asphalt mixtures, but there are still many problems in its implementation, evaluation and interpretation of results. Some of these concerns are related to the number of cycles, the value of stress level or the temperature of testing. Conventional test set is based on the standard EN 16659 [12] and indicates the stress levels (0.1 and 3.2 kPa), number of loading/unloading cycles at each stress level (10) and the geometry of the device (25 mm parallel plate with a 1 mm gap). The test temperature is not exactly defined, which remains still a questionable task for the future.

Some research works showed [8, 13, and 14] that in order to study the bitumen behaviour in the nonlinear domain, it is necessary to apply a higher stress level than 3.2 kPa, depending on the bitumen type. Moreover, according to studies [13, 14] the higher stress level is used, the better correlation coefficient for resistance to permanent deformation of mixtures (French rutting test) is received.

3. Used testing procedures

This paper compares different groups of bituminous binders due to the widest possible coverage of the different behavior of bituminous binders. Some variants are conventional paving grade bituminous binders, but in most cases modified binders (crumb rubber – labeled CR, polymer modified binders – PMB or low viscosity additives) are tested and analyzed. Attention has been given to assessing the impact on the selection of the applied stress level and temperature of testing. Furthermore, two different testing geometries were used on DSR device and finally calculations were compared using the average of 10 cycles with values calculated from the last 5 cycles.

3.1 Effect of test temperature and applied stress level

The attention in this part is paid to the use of different test temperatures and stress levels. In the technical standard EN 16659 [12] test temperature is not prescribed, but should be selected suitably in the range 50-80 °C. The exact prescription of one temperature is even not planned for the revised EN 14023 standard for PMB binders. Different polymer modified binders (PMB) and one neat bitumen were compared. PMB are used in highly loaded pavement structures and for this reason test was measured at normal test temperature (60 °C) and elevated test temperature (70 °C). On the contrary neat bitumen 50/70 was measured at normal test temperature (60 °C) and at a reduced test temperature (50 °C). Another parameter that can be determined during the MSCR test is using different stress levels. The standard EN 16659 prescribes stress level of 0.1 kPa and 3.2 kPa. But as mentioned above, values obtained at higher stress level correlate better with permanent deformations. In addition, the application of a higher stress can cause significant increase of J_{nr}. For this reason, the standard stress levels were completed with 8.0 kPa stress.

The following Table 1 presents the calculated values of elastic recovery R, non-recoverable creep compliance Jnr and stress sensitivity characteristics R_{diff} and $J_{nr,diff}$.

As seen in Table 1 it is obvious that the PMB binders have high value of elastic recovery and very low non-recoverable creep compliance. As the temperature increases the shear strain increases and that is associated with an increase of J_{nr} value. Increased susceptibility to permanent deformation is logically at higher temperature (higher J_{nr} value). This is connected with the reduction of elastic recovery, but surprisingly, slight increase of R was observed for two PMB binders at a stress level of 0.1 kPa. This increase is probably caused due to the variability of the measured values (detailed in section 3.3).

Table 1. Values of elastic recovery, non-recoverable creep compliance and sensitivity parameters

	0.1 kPa		3.2 kPa		8.0 kPa		0.1-3.2 kPa		3.2-8.0 kPa		0.1-8.0 kPa	
Binder type	R [%]	J_{nr} [kPa^{-1}]	R [%]	J_{nr} [kPa^{-1}]	R [%]	J_{nr} [[kPa^{-1}]	R_{diff} [%]	$J_{nr,diff}$ [%]	R_{diff} [%]	$J_{nr,diff}$ [%]	R_{diff} [%]	$J_{nr,diff}$ [%]
50/70_50 °C	2.5	1.59	1.2	1.09	0.5	1.08	53.7	-31.2	60.1	-0.9	81.5	-31.8
50/70_60 °C	0.5	4.93	0.1	4.41	0.0	4.57	87.3	-10.7	60.3	3.7	95	-7.4
45/80-65_60 °C	83.1	0.12	85.5	0.09	87.2	0.07	-2.9	-23.3	-2.0	-17.1	-5	-36.4
45/80-65_70 °C	89.8	0.19	85.6	0.20	57.4	0.64	0.1	-15.8	32.9	213.3	33	163.8
25/55-80_60 °C	69.4	0.15	66.6	0.16	66.4	0.17	4.1	3.8	0.3	5.1	4.3	9.1
25/55-80_70 °C	70.4	0.42	58.9	0.52	52.8	0.63	16.3	25.4	10.4	20.9	25	51.7
25/55-60_60 °C	76.8	0.13	38.5	0.49	29.7	0.57	49.9	267.0	22.9	14.7	61.4	320.8
25/55-60_70 °C	55.1	1.06	24.7	1.85	13.5	2.38	55.2	73.6	45.3	28.8	75.5	123.6

The paving grade bitumen 50/70 has high stress sensitivity R_{diff} (Figure 2), which is caused by very low elastic recovery values (Table 1.). This neat bitumen also experienced the most pronounced reduction in elastic recovery at a higher test temperature. The most significant change in J_{nr} value between test temperatures in case of PMBs was found for 25/55-60. This binder seems no longer to be in the linear viscoelastic range at 70 °C. Figure 2 presents that this binder has a high sensitivity R_{diff} (at both temperatures) and Figure 3 shows an extremely high sensitivity $J_{nr,diff}$ at 60 °C. An extreme increase is due to the fact that at stress 3.2 kPa, there was a significant increase of J_{nr} (compared with a stress 0.1 kPa), which leads to a high parameter $J_{nr,diff}$. This binder is the only one that has failed the US condition $J_{nr,diff} < 75$ % for stress levels 0.1 kPa and 3.2 kPa. The same binder has a lower $J_{nr,diff}$ (0.1-3.2 kPa) at temperature of 70 °C because it has a relatively high J_{nr} value at stress 0.1 kPa, (the binder is in nonlinear domain at higher temperature).

In contrary PMB 45/80-65 changed the determined values significantly between stress levels 3.2 kPa and 8.0 kPa at 70 °C (Figure 2.). Other binders can be described as very low stress sensitive. Generally, the increasing temperature increases deformation, reduces elastic recovery and increases non-recoverable creep compliance. Difference is more pronounced with higher stress level.

Additionally it shall be noted why the selected temperatures for the paving grade bitumen 50/70 and for the selected polymer modified binders are not the same. The determining factor in this case is the softening point. For the paving grade bitumen it was expected and proven by the ring & ball test that the softening point range around 50 °C, whereas polymer modified binders were expected to have softening point values ≥60°C. If all binders were tested for MSCR test at 50°C polymer modified binders would show very low values at the outline of measurability. On the contrary testing paving grade bitumen at 70°C would lead to uncertain results due to low viscosity the bitumen would have at this temperature. Therefore the temperatures used for paving grade and for modified binders differ.

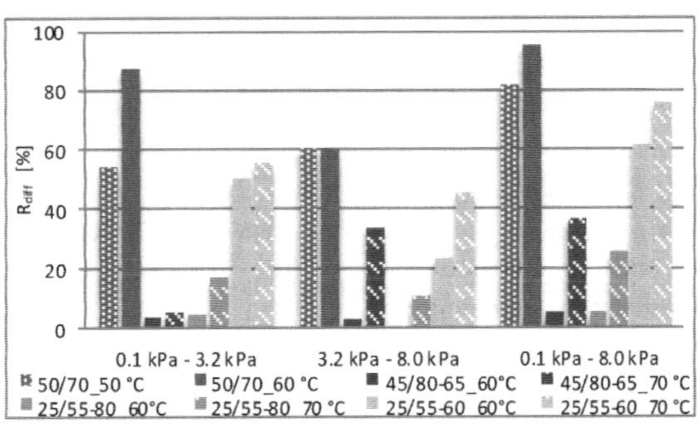

Figure 2. Stress sensitivity parameter R_{diff}

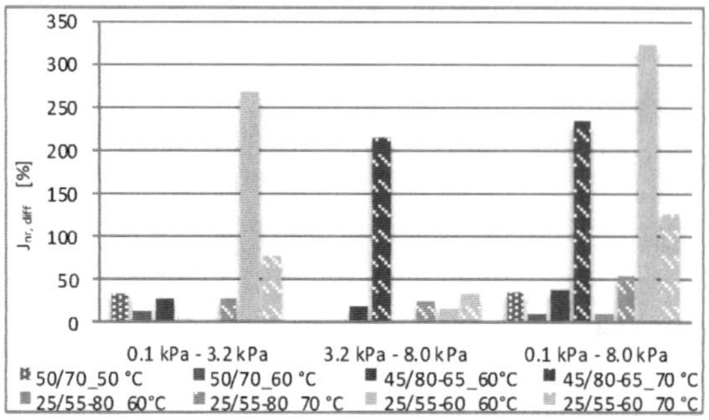

Figure 3. Stress sensitivity parameter $J_{nr,diff}$

Higher stress level 8.0 kPa was conducted also for the second group of binders containing variants with the basic paving grade bitumen 20/30. This binder was modified by selected low-viscosity waxes in an amount of 2-4 M% or by combination of crumb rubber CR (15 M %) with low viscosity waxes. Higher stress level was determined only for selected variants.

Table 2. Values of elastic recovery, non-recoverable creep compliance and sensitivity parameters

Binder type	0.1 kPa		3.2 kPa		8.0 kPa		0.1-3.2 kPa		3.2-8.0 kPa		0.1-8.0 kPa	
	R [%]	J_{nr} [kPa^{-1}]	R [%]	J_{nr} [kPa^{-1}]	R [%]	J_{nr} [[kPa^{-1}]	R_{diff} [%]	$J_{nr,diff}$ [%]	R_{diff} [%]	$J_{nr,diff}$ [%]	R_{diff} [%]	$J_{nr,diff}$ [%]
20/30	9.4	0.425	6.7	0.402	-	-	28.6	-5.3	-	-	-	-
2% CM	15.4	0.209	8.9	0.296	4.0	0.385	42.6	41.7	55.4	29.9	74.4	84.0
3% CM	18.1	0.222	14.1	0.236	-	-	22.0	6.4	-	-	-	-
3% BM	11.1	0.656	5.4	0.644	-	-	51.2	-1.7	-	-	-	-
4% BM	14.6	0.462	7.3	0.495	2.3	0.590	50.0	7.0	68.4	19.2	84.2	27.5
15% CR	73.5	0.015	67.9	0.018	-	-	7.6	20.9	-	-	-	-
15% CR+2% CM	81.7	0.006	74.6	0.008	73.3	0.009	8.7	31.8	1.8	6.3	10.3	40.2
15% CR+2% BM	78.6	0.007	75.7	0.008	73.3	0.009	3.7	12.5	3.2	11.4	6.8	25.3
15% CR+2% E10	72.7	0.011	70.6	0.012	67.6	0.013	2.9	1.8	4.2	9.1	7.0	11.1

Table 2 shows the fundamental values calculated from MSCR test. Crumb rubber (CR) modification resulted in a significant increase in the elastic recovery and decrease in J_{nr} in comparison with referential 20/30 bitumen. High J_{nr} values are expected to mean low resistance to permanent deformation. Application of low-viscosity additives (CM, BM and E10 synthetic waxes) does not significantly influence R or J_{nr} values. R values are very low and J_{nr} relatively high, when low-viscosity waxes are used with the bitumen 20/30. This leads logically to higher values of stress sensitivity characteristics R_{diff}. On contrary, CR binders have high elastic recovery, which helps to have lower values of R_{diff} (below 10 %) for all stress levels. This indicates good elastic properties even at a stress level of 8.0 kPa. When comparing the stress sensitivity parameter $J_{nr,diff}$, it is obvious that the US limit for stress level 0.1 kPa and 3.2 kPa ($J_{nr,diff} < 75$ %) would be fulfilled by all tested binders. But it is clear that a binder that has 2% CM is the most stress sensitive and especially $J_{nr,diff}$ is in this case high if comparing low (0.1 kPa) and high (8.0 kPa) stress level.

3.2 Effect of selected testing geometry used for DSR

In general, for more than two decades two testing geometries are used with the DSR for bitumen testing, namely 8 mm diameter plate-plate with a 2 mm gap (labelled PP08) or 25 mm diameter plate-plate and 1 mm gap (PP25). Geometry used for the testing device is based on the operating conditions. Geometry PP08 is generally used at low and medium temperatures (-5 to 30 °C) or for aged binders (frequency sweep test). Geometry PP25 is preferred for measurements at high temperatures (30 to 90 °C) and is therefore selected as a base for performing MSCR test. Different geometries were compared for PMBs and neat 50/70 after ageing, and in addition for bitumen modified by crumb rubber (CR).

Table 3. Values of elastic recovery, non-recoverable creep compliance and sensitivity parameters

Binder type	PP25				PP08			
	0.1 kPa		3.2 kPa		0.1 kPa		3.2 kPa	
	R [%]	J_{nr} [1/kPa]	R [%]	J_{nr} [1/kPa]	R [%]	J_{nr} [1/kPa]	R [%]	J_{nr} [1/kPa]
50/70+10% CR	35.5	0.50	7.8	0.81	33.7	0.42	7.5	0.70
50/70_R	5.6	1.59	1.1	1.57	8.1	0.54	5.1	0.58
25/55-60 _R	73.9	0.16	73.2	0.15	73.9	0.07	74.8	0.06
40/100-65_R	84.8	0.09	88.9	0.04	84.8	0.03	89.2	0.02
50/70_R+P	17.0	0.28	14.0	0.23	25.2	0.06	23.9	0.06
25/55-60 _R+P	64.5	0.07	64.2	0.06	69.0	0.02	65.1	0.03
40/100-65_R+P	88.6	0.02	88.2	0.02	91.7	0.01	90.1	0.01
50/70_3R	16.9	0.32	13.0	0.28	21.5	0.11	22.2	0.08
25/55-60 _3R	70.0	0.07	65.6	0.09	67.0	0.05	66.9	0.04
40/100-65_3R	92.1	0.02	91.8	0.02	82.7	0.02	90.4	0.01

Table 3 presents quite low difference of the percent recovery between the two geometries. The relative difference is up to 10 % with the exception of all variants of 50/70 bitumen, which have very low values of R and for this reason the percentage change is significant. The relative difference between the two geometries of J_{nr} value is shown in Figure 4. The J_{nr} value is always lower in the case of PP08 and even more than about half in most variants (with the exception of variant with CR). Overall, this implies that the selected testing geometry has a major influence on the J_{nr} values.

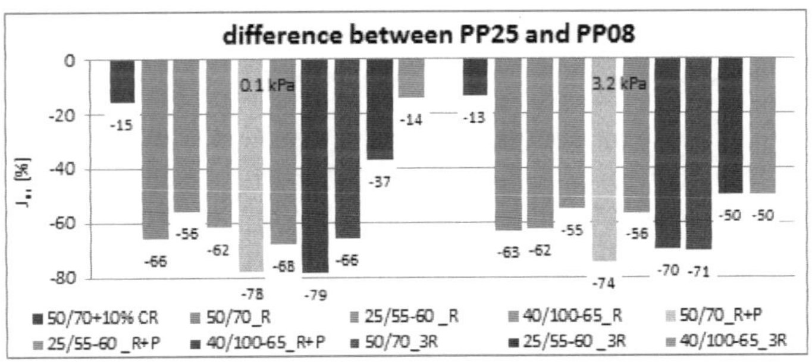

Figure 4. Percentage difference between two geometries

3.3 Effect of number of cycles for calculating characteristics of the MSCR test

Taking the average of the 10 cycles to calculate characteristics, as required in the MSCR standard procedure, could be misleading because the response is significantly changing with number of cycles. The following figures show the percentage difference for the calculation from the last 5 cycles compared to the calculation from all 10 cycles for each group earlier.

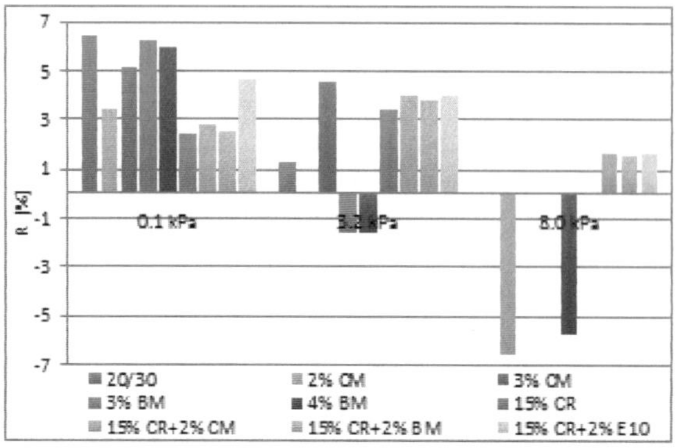

Figure 5. Percent change between two methods of calculating R

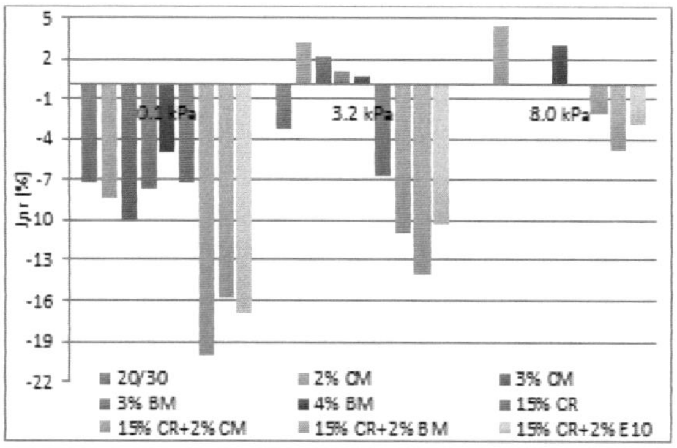

Figure 6. Percent change between two methods of calculating J_{nr}

Figure 5 presents the percent change of percent recovery between the values calculated from all 10 cycles, and only the last 5 cycles. The calculated differences are not significantly high at all stress levels (less than 7 %). Figure 6 shows the percentage difference of J_{nr} between two methods of calculating. The maximum difference is about 20 % at stress level of 0.1 kPa, which is quite significant change that may be a source of variability of the results. This change is lower at higher stress and it is about 5 % when the stress level is 8.0 kPa, which can be considered negligible. Higher difference of J_{nr} is evident for binders with CR.

Figures 7 and 8 show that applying higher stress level leads to reduction of result variability in the test. The dotted lines in the figures show the average values calculated from all 10 cycles. Individual values have larger fluctuations at a low stress level, while a higher stress level contributes to a better reproducibility of results. Modification with low-viscosity waxes (Figure 7a-b) have only a first calculated value significantly different from other values at the stress level 0.1 kPa. Conversely in case of modification with CR (Figure 8 a-b), there is not only the first value significantly different for all stress levels. For these binders (modified by CR with max. 15 M% content) would be therefore certainly appropriate to increase the number of cycles to obtain steady values (especially at lower stress level). This might by as hypothesis valid only for crumb rubber modified bituminous binders (CRmB) where the content of used crumb rubber would not exceed 15 M% and the maximum particle size would not exceed 1.0 mm.

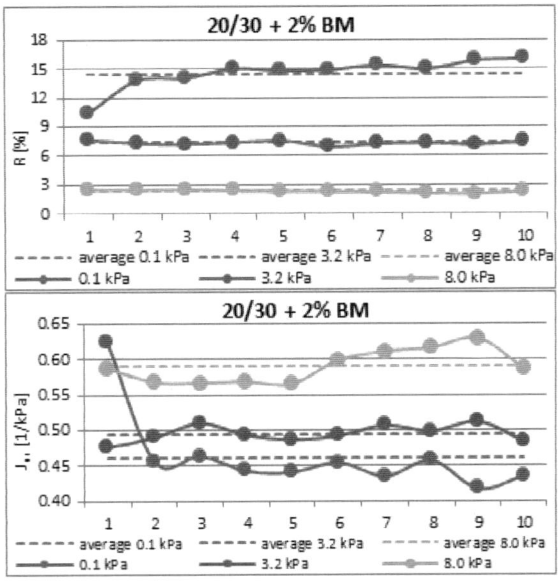

Figure 7. MSCR test values at each cycle, (a) elastic recovery R, (b) non-recoverable creep compliance J_{nr}

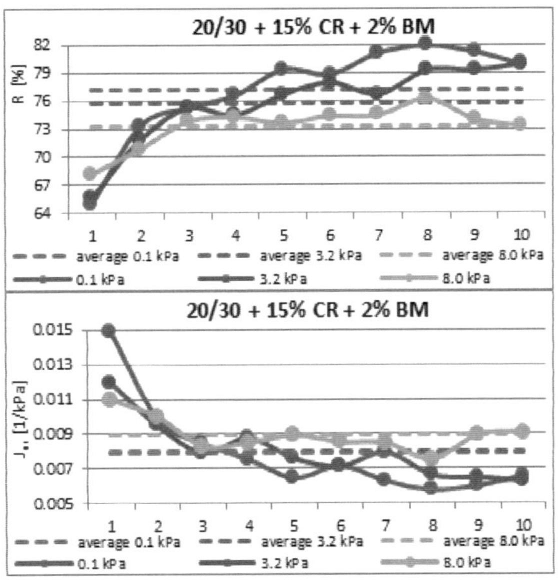

Figure 8. MSCR test values at each cycle, (a) elastic recovery R, (b) non-recoverable creep compliance J_{nr}

Figures 9 and 10 present the difference of the values calculated as the average of all 10 cycles in comparison with the average of last 5 cycles. The difference in elastic recovery ranges up to 12 %, but for certain binders are differences in the calculation of J_{nr} considerably higher. All PMB binders have similar trend which is seen in Figure 12 and 13, wherein the R values in the first 3-4 cycles gradually increase (J_{nr} values decrease) and in the following cycles are stabilized. This pattern is evident among stress levels 0.1 kPa and 3.2 kPa. Increase of relative change looks like a very low because of the relatively high values of R. Conversely, the variants that have very low J_{nr} have percent change significant between values calculated of 10 cycles and 5 cycles. Important change occurred in PMB

45/80-65 at 70 °C (Fig. 14), Jnr value is increasing in each cycle at stress 8.0 kPa, which is the opposite trend than at other stress levels. This likely indicates the fatigue of the material which occurs in the nonlinear viscoelastic region. The closest values of all cycles are evident for neat binders (represented by Fig. 11), only for the lowest stress level 0.1 kPa are apparent differences, but very low.

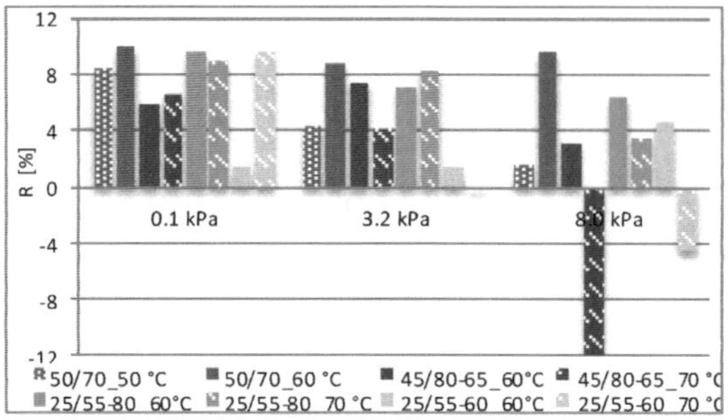

Figure 9. Relative change of two methods of calculating elastic recovery R

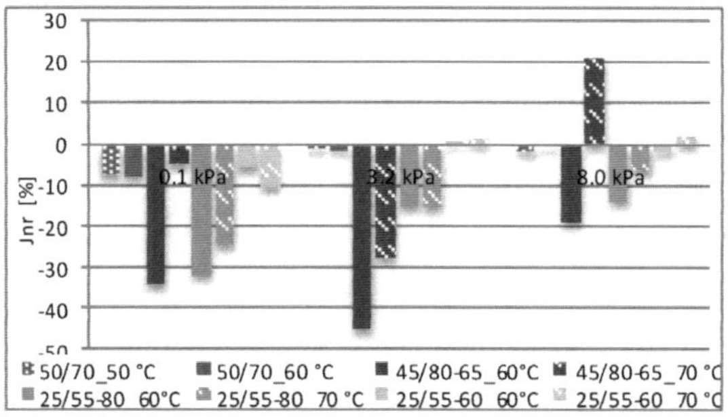

Figure 10. Relative change of two methods of calculating creep compliance J_{nr}

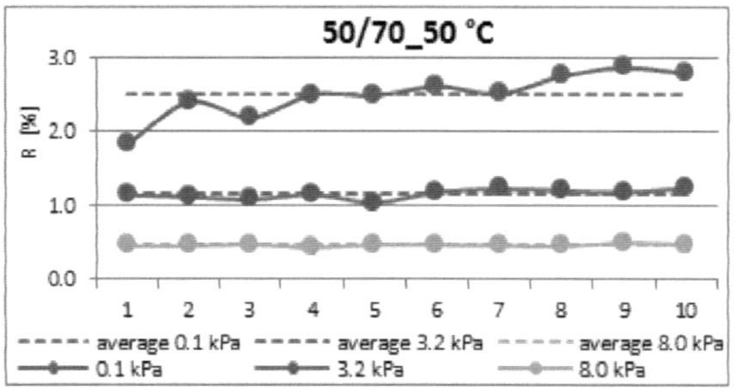

Figure 11. Elastic recovery values at each test cycle

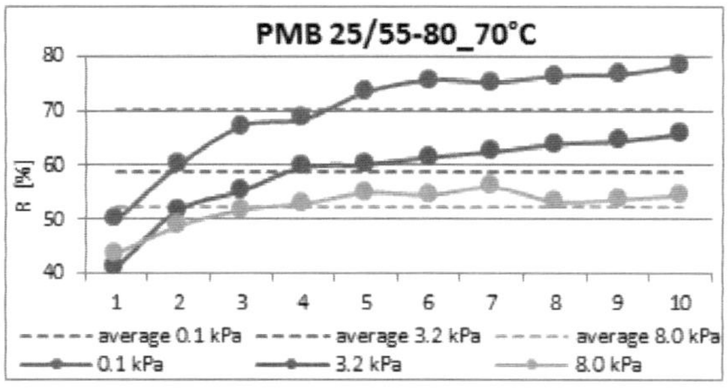

Figure 12. Elastic recovery values at each test cycle

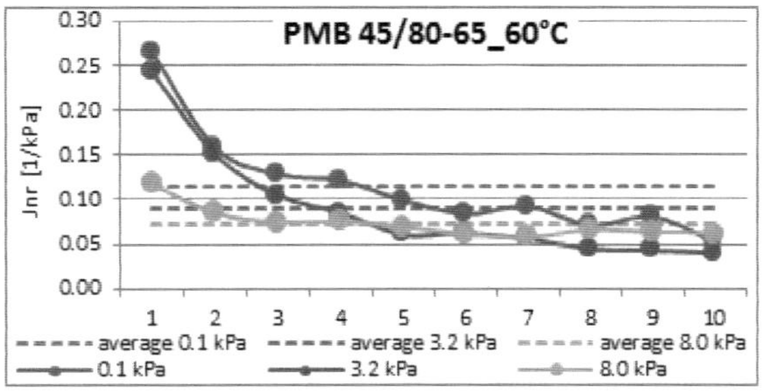

Figure 13. Non-recoverable creep compliance values at each test cycle

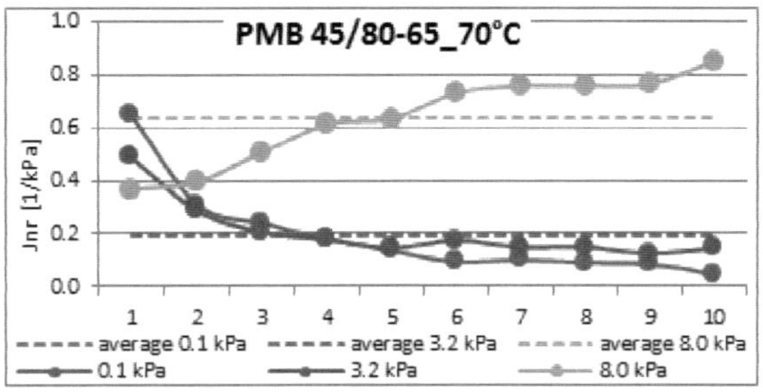

Figure 14. Non-recoverable creep compliance values at each test cycle

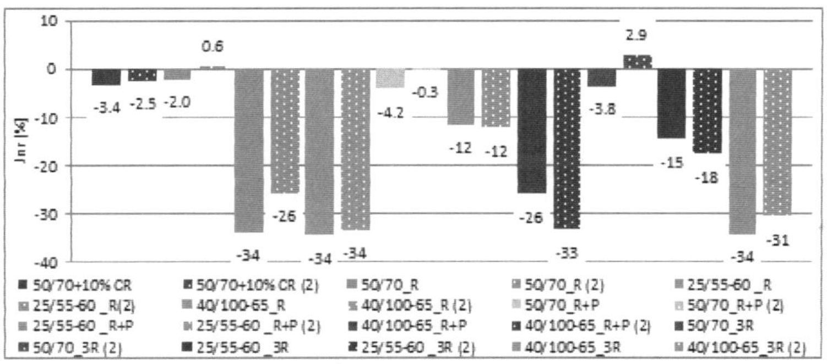

Figure 15. Percent change between two methods of calculating Jnr at stress level 3.2 kPa

Figure 15 shows the last group of bituminous binders with different geometry of the test device. Variants with geometry PP08 are labeled (2).While the change in the calculation of R, considering the last 5 cycles, is less than 10 % for all variants of this group (excluding 50/70 + 10% CR variant, which has a 15 % difference), the change of J_{nr} value is around 30 % for most PMBs, which generally corresponds to the previous group with PMB binders. Unmodified binders have the smallest difference of J_{nr}, but this is again mainly related to high J_{nr} value thanks to which, the percentage difference is very small between the J_{nr} value calculated from all 10 cycles and the last 5 cycles.

4. Conclusions

Currently, the MSCR test is promoted as the most appropriate method for assessing the resistance of bituminous binders to permanent deformation. The advantage of MSCR test compared to Frequency Sweep test is applying a larger load and deformation which reflects better the real situation in the pavement. Parameter G */sin δ characterizes bitumen in its linear domain (lower stress/strain) but permanent deformation occurs in nonlinear domain and it is necessary to describe the behavior in such domain by different parameters, e.g. J_{nr}.

This paper presented results of MSCR test comparing the influence of temperature, stress level, type of testing geometry and number of cycles for calculation individual values.

Performed measurements indicate that the method of averaging the response for 10 cycles at each stress level can lead to misleading values due to the changes during the cycles. Since the most significant differences are during the initial cycles. It would be therefore appropriate to calculate non-recoverable creep compliance and elastic recovery from the second half of the cycles at each stress level.

Research study showed that it would be preferable (especially for PMB) to increase the number of cycles since the 10 cycles for most of PMBs is according to presented findings not sufficient to achieve a stable secondary creep. An alternative option is to select a higher stress level, which leads to faster stabilization of test results.

In case the strain is very small (on the limit of the measuring device range) it is not possible to calculate sufficiently the accuracy of determined values. This problem occurs especially at lower test temperatures or at low stress level (0.1 kPa) when the divergence between cycles is high. Additionally, correlation with asphalt mixtures (resistance to permanent deformation) is lower when the stress level is 0.1 kPa [15]. For these reasons, it is preferable to use a higher stress level, e.g. 8.0 kPa.

Selected testing geometry primarily affects the J_{nr} value. Classical geometry PP25 is simpler for handling in DSR, as well as sample preparation (pouring) is easier. But in case of modification by crumb rubber with particle size of 1 mm or larger, it is necessary to use different geometry with a larger gap (PP08) or to analyze the possibility to use the standard geometry PP25 with a larger gap. Results show that the use of different geometry produces varying results and it is therefore necessary to specify the type of used testing geometry.

Performing the test at a higher temperature leads to a higher strain, thereby reducing the risk of inaccurate measurements due to the limits a DSR device has. Temperature of 60 °C would be an appropriate temperature for conducting MSCR test for many binders assuming that the test resistance to permanent deformation for asphalt mixtures would be performed at the same temperature (currently, the test is carried out at 50 °C), in order to better compare the measured values.

In general it is recommended to run the test always with the geometry plate-plate 25 mm (PP25) and consider proper test temperature in relation to softening point a bitumen has. It might be more difficult to test crumb rubber modified binders especially if the rubber particles are >1 mm. In such case the used test gap (bitumen sample thickness needs to be increased. Adding a higher stress level seems to be an interesting approach but this needs to be validated by additional test. In overall MSCR test confirms to be a suitable bitumen test which can help to indicate rutting resistance of the asphalt mixture.

5. References

[1] Neutag L, Zielke U, Beckedahl H J and Johannsen K 2009 Crumb Rubber Mmodified Binders for Flexible Pavements *Environmentally friendly roads - ENVIROAD 2009*, Warsaw 15-16 October 2009

[2] Chen C, Podolsky J H, Hernandez N, Hohmann A, Williams R Ch and Cochran E W 2016 Use of Bioadvantaged Materials for Use in Bituminous Modification *Transport Research Arena Conference* Warsaw, 18-21 April 2016

[3] Valentin J, Mondschein P and Miláčková K 2013 Selected Performance-Based Characteristics of Viscosity Improved Bituminous Binders *Ninth International Conference on the Bearing Capacity of Roads, Railways and Airfields* pp 593-605

[4] Bahia H U, et al 2001 *NCHRP Report 459, Characterization of Modified Asphalt Binders in Superpave Mix Design* (Washington DC: Transportation Research Board, National Research Council)

[5] Laukkanen O, Soenen H, Pellinen T, Heyrman S and Lemoine G 2014 Creep-Recovery Behaviour of Bituminous Binders and its Relation to Asphalt Mixture Rutting *Materials and Structures* **48**(12) pp 4039-53

[6] Stuart K, Mogawer W and Romero P 2000 Evaluation of the Superpave Asphalt Binder Specification for High-temperature Pavement Performance *Asphalt Paving Technologists Proceedings* pp 148–76

[7] D'Angelo R, Dongre R and Reinke G 2006 Evaluation of Repeated Creep and Recovery Test Method as an Alternative to SHRP+ Requirements for Polymer Modified Asphalt Binders *Proceedings Canadian Technical Asphalt Association*

[8] D'Angelo J, Kluttz R, Dongré R, Stephens K and Zanzotto L 2007 Revision of the 361 Superpave High Temperature Binder Specification: The Multiple Stress Creep 362 Recovery Test *Journal of the Association of Asphalt Paving Technologists, Proceedings of Technical Session* pp 123-62

[9] Domingos M D I and Faxina A L 2014 Creep-Recovery Behavior of Modified Asphalt Binders with Similar High-Temperature Performance Grades *Transportation Research Board (TRB)93rd Annual Meeting*

[10] Golalipour A 2011 *Modification of Multiple Stress Creep and Recovery Test Procedure and Usage in Specification* Master Thesis (Wisconsin: University of Wisconsin)

[11] AASHTO 2014 *Standard Specification for Performance-Graded Asphalt Binder Using Multiple Stress Creep Recovery (MSCR) Test* (American Association of State and Highway Transportation Officials)

[12] Czech Standardization Office 2016 *CSN EN 16659: Zkušební metody pro asfalty a asfaltové výrobky; Asfalty a asfaltová pojiva - Zkouška MSCR* (Czech: Czech Standardization Office)

[13] Wasage T L, Stastna J and Zanzotto L 2011 Rheological Analysis of Multiple Stress Creep and Recovery (MSCR) test *International Journal of Pavement Engineering* **12**(6)

[14] Dreessen S, Planche J P and Gandel V 2009 *Advanced Testing and Characterization of Bituminous Materials* (CRC Press)

[15] Dreessen S and Planche J P 2009 Seeking for a Relevant Binder Test Method for Rutting Prediction *Environmentally friendly roads - ENVIROAD* Warsaw, 15-16 October 2009

Acknowledgement
This paper was prepared under the funding scheme of Competence centre program by the Technology Agency of the Czech Republic within the project Centre for effective and sustainable transport infrastructure (CESTI), project no. TE01020168.

IOP Conf. Series: Materials Science and Engineering **512** (2019) 012030 doi:10.1088/1757-899X/512/1/012030

Optimization of mixing time for polymer modified asphalt

Ali M Babalghaith[1], Suhana Koting[1], Nor Hafizah Ramli Sulong[1] and Mohamed Rehan Karim[2]

[1]Center for Transportation Research, Department of Civil Engineering, Faculty of Engineering, University of Malaya, Kuala Lumpur, Malaysia
[2]Transportation Science Society of Malaysia (TSSM), Malaysia

E-mail: Bablgeath@hotmail.com

Abstract. Asphalt pavement made with neat binder is not sufficient to resist heavy traffic loads as well as harsh environmental condition. Therefore, there is a need to modify the binder with polymers in order to improve its properties. From literature, it was indicated that various blending process parameters, namely, blade shape, temperature, shear rate and mixing time influenced the properties of the produced polymer modified asphalt. In this study, the influence of mixing time on six different types of polymer with 60/70 asphalt binder was focused. The polymers were blended with asphalt binder and inspected at 30 minutes intervals during the mixing process. Then, all the samples were tested through Dynamic Shear Rheometer at 58ºC, 64ºC and 70ºC. It was shown that the complex modulus (G*) values at these temperature levels for all polymers modified asphalt indicated proper particles dispersion within the binder. The optimal blending time, which produces the best rheological properties of all modified asphalt is 30 minutes except for asphalt modified with Lucolast7010 and Anglomak2144, which is 60 minutes. In conclusion, mixing time has significant influence on the particle distribution of polymer within the binder to ensure the quality of asphalt-polymer blend.

1. Introduction

Since asphalt binder and polymers are of non-homogeneous nature, polymer is required to disperse and not to dissolve in the asphalt binder. Polymer modifiers are usually supplied indifferent forms. They are mainly supplied as a powder (fillers), pellets (granules), fibers, and liquid form. Each modifier has its own way of mixing and handling. It depends on the modifier type and the purpose of usage. For each modifier, the method of mixing and handling is known either from experience or from supplier recommendations [1]. The quality of asphalt-polymer blend is affected by the type of the asphalt, type and concentration of polymer and the shear rate applied by the mixer[2].

There are several ways of blending polymers with asphalt. In the field, one of the very useful techniques is in-line blending of the polymer into the asphalt feed line. Another technique is pre-blending polymer at the asphalt terminal, refinery, batch plant and drum plant operations[3, 4]. In laboratory, it is recommended that the blending machine is capable of providing high shear rate. The configuration of the blender head (i.e., the blade) and the speed of blending define the level of shear rate that the blender induces into the asphalt binder. The recommended speed should not be less than 2500 Rotation Per Minute (RPM)[5-8]. Optimum blending time is evaluated during the blending process at an interval of 30 minutes [7, 8],20 minutes [5], starting from zero until two hours. If a low shear mixer is used, longer mixing time is needed [9, 10].

Al-Dubabe et al.[5] modified Arab asphalt binders by different polymers to satisfy the performance requirements of the Gulf countries, in terms of rutting, fatigue, and low-temperature cracking. For identification of potential polymers, the polymers were blended with asphalt binder from Riyadh Refinery, which is the main supplier for asphalt binder in the Gulf countries. The polymers were blended at concentrations (2%, 4%, 6% and 8% by total binder weight) with 500 grams of asphalt binder to produce the Polymer Modified Asphalt (PMA). Visual inspection using an optical microscope and softening point testing were used to evaluate the homogeneity of the blend at 20-minute intervals from the beginning of the blending operation until 120 minutes.

Al-Abdul Wahhab et al.[7] studied the performance modification of Saudi-asphalt binders using five polymers produced by Saudi Basic Industries Company (SABIC). Polymers were blended with asphalt from Riyadh Refinery at various polymer concentrations (3%, 4% and 5% by total binder weight). Each polymer type was blended with 800 grams of asphalt binder to produce the PMA binder. A high shear mixer with maximum speed of 2500 RPM was used in the study. Optimum blending time was determined by means of complex shear modulus per sine phase angle ($G^*/\sin \delta$) at 76 °C and evaluated during the blending process at an interval of 30 minutes. The assumed maximum blending time was two hours.

Al-Soliman [8] studied the effect of using polymers for asphalt binder modification. Five grades of low-density polyethylene polymers (LDPE)(120W, 218W, M200024, R50035 and 2022J) and three grades of high- density polyethylene polymers (HDPE) (B4660, M200056, and F00952) were used. The polymer grades were blended at various polymer concentrations (3%, 5% and 7% by total binder weight) and visually inspected at 30 minutes intervals until 120 minutes in order to examine blend consistency and appearance of any unparsed particles. For each blend, a constant quantity of asphalt binder of 950 grams was mixed at a temperature of 180°C. A high shear mixer with maximum speed of 2500 RPM was used in the study.

The objectives of this study are, (1) to investigate appropriate mixing time for different polymer modified asphalt type, (2) to evaluate the effect of mixing time on binder rheological properties for different polymer modified asphalt type and (3) to propose optimum mixing time which satisfies the dispersion and consistency requirement of polymer modified asphalt.

2. Materials

The asphalt binder used in this study was a 60/70 penetration grade. The properties of the neat asphalt binder are tabulated in Table 1 [11, 12]. Six different polymers, namely, Lucolast7010, Anglomak2144, SBS KTR401, EE-2, Crumb Rubber and Pavflex140 were used in this study as shown in Figure 1. The physical and chemical properties of those polymers are tabulated in Table 2

Table 1: Properties of 60/70 asphalt binder[11, 12].

Properties	Reference	Value
Penetration @ 25°C (0.1mm)	ASTM D0005-05	68
Softening Point (°C)	ASTM D0036-95	48
Penetration Index	-------	- 0.99
Ductility (cm)	ASTM D0113-99	+ 100
Rotational Viscosity @ 135 °C, 20 rpm, (cps)	ASTM D4402-02	487
Specific Gravity	ASTM D0070-03	1.025
RTFO @ 163°C, for 85 minutes, (% wt loss)	ASTM D2872-04	0.07
High Temperature Grade (°C)	ASTM D7175-00	64
Low Temperature Grade (°C)	ASTM D6648-01	- 22

IOP Conf. Series: Materials Science and Engineering **512** (2019) 012030 doi:10.1088/1757-899X/512/1/012030

EE-2

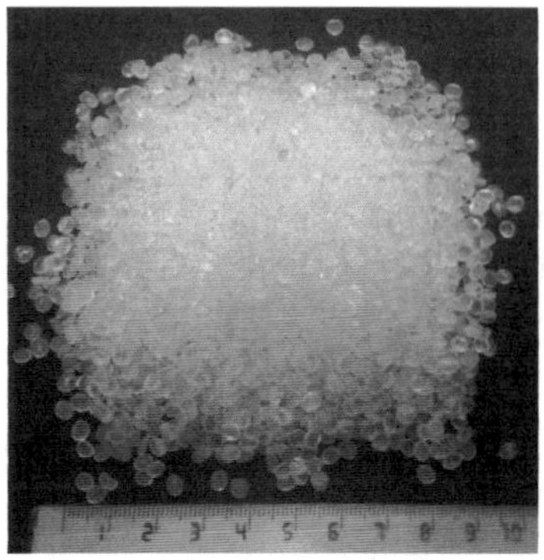
Lucolast7010

Anglomak2144

Crumb Rubber

Pavflex140

SBS- KTR401

Figure 1. Different polymers used in this study

Table 2: The physical and chemical properties of polymers

Polymers	Physical Form	Density (g/cm^3)	Melting Point (°C)	Melt Flow Index (g/10min)	Components
Lucolast7010	Pellet	0.924	95	3.9	Ethyllene and Butyl Acrylate with low crystallinity.
Anglomak2144	Pellet	0.930	96	3.5	Oxidised Polyethylene Homopolymer.
Paveflex140	Powder	-	212	-	Ethylene Vinyl Acetate Resins.
SBS KTR 401	Pellet	0.94	270	< 1	Styrene Butadiene Styrene.
EE-2	Pellet	0.96	-	-	Medium Density Oxidized Polyethylene.
Crumb Rubber	Pellet	1.2	200	-	Synthetic Rubber, Natural Rubber, Carbon Black, Anti-Oxidants, Fillers.

3. Preparation of polymer-modified asphalt

Polymer-modified asphalts were prepared in the laboratory by blending neat asphalt with polymers without curing. The blending procedure usually defines how the polymer particles disperse within asphalt particles. To get the required performance, the polymer particles should be properly dispersed within asphalt to produce a homogeneous blend. As shown in Figure 2, a Ross blender with a high shear blade was used to blend the polymers with the asphalt. The blending speed was controlled with a direct current motor which capable of producing up to 3000 RPM.

Figure 2. The Ross blending machine and its components

The temperature was controlled through an oil bath heating system. The blending temperature was determined based on asphalt plant blending temperature, as shown in Table 3. The asphalt binder was blended with different percentages of polymers in order to achieve high Performance Grade (PG), which was adopted by the Ministry of Transportation.PG76was selected since it represents the highest performance grading requirements in Saudi Arabia [5, 12]. The samples were tested using the Dynamic Shear Rheometer (DSR) test at 58 °C, 64 °C and 70 °C. 30 minutes intervals were taken during the blending process to examine blend consistency.

Table 3. Temperature and speed of blending

Polymers	Polymer (%)	Temperature (°C)	Speed (RPM)
Lucolast7010	3.6	170	2500
Anglomak 2144	3.2	170	2500
Pavflex140	5	245	2500
SBS KTR 401	3	180	2500
EE-2	4	170	1800
Crumb Rubber	8	180	2500

4. Result and Discussion

In this part, the process of polymer blending with asphalt binder was analyzed and discussed. The effect of mixing time was investigated in order to select the proper mixing time. Polymers were blended with asphalt binder and tested at 30 minutes intervals during the blending process to examine blend consistency and appearance. The samples were tested on dynamic shear rheological properties at different temperatures of 58 °C, 64 °C and 70 °C.

Dynamic shear rheometer test was conducted at various temperatures to evaluate the effect of mixing time on binder rheological properties. Dynamic complex moduli versus time at different temperatures are graphically presented in Figure 3 to Figure 8. It was indicated that there is variation in complex modulus (G*) for all PMA blends at 30 minutes of mixing time. The values of G* increased by increment of mixing time due to dispersion of polymer particles. The G* increment will continue until the polymer particles are completely dispersed. At that point, the G* is expected to stabilize at a certain limiting value. An example of this behavior is shown in Figure 3for asphalt modified with SBS KTR 401. It can be seen that G* values were increased by mixing time increment until 30 minutes. Then, the G* values were leveled off until 120 minutes of mixing time. This means that the minimum blending time to produce a homogeneous blend for the polymer SBS KTR 401 at a concentration of 3.0 % is 30 minutes. This concept was also used to identify the minimum blending times for polymer-modified asphalt binders as shown in Figure 4 to Figure 6.

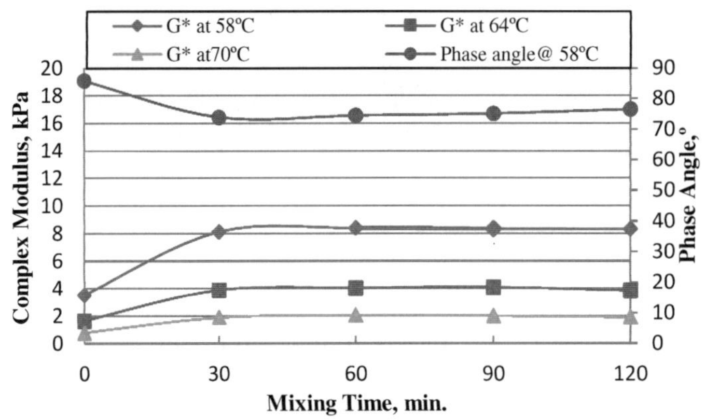

Figure 3. Dynamic modulus and phase angle with mixing time of SBS KTR 401

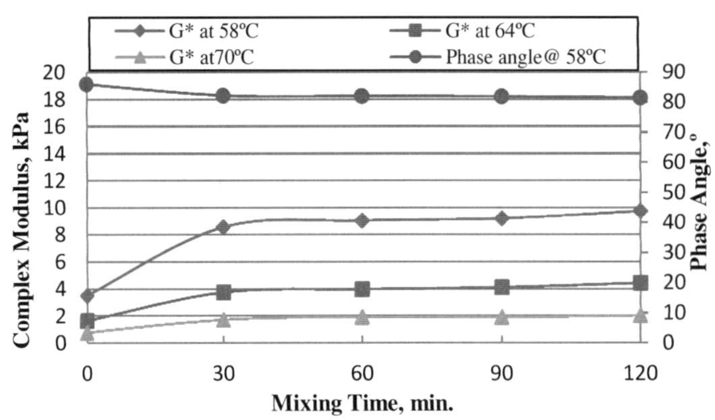

Figure 4. Dynamic modulus and phase angle with mixing time of Pavflex140

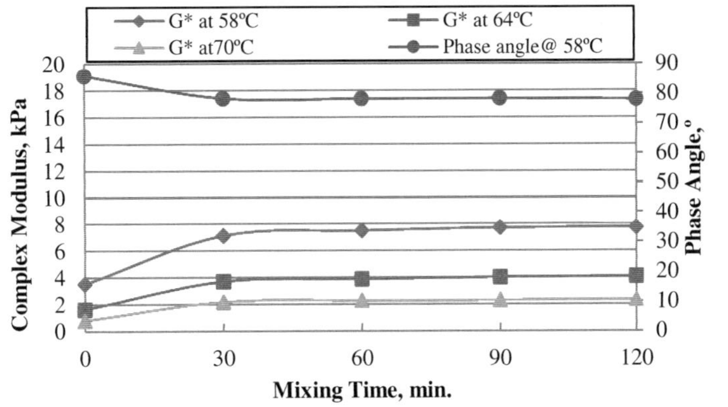

Figure 5. Dynamic modulus and phase angle with mixing time of EE-2

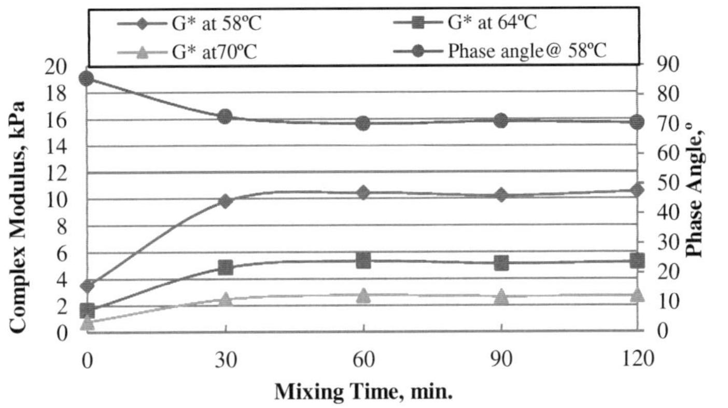

Figure 6. Dynamic modulus and phase angle with mixing time of Crumb Rubber

Lucolast7010 and Anglomak2144 show different trends, as shown in Figure 7 and Figure 8, respectively. Figure 7shows G* values versus mixing time behavior. In general, G* values increased with mixing time to a certain value, decreased slightly and then increased again. However, the amount of increment or decrement were depending on testing temperature. This behavior is probably due to the changes in the degree of polymer particle distribution within the binder as mixing time increases.

Figure 8displays that the value of G* increases with increasing mixing time due to more dispersion of polymer particles until the particle starts to become finely and completely dispersed. At that point, it starts to decrease until the end of fine dispersion. The same behavior was observed in the previous study which was conducted by Al-Soliman [8].

The variation of G* with temperature at specific mixing time can be considered as an indication of temperature sensitivity of the PMA blends. The G* at low temperature (58°C) showed higher variation with mixing time than at high temperature (70°C). The same phenomena can be seen for the phase angle, which is a measure of the elasticity part. As the mixing time increased, phase angle decreased due to polymer dispersion. When the dispersion starts to complete and fine, the phase angle stabilizes to a certain limiting value.

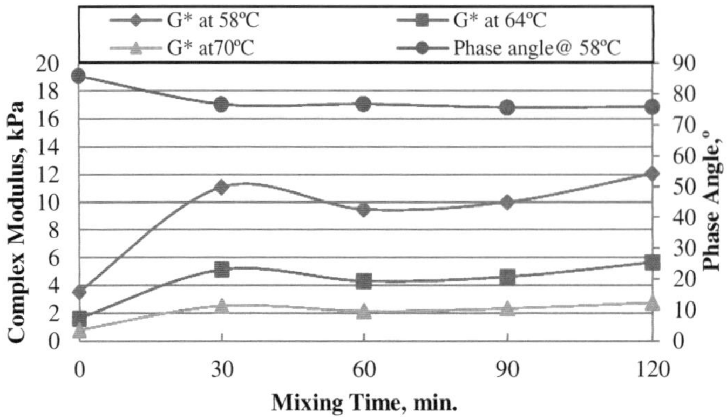

Figure 7. Dynamic modulus and phase angle with mixing time of Lucolast7010

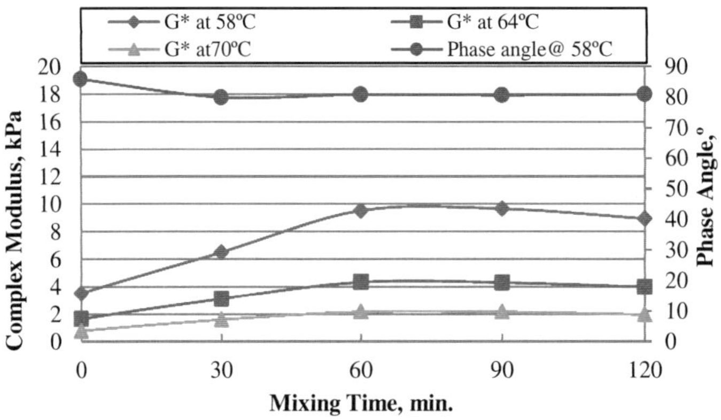

Figure 8. Dynamic modulus and phase angle with mixing time of Anglomak2144

5. Conclusion

Based on the results and analysis of this study, the following conclusions and recommendations can be obtained.

- Determination of mixing time for blending processes can best be made by measuring the changes in complex modulus at 30 minutes intervals during the blending process. Complex modulus at 58°C, 64°C, and 70°C were found to be good measures of proper dispersion and indicated the same trend.
- During the blending process of all polymers except Lucolast7010 and Anglomak2144, complex moduli tends to increase with mixing time until 30 minutes and then were leveled off until 120 minutes mixing time, indicating full dispersion of polymer particles and no change in consistency of blend.
- In this study, one shape of the mixer blade and shear rate were used to blend polymers with asphalt binder. For future work, it is recommended to study the effect of other shapes of mixer blade and different shear rates on blending time.
- 60-70 penetration grade asphalt binder was used in this study. It is recommended to investigate the effect of other asphalt binder grades on properties of polymer-modified asphalt.

6. References

[1] Terrel R and Epps J 1989 *Using Additives and Modifiers in Hot Mix Asphalt* (National Asphalt Pavement Association)

[2] Read J and Whiteoak D 2003 *The shell bitumen handbook* Thomas Telford

[3] May J and Wilkey T 1996 *Heating, mixing and storing modified asphalt* (Tennessee: Heatec Inc.)

[4] The Asphalt Institute 1989 *The Asphalt Handbook: Manual Series No. 4 (MS-4)* (Asphalt Institute)

[5] Al-Dubabe I A, Al-Abdul Wahhab H I, Asi I M and Ali M F 1998 Polymer modification of Arab asphalt *Journal of Materials in Civil Engineering* **10** 161-7

[6] Al-Abdul Wahhab H I 2012 Effect of modifiers and additives on fatigue behavior of asphalt concrete mixes in the gulf *International Journal of Pavement Research & Technology* **5**

[7] Al-Abdul Wahhab H I, Asi I M, Abaker I and Baig M G 2002 Performance modification of saudi-asphalt binders using sabic polymers *The 6th Saudi Engineering Conference* Dhahran pp 379-93

[8] Alsoliman H A 2010 Engineering characteristics of local polymer modified asphalt mixtures (Doctoral Dissertation, king Saud University)

[9] González O, Muñoz M E, Santamaría A, García-Morales M, Navarro F J and Partal P 2004 Rheology and stability of bitumen/EVA blends *European Polymer Journal* **40** 2365-72

[10] Lu X and Isacsson U 2001 Modification of road bitumens with thermoplastic polymers *Polymer testing* **20** 77-86

[11] Babalghaith A M 2017 Laboratory-based study of the characteristics of polymer-modified asphalt produced in Riyadh City (Saudi Arabia: King Saud University)

[12] Babalghaith A M, Alsoliman H A and Al-Suhaibani A S 2016 Comparison of rheological properties for polymer modified asphalt produced in Riyadh World Academy of Science, Engineering and Technology *International Journal of Civil, Environmental, Structural, Construction and Architectural Engineering* **10** 197-201

IOP Conf. Series: Materials Science and Engineering **512** (2019) 012031 doi:10.1088/1757-899X/512/1/012031

An overview of asphalt mix designs using various compactors

Hanani Mohamed Radzi, Ratnasamy Muniandy, Salihudin Hassim, T H Law and Fauzan Mohd Jakarni

Department of Civil Engineering, Universiti Putra Malaysia, 43400 Serdang, Selangor, Malaysia

E: mail: ratnas@upm.edu.my

Abstract. Generally, road network is a well-thought-out vital transportation medium that is in place throughout the world. As, such, utmost care is taken in providing the best possible safe and comfortable and yet cost-effective road transportation system. One of the areas that scrutinized is the performance of asphalt mixtures in the flexible pavements. Traditionally, laboratory asphalt mixture designs are based on Marshall Method whereby the asphalt mixtures are compacted using a 4.5kg drop weight hammer. Although the volumetric properties are reasonably accepted, the actual achievement of the roller compacted volumetric properties at the site is still farfetched. This is because the laboratory mix designs developed using a Marshall Drop weight compactor and in the field, the asphalt layer is compacted using a heavy-duty roller. In recent years, the Superpave Gyratory Compactor (SGC) have come into play with a static load and kneading approach. This objective of this paper is to look into the existing methods and processes carrying out the asphalt mix designs and review to see if there is or are any mismatch as compared with field compaction protocol.

1. Introduction

The existence of roads and highways are an undeniably a big thing that assist any country in its effort to nation building. Road engineering in general offers a bigger perspective, which is not only on the design and on construction of the roads but on the maintenance after the construction of roads and highways are completed. Typically, road transportation is a well-thought-out vital transportation medium throughout the world. Highway network is one of the major deeds in the history of the highway and transportation engineering and the last few decades' construction of highway has become an aggressive activity in most of the development countries (1).

Malaysia is not exceptional and the road authorities very stringently implement road standards and specifications. For reasons of logistic and travelling, road have become the pulse that generates the country's economic advancement. Road transportation has been a strong stake in supporting the nation's socio-economics, as it is a cheaper medium of transportation, as compared to water and air transportation. In Malaysia, based on the 'Statistik Jalan 2015' by the Road Facilities Maintenance Branch, PWD until up to year 2014, there was a total of 203,788.02 kilometers of roads which made up the whole road network of the country of which more than 80% are paved (2). However, with such a vast road network, the typical problems of rutting and cracking are still prevalent costing the road authorities to spend a lot of money on maintenance and rehabilitation.

Although several factors such as poor road materials, design and construction practises cited for the road failures, the mere fact of formulating mix designs in the lab in line with field procedures was not explored much. For instance, Marshall compactors are widely used in compacting asphalt mixtures while rollers used at the site. The closest equipment in simulating the field compaction is the super pave gyratory compactor but that it does not even come close to the real time road compaction. As such a mismatch is seen the formulation of asphalt mixtures and compacting them at the site.

1.1. Problem statement
Compaction is the simplest, most economical method of increasing road pavement life and improving its load-carrying capacity. The mix must be compacted to desired degree by increasing the density as well to reduce the porosity. Once the density is achieved with the needed degree of compaction then the air voids will complied (3). However, Hot Mix Asphalt (HMA) compaction may also affect by other subjective not only the method of compaction, but also the characteristics of the materials used in the materials or mix design, construction conditions, environmental conditions (4). Moreover, quality control from logistic or delivery from plant to site may contribute to result of thickness and in-situ density as a part after compaction process.

Occasionally, researchers found out that method of compaction especially in the laboratory resulted in final mix that delivered to production plant as well. Conventionally in the laboratory, asphalt mixture designs based on Marshall Method by 4.5kg drop weight hammer with certain number of blows. Although the volumetric properties acceptable within range, the impact compaction used in Marshall Method does not simulate mixture densification as it occurs in a real pavement using a heavy-duty roller. Over the time, the use of Superior Performing Asphalt Technology or Superpave profound in 90's is increasing around the world especially in United States (US). Another new mechanism of compaction been introduced with machine namely as Superpave Gyratory Compactor came up with static load and kneading action.

Even though SGC presented the volumetric properties consistently as compare to Marshall Method, yet still quite a mismatch procedure compaction at in-situ. Problem encountered on rutting developed a tiny groove produced in the asphalt concrete layer may resulted in riding quality on the pavement surface, as well serious risk in security (5). The number of vehicles in Malaysia increased over the last few decades. The trailers have been designed according to the logistics goods to be delivered (6) with the environmental factor such as aging and deterioration of pavement faster than the design life (7). The primary mode of failure in Malaysia is mainly cracking and rutting. (8).While Asphaltic concrete ACW 20 or AC14 roads failed predominantly through top-down and fatigue cracking (9).

1.2. Objective
The primary objective of this paper is to review the current practices in the designing of asphalt mixtures and the compacting procedures used. This could assist researchers to have a better understanding of the laboratory compaction procedures and the field compaction methods.

2. Various Compaction Methods
2.1. Compaction in laboratory
The purpose of compaction in asphalt mixtures is to even out and improve the physical aspects of asphalt mixtures. The influence of compaction in asphalt mixture by the degree of compaction was a dominant quality parameter in asphalt mixtures especially when the mixture is analytically calculated with a low asphalt range to convey resistance to perpetual deformation in mixtures such as dense asphalt concrete (1). It is fine proven that technique of compaction resulted in physical properties of compacted asphalt concrete samples. Asphalt concrete mixtures in the laboratory, is desired to formulate compacted specimens that narrowly identical to the characteristics of the real road pavement. The site parameters such as traffic loading conditions must be taken into the account in analyzing the best laboratory design that needs to be extended producing in mixing plant and laying it at site (1)(5). Each laboratory compaction method studied endeavors to separate specific engineering properties in order to support in predicting the reaction of a specific mixture to particular field

conditions. The four elementary types of compaction methods are Impact Compaction, Kneading Compaction, and Gyratory Compaction.

2.2. Marshall impact compaction

Earlier research done on a pavement that was compacted to only 95 percent of Marshall density would on this basis deteriorate twice as fast as if it had been compacted to the minimum requirement of 98 percent (4). The key benefit of the Marshall compaction method is the focus on air voids analysis in the compacted specimens. From the compaction process, obviously Marshall Hammer, create a straight load that results in limited deformation and a fewer even distribution of mixture ingredients (10).These principals to squeeze binder film on the aggregate surfaces and a more intense fatigue fracture. Four elementary types of compaction methods are Impact Compaction, Kneading Compaction, and Gyratory Compaction.

2.3. Kneading compaction

Hveem has come out in his mix design procedure with kneading method compaction by applying forces on one side of specimen through an irregularly triangular-shaped foot. It is comparable to the Marshall method of compaction, but the Hveem method is concentrated on voids, durability and strength. The primary test results determined that specimens produced via kneading compaction were impervious to permanent deformation, but most penetrating to aggregate characteristics (11). While specimens produced by rolling wheel exhibited less resistance to permanent deformation, and were most sensitive to asphalt characteristics. Samples prepared using the kneading compaction device as are more resistant to permanent deformation, due to expansion of a more through interparticle contact "structure" at least for densely graded aggregates mixtures especially in aggregate angularity and surface texture (5).

2.4. Gyratory compactor

The basis for gyratory compaction was a Texas gyratory compactor modified to use the compaction principles of a French gyratory compactor and found in the 1930s in Texas (12) and the enhancement was established and applied by the Army Corps of Engineers and the central lab of *Laboratoire central des ponts et chaussées* (LCPC) for bridges and roads in France. Later, Strategic Highway Research Program (SHRP) generating Superpave gyratory compactor. An underlying premise of the gyratory protocol selection is to gain the material characteristics from testing validation rather than the effects from the compactor itself.

2.5. Roller compactor

Laboratory half roller compactors compact asphalt slabs to the target mixture densities to simulate those of full-scale compaction equipment. Some of the common name is French roller compactor established by the LCPC (1). Major involvement with laboratory asphalt compaction since the 1950s, not only rolling wheel compaction, but also gyratory compaction, volumetric design, and failure mechanisms (13). Although the old rolling wheel compactor was a promising method of compaction, the entire set up was very bulky with some good property precision that most approximately simulate those of materials in the highway (14-17). Slabs can be prepared up to desired densities using loads that are comparable to those of site compaction machineries.

2.6. Compaction in roadworks

Compaction is needed to reduce air voids in the HMA mixture through the applied pressure or external forces. The discharge of air permits the mix to reduce volume, thereby increasing the density of the mass. Self-driven compactors deliver the external forces that convey into compaction energy (15). Usually a compaction train entails of more rollers with notable reason to achieve the required density to meet the specifications, as well to provide a smooth surface for riding quality.

2.6.1. Static compaction. Static compaction achieved from the deadweight of the roller. Tandem rollers and pneumatic tired rollers are used for this purpose. The effect of compaction somewhat low as compared to vibratory compaction influence. With tandem rollers, compaction is subjective by the static linear load (kg/cm) of the drum, with pneumatic tired rollers by the wheel load (t) and the tire inflation pressure (MPa) (3).The linear load can be achieved in range of 10-30 kg/cm. The low initial compaction will affected the function of static compaction. Whereby, the kneading and flexing effect from the wheels pneumatic tired rollers achieve a special quality in static compaction. Knowing that, the homogeneous distribution of the mix and closes the pores on the surface can be achieved.

2.7. Rolling wheel compactor

Previously, there were significant findings on the rolling wheel compactor that is practical for the production of asphalt concrete test specimens in a research laboratory with a large number of specimens of various geometries, which been produced on a daily basis. On top of that, the slab dimensions are easily varied and the equipment procedure can easily accommodate the compaction of pavement layers such as overlays (19). The major shortcomings of the rolling wheel compaction method required a big working area as well large amount of materials to be utilized. Typically, for beam fatigue purpose this the slab is compacted using rolling wheel compactor as well as vibratory compactor (20).

2.8. Vibratory compactor

The Vibratory Compactor was designed to simulate the action of vibratory rollers. The vibratory action obtained by compressed air causing a piston to vibrate vertically resulting in the acceleration of the compactor head The most common method of compacting beam specimens is by the Asphalt Vibratory Compactor(25)(26). Prior to research conducted the conclusions been made that the most common field practice for compacting granular soil was to use a smooth-drum vibratory compactor. On the enhancement part, the additional of addictive such as Sasobit® and Evotherm® (22).

2.9. French plate compactor

The French Plate Compactor compacts asphalt mixtures using either one or two smooth, reciprocating, pneumatic rubber tire (23-24). A costly compactor that leading to discontinued of the production of this machine worldwide.

3. Mix design

Generally, there are three types of mix design methods, which are commonly being used. All these three methods are common in the objective of developing an economical combination of aggregates and asphalt. These methods are Hveem, Marshall, and Superpave mix design. The Marshall method is very popular because of its relative simplicity, economical equipment and proven record. The Marshall method seeks to select the asphalt content at a desired density that satisfies minimum stability and range of flow values. The Superpave mix design method was designed to replace the Hveem and Marshall methods. The volumetric analysis common to the Hveem and Marshall methods. The Superpave system ties asphalt binder and aggregate selection into the mix design process and consider traffic and climate as well.

3.1. Gradation

Gradation is the particle size distribution in an aggregate mixture and is determined in terms of the percentage passing or retained on each of the sieves. Gradations are specified to ensure acceptable pavement performance (17).

For the purpose of this research, Hot Mix Asphalt been used and ACW14 mix design was selected as part of the work. The basis of ACW14 was taken from JKR Specification on Flexible Pavement (18). Five mix designs were formulated using five different gradations within the aggregate gradation envelop. Table 1 shows experimental matrix set up.

Table 1. Five selected gradations for ACW14

Sieve Size (mm)	LOWER LIMIT AC 14	UPPER LIMIT AC 14	% PASSING				
			(Median)	(Upper)	(Lower)	(Crossing 1)	(Crossing 2)
			G1	G2	G3	G4	G5
20	100	100	100	100	100	100	100
14	90	100	95	100	90	100	91
10	76	86	81	86	76	84	78
5	50	62	56	62	50	57.5	57
3.35	40	54	47	54	40	48	46
1.18	18	34	26	34	18	30	21
0.425	12	24	18	24	12	23	13
0.15	6	14	10	14	6	13.5	7.0
0.075	4	8	6	8	4	8	4
Pan	0	0	0	0	0	0	0

3.2. Superpave mixture design

Typically, for Superpave Mix design the determination is differ from the Marshall Method. Superpave mix design procedures depend on the traffic level of the pavement for which the HMA being designed. The compaction effort is determined from N_{inital}, N_{design} and N_{max} which respectively giving of 89%, 96% and 98% of Theoretical Maximum Density (TMD) value. The design is working parallel with number of gyrations for specified traffic levels and air temperature. With the same number of gyrations, the higher asphalt content the higher percentage of maximum density. Superpave is a height control concept based on the design required.

The compaction effort obviously determined from the design 7-day maximum air temperature as in Table 2. As stated in Arahan Teknik Jalan 5/85 the temperature is 38°C. Yet following the statistical from Cleveland, Ohio taking the standard deviation at 2°C for normal frequency distribution mean 7 day will exceeds 42°C.The selected air temperature will give 41-43 degree is taken at the designated ESALs at < 3 x 107 as shown in Table 2. The pattern on the asphalt content range normally plotted as Figure 1.

Table 2. Selection number of gyrations for SGC

Traffic (ESALs)	Design 7-day Maximum Air Temperature (°C)											
	< 39			39 - 41			41 - 43			43 - 45		
	N_{ini}	N_{des}	N_{max}	N_{ini}	N_{des}	N_{max}	N_{ini}	N_{des}	N_{max}	N_{ini}	N_{des}	N_{max}
$< 3 \times 10^5$	7	68	104	7	74	114	7	78	121	7	82	127
$< 1 \times 10^6$	7	76	117	7	83	129	7	88	138	8	93	146
$< 3 \times 10^6$	7	86	134	8	95	150	8	100	158	8	105	167
$< 1 \times 10^7$	8	96	152	8	106	169	8	113	181	9	119	192
$< 3 \times 10^7$	8	109	174	9	121	195	9	128	208	9	135	220
$< 1 \times 10^8$	9	126	204	9	139	228	9	146	240	10	153	253
$> 1 \times 10^8$	9	143	235	10	158	262	10	165	275	10	172	288

IOP Conf. Series: Materials Science and Engineering **512** (2019) 012031 doi:10.1088/1757-899X/512/1/012031

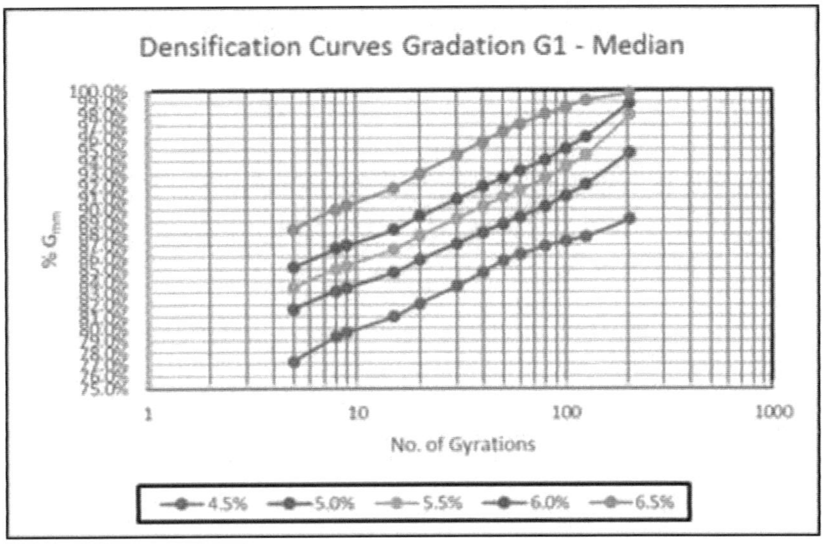

Figure 1. Typical pattern of SGC Optimum Asphalt Content

After the determination of Optimum Asphalt Content (OAC) at 96% of TMD, Superpave requirement was verified that the compaction and volumetric properties are satisfactory based on the 4% air voids, Voids in Mineral Aggregates (VMA) and as well Voids Filled with Asphalt (VFA) based on Table 3.

Table 3. Superpave Compaction Effort

Design ESALs (million)	Required Density (%Theoretical Maximum Specific Gravity)			Voids in the Mineral Aggregate (%) Minimum — Nominal Maximum Aggregate Size(mm)					Voids Filled with Asphalt (%)	Dust to binder Ratio
	N_{ini}	N_{des}	N_{max}	37.5	25.0	19.0	12.5	9.5		
< 0.3	≤91.5								70-80	
0.3 to < 3	≤90.5								65-78	0.6-1.2
3 to < 10										
10 to < 30	≤89.0	96	≤98	11.0	12.0	13.0	14.0	15.0	65-75	
≥ 30										

4. Overview on asphalt mix designs based on various compactors

A laboratory mix design supports in predicting the in-service performance of the asphalt mixtures through various performance evaluation tests. The compaction in the laboratory for specimen fabrication expected to simulate the properties of the pavement in the field. It is desirable that the laboratory compaction of specimens should be a true indicator of field performance in regards to particle orientation, air void content, permeability and mechanical properties (23).

A second method that is often used to specify compaction requires that the contractor compact the asphalt mixture to some minimum percentage of the theoretical maximum density (TMD). This is a direct method of specifying maximum in-place air voids and an indirect method for controlling compaction. This method involves taking a sample of the asphalt mixture during construction and conducting tests to measure TMD (ASTM D2041). The bulk density of the asphalt mixture is measured after the compaction and compared to TMD value. This comparison provides a direct measurement of in-place voids. For instance, a mixture compacted to 93 percent of TMD will have 7 percent air voids.

Regardless of material and time saving, the stability and flow from the Marshall applications being the prime variables in the performance of an asphalt sample as well for gradation and degree of compaction in whichever types of compactors. The OAC is traditionally estimated using Asphalt Institute Method or equivalent such as UPM's in-house method of averaging out the maximum binder contents based on Marshall Stability, Bulk Density and 4% VTM and the additional parameter of Resilient Modulus value.

It is observed that new mix designs should consider actual approaches in practice such as the roller compaction for mix designs. The current JKR (PWD) perspective only conforms and comply to using Marshall Compactor and Superpave Gyratory Compactors. New approaches in laboratory mix designs using roller compactor technology should be deliberated. It will become highly practical for the flexible road pavement construction.

5. References

[1] Banerji A K, Das A, Mondal A, and Biswas R 2014 A Comparative evaluation on the properties of HMA with variations in aggregate gradation of laboratory and field produced mixes *Int. J. Eng. Res. Technol.* 441–8
[2] Jabatan Kerja Raya 2014 *Statistik Jalan* (Kuala Lumpur: JKR)
[3] BOMAG 2009 *Basic Principles of Asphalt Compaction, 02/2009* (Germany: BOMAG GmbH)
[4] Mclaughlin A L 1977 *Field Compaction of Bituminous Mixes for Airport Pavements* (Washington)
[5] Button J W, Little D, Jagadam V and Pendleton O J 1994 Correlation of selected laboratory compaction methods with field compaction *Transp. Res. Rec.* **1454**
[6] Aziz Memon and Naeem 2006 *Comparison between superpave gyratory and Marshall Laboratory compaction methods* (Malaysia: Universiti Teknologi Malaysia)
[7] Jakarni F M, Muniandy R, Hassim S and Mahmud A R 2010 Analysis of stone mastic asphalt (SMA) slab dimensions for evaluation of the newly developed roller compactor (Turamesin) *Pertanika J. Sci. Technol.* **18** 13–22
[8] Jones C R, Ford W G and Hasim M S 1998 The maintenance of paved roads in Malaysia: performance of two full- scale experiments *Road Engineering Association of Asia and Australasia (REAAA) Conference* Wellington, New Zealand
[9] Hassan A and Sufian Z 2008 *Highway Materials and Construction* (Kuala Lumpur: Univision Press)
[10] Hartman A M, Gilchrist M D and Walsh G 2001 Effect of Mixture Compaction on Indirect Tensile Stiffness and Fatigue *J. Transp. Eng.* 370–8
[11] Sousa J B, Harvey J, Painter L, Deacon J A and Monismith C L 1991 *Evaluation of Laboratory Procedures for Compacting Asphalt Aggregate Mixtures* (Washington)
[12] Khosla N P and Sadasivam S 2002 *Evaluation of The Effects of Mixture Properties and Compaction Methods on The Predicted Performance of Superpave Mixtures* (North Carolina)
[13] Bahia H U, Coenen A and Tabatabaee N 2013 *Mix Design and Compaction* (Netherlands: Springer)
[14] Maupin J G W 1998 *Comparison of Several Asphalt Design Methods* (Virginia)
[15] ter Huerne H L 2004 *Compaction of Asphalt Road Pavements Using Finite Elements and Critical State Theory*
[16] Muniandy R, Jakarni F M, Hassim S and Mahmud A R2007 Development of criteria for slab compaction of laboratory roller compactor (Turamesin) *Am. J. Appl. Sci.* **4** 908–11
[17] Muraya P M 2007 *Permanent Deformation of Asphalt Mixtures* (University of Delft)
[18] JKR Malaysia 2008 *Standard Specification for Road Works, Section 4: Flexible Pavement* (Kuala Lumpur: JKR)
[19] Scholz T V, Allen W L, Terrel R L 1993 Preparation of Asphalt Concrete Test Specimens Using Rolling Wheel Compaction. *Transp Res Rec* **1417** 150–7

[20] Al-Qadi I L and Nassar W N 2003 Fatigue Shift Factors to Predict HMA Performance *Int. J. Pavement Eng.* **4**(2) 69–76
[21] Brown E R, Kandhal P S and Zhang J 2001 *Performance Testing for Hot Mix Asphalt*
[22] Kheradmand B, Muniandy R, Hua L T, Yunus R B and Solouki A 2014 An overview of the emerging warm mix asphalt technology *Int. J. Pavement Eng* **15**(1) 79–94
[23] Jakarni F M and Muniandy R 2006 *Properties of Stone Mastic Asphalt Slabs Compacted Using a Newly Developed Roller Compactor* (Kuala Lumpur: University of Putra Malaysia)
[24] Haghighi H and Muniandy R 2009 *Laboratory Asphalt Rotary Compactor* (Kuala Lumpur: University of Putra Malaysia)
[25] Kandhal P and Cooley L A 2003 *NCHRP Report 508 - Accelerated Laboratory Rutting Tests: Evaluation of the Asphalt Pavement Analyzer* (Washington: NCHRP)
[26] Wistuba MP 2015 The German segmented steel roller compaction method – state-of-the-art report *Int. J. Pavement Eng.* **17**(1) 81–6

Thermal, chemical and imaging analysis of Polyurethane/ Cecabase modified bitumen

F H Khairuddin[1,2], **N I Md Yusof**[2], **K Badri**[3], **H Ceylan**[4] **and S N Mohd Tawil**[5]

[1] Smart and Sustainable Township Research Centre (SUTRA), Universiti Kebangsaan Malaysia, Selangor, Malaysia
[2] Department of Civil Engineering, Universiti Pertahanan Nasional Malaysia, Kuala Lumpur, Malaysia
[3] School of Chemical Science & Food Technology, Universiti Kebangsaan Malaysia, Selangor, Malaysia
[4] Department of Civil, Construction and Environmental Engineering, IOWA State University, USA
[5] Department of Electrical & Electronic Engineering, Universiti Pertahanan Nasional Malaysia, Kuala Lumpur, Malaysia

E-mail: arename1004@gmail.com

Abstract. In this study, a 60/70 penetration grade bitumen was modified with polyurethane (PU) added with a Warm Mix Asphalt (WMA) additive, i.e. Cecabase, in the amount of 0.5 wt % with respect to bitumen. The thermal characterization was done using the Differential Scanning Calorimetry (DSC) while the chemical analysis was done using Fourier Transform Infrared (FTIR). Raman Spectroscopy was used to characterize the asphaltic materials in the bitumen, namely the D (Defect) and G (Graphite) bands. Test results showed that adding PU and Cecabase to the bitumen slightly shifted the T_g and T_m peaks of the thermal behaviour. The result of FTIR showed that there was no shift in the carbonyl and sulfoxide peaks, which indicate that no chemical interaction occurred in the unmodified and modified bitumen. Furthermore, the Raman Spectroscopy investigation revealed that PU and Cecabase reduced the wax content and decreased the bee-structure of the bitumen samples. However, the order structures of unmodified and modified bitumen were still within the boundary limit of the D and G bands of the Raman spectra. In conclusion, the use of PU and Cecabase as bitumen modifier are promising for alternative source for bitumen.

1. Introduction

In recent years, researchers have been making more attempts to modify bitumen with bio oils. Modification of bitumen is an alternative method to improve pavement performance in the attempt to deal with excessive traffic load and changes in the environmental conditions. Recent research were conducted utilizing various sources of bio-oils, such as waste cooking oil, swine waste, and waste wood as bio binder [1–3]. Sun et al. [1] reported the addition of waste cooking oil decreased the complex modulus and creep stiffness of the modified bitumen while at the same time increasing the m-value and phase angle of the modified bitumen. Julian et al. [2] investigated the feasibility of adding swine waste to base bitumen and observed a reduction in the viscosity of unage, short-term aged, and long term-aged modified bitumen. Furthermore, the results of rheological testing have shown that modified bitumen have lower complex moduli and phase angles in comparison to the base bitumen while the results of bending beam rheometer test indicated that the addition of swine waste have a potential to improve the thermal cracking performance of base bitumen by reducing creep stiffness and increasing m-value. Bitumen modified with bio-oil generated from waste wood sources requires lower mixing temperature and improve the performance of the bitumen [3]. Therefore, it can be concluded that bio-oil has a promising potential for use as a bitumen modifier.

The environmental problems brought about by the high temperature required for the manufacture of bitumen and road paving have prompted research to discover eco-friendly and sustainable materials. As a result, the Warm Mix Asphalt (WMA) technology was introduced and it has been widely utilised in the paving industry due to its ability to reduce the mixing and compacting temperatures by 30-50°C in comparison to Hot Mix Asphalt (HMA) [4,5]. Among the advantages of utilizing WMA are enhanced economic value, improved resistance and durability, less environmental pollution, and ease of construction. WMA reduces production temperatures and odour emission at the plant, extends paving seasons, allows for a longer hauling distance, allows for earlier traffic opening, reduces binder aging, reduces cracking, and minimises oxidative hardening since the mixtures are produced closer to operating temperatures. The example of WMA chemical additives are Evotherm, Rediset WMX and Cecabase RT. Rediset WMX and Cecabase RT are made of adhesion agents and surfactant that will improve the wetting condition of aggregates as well the adhesion behaviour without changing the actual bitumen performance [6,7]. Besides that, the mixing production temperature of asphalt mix Cecabase RT 945 reduced by 20 to 40°C and the mechanical properties are same as HMA [8,9].

Since the utilization of bio-oil and warm mix asphalt in the pavement industry are promising, this study is conducted to evaluate the feasibility of using polyurethane (PU) and WMA additive, i.e. Cecabase RT 975, as a modifier for base bitumen. The effects of this modifier were investigated through thermal, chemical, and image analysis and the modified bitumens were then compared with the base bitumen.

2. Experimental designs

2.1. Materials

This study used 60/70 penetration grade bitumen as the control sample. The physical properties of unmodified bitumen are presented in Table 1. A modifier, PU was produced through prepolymerisation process from palm kernel oil based polyol (PKO-p) and 2,4-diphenylmethane diisocyanate (MDI). A WMA chemical additive, Cecabase RT 975, was added to bitumen at a dosage rate of 0.5% by weight of bitumen, as recommended by supplier, in the attempt to achieve lower production and compaction temperatures of asphalt mixtures.

Table 1. Physical properties of 60/70 Penetration Grade Bitumen

Properties	Standard	Specification	Value
Penetration (dmm)	ASTM D5	60 - 70	66
Softening point (°C)	ASTM D36	47 - 52	49
Viscosity at 135°C (mPa.s)	ASTM D113	-	550
Ductility (cm)	ASTM D4402	> 100	100
Specific gravity	ASTM D70	1.0 – 1.06	1.03

2.2. Sample preparation

Three modified bitumen samples were prepared and coded as bitumen-PU (B-PU), bitumen-PU-Cecabase (B-PU-C), and bitumen-Cecabase (B-C). The base bitumen (B-B) was heated at 145°C until it turns into liquid. PKO-p was then added to the bitumen at 110°C and blended at 2000 revolutions per minute for 15 minutes using a mechanical shear mixer. MDI was then added and blended using the same mixing parameter. For the B-PU-C sample, Cecabase was added and blended for 10 minutes after the B-PU has been homogeneously blended.

IOP Conf. Series: Materials Science and Engineering **512** (2019) 012032 doi:10.1088/1757-899X/512/1/012032

3. Laboratory Testing

The effects of adding PU and Cecabase to bitumen were evaluated with regard to thermal condition by using the Differential Scanning Calorimetry (DSC) test. Shimadzu DSC-50 was used under Nitrogen atmosphere at flow rate of 10ml/min and heating rate of 10°C/min from room temperature to 600°C. The chemical characterization was done using Perkin Elmer-1650 thermogravimetric with wavenumbers ranging from 600 cm^{-1} to 4000 cm^{-1}. The topography of the bee-structures of unmodified and modified samples were determined using the inVia Confocal Raman equipped with a 532 nm solid state laser with a 20 µm x 20 µm area in-view laser beam size. The Raman spectra showed the structure of the D (Defect) and G (Graphite) bands of the bitumen.

4. Results and Discussion

4.1. Differential Scanning Calorimetry (DSC)

DSC tests were conducted to understand the thermal transitions as well as the initial wax crystallisation and melting temperatures of the unmodified and modified bitumen. Figure 1 shows the DSC curve for the transition glass temperature (T_g), melting temperature (T_m), and segregation temperature of the bitumen samples. It shows that the T_g, T_m, and segregation temperature of B-B are lower than those of the modified bitumen. The effects of low temperature on bitumen properties are typically indicated by the glass transition temperature (T_g). An exothermic peak was observed at a temperature of about 33°C for B-B, 31°C for B-PU, 33°C for B-PU-C and 30°C for B-C and this was interpreted as wax crystallisation or oxidation reaction [10]. The temperature required for melting transition, T_m to occur for B-B, B-PU, B-PU-C and B-C are 88°C, 97°C, 100°C and 90°C respectively. The temperature at which all wax completely melt is considerably higher than the temperature for initial crystallization. The DSC curve shows that the shape of the curve is very dependent on thermal history prior to recording the scans. B-C had the lowest T_g value, which makes it more resistance to low temperature cracking, while B-PU-C has the highest T_m value in comparison to B-B and B-PU. Segregation occurred at 447°C for B-B, 457°C for B-PU, 461°C for B-PU-C and 451°C for B-C. The endothermic peak was observed at 505°C, 513°C, 519°C and 520°C for B-B, B-PU, B-PU-C and B-C respectively.

Figure 1 shows that the addition of PU or Cecabase did not affect the T_g. B-B containing natural wax had a lower glass transition temperature and this can be attributed to the stiffening effect of natural wax when measurement of penetration was made at 25°C. At this temperature, natural wax, which is partly crystalline, stiffens the bitumen. Therefore, bitumen with a lower penetration grade is more likely to have a higher T_g [11,12]. These results indicate that the trends for T_g and T_m peaks are almost similar for the unmodified and modified bitumen.

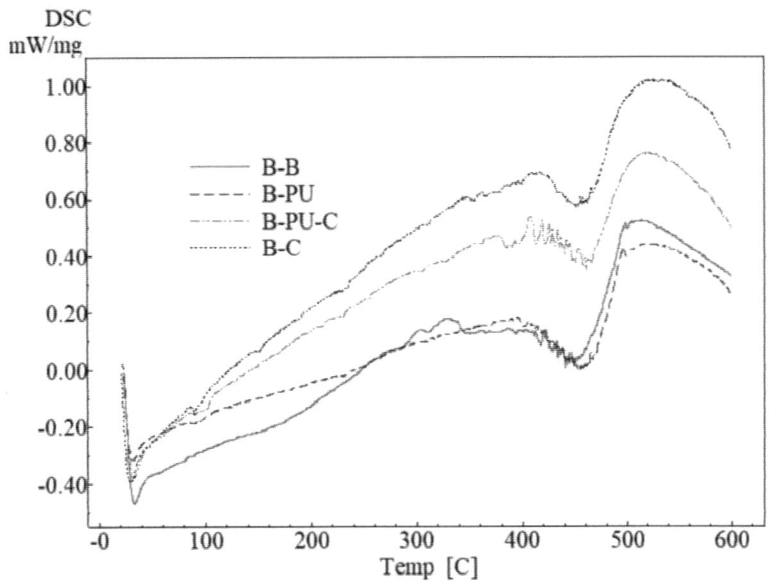

Figure 1. DSC curve for unmodified and modified bitumen

4.2. Fourier Transform Infrared (FTIR)

The chemical changes in modified bitumen were measured through FTIR analysis. The FTIR spectrum for unmodified and modified bitumen in figure 2 shows the peak position for possible functional group. The major peaks typically occur at similar wavenumber for all samples tested. The first major bond is the CH stretching vibrations of hydrocarbon at 2918-2850 cm^{-1} for B-B, 2918-2849 cm^{-1} for B-PU and B-PU-C and 2917-2850 cm^{-1} for B-C; this indicates that the possible peak position for this functional group is between 3055-2730 cm^{-1}. The next absorption peaks, which is due to C=O bond in the aromatic hydroxyl, was observed at similar wavenumber, i.e. 1653 cm^{-1}, 1655 cm^{-1}, 1664 cm^{-1}, and 1652 cm^{-1} for B-B, B-PU, B-PU-C and B-C respectively. The absorption peak at 1597cm^{-1} is due to C=C stretching bond, which was also observed at 1596cm^{-1} for B-B, 1598 cm^{-1} for B-PU, 1595 cm^{-1} for B-PU-C and 1597cm^{-1} for B-C.

The stretching vibrations of the sulfoxide bond S=O were observed at 1026cm^{-1} for B-B and B-PU, 1030 cm^{-1} for B-PU-C and 1028 cm^{-1} for B-C. Based on the IR spectra results, it can be concluded that the addition of PU and Cecabase did not cause any significant shift in the peaks of the modified bitumen [13,14]. Therefore, there was no chemical interaction between the unmodified bitumen and the modifier [15].

IOP Conf. Series: Materials Science and Engineering **512** (2019) 012032 doi:10.1088/1757-899X/512/1/012032

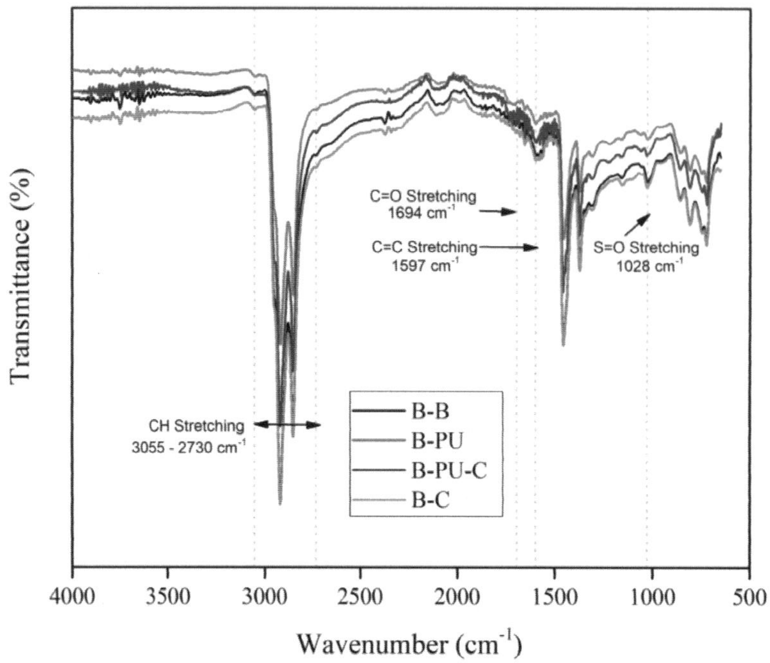

Figure 2. FTIR spectra for unmodified and modified bitumen

4.3. Raman Spectroscopy

4.3.1. Topography of the Bee-Structures. Bitumen is composed of several organic and inorganic compounds which have a complex microstructure. Loeber et al. [16] reported that bitumen structures are composed of wax and show a 'bee-like structure' or catana phase. Bitumen with wax content exhibits a more distinct bee-structure. Figures 3 (a-d) show the bee-structures of the unmodified and modified bitumen. However the number of bee-structures in B-PU and B-C are less than those in B-B and B-PU-C. This indicates that the addition of only PU or Cecabase to unmodified bitumen reduced the bitumen's wax content. The loss of wax could caused an increase in the temperature of the laser, which then caused the wax crystals to melt into the matrix phases of the bitumen. The size and shape of the structures decrease uniformly as a result of the melting process which begin outside the structures and move inwards [17].

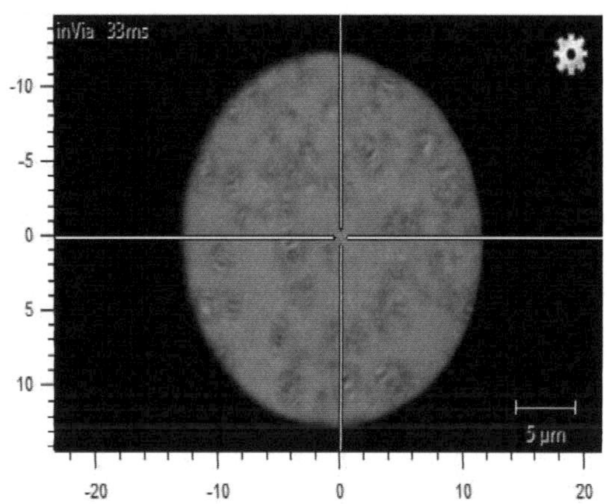

(a)

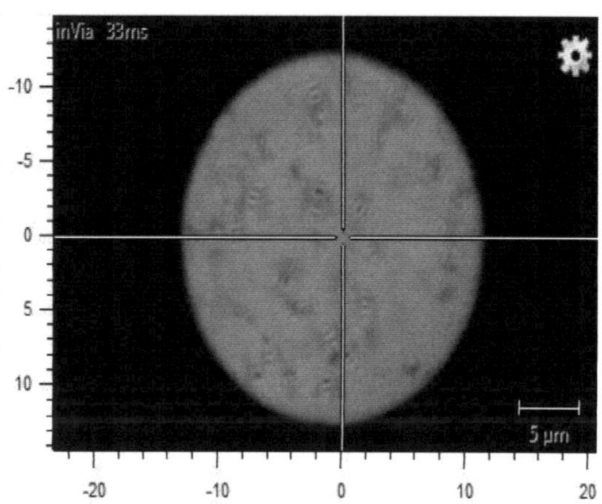

(b)

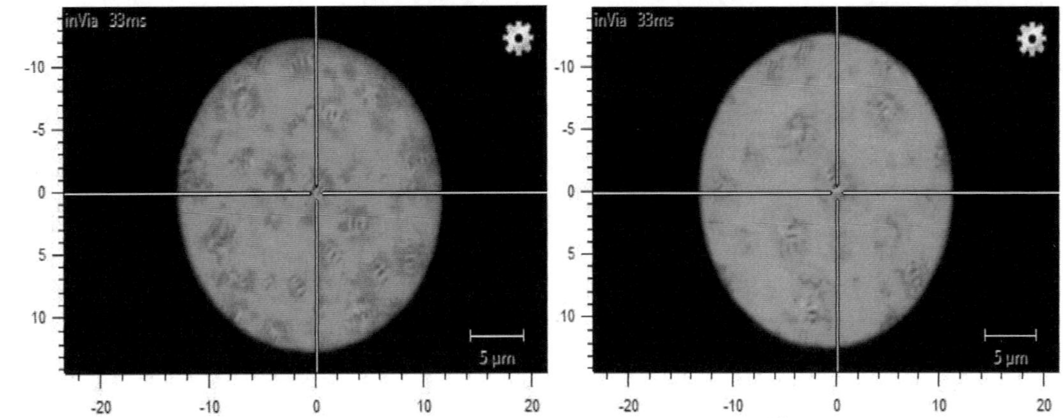

(c) (d)

Figure 3. Topography of bee-structure for (a) B-B (b) B-PU (c) B-PU-C and (d) B-C

4.3.2. Raman Spectra. Figures 4 (a-d) show the Raman peaks for unmodified and modified bitumen. Two types of band were analysed, the D band at a frequency of 1350 cm^{-1} and the G band at 1580 cm^{-1}. The D and G bands refer to the in-plane breathing vibration made possible by the defect in the ideal graphitic structure [18].

For B-B, the D band occurred at 1355 cm^{-1} and the G band at 1584 cm^{-1}.The addition of PU shifted both bands and the D band occurred at 1373 cm^{-1} while the G band at 1589 cm^{-1}. This is because the PU structure has a strong carbonyl compound. For B-PU-C and B-C, the D peaks occurred at 1352 cm^{-1} and 1372 cm^{-1} and the G peaks at 1582 cm^{-1} and 1586 cm^{-1}, respectively. These peaks show that the addition of Cecabase to B-PU slightly shifted both peaks. Hence, it can be concluded that the ordered structures for all modified bitumen are still within the boundary limits of the defect and graphite bands, which is consistent with the result obtained in previous study [19,20].

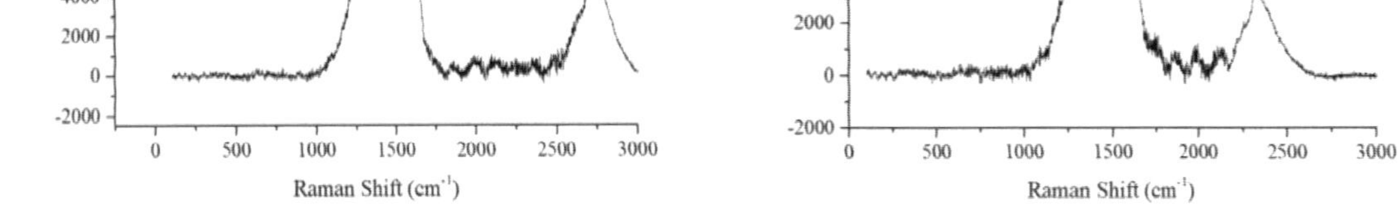

(a) (b)

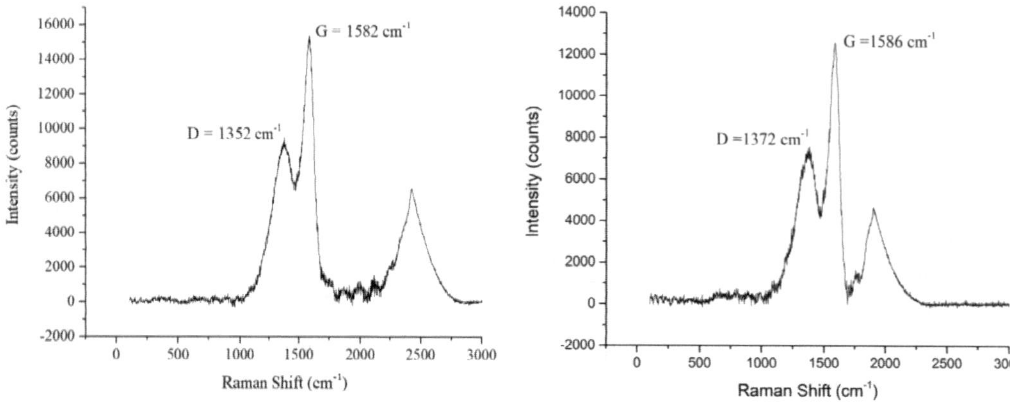

(c) (d)

Figure 4. Raman Spectra for (a) B-B (b) B-PU (c) B-PU-C and (d) B-C

5. Conclusion

The thermal behaviour of unmodified and modified bitumen were investigated using DSC. Results show similar trends for the tested bitumen, with B-B having the lowest T_g and T_m value. This indicates that addition of PU and Cecabase had little effect on the thermal transition of the B-B. Changes in chemical composition were determined through FTIR and result showed that there was no chemical interaction between the unmodified and modified bitumen since the carbonyl and sulfoxide bond remained almost unchanged. The morphology of the bee-structure from the Raman Spectroscopy revealed that there was a decrease in the wax content of B-PU and B-C, which reduced the catana phase, in comparison to B-B and B-PU-C. The ordered structures for all tested bitumen were still within the boundary limits of both the D and G bands. In conclusion, the use of PU and Cecabase as bitumen modifier are promising for alternative source for bitumen. It is recommended that the rheological and ageing conditions of PU/Cecabase modified bitumen to be conducted for further understanding of this new materials.

6. References

[1] Sun Z, Yi J, Huang Y, Feng D and Guo C 2016 Properties of asphalt binder modified by bio-oil derived from waste cooking oil *Constr. Build. Mater* **102** 496-504

[2] Mills-Beale J, You Z, Fini E, Zada B, Lee C H and Yap Y K 2014 Aging Influence on Rheology Properties of Petroleum Based Asphalt Modified with Biobinder *J. Mater. Civ. Eng.* **26** 358-66

[3] Yang X, You Z and Dai Q 2013 Performance evaluation of asphalt binder modified by bio-oil generated from waste wood resources *Int. J. Pavement. Res. Technol.* **6** 431-49

[4] D'Angelo J et al. 2008 Warm-Mix Asphalt:European Practice. *Technical Report Documentation*

[5] Jamshidi A, Hamzah M O and You Z 2013 Performance of Warm Mix Asphalt containing Sasobit : State-of-the-art *Constr. Build. Mater.* **38** 530-53

[6] Capitao S D, Picado-Santos L G and Martinho F 2012 Pavement engineering materials: Review on the use of warm-mix asphalt *Constr. Build. Mater.* **36** 1016-24

[7] Silva H M R D, Oliveira J R M and Peralta E J 2009 Evaluation of the rheological behaviour of Warm Mix Asphalt (WMA) modified binders *Advanced Testing and Characterization of Bituminous Materials* 661-74

[8] Abdullah M E, Zamhari K A, Buhari R, Abu Bakar S K, Mohd Kamaruddin N H, Nayan N, Hainin M R, Abdul Hassan N, Hassan S A and Md Yusoff N I 2014 Warm Mix Asphalt Technology : A Review *Jur. Teknologi.* **3** 39-52

[9] Sheth N M 2010 Evaluation of selected warm mix asphalt additives *Master Thesis*

[10] Lu X, Langton M, Olofsson P and Redelius P 2005 Wax morphology in bitumen *J. of. Materials. Scien.* **40** 1893-900
[11] Hamzaoui R, Guessasma S, and Bennabi A 2012 Penetration testing and thermal behavior of bitumen 35/50 and modified bitumen 13/40 *Eur. Phys. J. Appl. Phys.* **59** 10201
[12] Schmidt R J and Santucci L E 1966 A Practical method for determining the glass transition temperature of asphalts and calculation of their low temperature viscosities *Assoc Asphalt Paving Technol Proc* **35** 61-90
[13] Omar H A, Md Yusoff N I, Ceylan H, Abdul Rahman I, Sajuri Z, Mohd Jakarni F and Ismail A 2018 Determining the water damage resistance of nano-clay modified bitumens using the indirect tensile strength and surface free energy methods *Constr. Build. Mater.* **167** 391-402
[14] Al-Mansob R A, Ismail A, Rahmat R A O K, Borhan M N, Alsharef J M A, Albrka S I and Karim M R 2017 The performance of Epoxidised Natural Rubber modified asphalt using nano-alumina as additive *Constr. Build. Mater.* **155** 680-87
[15] Wei J, Li Y, Dong F, Fenf H and Zhang Y 2014 Study on the amorphous poly alpha olefin (APAO) modified asphalt binders *Constr. Build. Mater.* **66** 105-12
[16] Loeber L, Sutton O, Morel J, Valleton J m and Muller G 1996 New direct observations of asphalts and asphalt binders by scanning electron microscopy and atomic force microscopy *J. of Microsc.* **182** 32-39
[17] Mccarron B, Yu X, Tao M and Burnham N 2012 The Investigation of 'Bee -Structures' in Asphalt Binders *Major Qualifyng Project* Worcester Polytech Institute pp 1-48
[18] Baludikay B K et al. 2018 Raman microspectroscopy, bitumen reflectance and illite crystallinity scale: comparison of different geothermometry methods on fossiliferous Proterozoic sedimentary basins (DR Congo, Mauritania and Australia) *Int. J. Coal. Geol.* **191** 80–94
[19] Nciri N, Kim J, Kim N and Cho N 2016 An in-depth investigation into the physicochemical,thermal, microstructural, and rheological properties of petroleum and natural asphalts *Materials* **9**
[20] Jehlicka J, Urban O and Pokorny J 2003 Raman spectroscopy of carbon and solid bitumens in sedimentary and metamorphic rock *Spectrochim. Acta. - Part A* **59** 2341-52

Acknowledgments
The authors would like to express their gratitude to Universiti Kebangsaan Malaysia for the financial support for this work (DIP-2017-004).

IOP Conf. Series: Materials Science and Engineering **512** (2019) 012033 doi:10.1088/1757-899X/512/1/012033

Development of high workability grout on semi rigid wearing course

Nadiah Md Husain[1], Norfarah Nadia Ismail[1], Kamsiah Mohd Ismail[1], Deprizon Syamsunur[2] and Mariati Mohd Taib[3]

[1]Department of Civil Engineering, Kuliyyah of Engineering, IIUM, Gombak, Kuala Lumpur, Malaysia
[2]Department of Civil Engineering, Faculty of Engineering, Built and Technology, UCSI, Kuala Lumpur, Malaysia
[3]School of Housing, Building and Planning, Universiti Sains Malaysia, Penang, Malaysia

E-mail: drnadiah@iium.edu.my

Abstract. A composite pavement known as semi rigid wearing course is made by incorporating high workability fluid grout into a high porosity of porous asphalt skeleton. This fluid grout is particularly designed to have the consistency workability of water with high early and 28 – day strength. It is critical to maintain its high fluidity so it can flow through air voids while simultaneously generate high compressive strength with minimal porosity. The aim of this study is to investigate the influence of a known pozzolanic material towards the properties of fluid grout. A wide range of mix trials was formulated with various proportions of chemical and mineral admixtures. The fresh and hardened grout underwent flow cone and compressive test for its workability and strength determination respectively. The results obtained show that the combination of a known pozzolanic material and a known chemical admixture is possible to improve the fluidity of grout and strength when suitable and conducive mix proportions attained.

1. Introduction

A composite pavement known as semi rigid wearing course is made by incorporating high workability fluid grout into a highly porosity asphalt skeleton. This type of fluid grout is specially designed to have water-consistency workability with a high early and 28-day strength. This is in accordance to the standard set by Road Engineering Association of Malaysia (REAM). The properties of fresh fluid grout is crucial to ensure that the grout fills up small pores and voids in the porous asphalt skeleton, concurrently generate high compressive strength to effectively bond the two composition together with minimal porosity. High strength grout should also be able to withstand high compressive and static loads to maintain the stability of a structure. Besides that, the improvement of porous skeleton matrices also contributes to the function of impermeability to fluid grout, capable of giving it a better chemical resistance [1].

Research has shown that not only Portland cement is used to produce high compressive strength grout but also a mixture between Portland cement and supplementary cementitious materials (SCMs) is possible. SCMs that have been widely used in grout/concrete includes the pozzolanic materials such as fly ash (class F and class C), silica fume (SF), slag and metakaolin. These pozzolanic materials are

able to improve packing capacity at the interface, hence creating a tighter pore structure of the composite materials and produce high strength concrete/fluid grout of varying magnitude. This reflects the extent to which each addition is chemically active during the hydration process and therefore the extent to which they contribute to or modify the structure and properties of the hardened paste [2].
This paper aim to investigate the mechanical properties of fluid grout for the preparation of semi rigid wearing course. Workability and compressive strength analysis was performed on fresh and hardened grout using its specified testing machine.

2. Materials

2.1. Cementitious Binder

Cementitious binders used in this research were the ordinary Portland cement (OPC) type 1 and SF and in comply with BS EN196 and ASTM C1240-03a respectively. This study investigated the effect of SF replacement towards workability and strength of fluid grout within the range of 0% to 10% for the application in semi rigid pavement.

2.2. Chemical Admixture

Superplasticizer (SP) or also known as high-range-water-reducer (HRWR) has been introduced to the mineral admixtures to promote high-workability fluid grout. The selected SP was supplied by Fosroc and in comply with BS 5075 Part 3 and ASTM C494 Type F. The recommended dosages were in between 0.7% to 2.0% from the total mixture [3]. Excessive usage of SP will cause segregation and bleeding to the hardened grout [4][5].

2.3. Water/binder Ratio

Water to binder (w/b) ratio is known to be the weight of water to the weight of cement used in concrete mix. According to BS EN447 stated, concrete/grout complying with the standard will normally have a water w/b ratio below than 0.4. As such, the suggested w/b ratio were in the range of 0.30 to 0.36. With the aid of SP, which acts as a water removal agent, the final grout strength will help in the quality of concrete produced. This is also in agreement with Alsadey (2012) who claimed that SP is used to increase the workability without changing the w/b ratio.

3. Experimental Methods

3.1. Fluid Grout Mixture and Proportions

Table 1 shows the fluid grout mix design composition for various percentages of cement replacement and w/b ratio. The influence between various percentages of SF and w/b ratio towards the compressive strength and workability of fluid grout were studied and deliberated in accordance to the REAM Standard.

Table 1. Fluid Grout Mixture Proportions of Portland Cement and Silica Fume

	Mix Component			
Mix	OPC	SF	SP	w/b ratio
1	100%	0%		
2	95%	5%	1%	30%
3	90%	10%		
4	100%	0%		
5	95%	5%	1%	33%
6	90%	10%		
7	100%	0%		
8	95%	5%	1%	36%
9	90%	10%		

3.2. Grout Preparation and Curing

The fluid grout compositions were mixed in accordance to ASTM C305-99: Standard Practice for Mechanical Mixing of Hydraulic Cement Pastes and Mortars of Plastic Consistency. The fresh fluid grout will undergo the flow cone test which the dimension used is in comply with REAM standard.

The hardened cubes on the other hand were prepared and tested in accordance to BS 1881: Part 116:1983. Steel mold cubes with the size of 50 mm x 50 mm x 50 mm for compressive strength test should be prepared before casting. The hardened cubes were then removed from the mold and placed in the curing tank for 3,7 and 28 days.

4. Results and Discussion

4.1. Fluid Grout

Figure 1 shows the time taken for 3 different w/b ratios with 0% to 10% cement replacement respectively to empty the flow cone. The time taken to empty the cone varies from 11.5 s (high water consistency) up until 181.5 s (very cohesive). It was found that the increment of w/b ratios gave significant changes to the workability of fluid grout when measured using the flow cone test. The results exhibited the disadvantage of replacing OPC with higher percentage SF to the mixture where by revealed a longer time to empty the flow cone. It was found that by replacing OPC with a lower percentage of SF to the mixture generally gave shorter time for the fluid grout to empty the cone.

SP was introduced to the mix composition where it provides the possibility of a better dispersion of cement particles and improves stability through inter-particle repulsive force, thereby improving a higher fluidity [6]. Note that the SP used for this research is fixed to 1% as it could control the possibility of segregation and bleeding.

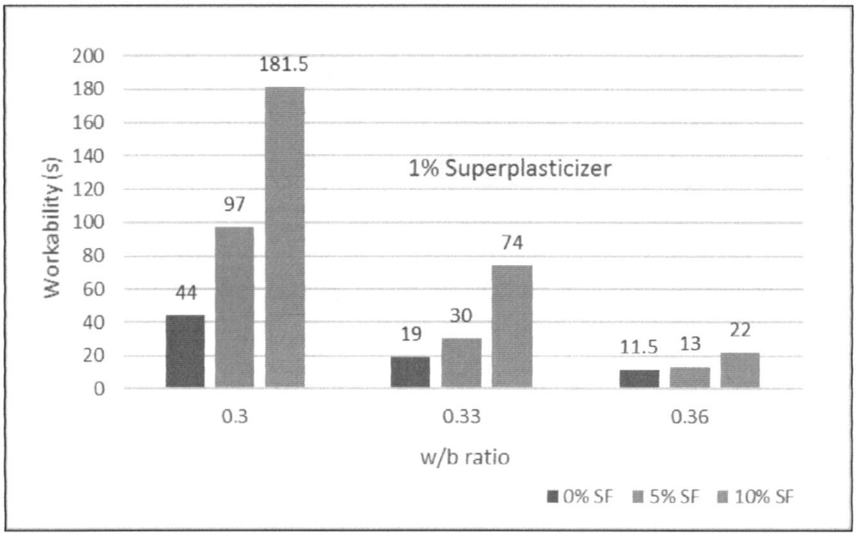

Figure 1. Workability of Fresh Fluid Grout at various w/b ratios and SF dosages.

4.2. Compressive Strength

The grout that is specified by REAM should comply a compressive strength on day 1 and day 28 of 45-50 MPa and 100-105 MPa respectively. This requirement is needed for the purpose of producing semi rigid wearing course.

Figure 2 illustrates the strength development at 0.3 w/b ratio over various percentage of SF replacement. The results achieved prove that higher dosage of SF incorporated into the mix possess higher strength especially on day 28 with 75 MPa compared to the controlled Mix 1 at only 68 MPa. The same pattern was also demonstrated for Mix 2 and Mix 3 from day 1 and maintained until day 28. The enhancement of compressive strength could be attributed to the pozzolanic reaction.

Figure 3 on the hand elucidates the strength development for mixes with 0%, 5% and 10% of SF replacement at 0.36 water to binder ratio on the 1st, 3rd, 7th and 28th day. It was observed that the results obtained for mixes with 0.36 w/b ratio are relatively low compared to those with 0.30 wa/b ratio. Lost of compressive strength from the hardened grout has been observed when the w/b ratio was increased to the mix particularly at 0.36. Bleeding and segregation were observed on the surface of the hardened grout. Although it gave good workability to the mixture however, it shows an underperformed orientation in strength. It can definitely be concluded that high dosage of water could affect the later days strength significantly and other related properties.

A noticeable finding can be highlighted from this study which includes the behaviour of cement replacement itself. The cement replacement by SF plays a vital role in the workability and strength of the fresh and hardened grout. Compared to the control mixes (0% SF), the increment of SF dosages managed to give higher compressive strength especially on the 28th day. It is generally accepted that SF improves pore structure by its small particle size that allow a filler effect to build the bridge between cement grains and the spaces between cement grains and aggregate [7]. This is also in agreement with Neville and Aitcin (1998) which stated that the very fine material was observed to reduce in the porosity of the grout matrix and provides a dense microstructure for microstructure properties improvement thus increasing the strength of the system.

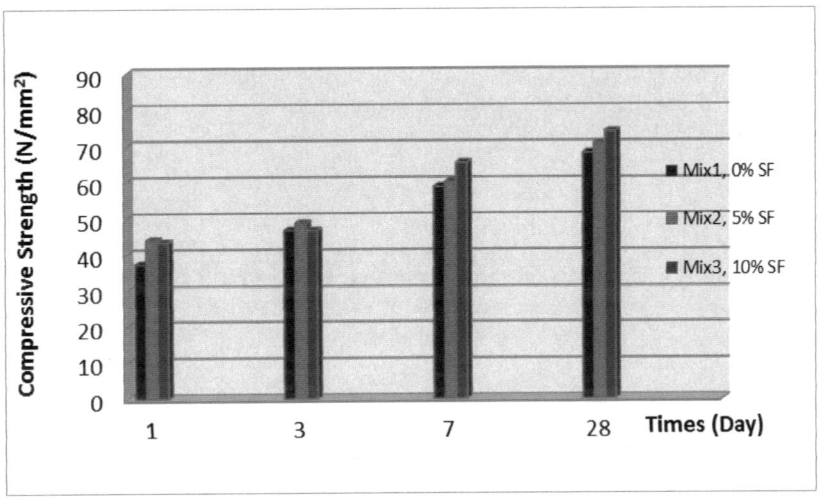

Figure 2. Strength Development at 0.3 w/b ratio over various percentage of SF replacement.

The introduction of SF will slightly lower the strength during the initial period (between day 1 and 3). The presence of SF in the mix composition shows lower strength development in the early days compared to the control mix.

Two types of reactions occur when dealing with pozzolanic materials – primary cement hydration, secondary pozzolanic reaction. The primary cement hydration produces an end product in the form of calcium silicate hydrate (C-S-H) and Calcium hydroxide ($Ca(OH)^2$). The C-S-H is known to determine the strength of the mixture. The secondary reaction focuses on the pozzolanic material, which is the SF. SF undergoes pozzolanic reaction with $Ca(OH)^2$ to produce more of C-S-H gel [9]. The C-S-H gel then leads to an additional reduction in capillary porosity. Eventually, the additional generated C-S-H allows for a further increase in strength to the hardened grout [10]. This explains the reason behind the decrement of strength at early days for mixes that has been incorporated with SF.

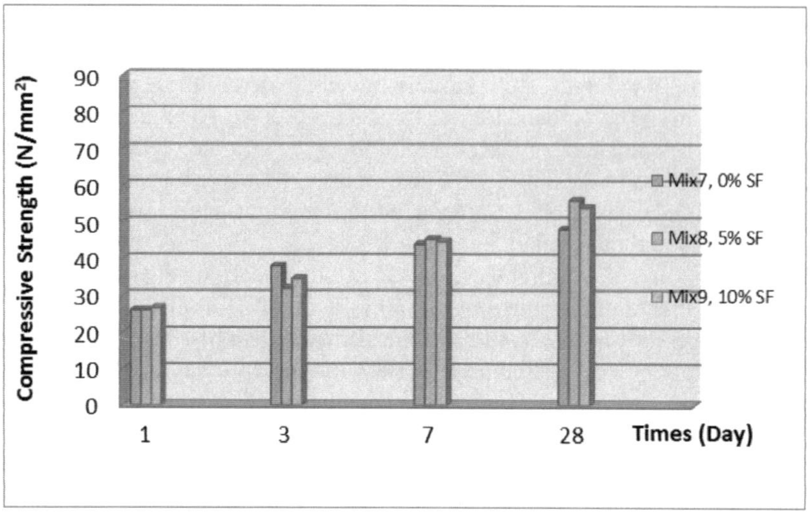

Figure 3. Strength Development at 0.36 w/b ratio over various percentage of SF replacement.

5. Conclusion

This study has concluded that workability of fresh fluid grout and strength of hardened grout is significantly reflected by the increment of w/b ratio and the incorporation of pozzolanic material especially at early and later days.

6. Recommendations

The recommended mixture for this particular set of data for the purpose of semi rigid wearing course production in compliance with REAM requirement probably would be Mix 2 from the 0.3w/b ratio. However

7. References

[1] Dumne S 2014 Effect of superplasticier on fresh and hardened properties of self-compacting concrete containing fly ash *American Journal of Engineering Research* 205-11

[2] Domone P and Illstone J 2010 *Construction materials, their nature and behavior 4th Ed.* (London and New York: Spon Press) pp 101-108.

[3] Fosroc 2011 *Conplast SP2000 (M) high performance super plasticizing admixture* (Shah Alam: Fosroc Technical Service).

[4] Alsadey S 2012 Effects of super plasticizing and retarding admixtures on properties of concrete *International Conf. on Innovations in Engineering and Technology* pp 25-6

[5] Suhana K, Karim MR, Mahmud H, Mashaan NS, Ibrahim MR, Katman H and Md Husain N 2014 Effects of Using Silica Fume and Polycarboxylate-Type Superplasticizer on Physical Properties of Cementitious Grout Mixtures for Semi Flexible Pavement Surfacing *The Scientific World Journal* pp 1-5

[6] Chandara S and Bjornstrom J 2002 Influence of cement and superplasticizers type and dosage on the fluidity of cement mortars - Part 1 *Cement and Concrete Research* pp 1605-11

[7] Bentz D and Stutzman P E 1994 Evolution of porosity and calcium hydroxide in laboratory concretes containing silica fume *Cement Concrete Research* pp 1044-50

[8] Neville A M and Aitcin P C 1998 High Performance Concrete-An Overview *Material Structures* **31** 111-117

[9] Siddique R and Khan MI 2011 *Silica Fume, Supplementary Cementing Materials* (Berlin: Springer).

[10] Shannag M J 2000 High Strength Concrete Containing Natural Pozzolan and Silica Fume. *Cement & Concrete Composites 22* 399-406

Acknowledgment

The authors would like to thank the Department of Civil Engineering of IIUM and to all colleagues whom directly and indirectly involved in this study.

Bridge loading in Malaysia : past, present and the future

Atikah Bt Zakaria@Ya, Fairul Zahri Mohamad Abas, Mohd Yunos Ali and Nadia Azmi

Bridge Design Division, PWD Malaysia.

Email: fairul.jkr@1govuc.gov.my

Abstract. A rapid growth of development in our country has demanded our road network to be expanded over the last few decades. To cope with rapid development, the transport industry has deployed larger vehicles with heavier axle loads. In many instances, the level of services of our road network governed by the bridge load capacity. Since pre-independence era, our bridges have been designed based on the British Code. Begin with British developed code of DSIR at early 20s to British Code (BS) and eventually to Eurocode, our bridge loadings have been evolved substantially. In the midst of foreign code, our National Axle Load study conducted in mid 80s has also marked a significant finding to our local bridge capacity in which three major groups of load carrying capacity have emerged from namely STAL, MTAL and LTAL. This paper presents the journey of our local bridge loading code of practice and the corresponding impact to the structural design and construction. Comparisons are made as early as in 20s at the time where the code was based on British Department of Scientific and Industrial Research (DSIR) to the early age of British Code and to the migration of Eurocode. A typical concrete bridge deck spanning of 20m to 60m are modeled and analyzed with respective live loading derived from these established code of practice. Discussion on the way forward of implementing the Eurocode into our local design practices are also included.

1. Introduction

Bridge structures as part of a road network has evolved since pre-historical time. The main function of the bridge has remained unaltered in defying the gravity force from the beginning of human civilization. Similarly in Malaysia, bridge structures as part of the road network played a significant roles particularly in promoting economic growth by connecting cities, allowing goods transportation and so on.

The understanding of the bridge mechanism evolved as the science and technology advanced. The foremost importance of the bridge structures is that it must be able to sustain the intended design load at that particular point of time. The locality, economy and geographical aspect of the country typically dictate the mode of transport thence the bridge capcity that need to be designed. As such, bridge and transport engineers have work hand to hand to ensure that the functionality of the bridge during service is not compromised.

The objective of this paper is to comprehend the evolution of the bridge loading in Malaysia and how is the intensity of load affects the bridge capacity in carrying the load.

2. From the past

Bridge design in Malaysia has always been influenced by the British Standard due to the colonization links between these two countries. The period between 1914 and 1918 marked a new era in the evolution of the highway bridge loading in United Kingdom. Malaysia has first adopted bridge design standard as early as 1922 from *British Standard Loading Train* as shown in Figure 1. This standard loading train consisted of a 20 tons tractor plus three trailer of 13 tonnes each respectively [1]. This configuration of loading was actually a direct representation of traction engine plus three trailer used on the road at that time as shown in Figure 2. This type of loading prevailed until late 1931 when British Ministry of Transport (MOT) adopted a new approach to design loading known as *MOT loading curve*. It consisted of a uniformly distributed load (UDL) plus with an invariable single transverse line load known as knife-edge load (KEL). The concept of loaded length was also introduced at this stage as part of the improvement from previous application of Standard Loading Train which has only specific dimensions. A reproduction of the curve of uniformly distributed load (UDL) against loaded length is as shown in Figure 3.

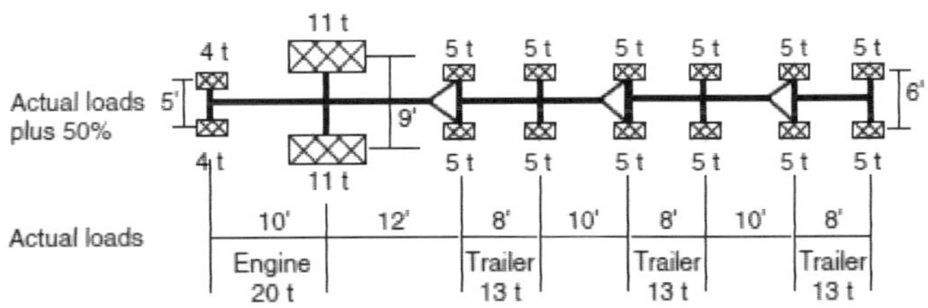

Figure 1: Standard Loading Train adopted in 1922 [1].

Figure 2: Traction engine plus three trailer c. 1910 (after ICE Manual)

At this point of time, the British Government through the MOT has also introduced the Construction and Use (C&U) Regulation for Lorries and Truck which indicated the legally allowed loads and dimensions for various types of vehicles.

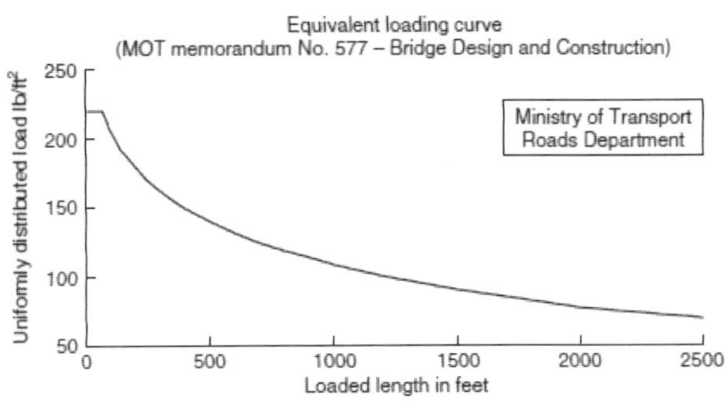

Figure 3: Original *MOT loading curve* of UDL against loaded length

In 1954, Henderson presented a paper suggesting reconsidering the load pattern of *British Standard Loading Train* and the *MOT Loading Curve* [2]. Henderson (1954) observed that in reality the loads in used were differed from the standard loading train or standard loading curve. It was found that there were those that could be describe as legal load that conform to C&U regulation while certain vehicle carrying abnormal indivisible heavy load, which were outside the regulation where special permission was required for transportation. Thus, his observations lead to the categorization of normal and *abnormal* indivisible load due to the fact that both ordinary traffic load and abnormal vehicle are dissimilar in weight and arrangement of wheels to those represented by the former loading trains. As a result, the MOT in conjunction with British Standard Institution (BSI) had come up with two kind of loading for design purpose, namely normal (HA) and abnormal (HB) loadings with the philosophy that the design should be made on the basis of normal loading and checked for abnormal traffic loading. This was embodied by the standard code document of BS 153. Several revisions have been made along the way as more precise information about traffic volumes and weight has been gathered and in 1978 the standard code of BS 153 was replaced by BS 5400.

Malaysia however did not completely adopt this type of loadings. Only normal HA loading was adopted for design purpose of major road while 2/3 of HA load intensity was used for designing minor roads up until year 1972 [3]. From late 1972 to 1990, our local bridge design practice has adopted to include HB in combination of HA type of loading. The HB loading vehicle load however was applied at the centreline of the bridge deck and was in coherent as per BS 5400.

In 1985, the Government of Malaysia has commissioned a National Axle Load Study for bridges in Peninsular Malaysia [3]. The study was completed in the late 1987. The details of the study are not presented in this paper. As the result, one of the major outcomes of the study lead to the formulation of Malaysian Design Bridge loads called Standard JKR Specification for Bridge Loading. The specification was read and applied in conjunction with BS5400 however. The only part that the specification covered was only the live load by replacing HA and HB loadings to LTAL and SV loadings respectively. The specification nevertheless valid for highway bridges with loaded not exceeding 50m.

3. At present

The LTAL load used as the specification of Bridge Design in Malaysia was not prolonged when it was found that the load effect of LTAL were higher than the load effect caused by the design load in UK, Canada and USA. In 2001, the British Standard Institution and Department of Transport has come to agreement that BD 37/01 will be used as the current code for Bridge Design in UK. BD 37/01 which is based on BS 5400 part 2 (BSI, 1978) was also adopted in Malaysia to replace the JKR Bridge Design Specification. Similar to BS 5400 part 2 (1978), the two most prominent load applications are defined as normal load (HA) only and HA in combination of abnormal loading (HB). The normal load consists of a uniformly distributed load (UDL) occupied on a designated notional lane plus a lane KEL. The intensity of UDL load curve is given as in Figure 4. It showed two-part curve in which each load curve W in kN/m is defined by equation (1) for one up to 50m loaded and equation (2) for the remainder up to 1600m respectively [4].

$$W=336(1/L)^{0.67} \quad (1)$$

$$W=36(1/L)^{0.1} \quad (2)$$

IOP Conf. Series: Materials Science and Engineering **512** (2019) 012034 doi:10.1088/1757-899X/512/1/012034

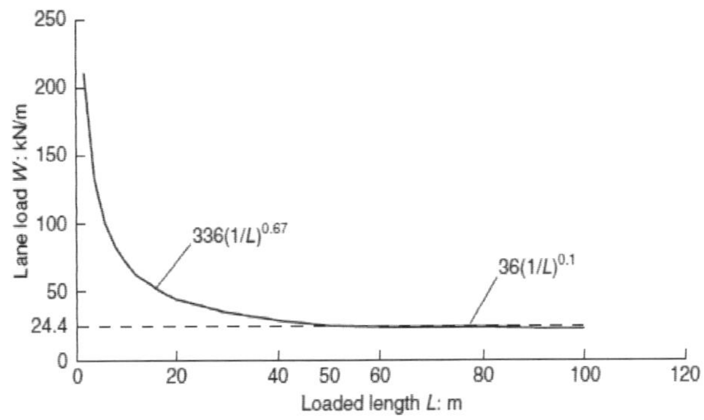

Figure 4: British Standard (BD37/01) normal loading curve for UDL HA type of loading

The application and intensity of normal load (HA-UDL) depends upon the carriageway width, the loaded length and the number of loaded lanes. The carriageway width will dictate how many numbers of notional lanes on a particular bridge deck. Notional lane in fact is not necessarily the same as the actual traffic lanes defined by the lane marking. It is given by:

$$\text{Int}\,[(b/3.65)+1] \quad (3)$$

where b is the total width of the bridge deck measured between kerbs on the deck.

The loading for abnormal vehicle in BD37/01 has a similar configuration as in BS 5400 part 2 (1978). It can be called as HB type of loading. This abnormal vehicle load is distributed on 16 wheels arranged on four axles as shown in Figure 5.The weight is measured by multiplying the unit per axle to the maximum number of 45. 1 unit of axle is taken as 10kN and produced up to 450kN per axle load for 45 unit of HB loading. It can be seen from Figure 5 that the configuration of HB vehicle can be adjusted to suit the influence line of the deck analysis interest. The inner axle spacing can vary to give the worst effect ranging from 6m to 25m in an increment of 5m.

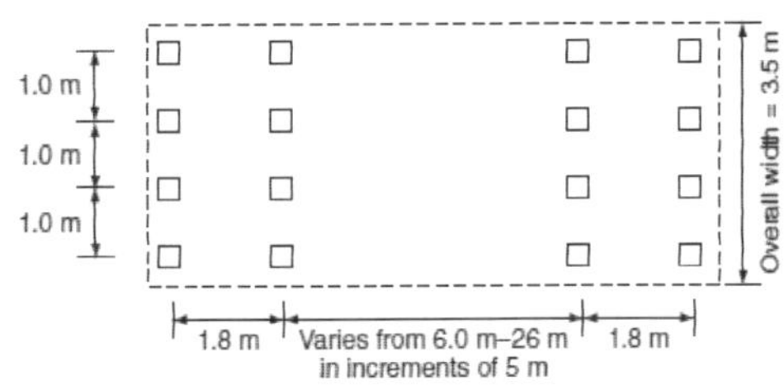

Figure 5: Abnormal HB vehicle configuration

BD 37/01 called that all road bridges shall be designed to carry HA loading. In addition to that, 30 unit of HB type of loading shall be taken in combination of HA loading for all road bridges except for accommodation bridge which allowed HA type of loading only. However, depending on the class of road carried by the structures, the number of unit of HB type of loading can be increased. For instance, 45 unit of HB type of loading for motorway and trunk road while 37.5 unit of HB loading for principal road. Other public road shall be taken as 30 of HB loading.

Similar to HA-UDL load application, the abnormal vehicle load of HB type is also applied onto the notional lane. However, there is an opportunity that the HB vehicle can be applied straddling onto two notional lane depending on the width of notional lane itself. HA-UDL type of loading is also applied on the remaining notional lane in the case of combination of HA and HB type of load application.

4. In the future

All the while Malaysia has adopted British Standard as its design code for civil works design and construction. In March 2010, British Standard Institute (BSI) withdrew 57 related and conflicting structural codes and had published 58 documents as the replacements. In view of this change, Top Management of JKR Malaysia in 2015 has given 00a directive to embark on the use of Eurocode for the design purpose including the bridge design loading and structural code. Initial point was raised on what loading shall be taken for the application of abnormal type of vehicle loading.

Eurocode for bridge loading compendium published in volume EN 1991-2. The load model for vehicle representation have been explicitly elucidate as to compared with previous BS code into 4 major type as listed in Table 1[5].

Table 1: Load model definitions in EN 1991-2 for bridge loading

Load Model	Definition
LM 1	Normal loading due to lorries plus cars
LM 2	A single axle for local effects
LM 3	Special vehicles for the transportation of exceptional loads
LM 4	Crowd loading

The normal loading represented by load model 1 (LM1) has a similar configuration as to British standard whereby it comprises of uniformly distributed load plus a double-axle tandem applied onto the respective notional lane [6]. The notional lane width is generally taken as 3m, and the number of notional lanes as *Int(w/3)* where *w* is the carriageway width. Instead of deriving from load curve, the UDL normal load of EN1991-2 together the tandem load are given by the characteristic values with respect to notional lane number as listed in Table 2.

Table 2: Characteristic value of normal load in EN 1991-2

Location (Not.lane)	Tandem System load TS	UDL System load
	Axle load (kN)	Q (kN/m2)
Lane Number 1	300	9
Lane Number 2	200	2.5
Lane Number 3	100	2.5
Other lanes	0	2.5
Remaining area	0	2.5

Abnormal loads are considered in a similar manner to the British Code, with a special abnormal load (model LM3) placed in one lane (or straddling two lanes) with a 25m clear space front and back and normal LM1loading placed in the other lanes. The vehicle may be specified by the particular load authority involved, or alternatively it may be as defined in EC1 which specifies eight load configurations with varying numbers of axles, and loads from 600kN to 3600 kN. Wheel areas are assumed to merge to form long areas of 1.2m x 0.15 m. Axle lines are spaced at 1.5m and may consist of two or three merged areas. A typical configuration for an 1800kN vehicle is shown in Figure 6.

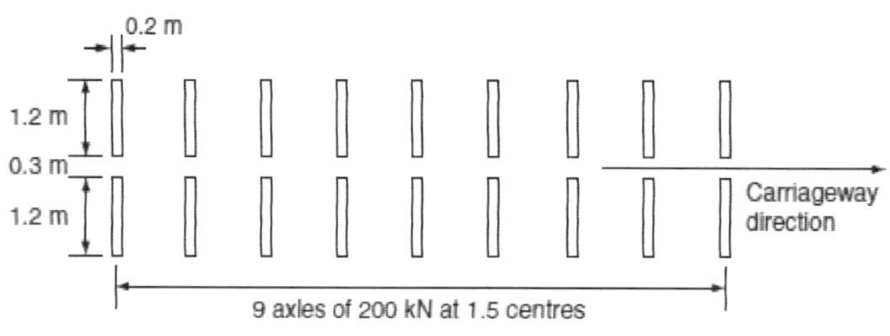

Figure 6: Typical LM 3 vehicle

5. BS and EC load comparison

To depict the advance of bridge live loading from British Standard to Eurocode, a simple comparison in term of bending moments for a single and multiple span bridges are carried out. A single span bridge of 30m, 50m and 100m length as shown in Figure 7 and 8 respectively have been modelled by using computer modelling package MIDAS. Similarly, a comparison on two-span simple beam bridge is modelled and the structures responses of bending moment are also compared [7].

Primary type of UDL loading which was derived from BS loading (HA –Type) and EN1 (Load Model 1 (LM 1)) were first compared in term of their load carrying capacity represented by value positive bending moment.The prerequisite data in term of number of notional lane, loaded length, partial factors that are required in deriving the UDL type of load in British Standard and Eurocode are the same for the purpose of comparison.

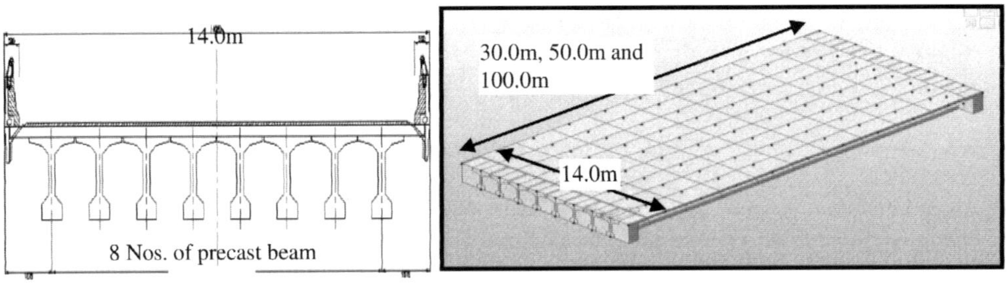

Figure 7: (a) Simple beam bridge deck of 14.0m wide and (b) the grillage model of the bridge deck

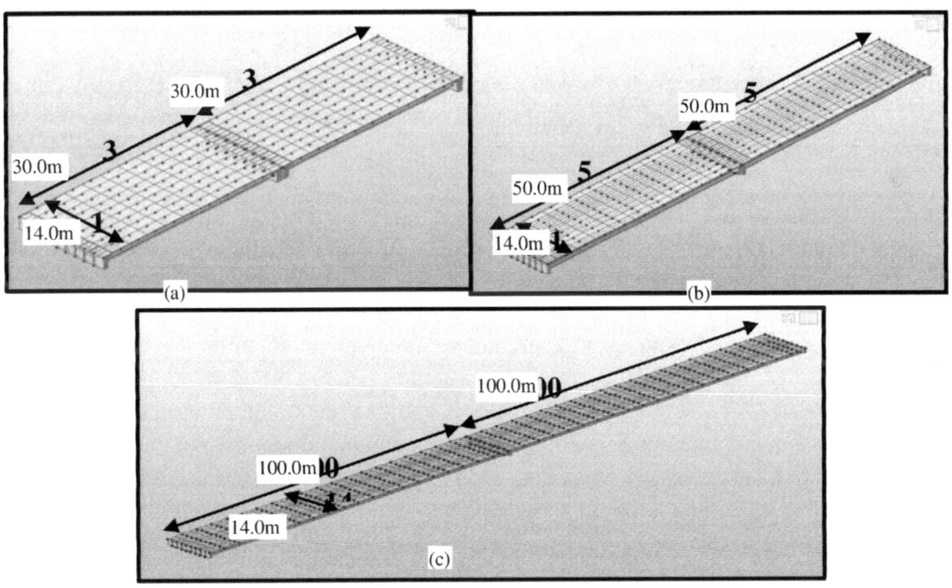

Figure 8: Simple two-span beam bridge deck of 14.0m wide and the grillage model of the bridge deck for (a) 30m each (b) 50m each and (c) 100m each span respectively

The positive bending moment values for the single span model with the span length of 30m, 50m and 100m respectively are shown in Table 3 while the results for two span span length are given in Table 4. In the case of simply-supported bridge model, the span length can be considered as the loaded length at which the traffic loading acted upon it. In this investigation the loaded length were made equal length of 30m, 50m and 100m each.

In order to see clear comparisons and relationships between the bending moment and its respective loaded length of two difference codes, the graph in Figure 9 and Figure 10 are plotted. From Figure 9, it can be seen that the positive bending moment values obtained by using the UDL type of traffic loading from BS code and EN code are almost identical. This demonstrates that the EN code can be replaceable to the BS code without any substantial deprivation of the existing load designed by using BS code. This is applicable when it deals with UDL type of bridge loading for a single span bridge. Similar pattern can also be seen when the UDL type of loadings of both codes were applied to two-span continuous bridge as shown in Figure 10. Agreements between these codes are prominent for a short loaded length whereas slight difference can be seen for a loaded length greater than 200m.

Although the EN code has omitted the influence of loaded length to the intensity of UDL loading it can be deduced that a characteristic values given for the LM 2 which consist of tandem (TS) and UDL systems respectively has produced a similar load effect as per BS. The main difference come from the substantial reduction in load applied to the location of traffic lane on the deck.

Table 3: Comparison of Positive Bending Moment Values for 30m, 50m and 100m Single-Span Bridge Model

Type	Loaded Length (m)								
	30			50			100		
	Unfactored	FOS	Factored	Unfactored	FOS	Factored	Unfactored	FOS	Factored
British Code (kNm)	2,342.59	1.50	3,513.89	4,647.81	1.50	6,971.72	12,711.11	1.50	19,066.67
Eurocode (kNm)	2,377.86	1.35	3,210.11	4,824.83	1.35	6,513.52	12,735.90	1.35	17,913.47

Table 4: Comparison of Positive Bending Moment Value for Continuous Two-Span Bridge Model of Equal Span of 30m, 50m and 100m respectively.

Type	Loaded Length (m)								
	2 Span @ 30m each			2 Span @ 50m each			2 Span @ 100m each		
	Unfactored	FOS	Factored	Unfactored	FOS	Factored	Unfactored	FOS	Factored
British Code (kNm)	1,365.28	1.50	2,047.92	2,879.42	1.50	4,319.13	7,543.06	1.50	11,314.59
Eurocode (kNm)	1,650.86	1.35	2,228.66	3,297.07	1.35	4,451.04	8,115.20	1.35	11,360.52

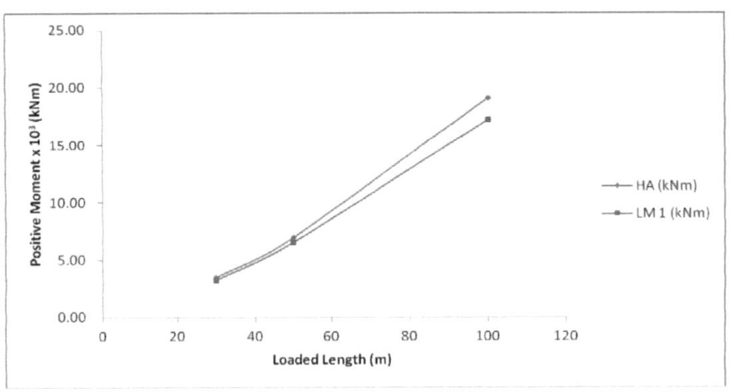

Figure 9: Comparison of Positive Bending Moment of UDL loading from British code BS (HA-UDL) and Eurocode EN (LM 1) for 30m, 50m and 100m Single Span

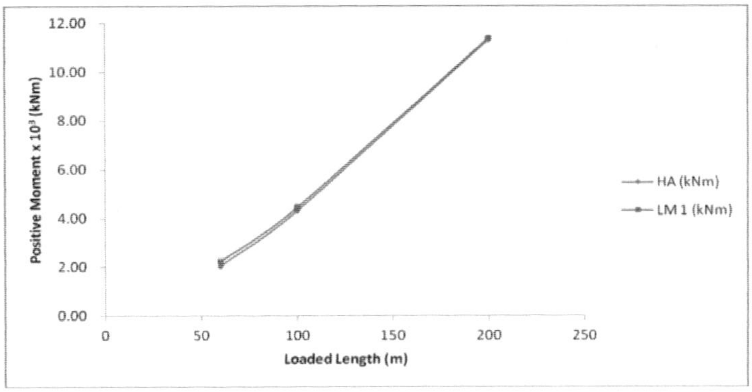

Figure 10: Comparison of Positive Bending Moment of UDL Loading from British code BS (HA-UDL) and Eurocode EN (LM 1) for Two Span Bridge Model.

The abnormal load between both BS and EC code were also compared. Comparison of current practice of British Code of abnormal load (HB 30) was made against Eurocode EN load of special vehicle (SV 100) while the abnormal loading of HB 45 in BS was compared against SV196 of EC code. These two sets are the common abnormal vehicle used for design loading for majority of principle road. The HB 30 and HB 45 were derived from the nature of exceptional industrial load with the total weight of 1200 kN and 1800kN respectively while SV 100 and SV 196 represent the effect of special type general order vehicle with the overall weight of 1000 kN and 1960kN respectively. Similar grillage models and span configurations as the primary UDL load was used to analyse these two type of loadings.

The positive bending moment values for the single span model with the span length of 30m, 50m and 100m respectively are given in Table 5 and 6 while the results for two span of varying span length are given in Table 7 and 8 respectively.

Table 5: Comparison of Positive Bending Moment Values for 30m, 50m and 100m Single-Span Bridge Model for HB 30 and SV 100 Loadings.

Type	Loaded Length (m)								
	30			50			100		
	Unfactored	FOS	Factored	Unfactored	FOS	Factored	Unfactored	FOS	Factored
British Code (kNm)	2,320.64	1.30	3,016.83	4,130.46	1.30	5,369.60	6,599.58	1.30	8,579.45
Eurocode (kNm)	2,691.07	1.35	3,632.94	4,416.05	1.35	5,961.67	6,702.12	1.35	9,047.86

Table 6: Comparison of Positive Bending Moment Values for 30m, 50m and 100m Single-Span Bridge Model for HB 45 and SV 196 loadings

Type	Length (m)								
	30			50			100		
	Unfactor	FOS	Factor	Unfactor	FOS	Factor	Unfactor	FOS	Factor
British Code (kNm)	3,480.96	1.30	4,525.25	6,195.69	1.30	8,054.40	9,899.36	1.30	12,869.17
Eurocode (kNm)	4,096.73	1.35	5,530.59	7,380.49	1.35	9,963.66	11,889.56	1.35	16,050.91

Table 7: Comparison of Positive Bending Moment for HB 30 and SV 100 Values for Continuous Two-Span Bridge Model of Equal Span of 30m, 50m and 100m Respectively

Type	Length (m)								
	2 Span @ 30m each			2 Span @ 50m each			2 Span @ 100m each		
	Unfactor	FOS	Factor	Unfactor	FOS	Factor	Unfactor	FOS	Factor
British Code (kNm)	1,748.60	1.30	2,273.18	3,185.63	1.30	4,141.32	5,504.20	1.30	7,155.46
Eurocode (kNm)	2,108.28	1.35	2,846.18	3,514.11	1.35	4,744.05	6,702.12	1.35	9,047.86

Table 8: Comparison of Positive Bending Moment for HB 45 and SV 196 Values for Continuous Two-Span Bridge Model of Equal Span of 30m, 50m and 100m Respectively

Type	Length (m)								
	2 Span @ 30m each			2 Span @ 50m each			2 Span @ 100m each		
	Unfactor	FOS	Factor	Unfactor	FOS	Factor	Unfactor	FOS	Factor
British Code (kNm)	2,622.90	1.30	3,409.77	4,778.44	1.30	6,211.97	7,997.56	1.30	10,396.83
Eurocode (kNm)	3,051.22	1.35	4,119.15	5,668.78	1.35	7,652.85	9,559.47	1.35	12,905.28

The comparison of HB 30 against SV 100 and HB45 against SV196 were also made by plotting the bending moment values against the loaded length under consideration as shown in Figures 11,12,13 and 14 respectively. From Figure 11, it can be seen that the maximum positive moment from HB 30 load is relatively in good agreement to the Eurocode load of SV 100 loading despite of the difference in overall weight of the application vehicles. A slight difference between these two load types is also worth mentioning at the loaded length of 30m. At this point, SV 100 produced a positive maximum bending moment of 20% higher than the HB 30 load which has the unfactored total load of 120kN. By referring to the methodology of this study, it can be deduced that the difference of the positive bending moment values is due the vehicle axle configuration at which the point loads acted upon. SV 100 has closer axle configuration as to compare with HB 30 axles configuration. Therefore, the intensity of the loading is higher, while HB 30 has the load that spreads apart within the span.

The argument of vehicle axle configuration is more prominent when it deals with two-span continuous bridge deck. At loaded length of 200m by which the vehicle is loaded at one span of 100m only, the difference in positive moment between BS code and EN code is as high as 26%. This is not so obvious in the case of short to medium span length. Let alone the effect of lateral bunching that may significantly affects to the short and medium span bridge.

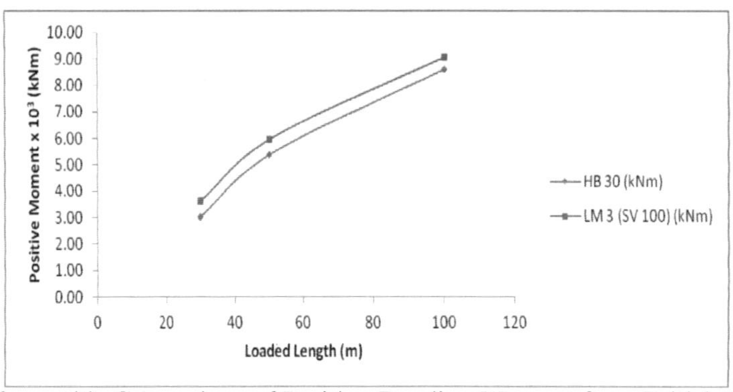

Figure 11: Comparison of Positive Bending Moment from British Code BS (HB 30) and Eurocode EN (SV 100) for 30m, 50m and 100m Single Span Models

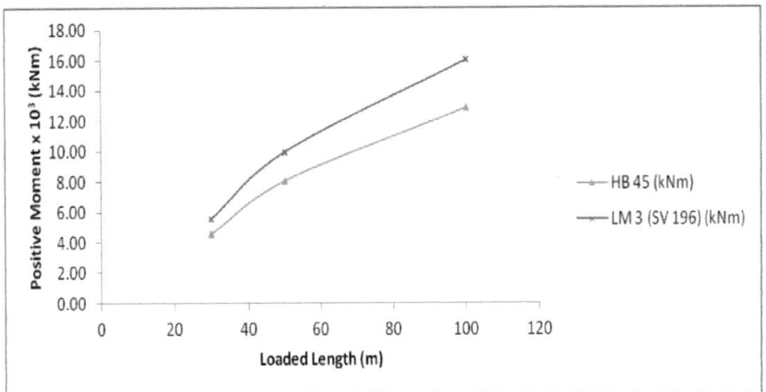

Figure 12: Comparison of Positive Bending Moment from British Code BS (HB 45) and Eurocode EN (SV 196) for 30m, 50m and 100m Single Span Model

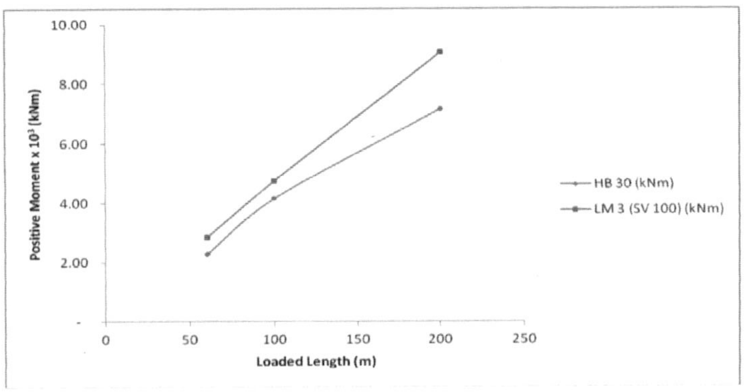

Figure 13: Comparison of Positive Bending Moment from British code BS (HB 30) and Eurocode EN (SV 100) for Two-Span Bridge Models

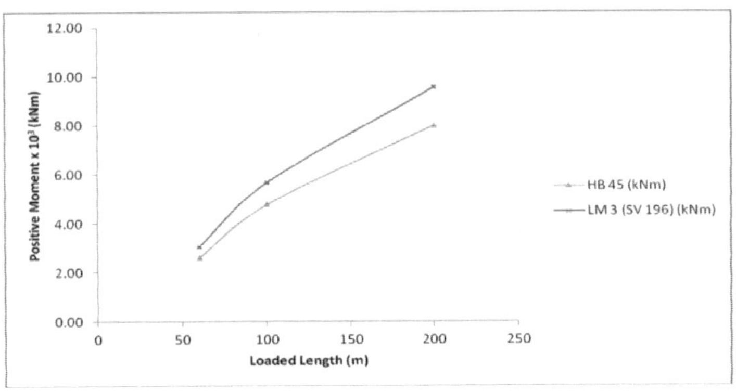

Figure 14: Comparison of Positive Bending Moment from British Code BS (HB45) and Eurocode EN (SV 196) for Two-Span Bridge Models

However, the comparison of HB 45 and SV 196 shows incompatible replacement of loads for medium to longer span. However, in the case of short span, the difference found from the analysis is to be 22%. Clearly that the vehicle axle configurations dictate significant load distribution on the bridge deck and thus affecting the bending moment response to the structure.

Similar trend of variation of moment against loaded length can be observed for two-span models. As the span length increased, the variation of bending moment produced by these two codes has becoming greater. At loaded length of 200m, the difference was found to be 24%.

A comparison between BS and EC against the JKR SV20 loading were also made. This is in view that the SV 20 has been used once as the bridge loading standard in Malaysia. It can be deduced from Table 9 that the SV20 loading has the highest bending moment value and yield to represent the highest vehicle model.

Table 9: Comparison of Positive Bending Moment for HB 45 and SV 196 against JKR SV20 loading for loaded length of single span 30m, 50m and 100m respectively.

Type	Loaded Length (m)		
	30	50	100
HB 45 (kNm)	2,642.29	5,490.29	7,498.78
SV 196 (kNm)	3,182.82	6,892.46	8,354.50
SV 20 (kNm)	3,317.12	9,489.52	15,491.70

6. Conclusion

Malaysia has always in line with British code of practice has been undergone many shift of bridge loadings since pre-independent period. The design of our local bridge has been very much influenced by the british code although there were attempts to instate our local bridge loading.

In the nutshell, the changeover bridge design guide from current British Code to the latest Eurocode would not significantly differ in terms of the structural response output except some decretions are required when deals with long-span bridge. The handling of the load intensity and application of the latest Eurocode design guides need to be clearly envisaged during the design process in line of current British Standard practice.

7. References

[1] Ryall M J 2008 *Loads and load distribution. ICE Manual of Bridge Engineering* (United Kingdom : Institute of Civil Engineers)

[2] Henderson W 1954 *British Highway bridge loading. Road Engineering Division Meeting, Road Paper No.44*

[3] Tharmabala T, Tautant D, Sani K M and Salleh R 2001 *Strength Evaluation and Load Testing of Bridges in Malaysia* (Kuala Lumpur : JKR)

[4] The Highway Agency and British Standard Institution 2001 *BD37/01 Design Manual for Road and Bridges*

[5] British Standard Institution 2003 *Eurocode 1: Actions on structures – Part 2: Traffic loads on bridges* (London : British Standard Institution)

[6] British Standard Institution 2003 *UK National Annex to Eurocode 1: Actions on structures – Part 2: Traffic loads on bridges* (London : British Standard Institution)

[7] Azmi N 2015 *Comparison of Bridge Loading Between BS5400/BD 37-01 and EN 1991-2* (MSc Thesis : University Putra Malaysia)

IOP Conf. Series: Materials Science and Engineering **512** (2019) 012035 doi:10.1088/1757-899X/512/1/012035

Threshold value of skid resistance and texture depth for Malaysia road

Aminuddin Suhaimi and Abdul Rahman Baharuddin

Road Branch, Public Works Department Malaysia, Kuala Lumpur, Malaysia

E-mail: aminuddins@jkr.gov.my & arahmanb@jkr.gov.my

Abstract. Skid resistance is an important pavement characteristic for roads, which determines the friction between the road surface and vehicle tyres. The level of friction is largely influenced by the pavement surface aggregates micro texture, macro texture, and drainage attributes. The objective of this paper is to present the findings of the research and analysis conducted on Skid Resistance Value (SRV), Mean Texture Depth (MTD) and Skid Number (SN) obtained from the data collected on federal roads over the years since 2008. This research also attempted to determine the appropriate limiting values for SRV (microtexture), MTD (macrotexture) and SN for the federal roads. The result of analysis indicated that the average value of SRV at the locations of wet surface accident is 52.31 $\pm$ 8.54. While for MTD, the accident risk is higher when MTD value is below the average of 0.45 $\pm$ 0.025. The SRV and MTD results were correlated to each other and it can be concluded that accident risk increases significantly when SN value is below the average value of 31.49 $\pm$ 7.87. Based on these findings, it is concluded that the minimum desirable values for SRV and SN for Malaysia's Federal roads to be 55 and 30 respectively.

1. Introduction

Road accident is a serious issue and a major concern to any road authorities and public at large. Accident statistics from the Malaysia Royal Police (PDRM) showed that the number of road accidents increased almost every year from 328,264 cases in year 2005 to 489,606 cases in 2015. About 1.3 - 1.8 % of the road accident are fatal accidents.

An important aspect of safety during travel using road transport is the interaction between vehicle tyres and the road surface. Two main factors that have high influence in this case are the skid resistance between the tyre and the road surface, and the texture depth of the pavement.

Skid resistance is the friction force developed when a tyre that is fully or partially prevented from rolling slides along a pavement surface under lubricated conditions. Friction is developed with the close contact between the tyre and the pavement aggregates.

On wet road surfaces, the capability to drain off the surface water between the tyre treads and pavement surface is another significant property required for the pavement wearing course. It is attributed to the aggregate size and amount of filler that filled up the spaces and voids within the aggregate matrix hence forming the residual texture depth of the pavement surface. Both skid resistance and texture depth have great impact over the tyre-pavement friction interaction mechanism and determined the safe vehicle operations on pavements [16].

IOP Conf. Series: Materials Science and Engineering **512** (2019) 012035 doi:10.1088/1757-899X/512/1/012035

1.1. Objective

The main objective of this study is to analyse the data on Skid Resistance Value (SRV) and Mean Texture Depth (MTD) collected for Federal Roads. The study will also review the effect on the Skid Number (SN) and the risk of skidding accident. This study also included the determination of appropriate threshold values of SRV and MTD. These threshold values shall provide some guide for key performance measures and intervention levels for Road Asset Management System. Hence, sufficient fund shall be allocated for the maintenance needs and thus, preserving the required safety level of the highway system.

1.2. Skid Resistance

Skid resistance is described as a movement resistance of two surfaces in at their contact location. It is characterized by the road pavement aggregates, pavement surface roughness and the friction forces developed when the pavement is exposed to the wheel load. Friction is expressed via friction coefficient which is a sum value of two acting forces, one is parallel to contact surface between two bodies and opposite to their movement direction (friction force), other acting force is perpendicular to the contact surface (normal force) [13].

The skid resistance properties also determine the distance travelled during emergency braking process to a stopping point. For safety requirements there is a minimum amount of skid resistance required based of the type of road locations. Transport and Road Research Laboratory (TRRL) (1969) has suggested the following requirements for skid resistance as shown in Table 1.

Table 1. TRRL (1969) requirements for skid resistance.

Category	Type of site	Minimum skid resistance
A	Roundabout, gradient, bends, approach	65
B	Heavily trafficked roads, urban	55
C	All other sites	45

Skid resistance has two major components: adhesion and hysteresis. Adhesion results from the shearing of molecular bonds formed when the tire rubber is pressed into close contact with pavement surface particles. Hysteresis results from energy dissipation when the tire rubber is deformed when passing across the asperities of a rough surface pavement. These two components of skid resistance are respectively related to the two key properties of asphalt pavement surfaces, that is micro-texture and macro-texture as presented in Figure 1.

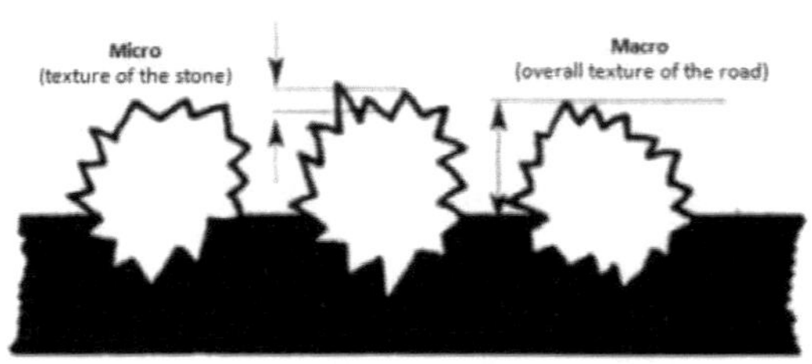

Figure 1. Illustration of Macro and Micro texture of a pavement surface [5].

1.3. Wet Weather Accidents

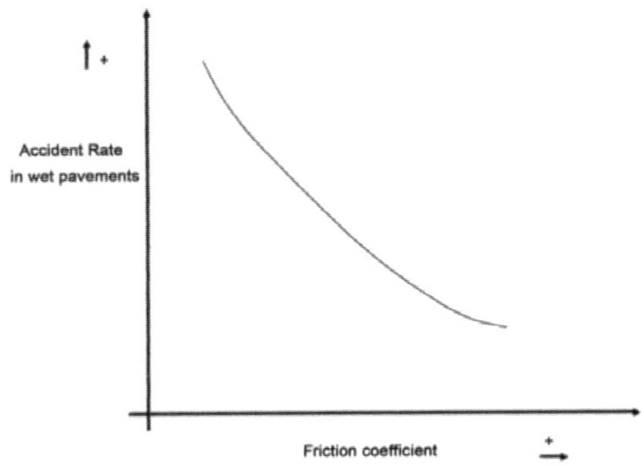

Figure 2. Relationship between accident rate in wet weather conditions and pavement surface friction [25].

Wambold et al. [27] reported that wet-weather accidents have a significant relationship with the skid numbers measured with a skid trailer. The effect of wet weather conditions on road safety was also demonstrated by a study conducted in Germany, where the proportion of wet crashes was compared to pavement surface friction, as shown in Figure 2. Friction in this study was measured at a speed of 80 km/hr. This figure clearly shows a significant decrease in wet pavement accidents as the pavement friction increases [25]. The relationship between pavement skid resistance and crash rates and the effect of pavement friction improvement on these crash rates is also demonstrated by several researchers [4, 10, 14, 17, 25].

1.4. Pavement texture

Road pavement texture can be divided into several categories. The Committee on Surface Characteristics of the World Road Association (Permanent International Association of Road Congresses (PIARC) 1995) has proposed a texture classification based on the wavelength range for each category as shown in Table 2.

Table 2. Texture classifications according to wavelength, PIARC (1995).

Texture classification	Relative wavelength
Microtexture	$\lambda < 0.5$ mm
Macrotexture	0.5 mm $< \lambda <$ 50 mm
Megatexture	50 mm $< \lambda <$ 500 mm
Roughness	0.5 m $< \lambda <$ 50 m

In the interaction between the vehicle tyres and the road surface, there are two main categories of surface texture that played very significant role in safety against skidding especially during wet conditions. They are the microtexture and macrotexture

With the wavelength of < 0.5mm, microtexture of individual aggregates contributed to the skid resistance at all speeds [6, 22, 24], and provides a greater contribution than macrotexture where vehicle operating speeds are low (<50km/hr) [11, 19, 23]. It contributes to the coefficient of friction through the creation of adhesion forces, which occur as a result of the tyre interlocking with the road surface [22] and molecular bonds being sheared as the rubber of tyres pass over the road surface [19]. Macrotexture refers to the overall texture of the road surface (0.5mm – 50mm) [7], that being the surface irregularities caused by the size, shape and spacing of stone chips in the pavement [2]. Macrotexture primarily contributes to the coefficient of friction through the dry contact with the aggregates and allowing water to drain through the texture spaces, thereby reducing the risk of aquaplaning especially at high speed.

2. Methodology

This research is based on evidence collected from most accident cases investigated by the JKR's Road Safety investigation team. The investigation procedures involved conducting several tests which included collecting data on skid resistance values and the texture depths of the pavement surface at the accident locations under investigation. Prior to the site investigation, basic information such as POL27 reports, accident and maintenance records are collected from the respective road authorities and agencies including the media.

2.1. British Pendulum Test (ASTM E303)

The British Pendulum Test (BPT) is a portable device that is widely used due to its simplicity, low cost and large historical database. It utilizes the loss in momentum of a swinging pendulum to measure pavement frictional properties. A rubber slider is attached to the end of the pendulum arm, which is raised parallel to the pavement surface prior to each test. Once released, the arm swings down and the slider travels across the pavement surface. The greater the friction of the surface, the greater the loss of momentum of the arm and the less amount of "backswing." A pointer follows the arm after the slider has left the pavement surface and measures the resulting Skid Resistance Value (SRV) or British Pendulum Number (BPN). This method is sometimes considered a surrogate for estimating effect of the microtexture.

Figure 3. British Pendulum Tester.

2.2. Sand Patch Method (ASTM E965)

One of the oldest, most often used and relied on methods for measuring macrotexture of pavement surface is the Sand Patch Method [1]. Even today most other tests are referenced to the sand patch method. This test method determines the average depth of pavement textured surface over a region by smoothing an area with a known quantity of sand. The method is designed to provide an average depth value of only the pavement macrotexture and is considered insensitive to pavement microtexture characteristics. The average depth between the bottom of the pavement surface voids and the top of surface aggregate particles can be calculated using the following:

$$MTD = \frac{4V}{\pi D^2} \quad (1)$$

Where: V = Sampling volume (cm3)
D = Average diameter of the area covered by the sand (mm)

Figure 4. Measuring texture depth using the Sand Patch Method.

3. Result and analysis

Data from accident investigation are being collected since 2008 and filtered for accident due to skidding of vehicles. Most of these accidents involved run-off vehicle cases but there are also other types of accidents that have been reported to be due to vehicles that are out-of-control. However, all these accidents have a common trend that is they had occurred on wet pavement surfaces. Table 3 shows the statistical analysis of data collected through the investigation on 37 wet surface accidents.

Table 3. Summary of the statistical analysis on wet surface accident data.

	Skid Resistance Value (SRV)	Mean Texture Depth (MTD)	Skid Number (SN)
Arithmetic Mean	52.31	0.45	31.49
Variance	72.87	0.0006	61.94
Standard Deviation	8.54	0.025	7.87

Based on the analysis, it can be concluded that the mean value of SRV and MTD are 52.31 $\pm$ 8.54 (43.77, 60.85) and 0.45 $\pm$ 0.025 (0.425, 0.475) respectively. By inserting these values in the Skid Number equation 2 [15], the mean value for SN is equal to 31.49 $\pm$ 7.87 (23.62, 39.36). SN is the Skid Number (Coefficient of Pavement Friction * 100) and SN0 is the Skid Number at zero speed. SN0 has been shown to be highly correlated to pavement microtexture which depends on the surface of aggregate asperities with its magnitude ranging from 1 to 500 mm (0.5 mm).

$$SN = SN_0 \exp\left[\frac{-(PNG)\,V}{100}\right] \quad (2)$$

Where: SN = Skid Number
$SN_0 = -31 + 1.38BPN$; Skid Number at zero speed
$PNG = 0.45\,(MTD)^{-0.47}$; Percent Normalized Gradient
V = Vehicle Speed (assumed constant urban street velocity of 60 km/hr)

3.1. MTD and SRV

According to the general principles and available literature, Microtexture and macrotexture are both necessary for the provision of adequate skid resistance. Thus, any measures available should be taken to increase the values of both parameters.

Increase in SRV may be achieved by reducing the aggregate particle size as well as the distance between particles, yet both of these adjustments to a mixture would create a decrease in MTD. Increase in MTD are beneficial to the overall skid resistance at high speeds, and this may be increased by coarsening the aggregate structure and extending the separation between particles. Greater MTD generates larger void spaces at the surface of the pavement, which reduces the effective contact area of the vehicle tire with the roadway – causing a reduction in SRV.

In Figure 5, MTD is plotted against SRV. The result shows weak correlation between MTD and SRV for wet surface accidents. Although a small amount of scatter was present in the data, the general trend of decreasing texture depth with increasing SRV was evident. Thus, maximizing both parameters is not advisable. A balanced approach is more appropriate.

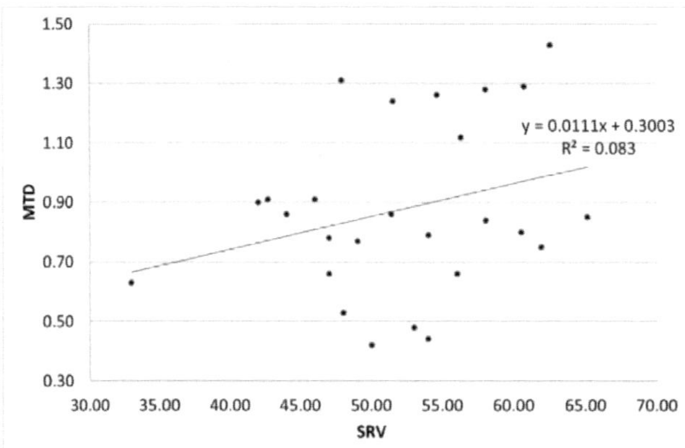

Figure 5. Mean Texture Depth vs. Skid Resistance Value.

3.2. Threshold value for SRV

Previously, in the absence of locally collected data, the minimum limiting value for SRV used by JKR is based on research conducted by TRRL on skid resistance in the United Kingdom (UK). A limiting value of 45 SRV (based on Category C, TRRL 1969) is being adopted as a guide for the minimum skid resistance value for all federal roads.

However, based on the results of the statistical analysis on the collected date, the mean value of SRV is 52.31 $\pm$ 8.54 (43.77, 60.85). It turns out that this finding is within the TRRL acceptable limits for roads in Malaysia (heavily trafficked roads). Hence, it is proposed that the minimum value of SRV for safe riding condition on Malaysian roads is 55 SRV.

3.3. Threshold value for SN

Page and Butas [21] claimed that the accident rate is low and uniform for pavements with SN of 26 and higher. However, the mean value computed from this analysis is 31.49 $\pm$ 7.87 (23.62, 39.36). Thus, it is proposed that the SN value of 30 to be used as a minimum threshold value.

4. Conclusions

The results of this analysis on the entire data relating to wet surface accidents are as follows:

- Minimum threshold value of SRV is 52.31 $\pm$ 8.54 (43.77, 60.85)
- Minimum threshold value of MTD is 0.45 $\pm$ 0.025 (0.425, 0.475)
- Risk of accident is higher when the SN value less than 31.49 $\pm$ 7.87 (23.62, 39.36)

This study recommended that the minimum SRV value for Malaysian Federal roads to be 55 SRV, minimum value of MTD to be 0.50 mm and minimum value of SN to be SN30.

5. References

[1] ASTM International 2015 *ASTM E965 Standard test method for measuring pavement macrotexture depth using a volumetric technique* (West Conshohocken: ASTM International).

[2] Austroads 2005 *Guidelines for the management of road surface skid resistance* (Australia: Austroads)

[3] Bonnot J and Ray M 1976 Rehabilitation of concrete pavements anti-skid properties *Transportation Research Record* **624** 72-81

[4] Bray J 2002 The role of crash surveillance and program evaluation: Nysdot's skid accident reduction program (SKARP) *28th International Forum on Traffic Records and Highway Information Systems* Orlando, 4-8 August 2002

[5] Bullas J C 2004 *Tyres, road surfaces and reducing accidents: A Review* (United Kingdom: AA Foundation for Road Safety Research).

[6] Cenek P D, Henderson R J, Loader M and Locke N 2002 *Texture depth requirements for state highway asset management purposes* (Lower Hutt, New Zealand: Opus International Consultants).

[7] Chelliah T, Stephanos P, Smith T and Kochen B 2002 *Developing a design policy to improve pavement surface characteristics* (Virginia, United States: Maryland State Highway Administration).

[8] Data Collection Ltd 2006 *Road measurement data acquisition system - Roughness surveys* Retrieved on 28th August 2006 from http://www.romdas.com/surveys/sur-rgh.htm.

[9] Giles C G, Sabey B E and Cardew K H F 1962 *Development and performance of the portable skid-Resistance tester* (United States: ASTM International)

[10] Gothie M 1996 Relationship between surface characteristics and accidents *International Symposium on Pavement Surface Characteristics Christchurch* 3-4 September 1996.

[11] Hall J W, Smith, K L, Titus-Glover L, Wambold J C, Yager T J and Rado Z 2009 *Guide for pavement friction* Available from http://onlinepubs.trb.org/onlinepubs/nchrp/nchrp_w108.pdf.

[12] Highway Agency 2004 *Design manual for roads and bridges-Pavement design and maintenance-Pavement maintenance assessment* (United Kingdom: Highways Agency)

[13] Kane M and Scharnigg K 2009 *Report on different parameters influencing skid resistance, rolling resistance and noise emission* TyroSafe Project Deliverable D10, 95

[14] Larson R 1999 *Consideration of tire/pavement friction/texture effects on pavement structural design and materials mix design* (HIPT: Office of Pavement Technology).

[15] Leu M C and Henry J J 1978 *Prediction of Skid Resistance as a Function of Speed from Pavement Texture-Transportation Research Record 666* (Washington DC: Transportation Research Board and National Research Council).

[16] Fuentes L G 2009 *Investigation of the factors influencing skid resistance and the international friction index* (Doctoral dissertation, University of South Florida).

[17] McLean J 1995 *The relationship between pavement conditions and road safety Load* - Pavement Interaction Workshop 1995 Newcastle, 5-6 July 1995

[18] Shafii M A 2009 *Skid resistance and the effect of temperature* (Master Thesis, University Teknologi Malaysia).

[19] Noyce D A, Bahia H U, Yambó J M and Kim G 2005 *Incorporating road safety into pavement management: Maximizing asphalt pavement surface friction for road safety improvements* (Wisconsin: Midwest Regional University Transportation Center - Traffic Operations and Safety (TOPS) Laboratory)

[20] Ong G P and Fwa T F 2007 Prediction of wet-pavement skid resistance and hydroplaning potential *Journal of the Transportation Research Board* 160-71

[21] Page B G and Butas L F 1986 *Evaluation of friction requirements for California State highways in terms of highways geometrics: Final report* (FHWA/CA-TL-86/01) (California Department of Transportation: Federal Highway Administration).

[22] Roe P G, Parry A R and Viner H E 1998 *High and low speed skidding resistance: The influence of texture depth* (Crowthorne, Berkshire: Transport Research Laboratory).

[23] Roe P G, Webster D C and West G 1991 *The relation between the surface texture of roads and accidents* (Crowthorne, Berkshire: Transport and Road Research Laboratory).

[24] Rogers MP and Gargett T 1991 A skidding resistance standard for the national road network *Highways and Transportation* **38** 10-16

[25] Schulze K, Gerbaldi A and Chavet J 1977 Skidding accidents, friction numbers, and the legal aspects involved report of the PIARC technical committee on slipperiness and evenness *Transportation Research Record* **623** 1-10

[26] Titus-Glover L and Tayabji S 1999 *Assessment of LTPP friction data (FHWA-RD-99-037)* (Mclean, Virginia: Federal Highway Administration).

[27] Wambold J C, Henry J J and Hegmon R R 1986 *Skid resistance of wet-weather accident sites. The Tire Pavement Interface (ASTM STP 929)* (West Conshohocken: ASTM International) http://dx.doi.org/10.1520/STP20001S.

Utilization of ultra - high performance concrete for bridge construction – a case study of Kg. Seberang Manong to Pekan Manong bridge

Norhidayu Kasim[1], Mohd Rozaiman Sulaiman[1], Kamarudin Abu Taib[2]
[1]Civil Engineer, Road Department, Public Works Department Malaysia, Kuala Lumpur, Malaysia
[2]Senior Lecturer, Faculty of Engineering and Build Environment, National University of Malaysia, Bangi, Malaysia

E-mail: Norhidayu.jkr@1govuc.gov.my

Abstract. The utilization of Ultra-High Performance Concrete (UHPC) in bridge construction is considered to be quite a new technology in Malaysia and its usage is still in the early stage of implementation as far as bridge design and construction in this country is concerned. The Bridge Design Division, Public Works Department (PWD) has taken the initiative to study and to begin utilising UHPC in designing a 300 m long bridge crossing the Perak River between Kg. Seberang Manong (Kg Menglembu) to Kuala Kangsar, Perak. The purpose of using UHPC is to reduce the number of intermediate piers that need to be constructed within the river channel in order to minimize the impact of debris flow disaster. This will allow free flow of logs thus avoiding them from being trapped and could possibly damage the piers during flooding. The bridge has 5 spans with the longest span of 70 m in length that uses UHPC in its construction. In this bridge construction, it is quite impossible to use typical normal concrete commonly utilize in conventional concrete bridge project that normally limits the span length to only 40 m. This project is also a pilot project in PWD that uses Malaysia Civil Engineering Standard Method of Measurement (MYCESMM), replacing the current Malaysian Standard Method for Measurement of Civil Engineering Construction. The MYCESMM covers a wider range of civil engineering works. This project is expected to contribute towards delivering cost effective civil engineering projects through the adoption of best practices in the preparation of the bills of quantities (BQ). The successful application and construction of this pilot project will have a great prospect in encouraging more projects to use UHPC and MYCESMM in the upcoming projects in Malaysia

1. Background of the project

Kg. Seberang Manong (Kg. Menglembu) is located on the western bank of the Perak River at a distance of 45 km from Kuala Kangsar and Pekan Manong (Figure 1). In the past 30 years, more than 5000 villagers were using ferry boat services to reach Kuala Kangsar and Pekan Manong to carry out their daily businesses. The proposed bridge, however, will provide basic infrastructure facilities to the villages from Kg. Seberang Manong and the surrounding area in order to cross the Perak River to Pekan Manong and Kuala Kangsar. The proposed bridge will also ensure the safety of villages who have been using ferry boat services for daily transportation. Moreover, the proposed bridge will also provide alternative routes and serve as economic stimulus to rural communities.

Figure 1. Location of Kampung Seberang Manong to Pekan Manong.

The government through Road Branch, Public Works Department (PWD) has agreed to implement a bridge project that links Kg. Seberang Manong (Kg Menglembu) to Pekan Manong and Kuala Kangsar, Perak. The scope of the project includes the construction of a 3.0 km single carriageway road and a 300 m bridge according to JKR standards. The 300 m bridge was designed in house under Bridge Design Division (BDD), Road Branch, PWD. BDD decided to design multiple spans bridge with a longest span of 70 m. In order to build a longer span bridge, BDD decided to use UHPC precast beam to solve the solution of having a span that is too long for a conventional beam preference. UHPC was found to be effective in fulfilling the design requirements due to the extent in span length of the bridge structure because of its high strength [1]. The purpose of the long span bridge is to provide ample space between piers in order to allow free flow of logs thus avoiding them from being trapped and could possibly cause damage to the piers during flooding.

The bridge project is considered to be a pilot project that uses Malaysia Civil Engineering Standard Method of Measurement (MYCESMM) in BDD. For tendering purposes, the preparation of the Bills of Quantities (BQ) for the bridge project that uses MYCESMM allows standardized measurements that provide a base to produce a good Bills of Quantities (BQ) used as procurement or contract documents in a project [2-4]. In other words, MYCESMM is needed in producing a sound BQ which constitutes as procurement or contractual agreement in the bridge project. The use of an appropriate standard BQ will allow consistent and better estimating of pricing in bidding, cost control and records by construction players during the difference stages and process of construction [2-3].

1.1 Overview and Characteristics of UHPC

Ultra-high performance concrete (UHPC) is an innovative material that show great potential to improve performance of infrastructure projects. It has been applied in civil engineering projects, especially those in bridge engineering [9]. UHPC needs to undergo the transition from research and development to actual construction uses [1, 10]. Since the application of UHPC to real-world projects has been sluggish, UHPC usage needs to be aggressively promoted in the construction industry. At the meantime, in Malaysia, very limited producers have experienced in the utilization of UHPC for precast or cast-in-place applications.

UHPC is a high-strength material created by combining Portland cement, silica fume, quartz flour, fine silica sand, high-range water reducer, water, and steel or organic fibres [1,10-13]. A typical UHPC material has a design compressive strength of up to 200 MPa while ductile tensile strength range of 10–15 MPa [13]. Table 1 summarizes the typical compositions of concrete materials at different levels of performance [5]. Ordinary and high performance concretes consist of cement, fine and coarse aggregates, and admixtures, compared to UHPC which is made up of cement, fine aggregates, admixtures, steel fibers, and nano-fillers. The coarse aggregate was replaced in UPHC with steel fibers and nano-fillers that enhance the mechanical strength of UHPC. The nano-filler increased not only the compressive strength but also the tensile strength [5, 11-12]. UHPC structures are capable of deforming and supporting flexural and tensile loads even after initial cracking. Figure 2 displays the compressive and flexural strength of UHPC in comparison to that of normal concrete [10]. With varieties of excellent material properties, UHPC applied to bridge structures can reduce the cross-sectional area, which decreases self-weight (dead load) and extension of service life due to outstanding durability [11-13].

Table 1. Compositions of Concrete Materials at Different Levels of Performance.

Ordinary Performance Concrete	High Performance Concrete	Ultra-High Performance Concrete
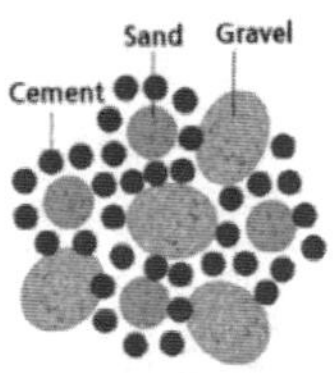	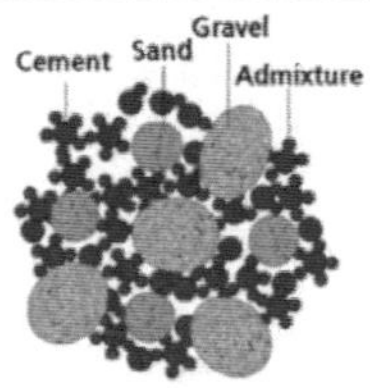	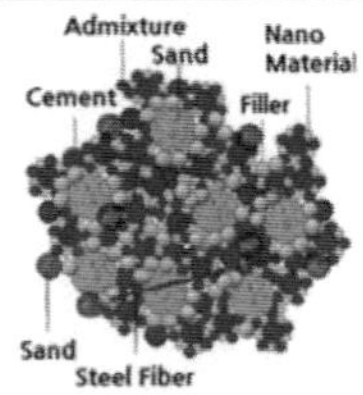
Strength : 20–40 MPa	Strength : 50–100 MPa	Strength : 100–200 MPa
Porosity : 10–20%	Porosity : 5–8%	Porosity : 2% below

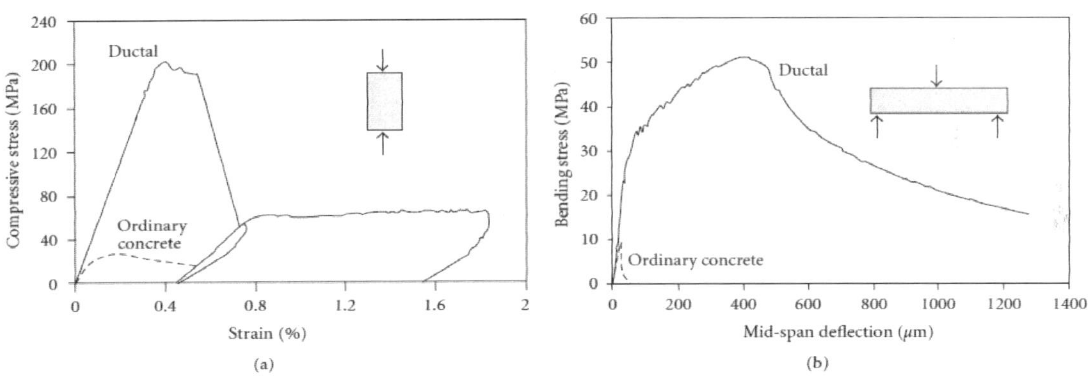

Figure 2. Behavior of Ductal in compression (a) and in bending (b) [10].

1.2 UHPC Segmental Beam - Deck System for Manong Bridge

The 300 m Manong Bridge crossing the Perak River was designed using Beam-Slab Bridge system with Multi-span continuous beam. The bridge was designed to withstand full traffic loading as per the Eurocode specification and comprises a total width of 11.5 metres that includes three spans of 70 m and two span of 45 m creating a total bridge length of 300 m (Figure 3). The loading of the bridge structures is designed according to BS EN 1990:2002[15] for permanent action and for the traffic loading according to BS EN1991-2 [16]. For traffic loading, combination of Load model 1 (LM1) which considers tandem system (TS) and uniformly distributed load (UDL) and Load model 3 (LM3) considering special vehicle (SV80) are used for the analysis in 3D Model MIDAS Civil. The global analysis of the bridge is carried out using a grillage model of the deck with the stiffness of the piers

IOP Conf. Series: Materials Science and Engineering **512** (2019) 012036 doi:10.1088/1757-899X/512/1/012036

and abutments represented by roller and pin. The superstructure and substructure are modelled together in a single 3D Model MIDAS Civil shown in Figure 4 and Figure 5. The superstructure of the bridge was designed and constructed using a post tension ultra-high performance concrete (UHPC) U-Girder semi integral beam-deck system, which was semi integral bridge with the substructure/foundation.

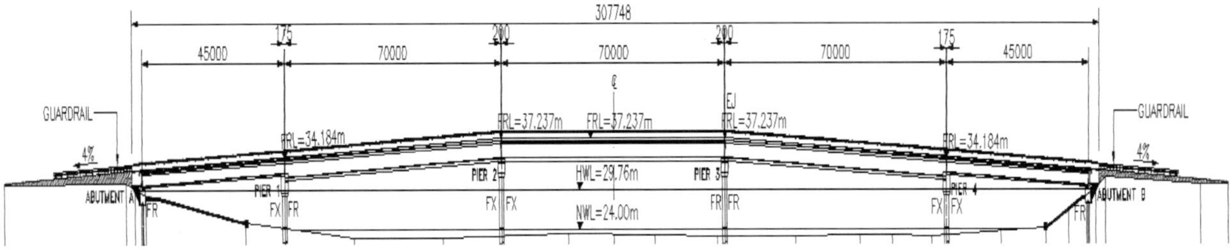

Figure 3. Longitudinal Section of the Bridge.

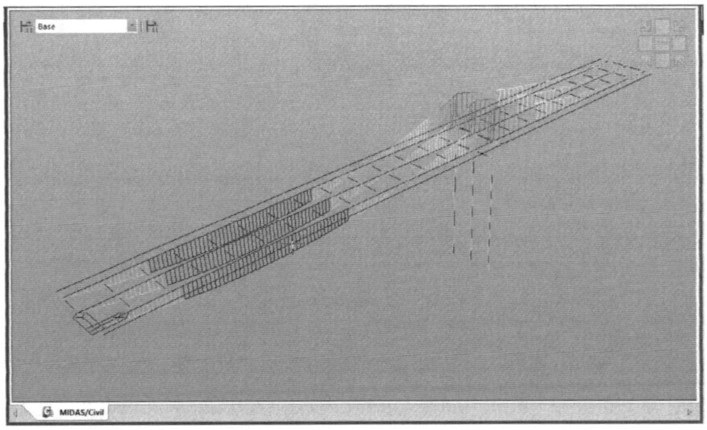

Figure 4. Bending Moment diagram using 3D Model MIDAS Civil for UHPC Girder beams.

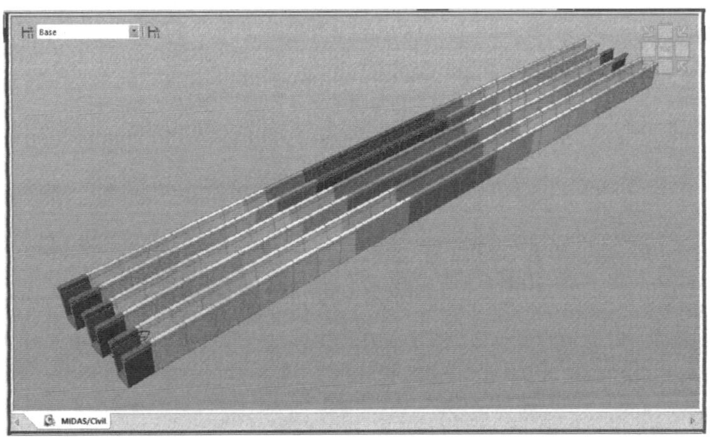

Figure 5. Stress analysis using 3D Model MIDAS Civil for UHPC Girder beams.

IOP Conf. Series: Materials Science and Engineering **512** (2019) 012036 doi:10.1088/1757-899X/512/1/012036

The foundation of the bridge abutment was seated on 750 mm diameter bored pile while bridge pier was seated on 1500 mm diameter bored pile bonded in 18.5 m in hard layer with approximate pile length of 37 metres. Figure 6 shows the typical Pier elevation of the bridge. The UHPC beams are spaced 3.25 metres c/c and each span consists of 3 beams. A total of 9 beams of 70 m and 6 beams of 45 m were used.

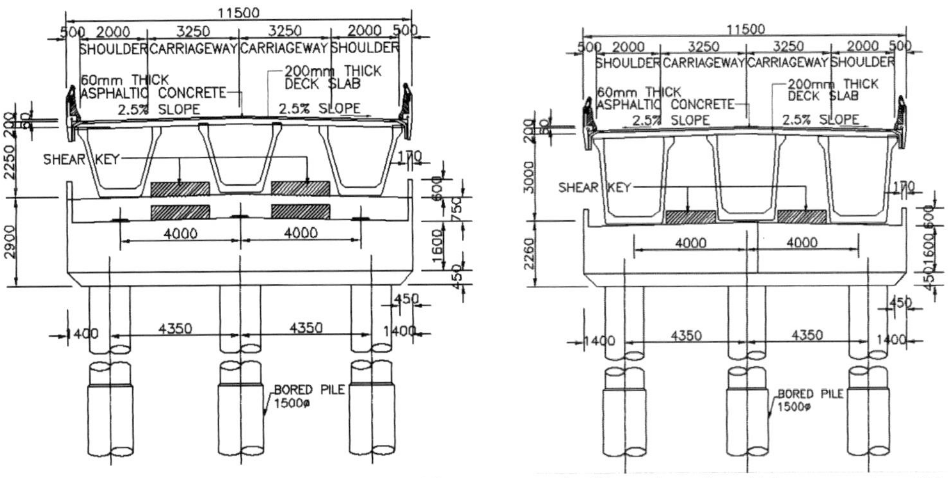

a. Pier elevation (Abutment A to P1, 45 metres span beam)

b. Pier elevation (Abutment A to P1, 70 metres span beam)

Figure 6. Typical Pier Elevation (a) Pier elevation (Abutment A to P1, 45 metres span beam), (b) Pier elevation (Abutment A to P1, 70 metres span beam)

Figure 7 shows the typical details of an UHPC U-Girder beam. A 45 m UHPC U-girder beam was designed with 2.25 m deep, 2.75 m wide at the top while 70 m UHPC U-girder beam was designed with 3.0 m deep, 2.818 m wide at the top (see Figure 7a, 7b, 7c and 7d). The beam is made from 2.5 m small panel box girder beam being assembled and become 70 m and 45 m U-girder beam (Figure 8).

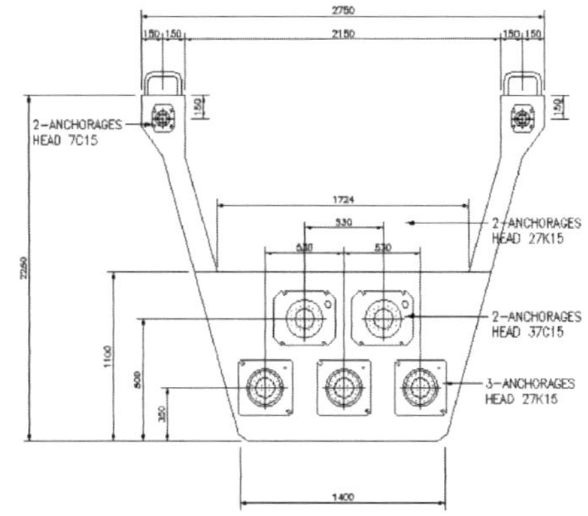

a. Anchorage Section for 45 metres UHPC U-Girder beam

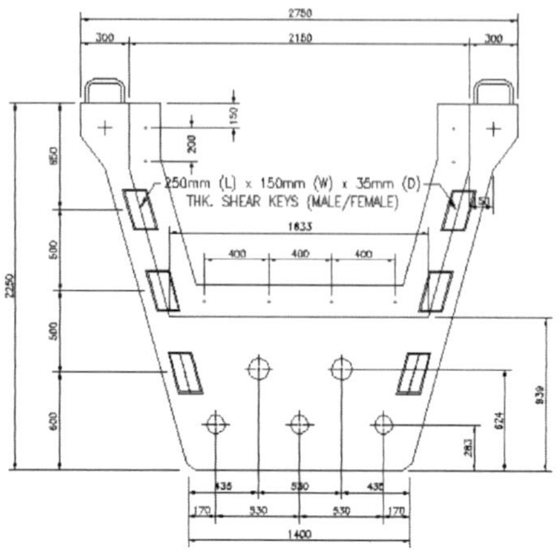

b. Joint Section for for 45 metres UHPC U-Girder beam

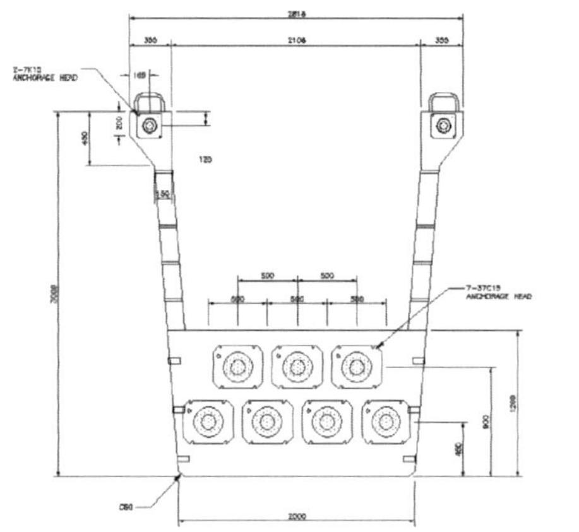

c. Anchorage Section for 70 metres UHPC U-Girder beam

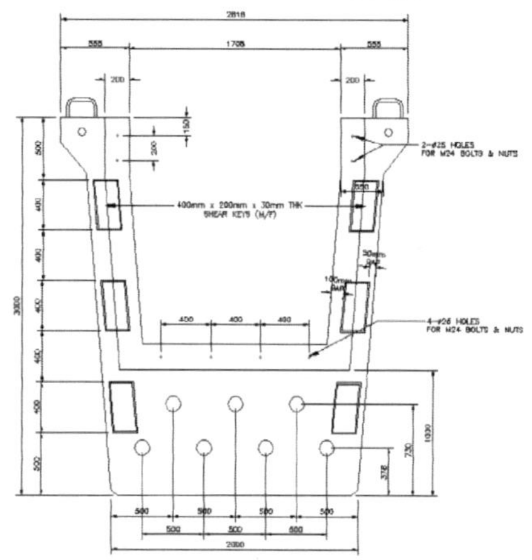

d. Joint Section for for 70 metres UHPC U-Girder beam

Figure 7. Typical details of U-Girdir Beam (a) Anchorage Section for 45metres UHPC U-Girder beam, (b) Joint Section for for 45metres UHPC U-Girder beam, (c) Anchorage Section for 70metres UHPC U-Girder beam (d) Joint Section for for 70metres UHPC U-Girder beam.

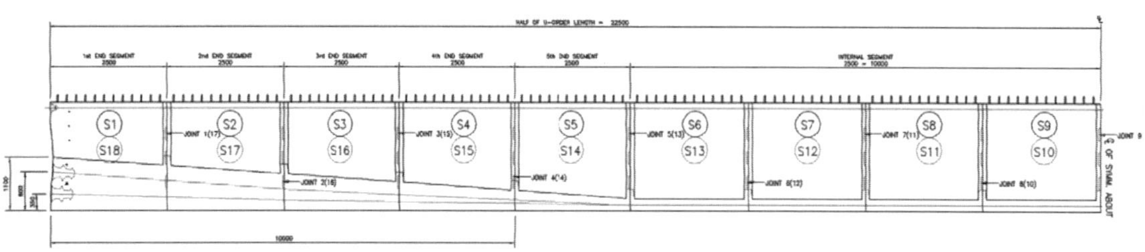

Figure 8. Elevation of full girder assembly.

1.3 Applications of Malaysia Civil Engineering Standard Method of Measurement (MYCESMM)
In January 2013, the Malaysian Government officially implemented MYCESMM in standardizing the method of measurement for civil engineering works in all public and private projects [17]. MYCESMM is intended to be a national document that features the rules of measurement, definition, coverage and description for civil engineering projects to be used as a basis in the preparation of the Bill of Quantities. Malaysia as a nation must seriously adopt the use of one standard method of measurement for civil engineering works. The adoption of one standard provides clarity in the preparation of the Bill of Quantities which leads to improved understanding of the work items hence enabling pricing of those items to be undertaken in a more accurately manner [3; 17-21]. Potential contractual claims and disputes due to ambiguities and discrepancies in the Bill of Quantities that can arise during the construction stage can thus be minimized which ultimately leading to a cost-effective in the delivery of civil engineering projects and smoother preparation of final certificates hence final payment [2-3].

This bridge project is a pilot project that uses MYCESMM within BDD in preparing the Bills of Quantities (BQ). At the moment, the bridge is in the construction phase and until present, there is no additional cost incurred due to the missing item in BQ as normally occured in conventional method of BQ. The construction cost of the Manong bridge was approximately RM23.0 million (US$5.65 million). This construction cost included the foundation/piling, substructure, superstructure, temporary works, and machinery and labor needed for the completion of the bridge. The cost, however, did not include road works, slope protection, and relocation of existing utilities and services. In terms of the cost rate per square meter, the structural component cost for the bridge was US$1650/m^2. This figure is quite comparable to the costs for conventional reinforced concrete construction in Malaysia which are approximately US$1,500/m^2 to $1,900/m^2, depending on the accessibility conditions and difficulties in constructing the bridge.

2. Conclusion
The purpose of this pilot project using UHPC superstructure is to accelerate the UHPC usage in Malaysia construction industry. While Manong bridge is still under construction, the successful construction of the Manong Bridge later on, could encourage the usage of UHPC bridge in Malaysia's construction industry. Besides that, to encourage the application of this material, more research needs to be carried out particularly on the types of projects that the UHPC is best for its application and how it could be used to attain its fullest potential. Therefore, in order to build a longer span bridge, UHPC has been found to be effective in fulfilling the design requirements due to the extent in span length of the bridge structure compared to the conventional reinforced concrete bridge design.

This bridge project is also a pilot project that uses MYCESMM in bridge construction within BDD. The purposes of preparing the Bills of Quantities (BQ) using MYCESMM are to allow the adoption of standardized measurements which will provide a base in producing a good BQ that shall be incorporated as procurement or contract documents in a project. The successful construction of Manong bridge without missing item in BQ will encourage the usage of MYCESMM in the future bridge construction industry in Malaysia.

3. References
[1] Graybeal B and Tanesi J 2007 Durability of an ultrahigh-performance concrete *ASCE Journal of Materials in Civil Engineering* **19**(10) 850-54

[2] Abd Rashid R, Mustapa M and Abd Wahid SN 2006 Bills of Quantities - Are they useful and relevant today? *International Conference on Construction Industry* 1–10

[3] Adnan H, Mohd Nawawi A H, Mohd Akhir S M, Supardi A and Chong H 2011 Bills of quantities: Perspectives of contractor in Malaysia *Australian Journal of Basic and Applied Sciences* **5**(11) 863–73

[4] Davis Peter R, Love P E D and Baccarini D 2009 Bills of quantities: Nemesis or Nirvana? *Emerald Journal* **27**(2) 99–108

[5] Kim SW 2007 *Development of ultra high performance concrete for hybrid stayed cable bridge* (Korea: Korea Institute of Construction Technology)

[6] Graybeal B 2006 *Material property characterization of ultra-high performance concrete (FHWA-HRT-06-103)* (McLean, VA: Federal Highway Administration)

[7] UHPC 2012 *Pathway to commercialization-Homeland security, science and technology* Retrieved from http://www.dhs.gov/xlibrary/assets/ultra-high-performance-concrete-roadmap.pdf.

[8] Yan P Y 2015 Application of UHPC in China *Proceedings of the 1st International Symposium of ACF on Ultra High Performance Concrete* Kolkata, India pp 79–82

[9] Acker P and Behloul M 2004 Ductal technology: A large spectrum of properties, a wide range of applications *Proceedings of the International Symposium on Ultra High Performance Concrete* Kassel, Germany.

[10] Graybeal B 2010 Design, fabrication, and testing of a 2nd generation ultrahigh performance concrete pi-girder *Proceedings of 3rd International Fib Congress* Washington DC

[11] Graybeal B 2006 *Material property characterization of ultra-high performance concrete (FHWA-HRT-06-103)* (McLean, VA: Federal Highway Administration)

[12] Brühwiler E and Denarié E 2008 Rehabilitation of concrete structures using ultra-high performance fiber-reinforced concrete *Proceedings of the 2nd International Symposium on Ultra-High Performance Concrete* Kassel, Germany.

[13] DM Engineering 2013 *Bridge harmony with human* Available from http://www.dmeng.com/img/main/DM_Engineering_Brochure.pdf.

[14] British Standards Institution 1991 *BS EN 1991-2:200 Eurocode 1: Actions on structures – Part 2: Traffic loads on bridges* (London, BS Institution)

[15] British Standards Institution 1990 *BS EN 1990:2002 Eurocode – Basis of structural design Includes Annex A2 – Application for bridges* (London, BS Institution)

[16] Che Mat McM 2010 The need to standardise the bill of quantities for civil engineering works *Seminar on Standard of Measurement in Construction Procurement* Kuala Lumpur pp 1-7

[17] Hassan S 2009 Role of SMM in Construction Industry *Seminar on SMM in Construction Procurement* Kuala Lumpur, 2nd July 2009

[18] Mohammad McF 2012 Standard method of measurement: A review of its importance and relevance in quantity surveying practice *International Seminar 2012: Responding to Global Challenges- The Need for Quantity Surveyor Professional Standard* Jakarta

[19] Rose TcM and Manley K 2012 Measurement constructs to explore innovation diffusion in construction *The Construction, Building and Real Estate Research Conference of the Royal Institution of Chartered Surveyors (COBRA)*

IOP Conf. Series: Materials Science and Engineering **512** (2019) 012037 doi:10.1088/1757-899X/512/1/012037

Deployment of GIS + BIM in the construction of Pan Borneo Highway Sarawak, Malaysia

Zohari Akob[1], Mohd Zaidee Abang Hipni[2] and Awang Abdullah Awang Abd Razak[3]

[1]Kementerian Kerja Raya, Tingkat 1-14, Kompleks Kerja Raya, Jalan Sultan Salahuddin, 50580 Kuala Lumpur, Malaysia
[2]Lebuhraya Borneo Utara Sdn Bhd, Level 23, Gateway Kuching, No.9, Jalan Bukit Mata, 93100 Kuching, Malaysia
[3]RS&GIS Consultancy Sdn Bhd, A4-2-3A Solaris Dutamas, Jalan Dutamas, Kuala Lumpur, 50480, Malaysia

Abstract. Construction of the Pan Borneo Highway across 786km of the Malaysian state of Sarawak is leveraging on technology to achieve greater efficiency and productivity throughout the project's life-cycle. With a tight construction delivery timeline and budget to keep in check, the ability to design and plan efficiently, minimize errors and discrepancies during construction, manage risks, visually monitor and track progress at the site, and finally to establish accurate 3D as-built models for future asset management are among the challenges faced. To address such challenges of a linear construction, Project Delivery Partner, Lebuhraya Borneo Utara (LBU), has deployed the use of Malaysia's first-ever Highway Information Modelling (HIM) as an integral part of its Integrated Project Management System. HIM is a combination of Geographical Information Systems (GIS) and Building Information Modelling (BIM) which incorporates three dimensional (3D) parametric highway models and optimal utilization of accurate and precise Geospatial Datasets. This paper discusses the application of GIS in development of the Pan Borneo Highway Sarawak project and its role as a foundation for design, planning, monitoring of construction and operation and maintenance of the future highway asset by providing precise data in a timely and cost-efficient manner.

1. Introduction

The long-awaited Pan Borneo Highway in the Malaysian state of Sarawak on the island of Borneo will span an estimated 1,060km. Works is on-going to develop and upgrade the current 2-lane single carriageway which connects the entire state into a 4-lane dual carriageway. The project is being undertaken by the Government of Malaysia and State Government of Sarawak through a Project Delivery Partner (PDP). A special purpose vehicle private entity, Lebuhraya Borneo Utara (LBU), has been appointed as the PDP.

Costing RM16.5 billion, Pan Borneo Highway Sarawak is deemed to be the single largest infrastructure project ever awarded by the Government in Sarawak. Implementation of the project involves construction of 786km of a new 4-lane highway in 11 works packages. Works began in January 2016, with the Highway targeted for completion in 2021. Operations and maintenance of the highway will follow upon completion of construction.

As PDP, LBU is in charge of the entire project lifecycle - from planning to design and construction, to operations and maintenance of the completed highway. LBU is committed to deliver the project on time and within budget.

There are numerous challenges in a linear construction project, particularly in highway construction, from challenges at project commencement, to engineering design stage, construction and also asset management. This paper highlights efforts to address issues that arise during construction stage and also anticipated obstacles post-construction by leveraging on technology, specifically with the use of Geographical Information System (GIS) and Building Information Modelling (BIM).

Project visibility is crucial for the project management team to monitor progress according to schedule and cost agreed. Visibility facilitates executive decision-making by providing accurate information and a realistic perspective of the project being executed, especially in construction stage. In the case of the Pan Borneo Highway Sarawak, it is burdensome to manually monitor progress of thousands of activities and tasks for the whole stretch of the highway. Physical progress monitoring through site visits is time consuming and employs many resources.

Post-construction, it is imperative for the asset owner to have an adequate geodatabase for maintenance of the asset. A complete geodatabase is important to develop a systematic plan for linear asset management and maintenance. Many road issues can be resolved faster with the availability of geo-information in a maintenance database.

This paper discusses the integration of GIS and BIM as a platform to support project management, construction management and asset management. In the case of the Pan Borneo Highway Sarawak, a dedicated task force was formed to manage the integration process of GIS and BIM to ensure the project is delivered on time and with minimal errors.

A foundation of the GIS and BIM project is geodesign which combines geography with design by providing designers with robust tools that support rapid evaluation of design alternatives against the impacts of those designs. Geodesign infuses design with a blend of science and value-based information to help designers, planners and stakeholders make better-informed decisions [6].

GIS plays a crucial role in linear projects, especially in infrastructure such as highway construction, pipeline and railroads. Meanwhile, BIM plays a major role in modern construction. By integrating GIS with BIM in one complete system, Malaysia's first-ever Highway Information Modelling (HIM) was developed for the Pan Borneo Highway Sarawak. GIS data provides the geolocation using real world coordinates, while BIM represents the physical and functional characteristics of the facility.

With technology, implementation of the Pan Borneo Highway Sarawak has been able to achieve increased efficiency of construction and project management with the aid of precise data provided in a timely and cost-efficient manner throughout the lifecycle of the project.

2. Methodology Overview

2.1. Design Stage

LiDAR enabled preliminary engineering design (PED) and detailed engineering design (DED) for the entire alignment to be completed according to the project implementation schedule as stipulated in the Project Delivery Partner Agreement with the Government.

In developing HIM for the Pan Borneo Highway Sarawak, all required data was captured using several methods and technologies. This included satellite imagery, ground survey data, LiDAR data, Unmanned Aerial Vehicle (UAV) imagery and Underground Detection Mapping (UDM). From there, all gathered data was processed to produce GIS layers to be used as base data for the designer to produce design drawings.

The integration of GIS and BIM has enabled better decision-making and improved communication. By using GIS and BIM together with time information, project teams now have a better view on the impacts of decisions made before, during and after construction of the project.

2.2. Construction Stage

Monthly UAV imagery at various locations along the 786km construction site are captured to assist in progress monitoring, towards ensuring the project is completed on time. Concurrently, GIS data together with UAV imagery is merged in a web-based monitoring platform designed for this project, namely HIM WebGIS. From WebGIS, key stakeholders, management, site personnel, contractors and consultants, have access to data to monitor progress of the project from anywhere in the world.

LiDAR will again be conducted upon completion of the highway to support the BIM/HIM requirements.

3. Geographical Information System (GIS)

Geographical Information System (GIS) is a framework for gathering, managing, and analysing data [7]. Rooted in the science of geography, GIS integrates many types of data. It analyses spatial location and organizes layers of information into visualizations using maps and 3D scenes. With this unique capability, GIS reveals deeper insight into data such as patterns, relationships and situations, where this would help the users to make the right decision [1].

In producing an effective solution for the Pan Borneo Highway Sarawak, certain methodologies have been conducted, such as topographical survey, LiDAR data feature extraction, Underground Detection Mapping (UDM), Unmanned Aerial System (UAS) and Web GIS as a common data environment to manage data.

- Ground Control Marker Survey Program: Prior to beginning the survey collection, a reconnaissance was done of existing controls in the project and surrounding areas. From the findings, controls to be included in the network were selected based on their location, horizontal and vertical orders, and accessibility. This program was introduced to establish the control network for the entire project area, towards ensuring a fixed common reference and an accurate and reliable survey control.
- Polygonal Control Network – Primary**:** A primary network of Global Positioning System (GPS) network control stations was established at certain intervals along the highway alignment. These polygonal networks (or closed traverse survey) allowed for the identification and correction of any 'erroneous' local datums or measurements, without compromising the accuracy of the total survey control network. A primary network of 75 GPS network control stations was established at approximately 15km intervals along the Pan Borneo Highway Sarawak alignment.
- Polygonal Control Network – Secondary: Temporary Bench Marks (TBM) support local site survey and investigation works (such as crossings and bridges), underground utility detection, as-built and hydrography / bathymetry surveys (as connections when these come on-shore) to ensure they match with the primary survey network. A secondary network of over 324 TBM was established, at around 2km spacing, interspersed between the primary Ground Control Marker (GCM) stations along the Pan Borneo Highway Sarawak alignment [3].

As part of the linear Pan Borneo Highway project, a complete geospatial engineering solution was implemented to manage and enhance the infrastructure lifecycle - from road design, construction, and maintenance. Throughout the infrastructure lifecycle, key processes were improved as the geospatial engineering solutions were deployed across the project team for data management, analysis, and coordination. As illustrated in Figure 1, this included data collection and as-built surveying, route planning and optimisation, pre-design analysis, infrastructure design, construction management, and operations and maintenance.

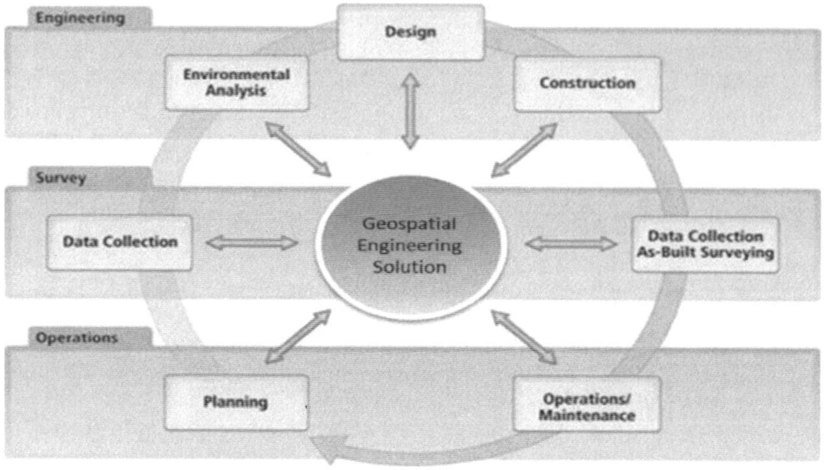

Figure 1. Geospatial Engineering approach for the infrastructure development lifecycle of the Pan-Borneo Highway Sarawak project

3.1. Data Acquisition

3.1.1. Satellite Imagery. Imagery is a type of data useful for many GIS applications and is defined as any type of photograph. Photographs used for GIS projects consist of images gathered from a satellite or aircraft. Printed topographic maps that are scanned as digital images also constitute imagery. In terms of the specific GIS data type, imagery is considered as raster data.

3.1.2. Light Ranging and Detection (LiDAR). LIDAR systems are varied and flexible. It should be specified to match the intended target area and user requirements. The GIS teams deployed leading-edge imaging systems, and applied the latest best-practice techniques and knowledge. This is to ensure a truly useful multi-disciplinary dataset is acquired to meet both the immediate and long-term requirements of this project.

Figure 2. LiDAR data from Pan Borneo Highway Sarawak Project

By utilizing LiDAR technology, multiple output could be generated in shorter time in comparison to ground survey data. A single flight LiDAR data acquisition was able to produce multiple resolution options to result in various outputs to meet both mapping and modelling needs. LiDAR data products

IOP Conf. Series: Materials Science and Engineering **512** (2019) 012037 doi:10.1088/1757-899X/512/1/012037

include Raw LiDAR geolocated data, processed point cloud data, Digital Surface Model (DSM), and Digital Terrain Model (DTM).

For Pan Borneo Highway Sarawak, the derived vector contour line mapping was acquired and utilized from this method as shown in Figure 2 above. In addition, GIS- and CAD-compatible digital surface models were generated to produce basemap layers for engineering, planning, and long-term monitoring.

In short, LiDAR data provided the Pan Borneo Highway Sarawak team with a broad coverage of high resolution data applications, such as high accuracy terrain/topography, road geometry, utilities, lane mapping, slope, structures, and many others. Figure 2 illustrates processed LiDAR data comprising approximately 23.6 billion LiDAR points captured throughout the 786km stretch of the Highway.

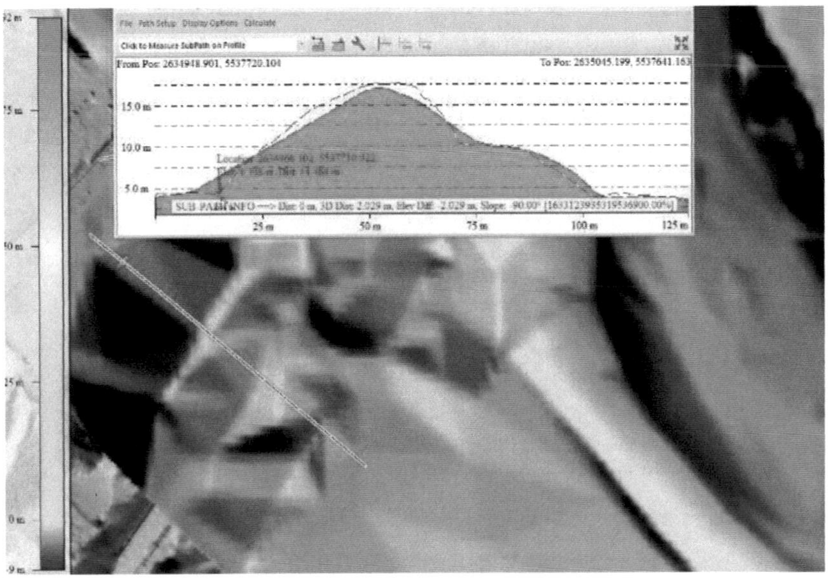

Figure 3. Ground Survey Data vs. LiDAR Survey Data

Reliability and accuracy of captured data (see Figure 3) is important. In comparing the precision of ground survey data versus LiDAR survey data, it was found that the LiDAR points were able to provide more information, thus reducing design errors and ensuring smooth progress during construction. With accurate elevation data and well defined DTM and DSM, such data can also be used as input into other powerful Alignment Optimisation Systems, such as Quantm System.

For the Pan Borneo Highway Sarawak, the LiDAR points density captured within one square meter range is between 20 to 30 points as shown in Figure 4.

LiDAR of the 786km highway alignment took 6 months to complete, as compared to 18 months if solely ground surveys were used. The LiDAR data thus enabled the preliminary engineering design (PED) and detailed engineering design (DED) for the entire alignment to be completed according to the project implementation schedule as stipulated the Project Delivery Partner Agreement (PDPA).

The final product of LiDAR is GIS. Among its applications was for land acquisition. In addition, GIS enabled the team to identify land usage and encumbrances, as well as resolve design conflicts which were not captured by conventional drawings.

Going further, as per the BIM/HIM requirement of the PDPA, LiDAR is proposed to be carried out upon completion of construction of the Highway replacing the normal conventional method for As-Built survey. In addition, around 28,000 aerial images were captured during the data acquisition flight. The resolution of images captured is 5cm, which is considered as very high quality and reliable aerial imagery for engineering usage.

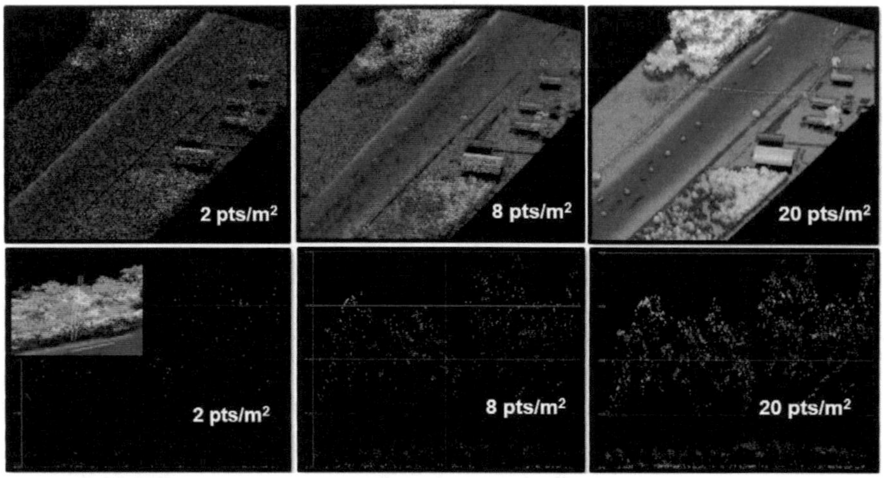

Figure 4. LiDAR points density captured in a square metre area

3.1.3. Underground Detection Mapping (UDM). The state of Sarawak is currently implementing major infrastructure developments. While doing so, the challenge lies in accurately detecting what lies underground - thousands of kilometres of water pipes, gas pipelines, electric cables and telecommunication cables. These underground utility cables, pipes and sewerage networking have to be detected using correct and proper devices to ensure accuracy, in order to avoid or minimise accidents by utility contractors doing excavation works.

Data collected from detection works were then surveyed and incorporated into the surveyor's overall reports and drawings. All such information has been studied, designed and consulted prior to excavation works.

Underground Utility Detection and Mapping is an important element at the planning level as shown in Figure 5. The application of Underground Utility Detection helps improve decision-making for relocating utilities, designing new proposed routes, quality survey, costing for project execution and data inventory enhancement [2].

Figure 5. Underground Utility Detection & Mapping in a high density area

IOP Conf. Series: Materials Science and Engineering **512** (2019) 012037 doi:10.1088/1757-899X/512/1/012037

3.1.4. Unmanned Aerial Vehicle (UAV). Unmanned aerial vehicle (UAV) is a type of aircraft that operates without a human pilot on-board. Recent technologies enable UAV to be used for mapping and site monitoring purposes [8].

For the Pan Borneo Highway Sarawak project, UAV are used to facilitate monitoring of progress. Monthly aerial imagery is captured using UAVs to see and track changes of progress made at the various construction sites. The data will then be processed into meshed imagery which is uploaded in the WebGIS (see Figure 6).

Data from the UAV can be compared monthly through time-slider feature in the WebGIS. The time-slider allows management and construction personnel to see the development and changes made in construction site on a single screen. Such technology has made project management more efficient and accountable.

Images captured through UAV are also meshed to generate 3D reality models for certain areas that require more detailed monitoring and amplification [5].

Figure 6. 3D Model from UAV photo mesh

3.2. Data Processing & Management

The GIS designer supports road designer to perform engineering related analysis by providing relevant geospatial data. For Pan Borneo Highway Sarawak, analysis is done by combining information from multiple independent layers and applying spatial operations to address a wide range of problems. This will allow users to visualize the results and gain deeper insights to make important decisions.

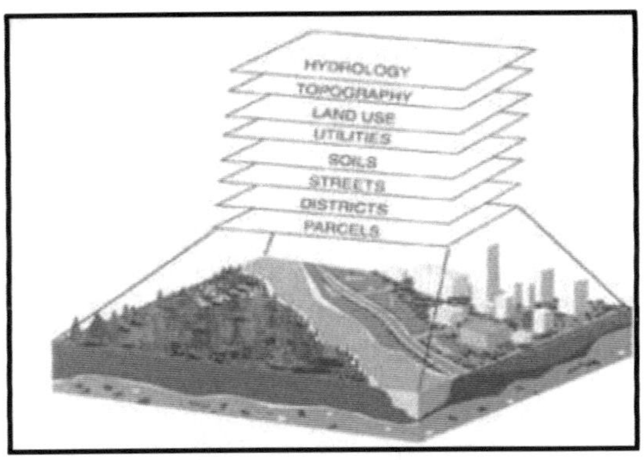

Figure 7: Geographic Data for Geoprocessing

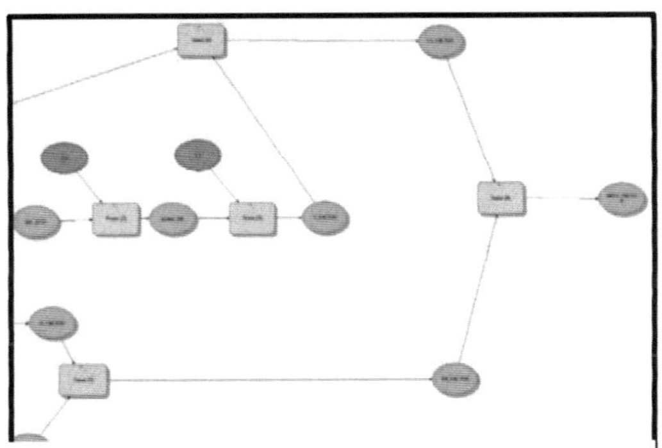

Figure 8. Example of Automation Analysis using ModelBuilder

Geoprocessing is a framework and set of tools for processing geographic and related data as illustrated in Figure 7 [6]. Various geoprocessing tools can be used to perform spatial analysis or to manage GIS data in an automated way. Geoprocessing is an integral part of the GIS analysis that creates new data from existing GIS data. Users are able to create a model to automate and keep track of the workflow.

Geoprocessing tools in ArcMap perform an operation on an input dataset and usually create a new output dataset. Geoprocessing tools perform analysis, data management, editing and other operations. ModelBuilder (Figure 8) is a graphical interface for making GIS models. It is one of the best environments for exploring ArcGIS tools, learning how inputs and outputs tools are used, and visually understanding how GIS modelling works. Figure 9 shows an example of output from geoprocessing.

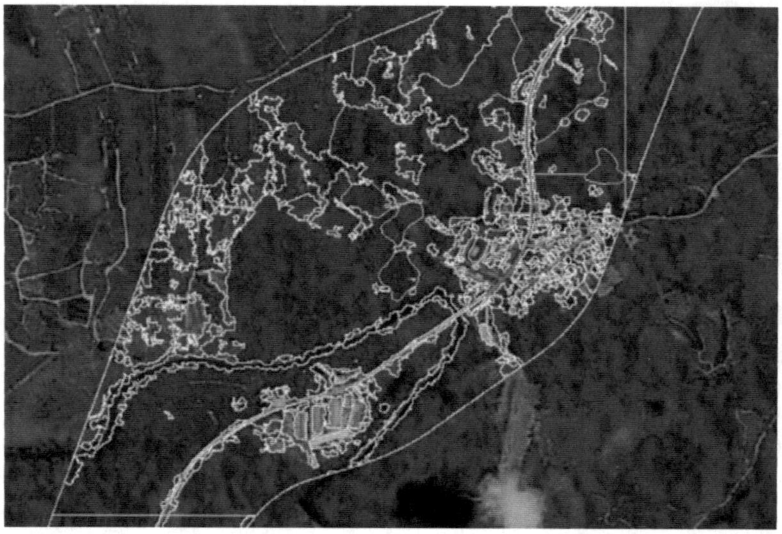

Figure 9. Example of Geoprocessing Output: Landcover Data

3.3. Data Visualization

3.3.1. WebGIS. Rapid deployment and implementation of a WebGIS-based Project Tracking and Monitoring System has helped Pan Borneo Highway Sarawak Project Management Team increase project delivery efficiency with access to tracking and monitoring functions on the computer and mobile devices.

Key datasets are integrated into the WebGIS server, and accessed by the Project management team and contractors as and when required. Database development is fast-tracked and real-time, as shown in Figure 10 whilst maintaining rigorous database design protocols to ensure compatibility for later conversion into a working Road Maintenance and Management System (RMMS), as illustrated in Figure 11, if required [4].

Figure 10. Time Slider feature in WebGIS

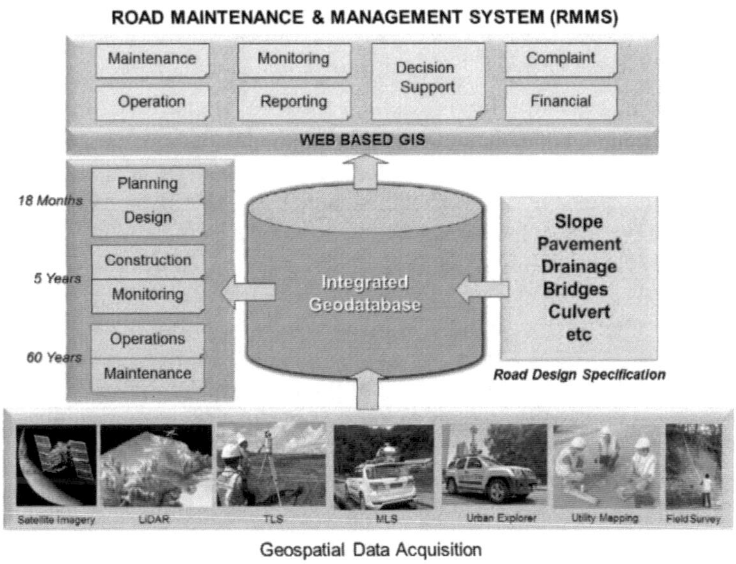

Figure 11. Geospatial Data Acquisition for RMMS

3.3.2. ArcGIS. ArcGIS 3D Analyst extension and ArcMap were used to provide a variety of ways to symbolize and display different types of GIS surface formats, such as rasters, Triangulated Irregular Networks (TIN), terrain and LAS datasets (point cloud format). Rasters, TINs and LAS datasets can be easily displayed inside ArcScene using different symbology options. Figure 12 and Figure 13 show different sets of LiDAR data displayed in ArcGIS.

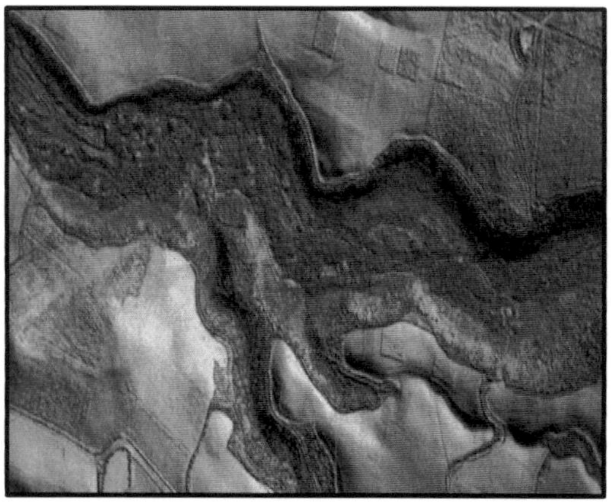

Figure 12. Example of LiDAR and Terrain

Figure 13. LiDAR point cloud of Pan Borneo Highway Sarawak

4. GIS + BIM Integration

GIS amplifies the value of Building Information Modelling (BIM) design data (2D drawings and 2D Models) through visualization and analysis of structures in the context of the natural and built environment.

The integration of GIS and BIM in the Pan Borneo Highway Sarawak to generate the first-ever Highway Information Modelling (HIM) has enabled better decision-making and communication. By using GIS and BIM together with time information, project teams now have a better view on the impacts of decisions before, during and after the construction of a project.

The process of integration was not without its challenges. The biggest obstacle to achieving GIS and BIM integration is the process of integrating BIM data, in the form of an assortment of proprietary files, web and database information stores, with GIS.

Engineers utilized information from BIM in GIS analytical and visualization tools. Converting CAD and BIM data into ArcGIS is the first step towards integration. This included converting individual BIM elements to GIS features, using a BIM file as a single complex GIS feature, or using BIM-derived data as a complex point marker symbol. The results of the integration can be viewed in WebGIS or any 3D model viewer as illustrated in Figure 14.

Figure 14. Integration of LiDAR 3D Point Clouds Data and BIM 3D Model for Batang Rajang Bridge, Pan Borneo Highway Sarawak

5. GIS in Linear Asset Management and Maintenance

Maintenance of roads is undertaken to ensure the safety of traffic and to sustain the serviceability and appearance of the road. Road maintenance involves both corrective maintenance and preventative maintenance. GIS and BIM elements, through HIM, will add greater value in the future Road Maintenance Management System employed by the Government by enhancing the system with a complete geodatabase and updated asset inventory.

6. Conclusion

As at October 2018, development and implementation of 786km of the Pan Borneo Highway Sarawak is nearly half-way in its construction timeline. Works on site is progressing satisfactorily at 11 works packages which is further broken down into 25 sections. With a tight timeline to meet and budget to keep towards delivery of the Project, efficient project management that leverages on technology is key.

An example is the efficient usage of LiDAR, completed in record time, thus enabled preliminary engineering design (PED) and detailed engineering design (DED) to be implemented according to schedule of the PDPA with the Government.

For Pan Borneo Highway Sarawak, an Integrated Project Management System (IPMS) is used to communicate with multiple stakeholders and manage risks across the project lifecycle on issues regarding planning, design, construction and later to prepare for asset management.

A key pillar of the IPMS is the Highway Information Modelling (HIM), which is a combination of GIS and BIM and WebGIS. Used daily by project teams at site across Sarawak, consultants, contractors, head office teams and Government stakeholders, the integration of GIS and BIM has successfully increased the efficiency of construction and project management by providing precise data in a timely and cost-efficient manner through geolocation which is coupled with the physical representation and functional characteristics of facility from BIM.

7. References

[1] Al-Sabhan W 2003 *Approaches to Developing a Web-based GIS Modeling Tool: For Application to Hydrological Nowcasting* (London: King's College)
[2] Smart Infra Resources 2013 *Proposal for a Geospatial Engineering Approach to Support the Planning, Design, Construction, and Maintenance of the Upgrading of Pan Borneo Highway, Sarawak - Underground Utility Mapping*
[3] Sumairi R 2018 *Lecture: Building Information Modelling (BIM) for Land Surveyor: The* Challenges & Opportunities (Kuala Lumpur: CIDB)
[4] Saad S 2016 Pavement Maintenance Management System: A Review *Trends in Transport Engineering and Applications* **3** pp 19-30.
[5] Nex F and Remondino F 2013 UAV for 3D mapping applications: A review *Applied Geomatics* **6**(1) 1-15
[6] Miller W R 2018 *Introducing Geodesign: The Concept* Available from https://www.esri.com/library/whitepapers/pdfs/introducing-geodesign.pdf.
[7] Dangermond J 2017 A framework for understanding, managing, and improving our world *GIS–the geographic approach. ArcNews* **29**(3) 1-3
[8] Remondino F, Barazzetti L, Nex F, Scaioni M and Sarazzi D 2011 *UAV Photogrammetry for Mapping and 3D Modeling - Current Status and Future Perspective* Available from https://is.muni.cz/el/

IOP Conf. Series: Materials Science and Engineering **512** (2019) 012038 doi:10.1088/1757-899X/512/1/012038

Role of BIM+GIS checker for improvement of technology deployment in infrastructure projects

Mohd Harris Ismail[1], Siti Salwa Mohd Ishak[2], and Mokhzal Osman[3]

[1] BIM Institute of Malaysia, Shah Alam, Malaysia
[2] Construction Management School of Housing Building & Planning (HBP), USM, Penang
[3] Ascension Technology Sdn. Bhd. Petaling Jaya, Malaysia

E-mail: mhim2u@gmail.com

Abstract. There has been a focus in construction industry, on creating, sharing and reusing digital project information, throughout the project lifecycle, which includes Building Information Modelling (BIM) and Geographical Information Systems (GIS). These co-existed technologies are said to have the ability to improve the efficiency and delivery of the infrastructure projects. Having said that, this can only be achieved through the right strategy, and proper collaborative environment, rather than relying on the cutting-edge technology alone. The involvement of 3rd party experts or "checker" is necessary to ensure the use of technology does not go off on a tangent from its original objectives. This paper will define what constitutes to successful process of infrastructure implementation, and how it helps to improve the use of technology with the correct methodology. Two ongoing mega infrastructure projects in Malaysia were selected as case studies. Comparison is being carried out on both projects to study the effectiveness of the technology that is being utilised during the construction works. Finally, the paper established a conclusion whether there is a need for a 3rd party "BIM+GIS checker" in order to ensure the right processes are being adhered to achieve efficiency in the delivery of linear construction projects.

1. Introduction

Virtual construction implies that it is possible to practice construction, to experiment and to make adjustments in the project before it is constructed [1]. Digital mistakes in the virtual construction generally do not have serious consequences provided that they are identified and addressed early enough so that they can be avoided in the actual construction of the project. When a project is planned and built digitally, most of its relevant aspects can be considered and communicated before the instructions for construction are finalized [2]. It is like running a simulation of a construction project by considering all aspects of the construction life cycle.

With Building Information Modelling (BIM) and Geographic Information Systems (GIS) technologies, one or more accurate digital information will be able to be captured, stored, manipulated, analysed and managed efficiently. GIS applications are tools that allow users to create interactive queries, analyse data, edit in maps and many other functions related to spatial and geographical information [9]. GIS can also be referred to as a number of different technologies, processes and methods deployed for the data capturing. It can be the foundation for a location-enabled horizontal construction which is attached to many operations and has many applications related to engineering, planning and management of the infrastructure works [3][4][5].

Meanwhile, BIM technology produces three dimensional (3D) virtual model of the construction components with the parametric information [2][3][7]. BIM support the whole construction phases from design, construction, up to the operation of the completed infrastructure. These computer-generated 3D parametric models contain precise geometry and data needed to support the construction, fabrication, and procurement activities through which the construction is realized [3]. BIM technology also accommodates many of the functions needed to model the lifecycle of a construction, providing the basis for ease of design review and efficient capabilities of error detection. BIM processes will optimize the success of project delivery by introducing changes in the overall methodologies, roles and relationships among project team. When adopted well, the application of BIM, GIS technology and the right processes will facilitate a more integrated design and construction that results in better quality at lower cost and reduced project duration [5].

In terms of the types of data and information that can be derived within BIM and GIS, generally they could provide a 2-D and/or a 3-D data with graphical or non-graphical information including geographical topography, specifications, cost data, scope data, and schedules [4][9][8]. On the other hand, categorised several types of information within Building Information Model based on the nature of the link between information and the model [1]. This pertains to all information that is part of, or connected to the components as well as the physical information inherent in the model itself such as size, location, etc. It is important that all information required in making an actual analysis be available from the BIM+GIS platform.

2. Technology Background

Dealing with a large-scale infrastructure project requires a proper methodology and workflow strategy. This is to ensure the smoothness of project delivery with better coordination and communication among the project team [6]. The deployment of BIM and GIS in the construction of infrastructure provides the tool for creating digital representation of the works for visualization, engineering analysis, code criteria checking, cost engineering, as-built product and many other functions [7]. Producing the proper BIM process could foster better integration, management and communication within the industry. The correct implementation of BIM+GIS processes within the lifecycle of project delivery is important to be strategize correctly since it affects the performance of the overall project. In the context of the whole construction life cycle, the Building Information Modelling processes can be used in every project phase, regardless of the type of project [3]. Hence, BIM methodologies had been utilized in most building projects and mega infrastructure projects in Malaysia.

2.1 Three Dimensional (3D) Model Development

BIM consists of two-facet, the 3D parametric modelling technology and the process of inputting information within the model [3][8]. Hence, at its basic core, deployment of BIM should begin with the production of 3D parametric model from the inception of the project lifecycle (model-based design), failing which, will jeopardize the efficiency and effectiveness of BIM implementation and reducing its benefits [6]. The approach of converting the 2D CAD drawings to 3D model is the lowest form of BIM methodology with the lowest potential benefits, while at the same time potentially creating new contractual disputes and disruption to the work progress [4]. It has been proven as inefficient approach since it was introduced in the early 2008. This method is also known as *PseudoBIM* and viewed by experts as a “first step to failure”, if it is not handled with due expertise.

The 3D parametric model that is produced from the 2D CAD model introduces some issues relating to its quality and integrity [6]. The fact that the conversion from 2D to 3D model is not carry out by consultant designers, but done by a 3rd party modeller with less project experience or non-AEC background, may result in the models being unsuitable for BIM uses, in fact, such 3D model may only be suitable for visualization and basic clash checking, which are only one-eighth (1/8) of the benefits reaped from BIM implementation. Further to that, the processes that are created from this approach are neither efficient nor productive thus not reducing the overall cost of the project (in contrary, adding more cost to the project), in which the reduction of cost and time are the essence of BIM. Nevertheless, the *PseudoBIM* approach is becoming a norm in Malaysia, where the involvement of a 3rd party "checker" or coach is inevitable in ensuring the use of BIM processes and methodology do not go off tangent from its original objectives. Hence, the first and basic criteria to ensure the success of BIM implementation and to reap the maximum BIM benefits is to have the 3D model developed from the inception of the project (model-based design) and use the information within the model for every aspect throughout the project lifecycle [2][3][6][7].

2.2 GIS Data Capture

Geographical Information Systems (GIS) application is a tool that allows users to analyse geospatial data, edit in maps and many other functions related to geographical information [4][9]. These highly accurate & precise Geospatial Datasets are acquired using the advanced surveying and remote sensing technologies such as GNSS, GPS, LiDAR and Photogrammetry [7]. The GIS base map is developed in 2D with the aerial data of the construction site captured using UAS methodology, and the terrestrial data is captured using field survey.

The UAS methodology for capturing GIS Data has been used extensively and proven professionally in many industrial sectors such as Property Development, Highway Construction, Plantation, Forestry, Agriculture, Engineering/Civil Infrastructure, as well as Oil & Gas [7]. Technically, these systems will deliver cost-effective services for the construction monitoring and Project Management of a linear construction. The drones/UAVs are capable of flying hours and covering distances up to 10km. Nevertheless, optimal flying specification must be properly identified for quality orthophoto captured during certain condition. For a focused area which require s high attention and details, the flying method is totally different from the way orthophoto data is captured, where it will require oblique and nadir data acquisition method. The flight altitude will be lesser while camera angle must be manually adjusted during data acquisition period to eliminate as many blind spots as possible [7]. The processed data can be opened using the given freeware viewer for offline viewing purpose.

The utilization of suitable drone platform and its operating procedures is crucial in order to achieve cost-effective operations. The flight path must be thoroughly planned and correctly configured prior to the work execution in order to cover the required area and obtain good set of aerial images. Additionally, potential areas for the ground station should be pre-determined, as it will serve as the take-off and landing location, as well as initial home reference. For the UAV flight planning for orthophoto generation of the linear area with minimal curves, such as road project with corridor width, the flying grid is to be properly determined to ease the image stitching and processing step later on.

In short, the development of 3D parametric model in BIM and capturing high quality GIS data can only be achieved through the right strategy, methodology, processes and proper planning rather than relying on the cutting-edge technology alone. In order to achieve efficiency and highest benefits from the 3D model and GIS data, the collaborative environment is inevitable.

2.3 Collaborative Environment

The collaborative environment in the construction industry originated from one segment of BIM methodology, namely Common Data Environment (CDE) [5]. The use of collaborative environment such as the CDE in BIM process is the segment that is crucial to achieve the highest benefits in deploying technologies [8]. It enables architects, engineers and contractors alike to work on a single platform from anywhere in the world. It condenses a plethora of information about every detail into a workable format.

It facilitates testing and analysis during the design phase and monitoring during construction phase, to find the best answer and solution to a problem [3]. Designing is made easier, and coordination is made simpler among team members, hence structure maintenance is made seamless too across the entire built environment. Without the proper deployment of CDE, the efficient and effective communication and coordination among the project team members will not be possible, defeating the original intention of BIM implementation [6]. Figure 1 shows the basic arrangement of CDE based on PAS1192-2. It provides a specific guidance for information management requirements associated with projects delivered using BIM process.

This information is to be managed in consistent and structured way to enable efficient and accurate information exchange. Specific details of the standards and processes should be properly adopted to deliver better outcomes. All types of information exchange and management standards between the entire project stakeholders need to be developed based on the specific project requirements. For the avoidance of doubt, all project information, either in GIS, BIM or conventional data formats should be shared using a single source of truth, i.e., the common data environment (CDE).

Nevertheless, not all information on a project will be originated, exchanged or managed in a graphical format. There should also include the non-graphical data, documentation and GIS information into the ecosystem of CDE [8]. Nevertheless, the data that is acquired through GIS acquisition is usually stored in its own geo-database. The geo-database data warehouse applies sophisticated rules and relationships to the data and has the ability to store a rich collection of spatial data in its own centralised location. All data stored in the geo-database is published to a GIS server and accessible via a web-based GIS application [7]. An example of the geo-database system architecture is shown in Figure 2. The geo-database in GIS also has the ability to define advanced geospatial relational models, integrate spatial data with other IT database and capable of leveraging spatial data to its full potential, where the CDE in BIM cannot provide these requirements within its platform. Hence, the interoperability between BIM CDE platform and GIS geo-database is another crucial aspect to be considered prior to implementing BIM+GIS.

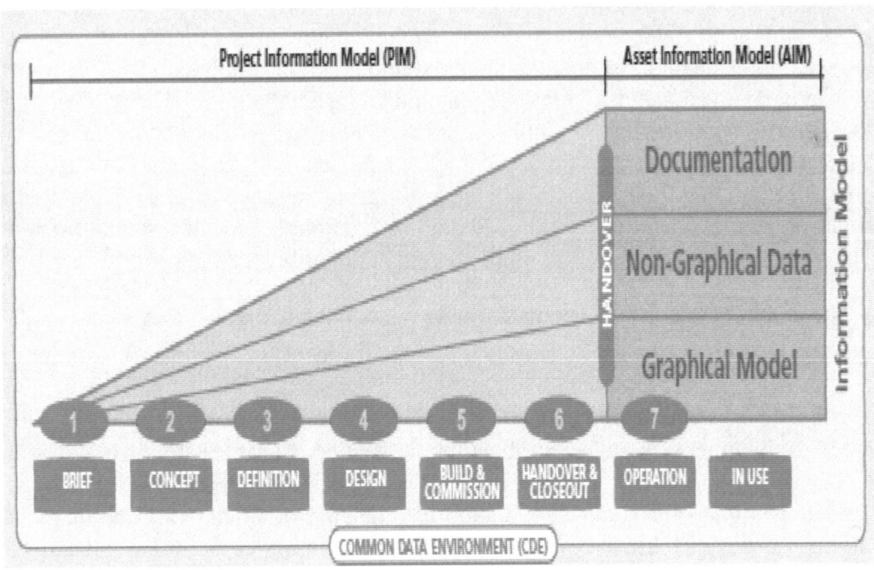

Figure 1. Common Data Environment based on PAS1192-2:2013

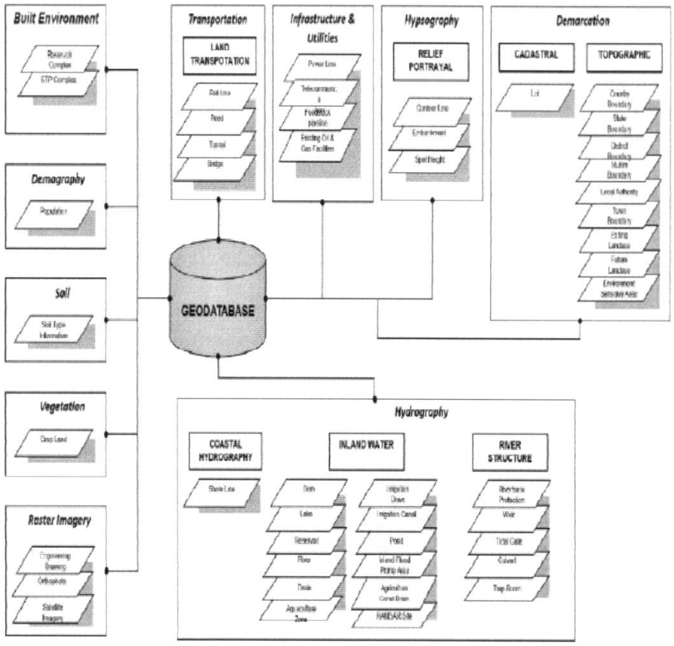

Figure 2. Example of Geo-Database

The key concept of Common Data Environment (CDE) is that the project team works together using the best collaborative tools at their disposal to ensure that the project will meet the owner's requirements at significantly reduced time and cost [3]. The owner needs to be either part of this team (to help manage the process) or a consultant, to represent the owner's interests. The trade-offs that are always part of the design process can best be evaluated using BIM—cost, energy, functionality, and constructability [2][3]. Therefore, BIM and project management go together and represent a clear break with current linear processes that are based on paper representation exchange of information. The effective use of BIM requires that changes be made to almost every aspect of project management approach (not just doing the same things in a new way) [3]. It requires some understanding of technology (BIM & GIS) and related processes and its implementation plan before the project may begin. Since, BIM and GIS having their own centralised data platform, they are required to be connected between each other, so that they are able to work as in a single assess point. The connection of two centralised platform (BIM+GIS) is known as the Connected Data Environment (CDE) for infrastructure projects.

3. Hierarchy of BIM Implementation (HBI)

The deployment of the cutting-edge technology alone do not guarantee benefits without a proper process [6]. There are proofs showing that technology can only be benefited so much, but with the correct methodologies and processes, technologies are able to provide up to 33% cost reduction and 50% of overall time savings. The lesser the standard processes being deployed; the lesser benefits will be gained from the technology deployment.

Figure 3 shows the hierarchy of BIM Implementation based on three (3) level of Classes, with 1st Class as the highest hierarchy while 3rd Class being the minimum BIM deployment with the least benefits [6].

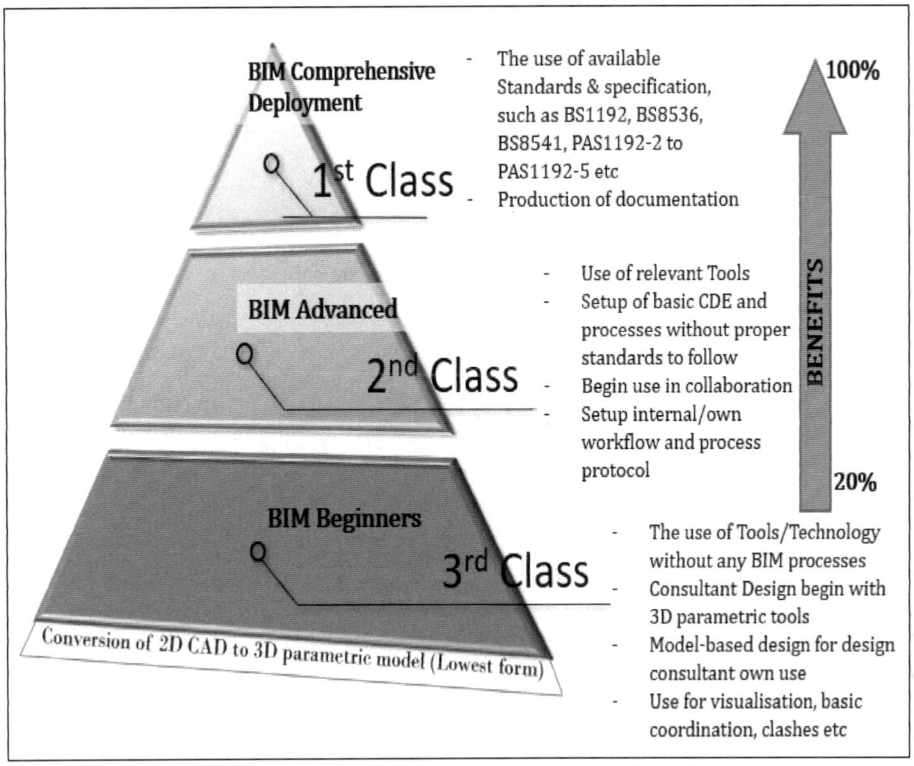

Figure 3. Hierarchy of BIM Implementation [6]

As earlier mentioned, conversion of 2D CAD drawings to 3D parametric models is the lowest form of BIM approach which potentially create other contractual disputes within the contract and may disrupts the entire work progress. At this stage, the appointment of *3rd party model checker* is inevitable to avoid the potential problems as described in item 2 above.

The 3rd Class implementation, is considered as the beginners of BIM adoption, it begins when the implementers deploying BIM with the use of BIM-related tools from the inception of the project, i.e. the design consultant deploys the BIM-related tools since day one of the project. However, the design consultant developing the models for the purpose of fulfilling their respective roles only (individual discipline) and production of drawings without the overall BIM process. At this condition, the use of 3D parametric models is for the basic purposes such as visualization or animation (3D), authority submission, basic estimation and discipline-based coordination–without the use of CDE while lacking of overall BIM work-process.

Implementation at 2nd Class acknowledges the implementers at higher BIM proficiency level, with the internal workflow being established and the use of some sort of CDE. Federated models and information are shared while collaboration and communication are carried out through this platform. However, the workflow established at this phase is produced based on isolated cases and suits the specific project or organisation only. Having said that, the benefits that are able to be gained at the 2nd Class implementation will be 40% to 60% higher than 3rd Class [6].

The most comprehensive implementation is marked as 1st Class BIM implementation, where the deployment of technology at this stage is by using the available accepted standards, specification and guidelines such as BS8541, BS8536 and the eight (8) pillars of BIM with high competency. The production of documentation is from the information developed through the

BIM process. A full BIM compliance to the 1st Class will potentially reap full benefits from the technology deployment. Hence, we need to move upward from current practices of 3rd Class BIM implementation with very low benefits towards the most efficient 1st Class practices.

4. Comparison of two (2) Mega Infrastructure Projects in Malaysia as Case Study

4.1 Methods

This research relies on an intrinsic case study approach. Two ongoing mega infrastructure projects in Malaysia were selected as the case studies. The first project is called Linear A, and it is a mega highway project in East Malaysia. The second project is a mega rail project in peninsular or West Malaysia, and is referred to as Linear B. These projects varied in size, phase, BIM execution approach and processes. The projects were selected from the authors' expert involvement.

Data of the study were collected through desk research (i.e., document analysis), observation and direct interviews with the BIM and GIS implementers, team members, and project managers. Comparisons of these two projects were developed to identify the evolving nature of BIM and GIS, and to create a common view of current BIM and GIS effective deployment practice and processes. The comparison constitutes seven (7) key criteria or checklists of the collaborative data environment, i.e., method of 3D model production, BIM modelling tools and coordination tools, BIM model checking tools, BIM CDE tools, GIS data capturing methods, GIS geodatabase accessibility, and availability of BIM Standards. Assessment tools developed by the BIM Institute of Malaysia are also utilized for the purpose of analysis and generalisation. The study period ended in May 2018.

4.2 Findings

Both, Linear A and Linear B projects implemented similar technologies, BIM and GIS, with two different sets of methodologies and processes. Comparison between both projects was carried out to identify the effectiveness of the technologies that are being utilised and the tangible benefits achieved during the construction works, since the GIS+BIM technology deployed in both projects are similar, hence the comparison is being made to their differences in methodology and processes.

The findings show that both projects have different outcomes and results, whereby, Linear B project is reaping more benefits from the technology deployment and achieving the significant saving in terms of time and cost including winning the international award of "Best BIM implementation" in 2017. Meanwhile, Linear A project is yet to show any significant benefits other than deploying a cutting-edge technology and having a detailed 3D parametric model and reality model of the infrastructure works with some clashes function (carried out by outsourced consultant) with a GIS geo-database accessible separately through the web-based application. The significant differences are on the organisational arrangement and CDE setup between both projects as shown in Figure 4 (Linear A) and Figure 5 (Linear B) respectively.

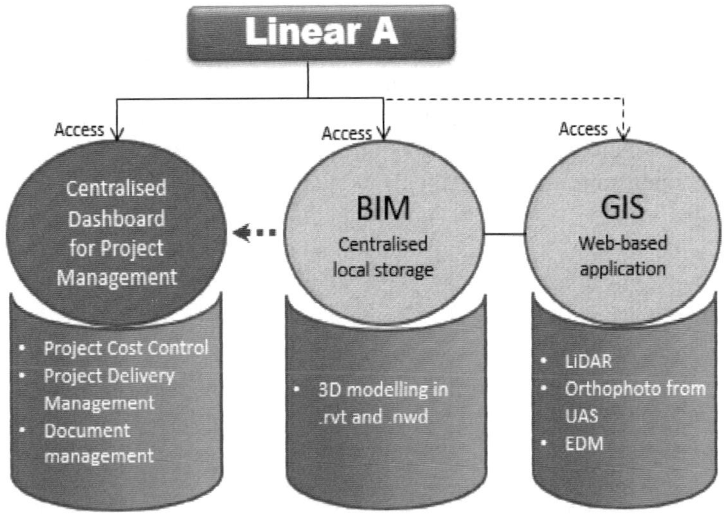

Figure 4. Centralised Data Storage for Linear A project

Linear A project providing separate centralised platform for Project and Information Management dashboard, BIM platform and GIS platform. The setting up of each platforms are being sub-contracted to two (2) separate service providers with separate contract to maintain each service. The GIS service, which includes a separate geo-database with a web-based application is a subset from BIM service. The project information management dashboard is a desktop and server application. The BIM centralised storage is in a local network server (Network Attached Server), while GIS is a web-based application. All three (3) platforms are able to be accessed separately with different access points. This creates disruption to the ease of operation and inconvenience to the users. The author did not find any properly established processes for the use of these three (3) storage areas which could streamline the use of these platforms.

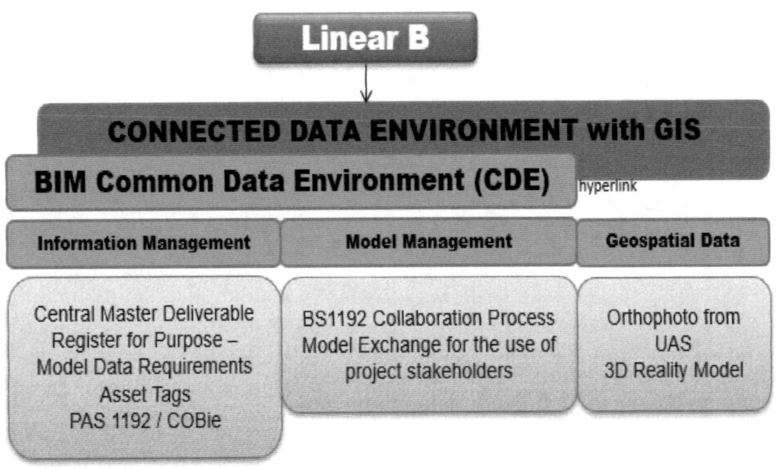

Figure 5. Connected Data Environment (CDE) for Linear B project

On the other hand, the setup of Linear B project is based on the single access point for BIM CDE, meanwhile the GIS web-based application is being connected through hyperlink in the GIS platform to the BIM CDE. Even though there are two (2) access points (BIM & GIS), it works as a single-entry platform, since both platforms are connected with each other through hyperlink (inter-connected). This approach makes the utilization of the platform much acceptable to site operation personnel. Meanwhile, the appointment of application setup is through software supplier while the operation and control of the system is through the internal experts with one single head of section to manage both aspects of technology. The process of information management for the entire project is established from the globally accepted standards and guidelines, i.e: BS1192: 2007; PAS1192-2:2013; PAS1192-3 and others, with some modification to suit local practice.

Table 1. Comparison between Linear A and Linear B project

Technology Deployment	Linear A	Linear B
1. Method of 3D model production	Modelling (conversion) from 2D CAD drawing	Model-based design from inception of project
2. BIM modelling tools and coordination tools	Autodesk Revit, Civil 3D, Infraworks and Naviswork	Autodesk Revit, Civil 3D, Naviswork and Bentley Aecosim
3. BIM model checking tools	N/A	Solibri Model Checker
4. BIM CDE	Bentley Projectwise	Bentley Projectwise
5. GIS data capturing	LiDAR, UAS	UAS
6. GIS geodatabase accessibility	WebGIS – view in plain view only	ArcGIS – view in plain and orbit view
7. Availability of Standards / Guides		
a. BEP	√	√
b. LOD definition specific for the project	x	√
c. Modelling guide	x	√
d. Establishment of internal Standards & operating Process for entire project	Minimal	Detail SOP for every single phase and operation
e. Globally accepted guides e.g: BS, PAS etc	some portion of Singapore BCA BIM Guide and BS1192:2007	BS1192:2007; PAS1192-2 to PAS1192-4

Referring to the aforesaid literatures in item 2.0 and the Hierarchy of BIM Implementation (HBI) at item 3.0, the setup of CDE and the processes used in both projects show that the implementation of Linear B project is at 1st Class implementation while Linear A could be at 2nd

Class or 3rd Class. In order to categorized accordingly, further comparison is made between Linear A and Linear B project as listed in Table 1.

From Table 1, it shows another major difference identified between Linear A and Linear B project, which is the method of 3D parametric model development. Linear A project develops the model from the design consultant's 2D CAD drawings. Which means, the project designers produced the design in accordance to the conventional method (2D CAD), hence, the client had sought the BIM consultant service to 'convert' the 2D CAD into 3D parametric model. As described in earlier literature and experiences, this method contributes further to the inefficiency of the project. According to HBI theory, this is the lowest form of BIM implementation approach. Having said that, there are also some elements of 2nd Class implementation that are available in Linear A project, hence the assessment of Linear A project may fall under the 3rd Class implementation.

Based on the assessment tool developed by BIM Institute of Malaysia, the results from the comparison of both projects against the Hierarchy of BIM Implementation (HBI) and Benefits of the technology deployment is shown in Figure 6.

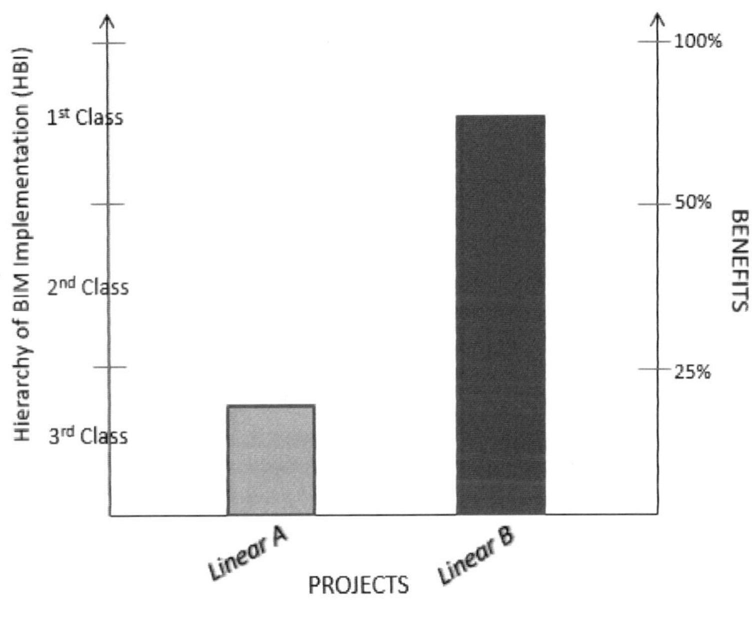

Figure 6. HBI and BIM benefits

5. BIM+GIS Process Checker

A comparison between the two (2) case studies clearly proves that deployment of technology alone may not provide optimum benefits if it is deployed without a proper process. The technical know how to use the technology will only benefit to certain level. The more important aspect is to have the knowledge to develop efficient and proper processes to suit the project requirements. The development of these processes must be in accordance to the globally accepted standards and guidelines. Therefore, even though having a BIM+GIS consultant is very helpful at certain level, there is still the need to counter check the BIM+GIS consultant. This is to ensure the appointed consultant is doing the correct approach to reap the maximum benefits from the technology implementation. The BIM+GIS checker can be a company or a person who is an

IOP Conf. Series: Materials Science and Engineering **512** (2019) 012038 doi:10.1088/1757-899X/512/1/012038

expert within the client's organisation or an external 3rd party BIM experienced and proven practitioners. Without the *Checker*, there will be no parties cross-checking the 'BIM consultant' whether they are doing the right BIM and giving benefits to the organization.

It is proven that by deploying technology alone will not reap the whole BIM benefits, hence, the role of the *Checker* is necessary in ensuring the process developed for the project is able to streamline the entire construction processes and the deployment of technology. In the event the project is forced to begin from the 2D CAD conversion to 3D model, the Checker is required to ensure the conversion is being done correctly and the developed models (from the 2D CAD) by the BIM consultant meet the minimum requirements for further BIM+GIS processes to take place. Therefore, the Checker (either internal or 3rd Party) must be fully equipped with knowledge and have sufficient experiences on what DO's and DON'Ts on the construction process, particularly relates to BIM implementation. The selection for BIM+GIS checker must be done thoroughly and strictly. They must be a construction-related background and possess both technical and process skills. They need to prove the competencies from the previous successful project and not simply *'involved'* in project or handled a *previously-failed* BIM related project. The BIM+GIS Checker can also act as a coach to the client's BIM+GIS team to enhance their skills to be a process checker and/or hand-hold the appointed BIM+GIS consultants through-out the project life-cycle.

6. Discussion

The findings show that both projects, Linear A and Linear B, have very distinct approaches, but to determine if the implementation is effective, it has been the subject of discussion and reference to the BIM, CDE, and HBI literature.

The significant differences in terms of approach are the strategy, execution, and model development. It can be said that the implementation of the Linear B project is considered state-of-the-art. Linear B was validated through extensive BIM processes and applications, making it one of the most robust and structured BIM-driven projects. Linear A is evolving as a pseudo-BIM project, and it is believed by most practitioners to be a less desirable approach to "disrupt" the BIM best practice.

This study observed the CDE execution, and the reflection is clustered around the processes and technologies used. The Linear B project CDE approach makes the utilization of all data platforms much more acceptable to the site operation personnel. Meanwhile, the CDE of Project A is ineffective, because the appointment of the application setup is through a software supplier while the operation and control of the system is through the internal experts with one single head of section to manage both aspects of the technology. There are also no established processes for the use of these three (3) storage areas, which could link and streamline the use of these platforms. This situation creates a disruption to the operation and an inconvenience to the users.

Client organizations are a complex network of sub-units with individuals with differing agendas and interests. Clients may have their own criteria on project success and perceived benefits. To improve the client's perceived value and benefits, and therefore the integrity in the BIM delivery, there is a need for client actors or "integrity agents" in the process of BIM. Based on observation, these agents are argued to possess the needed authority and influence to ensure the system integrity and process clarity by other project participants. There is clear merit to BIM and GIS consultants, as a means to deliver the BIM and enhance the project performance; however, alongside the workflow and process, the integrity check is deemed appropriate. This is also to ensure the appointed consultant is utilizing a suitable approach to reap the maximum benefits from the technology implementation as well as checks and balances in the process developed for the project and is able to streamline the entire construction process and the deployment of technology. In the event the project is constrained to take up the pseudoBIM approach, the "Checker" is required to ensure the conversion and data transformation is performed according to standards and the developed models (from the 2D CAD) meet the minimum requirements for further BIM and GIS integration processes to take place.

7. Conclusion

The paper present two realties of BIM+GIS-driven projects. BIM and GIS are cutting-edge technology and best-practice process, and by adopting these technologies, clients can manage mega infrastructure projects in a more effective, integrated, and structured way. The information, documents, and models are systematized in the entire timeline of the project from inception to operation and demolition. Since BIM and GIS are technologies that need specific competencies and experience, the appointment of the right expertise is crucial to ensure the achievement of the project objectives. A comparison of the case studies has proven that having cutting-edge technology alone will not provide much benefit to the project unless it is coupled with the appropriate processes and standards. It is clear to see that the Linear B project is reaping more benefits from the technology deployment and is achieving significant savings for time and cost. Meanwhile, the Linear A project is yet to show any significant benefits other than deploying a cutting-edge technology and having a detailed 3D parametric model and reality model of the infrastructure works with some clashes function

This paper shows that efforts to integrate BIM and GIS in Malaysia are evolving, and thus a "one-size-fits-all" approach is probably misguided and could hinder BIM diffusion in Malaysia. There is nothing wrong with pseudo-BIM, as despite being less desirable, it can however be improved and reoriented with an integrity checker role. The implementation and execution approach must be strategically planned and properly developed to avoid creating new disputes and conflicts. The Checker, either internal or third party, is appropriate to ensure the BIM execution meets the client's objectives and requirements.

8. References

[1] Kymmel W 2008 *Building Information Modelling. Building Information Modeling Planning and Managing Construction Projects with 4D CAD and Simulations* pp11, 47-50, 51-52

[2] Harris M, Irfan A, Haron T and Husairi A 2014 The Way Forward for Building Information Modeling (BIM) for Contractors in Malaysia *Malaysia Construction Research Journal* **15**. No.2

[3] Eastman C, Teicholz P, Sacks R and Liston K 2011 *2nd Edition BIM Handbook: A Guide to Building Information Modelling for Owner, Managers, Designers, Engineers, and Contractors* John Wiley and Sons., Inc. New Jersey

[4] Esri 2018 *What Is GIS* Corporate website of Esri https://www.esri.com/en-us/what-is-gis/overview

[5] McPartland R 2016 *What is the Common Data Environment (CDE)?* Corporate Website of NBS https://www.thenbs.com/knowledge/what-is-the-common-data-environment-cde

[6] Ismail M H 2018 *Improvement of BIM Deployment for Construction Project in MY* Cubicost, Glodon Ltd, China https://mp.weixin.qq.com/s/B4l-qKpkORqh5sU54AcZtA

[7] Razak A A 2018 *Provision of GIS + BIM Solutions to Support Utilities Relocation Works* Technical Proposal, RS&GIS Consultancy Sdn. Bhd.

[8] The British Standards Institution (BSI) 2013 Specification for information management for the capital/delivery phase of construction projects using building information modelling. *PAS 1192-2:2013 Incorporating Corrigendum No. 1*, 25

[9] Wikipedia,GeographicalInformationSystems, https://en.wikipedia.org/wiki/Geographic_information_system

A comparative study of monitoring methods in sustainable pavement management system

M J Chin, P Babashamsi and N I M Yusoff

Smart and Sustainable Township Research Centre, Faculty of Engineering and Built Environment, Universiti Kebangsaan Malaysia, Selangor, Malaysia

Email: cmjtheresse92@gmail.com

Abstract. Pavement management systems (PMS) have been receiving increasing attention in the attempt to ensure that roads are always in good condition and serving its purpose. Appropriate PMS must be implemented to gather valid and reliable pavement data. Previous studies which attempted to identify methods for monitoring pavements were limited by constraints such as cost, time, safety, data accuracy and sustainability. This study was conducted to review some of the pavement monitoring models introduced in previous studies. After completing a literature review, three models, i.e. manual survey, smart sensor, and optical image processing, were selected for a comparative study to determine which of these models is the most effective model in terms of cost, time, safety, accuracy and sustainability. A data quality guideline was modified to produce a rating system for ranking the models from the worst to the best. The results of this study shows that smart sensor has the highest score in the radar chart, and hence is the best model. In conclusion, the findings of this study could provide a guideline for the government and private sectors in determining the most effective pavement monitoring model to be used in their sustainable PMS strategy despite the constraining factors.

1. Introduction

Pavement management system (PMS) is established by highway agencies to ensure that roads are in good condition and serving its purpose. A pavement condition survey is an important element in the pavement management process. A survey of pavement condition will provide the valuable information needed for pavement performance analysis; it is also crucial in facilitating the forecast of pavement performance, anticipating maintenance and rehabilitation requirements, establishing maintenance priorities, and allocating funds [1]. The network-level pavement asset management system (PAMS) proposed by [2] has become an important tool in helping state highway agencies determine maintenance and rehabilitation schedules and allocating limited resources.

Pavement is one of the most important infrastructures which ensure a safe and comfortable journey. However, pavement is a type of consumable material and must be replaced at an appropriate interval since they deteriorate and are damaged by applied load and the effects of surrounding factors, such as heavy traffic, unpredictable climate, etc. [3]. Pavement maintenance is essential to ensure a safe travel as well as prevent traffic congestion, air pollution, and accidents [4]. Consequently, PMS is becoming increasingly important since it provides information on pavement condition, which in turn helps authorities to schedule pavement maintenance activities [1, 2, 5].

In order to ensure good road system, it is essential to continuously maintain and rehabilitate the existing road network [6]. PMS should be able to facilitate the decision making process regarding which segments of a pavement network should be preserved, maintained, and rehabilitated despite the budget constraints [7]. Before making a decision on road maintenance, it is important to know the actual condition of pavement in a particular area. Different types of pavement distresses would probably require different treatment [3, 8, 9]. Therefore, data on pavement distress is required before making a decision on the proper rehabilitation treatment to be carried out.

Several pavement monitoring methods can be employed to gather pavement distress data. Conventional method, such as walking survey in figure 1(a) and riding (also known as windshield) survey in figure 1(b) are used to monitor the conditions of pavement surface manually. However, these traditional methods are very costly, time consuming, labour-intensive and subjective, and evaluators conducting these surveys are exposed to hazards [8, 10-13]. These methods typically depend on individual evaluators' visual inspection, interpretation and judgement of the extent and severity of all distresses observed at each pavement section [14]. Consequently, the disparity in the distress data gathered through manual surveys has always been a contentious issue in PMS.

Figure 1. (a) Walking survey **(b)** Riding survey

On the other hand, data collection technology is one of the most rapidly evolving areas in asset management. The methods used to collect distress data range from manual surveys based on visual inspection to automated surveys, which utilized systems, based on photography, video cameras, or sensors. Since the mid-20th century, sensing technologies have been enhanced in terms of sensitivity, functionality, scale, survival rate, and resistance to harsh environment [11]. The development and application of ultrasonic, infrared, laser sensors, and high-speed computer processing have significantly improved the ability of transportation agencies to collect large volumes of pavement data quickly and efficiently. Sensor technologies can be used to gather data automatically, in addition to protecting their workers from exposure to hazards on the road. Technologies and procedures such as these allow agencies to gather data and report on pavement conditions more frequently and are often more cost-effective than the manual techniques [15]. Basically, sensors can be categorized into two types, i.e. wired and wireless. The present study used a wireless sensor which introduced by [16] called the Smart sensor. This sensor was designed with six main features: self-powered, non-volatile storage, small size, wireless communication, ability to withstand harsh loading and environmental conditions, and possibility of network deployment. Smart sensor can be used to automatically gather large amounts of pavement data via a moveable external radio frequency (RF) reader. This method can be implemented either manually or by driving a vehicle over the embedded sensors (figure 2). The use of sensors in PMS would reduce the length of time required to gather distress data and problems related to the safety of evaluators can be eliminated.

Many systems based on ground-penetrating radar or laser system have been proposed. However, the findings of recent research have shown that the optical image processing (OIP) method is more effective in conducting a non-invasive evaluation of surface degradation [17]. Since the 1990s, evaluation of pavement through digital images has become increasingly popular due to the significant advances in the sciences of computer vision and image processing [11]. OIP is performed using a mobile laboratory (figure 3), where a vehicle is equipped with special cameras to capture the image of pavement surface and record distress data. The data is based on high-speed digital acquisition system, which depends on the computed digital values and the images captured by the camera. The system is essentially a mobile laboratory, which gathers pavement surface data together with other supporting software such as global positioning system (GPS). Since the data is gathered using an OIP method, further analysis must be conducted, and this could be done using semi-automated or fully automated techniques. The data is processed to convert it into a usable format which provides the distress information of pavements. Fully automated data processing uses computers to interpret, reduce, and analyze the images without human intervention. Semi-automated data processing converts the gathered data using automated collection means, and the images will be viewed and interpreted by evaluators who to identify distress information [10].

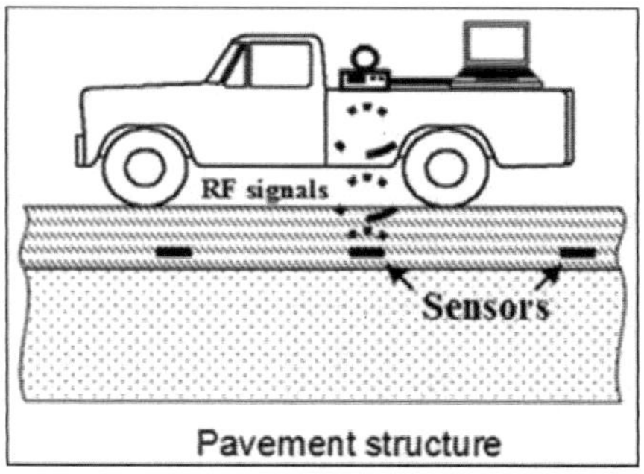

Figure 2. Smart sensor method

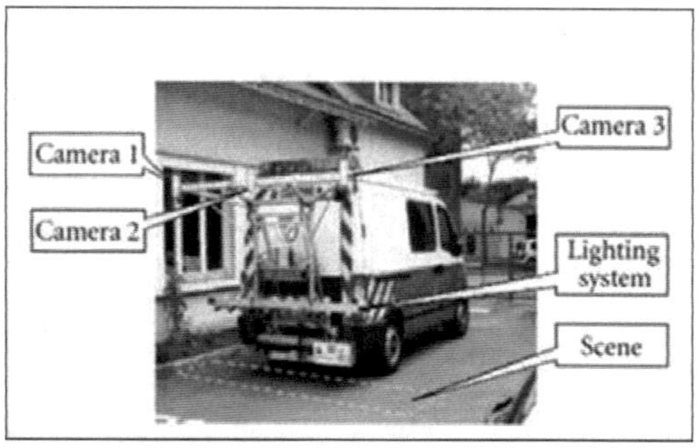

Figure 3. Optical image processing method

The objective of this study is to compare the frequently used monitoring methods, including manual survey, smart sensor study, and optical image processing in term of cost, time, safety, accuracy of data, and sustainability.

2. Methodology

After completing a review for three of the models in this study, the information gathered are summarised into a table (table 1). Table 1 presents an overall summary of the three methods. These three methods were chosen due to their outstanding performance in their respective category.

Manual survey is the first model to be considered in this comparative study due to fact that it is the very first pavement monitoring method used in PMS. Manual survey can be used as a basis for comparison of other models. Other automated methods were introduced subsequent to the implementation of the manual data collection survey. The use of sensor is important in an automated data collection in PMS. A wireless smart sensor is chosen for this comparative study due to its superior performance in comparison to that of a wired optical fiber sensor. Finding stated that normal wired sensor in PMS can only be used for short term purpose due to the need to use batteries, which have to be replaced when they run out of power [16]. This results in a higher operational cost. The use of solar power for other normal wireless sensor used to gather data in PMS is not feasible since the sensors are fully embedded under the pavement. Another model taken into account in this comparative study is the OIP model. Since this is an automated pavement image collection model, there is no concern of any data being missed at typical highway speeds.

This comparative study used a rating system table to assign average score to each model. The model with highest cumulative average score for all five indicators is regarded as the most effective pavement monitoring model. At the end of this study, the finding would assist the authorities choose the most effective pavement monitoring method by using valid and reliable data to implement a PMS strategy despite the constraining factors.

Table 1. Summary of comparison of the models

No.	Comparative Items		Manual Survey		Smart Sensor	Optical Image Processing
			Walking	Riding		
1	Cost	Equipment	Handheld tools or devices, data sheet	Computer, vehicle	Sensor, RF transponder and RF reader	High speed digital acquisition system, vehicle with camera, computer
		Labour & Training	- Trained evaluators - Trained for more than 6 months	- Experienced evaluators - Training required	- No labour cost - No training required	- Low labour cost - No training required
		Operation & Maintenance	- No operating cost - No maintenance fee	- Low operating cost - Low maintenance cost	- First time installation - Less maintenance due to high robustness	- Manual or software processes needed to convert the data into a usable format
2	Time		Very slow, 4 km per day	6000 km completed in 3 weeks	Very fast, continuous	Normal highway speed, 100 km/j
3	Safety		Dangerous, higher risk when gathering data	Safer, no risk	Safe, wireless communication	Safe to conduct on road
4	Accuracy		- Subjective, based on evaluator's experience - May be limited to certain span or segment of road	- May miss some data when the vehicle moves too fast	- Real time data, record and long term data storage	- More objective measurement - Repeatability
5	Sustainability		- Use of paper, record on data sheet - Walking will not causing pollution	- Emission of carbon monoxide gas by vehicle will cause air pollution	- Data can be read by RF reader either manually or by using a vehicle	- Emission of carbon monoxide gas by vehicle will cause air pollution

The components of the rating system are introduced through data quality guidelines [18 & 19]. The table was modified to suit the aims of this comparative study. The rank for each comparative item is arranged from worst to best. The worst rank is assigned a value of 1 and the best rank is assigned a value of 5. The rating system shown in Table 2 explains the comparative items in each rank. Each model will be given score based on the behaviour of the model. In the final stage, the score for each comparative item was summed up to compare the ranking of the models.

Table 2. Rating system for comparison items

WORSE TO BEST SCORE →

Compared Item	Rating System				
	1	2	3	4	5
Cost (For 1 km)	Very Expensive RM 400 and above	Expensive RM 301-RM 400	Moderate RM 201-RM 300	Cheap RM 101-RM 200	Very Cheap Below RM 100
Time (For 6000 km)	Very Slow More than 1 month	Slow 2 weeks to 1 month	Fair 1 to 2 weeks	Fast 1 to 6 days	Very Fast Less than 1 day
Safety	Very dangerous, may cause accident	Dangerous, exposure to hazard	Lower risk, safe	Safe to conduct on road	Very safe, no direct contact with traffic
Accuracy	Inaccurate or wrong data	Low accuracy, some unfitting and missing data	Accurate, data considered acceptable	Accurate, most of the required data is retrieved	High accuracy with real-time data
Sustainability	Not related to any sustainability practices	Does not implement sustainable practice, cause certain damage to environment	Minimal implementation of sustainable practice, potential risk to the environment	Sustainable practices, effort to reduce pollution	Support overall sustainable practice, environmentally friendly

The first compared item in the rating system (table 2) is cost indicator, which explained the budget rates for 1 km pavement are arranged from scale 1 to 5. The model that conducted with lowest cost will gain highest score of 5. Scores for the time indicator shows where the ranking is from the slowest to the fastest. The model which completes the monitoring of pavement condition in the shortest period of time is ranked as the best mode. In the rating system table also shows the ranking used for the danger and hazard exposure in the safety indicator. The fourth item compared in this study is accuracy of data collected. This is an important factor which is taken into account by most agencies when choosing the best pavement monitoring model. The availability of a more accurate data would allow agencies to make better decision with regard to their PMS strategy. In order to improve sustainability, pavement asset management systems should be able to account for sustainability indicators. Such as user time delay caused by preservation activities, additional fuel consumption caused by deterioration of pavement surface and other environmental impacts [2 & 20]. A compared item about sustainability is therefore also listed in the rating system table.

In the present study, a new rating system was introduced for the cost indicator to give a more specific and detail explanation about cost allocation. Similar with the concept in present rating system, the details in Table 3 is ranked from worst to best with the score ranging from 1 to 5. Since each model has been assigned with different types of cost, the cost indicator was divided into three categories, i.e. price of equipment, labour and training costs, and operating and maintenance cost.

Table 3. Rating system for cost indicator

Cost Indicator	Rating System				
	1	2	3	4	5
Equipment	Very expensive	Expensive	Affordable	Cheap	Very cheap
Labour + Training	High labour and training costs	Medium labour and training costs	Low labour and training costs	Low labour cost and no training cost	No labour and training costs
Operating + Maintenance	High conducting / operating and maintenance costs	Medium conducting / operating and maintenance costs	Low conducting / operating and maintenance costs	Low conducting / operating cost and no maintenance cost	No conducting / operating and maintenance costs

After prepared the rating system table as above (Table 2 & Table 3), the scoring process is able to proceed together with the finding listed in summary table.

IOP Conf. Series: Materials Science and Engineering **512** (2019) 012039 doi:10.1088/1757-899X/512/1/012039

3. Results and discussion

Following the tabulation of the rating system, the ranking process can be commenced by giving a score to each of the monitoring models. The scores are assigned based on the information provided by previous researches and studies, along with additional self-justification when the criteria are not considered in those studies. The comparative items begin with cost, time, safety, accuracy, and end with sustainability. Table 4 below shows the scores of average costs for all three pavement monitoring methods. Score 1 indicates the worst rank, which is given to the method that use the most expensive equipment; or highest labour charge and training; or highest operation and maintenance cost. For example, smart sensor study is having score 1 in the cost indicator of equipment because the material of sensor used is the highest among other methods.

Table 4. Average cost score ranking system

Models		Cost Indicator			Cost Average Score
		Equipment	Labour + Training	Operation + Maintenance	
Manual Survey	Walking	5	1	5	3.67
	Riding	3	2	3	2.67
Smart Sensor		1	5	1	2.33
Optical Image Processing		2	4	2	2.67

On the other hand, manual walking survey obtained score 5 in both equipment and operation and maintenance indicator. The reason was come from cheap handheld tools or devices used while conducting the survey and there are no any operation cost and maintenance fee needed.

Based on the summary of the costs and average score for all models in table 4, a model with the highest average cost score is considered as the best model. Manual walking survey has the highest average cost score of 3.67, which means that it has the lowest cost in comparison with other models in this study. The smart sensor model has the lowest cost average score of 2.33, which means that it is the most expensive method for PMS. The manual riding survey and optical image processing models have the same average cost score, which means that the cost for data collection in the two models are the same. Therefore, the models have to be compared based on other items to determine which the better model is. Table 5 presents the average cost scores for all models vis-à-vis other factors for a more comprehensive comparison.

Table 5. Average score of the five indicators

MODEL		Rating Comparison Item					Average Score
		Cost	Time	Safety	Accuracy	Sustainability	
Manual Survey	Walking	3.67	1	2	3	3	2.53
	Riding	2.67	2	3	2	4	2.73
Smart Sensor		2.33	5	4	5	5	4.27
Optical Image Processing		2.67	4	4	4	3	3.53

Through the time indicator's results shown above, manual survey apparently took the longest time to be completed, as have been shown in previous studies, such as the study conducted by the Arizona Department of Transportation (ADOT) [1]. But, the riding survey was able to cover a longer segment, in less time [1]. This survey was conducted using a very slow-moving vehicle [14]. Therefore, manual riding survey can obtain score of 2 in the time indicator.

For the safety indicator in this study, smart sensor model does not expose evaluators to any kind of hazard. This method does not require any human involvement subsequent to its installation. Smart sensors will automatically record and store pavement data independently and data is retrieved periodically via a wireless RF reader. Hence there is no need to gather pavement data on site. The scanning of RF transponder in the sensor to the reader can be done either manually or by mounting the reader on a moving vehicle. This process can be carried out fairly quickly without exposing anyone to any kind of hazard. Hence, the smart sensor model is given a score of 4 in safety indicator.

The fourth item compared in this study is accuracy. This is an important factor which is taken into account by most agencies when choosing the best pavement monitoring model. The availability of a more accurate data would allow agencies to make better decision with regard to their PMS strategy. [12] indicated that the data collected through manual walking survey tend to more subjective and less accurate since it is influenced by evaluators' experience. Even though evaluators must undergo training prior to working in the field, the data that produced by experienced and inexperienced evaluators will invariably be different. The researchers asserted that training programs can help evaluators to improve their evaluation skill, especially if focus is given to provide extra training to the types of distresses that are most likely to be interpreted differently. Thus, the accuracy of walking survey can actually be improved if evaluators are well-trained and are experienced. It should be noted that manual walking survey is sometimes limited to a certain segment of the road [15]. Humans have limited ability and it is not possible for any evaluator to survey the entire route if the distance is too long. Hence, walking survey is given a score of 3, which means that the data is considered acceptable.

There are two factors in manual walking survey could have positive/negative influence on sustainability. Since this is the only model which requires walking in the data collection stage, it is considered as an environmentally friendly method. It does not cause air pollution or release harmful carbon emission into the environment. On the other hand, the use of paper data sheets (to record the pavement distress data) makes this method less sustainable when compared with digital data storage. Unlike the walking survey, the other three method in this study do required the usage of vehicle during collect the pavement data. Manual riding survey and OIP methods both are marked with lower score in term of sustainability compare to smart sensor. The reason behind these scores is relied on the repeatability and frequently to the operation of vehicle on each method. The use of automatic and self-powered from smart sensor has eliminated the need to use vehicles during the data collection process. This also means that there is no gas emission. Vehicles are only used periodically to retrieve data via wireless communication from the RF reader. Data retrieval can also be done manually if the distance for data retrieval is not too far.

Once the scoring process for each of the models in five comparison items has been completed, the average scores were computed, and the results are formed. The model with the highest average score is considered as the most effective pavement monitoring model. The smart sensor model has the highest average score of 4.27. The optical image processing model has the next highest score of 3.53. The scores for walking and riding methods are almost similar, with the riding survey showing a slightly better score than walking survey. Both manual surveys (walking and riding) are ranked the lowest, with the walking survey having a score of 2.73 and riding survey a score of 2.53. In addition, a radar chart (Figure 4) was produced to summarize the results of the comparison.

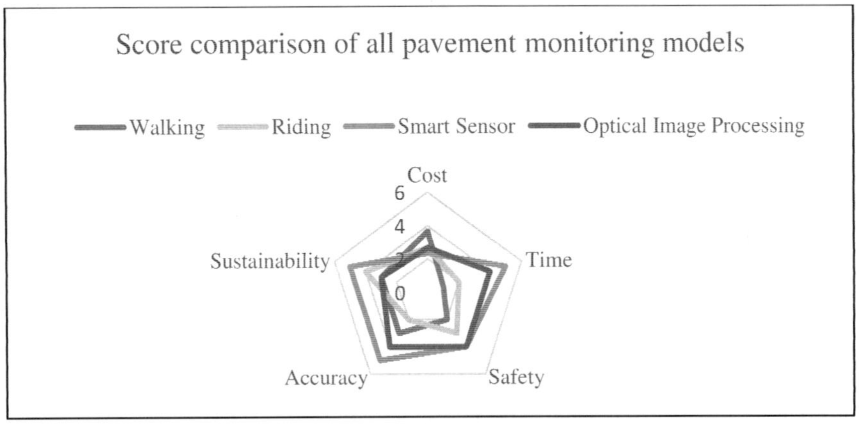

Figure 4. Radar chart for the comparative study

The radar chart above presents the results of the study in a graphical form. There are four polygons with each polygon representing one model in this comparative study. The red polygon represents the manual walking survey model, the yellow represents the manual riding survey model, the green represents the smart sensor model, and the blue represents the optical image processing model. The score value of the models expands from center of chart to the outer range. The maximum value for this rating system is 5. In this study, the polygon with the largest area is ranked as the best monitoring model. It can be seen that the smart sensor model has the largest polygon area; it has a very high score for four indicators, i.e. time, safety, accuracy, and sustainability. The manual walking survey has the highest score for cost indicator although its scores for other indicators are extremely weak in comparison with other models. Although the polygon for the optical image processing model is quite balanced, it does not have a good overall score value when compared with the smart sensor model. It should be noted that although the manual riding survey has a weak accuracy indicator, it is still better than the walking survey in terms of time, safety and sustainability. The radar chart in Figure 4 above has presents a clear picture of the overall findings of the study.

4. Conclusions

Three models were chosen for a comparative study, i.e. manual survey (comprising walking and riding surveys), smart sensor, and optical image processing. Comparison was done base on five indicators, i.e. cost, time, safety, accuracy and sustainability, to determine which model is the most effective in performing pavement monitoring functions. Results show that the smart sensor method is the best overall model for pavement monitoring. As a conclusion, it is hoped that the findings of this study would help relevant agencies make a decision with regard to the most effective pavement monitoring model for their PMS strategy relative to the constraining factors they may have.

5. References

[1] McQueen J M and Timm D H 2005 Statistical analysis of automated versus manual pavement condition surveys *Transportation Research Record: Journal of The Transportation Research Board* **1940**(1) 54–62

[2] Zhang H, Keoleian G A and Lepech M D 2013 Network-level pavement asset management system integrated with life-cycle analysis and life-cycle optimization *Journal of Infrastructure Systems ASCE* **19**(1) 99–107

[3] Salpisoth H 2014 Simple evaluation methods for road pavement management in developing country

[4] Chan C Y, Huang B, Yan X and Richards S 2010 Investigating effects of asphalt pavement conditions on traffic accidents in Tennessee based on the pavement management system (PMS) *Journal of Advanced Transportation* **44**(3) 150–61

[5] Santos J and Ferreira A 2013 Life-cycle cost analysis system for pavement management at project level *International Journal of Pavement Engineering* **14**(1) 71–84
[6] Babashamsi P, Golzadfar A, Yusoff N I M, Ceylan H and Nor N G M 2016 Integrated fuzzy analytic hierarchy process and VIKOR method in the prioritization of pavement maintenance activities *International Journal of Pavement Research and Technology* **9**(2) 112–20
[7] Torres-Machí C, Chamorro A, Videla C, Pellicer E and Yepes V 2014 An iterative approach for the optimization of pavement maintenance management at the network level *The Scientific World Journal* **201** 11
[8] Cafiso S, Graziano A D and Battiato S 2006 Evaluation of pavement surface distress using digital image collection and analysis *Seventh International Congress on Advances in Civil Engineering* pp1–10
[9] Nikolaou A 2016 Study of asphalt pavement deterioration through remote sensing
[10] Wolters A, Zimmerman K, Schattler K and Rietgraf A 2011 *Implementing pavement management systems for local agencies: implementation guide No. ICT-11-094-1*
[11] Xue W, Wang D and Wang L 2012 A review and perspective about pavement monitoring *International Journal of Pavement Research and Technology* **5(5)** 295–302
[12] Attoh-Okine N and Adarkwa O 2013 *Pavement condition surveys-Overview of current practices* (Newark DE USA: Delaware Center for Transportation University of Delaware)
[13] Schnebele E, Tanyu B F, Cervone G and Waters N 2015 Review of remote sensing methodologies for pavement management and assessment *European Transport Research Review* **7**(2) 7
[14] Bogus S M, Song J, Waggerman R and Lenke L R 2010 Rank correlation method for evaluating manual pavement distress data variability *Journal of Infrastructure Systems (ASCE)* **16**(1) 66–72
[15] Pierce L M, McGovern G and Zimmerman K A 2013 *Practical guide for quality management of pavement condition data collection* 170
[16] Lajnef N, Rhimi M, Chatti K, Mhamdi L and Faridazar F 2011 Toward an integrated smart sensing system and data interpretation techniques for pavement fatigue monitoring *Computer-Aided Civil and Infrastructure Engineering* **26**(7) 513–23
[17] Chambon S and Moliard J M 2011 Automatic road pavement assessment with image processing: review and comparison *International Journal of Geophysics* **2011** 1–20
[18] Cooper J S and Kahn E 2012 Commentary on issues in data quality analysis in life cycle assessment *The International Journal of Life Cycle Assessment* **17**(4) 499–503
[19] Babashamsi P 2017 *PAVECO: An integrated life cycle cost analysis and life cycle assessment toward sustainable airport pavement management system* (Doctoral dissertation, Universiti Kebangsaan Malaysia)
[20] Babashamsi P, Yusoff N I M, Ceylan H, Nor N G M and Salarzadeh J H 2016 Sustainable development factors in pavement life-cycle: highway/airport review *Sustainability* **8**(3) 248

Location definition for road asset management systems: case of Sarawak Public Works Department

A Behnam[1] and R Kenley[1, 2]

[1] Swinburne University of Technology, Hawthorn, Victoria, Australia
[2] Unitec Institute of Technology, Mount Albert, Auckland, New Zealand

abehnammanesh@swin.edu.au

Abstract. Sarawak Public Works Department will be expected, in the near future, to change the way their road assets are designed, constructed and managed. Currently, there is a need to identify how road asset information is being collected, stored and retrieved. Location definition is an attribute that provides a simple framework for the identification of current technical requirements to exchange information within the relevant road asset management system. The balance of this paper presents an overview of the current road asset management systems and their location definition. The findings are results of the analysis of asset management systems and standards as well as the verbal engagement with PWD Sarawak experts in this field. The findings suggest that emerging systems and processes are moving from solely linear referenced models towards incorporating geospatial location data at different dimensional levels. Given this evolution, in this paper, road asset management systems are classified into three location definition categories considering their location referencing methods and their levels of dimensional detail. However, further research is required within PWD Sarawak to identify required location referencing in order to improve automation of the information exchange between the emerging BIM models, geospatial systems and the automated data collection tools.

1. Introduction

Under the Sarawak Socio-Economic Transformation Plan (SETP), road construction in Sarawak is of great importance [1]. Sarawak has set a target of a highly connected road network by 2030 [2]. While many initiatives have been implemented to accelerate the state growth, more are in the planning stage. Examples include several road developments such as the Pan Borneo Highway Phase 1/2 and Sarawak Coastal Highways, bridges including ASEAN, Batang Lupar and Batang Rambungan as well as future rail developments such as Kuching Light Rail Transit (LRT). To assist in meeting this target, a more complex integrated digital engineering system is being considered to improve the information exchange and interoperability between asset management systems, models and tools [3].

The adoption of Building Information Modelling (BIM) in Malaysia was primarily promoted by the private sector in early 2000 [4]. However, BIM has only become a popular term when Public Works Department (PWD) introduced BIM in construction planning for public buildings in 2007 through their BIM Manual and Guidelines [5]. PWD which is the largest technical department under the Ministry of Works Malaysia (MOW), is responsible for all design, planning, construction, development and maintenance of public infrastructure and government buildings in Malaysia.

IOP Conf. Series: Materials Science and Engineering **512** (2019) 012040 doi:10.1088/1757-899X/512/1/012040

In the past years, PWD together with the Construction Industry Development Board (CIDB) has formed the BIM Steering Committee, BIM Roadmap, and BIM Portal, involving policy makers, practitioners, and academicians [6]. While, there has been an increasingly demand for adoption of BIM for vertical construction, little emphasis has been placed on the adoption of digital engineering practices for horizontal infrastructure assets such as roads and rail.

1.1. Problem statement

Sarawak Public Works Department will be expected, in the near future, to change the way their road assets are designed, constructed and managed. Digital engineering practices in construction sector, such as Building Information Modelling (BIM) and Geographic Information System (GIS) will provide a new method for capturing and managing road asset information [7, 8].

The lifecycle information of road assets has generally been categorised differently within each functional unit [9]. Globally, road authorities have developed multiple units with multiple functional and technical information requirements. This silo of organisational structures within road authorities and their relevant stakeholders has been identified as a barrier to the integration and exchange of information between the lifecycle functions of road asset [10]. This is currently the case for Sarawak Public Works Department. Thus, the Sarawak Public Works Department has entered into a collaborative research project with Swinburne University of Technology, to identify current road asset lifecycle information requirements that could be changed to suit an integrated digital asset management system.

1.2. Research objective

Currently, there is a need to identify how road asset information is being collected, stored and retrieved. The first step is to gather information on current digital asset management systems and processes related to the road asset lifecycle information. This can be done by focusing on an attribute or a definition that could provide an indication of what lifecycle asset information has been or has not been exchanged or integrated into asset management systems.

Location definition is an attribute that provides a simple framework for the identification of the underlying information that is currently recorded and managed into the relevant asset management system. The aim of this research is to look into the digital engineering practices in road asset management systems of Sarawak Public Works Department from the lens of location definition. The findings of this exploration aim to present a location definition model that identify current technical information requirements that Sarawak Public Works Department needs to consider for having an integrated digital asset management system. Consequently, this emergent exploratory research revolves around the following question: to explore the location definition for management and integration of information in road asset management systems and processes of Sarawak Public Works Department

There has been an array of definitions within academic articles and industry documents for concepts and terms that are relevant to this research. However, the following definitions in the Table 1 are the most appropriate definitions for this research.

Table 1. Definition of Key Terms and Concepts

Term	Definition	Source
Digital Engineering	A collaborative way of working, using digital processes, to enable more productive methods of planning, designing, constructing, operating and maintaining built assets.	[11]
BIM	Building information models (note models not modelling) are parametric and 3D objects that can be used as a shared knowledge resource by different stakeholders involved in lifecycle of a building or infrastructure asset.	[12]

Table 1. Definition of Key Terms and Concepts (continued)

GIS	A digital information system that is designed to work with data referenced by geographical coordinated-based features	[13]
Object	Any kind of object or structure which can reasonably be related to a geographical location, such as points of interest (POIs), roads, places, bridges, buildings or agricultural areas.	[14]
Location	A virtual object that is temporal, graphical and model-independent. The virtual object can be a point, line, path or area, representing a physical place on a road network.	[8]
Location Definition	Identification of a location in terms of its level of dimension, position and geometry.	[15]

2. Research Approach

The methodology adopted in this qualitative research has been exploratory and emergent [16, 17]. At the core of this methodology, it was aimed to actively search for documents of digital *systems* or *processes* related to road asset lifecycle information [18]. The documents were analysed to identify their location definition and types of information which they contain of [19]. The documents were both in electronic and non-electronic format and included systems databases, guidelines, policies, design models, etc. The asset information were classified into structured data (e.g. relational databases), semi-structured data (e.g. spreadsheets), or non-structured data (e.g. images/sketches or maps) [20].

The analytical process has been inductive, with the aim of exploring location definition for various asset management system or process. This has involved in identifying the level of location dimension (e.g. point, polyline), position of a location with respect to other locations in the system/process as well as the geometry of location (e.g. straight line or horizontal/vertical curve). Table 2 shows an overview of the data collection methods and framework for data analysis in this study.

Table 2: Overview of Data Collection Methods

Verbal Engagement			Written Documents		
Verbal engagement with 25 expert informants in the area of Asset Management (AM), BIM and GIS			24 Documents (e.g. manuals, standards, progress reports) Asset Management *systems* or *processes*		
Sarawak Public Works Department-owned road assets			Location Definition		
6 internal branches	2 consultants BIM/GIS/AM	1 project manager and 3 contractors	Level of Dimension	Position/Topology	Geometry

3. Road Asset Management Systems

This section provides a summary of asset management systems and processes related to Sarawak PWD-owned road assets. These systems and processes are analysed for identifying their location definition and hence types of information which they record.

3.1. RMMS and MARRIS

The Road Maintenance Management System (RMMS) developed in 1987, is a general road maintenance system for Sarawak State roads. It is designed to manage and monitor concessionaires' maintenance operations by implementing work methods and controlled maintenance expenditures. In the current version of Road Maintenance Management System (RMMS), pavement locations are recorded at every 100m chainage to report pavement attributes such as material, condition or any assigned maintenance activities by a road party. However, majority of non-movable assets such as road furniture, culvert, bridges, riverine infrastructure, drainage and irrigation and others are still recorded using manual registers or spreadsheets. Several improvements have been considered for the new version of RMMS. Figure 1 shows some of the previous, current and proposed features.

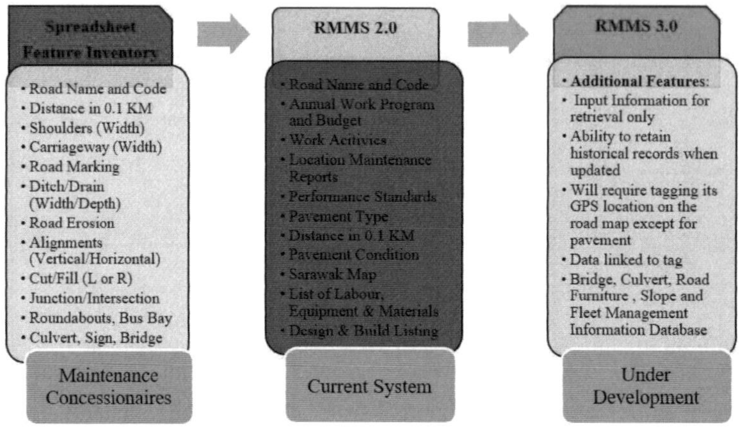

Figure 1. The evolution of PWD Sarawak RMMS

In the current discussions within PWD Sarawak, it is evident that there is an increasing interest towards the use of Global Positioning System (GPS) coordinates tagging of bridges, culverts and road furniture on the road map, which can populate the existing GIS database. However, current or future systems do not yet capture geospatial polylines for pavements and other polygon assets such as drainage and slopes. On the other hand, the GIS unit in the organisation captures capture and record polyline and polygon geospatial locations for populating the condition data. The GIS database is also used for the purpose of reviewing the wayleave application from other infrastructure departments in Sarawak. Nevertheless, according to the expert informants, there is little integration between road asset management data and the geospatial system. Also, there is still no integration or information exchange between the GIS databases managed by Sarawak Land & Survey Department and other infrastructure departments.

Another system containing some pavement data is the Malaysian Road Record Information System (MARRIS) that is designed for accurate road length measurement and for eliminating double registration of State Roads, the Municipal Roads, the Rural Roads, Island Roads and the Village Roads. The system records the GPS coordinates of the start and ending points as well as the width of road carriageways, shoulders, drainage and roundabouts. The manual published by the Ministry of Finance specifies the reference points for the length and width measurements of single and dual carriageways, shoulders as well as the circumference of roundabout circles.

Considering the increasing capability of GPS and other remote sensing technologies and tools, road asset management systems have realized the benefits of GIS mapping of road assets. An example is a pilot Sarawak Integrated Asset Management System (IAMS) for 120km of Federal and State roads in Sarawak.

3.2. Pilot road asset management system

The pilot IAMS is designed to record and locate various features such as road pavement and road furniture using GPS coordinates via Radio Frequency Identification (RFID) System for road asset tracking. In addition, a Road Asset Condition Assessment System (RACAS), a data capturing system (HD camera, GPS, G-sensor) is adopted for capturing the pavement condition data which has been populated on a web-based GIS database management system (shown in Figure 2). Roads centreline polylines and road furniture point assets are located with GPS coordinates which are then converted to PWD road chainage and road coding system.

Figure 2. Road Furniture GIS Mapping & Pavement Condition Captured by RACAS

IAMS and other individual systems, managed by road maintenance concessionaire in Sarawak, are influenced by digital systems such as RAMS (Road Asset Management System) developed for Toll Expressways in West Malaysia. RAMS is designed for creating a complete register of asset inventory, condition and work information for both discrete assets (e.g. bridge, slopes) and continuous assets (e.g. guardrails, pavements). Unlike IAMS, RAMS provides the geospatial polygon location of certain assets which are linearly referenced and uniquely identified by an asset ID. Nevertheless, BIM or so called Highway Information Modelling (HIM) is gaining momentum in the design and construction phase of the Pan Borneo Highway project in Sarawak. These processes and the resulting models provide a significant source for capturing the geometric and geospatial location information that can be accommodated into current and future road asset systems.

4. Highway Information Modelling and Asset Tagging

The term Highway Information Modelling (HIM) has been an important in discussions within Sarawak and Federal PWD over the past two years. HIM reflects some of the digital engineering concepts and practices that are implemented through the application of both BIM GIS mapping and UAV technology in planning, designing and monitoring the construction of the on-going Pan Borneo Highway project. In this exercise, 2D CAD drawings are converted into construction 3D models and ultimately as-built models (shown in Figure 3). The 3D models provide simulation of construction objects (e.g. useful for clash detections) and also facilitate the extraction of quantities for construction of pavement, drainage, and sewage. Considering the lack of open data exchange standards for roads (e.g. Industry Foundation Classes or IFC), the 3D models are converted to i-Models to be used within ProjectWise and Navigator environments.

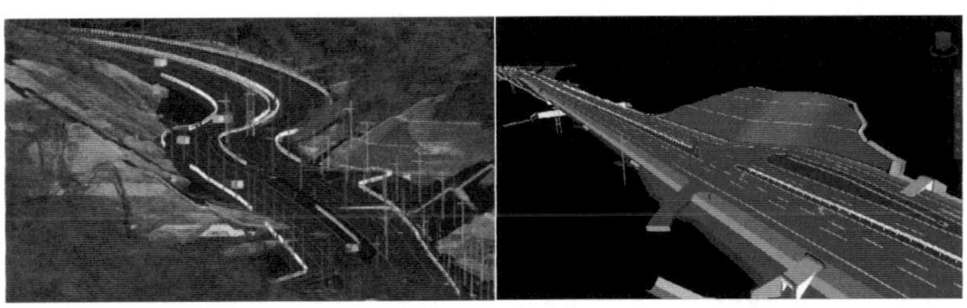

Figure 3. 3D Modelling using Autodesk + Bentley Microstation + Bentley i-Models

In the current discussion, the AssetWise network database and linear referencing engine is to be used to provide linear referencing capability to ProjectWise data (Figure 4). Asset or project location records can be stored and maintained in the AssetWise database in the form of network locations. Asset or project property attribute records and associated documents are stored and maintained the ProjectWise database. The special features in AssetWise can be used to display the ProjectWise record on the AssetWise map. A link back to the ProjectWise database means that the map interface can be used to query attributes stored in ProjectWise.

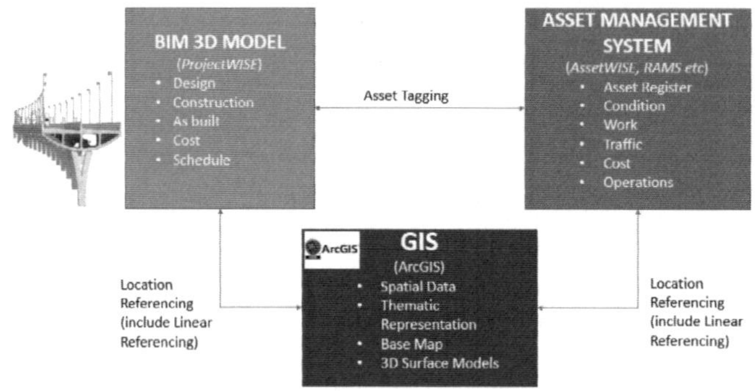

Figure 4. Integration of BIM Asset Tagging, GIS, and Asset Management System

An important requirement is to move design and construction data into the operation system. Design data will include alignments, asset inventory, bridge details, and drawings, 3D models and other data valuable to operations. The geometric alignment data is transferred by tracing function in the Linear Referencing System (LRS) tool. Assert inventory is also entered into the forms, or bulks loaded from files. Bridge detail and drawings, 3D models and other data valuable to operations can be accessed from ProjectWise. AssetWise supports data to be stored at the lane level as required to describe the lane, its construction history, condition, repairs from maintenance through rehabilitation/reconstruction and other affiliated data as identified by user. Data collected against the network using either linear referencing or sections can be retrieved against the chosen linear reference or section definition.

Asset tagging of HIM needs to be compatible with a framework of a federal Public Works Departments asset management system called MySPATA system (Government Asset Management System, in Malay). MySPATA is developed for registering necessary information about the federal government immovable assets including building and infrastructure assets (Figure 5 and 6). MySPATA system is supported by an asset tagging system called SKATA (Codes of Asset Tagging System, in Malay). Prior to this standard, there were various data collection methods used by agencies and there was no specific guideline in collecting assets location data. This caused the data collection being inaccurate, inconsistent and unsystematic.

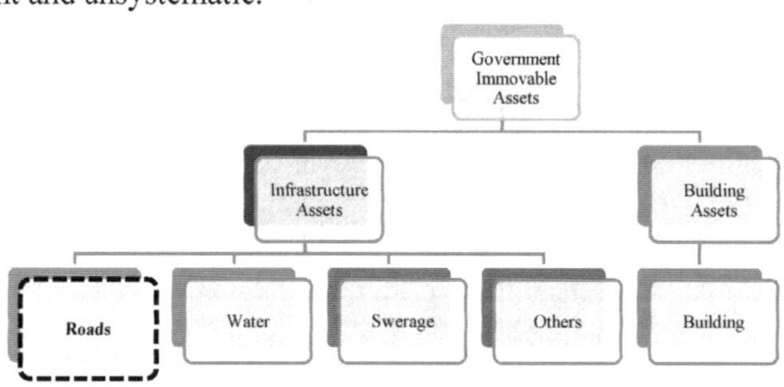

Figure 5. Federal Government Immovable Assets

Assets tags for physical road assets is divided into two codes 1) DPA which is the first level code structure and contains basic information about the owner and location of the asset 2) DKA which is the second level code structure and contains more detailed information about the road assets/components and their exact location. DPA tag is located at the start and final locations of each road and DKA tag is located at every 1 km. DKA represents the location of each asset and its components in terms of each road section number, subsection number, position (e.g. Left or Right), category and component code (shown in Figure 7).

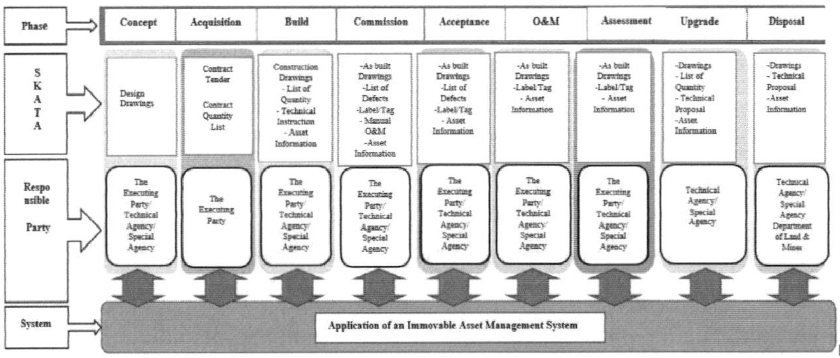

Figure 6. MySPATA Application and Lifecycle Information Requirements

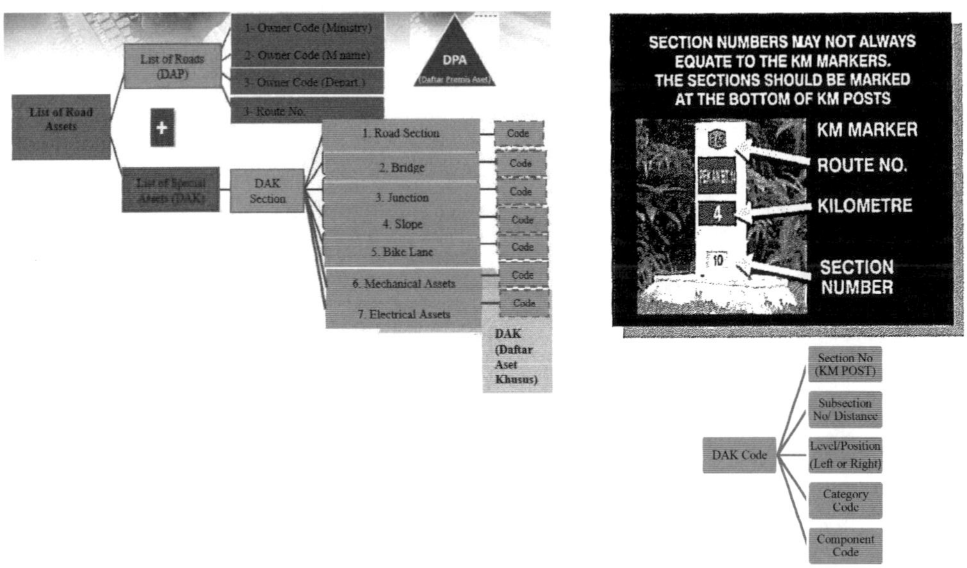

Figure 7. DPA and DKA Tags for Asset Types

5. Definition Categories and Sophistication Levels

The overview of road asset management systems and processes for Sarawak PWD road assets suggest that the linear referencing and GPS coordinates are the two main location referencing methods used for referring to the position of a location. On the other hand, the dimension of representation of assets vary from zero-dimension (point locations) to 2D dimension (polygon representations) and 3D dimension objects produced from BIM models. In addition to GPS coordinates and linear referencing, geometric location definition is also found in the as-built BIM models. Table 3 shows various categories of location models which result from different location referencing types (providing positional and geometrical information) and level of dimensional details for locations representations.

Table 3: Location Definition for Road Asset Management System

Road Asset Management Systems		Level of Detail for Asset Representation			
Location Definition Categories		Zero Dimension Point	One-Dimension Polyline	Two-Dimension Polygon	Three-Dimension 3D object
Location Referencing	Geospatial Coordinates	– 1 –	– 4	– 7 –	– 10 –
	Topological Linear Ref.	– 2 –	– 5 –	– 8 –	– 11 –
	Geometric Alignment	– 3 –	– 6–	– 9 –	– 12 –

It can be argued that current road asset management systems which are either fully adopted or considered, can be classified into three location definition sophistication categories depending on their level of dimensional details, and positional referencing (no geometric yet incorporated into the system). These categories include 1) Zero-dimensional linear referenced information 2) One dimensional geospatial polylines asset representation with zero-dimensional linear referenced information and finally 3) Two dimensional geospatial polygons asset representation with zero-dimensional linear referenced information. This means while some locations are captured and visualised using GPS coordinates, the information associated to assets are still tagged and located using zero-dimensional linear referencing.

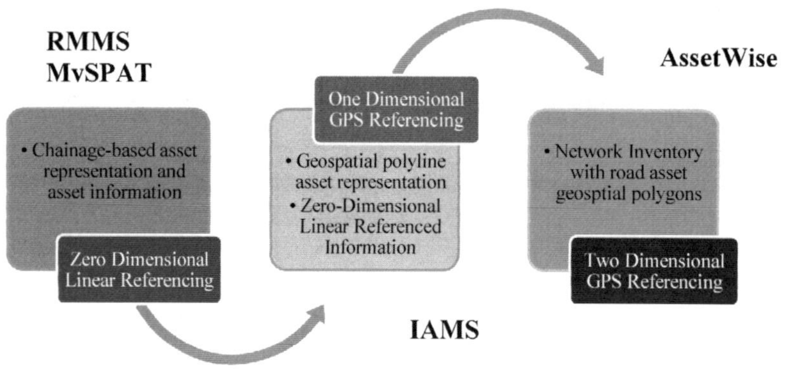

Figure 8: Location Definition Sophistication Levels in Current PWD Asset Management Systems

In order to have an integrated asset management system which records various types of construction and maintenance data, it is not possible to rely on only one type of location referencing method and a specific level of dimension. Therefore, the integration of different location referencing methods and level of dimension is essential for creating an integrated asset management system. Higher level of sophistication means that systems can accommodate higher level of location dimension and also are potentially capable of capturing geospatial information. Having said that, none of the systems except BIM models are storing geometric data. The transfer of geometric information is currently done by tracing the geometric alignment and GPS tagging of road centreline and maybe other polyline/polygon assets.

6. Conclusions

The findings highlight that considering the increasing use of 3D geometric BIM models and geospatial information in individual asset management processes in Sarawak, more detailed and sophisticated location definitions models for current and future asset management systems should be considered. In particular, there is a need for incorporating higher level of dimensional details and higher variety of location referencing methods for collecting and recording information related to road assets lifecycle. Furthermore, it is essential to explore the ways to improve the automation for two-ways information exchange between the as-built HIM models and the geospatial systems. It is also recommended that future research explore the location definition of information that is captured by the automated data collection technologies and tools currently adopted in Sarawak road projects.

7. References

[1] Khan R, Liew M and Ghazali Z 2014 Malaysian Construction Sector and Malaysia Vision 2020: Developed Nation Status *Procedia-Social and Behavioural Sciences* **109** 507-13
[2] Koen V 2012 *Malaysia's Economic Success Story and Challenges Economics Department Working Papers* (Paris: France Organisation for Economic Co-operation and Development)
[3] Kenley R, Harfield T and Behnam A 2016 BIM Interoperability Limitations: Australian and Malaysian Rail Projects *MATEC Web of Conferences* 001021-26
[4] Harun A 2016 Existing Practices of Building Information Modeling (BIM) Implementation in the Public Sector *Supply Chain Manag.* **5** 166-77
[5] Latiffi A, Brahim J and Fathi M 2016 Transformation of Malaysian Construction Industry with Building Information Modelling (BIM) *MATEC Web Conference* **66** 22-30
[6] Zainon N, Mohd-Rahim F and Salleh H 2016 The Rise of BIM in Malaysia and Its Impact towards Quantity Surveying Practice *MATEC Web Conference* **66** 60-68
[7] Bradley A 2016 BIM for Infrastructure: An Overall Review and Constructor Perspective *Automat constr.* **71** 139-52
[8] Kenley R and Harfield T 2016 *Scoping Study for a Location Referencing Model to Support the BIM Environment AP-R568-18* (Sydney: Austroads Publication)
[9] Van der Velde J, Klatter L and Bakker J 2013 A Holistic Approach to Asset Management in the Netherlands. *J Struct. Infrastruct. E.* **9** 340-48
[10] Bennett P 2013 *Emerging Digital Mapping Requirements for C-ITS AP-R432-13* (Sydney: Austroads Publication)
[11] Sparrow M, Gallagher A and Patis M 2017 Digital Engineering for Bridge Infrastructure Projects *Austroads Bridge Conference* Melbourne, Australia
[12] Aram S and Eastman C 2013 Integration of PLM Solutions and BIM Systems for the AEC Industry *Proceedings of 30th International Symposium of Automation and Robotics in Construction and Mining in Montreal* 1046-55
[13] Shaw S 2013 Geographic information systems for transportation: from a static past to a dynamic future *Ann. GIS.* **16** (3) 129-40
[14] Hackeloeer A 2014 Geo-referencing: A Review of Methods and Applications *Ann. GIS.* **20**(1) 61-69
[15] Rehrl K 2017 Generation of Meaningful Location References for Referencing Traffic Information to Road Networks Using Qualitative Spatial Concepts *Progress in Location-Based Services* ed G Gartner and H Huang (Heidelberg: Springer) pp 173-91
[16] Saunders M, Lewis P and Thornhill A 2007 *Research Methods for Business Students* (Upper Saddle River: Financial Times/Prentice Hall)
[17] Hesse-Biber H and Leavy P 2010 *Handbook of Emergent Methods* (New York: Guildford Press)
[18] Sinha R and Hassan A 2014 Respondents versus Informants Method of Data Collection: Implications for Business Research *MAJESS* **2** 1-13

[19] Transportation Research Board 2016 *Civil Integrated Management (CIM) for Departments of Transportation* ed W O'Brien (Washington DC: the National Academies Press)
[20] Vossebeld N and Hartmann T 2016 Modeling Information for Maintenance and Safety along the Lifecycle of Road Tunnels *J Comput. Civil Eng.* **30** 4-13

Acknowledgments
The authors would like to thank Swinburne University of Technology Melbourne Sarawak Research Collaboration Scheme ID number 3000004617 for providing funding for this research. Also, special thanks to Sarawak Public Works Department for facilitating the data collection process for this research. Last but

IOP Conf. Series: Materials Science and Engineering **512** (2019) 012041 doi:10.1088/1757-899X/512/1/012041

CONie is not COBie: information exchange differences between network and building handover to asset management

R Perumpilly[1], R Kenley[1,2] and T Harfield[1]

[1]Swinburne Business School, Faculty of Business and Law Engineering, Swinburne University of Technology, John St, Hawthorn VIC 3122, Australia
[2] Department of Construction and Services, Unitec Institute of Technology, Carrington Road, Mount Albert, Auckland, New Zealand

tharfield@swin.edu.au

Abstract. This desktop review of road asset management information requirements at the handover phase of construction studies the different needs of vertical and horizontal infrastructure. The paper considers three specifications for construction information at handover to asset management. COBie (*Construction to Operations for Building information exchange*), and the proposed CONie (*Construction to Operations for Network information exchange*). COBie has been successfully used for vertical construction since 2007, but the specification was not intended for use in road construction. However, the UK drive to introduce digital information exchange for all new government projects resulted in in an attempt to development a hybrid COBie-for-All specification to include both vertical and horizontal infrastructure. However, differences in definitions and utilisation of data linked to specific attributes, such as location, cannot be force fitted into a dual-use information exchange specification. Thus, the development of a single-use specification intended for horizontal network assets is deemed the only logical option. That is the purpose of the CONie research project into the development of a handover specification to meet the needs of asset management of a number of road authorities from Australia and New Zealand.

1. Introduction

As part of the drive for governmental efficiency, the use of building information modelling (BIM) has been mandated for public works, for example, Singapore, Japan, UK and Malaysia [1, 2, 3]. This is because the concept of Integrated Project Management suggests value-added is gained by sharing and exchanging information utilising a BIM environment [4].

Integrated Project Management could be applied to the obvious bottleneck in the construction process, handover. Currently building owners globally continue to allocate resources to both finding and converting construction information at handover [5]. Their major problem is the need to reduce the project information for useful input into their facilities management systems. COBie (Constructions to Operations Building information exchange) was developed to simplify this process [6]. COBie, as the name suggests, was specifically designed to focus on the asset management for buildings.

COBie has been successfully used in vertical construction since 2007. The specification was developed in the USA as a part of their contribution of the expansion of the global effort to find an international standard for Integrated Project Delivery (IPD) [7].

One of the reasons that COBie has been successful is because during development, facilities managers (the expected end-users) were asked what information they needed [6]. The end-user information supplied the constraints for the specification, which in turn led to functionality in BIM enabling systems [2].

Obviously, a handover specification that was developed specifically for road asset management could provide horizontal infrastructure the same value-add as COBie does for vertical infrastructure. Thus, the industry need to conceptualise a handover specification is the aim of a government funded research project with road authorities in Australia and New Zealand [8, 9].

This paper outlines the fundamental differences between construction information required at handover for COBie and a proposed CONie. The structure of this paper begins with details of the research method in Section two. Section three explains some COBie specification attributes for buildings. Section four focuses on the lack of development of a proposed COBie-for-All. Section five discusses the need for an information exchange for horizontal infrastructure. CONie, a proposed information exchange for road networks, is outline in Section six, followed by the conclusion.

2. Research design

The purpose of this historical desktop research is to explore the development of COBie and the rationale for the proposed CONie. Thus, a three stage method was used [10].

2.1. Stage one

Online (open access) and university library (limited access) resources were used as the datasets. Documents found included: academic journal articles, academic conference papers, government reports, international standards and industry research publications (presentations or talks were excluded).

The first search terms were: COBie and BIM. The search was initially based on two timeframes 1990-1998 and 1999-2007 because of the large number of digitalised documents that became available early in this century.

The development of the BIM technology is the topic of a large range of documents from a variety of points of view. For this project, a limited number of seminal references and those that feature COBie were considered relevant. Although COBie claims to be integrated into 170 software programs, little academic literature was found [11, 12]. However, industry research reports and guidelines on BIM implementation often includes COBie usage and/or recommended modifications [1, 2, 13].

2.2. Stage two

The search was extended to include 2008-2018 in order to find additional documents for COBie. This also included a new search terms COBie for All and CONie. However, only a small subset of items were found.

2.3. Stage three

Stage three was a systematic review of a documents of a select dataset containing text related to COBie, COBie for All, and CONie. The review was only to verify that the documents used for this report actually discussed in some detail these three types of digital information exchange.

For example, topics such as: COBie as part of the Industry Foundation Classes (IFC), the importance of appropriate definitions of location in the different types of models, and difference in the needs of asset management for vertical and horizontal infrastructure.

Little was found about CONie. This is explained because this information exchange open specifications is still in the early stages of development, and should be considered a 'concept' rather than a specification.

3. COBie: handover for vertical infrastructure

SMART international has developed definitions for BIM based on object libraries and IFC standards [14]. The international standard, ISO16739, specifies a conceptual data schema and an exchange file format for BIM data [15]. Increasingly, BIM environments are being used to design 3D models for construction projects. IFC was initially developed specifically for the vertical infrastructure.

Traditionally, the facility management information was created after construction. On completion of the project all project materials (drawings, manuals, schedules, etc.) were delivered to the facility owner/operator [5]. However, with the increase in the use of digital modelling, devising a new method was obviously necessary and resulted in the development of COBie [16].

COBie *(*Constructions to Operations Building information exchange*)* is a subset of IFC2. COBie defines a precise set of information needed to solve a specific problem at a specific point in the construction life cycle: handover [17].The required information for building facilities management is small subset of all construction information. It is needed to support the operational requirements and locate assets in a building. COBie information exchange format has been achieved by extracting the required information from the BIM file or an IFC schema and transporting it into a standard COBie schema, ready to import into a facility management system.

4. COBie-for-All: handover for buildings and roads

In May 2011, the UK government published the *Government Construction Strategy* paper, announcing the intention of requiring Level 2 BIM by 2016 (collaborative 3D BIM with all project and asset information, documentation, and data being electronic) for all government infrastructure projects.

Both the UK drive to introduce digital information exchange for all new government projects and the lack of interoperability between object-based design (vertical infrastructure) and string-based design (horizontal infrastructure) was the business case for the development of a new COBie-for-All specification.

Based on the initial studies conducted between 2011 and 2013, a draft report was released for public comment on 15 October 2013 recommending essentials be included in COBie for use in Civil and Infrastructure projects. The title of the document provides a simple explanation of the contents:

COBie for All*:* Required Information for Facility Ownership Buildings & Civil/Infrastructure, Understanding How COBie Works in the UK Infrastructure Market. Currently, this document does not appear to be available from an open access repository online.

Clearly the 60-page manual of 'problems' and 'solutions' aims to stretch the initial COBie specification for buildings to include roads. Three of the 12 problems are directly related to issues of location for road asset management.

The BIM Task Force that developed strategies to continue British development of the COBie-for-All information exchange has since been disbanded [13]. In addition, personal communication in 2018 with British contractors tendering for government infrastructure, indicate they have not been required to use COBie-for-All. Therefore, it is assumed that the development of a vertical/horizontal infrastructure information exchange specification is no longer being developed.

However, the global drive to increase the implementation of BIM platforms continues, thus, the need for a horizontal infrastructure specific digital information exchange specification.

5. Why we need a network not a building information exchange

A digital information exchange specification for roads requires a horizontal infrastructure perspective. One example of this type of perspective is the different ways that location is defined in models of building or roads. Therefore, in the development of an information exchange specification for horizontal infrastructure, the specification elements for location must be based on appropriate definitions of location [9].

5.1. Location information in buildings

The IFC schema defines a set of consistent data representations of building information exchange between the various software applications. The IFC vertical infrastructure model includes location information as an essential part of the definition. An object-view of assets requires precise geometrical information about the spaces and equipment required in the detailed design stage that are essential for construction.

Continuous Objects: Building (Floors and Spaces) components tend to be discrete entities. In COBie there is an assumption that all components are discreet elements because discrete elements can be identified by their physical location limits.

Geospatial Location: Containment is achieved by the assignment of a component to a space where spaces or rooms are contained on the floors without defining their precise geometry [13]. Therefore, building location is only an area, not a line or a point.

5.2. Location for networks

Networks of pipes, culverts or roads have a linear aspect that requires location to be very specific over a long distance. Therefore, mechanisms for network location integration into BIM models remains problematic.

Continuous Objects: In continuous linear entities, such as roads, the variation of horizontal infrastructure elements of the continuous assets needs to be segmented with defined limits [18].

Geospatial Location: Obviously, road projects have larger spatial extents than any individual building. Therefore, Geographical Information Systems (GIS) play a more significant role in defining asset management needs. However, it is important not to confuse the concept of 'spaces' with the concept of 'location'. It is therefore, important to understand the network information context with the multiple location definitions of points, lines or areas and how they can be combined into identified segments [8].

6. Proposed CONie: handover for horizontal infrastructure

Currently there is no handover information exchange specification that can be used in road asset management systems. However, it is obvious that the same problem for vertical infrastructure of too much information, in too many formats, is the same for horizontal infrastructure. Thus, the development of an information exchange specification that can limit the amount of information required, must begin with identifying what varieties of information road asset managers currently access.

6.1. Requirements for a network information exchange model

Road asset management includes several different types of systems. These systems are used for different types of ongoing projects that impact on specific parts of the road network. To ensure best practice and code compliance, road authorities have access to extensive technical libraries of standards and references for regional or national standards. This means that road asset systems utilise several different types of management resources.

This complexity must be captured in the development of an information exchange specification. Therefore, a model of construction to network information exchange needs to consider five types of information:

- Systems: An interconnected network of road resources of specific types, typically defined by number of lanes and median type, and managing jurisdiction.
- Projects: A set of work with a defined start and end. Projects require engineering plans and specifications and are executed by either road authorities or contractors.
- Jobs: A recurring set of projects needed to keep the network operating at appropriate levels of performance. Jobs typically do not require engineering plans and specifications because they are based on standards-of-practice.
- Resources: A set of tools, materials, labour, and training needed to perform jobs that can be internal or external to a road authority thus, requiring knowledge of how to transfer the information that might be useful for asset management functionality.
- Standards: A set of templates for roadway profiles (and associated assets) and jobs, as well as engineering details necessary to insure consistency of projects within the network. Specific

departments within a road authority usually can develop standards for processes, products and information. In addition, national or international standards can in applied.

6.2. What do asset managers need at handover?

Handover is considered to be the set of all documents that must be provided by the contractor [16]. However, even with IPD, by the end of construction many types of documents will have accumulated, and not all of them will have information useful to road asset managers.

For example, pavement information will be accurate to the sub-metre. Other structures such as bridges and tunnels will be accurate to the millimetre. This is usually much more data than is necessary for asset management functions.

Therefore, the information that is exchanged should be based on current and foreseeable future data requirements for performance-based asset management and operations of the road network [17]. This will ensure that handover is implemented in a reasonable amount of time, and with the least amount of disruption to network Maintenance and Asset Management information and systems.

7. Conclusion and future research

The use of building information modelling (BIM) has been mandated for public works as part of the drive for governmental efficiency. The concept of Integrated Project Management suggests value-added is gained by sharing and exchanging information utilising a BIM environment. This would be a means to eliminate the information exchange bottleneck in the construction process, handover to asset management.

Therefore, this desktop review has provided a comparison of three related information exchange specifications for construction information at handover to asset management. COBie (*Construction to Operations for Building information exchange*), COBie-for-All and the proposed CONie (*Construction to Operations for Network information exchange*).

Of the many governments mandating the use of BIM, only the UK has attempted to modify the original use of COBie to include both vertical and horizontal infrastructure. However, development of this hybrid, COBie-for-All specification, appears to have been abandoned.

Proactive road authorities in Australia and New Zealand are currently working on a solution through the development of a specification intended for use in road asset management systems.

8. References

[1] Kaneta T, Furusaka S, Tamura T and Deng N 2016 Overview of BIM implementation in Singapore and Japan *Journal of Civil Engineering and Architecture* **10** 1305-12

[2] NBS 2017 *National BIM Report National Building Specification Report* Available from https://www.thenbs.com/knowledge/nbs-national-bim-report

[3] Latiffi AA, Mohd S, Kasim N, Fathi MS 2013 Building Information Modeling (BIM) application in Malaysian construction industry *International Journal of Construction Engineering and Management* **2** 1-6

[4] Kent D C and Becerik-Gerber B 2010 Understanding construction industry experience and attitudes toward integrated project delivery *J. Constr. Eng. Manage.* **136** 815-25

[5] East B, Kenley R and. Harfield T 2015 CONie: lessons to be learned from COBie specifications and standards *SBEnrc Project 2.33 New Project Management Structures Briefing Paper* (Perth: Sustainable Built Environment National Research Centre)

[6] East EW (2007) *Construction operations building information exchange (Cobie): requirements definition and pilot implementation standard, Construction Engineering Research Laboratory TR-07-30 ERDC/CERL* (Champaign IL: US Army Corps of Engineers, Engineer Research and Development Center)

[7] Cerovsek T 2011 A review and outlook for a 'Building Information Model' (BIM): a multi-standpoint framework for technological development *Adv.Eng. Inform.* **25** 224-44

[8] Kenley R and Harfield T 2018 *Scoping Study for a Location Referencing Model to Support the BIM Environment* (Sydney: Austroads)
[9] Kenley R and Harfield T 2016 *Location Importance in Infrastructure: New Project Management Structures SBEnrc Project 2.33* (Perth: Sustainable Built Environment National Research Centre)
[10] Fielding NG 2008 *The SAGE Handbook of Online Research Methods* (ebook: Sage Publications)
[11] Becerik-Gerber B, Jazizadeh F, Li N and Calis G 2011 Application areas and data requirements for BIM-enabled facilities management *J. Constr. Eng. Manag.* **138** 431-42
[12] Zhao X 2017 A scientometric review of global BIM research: analysis and visualization *Automat.Constr.* **80** 37-47
[13] Scarponcini P, Nisbet N and Bew M 2013 *COBie for All Required Information for Facility Ownership Buildings & Civil/Infrastructure: Understanding how COBie works in the UK infrastructure market* (London: BIM Task Group)
[14] Eastman C, Teicholz PR, Sacks R and Liston K 2008 *BIM Handbook: A Guide to Building Information Modelling for Owner, Managers, Designers, Engineers, and Contractors* (New Jersey: John Wiley and Sons)
[15] ISO 2013 *Industry Foundation Classes (IFC) for Data Sharing in the Construction and Facility Management Industries* (Online: International Standards Organization)
[16] East B, Construction-Operations Building information exchange (COBie) *Whole Building Design Guide* Available from http://www.wbdg.org/resources/construction-operations-building-information-exchange-cobie
[17] East EW, Nisbet N and Liebich T 2013 Facility management handover model view *J. Comput. Civ. Eng.* **27** 61-67
[18] ISO 2012 *Geographic Based Information-Linear Referencing* (Online: International Organization for Standardization)

Acknowledgements
This research has been supported by the Australian Research Council Linkage Project 116100524 and the Austroads ASP2112 Asset Management Harmonisation Project. Appreciation to Bill East for his contribution.

IOP Conf. Series: Materials Science and Engineering **512** (2019) 012042 doi:10.1088/1757-899X/512/1/012042

Asset management innovation for rigid pavements

F Varela and R Pacheco-Torres

Department of Civil Engineering: Construction, Infrastructure and Transport, Faculty of Civil Engineering, Polytechnic University of Madrid, Madrid, Spain

E-mail: fernando.varela@upm.es

Abstract. Today, the new pavement management tools encourage preventive action, whereas more traditional approaches focus on acting on observed damage. An adequate initial evaluation is essential to properly manage pavement maintenance. In this regard, non-destructive tests such as deflection tests make it possible to obtain first-hand information without altering the condition of the pavement, and their use is becoming increasingly more common. This paper presents a study of the structural evaluation of rigid pavements based on measured deflections. Several aspects are identified which will make it possible to improve these procedures. A 62-kilometre stretch a motorway in Mexico has been used as a case study. A campaign of on-site measurements was carried out using a Falling Weight Deflectometer (FWD). The deflection values were used in several iterative methods with calculation programs and manual analytical methods, comparing the results obtained. The elastic moduli of the pavement layers were determined. These are the main parameters for calculating the remaining life of the structure, which facilitate decision-making around pavement management. The results obtained in this study demonstrate the usefulness of back-calculation methods for analysing the structural condition of rigid pavements as an initial stage in the development of pavement management systems. The conclusions validate the use of back-calculation programs in the structural analysis of rigid pavements, by comparing the results obtained with manual analytical methods.

1. Introduction

The current approach to road maintenance involves anticipating damage by monitoring and periodically evaluating the road surface and creating evolutionary models, thus identifying the areas that require action, scheduling their maintenance and assigning resources efficiently. Non-destructive tests make it possible to gather information about the condition of the pavement and process it using suitable software, in order to determine the actions to be taken to keep the infrastructure in optimum condition for use. These tests are generally faster and less expensive, and cause less inconvenience to users, compared to traditional techniques.

Among non-destructive tests, deflection analysis using a Falling Weight Deflectometer (FWD) has become popular due to its ability to estimate the properties of the pavement materials by measuring the deflections caused by an applied load. The data obtained using a FWD can be used in the structural evaluation of the concrete slab and support layers, respectively. These are required inputs for designing reinforcements during rehabilitation.

The main objective of this paper is to conduct a study on the rigid pavement management within a preventive approach, based on the current techniques for structural evaluation. The specific objectives are to: (i) identify the main parameters that allow an improvement on the current calculation procedures; (ii) validate the results of automatic calculation programs by comparing them with those obtained by analytical methods.

Data obtained from a test campaign with a FWD is used to analyse the current condition of the road on a 62-kilometre stretch of motorway in Mexico. Based on these data, various methods were used to obtain the modulus of subgrade reaction and the elastic modulus of the concrete slab. The results obtained from several commonly used calculation programs and those obtained by analytical methods are then compared.

2. State of the art

An analysis of the state of the art has been made in terms of maintenance of rigid pavements, which is gathered below.

2.1. Management systems for road pavement maintenance

The basic aim of pavement management is to use reliable and consistent information to develop decision-making criteria, providing realistic investment alternatives and contributing to efficient decision-making [1]. Until not long ago, management was based on experience, assigning the resources available for network maintenance in the most effective manner possible. With current technology, more advanced road pavement management systems have been developed. These modern systems make it possible to classify and prioritise the actions to be taken, by studying changes in the structural and functional characteristics of the pavements.

In general, maintenance programmes should improve the long-term performance of the pavement, fulfilling the expectations of drivers. This goal is achieved through preventive maintenance, non-structural rehabilitation and routine maintenance. A management system must include road pavement performance models, making it possible to predict real changes and the condition of the road pavement at any given time.

Most efforts in this area have focused on developing management systems for flexible pavements. For this purpose, HDM-4 is the tool most commonly used for analysis, planning, management, maintenance assessment and investment decisions regarding roads. This tool can also be used to evaluate flexible road pavement. However, a recent study by the Washington State Department of Transportation (WSDOT) indicates that version 1.3 of HDM-4 does not provide meaningful analysis output for the rigid pavements studied by WSDOT [2].

2.2. Non-destructive tests in pavement evaluation

Evaluating pavement condition is an essential step in assigning suitable maintenance actions and must be carried out from a functional and structural perspective. This can be done using destructive or non-destructive tests. Among non-destructive testing techniques, pavement deflection tests represent an extremely valuable engineering tool for evaluating the uniformity and structural condition of the existing pavement. Deflection is the structural response to the application of an extreme vertical load produced by vehicle traffic and can be related to the structural capacity of the pavement [3].

In rigid pavements, the deflections results obtained provide a great deal of information about the structure of the existing pavement, including variability in deflections, the efficiency of load transfer across joints and cracks, and the presence of voids under the slab. The information is also useful for back-calculating the main properties of the material, specifically the elastic modulus of the concrete and the modulus of subgrade reaction, thus making it possible to evaluate the structural condition of the pavement.

Currently, the most common device for measuring deflections is the Falling Weight Deflectometer (FWD). This device releases a known weight from a given height onto a plate resting on the structure of the pavement, producing a load similar in magnitude and duration to that of the wheel of a heavy vehicle [4]. The impact of the falling mass triggers a load wave which is transmitted to the pavement through a system of elastic buffers.

2.3. Analytical methods

There are several methods of analysing pavement condition based on deflections: the AASHTO 1993 Method, Lukanen's Method, YONAPAVE Method and Rohde's Method [3]. These methods make it possible to calculate the elastic modulus of the concrete slab and the modulus of subgrade reaction (k value), with which it is possible to determine the value for the effective structural number of an existing structure in order to design and evaluate stresses or rehabilitation based on the modulus of the subgrade and the concrete slab.

The recommended procedure for calculating the effective k value of non-destructive tests is based on the AREA concept, which was originally proposed by Hoffman and Thompson [5]. It consists of calculating the AREA parameter using the following expression, which describes the deflection basin (Eq. 1):

$$AREA=150*[1+2*D_{30}/D_0+2*D_{60}/D_0+D_{90}/D_0] \quad (1)$$

where D_0 is the surface deflection at the point of impact and D_{30}, D_{60} and D_{90} are the deflections at a distance of 30, 60 and 90 cm from that point. The AREA value is a good indicator of the pavement modulus, with high values corresponding to road pavements with good structural characteristics.

For a given load radius and sensor configuration, there is a unique relationship between the AREA parameter and the radius of relative stiffness in the pavement system. In the case of a load plate with a radius of 150 mm (5.9 inches), it is possible to apply the following equation by Hall (1991) (Eq.2):

$$l_k=[\ln((36*AREA)/1812.279133)/(-2.559340)]^{4.387009} \quad (2)$$

With the AREA parameter calculated based on deflections measured by the FWD and the radius of relative stiffness, the effective k modulus can be obtained using Westergaard's deflection equation (Eq.3):

$$k=[P/(8*d_o*l_k^2)]*\{1+(1/2\pi)\{\ln[a/(2*l_k)]+(\gamma-1.25)\}(a/l_k)^2\} \quad (3)$$

where d_o represents maximum deflection, P is the load, γ is Euler's constant and a is the load radius. Finally, the k value of the subgrade can be used to determine the elastic modulus of the concrete, using Eq.4:

$$E_{PCC}=[12*l_k^4*(1-\mu_{Pcc}^2)*k]/h^3 \quad (4)$$

where E_{PCC} is the elastic modulus of the concrete slab, l_k is the radius of relative stiffness and μ_{pcc} is Poisson's modulus for the concrete.

2.4. Back-calculation process

In pavement design, stresses, deformations and deflections are calculated based on defined traffic loads, layer thicknesses and the structural contribution of the materials. Back-calculation is the opposite procedure, making it possible to describe the mechanical resistance of the layers that make up the structure of an existing pavement, based on the deflections measured on the surface [6]. This enables us to calculate the elastic modulus of the concrete that is used to evaluate the structural condition of the slab, and the effective modulus of subgrade reaction (k value) used to evaluate the support layer [7]. The ASTM D5858 standard provides an approach to back-calculation based on deflection measurements [8]. Figure 1 shows the flow diagram for the back-calculation process.

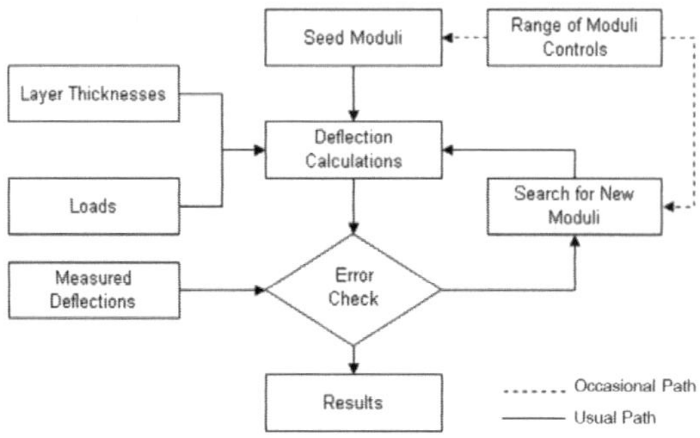

Figure 1. Flow diagram of the back-calculation process.

The back-calculation process is sensitive to the seed moduli and the more realistic these are, the better the results obtained in the process of iterations. For each deflection basin measured, the required input data are entered. These include the load characteristics of the FWD device, Poisson's coefficient, layer thicknesses, deflections and locations, and possibly, the initial estimates of the layer moduli (seed moduli). The materials classification and approximate layer thicknesses can be obtained from construction records, non-destructive tests or a core extraction programme.

3. Methodology

The following describes the methodology applied to the analysis of a specific road in which deflection measurements have been taken in recent years.

3.1. Case study

A 62-kilometre stretch of motorway in Mexico was used as a case study. Each year, goods with a transported value on the order of 25% of the country's gross domestic product travel along the section which the object of study.

3.1.1. Initial data

The current pavement structure was put into service in 1995. It is made up of concrete slabs resting on an asphalt layer, which in turn rests on a base with an average thickness of 20 cm. In order to obtain the thicknesses of the layers that make up the pavement along the lane under study, a field exploration programme was conducted, which involved taking test borings or bore holes every kilometre. This made it possible to determine the stratigraphic profile of the pavement, which is shown in figure 2. Two different structures can be identified. The first is made up of a layer of concrete, followed by asphalt mix and a layer of hydraulic base. The second type of structure identified consists of a layer of concrete, asphalt mix and a cement-stabilised base layer.

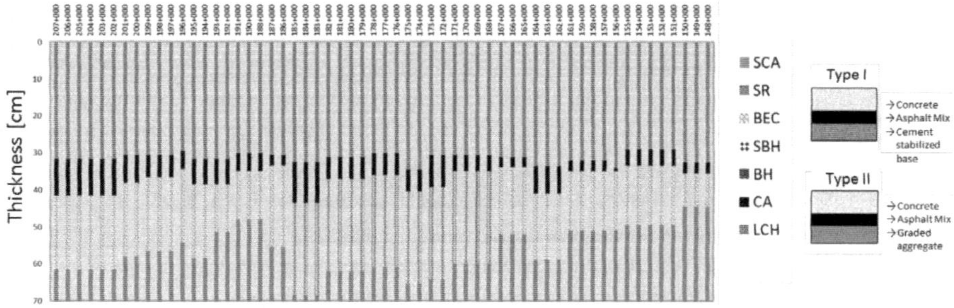

Figure 2. Stratigraphic profile of the pavement (KP 207+000 to KP 148+000) and types of structure identified.

3.1.2. Deflection measurement

Pavement deflection measurement was carried out in March 2017, using a Dynatest 8082 Falling Weight Deflectometer. The equipment sensors are located at a distance of -30, 0, 30, 60, 90, 120, 150 and 180 cm, measured between the system's load axis and the location of each sensor. The radius of the load plate is 15 cm (5.9"). The deflection measurements were taken every one hundred metres, in the centre and at the edges of the slab, recording the temperature of the pavement surface and the air at the time of measurement. The distance travelled and location of each measurement point were also recorded using GPS. Three blows were used at each point, gradually increasing the load. Lastly, an output file was generated, containing this information and the deflection measurements. Figure 3 shows the measurement scheme of an FWD and the arrangement of the geophones.

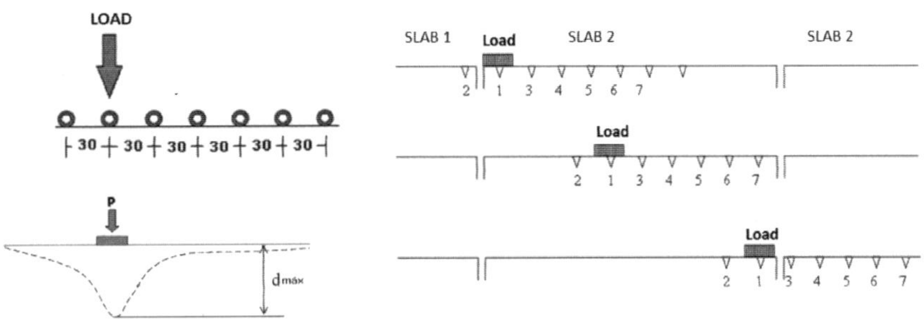

Figure 3. Measurement scheme of an FWD and arrangement of the geophones.

3.2. Analysis

This section studies the deflections from the stretch of motorway between KP 148 and KP 210 using the Elmod and Rubicon programs and analytical methods (AREA Method and YONAPAVE Method). The purpose is to obtain values for the elastic modulus of the concrete slab and the subgrade, the main parameters in the structural analysis of rigid pavements, and to determine its remaining structural capacity. Only the deflections measured at the centre of the slab were used.

3.2.1. Using software

Today there is a range of software programs available to perform back-calculation. However, there is no standardised criterion for choosing the most suitable in each case study [9]. Of the available programs, in this study the deflection analysis was performed using Elmod software and Rubicon software.

Elmod version 6.1.8.2, developed by Dynatest Consulting, was used for the analysis. This program is capable of analysing large batches of FWD data, a valuable characteristic when analysing long stretches of road. The Elmod program standardises the deflections for a pressure of 700 kPa. Lastly, it provides the elastic modulus of each layer, including the subgrade. Calculations made with the Rubicon Toolbox program used version 2.4.1., developed by the company Rubicon Solutions. The program has a back-calculation module that makes it possible to perform an automated back-calculation of the elastic moduli of the layers, based on FWD data. It uses an effective search method to find the rigidity of the layer which will produce a calculated deflection basin which is close to the FWD deflection basin measured. This program allows for a maximum of 50 iterations. The initial elastic moduli for the calculation were established based on the ASTM D5858 - 96 Standard, as indicated in table 1, depending on the material in each layer.

IOP Conf. Series: Materials Science and Engineering **512** (2019) 012042 doi:10.1088/1757-899X/512/1/012042

Table 1. Elastic modulus and Poisson's Modulus for different types of material.

Material	Elastic Modulus [MPa]	Poisson's Coefficient
Asphalt concrete	3500	0.30–0.40
Hydraulic cement	35000	0.10–0.20
Cement-treated bases	4100	0.10–0.30
Granular bases	350	0.20–0.40
Lime-stabilised soils	200	0.10–0.30
Granular sub-bases	140	0.20–0.40

3.2.2. Analysis of the deflections using the AREA Method

In the second part of this methodology, the AREA Method is used to perform a numerical analysis of the deflections in order to obtain the elastic modulus of the concrete slab and compare these values with those generated by the software programs. The deflections were not corrected for temperature, as the existing literature does not indicate how to apply this correction to rigid pavements.

Using Eq. 1, the AREA parameter was calculated for each blow, and with this, the radius of relative stiffness, utilising Eq. 2. With these parameters and knowing the maximum deflection, the applied load and the radius of the load plate, the effective k modulus was calculated using Westergaard's equation (Eq. 3). Lastly, the elastic modulus of the concrete slab was determined using Eq. 4. The YONAPAVE Method was used to estimate the elastic modulus of the subgrade.

4. Results

Figure 4 shows the standardised deflections for the project. It is possible to see that along some sections, the deflections are highly variable, with high values. In contrast, others are uniform and remain within the same range. Uniform and low deflections indicate good structural capacity, while very high and variable values indicate weaker pavement with the presence of damage.

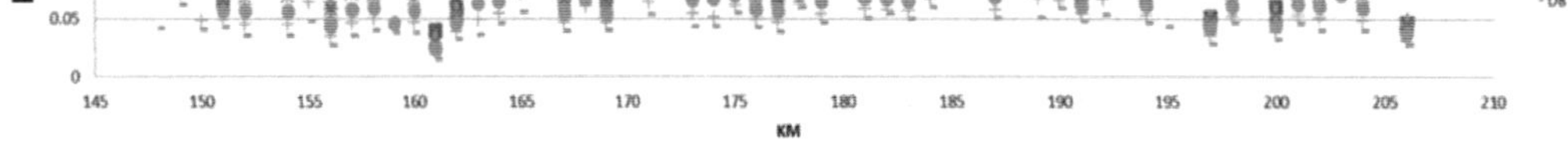

Figure 4. Variation in standardised deflections in the project.

The results of deflections back-calculated by the Elmod software are shown in figure 5. It can be observed that at KP 179 the deflections are mostly uniform in all the measurements taken, and that the elastic moduli also show little variation.

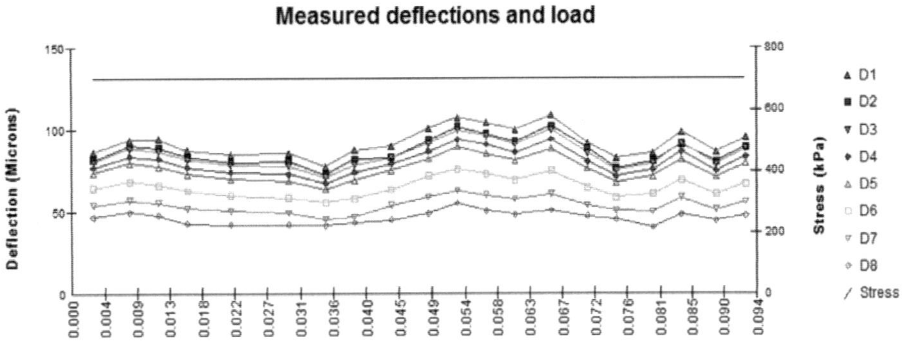

Figure 5. Deflections plotted by Elmod and standardised to 700 KPa, KP 179.

The elastic modulus of concrete slab E1 was calculated using the AREA Method. The resulted are compared to those obtained using Elmod and Rubicon in figure 6. It is possible to see a similarity between the values obtained by Elmod and the AREA Method, demonstrating the same behaviour throughout the route. In contrast, the values obtained with Rubicon show greater variations and smaller magnitudes along some sections, compared to those found using the AREA Method.

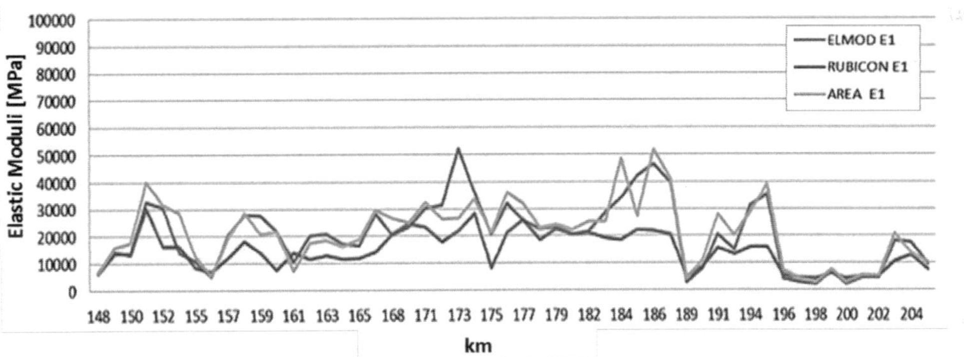

Figure 6. Comparison of the elastic modulus of the concrete subgrade according to calculation method.

Figure 7 compares the results obtained when calculating the modulus of subgrade E4 using the YONAPAVE Method with those calculated with Elmod and Rubicon. It can be seen that there is a relationship between the values obtained with Elmod and the AREA Method. The values obtained demonstrate similar behaviour in both cases. In contrast, the values obtained with Rubicon are lower than the others and almost constant throughout the project, with few and slight variations.

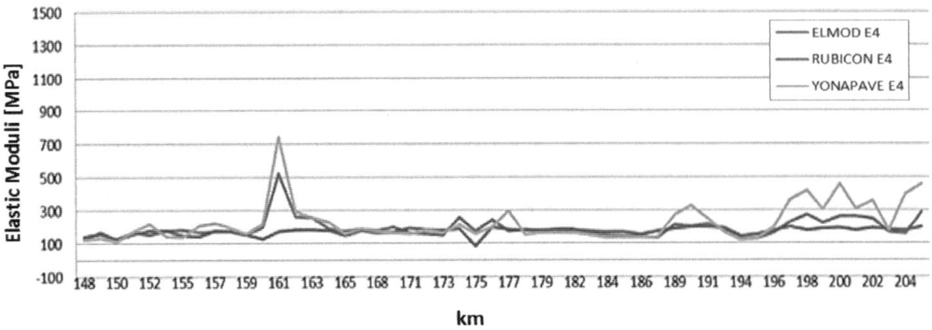

Figure 7. Comparison of the elastic modulus of concrete subgrade E4 for each calculation method used.

The residual deflection values were calculated as the y-intercept for the linear regression line in a graph of deflection vs. load. Figure 8 shows the graph of residual deflections to detect voids for KP 148, calculated using Elmod.

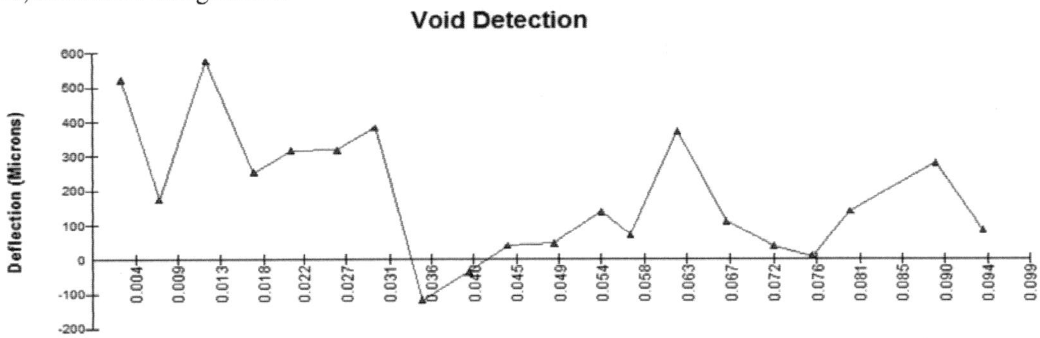

Figure 8. Graph of residual deflections for detecting voids using Elmod, KP 148.

The residual deflection values are plotted for the entire project (see figure 9). The red line indicates the limit of 50 mm^{-3}. According to this, there are a large number of sections in which there are indications of the presence of voids under the slab.

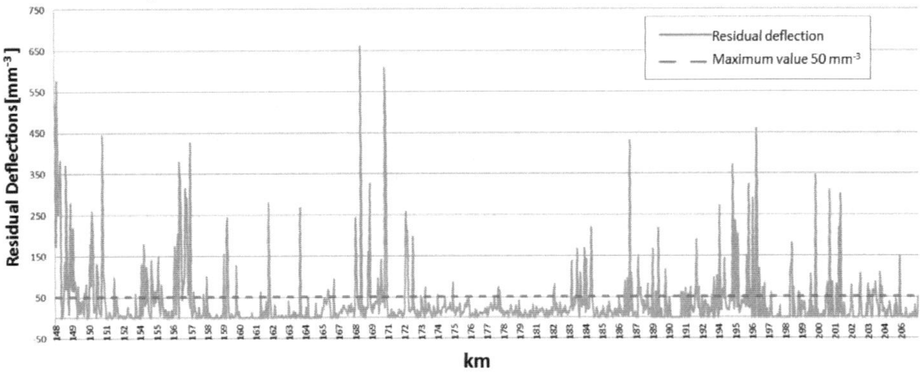

Figure 9. Graph of residual deflections for the project.

5. Discussion of results

The Elmod and Rubicon results differ with regard to the elastic moduli. In the case of Elmod, an iterative method based on the seed moduli entered is used, with the iterations ending when the maximum permissible error is reached. In the case of Rubicon, a database search method is used, whose result is also limited by the seed moduli and restricted by the maximum number of iterations permitted by the program (50 iterations).

When the results from the Elmod and Rubicon Toolbox programs were compared with analytical methods for estimating the elastic modulus of the concrete slab and the subgrade, Elmod came closer in both cases. The Rubicon results were less similar to those of the analytical methods and it is likely that the limited number of iterations may have affected these results.

The deflections obtained by the FWD were highly varied, indicating low capacity and/or the presence of damage. When the elastic moduli of the concrete slab and subgrade were back-calculated, abnormal values were found. This fact can be attributed to three main factors: the presence of voids, warping of the slab due to thermal gradient, and damage (primarily in the form of cracks).

It is reasonable to infer that the estimated values of the moduli are affected by the condition of the surface, producing lower values where the deflections are higher or close to the areas with more severe damage. Consequently, the low moduli imply damaged areas which warrant special attention in maintenance programmes. In the case of the data analysed, it is possible that the majority may be affected by the surface degradation of the slabs, based on the visual inspection performed using photographic records.

6. Conclusions

Non-destructive tests on pavements offer advantages over traditional methods. These include greater speed and ease in performing them, cost savings and fewer requirements, and an increase in the amount of data obtained, with the possibility of evaluating the project as a whole and making fewer changes in the infrastructure. Such tests make it possible to identify structural deficiencies before these manifest in the form of damage.

This research has compared the results obtained by analytical methods and calculation programs in order to estimate the elastic modulus of the concrete slab and the subgrade based on the deflections obtained. The following specific conclusions are noteworthy:

- The Elmod program came closest in calculating the elastic moduli.
- The results of the Rubicon program were less similar to those of the analytical methods, with these differences being primarily attributed to the limited number of iterations allowed by this program.

The deflections obtained in the FWD tests must be carefully analysed, as these may be affected by a range of factors and alter the results. The person performing the analysis must be knowledgeable about the typical values of each material in order to make reasonable suppositions and use their judgment in the back-calculation process.

When tests are performed using the FWD, under ideal conditions, the slabs to be tested would not have any cracks. Additionally, it is advisable refrain from performing the tests in the middle of the day in order to avoid warping due to thermal gradient.

It was determined that in areas with very limited layer thicknesses, the back-calculation of the elastic moduli is affected. Estimating the moduli for layers of pavement less than five centimetres thick is a common difficulty in back-calculation.

The loss of support from the soil under rigid pavements associated with voids entails an increase in stresses and deflections, causing a significant reduction in pavement life and contributing to its premature breakage. Any maintenance programme for rigid pavements must include measuring deflections for the purpose of detecting early voids and planning injections, saving on later rehabilitation and reconstruction costs.

IOP Conf. Series: Materials Science and Engineering **512** (2019) 012042 doi:10.1088/1757-899X/512/1/012042

7. References

[1] Solminihac H 2001 *Gestión de infraestructura vial* (Chile: Pontificia Universidad Católica)

[2] Washington State Department of Transportation 2004 *Application of HDM-4 in the WSDOT highway system* (Washington: Washington State Department of Transportation)

[3] Higuera C 2009 Caracterización de la resistencia de la subrasante con la información del deflectómetro de impacto *Revista Facultad de Ingeniería UPTC* 19

[4] Smith K, Harrington D, Pierce L, Ram P and Smith K 2014 *Concrete Pavement Preservation Guide* (USA: National Concrete Pavement Technology Center)

[5] Hoffman MS and Thompson MR 1981 *Mechanistic Interpretation of Nondestructive Pavement Testing Deflections* (Illinois: University of Illinois at Urbana-Champaign)

[6] Noriega MH 2011 *Relación entre el módulo resiliente hallado por retrocálculo y el encontrado en ensayos de laboratorio* (PhD dissertation, Universidad Nacional de Colombia)

[7] American Association of State Highway and Transportation Officials (AASHTO) 1993 *Guide for Design of Pavement Structures* (United States: AASHTO)

[8] ASTM International 2015 *ASTM D5858 – 96. Standard Guide for Calculating In Situ Equivalent Elastic Moduli of Pavement Materials Using Layered Elastic Theory* (ASTM International)

[9] Tarefder R and Ahmed M 2013 Consistency and accuracy of selected FWD back calculation software for computing layer modulus of airport pavements *Int. J. Geotech. Eng.* **7** 21–35

IOP Conf. Series: Materials Science and Engineering **512** (2019) 012043 doi:10.1088/1757-899X/512/1/012043

Maintenance management of pavements for expressways in Malaysia

M A Shahid

Principal Pavement Specialist, Opus International (M) Berhad, UEM Edgenta, Kuala Lumpur, Malaysia

drshahid@opusbhd.uemnet.com

Abstract. Pavements deteriorate over time due to traffic loading and environmental effects. The maintenance of pavements to improve the riding quality and level of service on major expressways, including the North-South Expressway (850 km in length), is a key challenge for the maintaining authorities and highway concessionaires. Pavement maintenance works not only incur high costs but also cause interruption to traffic flow resulting in complaints from the road users and the authorities. To maintain satisfactory condition of existing pavements, timely intervention is necessary which helps to minimise pavement deterioration and maintenance costs. Pavement condition assessment is required to determine the functional and structural condition which translate into service life. This paper describes an approach with case studies of pavement maintenance management for expressways in Malaysia.

1. Introduction

Pavement maintenance works form a major component of the annual expenditure. Appropriate intervention for maintenance and rehabilitation of existing pavements is necessary for asset preservation when signs of surface defects are observed. Such defects otherwise can affect the structural integrity of pavements and the level of service to the road users.

The maintenance management requires pavement condition assessment to be carried out at regular intervals. The data is used to determine the functional and structural condition. The data is also used to identify areas of poor condition pavements which are then followed by pavement treatment to be applied at the most appropriate time. The pavement treatment is determined using mechanistic design. The benefits of this design approach over empirical methods are that the in-situ material properties of pavement layers from a specific site can be modelled. The engineering properties of innovative materials can be incorporated into the design procedure. The mechanistic design also helps to ensure that pavements in different sections along the expressway are not over-designed or under-designed. This design approach therefore provides optimal and cost-effective maintenance solutions.

2. Pavement serviceability

Figure 1 shows conceptually how pavement serviceability changes with traffic loading over time [1]. Pavement deterioration starts off at an accelerated rate at a point known as the "critical or intervention level." It is important to ensure that pavement strengthening is carried out at or before the critical/intervention level so that the structural integrity is largely retained. Before the intervention level, strengthening of pavements requires minimum treatment such as resurfacing or mill and pave in localised areas followed by thin overlay to restore the pavement condition. This helps to minimise the maintenance costs. The serviceability of existing pavements is determined through pavement condition assessment which is discussed in the following sections.

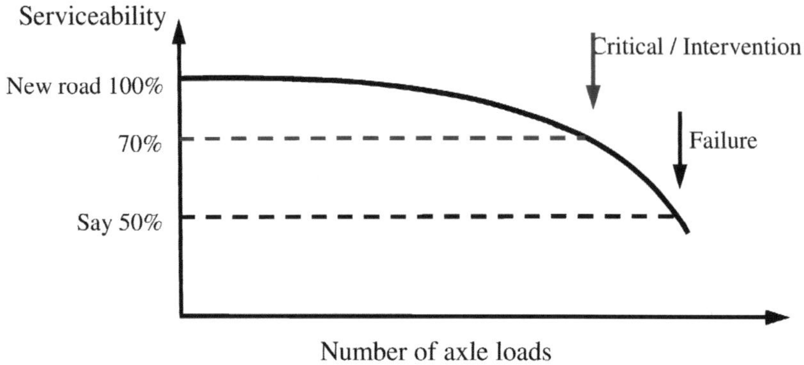

Figure 1. Pavement serviceability with traffic loading [1]

3. Pavement condition assessment

Pavement condition assessment (PCA) is carried out periodically for the whole expressway network. The purpose is to determine the functional and structural condition of existing pavements. A two-stage approach, termed as the network level PCA and project level investigation, is considered. The network level PCA provides an overall assessment of the performance condition of pavements which helps to identify sections of poor condition pavement. These sections are considered for project level pavement investigation. The PCA involves non-destructive testing techniques as described below.

3.1. Multi laser profiler (MLP) survey

The MLP survey is carried out annually to measure the surface properties which are directly related to the riding quality of pavements (figure 2). The survey is carried out at traffic speed of 70-90 km/hr. The data is recorded for every 25mm spacing and reported at 20m to 100m intervals based on the project requirements. The data is analyzed and processed to determine the following:

- Riding quality in terms of International Roughness Index (IRI) expressed in units of m/km.
- Transverse profile (rutting).
- Surface texture of pavements (macro texture).

3.2. Falling weight deflectometer (FWD) testing

The FWD testing is carried out to determine the structural condition of pavement (figure 3). The testing is carried out along left wheel-path of slow lane and right wheel-path of fast lane. The test spacing is generally 200m for network level PCA depending on the size of road network and 25m to 50m for project level investigation.

The FWD contact pressure is around 0.7 MPa for flexible pavement and 1.0 MPa for rigid pavement which produces a load pulse similar to that from an axle load of heavy vehicle. Pavement and air temperatures are recorded during testing and used in the subsequent analysis. The FWD data is analysed to determine the stiffness moduli of pavement layers and subgrade.

IOP Conf. Series: Materials Science and Engineering **512** (2019) 012043 doi:10.1088/1757-899X/512/1/012043

Figure 2. Multi Laser Profiler

Figure 3. Falling Weight Deflectometer

3.3. Automated crack detection (ACD)

The ACD survey is carried out using the Hawkeye (2000) system developed by ARRB, Australia. The Hawkeye system comes with a software which allows reporting of the type, severity and extent of cracks. It also allows reporting of wide cracks and percentage of all cracks over a specified length. The system uses the latest digital camera technology to record high resolution video images of the pavement surface condition.

All images are recorded using four channel digital cameras from different views such as vertical (down-facing to pavement surface), front view, right side and left side views. The cameras can log these images against other parameters such as distance and GPS coordinates. The cameras can also record roadside furniture such as guardrails, kilometer markers, signboards and road marking.

3.4. Visual condition survey

A detailed visual condition survey is carried out for project level investigation. The sections with different types of surface defects are identified and recorded following the procedures given in the JKR Guidelines on visual assessment of flexible pavement surface condition [2] and Transport Research Laboratory (TRL) Overseas Road Note 18 [3].

All types of pavement defects such as cracking, rutting, potholes, patching, bleeding and raveling as applicable are recorded. The data is reported at 100m intervals. The defects are characterised in terms of their type, severity and intensity. The severity of longitudinal, transverse, crocodile and/or block cracks is recorded as follows:

- Hairline crack - Very Low (VL)
- Crack width up to 1mm - Low (L)
- Crack width 1mm to 3mm - Medium (M)
- Crack width greater than 3mm - High (H)
- Crack width greater than 3mm with spalling - Very High (VH)

Rutting along wheel-path is measured at regular intervals using a 3m straight-edge. The severity of rutting is reported as follows:

- Rutting < 5mm: Low (L)
- Rutting 5mm to 15mm: Medium (M)
- Rutting 15mm to 20mm: High (H)
- Rutting > 20mm: Very High (VH)

3.5. Coring and dynamic cone penetrometer test

Cores (150mm diameter) are taken through asphalt layers to determine the thickness of asphalt and assess materials condition. Cores are taken from areas of good and poor pavement condition. A dynamic cone penetrometer (DCP) test is carried out through coreholes. The coreholes are then reinstated using suitable materials. The DCP data is analysed and interpreted to determine the thickness of granular materials for roadbase, sub-base and capping (if any) and the in-situ CBR of subgrade.

3.6. Skid resistance survey
Skid resistance survey is carried out using a Grip-Tester. It involves measurements which represent friction force on a test wheel operated at constant load and longitudinal reference slip. Water is sprayed ahead of the reference tyre by a self-watering system to produce 0.25mm thick water film and the test wheel is brought to the road surface by means of a hydraulic actuation system. The data is analysed to derive a Grip-Number (GN) which represents the skid resistance of pavement surface. The GN values are then used to determine pavement sections which fall under acceptable or investigatory level as per UK Highways Agency guidelines [4].

3.7. Traffic volume and axle load survey
Axle load survey is carried out using portable weigh pads or Weigh-in-Motion (WIM) system. The survey sites are selected to represent the axle load spectrum from commercial vehicles along the expressway network. The axle load data is used to determine Equivalence Factor (EF) values which are then used to calculate the design traffic loading in millions of equivalent standard axles [5].

3.8. Trial pit investigation
Trial pit investigation is carried out only in selected areas of existing pavements. The trial pit locations are based on the analysis of FWD deflection data and site conditions. Trial pits allow a more detailed visual inspection of pavement layer thicknesses and materials condition including the presence of water. It also allows to take bulk samples of granular materials and subgrade for laboratory testing.

3.9. Equipment calibration
The calibration of all survey equipment is necessary to ensure the accuracy of PCA data. The equipment shall be capable of providing satisfactory data with repeatability. All equipment is calibrated as per manufacturer's recommendation.

The FWD and MLP require annual calibration carried out by the manufacturer and a relative or reference calibration prior to the commencement of any survey. The FWD relative calibration involves stacking tower test for deflection sensors. This helps to compare and verify deflection measurements from all deflection sensors. Any sensors with inconsistent readings are removed and replaced prior to commencement of data collection.

4. Analysis and interpretation

4.1. Functional and structural condition
The PCA data is analysed to determine the functional and structural condition of pavements. The pavement condition is classified as good, fair, poor or bad with respect to the Key Pavement Performance Indicators (KPPI) established by the Malaysian Highway Authority. The structural condition is represented by Pavement Condition Index (PCI) and FWD deflections. The pavement assessment criteria with respect to roughness, rutting, surface texture and PCI are shown in table 1.

Table 1. Pavement Condition Assessment Criteria

Pavement Condition	Roughness IRI (m/km)	Rutting (mm)	Texture Depth (mm)	Pavement Condition Index (PCI)
Good	< 2	< 5	> 0.5	< 2
Fair	2 - 3	5 - 10	0.3 - 0.5	2 - 3.9
Poor	3 - 3.8	10 - 20	< 0.3	≥ 4
Bad	≥ 3.8	≥ 20	-	-

IOP Conf. Series: Materials Science and Engineering **512** (2019) 012043 doi:10.1088/1757-899X/512/1/012043

Pavement Condition Index (PCI) = CI + RI, where:

CI = Crack Index	RI = Rut Index (derived from the MLP survey)
0 = No crack	0 = 0 - 5 mm
1 = Single crack	1 = 6 - 10 mm
2 = More than one crack (not connected)	2 = 11 - 15 mm
3 = Inter-connected cracks	3 = 16 - 20 mm
4 = Crocodile cracks/block cracking	4 = > 20 mm
5 = Complete disintegrating or potholing	

Figures 4 and 5 show typical example of the structural and functional condition of pavements with respect to PCI and surface roughness (IRI) respectively. The percentage of good, fair and poor condition areas indicate performance of pavements over the expressway network.

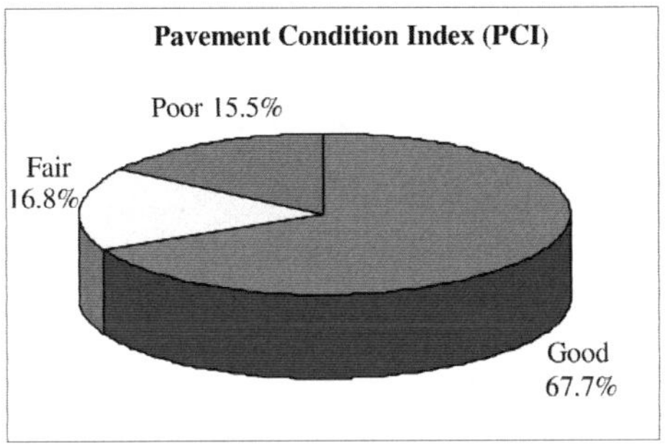

Figure 4. Pavement Condition based on PCI

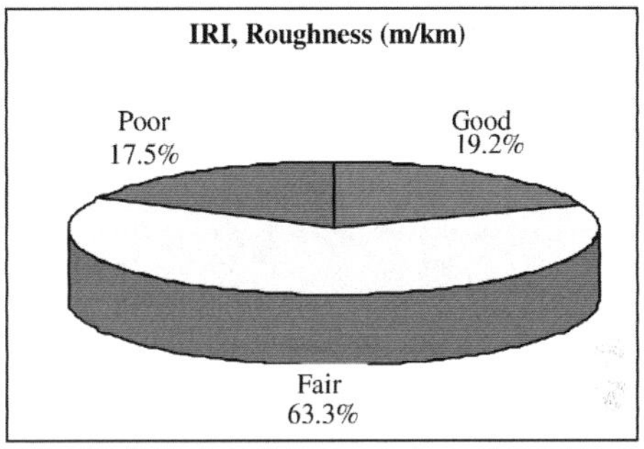

Figure 5. Pavement Condition based on IRI

4.2. HDM4 analysis

The highway development and management model (HDM4) has been used for planning of maintenance budgets for road projects in many countries. The HDM4 analysis is carried out to provide maintenance strategies which allow monitoring of pavement performance with respect to the level of service based on different financial budgets. The HDM4 model has been calibrated for local conditions in Malaysia. This had involved setting up and performance monitoring of long-term pavement performance (LTPP) sites over 10 years. The input data for HDM4 analysis includes the following:

- Pavement condition assessment data
- Pavement construction details and age
- Traffic volume data
- Axle load data with Equivalence Factors (E.F.) of commercial vehicles
- Pavement maintenance record and treatment options
- Unit rates for the treatment options

HDM4 analysis produces a long-term rolling programme (LTRP) which shows medium to long-term maintenance works and expenditure profile. It provides different maintenance scenarios which can be analysed over a period of 5, 10, 20 or 30 years. This scenario analysis is particularly useful for the financial planning of maintenance budgets in order to achieve target key performance indicators for the expressway network. Figure 6 shows a flow chart of HDM4 analysis.

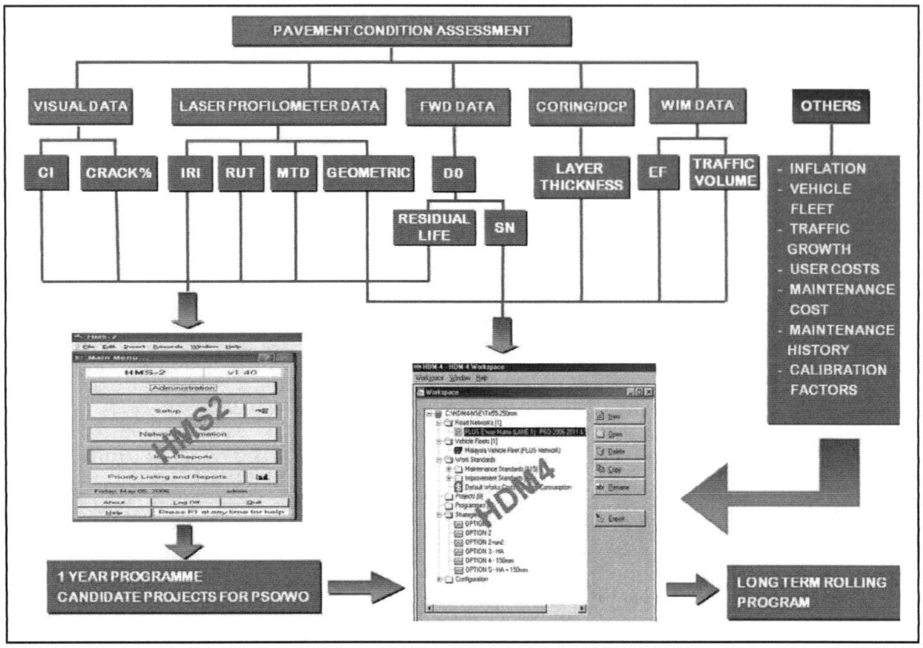

Figure 6. Flow chart of input data for HDM4 Analysis

4.3. Pavement sections for maintenance

The PCA results in terms of surface rutting and cracking can be presented on a strip-map as shown in figure 7. This gives an idea of the extent and location of good, fair and poor condition pavements along the expressway network. The strip-map is used to identify sections of poor pavement condition which require attention for pavement maintenance and rehabilitation works.

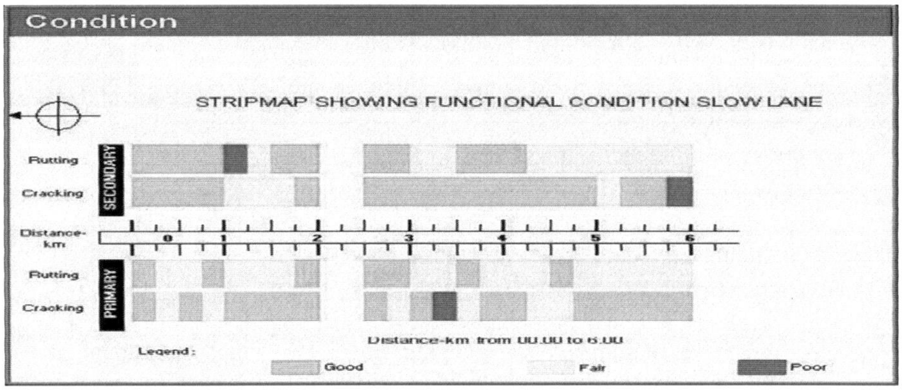

Figure 7. Strip-map of pavement surface condition

5. Pavement rehabilitation design

5.1. Design approach

Pavement rehabilitation design is carried out using a mechanistic approach [6, 7]. The mechanistic design offers several advantages over empirical methods such as:

- The engineering properties of each pavement layer and subgrade can be modelled.
- Temperature and environmental data specific to a particular site can be considered.

- Sensitivity of design thickness to changes in data can be assessed.
- Pavement design for extra heavy and unusual vehicles loading can be carried out.
- Design options for alternative pavement structures can be determined.
- Innovative materials can be incorporated into the design procedure.

Pavement rehabilitation design based on mechanistic approach has been applied to major expressways in Malaysia over several years. The objective of the design procedure has been to retain as far as possible the existing pavement materials and obtain optimal maintenance solutions.

5.2. Design procedure

The design procedure requires computation of strains at the base of asphaltic layers and on top of subgrade [8]. These strains are related to fatigue cracking in asphalt and permanent deformation (rutting) of subgrade. Computation of tensile stresses in cement treated base (where applicable) is also carried out. The critical strains in the asphaltic layers and subgrade are compared to the permissible strains based on traffic loading. Figure 8 shows a flow chart of the design procedure for fatigue cracking. A similar approach is adopted to determine strains for subgrade permanent deformation.

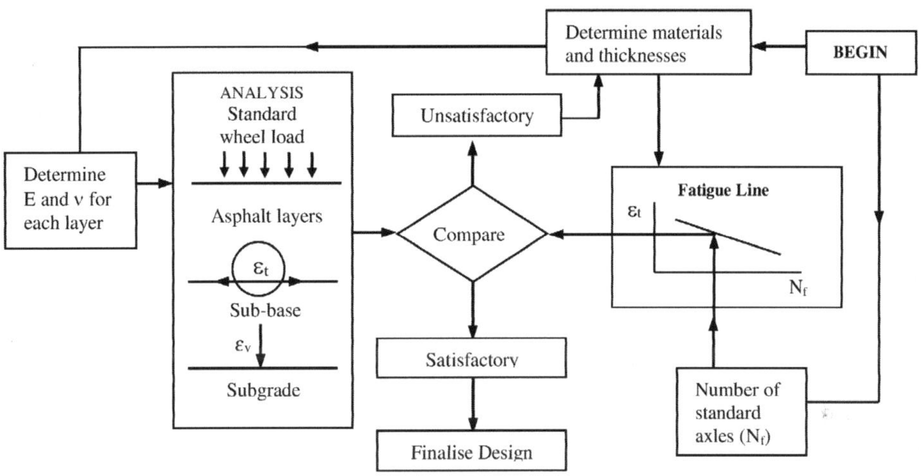

Figure 8. Flow chart of design procedure for fatigue cracking

The above design process considers conventional failure mechanism of asphalt cracking and rutting in flexible pavement. Other failure modes such as reflection cracking, surface initiated (top-down) cracking, poor drainage, poor asphalt mix and workmanship are also considered. Depending on the structural condition of existing pavement layers (i.e. granular roadbase, sub-base and subgrade), pavement rehabilitation design would generally involve asphalt overlay with partial reconstruction of slow lane. Full reconstruction may also be required in some localised areas with contaminated granular layers and poor subgrade. The overall objective of the design procedure is to preserve the asset with optimised treatment. This helps to minimise the maintenance costs.

5.3. Case studies

Pavement maintenance and rehabilitation works along the North-South Expressway (NSE) have been carried out on a yearly basis. The works have been completed in several areas along the NSE. These areas were identified with poor functional and structural condition. Pavement condition has been monitored based on the PCA data. The results have indicated improved functional and structural condition of pavements where rehabilitation works have been completed. Case studies for two sites along the NSE, referred to as Site A and Site B, are discussed below.

5.3.1. Site A - NSE Section S3 (KM 72.0 - 79.13) Northbound. Total length of this site was 7.13 km comprising a two-lane carriageway plus hard shoulder. This site involved flexible pavement with 50% of the pavements constructed on embankment (fill areas) and 50% in cut sections. PCA 2008 results showed poor functional and structural condition with poor IRI, high defections and high crack index (Figure 9a).

Project level pavement investigation was carried out between Nov 2008 and Feb 2009. This involved FWD testing, detailed visual condition survey, coring through asphaltic layers, in-situ DCP test, trial pit investigation and laboratory tests. The visual condition survey showed medium to high severity cracks on 70% of the slow lane. Rutting of medium to high severity (15-20 mm) along wheel-paths was observed over the entire length of slow lane. The fast lane pavements were generally in good condition with low severity cracks and low severity rutting (less than 5mm). The asphalt thickness was around 180mm to 220mm. A number of cores taken from slow lane showed cracked and broken asphalt material. The average thickness of granular materials (wet mix roadbase and sub-base) was 400-500 mm. A capping layer of 600-800 mm thickness was found underneath of sub-base. The in-situ subgrade CBR was generally 6% and higher.

The design traffic was calculated based on the traffic volume and axle load data from Weigh-in-Motion stations. The 10-year design traffic was found to be 48.6msa for slow lane and 12.2msa for fast lane. The pavement residual life was found to be less than 1 year for slow lane and 4-6 years for fast lane.

5.3.2. Site B - NSE Section N2 (KM 90.0 - 93.0) Northbound. Total length of this site was 3.0 km comprising a two-lane carriageway plus hard shoulder. This site involved flexible pavement with 10% of the pavements constructed on embankment (fill areas) and 90% in cut sections. PCA 2010 results showed poor functional and structural condition of pavements with poor IRI, high crack index and high rut index (Figure 10a).

Project level pavement investigation was carried out between Oct 2010 and Jan 2011. This involved FWD testing, detailed visual condition survey, coring through asphaltic layers, in-situ DCP test, trial pit investigation and laboratory tests. The visual condition survey showed medium to high severity cracks on 80% of the slow lane. Rutting of medium severity (5-15 mm) along wheel-paths was observed on slow lane. The fast lane pavements were generally in good condition with low severity cracks and low severity rutting (less than 5mm). The asphalt thickness was around 180mm to 200mm. The average thickness of granular materials (wet mix roadbase and sub-base) was generally between 500-600 mm. A capping layer of 400-600 mm thickness was found underneath of sub-base. The in-situ subgrade CBR was generally 5% and higher.

The design traffic was calculated based on the traffic volume and axle load data from Weigh-in-Motion stations. The 10-year design traffic was found to be 38msa for slow lane and 10msa for fast lane. The pavement residual life was found to be less than 1 year for slow lane and 3-5 years for fast lane.

5.3.3. Pavement performance. Pavement rehabilitation designs were carried out using mechanistic approach for 10-year design life. The treatment involved mill and pave existing pavement on slow lane (to varying depth) then followed by asphalt overlay on full width of carriageway. The pavement rehabilitation works were implemented and completed in Year 2009 for Site A and in Year 2011 for Site B. The performance of these sites has been monitored since completion of pavement works and is described below.

Site A - NSE Section S3 (KM 72.0 - 79.13) Northbound

- Figure 9a shows PCA 2008 data which is before the rehabilitation works. The results indicate poor condition of pavements with poor IRI, high defections and high crack index.
- Pavement rehabilitation works were completed in Year 2009.
- Figure 9b shows PCA 2017 data which is 9 years after completion of pavement works. The results indicate satisfactory pavement condition.

Site B - NSE Section N2 (KM 90.0 - 93.0) Northbound

- Figure 10a shows PCA 2010 data which is before the rehabilitation works. The results indicate poor condition of pavements with poor IRI, high defections and high crack index.
- Pavement rehabilitation works were completed in Year 2011.
- Figure 10b shows PCA 2017 data which is 7 years after completion of pavement works. The results indicate satisfactory pavement condition.

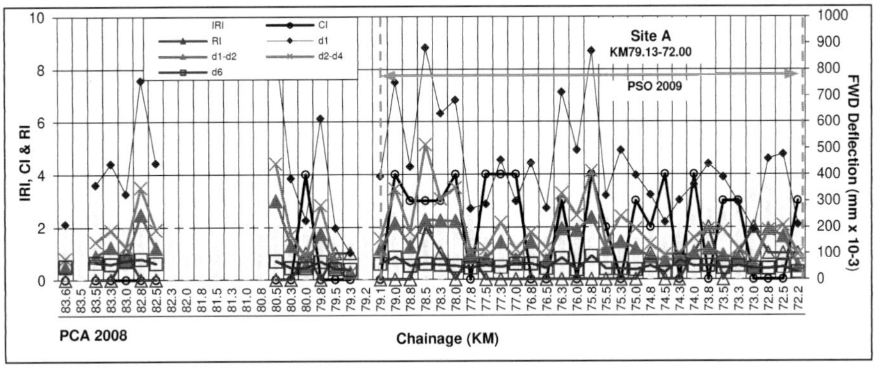

Figure 9a. Site A - Pavement condition of slow lane based on PCA 2008

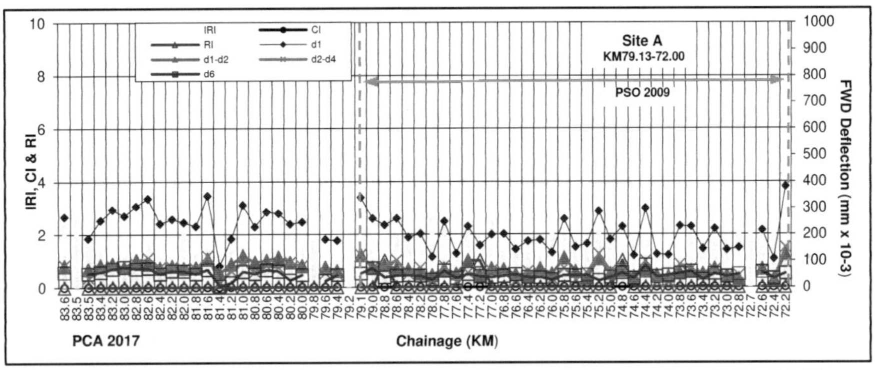

Figure 9b. Site A - Pavement condition of slow lane based on PCA 2017

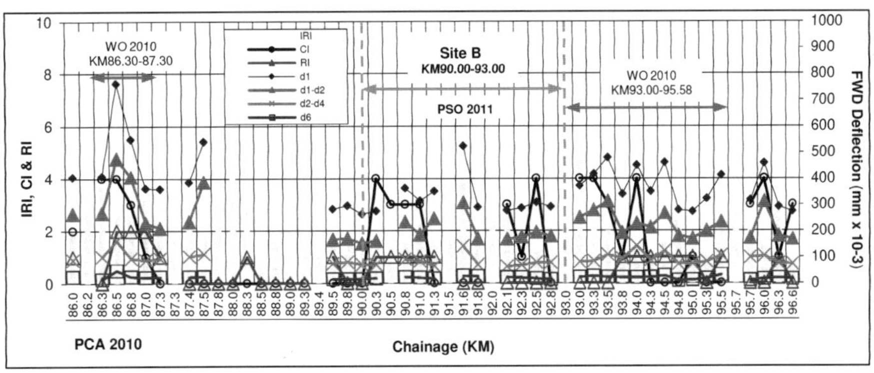

Figure 10a. Site B - Pavement condition of slow lane based on PCA 2010

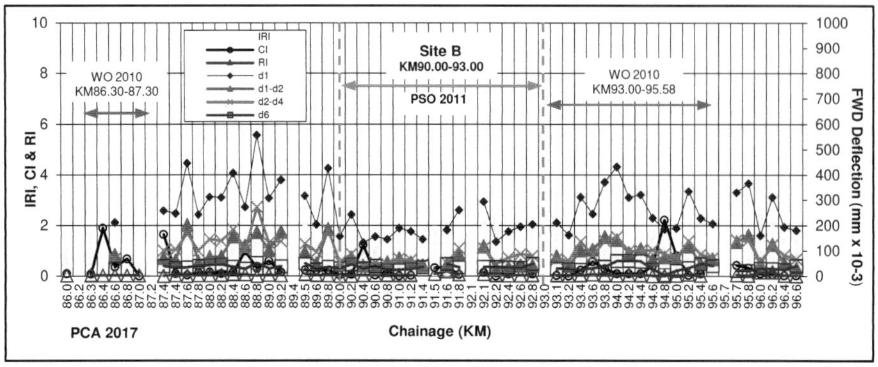

Figure 10b. Site B - Pavement condition of slow lane based on PCA 2017

The above shows improved functional and structural condition of pavements at sites A and B after the rehabilitation works.

6. Conclusions

Pavements deteriorate over time due to traffic loading and environmental effects. Timely intervention is necessary to maintain the existing pavements in satisfactory condition which helps to minimise pavement deterioration and maintenance costs with reduced interruption to the road users.

Pavement condition assessment (PCA) is carried out to determine the functional and structural condition. The network level PCA is carried out periodically and provides an overall assessment of the performance condition of pavements over the expressway network. It also helps to identify pavement sections with poor structural condition along the expressway.

Project level investigation is carried out for selected pavement sections which require maintenance. The data is used to determine the rehabilitation design options using mechanistic approach. The design procedure allows material properties of pavement layers to be incorporated. This approach helps to optimise the design options based on the structural condition of existing pavement.

HDM4 analysis is carried out to determine maintenance strategies which allow monitoring of pavement performance. It also provides maintenance scenarios and financial forecast over a certain period. This scenario analysis helps to determine the most optimal maintenance strategy and financial budget to maintain satisfactory pavement condition along the expressway network.

Pavement maintenance and rehabilitation works along the North-South Expressway (NSE) have been carried out on a yearly basis. Pavement rehabilitation works have been completed in several sections along the NSE. The performance of completed works has been monitored on a yearly basis. The results have shown improved functional and structural condition of pavements.

7. References

[1] Brown S F 1998 *Introduction to pavement design, Residential course on bituminous pavements - materials, design and evaluation* (UK: University of Nottingham)

[2] JKR 1992 Visual assessment of flexible pavement surface condition (Kuala Lumpur: Jabatan Kerja Raya)

[3] TRL Overseas Road Note 18 1999 *A guide to the pavement evaluation and maintenance of bitumen-surfaced roads in tropical and sub-tropical countries* (Crowthorne, UK)

[4] The Highways Agency (HD28/04) 2004 *Skid resistance, Design Manual for Roads and Bridges, Volume 7, Pavement Design and Maintenance* (United Kingdom)

[5] JKR 1985 *Arahan Teknik (Jalan) 5/85 Manual on pavement design* (Kuala Lumpur: Jabatan Kerja Raya)

[6] Brown S F and Brunton J M 1986 *An introduction to the analytical design of bituminous pavements, Third Edition* (University of Nottingham)
[7] Brown S F 1998 Developments in pavement structural design and maintenance *Proc. Institution of Civil Engrs & Transp (London)* **129** pp 201-6
[8] Powell W D, Potter J F, Mayhew H R and Nunn M E 1984 *The structural design of bituminous roads Laboratory Report 1132, Transport and Road Research Laboratory* (Crowthorne, UK)
[9] Shahid M A and Chan F W K 1998 Use of FWD for quality assessment during pavement construction *Proc. of the 3rd International Conference on Road and Airfield Pavement Technology (Beijing)* **2** pp 1239-48
[10] Shahid M A and Chan F W K 1999 Full scale site trials of in-situ recycled cement bound roadbase in Malaysia *Proc. of the 21st World Road Congress, PIARC* Kuala Lumpur

Acknowledgments
The author would like to acknowledge the support and encouragement from the management of UEM Edgenta and Opus International (M) Berhad during preparation of this paper. The author would also like to thank for the cooperation from the management of Projek Lebuhraya Usahasama Berhad who is the concessionaire for major expressways including the North-South Expressway in Malaysia.

IOP Conf. Series: Materials Science and Engineering **512** (2019) 012044 doi:10.1088/1757-899X/512/1/012044

Sustainable development (urban transport and mobility) - "sharpening the saw'' in shaping liveable cities towards quality of life experiences

Alan T L Wong

Transport & Traffic Branch, JKR Sabah, Malaysia

Email: alan.wong@sabah.gov.my

Abstract. "Sharpening the saw" attempts to refresh and re-energise in enabling us to understand better through a brief insight to the topics on: Liveability, Urban Transport, Quality of Life, and Sustainability associate to Sustainable Development (Urban Transport and Mobility) as the urban challenges through the "Mind – paradigm shifting" principle that creates new way of thinking: Strategic Thinking in "Doing the Right Things", and Tactical Thinking in "Doing the Things Right", which can lead to shift in behaviour that transform ourselves with added values in four dimensions of focus towards developing the "Arts of Sustainability - Sustainable Development quadrant (urban transport and mobility) ". The SD Quadrant, a loop-framework with the four dimensions of focus: Leadership: championing "think global, act local"; Stakeholder Engagement: aligning purpose, wants and needs; Smart Growth: innovating the future we want and; Transformational Edge: shaping liveable city, enabled and direct us to kick-starting the sustainable development initiatives through individual and collective participation for mutual benefits with urban transport and mobility focus. This is a way forward in shaping Liveable City together towards QoL experiences within the concept of sustainability that not only benefit us now but our future generation to come – a responsibility that we own and owe.

1. Sharpening the saw

1.1. Re-energising ourselves

"Sharpening the saw ", a catch-phrase and life-principle popularised by the late Stephen Covey through his much-acclaimed book, "The 7 Habits of Highly Effective People" attracted my curiosity with excitement when read 20 over years ago in a continuing effort to continue self-development through upward spiral of "Commit, Learn, Practice" life-skill. Lessons learnt include cultivating oneself over times with renewing to balance the concept across the four vital dimensions of lives for personal change towards effectiveness**:** the body, the mind and the emotional and spiritual [1].

1.2.The Mind – paradigm shifting; the objective

The Mind is the one life dimensions of "Sharpening the Saw" that we are sharing @ this moment. It is about knowledge updating, learning and un-learnt, sharing together through the journey of writing and presenting this Paper; through your reading of this Paper somewhere and/or together we are out for a date - right here at the 10MRC 2018; and through opportunities in engaging during the conference and possibly networking thereafter.

Urban transport often associates to private vehicles for mobility because this individualised transport has many superior traits (flexible, greater comfort; convenient). This transport trend if continues can effect in three key urban mobility concerns: traffic congestion, road crashes, environmental degrading that affects Quality of Life (QoL) experiences.

This "sharpening the saw" session attempts to refresh and re-energise in enabling us to understand better on the topics of: Liveability, Quality of Life, Sustainability associate to Sustainable Development (Urban Transport and Mobility focus) as the urban challenges through the "Mind – paradigm shifting" principle that creates new way of thinking (Strategic and Tactical) which can lead to shift in behaviour that transform ourselves with added values towards developing the "Arts of Sustainability" - Sustainable Development quadrant (urban transport and mobility focus), a loop-framework to guide and set direction in kickstarting the sustainability initiatives through individual and collective participation for mutual benefits [2, 3, 4, 5]:

Strategic Thinking: Do the Right Things

Strategic thinking is typically leadership focus with "doing the right things" for effectiveness in delivering the desired values based on aligned strategies and goals with commitment in (right sized) human and financial resources as well as in engaging stakeholders [4].

- Desire to "Think Global" with sustainability in Mind motivates us to "Act Local" that inspired leadership who shall champion in sustainable development locally with involvement of the others (in a team) through the leadership traits of being dynamic, excited, creative, trustworthy, absolutely committed and with the ability to inspire/influence/drive the team to create mutual values, taking the bold move in making great things happen and deliver the change or transformation in shaping liveable city [5].
- Forward thinking leadership direct oneself/the-champion (with others in the team) to engage stakeholders in developing mutual respect to build trust with shared passion-driven commitment and accountability-together interact to better understand (sustainable development, urban transport and motility focus) to close knowledge gap, encourage in the pursuits of localised sustainable development goals and associate targets guided by aligned purpose, wants and needs in serving for greater good towards innovating the future we want [2].

Tactical Thinking: Do the Things Right

Tactical thinking is management focus with "doing things right" collectively in executing priority activities with high performance or efficiency to produce or achieve the expectations [4].

- Learnt or un-learnt Knowledge updates and leads us with the knowhow and ability to being innovative in the approach to smart growth development for the future we want, collectively.
- Executing transformation in shaping liveable city needs a business plan and a roadmap with goals that align strategies, and activities with time line for performance measurement and in guiding towards high performance and hence greater success [3]. High performance or efficiency can be achieved with clear individual or team shared responsibility and accountability that creates personal and team importance, these can positively affect commitment, progress, practice follow through with what matters most first, and continuous self/team-development in delivering the expectations.

"Sharpening the saw" together through this Paper or Presentation can lead us to a paradigm shift in the new way of thinking about urban transport and mobility with sustainability in mind. The ultimate reach-out (through words of mouth, social medias and other accessible means) shall be to our city stakeholders: being road users; engineering-architectural-planning professionals as well as academicians, leaders, politicians, policies makers, and decision makers; interest groups, communities and individuals alike. A way forward in shaping liveable city together towards Quality of Life experiences within the concept of sustainability (urban transport and mobility focus) that not only benefit us but our future generation to come – a responsibility that we own and owe.

2. Liveability

Liveability emerged as an academic topic in the 1980s. It was briefly introduced in the town planning module as part of my UK postgraduate study in Transport & Traffic Engineering. The concept of liveable cities surfaced as part of urban planning efforts to 're-verse' the declining urban centres due to the rapid growing urban sprawl built-environment at out of towns suburban areas, characterised mostly by single-use residential development with private car-dependent communities [6]. Urban Sprawl increases traffic congestion level and exposes road users to conflicting movements that can be hazardous at the urban road network due to residents having to travel great distances (same time periods and routes simultaneously that creates rush hours, and possible aggressive driving with high risks taking situations) specially to reach for schools, works, and business in the urban or city areas. On top of congestion that effect in loss of productivities for the communities and road crash that effect in damaged or loss of life, the greater use of private or the usually low-occupancy vehicle during peak hours for the travel needs, increases local carbon footprint - an environment concern with the effects on global climate change and deterioration in urban air quality with it associated medical and economic costs.

Liveability initiatives include regenerating and planning city or urban centre as a place to "Live, Work and Play" in an environment that is conducive and sustainable with the presence or support of good infrastructure, public facilities and services, wide and accessible transport choices, affordable housing, abundance of job opportunities, and pleasing built or natural amenities for Quality of Life experiences by the local communities and/or visitors alike. Not forgetting for the future generation to have similar liveability dimensions, complimenting with feel-good or desirable QOL experiences if not better through subscribing to the sustainable development concepts (integrating environmental protection/issues into socio-economic planning and development) made known in the world 1st UN earth summit, the Rio Earth Summit 1992, and then popularised in evolving into high profile Global and Local Agendas ever since.

Liveability through investing urban transport for a city, such as programs in: making active mobility (walk; pedal-cycle) an attractive form of basic transport for urban mobility; providing affordable and widely accessible alternative "green" public transport choices for all walk of life; balancing (act of) the private-public transport mode-share and personal trip-share that harmonises with land development can help shape our urban form (and culture) that influences our lifestyles towards QoL experiences. This liveability commitment on urban transport shall enable a city in achieving its localised Sustainable Development Goals that are linked to mobility, referencing to the 3 key dimensions of Sustainability:

- Social or Community development – harmonised land development and transport, coupled with accessible to affordable transport choices and safe mobility enhances sense of place to work, live and play;
- Economic – city with active development and economic productivity facilitate through urban transport and mobility, enabled it to evolve and prosper as centres for employment, trading, shopping, education, and other services;
- Environment – reduce greenhouse gas GHG emissions from motorised vehicles; clean air quality through non-motorised transport enhances public health.

Liveability in a nut-shell refers to the presence with quality amenities of the built and natural environments as well as of community services. The term liveability has gained much attention especially with the annual liveable cities ranking surveys (US-Mercer, EU-EIU, UK-Monocle) that makes for a great news story to attract extensive media coverage [7, 8, 9]. Being high on a world liveable ranking not only bring pride to the city but can be a valuable marketing tool to lure prospective new talent, investment as well as attracting people to visit, study and do business in the city. World liveable cities rankings seem to have developed into coveted symbols of urban planning success among cities around the world – however, a city not only needs successful planning, it needs to be managed well too with support of love and care from the local communities or visitors alike so that everyone in the same city now and future can continue enjoy the QoL experience towards achieving or up keeping the liveable city image with sustainability in mind. The annual liveable cities ranking surveys base on quantitative and qualitative measures to assess each city chosen, typically on factors such as political stability, housing, safety, healthcare, education, recreation, public services, environmental quality, infrastructure and transportation. These surveys measure QoL experiences for wellbeing benefits associate to the above factors.

3 Quality of life experience

3.1 Urban public transport: Kota Kinabalu bus system

Quality of Life refers to user experience of the built and/or natural environment amenities as well as of community services that bring or associate to wellbeing benefits. Urban transport system shapes our urban form, influences our lifestyles, and determines the QoL that affect liveable image of a city like Kota Kinabalu [10].

Public transport such as buses, being larger in size than private vehicles can carry higher number of passengers. Thus, will free the urban road space from the presence of large number of private vehicles if more buses are plying along the urban routes to carry commuters for the work, school, leisure, shopping trips - instead of private vehicle which on average carrying less than 2 persons/vehicle in Kota Kinabalu [11]. Urban road network will be less congested when more high capacity public transport vehicles instead of the low-capacity/occupancy carrying private vehicles are plying along the same routes and at the same period within the same road network. The current Kota Kinabalu urban public transport (buses) has rather poor image, viewed as an inferior urban transport system provision as well as of low quality service in-terms of no scheduled service or un-reliable and no services after 9pm or earlier; less comfort in-vehicles facilities and lacking good urban design features of the road-side facilities while waiting for bus as well as when walking towards or away from the bus stops. Perhaps only captive passengers will remain as the patronage for their trips purpose without the attractiveness to capture bus usage from those who have the choice of travel by private vehicles. Vehicles used as buses are often small and mid-sized vans with tight leg-room, fares based on distance/stage are cash payment on board to conductor. Smart card is not used as payment method, hence crucial patronage data is not captured. No on-board security cameras and GPS system that facilitate both security and information on bus tracking for intend passengers. Even basic information such as bus and route maps not easily available, what more is the bus time table – there is none.

Kota Kinabalu's public transport business model of individual bus operators competing for passengers along the same catchment-corridors is creating highly competitive business environment that leads to operators finding "self-survival" practices, such as depart only when bus is having 'enough passengers', not plying those routes that have passengers but low volumes that are considered 'non-profitable' routes, less focus on up-keeping of vehicles in-terms of cleanliness, fitness etc to reduce operating costs. Such current practices are not conducive in making public transport (bus) as an attractive mode of transport to move people especial those who have the choice of private vehicle use.

The current unattractive Kota Kinabalu urban public transport (bus) system does not create quality of life experience to benefit much on our well-beings especial for those who are not captive passengers. As of current level of comfort (or discomfort actually), it will be a huge-huge challenge to attract local communities to commute with buses in associate to their travel needs to work, school, shop and recreation.

The challenge to balance the private and public transport in mode share and personal trip shares highlighted earlier needs a serious though.

3.2. Better data-better information, better life

Private vehicles whose average occupancy is at a low of 1.4-1.5 person per vehicle for the work and home bound traffic form majority of the road usage that often congests the Kota Kinabalu urban road network at peak periods as private vehicles, being less effective in moving more people occupy most of the road space [12]. The 2015 'onion' pilot transport survey by T&P/JKR Sabah at 4 key routes of the urban radial road network indicates a range of 22-24% bus ridership share at "work and home bound" peak hours, with a low 2-7% of the total vehicles being buses [12]. The low volume of bus plying along the urban routes but with a medium double-digit ridership shares (almost 23:77 in favour of private vehicle usage) indicate well that Kota Kinabalu buses indeed "move more people with less". But, dominant private vehicles in low occupancy scenario results in less effective use of road space, and this effect in traffic congestion at peak demand periods.

Traveling speed is a measure of traffic congestion level – the lower the speed the higher is the congestion level. The 2014 GPS travel speed measurements by T&P/JKR Sabah along the 7 key radial routes in and out of Kota Kinabalu city centre, indicates that travel speed for the commuter journey during peak hours along the non-CBD routes within the urban areas is approximately ½ less than the average journey speed during the non-peak lunch hour period [12]. Average journey speed during the week day lunch hour is around 45km/h – a speed that is still fast enough for some of us in Kota Kinabalu within the 10km radius to go home for lunch and on time back to Office after the short break with family - a QoL experience that benefits culture value. Can this better life "privilege" be sustained for the next generation to benefit [11]?

Necessary for us to revisit the 1960s or 70s when our towns are small-medium size with compact mixed developments concepts/trends such that one can live (near) where one work or work where (near) one live – thus reduce the needs to travel long distance by motorised vehicles or avoid, where walk and pedal cycle are the conveyance means for the short-medium trips on the road? – Road is a share space where all road users should be able to use safely and efficiently for mobility in an environment that should have been planned, designed and developed for both the motorise and non-motorise urban transport system.

Travel time reliability can often be affected by incidence (such as stalled vehicles, road crashes) at the urban road network in Kota Kinabalu where lane blockage affects road capacity that in-turns make commuters in both private and public transport having uncomfortable QoL experiences with the extra travel time needed especially during peak hours travel. With the availability of the online mobile "Waze" and "Google Traffic" Apps, this seem to be a step forward in alerting road users and authority of routes with slow speed and/or with incidence for road users to managing their mobility encounter/experience, and for possible quick response by the authority, such as in removing the lane-blockage. However, availability and usage of such Apps for crucial mobility information needs to be reliable and up to date to facilitate mobility QoL experience in real time.

Transport & Traffic data and/or information when available and after being processed and analysed can increase knowledge not only for urban planners but stakeholders to familiarise with the situations so to facilitate discussion in drafting strategies and planning towards good practice. In-addition to that, when available on-line with live feeds, these data and/or information enable fast response with good decisions in managing urban traffic for better flow, and in operation and/or usage of public transport system towards QoL experience. The catch-phrase "better data-better information, better life" complements the QoL experience needs with Sustainability in mind.

4. Sustainability

4.1. Sustainable urban transport and mobility development

The Rio +20 Conference in 2012 through its "the future we want" document refreshed its global supports of voluntarily commitments to the development of sustainable urban transport system. A sustainability concept introduced in the Rio Earth Summit 1992, but with a new sustainability paradigm: "Avoid; Shift; and Improve" [13]:

- Avoid the needs for travel through smart growth concepts or better physical planning and 'tele-works' (complementing smart city concept);
- Shift trips to more sustainable green modes of transport;
- Improve vehicle energy efficiency through use of better technologies

The Rio Earth Summit 1992 popularizes the 3 pillars of Sustainable Development concept: the triple bottom line of Social (People), Economic (Profit) and Environment (Planet) sustainability as well as promotes the development of Local Agenda 21. LA 21 is the global blueprint for sustainability, i.e. think global, plan-decide-implement local actions or doing things today that will not prevent future generations from having the same opportunities locally – the LA21 ideology of 'think globally, act locally'.

"Transforming our world": 2030 Agenda, the new agenda for Sustainable Development, adopted at the 2015 UN Summit is a roadmap, document in guiding and stimulating actions to shift the world onto a resilient path towards sustainability through a set of aspiring global Sustainable Development Goals (SDG) and associated targets that can be localised for the benefits of all – generations now and future. Urban Sustainable transport is not represented by a standalone SDG in this new Agenda, but it is being integrated directly or indirectly to facilitate targets achievement of the SDG, such as Goal [14]:

1. Sustainable Cities and Communities (e.g. targets: urban access and affordable for all);
2. Good Health and Wellbeing (e.g. targets: safer road, cleaner air-less vehicle emissions);
3. Industrial innovation and infrastructure (e.g. targets: quality, reliable, sustainable and resilient transport infrastructure);
4. Affordable and clean energy (e.g. target: energy efficient transport);
5. Responsible consumption and protection (e.g. target: review fossil fuel subsidy);
6. Climate Action (e.g. targets: climate change – mitigation and adaptation);

Sustainable Urban Transport and Mobility development in a nutshell is the volunteer initiative to think global (sustainability concerns), and the commitment or broad move to executive local actions to achieve localized targets of the global SDG for a safe, efficient, accessible, and affordable urban transport system to move people, goods and services associate to the land development in ways that benefits all on the Environment, the Economy, and the Society. The sustainable urban transport and mobility development shall, among other benefits enable local communities and visitors alike to travel within an acceptable level of traffic congestion, say target of "35km/h" with reliable journey time during peak hours towards feel good experience for their mobility needs in associate to the built or natural environment and community services of the city, without compromising the QoL of future generations – that's sustainable development concept.

4.2. Sustainable urban transport and mobility index-ranking

Development creates asset - for continuous sustainability, this asset needs to be managed well as part of the sustainable life cycle in continuous development for better with each review of the overall state of the urban transport system. However, managing for continuous system and service improvement is not effective without the availability of data through performance measurements or assessment of sustainability indicators and/or without the dare to be benchmarked against other cities (globally or locally) on urban transport and mobility index for competitive edge.

Sustainable urban transport and mobility has gained much attention especially with the global city index-ranking (UNESCAP-SUTI; Arcadis-SCMI; Little_UITP-UMI; Deloitte- CMI) of system and service performance measurement which enable city stakeholders and the public to understand their current urban transport and mobility situation through a point -based results framework [15, 16, 17, 18]. This assessment provides cities with a yardstick to independently measure their performance and benchmark their progress against other cities that shall motivate continuous improvement or development of urban transport and mobility.

IOP Conf. Series: Materials Science and Engineering **512** (2019) 012044 doi:10.1088/1757-899X/512/1/012044

The index-ranking surveys measure and assess sustainability dimensions with performance indicators on the state of sustainable urban transport (system) and mobility (service), typically are based on the triple bottom-line approach with the three pillars of sustainability, covering the dimensions:

- *People* measures the human and social implications of the mobility systems in place (e.g. access, road safety) and quality of life (e.g. affordability, promote active mobility - walking and cycling);
- *Planet* measures the environmental impacts (e.g. energy, pollution, and emissions) and the transport mode (e.g. modal split);
- *Profit* assesses the transport initiatives (e.g. improve mobility), efficiency and reliability of the system (e.g. less congested routes and travel time to/from work, frequent on-time public transport service) to facilitate economic growth.

Vienna and Zurich, not only are in the top 10 of the sustainable urban transport and mobility index-ranking but these two European cities have been highly ranked in the liveable city ranking for QoL experience too in the consecutive few years – what are the strategies for such sustainable achievement in serving the communities and visitors alike that continue having QoL experiences associate to urban transport and mobility? Perhaps, our 'sharpening the saw' session needs 2nd round when we next meet to share-learn from the best by familiarising with sustainable urban transport and mobility, the Vienna and Zurich experiences.

5. Way forward

5.1. Compass & direction

Having "sharpened the saw" (or still sharpening?) together with the renewed self to the new way of thinking (strategic and tactical) with sustainability in mind regarding urban transport and mobility challenges in our city, in recognizing ownership to these concerns and taking responsibility as being stakeholders to our city, it is an encouraging "baby-step" but can "leap-frog" together and synergise to developing the "Arts of Sustainability" in shaping liveable city towards QoL experiences in the place where we live, work and play – to be proud of.

5.2. The arts of sustainability - sustainable development quadrant (urban transport and mobility focus)

A loop-framework which encourages iterative process where continuous improvement is facilitated by lessons learnt from previous experience that promote sustainable development, and work towards reducing, or perhaps minimising or solving land development & transport related urban challenges (such as traffic congestion, road crash, and environment concerns) will increase vitality of a city, as well as encouraging it to grow, prosper, and enhance liveable image of the city that is developing towards sustainability for QoL experiences of generations now and future.

The Sustainable Development Quadrant has four element of focus that encourages practice in:

"Do the Right Things" -

✓ *Quadrant 1 Leadership focus: championing "Think Global, and Act Local"*

1. Leaders of change and leaders for change: visionary, forward-thinking leadership and political-will that subscribe to sustainable development concepts (urban transport and mobility focus) and commit to engage stakeholders and take the bold moves in leading to change based on the three pillars of sustainability: People, Planet, Profit/Prosperity;
2. Learn from the best: familiarise with cities that are top in the urban transport and mobility and/or liveable city index-ranking. However, needs caution when looking at other successful cities for inspiration in direct transfer of solutions in terms of the appropriateness and effectiveness as there can be differences in culture and having owns distinctive local 'flavour';
3. Commit in investing and the running of sustainable public transport business: create alternative revenues and make available financial resources for right -sized funding;

develop human resources who are passion-driven towards being knowledgeable and skilful with high performance in the business of sustainable urban transport and mobility.

✓ *Quadrant 2 Stakeholders Engagement focus: aligning "Purpose, Wants and Needs"*

4. Establish the Wishlist program: Commit to engage with stakeholders of the city to generate ideas together with aligned purpose, wants and needs onto the type of sustainable urban transport and mobility that everyone wishes for, and support towards shaping liveable city;
5. Introduce awareness program: Closing knowledge gaps - share to city stakeholders, the trends and milestones of sustainable development (urban transport and mobility focus) in other cities of developed and developing countries as well as own for lessons to be learnt or un-learnt;
6. Overcome cultural barrier such as social attitudes towards public transport as second class option or inferior mode of urban transport;
7. Integrated decision-making: stakeholders to share vision with focus on what the future city would look like; adopt common goals and localised targets to integrate decision-making on economically viable, affordable, people-oriented, environmental-friendly sustainable urban transport and mobility development. Decisions made today will shape liveable city and influence QoL experiences for now and the future generations to come - a responsibility that we own and owe, for sustainability.

✓ *Quadrant 3 Smart Growth focus: innovating "the Future We Want"*

8. Promote compact mixed-used development: easy reach, accessible, convenience - work (near) where you live, live (near) where you work to avoid motorised car travel;
9. Recognise road as a share space, and adopt strategic road planning concept with alternative new routes or road upgrading that are solution for all road users as a relieve route or choice to promote sustainable growth;
10. Shift from cars (with targets): Public transport corridor-ready road along new or existing routes to facilitate development of fast, safe and reliable public transport services; encourage transit-oriented site developments concept to attract greater public transport usage; car-light city;
11. Enhance travel choices: attractive bus services and facilities, include ride-share call on demand service – "bus goes to the people/potential passengers" concept;
12. Urban street design innovations to encourage basic or active mobility (walk; pedal-cycle), electric "hoverboard" or scooter and public transport usage: conducive environment - safe and comfortable;
13. Integrated urban transport: commuters or visitors can plan with the convenience of mobile devices, and complete their journey with an easy to use smart-card/device for payment through integrating various urban transport modes in the city;
14. Create a good public transport image for a realistic and quality alternative mode of transport that symbolize modern lifestyle/icon to attract new rider-ship as well as retain existing;

✓ *Quadrant 4 Transformation Edge focus: shaping "liveable city"*

15. Revamp existing un-attractive basic bus system to include review of bus business model: business ownerships/authority, bus fleets, services, performance measurements and bus operators as well as innovate alternative revenues for business sustainability towards QoL experiences;
16. Sustainable public transport (bus) business model needs to be developed for the likely profitable and non-profitable routes with area-wide coverage so not to create

in-equality in the city on bus service availability, and avoid direct competing for same passengers: one operator one complete route (some short overlapping segment of route may occur);

17. Start planning now: alternative high capacity mode of transport such as BRT or Tram/LRT along development catchment areas;
18. Integrated planning: allocate road space for the basic or active mobility (walking, cycling), public transport, motor vehicles; integrate their connectivity and balance the private-public transport mode and/or personal trips shares to optimise space usage and reduce private car use;
19. Bus system to attract usage with: low floor, clean-air, high capacity buses; in bus free Wi-Fi and, USB charger port; mobile Apps for bus live tracker information; smart travel card and security cameras in bus and at bus stops;
20. Enhance travel reliability: embrace technology and on-line live feeds on traffic flow and/or incidences for information to manage self-mobility and/or for response by authority to manage traffic flow to minimise travel disruption or delay;
21. Establish data warehousing for on-line share use of data and information that are relevant to the planning, executing and managing of urban transport: Data on patronage associated to demand, time and day needs to be part of the smart ticketing system to ease collection and retrieval for use as performance database and managing operation for better; Buses to equip with on board GPS unit to track routes, stoppage/delay, travel time and distances travelled for trip database that can be used for analysis planning, monitoring and quick decision making to manage successful bus system;
22. Competitive edge: anticipate new trends (include disruptive technologies), and innovate to embark on efficient transformation or evolving in delivering values and uniqueness, or expectations, supported by personal and team importance where their talent, skills, and high-performance are crucial towards developing sustainable urban transport and mobility for QoL experience;
23. Publish Local Sustainable Plan (Urban Transport and Mobility focus): a 5-year action plan (a living document with yearly review), subscribed to sustainability development concepts and guided by a business roadmap with localised Sustainable Development Goals, Targets and Actions aligned to strategies to be achieved with time-line, human and financial resources;
24. Dare to be benchmarked against sustainable urban transport and mobility index-ranking (national/regional/global) in energising ourselves to continually “Commit, Learn, Practice” for transformational edge in evolving and shaping liveable city towards Quality of Life experiences

5.3. Taking to the next level

Have we sharpened our saw with understand better of the urban living challenges and re-energise ourselves with the new way of thinking (strategic and tactical) towards sustainable development initiatives - urban transport and mobility focus, ready to take to the next level?

Perhaps, take the challenge to head-hunt a champion through open invitation to an opening engagement session in each city, and to select, lead a mid-size team of 10-15 city stakeholders to kickstarting the sustainable development initiatives by aligning the purpose, wants and needs with follow through in innovating the future we want and transformation to shaping our liveable city towards QoL experience, guided by the Sustainable Development Quadrant, a loop-framework in setting direction for kickstarting the sustainable development initiatives through individual and collective participation for mutual benefits with urban transport and mobility focus, created and illustrated in Figure 1.

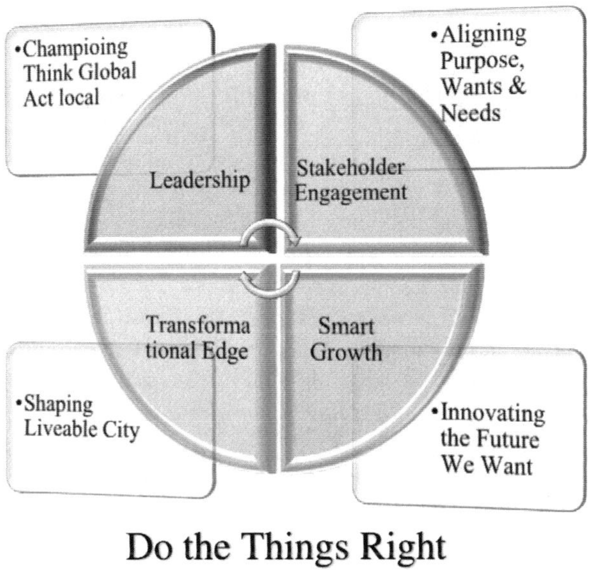

Figure 1. Sustainable Development Quadrant (urban transport and mobility focus), loop framework
"Urban transport system shapes our urban form and influences our lifestyle;
Mobility service determines our QoL experiences for liveable image of our city"- AW

6. Conclusion

Sharpening our saw to understand better the topics of: Liveability, Quality of Life, Sustainability associate to Sustainable Development (Urban Transport and Mobility focus) as the urban challenges through the "Mind – paradigm shifting" principle that creates new way of thinking: Strategic thinking in "Doing the Right Things", and Tactical thinking in "Doing the Things Right", illustrated in "the Art of Sustainability - Sustainable Development Quadrant" shall energise and direct us to kickstarting the sustainable development initiatives – urban transport and mobility focus. A way forward in shaping liveable city together towards Quality of Life experiences within the concept of sustainability that not only benefit us now but our future generation to come – a responsibility that we own and owe.

7. References

[1] Covey S R 1989 *The 7 Habits of Highly Effective People: Powerful Lessons in Personal Change. First Edition* (New York: Free Press)

[2] Jeffery N 2009 *Stakeholder Engagement: A Road Map to Meaningful Engagement #2 in the Doughty Centre 'How to do Corporate Responsibility' Series* (Doughty Centre: Cranfield School of Management)

[3] McChesney C, Covey S and Huiling J 2013 *The 4 Disciplines of Execution: Achieving Your Wildly Important Goals* (Riverside: Simon & Schuster)

[4] Stack L 2016 *Doing The Right Things Right: How the Effective Executive Spends Time. First Edition* (Oakland: Berrett-Koehler Publishers)

[5] Thompson M 2009 *The Organizational Champion: How to Develop Passionate Change Agents at Entry Level. First Edition* (USA: McGraw-Hill Education)

[6] Wong T L A 2017 *Raising the Bar towards Best Practice in Urban Planning: urban development & transport system*

[7] Brinded L and Martin, W 2016, 2017, 2018, *Mercer Quality of Living index: cities with the best quality of life in the world* (UK: Business Insider)
[8] Monocle Magazine 2016, 2017, 2018, *Quality of Life Annual Survey: Top 25 Cities*
[9] The Economist Intelligence Unit 2015, 2016, 2017, *The Annual Global Liveability Index Report*
[10] Wong T L A 2016 *Urban Transport vs Quality of Life*
[11] Wong T L A 2014 *Tracking Urban Traffic Congestion: Synergizing Towards Integrated Strategies*
[12] T&P/JKR Sabah 2018 *Transport & Traffic Source Book* (under construction)
[13] UN General Assembly 2012 *Resolution adopted by the General Assembly on 27 July 2012: The Future We Want* (United Nations)
[14] UN General Assembly 2015 *Resolution adopted by the General Assembly on 25 September 2015: Transforming our world: the 2030 Agenda for Sustainable Development* (United Nations)
[15] Arcadis 2017 *Sustainable Cities Mobility Index 2017 Bold Moves* (Arcadis)
[16] Dixon S, Irshad H, Pankratz D M and Bornstein J 2018 The Deloitte City Mobility Index: Gauging Global Readiness for the Future of Mobility *Deloitte Review* **23**
[17] Arthur D L 2018 *The Future of Mobility 3.0: Reinventing Mobility in the Era of Disruption and Creativity* March 2018
[18] *UNESCAP* 2018 S*ustainable Urban Transport Index (SUTI) Intro Brochure May 2018* (United Nations Economic and Social Commission for Asia and the Pacific)

IOP Conf. Series: Materials Science and Engineering **512** (2019) 012045 doi:10.1088/1757-899X/512/1/012045

Road infrastructure development and economic growth

C P Ng[1], T H Law[2], F M Jakarni[2] and S Kulanthayan[3]

[1]Department of Civil Engineering, Faculty of Engineering, Universiti Pertahanan Nasional Malaysia, Kuala Lumpur, Malaysia
[2]Department of Civil Engineering, Faculty of Engineering, Universiti Putra Malaysia, Serdang, Selangor, Malaysia
[3]Department of Community Health, Faculty of Medicine and Health Science, Universiti Putra Malaysia, Serdang, Selangor, Malaysia

E-mail: cpng@upnm.edu.my

Abstract. Stimulating economic growth and development of road infrastructure in economical lagging regions is the goal of many countries. This is because road infrastructure plays a crucial role by providing mobility for the efficient movements of people and goods, as well as providing accessibility to a wide variety of commercial and social activities. However, to achieve a sustainable economic growth, focusing on road infrastructure development alone would not be sufficient. Thus, this study analyse the contribution of road infrastructure development and other socio-economic factors that contributed to economic growth. To shed light on this issue, fixed-effects panel linear regression analysis was conducted using time-series cross-sectional data for 60 countries over the period of 3 decades from 1980 to 2010. The key finding of this study demonstrated that the growth in road length per thousand population, per capita export, per capita education expenditure and physical capital stock per worker contributed positively to economic growth. It was observed that there is an inverted U-shaped dependency relationship between urbanization and economic growth. That is, the economic growth increases at low urbanization levels but decreases once urbanization exceeds a threshold level. Moreover, it was also observed that the growth in road length per thousand population would facilitate export growth. In summary, this study suggest that policies focused on road infrastructure development should be implemented hand-in hand with other socio-economic and urban growth policies, in order to realize a sustainable economic growth.

1. Introduction

A breadth of research [1-6] revealed that the road infrastructure development were positively associated with economic growth. Generally, road infrastructure plays a crucial role by providing mobility for the efficient movements of people, goods and services as well as providing accessibility to land and a wide variety of commercial and social activities [7]. The provision of road infrastructure not only lower the physical barrier by stimulating the movements of people, goods [8] and services but also improve access to markets, social services and employment by reducing the overall transportation times and costs. The development or provision of high mobility road infrastructure such as expressway can increase the speed and improve the efficiency of domestic and international trades by reducing the transportation times and costs; whilst the development or provision of high accessibility road infrastructure such as local road allow easy land access and promote commercial and social activities at local level [9-10].

Empirical studies and scientific evidence show that the investment in paved roads, especially in countries with shortages of road infrastructure have been proven to provide an impressive return in economy [1]. In Sri Lanka, the investment in highway infrastructure has led to increases in the industries' output by more than 60% [2]. The road construction in China, especially rural roads has raised the country's national gross domestic product (GDP) four times higher than high-grade roads due to greater benefits than costs on the investments of rural roads [3]. Similarly, improvements in urban roads and major regional roads had increased the GDP share for both manufacturing and service industry in China [4]. Nevertheless, several scientific evidence showed that the growth of road infrastructure has no significant impact on economic growth [11-12]. For example, the improvements in highway infrastructure in the North Carolina counties in the United States between 1985 to 1997 has no significant impact on employment growth [11] and improvements in paved roads and highways in China from 1986 to 2003 has no significant effect on per capita GDP growth [12].

As shown above, the evidence on the role of road infrastructure development in economic growth is indecisive. However, historical development of road infrastructure indicated that countries that are more affluent generally have better road infrastructure, thus promoting agriculture, trade, industry and commerce, thereby helping these countries to sustain higher economic growth. In contrast, impoverished countries generally lack of road infrastructure and agriculture is the main economic source of income for most of the population, thus, their economic growth is limited. An efficient road infrastructure or a good road network created a competitive edge in moving goods economically. Conversely, lack of road infrastructure or poor road network systems are barriers to agriculture, industry and trade, and may hinder the process of urbanization and socioeconomic development. In view of this, this study intended to investigate the role of road infrastructure development on economic growth by utilizing a larger sample size with longer time period. Apart from this, empirical research also demonstrated that the positive effect of road infrastructure development on economic growth declines if the development of roads is increased in isolation from other socioeconomic development factors such as physical capital, human capital, health and education [1, 13]. Thus, the second objective of this study was to determine the effect of other socioeconomic factors (exports, education, physical capital stock and urbanization) that contributed to economic growth.

The rest of the paper was organized as follows. First, in Section 2, the literature of the socioeconomic factors on economic growth was discussed. Following this, in Section 3, the data and variables use in the study was described. Then, in Section 4, the methodology used to estimate the economic growth was discussed. Next, in Section 5, results and discussions were presented. Finally, the conclusion of the study, policy implications, limitations and the directions of the future research was discussed in Section 6.

2. Literature Review

The Solow growth model indicated that the human capital stock and the physical capital play a significant role in economic growth [14-16]. The human capital stock refers to the level of education and the life-long learning or the earning ability in the labour force, which contributed directly and indirectly, to economic growth. Empirical scientific evidence suggested that the investment in education is essential in accumulating human capital required to facilitate and sustain higher economic growth [17-18].

The physical capital stock is a measure of non-residential capital stock per worker and it is an integral component of economic growth. It is defined as the stock of capital goods, such as machines, equipment and technology required for production. It could not be measured directly but it can be estimated using the perpetual inventory method [19-22].

Economy theory shows that economic impulse based on expansion of domestic market is bound to be exhausted quickly. However, economic driven by export market promote a positive long run effects on economic growth [23-29]. This is because export market has no growth restriction on the demand for goods [29-32] and export expansion could increase the total factor productivity [30, 33].

Inevitably, development in road infrastructure would facilitate the export growth and consequently enhance the economic growth.

Urbanization refers to a transition process, in which, people move from rural to urban areas, which can have profound impacts on development and it is an inherent part of economic growth [34-36]. This is because urban areas provide higher economic opportunity and other opportunity such as education centre, health care services and credit facilities. The urbanization and economic growth theory is supported by Kuznets, where the population migration from rural to urban areas, is afforded by population growth and structural changes that generate productivity and wage gains per worker and the per capita income increase is associated with economic growth [37]. Scientific evidence also showed that there is an inverted U-shaped dependency relationship between urbanization and economic growth [38-40] that is economic growth increases at low urbanization levels but decreases once urbanization exceeds a threshold level.

In view of this, this paper attempted to investigate how road infrastructure development and other socioeconomic factors (export, education, physical capital stock and urbanization) facilitates economic growth.

3. Data and Variables

This section describes the data and variables used in this study. All data and variables used were secondary data obtained from various sources, such as the World Development Indicator (WDI) and Education Statistics (ED) from World Bank, Penn World Table 7.1 (PWT 7.1) and International Road Federations (IRF). The data used were annual observations on 'purchasing power parity converted gross domestic products (GDP) per capita (chain series) at 2005 constant price' or 'rgdpch', 'road length per thousand population' or 'roadpp', 'per capita exports of goods and services' or 'exp', 'per capita government expenditure on education' or 'edu', 'physical capital stock per worker' or 'kpw' and 'ratio of urban population to rural population' or 'urb' of 60 countries over a period of three decades from 1980 to 2010. The description of variables and data sources were given in Table 1.

Table 1. Description of variables and data sources.

Variable	Definition	Data Sources
rgdpch	Purchasing power parity converted gross domestic products per capita (chain series) at 2005 constant prices	PWT 7.1
roadpp	Road length per thousand population	IRF, WDI
exp	Per capita export of goods and services	WDI, PWT 7.1
edu	Per capita government expenditure on education	ES, WDI, PWT 7.1
kpw	Physical capital stock per worker	PWT 7.1, WDI
urb	Ratio of urban population to total population	WDI

The total sample used in the panel linear regression analysis was 1174. The data in each country has at least a minimum of 5 years observations. The list of countries included in this study was stated in Table 2.

The dependent variable in this study was the 'purchasing power parity converted gross domestic products per capita (chain series) at 2005 constant prices' or 'rgdpch'. This data was obtained from PWT 7.1 and was utilized as a proxy to describe the economic growth of a country.
The independent variable - the 'road length per thousand population' or 'roadpp', was derived by dividing the total length of roads by thousand population for a particular country in a particular year. The road length data was obtained from the World Road Statistic annual yearbooks published by the IRF while the population data was taken from WDI. The 'roadpp' variable was used as a proxy to describe the road infrastructure development of a country.

Other independent variables, such as 'per capita exports of goods and services' or 'exp', was derived by multiplying exports of goods and services as percent of GDP drawn from WDI with 'rgdpch' taken from PWT 7.1. The 'exp' was used as a proxy to describe the countries' export level.

The 'per capita government expenditure on education' or 'edu' was used as a proxy for describing human capital as indicated in the Solow growth model. This variable was computed by multiplying government expenditure on education as percent of GDP with 'rgdpch'. The per capita government expenditure on education was drawn from ES and WDI.

Data required to estimate the 'physical capital stock per worker' or 'kpw' was taken from PWT 7.1 and WDI. It was computed following the perpetual inventory method [19-22]. This data was transformed to natural log prior to the panel regression analysis to improve the interpretation and to reduce the heteroskedasticity in the data.

The urbanization ratio variable was used as a proxy to describe the urbanization level of a country. It was computed by dividing the total urban population by the total rural population. Population data was obtained from WDI.

Table 2. List of countries.

No.	Country	No.	Country	No.	Country
1	Argentina	21	Georgia	41	Netherlands
2	Austria	22	Germany	42	Norway
3	Azerbaijan	23	Greece	43	Pakistan
4	Bangladesh	24	Hungary	44	Panama
5	Belgium	25	India	45	Peru
6	Bolivia	26	Indonesia	46	Philippines
7	Botswana	27	Iran	47	Poland
8	Bulgaria	28	Ireland	48	Romania
9	Canada	29	Israel	49	Slovak Republic
10	Chile	30	Italy	50	Slovenia
11	China	31	Jordan	51	Spain
12	Costa Rica	32	Kazakhstan	52	Sweden
13	Croatia	33	Latvia	53	Switzerland
14	Czech Republic	34	Lithuania	54	Tanzania
15	Denmark	35	Macedonia	55	Thailand
16	Ecuador	36	Malawi	56	Tunisia
17	Egypt	37	Malaysia	57	Turkey
18	Estonia	38	Mexico	58	Ukraine
19	Finland	39	Moldova	59	United Kingdom
20	France	40	Morocco	60	United States

4. Methodology

This study utilize the panel linear regression with exogenous covariates to examine the relationship between the economic growth and the determinants such as road development, export level, education level, physical capital stock and urbanization level. In the panel linear regression, either a fixed-effects (FE) or random-effects (RE) model is employed to control for heterogeneity, as well as offered consistent and efficient estimates of model parameters in the existence of heterogeneity [41]. The FE model supposes that the country-specific intercept is correlated with the independent variables, on the other hand the RE model supposes that the country-specific intercept is a component of error term and not correlated with the independent variables [42]. The Hausman test was used to examine whether the FE model or RE model is more suitable. The null hypothesis is the country-specific intercept is not correlated with the independent variables and, therefore, rejection of null hypothesis indicates FE model is more suitable than RE model [41, 43-44].

As discussed earlier, road infrastructure development and other socioeconomic factors (export, education, physical capital stock and urbanization) facilitates economic growth. To examine these effects, a modified version of the framework developed by Fan and Chan-Kang [3] based on the Cobb-Douglas production function was adopted and the FE models were found to be more appropriate to describe the relationship between various factors and economic growth. The FE model for the panel linear regression can be written as follows.

$$rgdpch_{it} = \alpha_i + \mu_1 roadpp_{it} + \mu_2 roadpp \cdot exp_{it} + \lambda_1 urb_{it}^2 + \lambda_2 urb_{it} + \beta_i X_{it} + \varepsilon_{it} \quad (1)$$

where the dependent variable 'rgdpch' is the proxy for economic growth, sub-index i denotes country, t denotes period of time; α_i is the country-specific intercept (FE); μ, λ and β are the model parameters; ε is the error term; 'roadpp' is the proxy to measure the road infrastructure development, roadpp·exp is the interaction variable between the 'roadpp' and 'exp'; 'urb' is the proxy to measure the urbanization level, X are other independent variables which includes, 'exp', 'edu' and 'kpw' that act as proxies for measuring country's export level, education level and physical capital stock per worker, respectively.

5. Results and Discussions

Table 3 shows the descriptive statistics for the dependent and all the independent variables used in the analysis.

Table 3. Descriptive statistics.

Variable	Obs	Mean	Std. Dev	Min	Max
rgdpch	1174	15316.73	12187.87	431.42	51798.08
roadpp	1174	9.78	9.38	0.58	62.02
exp	1174	6128.50	6361.51	44.86	36060.10
edu	1174	791.30	743.34	11.05	3575.11
kpw	1174	1.46e+14	3.55e+14	2.14e+11	3.32e+15
urb	1174	3.18	4.87	0.16	41.39

Table 4 reports correlation matrix between all independent variables used in the panel linear regression analysis. The highest correlation observed was 0.7802 between 'edu' and 'exp'. Multicollinearity did not appear to be a major issue because the correlation coefficient is not greater than 0.8 [42]. Overall, the Hausman test results show that the FE models are more suitable than RE model for the panel data set.

Table 4. Correlation matrix.

	roadpp	exp	edu	ln(kpw)	urb
roadpp	1.0000				
exp	0.5011	1.0000			
edu	0.5958	0.7802	1.0000		
ln(kpw)	0.1031	0.1441	0.3185	1.0000	
urb	0.1402	0.4080	0.3411	0.0903	1.0000

In Table 5, five models were used to describe the individual relationship between each independent variable and economic growth for all countries. Model A, B, C, D, and E investigated the relationship between economic growth with road length per population, export level, education level, physical capital stock per worker and urbanization level, respectively.

Table 5. Estimates of economic growth.

DV = rgdpch	Model				
IV	A	B	C	D	E
roadpp	293.35**				
exp		1.06**			
edu			12.63**		
ln(kpw)				4491.90**	
urb^2					-62.80**
urb					4844.45**
constant	12449.15**	8835.70**	5323.30**	-125581.9**	2051.55**
no. of observations	1174	1174	1174	1174	1174
no. of groups	60	60	60	60	60
R^2 (within) sample	0.10	0.78	0.72	0.49	0.34
Overall model significance	177.61	417.57	66.09	435.3	302.38
Hausman test (chi-square)	13.44**	17.59**	42.76**	9.25*	24.50**
Turning point					38.57

*Remarks: ** significant at 1%, * significant at 5%*

In Table 6, Model F was used to investigate the relationship between economic growth and all the independent variables. The road length per population was interacted with export level to measure the interaction effect of these variables on economic growth as shown in Models G, H and I. As the interaction variable was the combination of the independent variables, the presence of it together with the independent variables in the regression can lead to multicollinearity. To address this issue, the independent variables 'roadpp' and 'exp' were removed from the model when the effects of the interaction variable was tested in Model G. In models H and I, the independent variable 'roadpp" and 'exp' was added one-by-one to check the multicollinearity between the interaction variable with the independent variable. Model J is the full model consisted all the independent variables, including the interaction variables. The presence of the interaction variable 'roadpp*exp' and 'roadpp' has a high correlation of 0.8151 that caused multicollinearity in models H and J.

Table 6. Estimates of economic growth.

DV = rgdpch	Model				
IV	F	G	H	I	J
roadpp	35.36**		-269.40**		-44.49*
roadpp*exp		0.012**	0.023**	0.0033**	0.0059**
exp	0.60**			0.54**	0.49**
edu	5.39**	7.58**	6.40**	5.34**	5.33**
ln(kpw)	991.67**	1449.29**	1563.50**	1035.70**	1088.54**
urb^2	-13.33**	-15.48**	-15.58**	-13.76**	-13.91**
urb	634.32**	931.54	885.95**	674.57**	688.15**
constant	-25630.65**	-39675.47**	-40536.94**	-26656.26**	-27868.16**
no. of obs.	1174	1174	1174	1174	1174
no. of groups	60	60	60	60	60
R^2	0.89	0.84	0.8706	0.89	0.90
Overall model significance	264.79	105.05	132.93	139.19	122.16
Hausman test (chi-square)	754.86**	215.09**	511.94	93.29**	1811.89**
Turning point	23.80	30.09	28.44	24.52	24.73

*Remarks: ** significant at 1% and * significant at 5%*

Models A and F showed that road length per thousand population was positive and statistically significant to economic growth while Models H and J showed that the relationship is negative but still statistically significant. The negative and statistically significant relationship in Models H and J was attributed to multicollinearity. In general, the growth of road length per thousand population would increase the economic growth. This is in line with previous research [1-6] as development of road infrastructure would contributed substantially to economic growth.

Models B, F, I and J showed that export was positive and statistically significant to economic growth. This result is consistent with the results implied by previous research in which export growth promotes economic growth and in turns, creating more opportunities in employment and improve productive capacity within a country [23-27].

In common with previous studies [17-18], education level was positively significant with economic growth as seen in Models C, F, G, H, I and J. Thus, these results implies that education is an important socioeconomic factor that contributed substantially to economic growth. The physical capital stock per worker was also positively related to economic growth, as shown in Models D, F, G, H, I and J which is in accordance with previous studies [14-16].

Consistent with the findings of previous studies [38-40], the results confirmed a statistically significant inverted U-shaped dependent relationship between urbanization and economic growth. The estimated turning point of 'urb' in Model I was 24.52, which was at about 97th percentile of the urbanization data.

The estimations in Table 6 also demonstrated that the coefficient for the interaction variable between the 'roadpp' and 'exp' in Models G, H, I and J was positively significant with economic growth. This shows that the road infrastructure development enhance the effect of export on economic growth. This is particularly true as the development in road infrastructure could improve the transport efficiency for delivering export goods.

6. Conclusion and Recommendations

This study discussed the effect of road infrastructure development (i.e. the growth of road length per thousand population) and other socioeconomic factors (i.e. exports, education, physical capital stock and urbanization) on economic growth. The results indicated that road infrastructure development, export, education and physical capital stock per worker contributed substantially to economic growth. This implies that policies to improve road infrastructure development, export, education and physical capital stock should be carried out hand-in-hand in order to sustain higher economic growth. Nevertheless, the inverted U-shape dependency relationship between urbanization and economic growth suggested that over urbanization (when urb > 24.52) would lead to a decrease in economic growth. As such, special attention should be paid when implementing the urban growth policy together with other socioeconomic development policies. Another key finding of this study was that the development in road infrastructure would leads to an increase in export growth that would also contributes to economic growth.

Several limitations of this study should be noted. First, the road length per thousand population was used in this study to describe the road infrastructure development without considering the road network capacity and connectivity. It is no doubt that countries with greater road network capacity and connectivity would certainly has greater impact on economic growth. Second, besides socioeconomic factors describe in this study, empirical research also shows that improvements in other transport facilities such as railways [45-46] and infrastructures such as telecommunications [47-48] contributed substantially to economic growth. Thus, including these factors in further study could definitely provide a more comprehensive overview on the factors affecting a country's economic growth.

7. References

[1] Canning D and Bennathan E 2000 The social rate of return on infrastructure investments, *Development Research Group, Public Economics and Private Sector Development and Infrastructure Group, World Bank*

[2] Gunasekera K, Anderson W and Lakshmanan T R 2008 Highway-induced development: evidence from Sri Lanka *World Development* **36(11)** 2371-89

[3] Fan S and Chan-Kang C 2008 Regional road development, rural and urban poverty: Evidence

from China *Transport Policy* **15** 305-14

[4] Ding C 2013 Transport development, regional concentration and economic growth *Urban Studies* **50**(2) 312-28

[5] Zhang X 2013 Has transport infrastructure promoted regional economic growth? – With an analysis of the spatial spillover effects of transport infrastructure *Social Sciences in China* **34**(2) 24-47

[6] Jiwattanakulpaisarn P, Noland R B, Graham D J and Polak J W 2009a Highway infrastructure and state-level employment: a causal spatial analysis *Papers in Regional Science* **88(1)** 133-60

[7] Meyer M D and Miller E J 2001 *Urban transportation planning* 2nd edition (New York: McGraw Hill)

[8] Motamed M J, Florax R J G M and Masters W A 2014 Agriculture, transportation and the timing of urbanization: global analysis at the grid cell level *Journal of Economic Growth* **19** 339-68

[9] Saunders S C, Mislivets M R, Chen J Q and Cleland D T 2002 Effects of roads on landscape structure within nested ecological units of the Northern Great Lakes Region, USA *Biological Conservation* **103**(2) 209-25

[10] Aljoufie M, Brussel M, Zuidgeest M and van Maarseveen M 2013 Urban growth and transport infrastructure interaction in Jeddah between 1980 and 2007 *International Journal of Applied Earth Observation and Geoinformation* **21** 493-505

[11] Jiwattanakulpaisarn P, Noland R B, Graham D J and Polak J W 2009b Highway infrastructure investment and county employment growth: a dynamic panel regression analysis *Journal of Regional Science* **49**(2) 263-86

[12] Banerjee A, Duflo E and Qian N 2012 On the road: access to transportation infrastructure and economic growth in China *National Bureau of Economic Research* 1-50

[13] Gannon C and Zhi L 1997 *Poverty and transport: Discussion Paper TWU-30* (Washington: World Bank)

[14] Solow R 1956 A contribution to the theory of economic growth *Quarterly Journal of Economics* **70** 65–94

[15] Barossi-Filho M and Silva RG 2005 The empirics of the Solow growth model: longterm evidence *Journal of Applied Economics* **8** 31-51

[16] Doucouiagos C and Ulubasoglu M A 2006 Economic freedom and economic growth: Does specification make a difference? *European Journal of Political Economy* **22** 60-81

[17] Dauda ROS 2010 Investment in education and economic growth in Nigeria: An empirical evidence, *International Research Journal of Finance and Economics* **55** 158-69

[18] Mercan M and Sezer S 2014 The effect of education expenditure n economic growth: the case of Turkey *Procedia – Social and Behavioural Sciences* **109** 925-30

[19] Hall B H, Cummins C, Laderman E S and Mundy J 1988 The R&D master file documentation. *NBER Technical Working Paper No. 72*

[20] Hall R E and Jones C I 1999 Why do some countries produce so much more output per worker than others? *Quarterly Journal of Economics* **114**(1) 83-116

[21] Law T H, Noland R B and Evans A W 2010 The direct and indirect effects of corruption on motor vehicle crash deaths *Accident Analysis and Prevention* **42** 1934-42

[22] Ng C P, Law T H, Wong S V and Kulanthayan S 2017 Relative improvements in road mobility as compared to improvements in road accessibility and economic growth: a cross-country analysis *Transport Policy* **60** 24-33

[23] Balassa B 1978 Exports and economic growth *Journal of Development Economics* **5** 181-9

[24] Bernard A B and Jensen J B 1999 Exceptional exporter performance: cause, effect or both? *Journal of International Economics* **47**(1) 1-25

[25] Aw B Y, Chung S and Roberts M 2000 Productivity and turnover in the export market: micro evidence from Taiwan and South Korea *The World Bank Economic Review* **14**(1) 1-65

[26] van Biesebroeck J 2006 Exporting raises productivity in Sub-Saharan African manufacturing firms *Journal of International Economics* **67(2)** 373-91

[27] De Loecker J 2007 Does export generate higher productivity? Evidence from Slovenia *Journal of International Economics* **73** 69-98

[28] Jun S 2007 Bi-directional relationships between exports and growth: a panel cointegration analysis *Journal of Economic Research* **12** 133-71

[29] Siliverstovs B and Herzer D 2007 Manufacturing exports, mining exports, and growth: cointegration and causality analysis for Chile (1960-2001) *Applied Economics* **39** 153-67

[30] Helpman E and Krugman P R 1885 Market structure and foreign trade *MIT Press, Cambridge, MA*

[31] Esfahani HS 1991 Exports, imports, and economic growth in semi-industrialized countries *Journal of Development Economics* **35** 93-116

[32] Easterly WR 2007 Free market and economic growth *International Symposium on Poverty Reduction and Beyond Development Strategies for Low Income Countries*

[33] Feder G 1983 On exports and economic growth *Journal of Development Economics* **12** 59-73

[34] Bertinelli L and Black D 2004 Urbanization and growth *Journal of Urban Economics* **56** 80-96

[35] Brülhart M and Sbergami F 2009 Agglomeration and growth: cross-country evidence *Journal of Urban Economics* **65**(1) 48-63

[36] Han X, Wu P L and Dong W L 2012 An analysis on interaction mechanism of urbanization and industrial structure evolution in Shandong, China *Procedia Environmental Sciences* **13**(1) 1291-300

[37] Kuznets S 1966 *Modern economic growth* (New Haven, CT: Yale University Press)

[38] Junius K 1999 Primacy and economic development: bell shaped or parallel growth of cities *Journal of Economic Development* **24**(1) 1-22

[39] Davis J and Henderson J V 2001 Evidence on the political economy of the urbanization process *Journal of Urban Economics* **53**(1) 98-125

[40] Henderson J V 2003 The urbanization process and economic growth: the so-what question *Journal of Economy Growth* **8** 47-71

[41] Wooldridge J M 2002 Econometric analysis of cross section and panel data *MIT Press, Cambridge, MA*

[42] Judge GG, Hill RC, Griffiths WE, Lutkepohl H and Lee TC 1988 Introduction to the theory and practice of econometrics 2nd Ed. (New York: John Wiley and Sons)

[43] Greene W H 2003 *Econometric analysis* (New Jersey: Prentice Education, Upper Saddle River)

[44] Baltagi B H 2005 *Econometric analysis of panel data* (New York: John Wiley and Sons Inc)

[45] Hou Q and Li S 2011 Transport infrastructure development and changing spatial accessibility in the Greater Pearl River Delta, China, 1990-2010 *Journal of Transport Geography* **19** 1350-60

[46] Pradhan R P and Bagchi T P 2013 Effect of transportation infrastructure on economic growth in India: the VECM approach *Research in Transportation Economics* **38** 139-48

[47] D´emurger S 2001 Infrastructure development and economic growth: an explanation for regional disparities in China *Journal of Comparative Economics* **29** 95-117

[48] Pradhan R P, Arvin M B, Norman N R and Bele S K 2014 Economic growth and the development of telecommunications infrastructure in the G-20 countries: a panel-VAR approach *Telecommunications Policy* **38**(7) 634-49

Value management / value engineering (VM/VE) application in Malaysian public construction projects: application of VM/VE study improved roads project sustainability

Mukhzani Abd Latif and Zawidatul Asma Ghazali

Value Management Unit, Public Works Department Malaysia, 50400, Kuala Lumpur, Malaysia

E-mail: mukhzani.jkr@1govuc.gov.my

Abstract. The application of Value Management/Value Engineering (VM/VE) in project development has a significant impact on the sustainability of a project. There is a relationship established between VM/VE methodology/process with the sustainability concept, which enables to enhance and achieve the intended project values. As such, to ensure value for money projects, the Malaysian Government has introduced the application of VM/VE in the development of public projects. Public Works Department (PWD) Malaysia is the organization responsible for VE application in public project development. In relation to sustainability, PWD has developed a sustainable rating tool known as Penarafan Hijau JKR (pH JKR) as a mean to determine the sustainability level of projects. This paper elucidates the relationship of the application of VM/VE towards sustainability of road projects by using selected VE case studies. Verification was done between the findings of VE for the selected case studies with the sustainability criteria of pH. It shows a significant result whereby VE findings improved the sustainability of road projects. It also provides evidence that the application of VE contributes to the sustainability of the project.

1. Introduction

1.1. Value Management/Value Engineering (VM/VE) in Malaysia

VM in Malaysia is defined as a structured multidisciplinary team approach to seek functional requirements of programmes/ projects/products/services to achieve optimization in delivering the client's perceived value [1]. This is consistent with other established definitions, such as by SAVE International [2], which defines VM as a systematic process that follows a Job Plan, involving a multidisciplinary team to improve the value of a project through the analysis of functions.

To achieve value for money in public projects, the Malaysian Government has ordered implementation of VM in public programs and projects valued at MYR 50 million and above through a directive circular by The Economic Planning Unit (EPU), Prime Minister Department in December 2009 [3] and updated circular in November 2015 [4]. The requirement is mandatory and failure to implement it may result in the rejection of the programs and projects.

To further enhance VM implementation in public projects, EPU in 2011 has launched Value Management Implementation Guide in Government Programs and Projects [5] with the intention to introduce the major VM intervention points in the project lifecycle and stipulate the process of VM implementation in public programs and projects. The VM study interventions are known as Value

Assessment (VA) at Strategic Planning Stage, Value Engineering (VE) at Design Stage and Value Review (VR) at Operation Stage which applied at various stages of the project's lifecycle

Aligned with this, PWD has established VE Application Guidelines for Public Projects which outlines the policies on VE governance, framework and study interventions focusing on PWD work process, and as well as to set standards of VE practice in public projects [1]. Additionally, PWD has also developed Value Engineering Facilitation Kit to provide quality information and practical assistance to facilitators in conducting VM/VE studies for public projects [6]. The kit outlines the concept of facilitation, the methodology of facilitating of VM/VE study and application of tools and techniques commonly used in facilitation VM/VE studies, which also known as VEMM. Both documents are in line and complement the circulars and VM Guide published by EPU.

1.2. Sustainable Development & Sustainable Construction

Sustainable development is defined as development that meets the needs of the present without compromising the ability of the future generations to meet their own needs [8]. To ensure sustainable development is achieved, it is crucial to integrate and harmonize the three pillars of sustainability, which are the economic, environmental and social factors. Balancing these factors are important and crucial; therefore, it needs to be sustained for the future survival and well-being of individuals and societies [9].

Sustainable development implicates improvement in the quality of life of humanity, the transformation of natural resources into the productive output, environmental conservation and economic progress. Consequently, United Nation in 2016 has outlined seventeen (17) goals of sustainable development to improve the lives of people as well as protecting the planet [9], which universally it applies to all countries/governments.

Meanwhile, sustainable construction is described as the subset of sustainable development philosophy. It explores the responsibilities of the construction industry to attain sustainability [10]. Therefore, to ensure sustainable construction delivered in the construction industry, cooperation is needed from the industry players such as the policymaker, clients, implementers, designers, contractors, and suppliers. These industry players are required to take necessary actions and play their roles towards achieving sustainability in the planning, design, construction, and operation of the building and infrastructure. Willingness and participation of all parties to explore new approaches in related fields in construction are required and crucial to ensuring the success of sustainable construction [11].

1.3. VM/VE and Sustainability

The integration of VM/VE with sustainability does not alter the VM/VE process. However, the study only emphasizes and focuses on sustainability [13]. Therefore, it is crucial to determine the sustainability of projects during the project development stage as well as other related factors involved (sustainability cost, planning, etc.). Well defined sustainability goals ensure sustainability issues/elements are addressed accordingly in the study in addition to other client values such as cost, functionality, quality, to achieve the best possible solutions or proposal for the projects. A structured process of VM/VE (stages and phases) provides guidance to the project team in achieving sustainability in the projects at the same time fulfilling the objectives and outcomes of the project.

1.4. pH JKR

PWD as the technical agency to the Malaysian Government has introduced several sustainability initiatives in the implementation of public projects. Recently in May 2017, PWD Sustainability and Green Mission 2.0 [12] was established to outline the activities to be carried out by PWD to ensure all government / public projects are executed sustainably. One of the activities is to utilize pH JKR as a green rating tool to determine sustainability the level of a project, which comprises of building and roads projects. pH JKR provides guidance to the road designers and project managers in considering sustainable elements in design and construction, further ensuring projects are delivered to the desired level of sustainability [7]. Seven criteria are used as rating criteria in pH JKR as listed in Table 1. Meanwhile Table 2, describes the classification of pH JKR rating specifically for road projects.

Table 1. Criteria of pH JKR.

Abb.	Criteria
SM	Sustainable Site Planning & Management
EW	Environment & Water
AE	Access & Equity
CA	Construction Activities
MR	Material & Resources
PT	Pavement Technologies
IN	Innovation

Table 2. pH JKR rating Classification.

Percentage (%)	Star	pH JKR Ratings
40 - 49	☆☆	Potential recognition
50 - 69	☆☆☆	Best management practices
70 - 84	☆☆☆☆	National excellence
85 - 100	☆☆☆☆☆	Global excellence

2. Improving Road Project Sustainability Through Application of VM/VE

The methodology of this study is outlined in Figure 1:

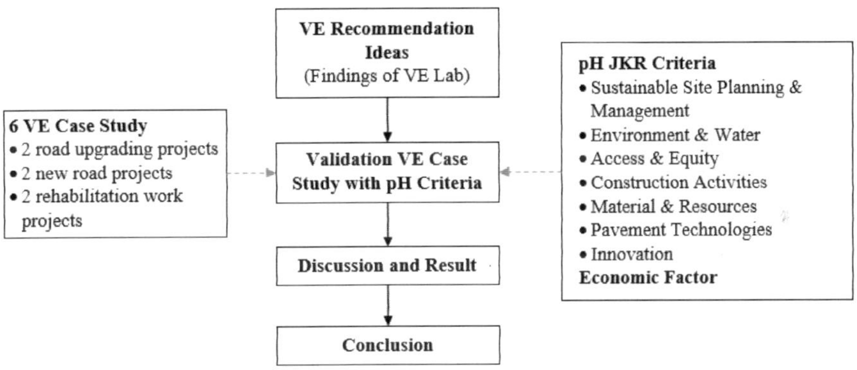

Figure 1. Study Methodology.

2.1 Relationship VEMM and Sustainability

This study investigates the relationship between PWD VM/VE methodology or known as Value Engineering Methodology Matrix (VEMM) in relation to sustainability for road projects. From the study, a relationship was established between VEMM and sustainability of roads projects. During the earlier stages of the VM/VE study (pre-lab stage, information phase and function analyses in the lab stage), VEMM activities ensured that sustainability issues are discussed and addressed accordingly. The application of appropriate tools and techniques as outlined in VEMM helped to identify issues, concerns, details, directions, and others on the sustainability of the road projects. Meanwhile, the later stage of the VM/VE study (creative phase, evaluation phase, development phase, presentation phase and post-lab stage), focusses on resolving the sustainability issues by generating new/alternative ideas, evaluate and further develop the ideas into workable solutions. Table 3 summarizes the relationship between VEMM and sustainability.

Table 3. The relationship between VEMM with sustainability.

Study stages / Lab phases	Key VEMM activities	Sustainability Focus
Pre Lab Stage • **Check project readiness for VE Study** • **Gather and synthesize project information** • **Plan and prepare for VE Lab activities**	• Interface with project team • Collate relevant and sufficient project information • Commence the initial study • Develop lab program and arrange logistics for VE Lab execution	• Client direction on project sustainability • Compliance status of authorities' requirements (EIA, EMP, TIA, RSA. pH JKR Rating) • Compliance status of project objectives, eg: road user safety, road user comfort, impacts to the environment and local surrounding • Design concept and alignment selection – impact to sustainability • Economy aspect of the latest project cost • Identification of Subject Matter Expert (SME) to advise on project sustainability
Lab Stage - Information Phase • **Develop an understanding of the project and share the required details** • **Gather, synthesize and verify project information**	• VE study briefing • Project information briefing • Confirmation VE study model(s) - as baseline for optimization / improvement. • Confirmation of Client Value System (value criteria to be incorporated into the design solutions and project deliverables) • Confirmation Lab working groups and VE study scopes • Establishment of project parameters to be studied.	• Client direction on sustainable elements made explicitly • Confirmation / verification status on authorities' requirements • Confirmation / verification status on the economic status of the latest project cost • Confirmation / verification status on project objective's performance • Relevant sustainability cost allocation, i.e., OSHA, EPW, etc. • Issues, risks, and implication arise on project sustainability • SME input/views on project sustainability
Lab Stage - Function Analysis Phase • **Understand the functions of the project/ elements / components /system** • **Identify mismatches and potential value improvement on functional requirement**	• Verification project functions • Analyse functions of elements / components / systems • Identification of value mismatches	• Identification and analysis of project / elements / components / systems functions related to sustainability
Lab Stage -Creative Phase • **Generate a broad list of creative and innovative ideas as alternative ways to enhance the value of the project.**	• Generation of alternative ideas to the original design proposal that may improve the value of the project.	• Generation of alternate ideas to improve/overcome sustainability issues/problems arise during the information phase and function analysis phase

IOP Conf. Series: Materials Science and Engineering **512** (2019) 012046 doi:10.1088/1757-899X/512/1/012046

Table 3. The relationship between VEMM with sustainability (continued)

Lab Stage - Evaluation Phase • **Shortlist the generated ideas to potential ideas for value improvement**	Evaluation of generated ideas by: • Judgment of ideas • Categorization of the judged ideas	• Evaluation of ideas related to sustainability for evaluation • Potential ideas will be further evaluated in the development phase
Lab Stage - Development Phase • **Analyze further the viability of ideas (developing it into workable solutions) for recommending best or preferred options**	• Development of shortlisted ideas into workable options • Consensus on the best solutions as VE Study recommendations (through a plenary session) • Summarization of all recommended ideas and findings • Development of an action plan for post-lab activities • VE Study achievement review and identification of lessons learned.	• Development of shortlisted sustainability ideas into workable options and determine the best solutions • Consensus on best solutions selected ideas • Registration of further action required on project sustainability in post lab action plan • Documentation of any lesson learned related to sustainability in the report
Lab Stage - Presentation Phase • **Present and gain acceptance from stakeholder**	• Presentation of VE study recommendations and findings • The consensus from lab team members/stakeholders on VE Study recommendations and findings.	Highlighted ideas/achievement related to sustainability
Post Lab Stage • **Prepare and submit VE report** • **Monitor and assess implementation of VE Study recommendation and finding** • **Assess performance of VE Study implementation**	• Documentation of VE Study outputs. • Dissemination of VE Study Report • Follow up on agreed Action Plan • Monitoring and assessment of the implementation of VE recommendations and findings • Assessment of performance VE Study implementation	• Documentation of recommendation and findings related to sustainability elements. • Monitoring and assessing the implementation and performance of sustainability elements

2.2 VE Case Studies Selection

Six (6) VE case studies were selected for the validation process. To represent the uniqueness of road projects, various types of projects such as upgrading works, new construction works, and rehabilitation works have been considered for this study. For all the selected VE case studies, the VE intervention level is set at the development of details design and development of a bill of quantities. The VE study scope involved are elements and components of the road design. Listed below are the details of the VE case studies selected and Table 4 summarizes the findings and achievements VE case studies.

VE Case Study 1

The proposed project involves the upgrading works of the existing main trunk road to the city center which experiences severe congestions in the morning and evening. It is located in an urban area and the major scope of works involves widening works of existing 4 lanes dual carriageway to 6 lanes dual carriageway (U5 standard) for almost 8.1 km, construction of 4 interchanges (flyover) and 2 new bridges along the road. The proposed project is expected to improve the level of services (LOS) of the road segments as well as the relevant intersections. The approved project cost is approximately RM 420 million. In the VE study, 74 ideas for improvement have been generated whereby 43 ideas have been recommended for implementation. Generally, VE study manages to optimize project design and documentation with a cost saving of 0.23%.

VE Case Study 2

The proposed road project involves the upgrading works of existing rural coastal road to the Standard of R5. The existing road is experiencing road safety and congestion issues as there is no median provided along the 4 lanes roads and overcapacity of the existing intersections. The project scopes involve upgrading works (inclusive of geometric improvements, pavement rehabilitation) of 50 km road which comprises of 45 km of 4 lanes single carriageway to 4 lanes dual carriageway (with median) and 5 km of 2 lanes single carriageway to 4 lanes dual carriageway. It also involves upgrading 5 grade junctions to interchanges (flyover), construction of U-turns, single span bridge and pedestrian bridges for road crossing. The total project cost is estimated about RM 630 million. 99 ideas have been generated in the study and 59 numbers are recommended to be implemented. Generally, design and project cost have been optimized during the VE study with improvements to the design, item rates, and quantities as well as engineering judgment made (due to insufficient data). A cost saving of 0.87% has been recorded during the study.

VE Case Study 3

The proposed road project is to complete the remaining work of abandoned road alignment as well as damages that have occurred since 2011. It involves rehabilitation works and construction of a new road for a length of 20 km which include incomplete road alignment inclusive of road works and structure works. The road is categorized as the major road linking the west coast and east coast of Malaysia and has been designed to R5 Standard. The scope of works consist of construction of new road and structures (for incomplete road), repair works (for the road, structures, and drainage) and other related works. The cost is approximately RM290 million. In the VE study, 25 ideas have been generated and 19 are accepted for recommendations. 0.45% of cost saving has been achieved, and project/design elements are also reviewed during the VE study.

VE Case Study 4

The proposed road project involves the rehabilitation of pavement works of a 40km existing road inclusive of bridge replacement and drainage improvement along the segment. This project will improve the safety and comfort of the road users as the condition of the existing road experienced severe defect and damages. Furthermore, the existing bridges are narrow and the drainage system are improper. The approved project cost is RM 85 million. In the VE study, 39 ideas for improvement have been generated and 26 ideas have been recommended for implementation. Generally, some improvements on sustainability have been achieved during the VE study and 1.05% of cost have been saved. The VE study also managed to review the construction and project elements as well as the project documentation.

VE Case Study 5

The proposed project is to construct a 25 km new bypass road which conforms with JKR R5 standard and is 4 lanes dual carriageway. The construction of this new road will ease the congestion of the existing road and provide an alternative road for the locals to access the town. The total project cost is estimated at around RM 370 million. 75 ideas have been generated to improve and optimise the project value. However, only 33 are accepted for recommendations. Generally, VE study manages to optimize the project design and cost with a cost saving of 0.95%.

VE Case Study 6

The proposed road project involves the construction of a new alternative road for approximately 7 km with the configuration of 4 lanes dual carriageway. The project is a part of the overall alignment of a new road which consists about 75 km. However, due to cost constraint, only few packages have been approved and VE case study 6 is one of the approved package. The proposed new road will be connect the existing road networks to the surrounding area. The approved project cost is about RM 190 million. In the VE study, 48 ideas for improvement have been generated with 23 ideas have been recommended for implementation. Generally, VE study manages to optimize the project design and cost with a cost saving of 0.03%.

Table 4. VE case studies findings

VE Case Study	Type of Projects	Sustainability Achievements
1	Upgrading Works	• Cost optimization (cost saving 0.23%) • Optimization of land acquisition area and cost • Improvement of pedestrian safety (improve the location of bus layby facilities, provision of pedestrian bridge) • Provision and improvement of Occupational Safety & Health (OSHA) items and rates • Provision and improvement of the Traffic Management Plan (TMP) items • Revision of earthwork quantities (cutting/filling) and omitting excess import material • Revision of design concept of structure (from slip in slip out bridge to box culvert)
2	Upgrading Works	• Cost optimization (cost saving 0.87%) • Revision quantities for site clearing and related works to minimize affected area • Improvement of road safety features (provision of road lighting) • Revision of access and junctions design (improve the length of auxiliary lanes) • Reuse and recycle existing material for new construction (existing road base as new subbase material, milling waste as new road base material) • Provision of maintenance items after construction to ensure effective maintenance works (maintenance of bridges and culverts) • Provision of pavement performance monitoring after construction • Provision of an alternative method for pavement rehabilitation (partial reconstruction) • Using the latest pavement guidelines as design reference Arahan Teknik Jalan (ATJ) 5/85 (Pindaan 2013) – improve pavement performance • Revision of design concept of the bridge (bridges re-profiling proposed to improve safety and comfort)
3	Rehabilitation works	• Cost optimization (cost saving 0.45%) • Revision of project length to improve safety and accessibility of the road user (to include existing junction improvement) • Provision of temporary protection for slopes (eroded & bare slope) • Provision and improvement of EPW design and quantities (for the physical works) • Improvement of road safety features (provision of road lighting at bridges area and increase quantities for NJB to accommodate substandard curves) • Provision of Occupational Safety & Health (OSHA) items and rates • Improvement of Traffic Management Plan (TMP) items and rates • Revision of platform level in design to minimize earthwork quantities (cutting/filling) and right of ways (ROW) affected • Provision of building information modeling (BIM)

Table 4. VE case studies findings (continued)

4	Rehabilitation works	• Cost optimization (cost saving 1.05%) • Improvement of drainage system (earth drain to U drain – improves flow and maintainability) • Improvement of road safety features (traffic light system-vehicle actuated system provided) • Revision of access (closing conflict access near to the bridge) • Improvement of Traffic Management Plan (TMP) items • Provision items for pavement rehabilitation (milling, regulation, etc)
5	Construction of a new road	• Cost optimization (cost saving 0.95%) • Revision of bridge geometric design (for uniformity and comfort) • Provision and improvement of EPW design and quantities • Improvement of drainage system (provision of subsoil drain at cut areas) • Provision and improvement of Road Safety Audit items (Stage 1-5) • Review of access and junction design (provision of U-turns, from the roundabout to signalized junction) • Improvement of pedestrian safety (provision of bus layby facilities) • Provision of a Traffic Management Plan (TMP) items • Provision of maintenance items after construction to ensure effective maintenance works • Road lighting using LED technology for energy efficiencies • Provision of International Roughness Index (IRI) testing
6	Construction of a new road	• Cost optimization (cost saving 0.03%) • Optimization of land acquisition area and cost • Provision of slopes protection (turfing) • Improvement of drainage system (revision of culverts numbers) • Improvement of Road Safety Audit items (Stage 4-5) • Improvement of EPW design and quantities • Revision of design concept of the box culverts (from 2 numbers adjacent box culvert to single span bridge structure)

2.3 Result Verification

Verification was made between VE recommendation ideas of the VE case studies with the pH JKR criteria and evaluated for relevant sustainable VE recommendation ideas. Figure 2, Figure 3 and Figure 4 show the verification results of this study.

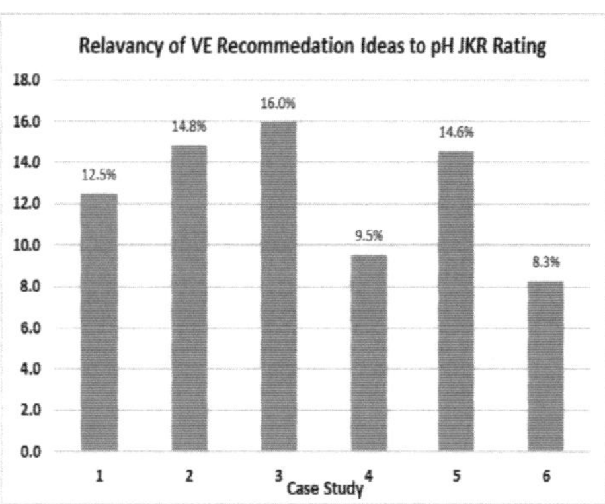

Figure 2. Relevancy of VE Case Studies against Sustainability.

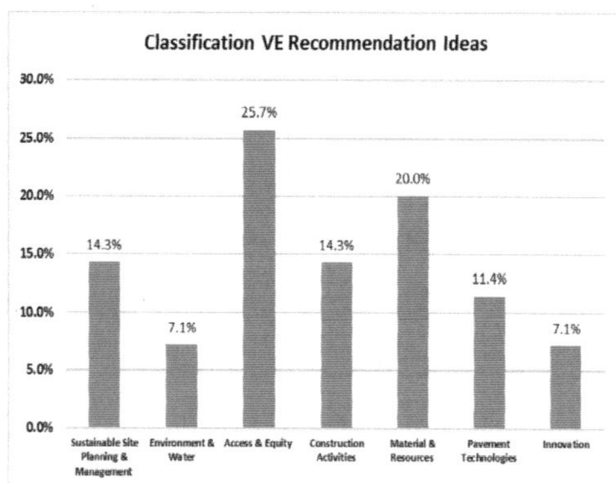

Figure 3. Classification VE Recommendation Ideas.

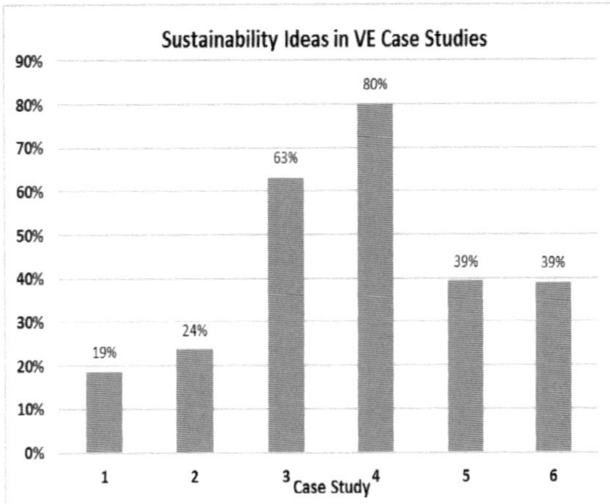

Figure 4. Percentage of Sustainable Ideas in VE Case Studies.

2.4 Discussion of Result

The detailed findings of this study are described as follows:

1. VE application contributes to the improvement of the sustainability of road projects regardless of the type and scope of the project. Additionally, implementation of VE ensures sustainability issues/elements of road projects are discussed and addressed accordingly. Consequently, the sustainability of road projects is improved as shown in Figure 2
2. VE improves the three pillars of the sustainability of road projects which are economic factor, social factor and environment factor. VE application has contributed significantly to the pH criteria representing social and economic factor as shown in Figure 3. While, for the economic factor, VE case studies have indicated cost avoidance/saving from the project costs ranging from 0.03% to 1.05%, which ensures that the final projects cost is within the approved project cost.
3. The total VE recommendation ideas related to sustainability was recorded as high as 80% as shown in Figure 4. Hence, it verifies that VE application to road projects contributes to sustainability.

In general, VE application using the VEMM approach shows a significant impact on the improvement of sustainability in road projects supported with appropriate VE tools and techniques. Table 5 highlights the area of VE recommendation ideas which contributes to the improvement of sustainability in road projects

Table 5. Highlighted VE Recommendation Ideas Contributes to Sustainability.

pH JKR Criteria	Area of VE Recommendation Ideas
Sustainable Site Planning & Management	• Optimization of land acquisition area • Provision of temporary protection for slopes (eroded & bare slope) • Revision of design and quantities for site clearing and related works to minimize affected area • Revision of bridge geometric design (for uniformity and comfort) • Revision of project length to improve safety and accessibility of the road user
Environment & Water	• Improvement of drainage system (culverts, surface drainage) • Provision and improvement of EPW design and quantities
Access & Equity	• Improvement of road safety features (signboards, barriers, road marking, road lighting, traffic light system) • Improvement of pedestrian safety (bus layby facilities, pedestrian crossing) • Provision and improvement of Road Safety Audit items (Stage 1-5) • Revision of access and junctions design
Construction Activities	• Provision and improvement of Occupational Safety & Health (OSHA) items (Safety officer, personal protective equipment, site safety, reporting, etc) • Provision and improvement of Traffic Management Plan (TMP) items (design and quantities, for temporary diversion, temporary works)

Table 5. Highlighted VE Recommendation Ideas Contributes to Sustainability (continued)

Material & Resources	• Revision of platform level in design to minimize earthwork quantities (cutting/filling) and right of ways (ROW) affected • Reuse and recycle existing material for new construction (topsoil, road base material, milling waste, etc) • Provision and improvement of maintenance items after construction to ensure effective maintenance works • Road lighting using LED technology for energy efficiencies
Pavement Technologies	• Provision of pavement performance monitoring after construction • Provision of an alternative method for pavement rehabilitation (partial reconstruction, CIPR, etc) • Using the latest pavement guidelines as design reference Arahan Teknik Jalan (ATJ) 5/85 (Pindaan 2013) – improve pavement performance • Provision of International Roughness Index (IRI) testing
Innovation	• Provision of building information modeling (BIM) • Revision of design concept (bridges re-profiling, a combination of structures, changes to the type of the structure, etc)

3. Conclusion
This paper has demonstrated the VE application using the VEMM approach contributes to the improvement of sustainability in road projects. VE case studies validation provides evidence and quantifies the contribution of VE application in improving sustainability (economic factor, social factor and environment factor). Furthermore, VE application facilitates the designer to improve the sustainability elements in the road design although it has been considered at the earliest stage. This indirectly enhances and compliments the application of pH JKR as a sustainable rating tool for public projects although economic factor is excluded from the evaluation criteria in pH JKR. However, the result shown is subjective and may differ subjected to the comprehensiveness of sustainability consideration in road design by the designer prior to VE study and changes in project scenarios due to project issues, problems, information availability, new requirements, etc.

4. References
[1] Public Works Department (PWD) 2013 *Value Engineering Application Guidelines for Public Projects* (Kuala Lumpur: Public Works Department)

[2] SAVE International® 2007 *The International Value Standard and Body of Knowledge* (SAVE International®)

[3] Economic Planning Unit (EPU) 2009 *Value Management Implementation Guideline No 3/2009* (Putrajaya: Malaysia Prime Minister's Department)

[4] Economic Planning Unit (EPU) 2015 *Improvements in Value Management Implementation and Guidelines & Rules for Building Planning in Federal Government Programs / Projects No 1/2015* (Putrajaya: Malaysia Prime Minister's Department)

[5] Economic Planning Unit (EPU) 2011 *Value Management Implementation Guide in Government Programmes / Projects* (Putrajaya: Malaysia Prime Minister's Department)

[6] Public Works Department (PWD) 2015 *Value Engineering Facilitation Kit, Malaysian Public Works Department, Kuala Lumpur* (Kuala Lumpur: Malaysian Public Works Department)

[7] Public Works Department (PWD) 2015 *JKR Green Rating Manual (Road Sector)* (Kuala Lumpur: Public Works Department)

[8] United Nations 1987 *Report of the World Commission on Environment and Development (WCED) Our Common Future* (United Nations)

[9] United Nations *The Sustainable Development Goals, 17 Goals to Transform Our World* Available from http://www.un.org/sustainabledevelopment/sustainable-development-goals/.

[10] Nazirah ZA 2009 Sustainable construction in Malaysia-developers Awareness *Proceedings of World Academy of Science, Engineering and Technology* **41** pp 807-13

[11] CIDB Malaysia 2013 *Guidebook on Planning and Implementing Green Practices for Building Construction Works* (Construction Industry Development Board Malaysia).

[12] Public Works Department (PWD) 2017 *Sustainability & Green Mission 2.0*(Kuala Lumpur: Malaysian Public Works Department).

[13] Abd Karim SB et al 2017 Advocating Sustainable Building Paradigm Through Value Management *Journal of Design and Built Environment Special Issue* pp. 95-106.

Acknowledgement

Sincere appreciation extended to the co-author- Mdm Zawidatul Asma, Ir. Ab Hamid Mohd Daud, Senior Director of CPAB, JKR. Greatest thanks to all VM Unit team members for continuous support; friends and families, and to all the people who are directly and indirectly involved in this study.

IOP Conf. Series: Materials Science and Engineering **512** (2019) 012047 doi:10.1088/1757-899X/512/1/012047

Highway infrastructure: visions & challenges in the next decades

Harun Sazali and Ismail Mohd Firdaus

Planning Division, Malaysian Highway Authority, KM-6, Jalan Serdang-Kajang, 43000 Kajang, Selangor Darul Ehsan, Malaysia.

E-mail: Sazali@llm.gov.my & firdaus.ismail@llm,.gov.my

Abstract. Malaysia has achieved a significant economic and social progress over the past several decades through the implementation of First Malaysia Plan (1966-1970) to Ninth Malaysia Plan (2006-2010). Further the Government of Malaysia aims at achieving high income nation status by 2020 through the introduction of Malaysia Vision 2020. The vision is crafted by the Fourth Malaysian Prime Minister to shape Malaysia as a developed nation not only in an economic sense but also in terms of social justice, political stability, government system, quality of life, social and spiritual values, national pride and confidence. The initiation of Greater Kuala Lumpur/Klang Valley (KL/KV) has resulted in rapid growth in urbanisation, transportation, infrastructure and construction industry sectors. Malaysian infrastructure development especially the transportation sector is going through rapid changes in terms of adapting state of the art construction technology, adjusting to industry evaluation, and use of intelligent transportation system to achieve the set goals for the future and equip the nation to brace the Industry Revolution (Industry 4.0). To stay relevant, the highway infrastructure need to be to be smart and green. The smart highway will provide better, safer and greener highway through digital and green technology through the introduction of Intelligent Transportation System (ITS). The idea of smart highway is not only limited to the application of technology but also to achieve the goal of sustainable and green environment. It is in line with the Green Technology Master Plan Malaysia (GTMP) 2017-2030 that outlines the strategic plans for green technology development to create a low-carbon and resource efficient economy. The challenges for smart highway implementation is waiting ahead. Issues related to technology acceptance, political will and policies, creativity and innovations, construction methods, ethics and changing in human values need to be discuss in depth.

1. Introduction

Malaysia has achieved a significant economic and social progress over the past several decades through implementation of various economic plans. Presently with more competitive and challenging global economic environment, Malaysia has reached a defining moment in its development path.

Malaysia has embarked on comprehensive economic agenda to achieve the developed nation status and among top 20 countries in the world through various initiatives such as Vision 2020, Government Transformation Programme (GTP), New Economic Model (NEM), Economic Transformation Programme (ETP) and 10th Malaysia Plan [1].

The Malaysian infrastructure development especially the transportation sector is going through rapid development to equip the nation to brace the Industry Revolution (Industry 4.0) in terms of technology and innovation.

This paper aims at presenting the various strategies to overcome the challenges and key factors to be amongst the top countries in economic development, citizen well-being and innovation.

2. Malaysia Economic Planning

Traditionally, the progress in Malaysian economic and social development over the past several decades is through the implementation of First Malaysia Plan (1966-1970) to Ninth Malaysia Plan (2006-2010). Besides the Government of Malaysia also aims at achieving high-income nation status by 2020 (Malaysia Vision 2020).

Malaysian Vision 2020 was unveiled by the fourth Prime Minister of Malaysia, Tun Dr. Mahathir bin Mohamad at the inaugural meeting of the Malaysian Business Council on 28 February 1991. The Malaysian cabinet approved the Vision unanimously and mandated for an immediate national conference, which took place in December 1991 [2].

The vision is crafted to shape Malaysia as a developed nation not only in an economic sense but also in terms of social justice, political stability, government system, quality of life, social and spiritual values, national pride and confidence [3].

It is also a Malaysian Government desire to become one of the top 20 countries in the world in terms of economic development, citizen well- being and innovation. The Greater KL/KV is already on the global map as one of the iconic cities in Southeast Asia. It boasts world-renowned landmarks such as Petronas Twin Towers, a unique blend of diverse cultures and heritage, an extensive road network and high quality basic services such as water and electricity. However, it now faces fierce competition from neighbouring cities in development progress and attracting multinational talent. Its liveability lags many other Asian Cities, inadequate public transport system and many other natural assets remain untapped.

3. Malaysian Transport Infrastructure

The initiation of greater KL/KV has resulted in rapid growth in urbanisation, transportation, infrastructure, and construction industry sectors beside others. Availability of space in the urban environments is very scarce and hence the need for integration of infrastructure facilities and their coexistence.

Greater KL/KV regional connectivity will be enhanced and at the same time, intra city connectivity will be improved with light rail, mass rapid transit systems and upgraded pedestrian facilities.

The Malaysian transport infrastructure mainly comprise public transport systems like Light Rail Transit (LRT)/ Mass Rapid Transit (MRT)/ monorail/ inter urban bus and rail services and extensive highway network system besides other modes of transport like airways.

Since the above existing transport system is still inadequate to meet the current demand as well the future, additional public transport systems are being planned and developed to enhance the capacity and meet the future demand such as Light Rail Transit (LRT 3), Mass Rapid Transit (MRT Line 2), High Speed Rail (HSR) and East Coast Rail Link (ECRL).

In the aspect of Malaysian highway infrastructure, Malaysia can be considered to be one of the best among developed nations. In Malaysia currently, there are about 31 inter urban & intra urban highways/expressways in operation, 6 are under construction phase and another 5 are in planning stage besides several other major arterial roads. This highway network system has been contributing significantly towards the country's economic growth from time to time.

The North-South Expressway which is a brainchild of Malaysian fourth Prime Minister, is the Malaysia's longest access controlled highway with the total length of about 823km which connects Bukit Kayu Hitam near Malaysia - Thailand border on northern side and Johor Bahru at the southern portion of Peninsular Malaysia and to Singapore [4].

This expressway was fully operational in the year 1994, acts as a backbone connecting major cities and towns in Western Peninsular Malaysia and has spurred the economic growth significantly in the adjoining regions all along its route.

4. Challenges for Highway Infrastructure

Malaysian infrastructure development especially the transportation sector with the above large scale ongoing projects is presently going through rapid changes in terms of adapting state of the art construction technology, adjusting to industry evaluation, adapting to internet of things, and use of integrated transport information system to achieve the set goals for the future and equip the nation to brace the Industry Revolution 4.0 (Industry 4.0).

In line with the Industry 4.0, the concept of Smart City has been promotes through the globe even though the absolute definitions of "smart' is still questionable. However, the Smart City and Digital City are the most used terminologies in literature to indicate the smartness of a city.

Citing the definition by Washburn et. Al (1980) [5], Smart City is stated as "the use of Smart Computing technologies to make the critical infrastructure components and services of a city-which include city administration, education, healthcare, public safety, real estate, transportation and utilities-more intelligent, interconnected and efficient". It can be imagined that smart city will be equipped with electronics data collections sensors to ease the traffic flow, made parking easier, improved the energy-efficient and comfort in building as well as the efficiency of public lighting.

In the coming decades, the Malaysian highway infrastructure shall face mainly the following challenges but not limited to:

a) Space availability for development and sharing of space for coexistence;
b) Utilisation of existing highway infrastructure to its maximum time & capacity
c) Upgrading to latest construction technology;
d) Adoption of intelligent transportation system;
e) Inclusion of Creativity & Innovation;
f) Understanding of internet of things and its application;
g) Integration of existing and planned public transport system;
h) Changing value and lifestyle;
i) Legal issues and risk factors; and
j) Policy framing and political will

The introduction of smart city concept also need to be considered by the highway industry in order for the highway infrastructure to be smart and green to complement the smart city. The smart highway will served as a connection from one smart city to the other smart city with the mission to provide better, safer and greener highway through digital and green technology.

5. Smart Highways

In relation to Smart City concept, there is no absolute definition on 'smart' when it comes to highway. Nevertheless, the introduction of Intelligent Transportation System (ITS) has given some light toward the implementation of a smart highway.

The terms of Intelligent Transportation System (ITS) has been widely promoted and recognized globally. ITS refer to integrated applications, employing combinations of information, communications, computing, sensor and control technologies and can be applied to various sectors such as transportation, building and vehicles. These technologies are already revolutionizing the way we live and work and they can equally well revolutionize the way that we travel.

In the transportation sectors, Intelligent Transportation Systems (ITS) aim to improve transport safety and mobility and reduce vehicle emissions [6]. Many such technologies have been developed to enhance vehicle safety, to prevent crashes, reduce trauma during a crash or to reduce trauma following a crash and used in mitigating traffic congestion. The ITS solution will enhance transport performance in cities which at the same time would not exclude further construction of roads and other transport infrastructure. Figure 1 shows the typical application of ITS.

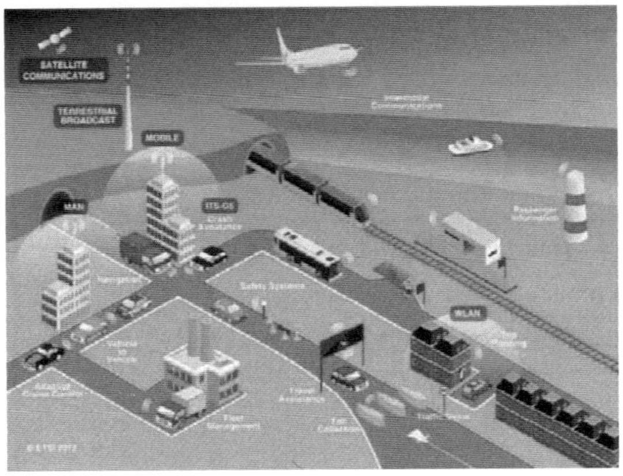

Figure 1. Typical application of ITS.

Some ITS applications are already existence in Malaysia and could be mostly seen on the expressway. The applications are listed as follows:

- Use of state of the art Intelligent Transport Management Systems, Traffic Control & Surveillance System, Toll Collection System, and Emergency Response Plans.
- Progression of Malaysian highways to migrate from Touch & Go/Smart TAG systems to Radio Frequency Identification (RFID) in 2018 & Multi Lane Free-Flow (MLFF) in 2025. Figure 2 shows the RFID implementation and Figure 3 shows the MLFF gantry in testing phase;

Figure 2. ETC System - RFID implementation.

Figure 3. ETC System - MLFF Gantry in Testing.

Efficient Use of Internet of Things, Big Data & Cloud Computing to make use in real time traffic management & surveillance. This includes traveller information system, advanced traffic information system, connecting vehicles, fleet management, congestion management, congestion pricing, real time traffic signal control system, parking management system, vehicle weigh-in-motion, vehicle entrance permission, vehicle on-board units (OBU), automatic enforcement system and efficient transport planning by public authorities (Refer to Figure 4)

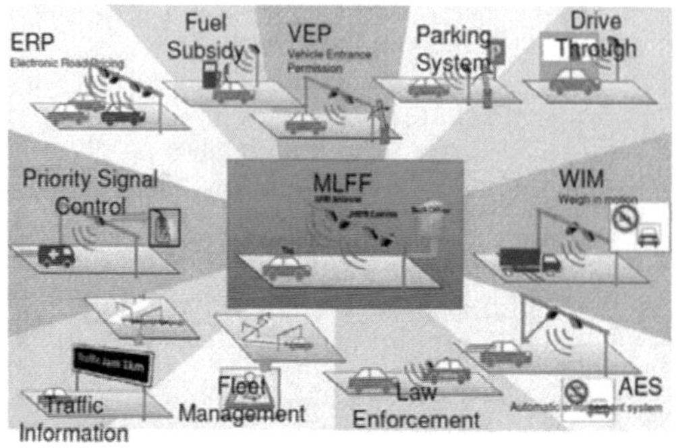

Figure 4. Various Applications of ITS.

The implementation of Intelligent Transportation System (ITS) are now globally accepted as the modern approach to solve issues related to highways infrastructure and traffic operations such as traffic congestion [7], traffic signals [8] and traffic lighting [9]. The ITS implementation are also made to meet the purposes to improved safety, higher productivity and efficiency, more environmental friendliness and enhancing for a better life quality [10].

The technologies that can be seen today has gone through a long journey to be developed. Definitely many *'try and error'* experiments has been conducted for improvements. Looking back at the history, Saunders, N. as recited by A Martin, H Marini, S Tosunoglu (1999) stated that Japan in the year 1973 has already started a project related to Intelligent Vehicle/highway System (IVHS). The Japan's Government through the Ministry of International Trade and Industry has been conducting an activity to develop and evaluate an on-board dynamic route guidance system, working in junction with the private sectors in different areas.

The effort to develop and evaluate IVHS continues with a strong support from the Japanese government. All this work was then unified in 1990 in a project called the Vehicle Information and Communication System (VICS). With all these projects under way, Japan has moved ahead in a number of areas. In 1994, Japan formed its Vehicle, Road, and Traffic Intelligence Society (VERTIS) [7].

Europe started the activity related to the Intelligent Vehicle/highway System (IVHS) back in the year 1988 with DRIVE (Dedicated Road Infrastructure for Vehicle Safety in Europe) followed by another programs called European PROMETHEUS Project (Program for European Traffic with Highest Efficiency and Unprecedented Safety) [7], These two programs is reported have been working for about 12 years to demonstrate the necessity of viewing vehicles and highways as an integrated system.

A Martin et. al (1999) also recorded in their study that in the United States of America., the first efforts in the IVHS were undertaken by individual states and cities back in 1987 where 'Mobility 2000' was created to promote the creation of national programs in the field of IVHS. 'Mobility 2000'was a main factor in the completion of a set of planning documents to guide national support for the development of an IVHS program. In 1990 'Mobility 2000'sponsored the National Workshop in IVHS in which representatives of federal and state government agencies, industry and the academic community participated. There, the U.S. Department of Transportation (USDOT), through its Federal Highway Administration (FHWA), mobilized federal support for a nationally coordinated program. Representatives of the major automobiles, communications, information systems, transportation equipment, and consulting companies in the United States also attended the workshop. Mobility 2000 promoted the formation of the Intelligent Vehicle Highway Society of America (IVHS America).

Continuous research, development, and implementation of ITS in Japan, Europe and United States of America has given a big contribution to the world. It is believed that these developed countries are now adopting more and more Intelligent Traffic System on their highway infrastructure which will be as a model of reference by others.

It has to be noted that the idea of smart highway is not only limited to the application of technology through Intelligent Transportation System (ITS) but also to achieve the goal of sustainable and green environment. It is in line with the Green Technology Master Plan Malaysia (GTMP) 2017-2030 [11] that outlines the strategic plans for green technology development to create a low-carbon and resource efficient economy. In the first edition of the GTMP, six key sectors has been choose which are Energy, Manufacturing, Transportation, Building, Waste and Water. In the transportation sector, several approach has be taken such as:

a) Energy Efficient Vehicle (EEV)
In the National Automotive Policy (NAP) 2014, Malaysian Government has set Energy Efficient Vehicle (EEV) as the way forward for the national automotive industry. According to NAP 2014, EEV definition is applicable to any vehicle that fits the set of specifications in terms of carbon emission level and fuel consumption, and encompasses fuel-efficient vehicles, hybrid electric vehicles, full electric vehicles, alternative fuel vehicles etc. Driven by the EEV initiative, local car manufacturers Proton and Perodua have EEV models, such as Ertiga and Axia respectively (Figure 5). R&D&I on EEV should be continued to allow local automotive industry to move up the value chain in transport GT while positioning Malaysia as the local EEV hub.

Figure 5. Perodua AXIA (EEV).

b) Electric Vehicle (EV) Technology
For personal transport, there are several types of EVs available in the Malaysian market, ranging from Hybrid Electric Vehicle (HEV, which runs on both fossil fuel and battery, without external charging function, such as Toyota Prius and Honda Jazz), Plug-in Hybrid Electric Vehicle (PHEV, which runs on both fossil fuel and battery while allowing the battery to be charged externally, such as BMW 330e), Battery Electric Vehicle (BEV, which runs solely on battery, such as Nissan Leaf and Tesla Roadster) are also in the market. Figure 6 show the example of electric vehicle.

Figure 6. Electric vehicle.

c) Electric Motorbikes
In terms of electric motorbikes, local company Eclimo Sdn Bhd has been developing electric scooter that can cater local as well as regional markets. Moreover, Eclimo has been working with the Research Centre for Applied Electromagnetics (UTHM) and UM Power Energy Dedicated Advanced Centre (UMPEDAC, UM) to develop electric drive train for its electric scooter, as a move to localise the EV technology.

d) Ride Sharing and E-Hailing
The rise of ride sharing and e-hailing services such as Grab [12] and Uber [13] have brought a new mode of efficient "public transport" which substantially changed the way urbanites commutes. Ride sharing and e-hailing services are seen as the future trend in mobility and the services could potentially increase transport efficiency and mitigate congestion issues.

e) Bike Sharing
The smart bike sharing green system implemented recently viz., oBike [14] (Figure 7) is Malaysia's first and largest dockless system which uses technology. It allows commuters to travel – via bicycles located all over the city and across country to reduce congestion and pollution.

Figure 7. oBike sharing.

6. Future Trends for Smart Highways

Going into the future, transportation system and operations will be fully mechanized, automated and sustainable with the use of advanced sensors, user-friendly digital technologies as well as green technologies. The goal of the smart highway is to provide better, safer and greener highway through digital and green technology.
Change of mind-set is crucial among players in the highway industry in order to stay relevant in the coming decades. Continuous research is needed to invent new technologies. Some of the future technologies for use on road are already explored. The examples are elaborated as follows:

a) Interactive lights
Smart road lighting using relevant automated system and data connectivity offers lighting asset management (analyse, plan & maintain), remotelighting management (monitor, manage & measure) that is safe and green. Interactive lights works in the sense that when a car approaches a particular stretch of a road, the motion sensors will light up only that section of the road [15]. The lights will grow brighter as the car comes closer and will slowly dim away as it passes. Figure 8 shows the concept of interactive light.

Figure 8. Interactive lights.

b) Glowing lines

Created by Daan Roosegaarde's and Heijmans Infrastructure and introduced on a highway in Oss, Netherlands. Smart Highway Project combines green energy with safer driving and an art installation experience. It consists of glowing lines that are charged during day-time and glow at night for eight hours. The first pilot has been realised, and will further be developed and launched internationally. Figure 9 shows the concept of glowing lines.

Figure 9. Glowing lines.

c) Electric priority lane

Smart Highway also provides an extra Electric Priority Lane. Underneath the road surface, this priority lane is equipped with induction coils which enables electric cars to recharge as they drive over. Figure 10 shows the concept of electric priority lane.

Figure 10. Electric priority lane.

d) Intelligent (Networked) highway

Deployment of fibre optic cables along major road networks with high transmission capacities will contributes to future ITS systems development that requires rapid high-volume transmission and huge traffic data aggregation to regional or national data centres. (Eg. Asia Pacific Information Superhighway initiative). The concept may also be referred to roadside 'listening' stations that will link up with GPS receivers in cars to monitor traffic patterns and accidents.

e) Piezo-electric energy road

Piezo-electric crystal can generate energy from the vibrations generated by the vehicles. Focusing on the transition of energy to piezoelectric integrated roads, a renewable energy harvesting method, will lead the next power generation into a feasible and more reliable source of energy. Figure 11 shows the concept of Piezo-electric energy.

Figure. 11. Piezo-electric energy.

f) Illumination Traffic Sign/Bollard

Development of light-collecting illumination signs/bollard or visibility during night time and inclement weather. Figure 12 shows an example of illumination traffic bollard.

Figure. 12. Illumination traffic bollard.

These innovations has to be cultured to ensure the highway infrastructure are moving together with the time. The implementation of new technology to highway infrastructure will result in benefits such as improving safety, reduce congestion, reduce emissions and use of fossil fuels, improve access to jobs and services, reduce transportation cost and improve accessibility and mobility. Thus the effort made has to be fully supports to cultivate many more innovators to contribute ideas for mutual prosperity.

IOP Conf. Series: Materials Science and Engineering **512** (2019) 012047 doi:10.1088/1757-899X/512/1/012047

7. Conclusion

With the desire and strong determination, Malaysia is in the right path to achieve the set goal to become developed nation status and to brace the challenges of Industry 4.0 in the coming decades. The Malaysian highways will surely emerge as a world class, fast, safer, convenient and intelligent system with fully mechanised, automated operations and sustainable with the use of user friendly intelligent transportation systems, digital and green technologies in meeting the present needs and without compromising those of future generations.

The steps towards the smart highway are already implemented phase by phase and at the mean time we are preparing for the future to come in. The technology bloom can't be avoided thus well preparation has to be made to welcome it to be a part of us. The challenges for smart highway implementation is waiting ahead. Issues related to technology acceptance, political will and policies, creativity and innovations, construction methods, ethics and changing in human values need to be discuss in depth. Government roles is not only limited to provide infrastructure and access to technology but further extended to bond collaboration with the next generation of industry players, stakeholders and academicians in order to develop policies, clear frameworks and proper education to create a shared vision of the future without compromising the basic value of humanity.

8. References

[1] Executive Summary 2010 *Economic Transformation Programme - A Road Map for Malaysia.*
[2] Rafikul Islam 2010 Critical success factors of the nine challenges in Malaysia's vision 2020 *Socio-Economic Planning Sciences*
[3] Khan RA, Liew MS and Ghazali Z 2013 Malaysian Construction Sector and Malaysia Vision 2020: Developed Nation Status *2nd World Conference on Business, Economics And Management*
[4] Retrieved from www.plus.com.my
[5] Washburn D, Sindhu U, Balaouras S, Dines RA, Hayes NM and Nelson, LE 2010 *Helping CIOs understand smart city initiatives: Defining the smart city, its drivers, and the role of the CIO* (Cambridge, MA: Forrester Research, Inc.)
[6] Ambak K, Atiq R and Ismail R 2009 Intelligent Transport System for Motorcycle Safety and Issues *European Journal of Scientific Research*
[7] Alberto M, Hector, M and Sabri T 2018 *Intelligent vehicle / highway system: A survey Part 1*
[8] Ziarmand A 2013 Smart Road Infrastructure *East-West Design & Test Symposium 2013*
[9] Mustafa O, Abubakr A, Derbala E, Ahmed and Mokhtar B 2016 Towards a smart highway lighting system based on road occupancy: Model design and simulation *Lecture Notes of The Institute for Computer Sciences, Social Informatics And Telecommunications Engineering (LNICST)* 179
[10] Huawei Technologies Co. Ltd. 2016 *Smart transportation: Connecting to the smart future.*
[11] Ministry of Energy, Green Technology and Water Malaysia 2017 *Green Technology Master Plan Malaysia 2017-2030* (Putrajaya)
[12] Retrieved from https://www.grab.com/my/
[13] Retrieved from https://www.uber.com/en-MY/
[14] Retrieved from https://www.o.bike/my/
[15] Retrieved from https;//www.studioroosegaarde.net/project

Moisture damage evaluation of asphalt-aggregate constituents and warm mix asphalt using imaging techniques and tensile tests

S Y Teh[1] and M O Hamzah[2]

[1]UEM Edgenta Berhad, Bangsar South, 59200 Kuala Lumpur, Malaysia
[2]School of Civil Engineering, Universiti Sains Malaysia, Penang, Malaysia

E-mail: cemeor@yahoo.com

Abstract. Warm Mix Asphalt (WMA) is more susceptible to adhesive failures at the asphalt-aggregate interface than Hot Mix Asphalt (HMA) due to its lower production temperature. This paper presents an innovative approach to quantify the adhesive failures of asphalt-aggregate constituents and mixtures in direct and indirect tensions, respectively. Advanced imaging techniques were utilized to compute the specimens percentage adhesive failure. A PG-64 bitumen was the conventional binder used. Hydrated Lime (HL) and Pavement Modifier (PMD) were used as anti-stripping fillers, while Evotherm (EV) was incorporated as the surfactant-WMA additive. Three moisture conditioning levels were performed namely; unconditioned, 1 freeze-thaw (F-T) and 3 F-T cycles prior to testing. The specimens percentage adhesive failure increased, while tensile strength reduced with F-T cycle. PMD specimens exhibited slightly lower percentage adhesive failure than HL specimens. The percentage failure further reduced with the incorporation of EV. The effects of incorporating EV were more pronounced when combined with PMD as shown by the higher tensile strength and lower percentage failure than HL-EV. Limestone specimens showed superior moisture resistance than granite in moist condition. The high percentage broken aggregate on the fractured surfaces of tested asphalt mixtures was found to be more related to aggregate orientation than shape.

1. Introduction

The destructive effects of moisture can adversely affect mixture performance and functionality. This is particularly true when granite aggregates are used in the mixture since such siliceous aggregates are known to be hydrophilic. Inadequate drying of aggregates due to the lower production temperature of WMA contributes to adhesive failures at the binder-aggregate interface [1]. Infiltration of moisture into asphalt pavements results in two failure mechanisms, namely adhesive and cohesive failures. Adhesive failure takes place when binder film strips away from the aggregate surfaces, while cohesive failure is caused by softening of the asphalt mastics [2]. The asphalt binder and aggregate bond strength is an important parameter to evaluate the binder ability to resist moisture damage. A test method that can effectively assess the effects of water in both adhesive and cohesive failures in an asphalt-aggregate system can lead to a better comprehension of the mixtures moisture sensitivity [3]. According to Canestrari et al. (2010), a repeatable, reliable and practical method to investigate the adhesive and cohesive properties of asphalt-aggregate systems is based on the Pneumatic Adhesion Tensile Testing Instrument (PATTI) [4]. However, a standard procedure for evaluating moisture damage at the asphalt-aggregate interface is still not available.

This study investigated the adhesive strength of various asphalt constituents and mixtures in relation to its mechanical strength using direct and indirect tensions, respectively. Adhesive failures attributed to moisture conditionings were investigated. Compacted asphalt mixtures were first subjected to indirect tensile strength (ITS) test, followed by examination of the fractured surface using 3-D image analysis. On the other hand, asphalt constituent test specimens were tested using the PATTI test and specimen fractured surface was examined using 2-D image analysis. The 2-D imaging technique used was strictly monoscopic with images of specimens taken from merely one viewpoint. This study attempts to overcome this shortcoming by instead employing the 3-D imaging technique. This gives a more precise quantification of adhesive failure. In Malaysia, qualitative measurement is still carried out based on visual inspection using the naked eye. Such quantification is very subjective in nature and varies with different observers. With the introduction of 3-D imaging technique, it may be incorporated into the Malaysian Public Works Department (PWD) specifications to evaluate stripping if it is made mandatory in the near future. Furthermore, preliminary studies were also carried out to evaluate the effects of aggregate shape and orientation upon mixtures fractured surfaces.

2. Methodology

2.1. Materials

The unmodified PG-64 was the binder type used in this investigation. Granite aggregates used in the preparation of all mixtures were supplied by Kuad Sdn. Bhd. in Penang. Limestone boulders were sourced from Pens Industries Sdn. Bhd., Perlis, for aggregate substrate preparation. The adopted median aggregate gradation was in accordance with the Malaysian PWD gradation limits for asphaltic concrete mixture type AC-14 [5]. All mixtures incorporated Pavement Modifier (PMD) as the filler, while asphalt mastics used PMD and Hydrated Lime (HL). The adopted filler to asphalt ratio was F/A = 1. Hence, base binders were added with the same amount of fillers by mass of binder. PMD is a grayish-black powder mineral filler that has been used as anti-stripping additive for asphalt mixtures. According to Hill et al. (2012), the surfactant-WMA (EV) was added at an optimum dosage rate of 0.5% by mass of PG-64 binder [6].

2.2. Specimen preparation

Asphalt mixtures prepared for this investigation included three conditioning types, four mixing and compaction temperatures and one anti-stripping agent as filler. The specimen dimensions were 100 mm in diameter and 63.5 mm in height. The Servopac Gyratory Compactor which better simulates mixture densification in the field was used to compact specimens to 7±1% air voids. The Marshall mixture design ASTM D1559 [7] procedures conforming to PWD specification [5] for mixture type AC-14, were applied to determine the mixtures optimum binder content. The optimum binder contents of base PG-64 binder were 5.0% and 5.2% for HMA and WMA, respectively. The mixing and compaction temperatures of all asphalt mixtures are listed in table 1. A total of 48 asphalt-aggregate test specimens were also prepared, including three conditioning types, two aggregate and filler types. For preparation of asphalt mastics, binders and fillers were preheated to 160°C prior to blending using the high shear mixer for 30 min.

Table 1. Mixing and compaction temperatures of asphalt mixtures.

Mixture	Mixing Temperature (°C)	Compaction Temperature (°C)
HMA	160	150
WMA Control (0% Additive)	140	130
WMA (0.5% Additive)	120-140	110-130

2.3. Laboratory moisture conditioning

Moisture conditioning was conducted in accordance with ASTM D4867 [8] procedures. The only modification made was the addition of Sodium Carbonate (Na_2CO_3) to the distilled water at a concentration of 6.62 gm per liter to accelerate stripping within the asphalt specimens [9]. Investigation was carried out on three sets of specimens; unconditioned, 1 freeze-thaw (1 F-T) cycle and 3 F-T cycles. Prior to heating the asphalt mastics and aggregates for specimen preparation, silicon cuts as shown in figure 1 were glued to the substrate surface. The use of silicon cut provided lateral confinement to the asphalt mastics during preparation and thawing at 60°C. Moisture conditioned specimens were placed in water inside the ALVS chamber for 30 min at 35°C prior to F-T cycle.

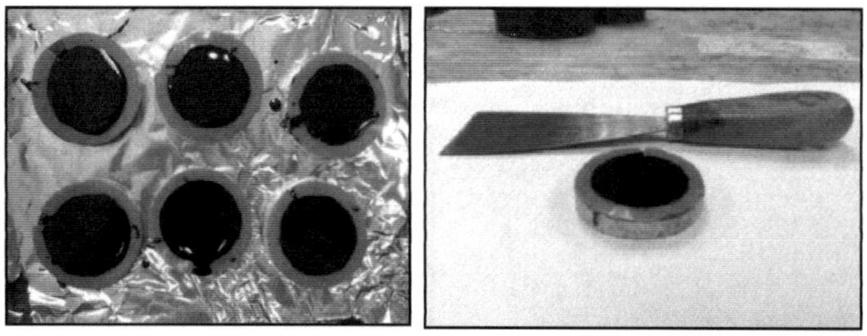

Figure 1. Silicon cuts.

Figures 2(a) and (b) show the asphalt binder and mastics specimens, respectively after exposure to F-T cycle. An interesting observation pertaining to the moisture-damage mechanism was made in this study where moisture penetrated into the asphalt-aggregate interface via holes on the asphalt film. The generated holes started from the top, open surface of asphalt specimens during conditioning at 60°C in a submerged water condition. The holes were observed to be cylindrical and randomly distributed. The holes became visible within approximately 2 h after conditioning and the holes diameters varied from microscopic to more than 1 mm. The holes eventually reached the bottom of asphalt film. After reaching the interface through holes, moisture started to replace the asphalt film from the substrate surface, and adhesive failure or stripping began to take place. The process started with dewetting of asphalt film from substrate, followed by spreading of moisture on the substrate. Figure 2 also shows that more holes were developed on the binder specimen compared to asphalt mastics.

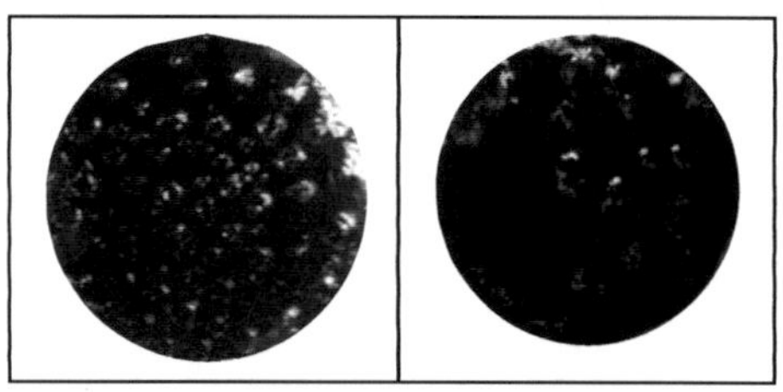

(a) Binder (b) Mastics

Figure 2. Formation of holes on asphalt-aggregate specimens after moisture conditioning.

2.4. Test methods

The Indirect Tensile Strength (ITS) test was employed to evaluate mixtures moisture susceptibility according to ASTM D4867 [8] procedures. Test temperature for all mixtures was set at 15°C. Such a low temperature stiffened the specimens and easily disintegrated when tested until failure [10]. The PATTI with modified pull stub as shown in figure 3 was used to record the maximum tensile stress required to separate the asphalt mastics from aggregate substrate. The asphalt film thickness is the space between the pull-off stub and aggregate surface, in this case equivalent to 0.8 mm. Failure within specimen occurred once tensile strength exceeded the bond strength at the pull-off stub and substrate interface, or the asphalt mastics cohesive strength.

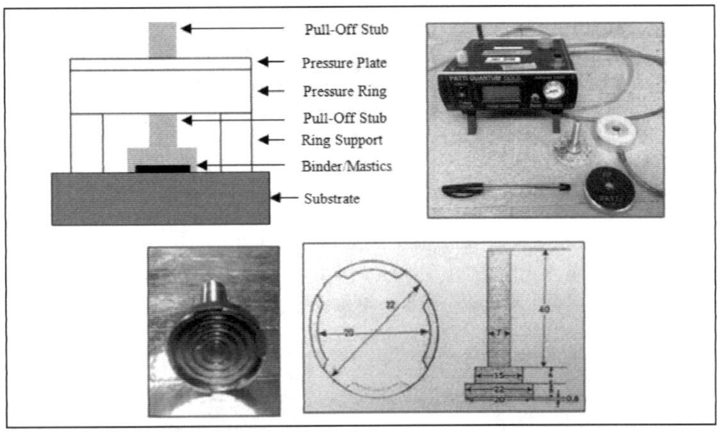

Figure 3. PATTI instrument with modified stub [11].

3. Imaging technique

Images of specimen fractured surfaces were taken after the test for image analysis. Imaging technique in 3-D was used to compute the percentages of adhesive failure and broken aggregates of asphalt mixtures, while 2-D imaging technique was carried out on fractured surfaces of asphalt constituent test specimens.

Imaging technique in 3-D was introduced to more precisely quantify the amount of binder stripping from the aggregate surfaces. Prior to 3-D image analysis, ordinary 2-D photos had to be first captured, merged and converted to a 3-D model using the 123-D Catch software. Autodesk 123-D Catch software was used to merge the ordinary 2-D images into a 3-D model. A minimum of 20 photos were captured for each specimen at various angles from the horizontal axis. Figure 4 shows the transformation of 2-D images, taken at different angles, into a 3-D model. The final assembled 3-D models were then saved and exported into CloudCompare for 3-D image analysis.

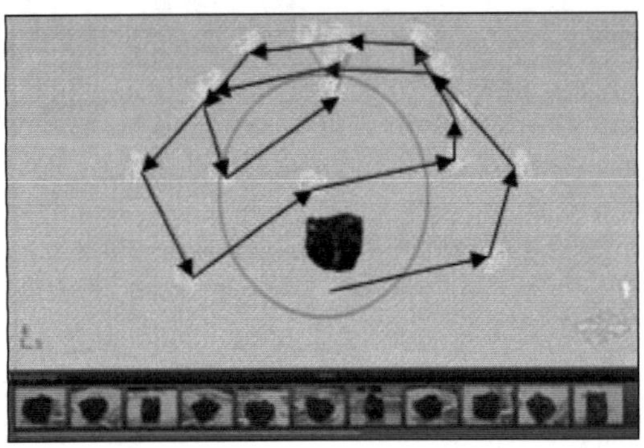

Figure 4. Conversion of ordinary 2-D images into a 3-D model.

Figure 5(a) demonstrates one of the 3-D models imported from 123-D Catch. Prior to analysis, this model had to be first cropped or segmented, leaving only the fractured surface of specimen as depicted in figure 5(b). This model was then converted to scalar fields for statistical computations. Similar to 2-D image analysis, the multidimensional domain of the model, in red, green and blue (RGB) was reduced to lower dimensions, consisting of only one or two component axes. Conversion to grayscales yielded models with merely one band or intensity. Binder covered areas were represented by low gray values, while higher intensities were associated with stripped and broken aggregate surfaces. Figures 6(a) and (b) show the conversion of the segmented fractured surface of the original 3-D model into grayscales. Classification of 2-D images and 3-D models was carried out by means of thresholding. The number of classes used in this software was by default, 255. Threshold range values for stripped and broken aggregate surfaces were obtained via the trial and error method by narrowing the number of classes between 0 and 255. Figure 6(c) shows the classification of the model into stripped and broken aggregate areas. Figure 7 shows the original images of PG-64 asphalt mastics over granite aggregate substrate taken after testing specimens using PATTI test in 3 F-T condition. These images were analyzed using the Earth Resource Development Assessment System (ERDAS) software. Figure 8 presents the classified images that clearly show the adhesive failure over aggregate substrate. The statistical tool included in this software will conveniently display the amount of pixels that corresponded to each of the threshold range value selected. These values were then used to quantify the percentages adhesive failure and broken aggregates.

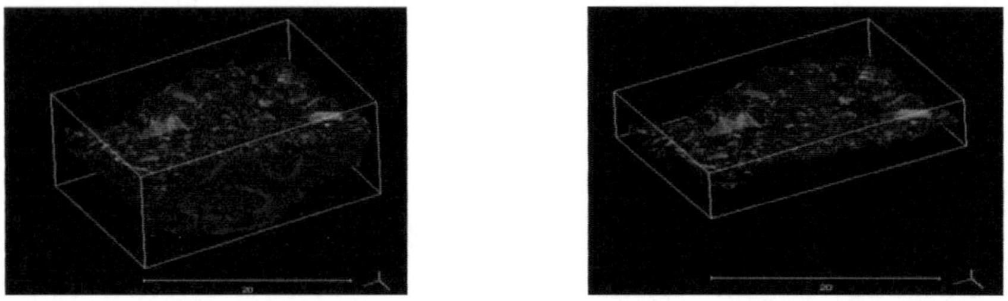

(a) A 3-D model of WMA 120°C (1 F-T) (b) Segmented fractured surface of the 3-D model

Figure 5. A 3-D model imported from 123-D catch.

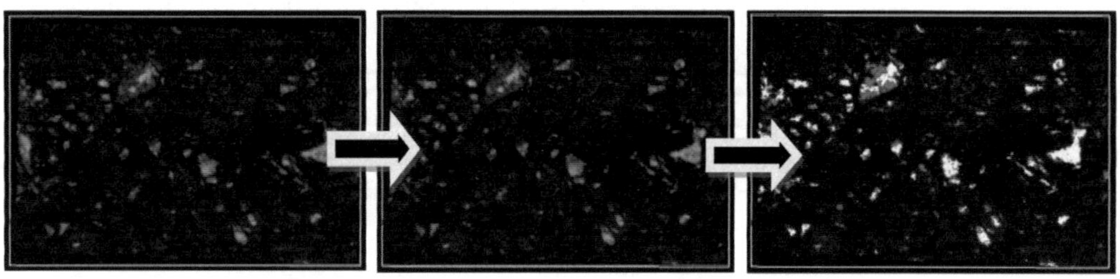

(a) Original model (b) Grayscales (c) Classification

Figure 6. Classification of the segmented fractured surface of 3-D model.

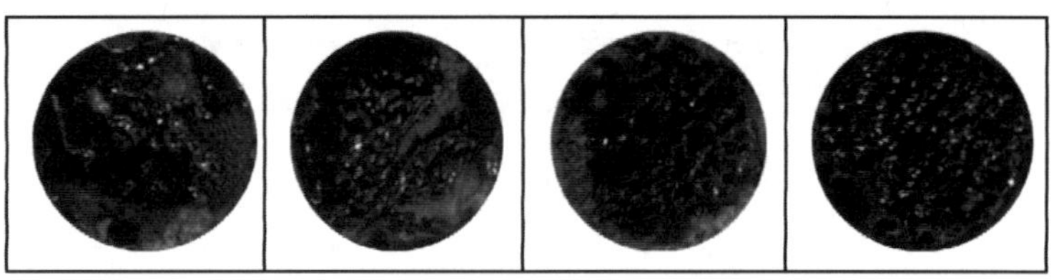

(a) HL-Control (b) HL-EV (c) PMD-Control (d) PMD-EV

Figure 7. Original images of PG-64 asphalt mastics on granite substrates after 3 F-T cycles.

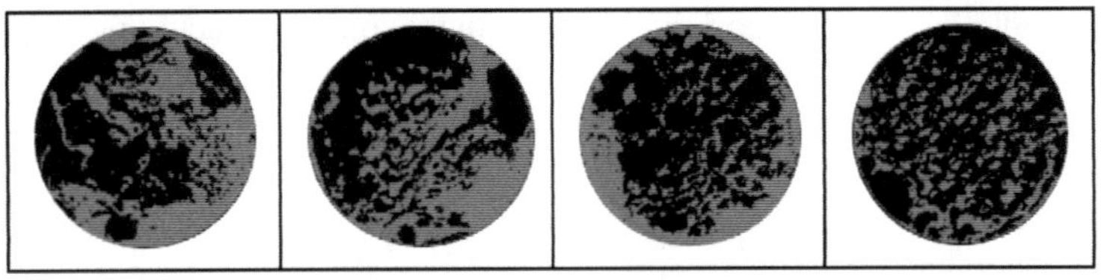

Figure 8. Classified images of PG-64 asphalt mastics on granite substrates after 3 F-T cycles.

4. Studies on broken aggregates

All tested ITS test specimens were preheated in the oven at 80°C for two hours. Broken aggregates observed on the specimen fractured surfaces were extracted. Each tested ITS test specimen consisted of two broken parts with symmetrical fractured surfaces. Photographs of broken aggregate orientation were immediately taken to measure the orientation angle between aggregate longitudinal axis and fractured plane. The ImageJ software was utilized for this purpose. Only aggregates with orientation angle equal to or less than 80° were regarded as having orientation issues. Figure 9 shows the schematic diagram of broken aggregate orientation with respect to fractured plane. All extracted aggregates were stored in individual plastic bags and left to cool at room temperature. The aggregates were glued together and weighed to obtain the total extracted aggregate mass. Aggregate shape was examined using the proportional calliper according to ASTM D4791 [12] procedures. The mass and percentage of elongated and flat aggregates were determined.

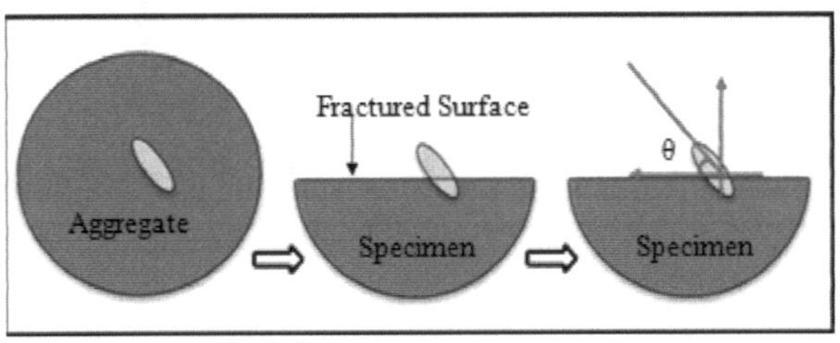

Figure 9. Schematic diagram of broken aggregate orientation with respect to fractured surface.

5. Results and discussion

The image analysis results are used to determine the amount of pixels that corresponds to the total fractured surface, stripped and broken aggregate areas. For computation of results, failures attributed to adhesion and broken aggregates are expressed in percentage, while ITS is in MPa. The pull-off strength of PATTI test is expressed in kPa.

5.1. Effects of moisture conditioning on asphalt mastics adhesive failure

Figure 10(a) shows the percentage adhesive failure of unconditioned and conditioned asphalt mastics on granite and limestone substrates using the pull-off tensile test. In unconditioned state, there is no clear relationship in terms of percentage adhesive failure between specimens of different aggregate types, filler types and EV contents. The aggregate type shows insignificant effects on percentage adhesive failure of unconditioned specimens. In addition, cohesive failure is generally observed in tested unconditioned specimens. In the absence of water, the asphalt-aggregate adhesive bond is higher than the asphalt film inner cohesion. From figure 10(a), the percentage adhesive failure of all PG-64 specimens escalates after 1 F-T cycle, regardless of EV content, aggregate and filler types. The highest percentage failure after 1 F-T cycle is recorded by HL-Control with 29.0% adhesive failure, followed by PMD Control with 24.2%. The corresponding HL and PMD specimens incorporating

0.5% EV exhibit lower percentage adhesive failure than control specimens. This implicates the beneficial effects of adding EV to further enhance the stripping resistance of WMA specimens when subjected to severe moisture conditionings.

Similar trend is also found in specimens using limestone aggregate where specimens added with WMA additive demonstrate lower percentage adhesive failure than control specimens. The lowest percentage failure is shown by PMD-EV with only 18.0% failure, followed by HL-EV with 20.5% failure. After 3 F-T cycles, the highest failure is still recorded by HL-Control, followed by HL-EV, PMD-Control and the lowest by PMD-EV. Specimen HL-Control registers 47.6% and 44.8% failures using granite and limestone substrates, respectively. Specimens incorporating EV still exhibit lower percentage adhesive failure than the corresponding control specimens after 3 F-T cycles. However, when comparing between filler types, asphalt mastics using PMD exhibit higher resistance to moisture damage than HL shown by the lower percentage adhesive failure. In addition, the effects of using EV are more significant when used in combination with PMD. This is evidenced from the lower percentage adhesive failure of PMD-EV specimens compared to HL-EV. Limestone specimens exhibit lower percentage adhesive failure after moisture conditionings compared to specimens using granite aggregate. These findings are in agreement with Kanitpong and Bahia (2005) [13].

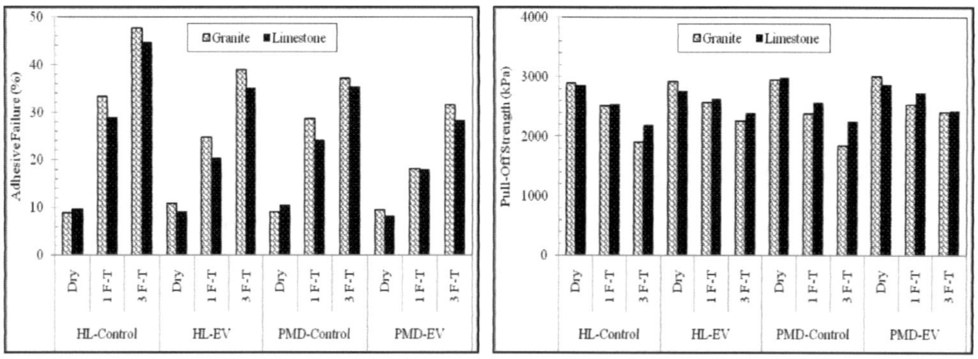

(a) Adhesive failure (b) Pull-off strength

Figure 10. Adhesive and strength properties of unconditioned and conditioned asphalt mastics specimens.

5.2. Effects of moisture conditioning on asphalt mastics bond strength

Figure 10(b) shows the pull-off strength of PG-64 mastics using both aggregate types. The results show no obvious trend in terms of dry specimen bond strength using different aggregate and filler types. The pull-off strength of PG-64 specimens reduces more significantly after 3 F-T cycles than after 1 F-T cycle. Specimens using PMD exhibit slightly higher pull-off strength than HL specimens after moisture conditionings. Moreover, the effects of incorporating EV are more significant when used in conjunction with PMD, as shown by the higher strength in PMD-EV than HL-EV. Specimens incorporating EV demonstrate higher pull-off strength compared to control specimens, regardless of filler type. After 3 F-T cycles, the PMD-EV and HL-EV specimens exhibit the highest strength values. Similarly, limestone which has better affinity towards conditioned asphalt mastics shows higher pull-off strength than specimens using granite aggregate. The percentage adhesive failure results described in Section 5.1 show that adhesive failure increases with increasing freeze-thaw cycle. Hence, the specimen pull-off strength is inversely proportional to the percentage adhesive failure in moist conditions.

5.3. Effects of moisture conditioning on mixture indirect tensile strength

Figure 11(a) shows the ITS test results of PG-64 mixtures. The ITS of all mixtures decreases with F-T cycle. The reduction in ITS is more significant after 3 F-T cycles compared to after 1 F-T cycle and in dry condition. The ITS of all WMA is found to be lower than HMA, regardless of conditioning level. This implicates the higher moisture damage susceptibility of WMA due to their lower production temperatures. When comparing all WMA, WMA with 0% additive exhibits the lowest ITS. This finding can again be substantiated by the absence of warm mix additive to help improve its adhesive strength. Similar to the PATTI test results, the ITS is also found to be inversely proportional to percentage adhesive failure. This finding is also in agreement with Hamzah et al. (2014) [14]. From figure 11(b), almost all PG-64 mixtures meet the minimum 80% indirect tensile strength ratio (ITSR) limit except for WMA produced at the lowest mixing temperature of 120°C after 1 F-T cycle. After 3 F-T cycles, all PG-64 mixtures including HMA fail the test.

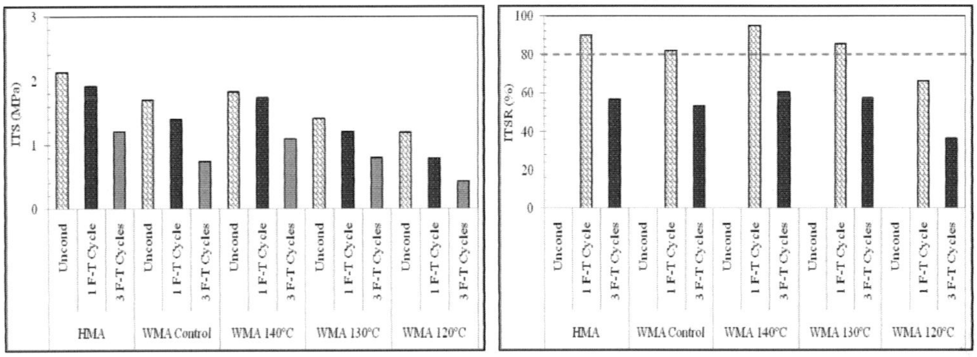

(a) ITS (b) ITSR

Figure 11. ITS results of fractured specimens (quantitative measurement).

5.4. Effects of moisture conditioning on mixture adhesive failure

Figure 12(a) represents the percentage adhesive failure of PG-64 mixtures, quantified using 3-D imaging technique. The percentage adhesive failure escalates with the number of F-T cycles. The percentage adhesive failure of dry specimens for all PG-64 mixtures, produced at different mixing temperatures, is similar. However, WMA prepared at 140°C with 0% additive, 130°C and 120°C, exhibit the highest failure with respect to adhesion after 1 F-T cycle. This may be attributed to the effects of lower production temperature, resulting in WMA more susceptible to adhesive failure due to water infiltration.

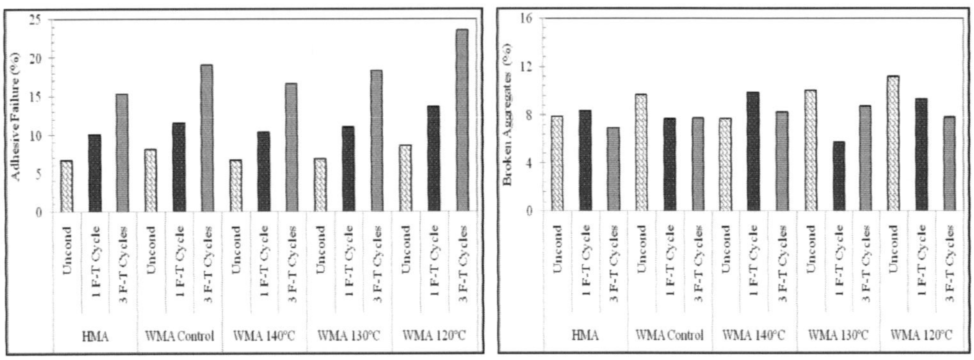

(a) Adhesive failure (b) Broken aggregates

Figure 12. Percentage failure results of fractured surfaces (quantitative measurement).

As for WMA with 0% additive (WMA Control), this may be caused by the absence of warm mix additive to further enhance the mixture adhesive properties. The WMA (with additive) produced at 140°C shows comparable performance with HMA even after exposure to 1 F-T cycle. After 3 F-T cycles, the percentage adhesive failure of HMA using PG-64 binder is still lower than the corresponding WMA produced at different mixing temperatures. As for WMA, the percentage adhesive failure is relatively equivalent to one another after 3 F-T cycles.

5.5. Shape and orientation analysis on broken aggregates

Figure 12(b) shows the percentage broken aggregate quantified using 3-D imaging technique. These broken aggregates are assumed to be caused by fractures during compaction or due to the indirect tensile force applied onto the specimen. The percentage broken aggregate failures show no obvious trends when mixtures were moisture conditioned. From the investigation carried out in Section 4, the high percentage broken aggregate in tested specimens is found to be more related to aggregate orientation rather than shape. Aggregates with orientation angles equal to or less than 80° account for more than 90% of the total extracted aggregate mass, while elongated and flat aggregates represent more than 40% of the total mass. Aggregates with orientation issue are subjected to flexures or bending stresses during testing and may break easily. In addition, elongated and flat aggregates are less desirable as they could affect pavement durability.

6. Conclusions

The percentage adhesive failure of asphalt-aggregate test specimens escalated, while pull-off strength reduced with F-T cycle. Specimens incorporating PMD exhibited higher moisture resistance than HL specimens. Incorporation of EV further reduced the percentage adhesive failure of all specimens. Limestone which showed better affinity towards asphalt mastics in moist condition exhibited lower percentage adhesive failure than granite specimens. The use of 3-D imaging technique was necessary for a more precise quantification of adhesive failure. The mixture ITS reduced with the number of F-T cycles. The asphalt constituent and mixture test specimens incorporating EV exhibited comparable, if not better, moisture resistance than the control specimen. The high percentage broken aggregate in tested specimens was found to be more related to aggregate orientation rather than shape.

7. References

[1] Mogawer W, Austerman A and Bahia H U 2011 Evaluating the effect of warm-mix asphalt technologies on moisture characteristics of asphalt binders and mixtures *Transp. Res. Rec.: J. Transp. Res. B.* **2209** 52-60

[2] Copeland A and Kringos N 2006 Determination of bond strength as a function of moisture content at the aggregate-mastic interface *In 10th Int. Conf. on Asphalt Pavements* Quebec, Canada, 12-17 August 2006 pp 709-18

[3] Kanitpong K and Bahia H U 2003 Role of adhesion and thin film tackiness of asphalt binders in moisture damage of HMA *In Assoc. of Asphalt Paving Technologists Technical Sessions* **72** Kentucky, USA

[4] Canestrari F, Cardone F, Graziani A, Santagata F A and Bahia H U 2010 Adhesive and cohesive properties of asphalt-aggregate systems subjected to moisture damage *Road Mater Pavement* **11(sup1)** 11-32

[5] PWD 2008 *Standard specification for road works in Malaysia, section four: Flexible pavements* (Kuala Lumpur: PWD)

[6] Hill B, Behnia B, Buttlar WG and Reis H 2012 Evaluation of warm mix asphalt mixtures containing reclaimed asphalt pavement through mechanical performance tests and an acoustic emission approach *J. Mater. Civ. Eng.* **25**(12) 1887-97

[7] ASTM D1559 2006a *Test method for resistance of plastic flow of bituminous mixtures using marshall apparatus* (West Conshohocken: ASTM)

[8] ASTM D4867 2006b *Standard test method for effect of moisture on asphalt concrete paving mixtures* (West Conshohocken: ASTM)

[9] Hicks RG, Santucci L and Aschenbrener T 2003 Moisture sensitivity of asphalt pavements *A National Seminar* (San Diego, California: Transportation Research Board)

[10] Xie J, Xiao Y, Wu S and Huang J 2012 Research on fracture characteristic of gneiss prepared asphalt mixture with direct tensile test *Construction and Building Materials* **28**(1) 476-81

[11] Moraes R, Velasquez R and Bahia H U 2011 Measuring the effect of moisture on asphalt-aggregate bond with the bitumen bond strength test *Transportation Research Record: Journal of Transportation Research Board* **2209** 70-81

[12] ASTM D4791 2010 *Standard test method for flat particles, elongated particles, or flat and elongated particles in coarse aggregate* (West Conshohocken: ASTM)

[13] Kanitpong K and Bahia H U 2005 Relating adhesion and cohesion of asphalts to the effect of moisture on laboratory performance of asphalt mixtures *Transportation Research Record: Journal of Transportation Research Board* **1901** 33-43

[14] Hamzah M O, Kakar M R, Quadri S A and Valentin J 2014 Quantification of moisture sensitivity of warm mix asphalt using image analysis technique *Journal of Cleaner Production* **68** 200-08

Acknowledgements

The authors would like to acknowledge the Malaysian Ministry of Higher Education that has funded this research grant through the Fundamental Research Grant Scheme (Grant Number 203/PAWAM/6071277) which enables this paper to be written. Special thanks goes to Mr. Chet Arunyakasem from Ingevity Corporation for the supply of surfactant-WMA additive.

IOP Conf. Series: Materials Science and Engineering **512** (2019) 012049 doi:10.1088/1757-899X/512/1/012049

Performance of pavement preservation with Ralumac Micro surfacing at LATAR highway

Shamsuni Salleh[1,3] Rahimah Muhamad[1], Mohd Haryazi Abdillah[2] and Ahmad Fuad Ahmad Shahimi[3]

[1]Razak Faculty of Technology and Informatics, Universiti Teknologi Malaysia, Kuala Lumpur, Malaysia
[2]Terratech Consultants (M) Sdn Bhd, Bt Caves, Selangor, Malaysia
[3]KL-Kuala Selangor Expressway Berhad, Bestari Jaya, Selangor, Malaysia

E-mail: shamsuni@latar.com.my

Abstract: Maintaining pavement to desired level of serviceability as required by Malaysian Highway Authority (MHA) give immense pressure to highways operators. Pavement distress such as roughness, rutting and cracking are common problems that occur after the highway has been operational. Pavement preservation has been sought as a potential alternative for managing all the pavement distress. This paper evaluates the performance of pavement preservation with Ralumac Micro surfacing for slow lane and fast lane from KM23.70 to KM24.10 Eastbound at LATAR highway. The evaluation comprises of International Roughness Index (IRI), texture depth and skid resistance. The result revealed that micro surfacing treatment has significant improvement on roughness, rutting, and skid resistance even after 36 months operational. The performance of Ralumac Micro surfacing at slow lane is slightly lower than Ralumac Micro surfacing at fast lane as it affected by heavy vehicles.

1. Introduction

Pavement maintenance treatment can be categorized into three main categories: routine maintenance, preservation, and rehabilitation. Routine maintenance is reactive and often comprises relatively inexpensive and corrective actions to address specific problems that may compromise safety, such as localised potholes. Preservation treatments are proactive and consist of well-timed and executed activities to prevent premature distresses and to slow the rate of deterioration. Finally, rehabilitation consists of structural enhancements that renew the service life of an existing pavement and improve its load carrying capacity.[1]

The current practices of most highway agencies concentrate on the corrective maintenance rather than preservation treatment which are either ignored or treated as a low priority [2,3]. Most current maintenance activities have budget issues and only focus on removing deteriorated pavement sections or to defer the repair works. This approach has impacts on the environment and on safety to road user, and may thus affect the costs associated with maintenance [3].

Pavement preventive maintenance is a proactive, planned strategy that extends the life of the pavement and provides a cost-effective solution for pavement management. [4]. Its include fog seal, chip seal, slurry seal, micro surfacing, crack treatment, and thin hotmix overlays. The selection of a pavement preventive maintenance should be based on the condition of the existing pavement, traffic volume, and environmental conditions. Other factors include experience, budget constraints, and availability of the technology or materials [5].

Micro surfacing is the important method in the preventive maintenance applied by highway agencies. Even more importantly, Micro surfacing is poised above all other treatments to comply strict environmental regulations and to optimize budget availability [6]. According to Koichi Takamura [7] Micro surfacing is more "eco-efficient" than hot mix overlay because of less material being used, less transportation of materials and lower overall emissions during the life of the treatment. Micro surfacing performs best when applied to correct surface friction, oxidation, raveling and rutting on pavements that have adequate structural capacity [8] and cost-effective maintenance technique for higher severity cracks for at least a year [9].

Micro surfacing is a cold-applied paving mixture composed of polymer-modified asphalt emulsion, crushed aggregate, mineral filler, water and a hardening-controlling additive, proportioned, mixed and uniformly spread over a properly prepared surface [10]. It can be used a blanket cover on pavements suffering from loss of skid resistance, oxidation, raveling and surface permeability. In addition, Micro surfacing can be used to fill ruts and improve rideability by removing minor surface irregularities. This treatment can last on average 8 to 9 years [11]. This study was carried out to evaluate performance of Micro surfacing from KM23.70 to KM24.10 Eastbound LATAR after 36 months operational. Evaluation of surface roughness, rutting, texture depth, and skid resistance are carried out at slow lane and fast lane to determine effectiveness micro surfacing.

2. Objective

This study evaluates the performance of pavement preservation with Ralumac Micro surfacing for slow lane and fast lane from KM23.70 to KM24.10 Eastbound at LATAR highway after 36 months operational. The evaluation comprises of International Roughness Index (IRI), pavement rutting, texture depth and skid resistance.

3. Specification and Mix Design

The aggregate gradation for Micro surfacing according to JKR specification as illustrated in Table 1. In order to improve more on skid resistance for Ralumac Micro surfacing at LATAR highway, the ASTM sieve size 4.75mm and 3.35mm were changes to 5.0mm and 3.15mm as per Table 2.

Table 1. Aggregate gradation JKR/SPJ/2008-S4.

ASTM Sieve Size (mm)	% Passing by weight
8.0	90 -100
4.75	60 – 74
3.35	45 – 58
2.0	36 – 50
1.0	23 – 38
0.710	20 – 33
0.090	5 – 12

Table 2. Aggregate gradation of Ralumac Micro surfacing at LATAR.

ASTM Sieve Size (mm)	% Passing by weight	Quarry Dust	Chipping	Combined (dust & chipping)
8.0	90 -100	99.29	75.06	94.54
5.0	60 – 74	82.89	16.19	70.37
3.15	46 – 58	61.36	5.78	50.54
2.0	36 – 50	45.28	3.94	37.97
1.0	23 – 38	31.11	3.05	26.48
0.710	20 – 33	26.41	2.74	22.64
0.090	5 - 12	8.51	3.01	7.52

IOP Conf. Series: Materials Science and Engineering **512** (2019) 012049 doi:10.1088/1757-899X/512/1/012049

The sieve analyses were carried out by taking representative samples of each individual coarse and fine aggregates. The combined aggregate used for Ralumac Micro surfacing was blended in the proportion of quary dust:chipping = 3:1.

The bitumen used for the manufacturing of the Ralumac is from Shell pen. Grade 80/100. The average penetration is 86.82 while average softening point is 46.43 °C. The Ralumac emulsion content included Bitumen 80/100, Peral 416, and Latex.

The average test result for Ralumac emulsion used for this project is Binder residue test is 67.40%, penetration (0.1mm) is 50.44mm and softening point is 58.5°C. The laboratory mix for Ralumac Micro surfacing is shown in Table 3.

Table 3. Laboratory Mix for Ralumac Micro surfacing.

Item Description	Requirement / Results
Combined aggregates	100%
Ralumac emulsion	10 – 12% by the weight of aggregate
Cement	2% by the weight of aggregate
Water	8 – 10% by the weight of aggregate
Additive	0.05% by the weight of aggregate
Cohesion test	
30 minutes	16.28 kg-cm
60 minutes	22.67 kg-cm
Wet track abrasion test	38.67 g/ft^2
Mix time	60 ec

4. Methodology

This study has been carried out to determine the performance of Ralumac Micro surfacing system from KM23.70 to KM24.10 Eastbound for both lanes; slow lane and fast lane at LATAR highway as shown in Figure 1. These performance tests are International Roughness Index (IRI), rutting, texture depth and skid resistance have been measured on the Ralumac Micro surfacing.

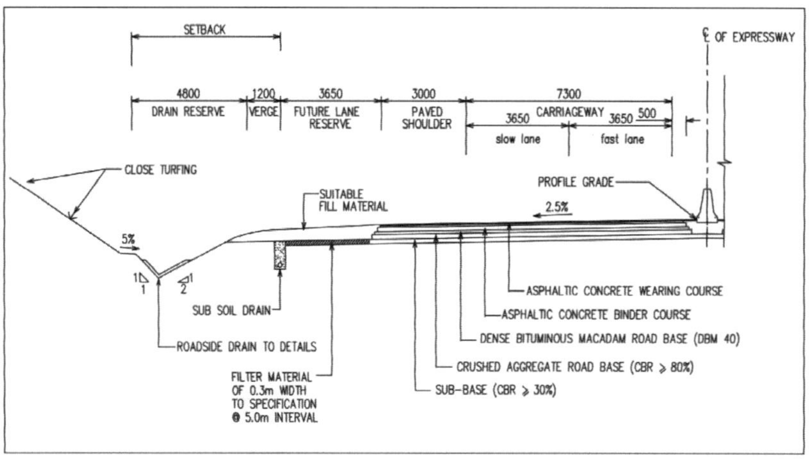

Figure 1. Pavement cross section at KM23.70 to KM24.10 Eastbound.

IKRAM Road Scanner (IRS) were used to capture pavement functional condition data, survey mapping information and roadside asset details whilst travelling at highway speed. This fully-featured vehicle contains a compact workstation capable of capturing and storing individual data elements for roughness, rutting, texture and surface condition, as well as providing high-detail video images of road and road side assets along the LATAR highway.

Meanwhile skid resistance was performed using the Grip Tester (GT) along the left wheel path of the road. The GT has one test wheel, fitted with apparatus containing transducers, instrumentation, actuation controls, a water supply and automatic control system. It uses measurements representing the friction force on a test wheel operated at a constant load and longitudinal reference slip. During a test run, water was sprayed ahead of the measuring wheel by a self-watering system to produce 0.25 mm thick water film and the test wheel is brought to the road surface by means of the actuation system. The information was process and the Grip Number was converted to SCRIM (Sideway-force Coefficient Routine Investigation Machine) value by the following relationship:

$$\text{SCRIM value} = 0.85 \times \text{Grip Number} \quad (1)$$

Figure 2. IKRAM Road Scanner (IRS) and Grip Tester (GT).

The data performance of Ralumac Micro surfacing has been collected since 2014 and has been evaluated for every 100m. The test results have been compared to MHA advisory condition criteria as illustrated in Table 4 and Table 5

Table 4. MHA Advisory Condition Criteria for Roughness, Rutting, and Texture Depth.

Parameter / Condition Criteria	Roughness (IRI)	Mean Rut Depth (mm)	Texture Depth (mm)
Good	< 2.0	< 5.0	> 0.5
Fair	2.0 – 3.0	5.0 – 10.0	0.3 – 0.5
Poor	3.0 – 3.8	10.0 – 20.0	< 0.3
Bad	≥ 3.8	≥ 20.0	N/A

Table 5. Skid Resistance Indicative Condition Criteria.

Condition Criteria	SCRIM Value
Acceptable Level	≥ 0.38
Investigatory Level	< 0.38

5. Result and Discussion

The data performance of roughness, rutting, texture depth and skid resistance on Ralumac Micro surfacing has been calculated and discussed in the next subsection.

5.1. Roughness

Table 6 and Figure 3 indicate the pavement roughness for Ralumac Micro surfacing after 36 months operational at fast lane is improved up to 34% at KM23.90 while at slow lane it is increased up to 25% at KM23.80 from original pavement condition. Furthermore, overall pavement condition at slow lane is more 5% roughness compare to roughness at fast lane as surveyed at KM23.80. This occurred due to the slow lane receive more traffic loadings as it was designed mainly for heavy vehicles. However, overall roughness for the Ralumac Micro surfacing after 36 months operational are still less than 2.0mm/km which are shown as a good rating as per Table 4 although Annual Average Daily Traffic (AADT) had increased by 24%.

Table 6. Pavement roughness for Ralumac Micro surfacing.

KM	Lane	0 mth (AADT 56,031)	12 mth (AADT 66,520)	24 mth (AADT 67,616)	36 mth (AADT 69,530)	Performance after 36 mth	Performance at Slow Lane vs Fast Lane
23.70	Fast Lane	2.25	1.98	1.79	2.04	+10%	+13%
	Slow Lane	1.99	1.76	1.81	1.80	+10%	
23.80	Fast Lane	1.77	1.61	1.41	1.75	+1%	-5%
	Slow Lane	1.47	1.69	1.32	1.84	-25%	
23.90	Fast Lane	2.20	1.33	1.47	1.45	+34%	+18%
	Slow Lane	1.49	1.13	1.81	1.22	+18%	
24.00	Fast Lane	1.77	1.33	1.54	1.40	+21%	-18%
	Slow Lane	1.72	1.58	1.83	1.72	0%	
24.10	Fast Lane	1.65	1.30	1.80	1.39	+16%	-24%
	Slow Lane	1.63	1.77	1.62	1.83	-13%	

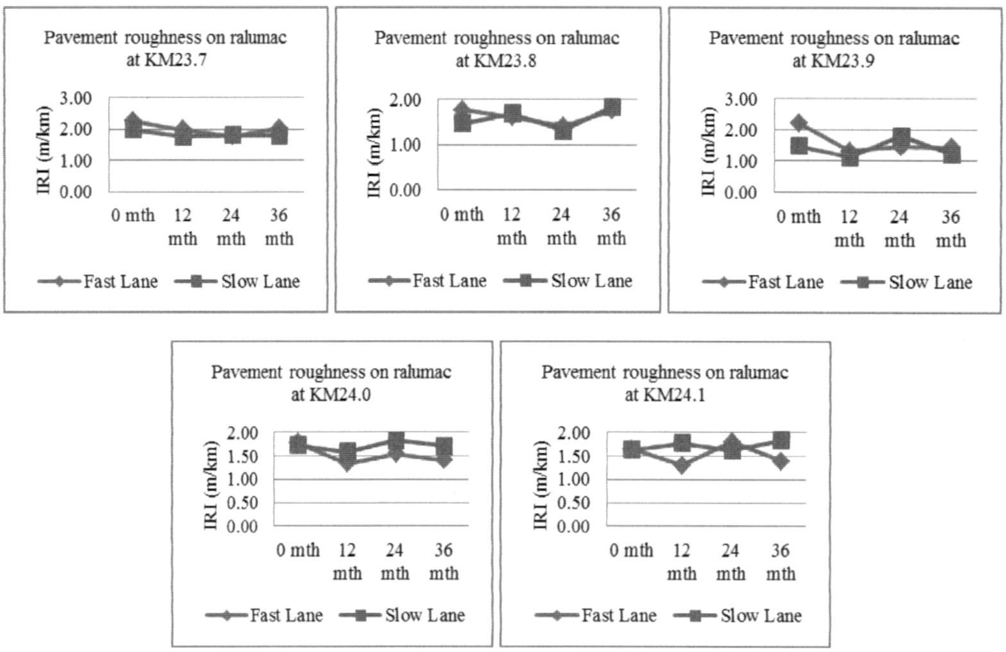

Figure 3. Pavement roughness at fast lane and slow lane.

5.2. Rutting

Table 7 and Figure 4 indicate the overall pavement rutting for Ralumac Micro surfacing after 36 months operational at fast lane is increased up to 63% as surveyed at KM23.70. Meanwhile pavement rutting at slow lane is increased up to 117% as surveyed at KM24.10 from original pavement condition. Furthermore, overall pavement condition at slow lane is more 36% rutting compare to rutting at fast lane as surveyed at KM23.70 and KM23.80. This occurred due to the slow lane receives more traffic loadings as it was designed mainly for heavy vehicles. However, overall pavement rutting on Ralumac

Micro surfacing after 36 months operational are still less than 5.0mm which are shown as a good rating as per Table 4 although Annual Average Daily Traffic (AADT) had increased by 24%.

Table 7. Pavement rutting for Ralumac Micro surfacing.

KM	Lane	0 mth (AADT 56,031)	12 mth (AADT 66,520)	24 mth (AADT 67,616)	36 mth (AADT 69,530)	Performance after 36 mth	Performance at Slow Lane vs Fast Lane
23.70	Fast Lane	1.60	2.81	2.50	2.60	-63%	-36%
	Slow Lane	2.70	3.99	3.20	4.05	-50%	
23.80	Fast Lane	2.00	2.52	2.90	2.36	-18%	-36%
	Slow Lane	2.20	3.47	2.40	3.71	-69%	
23.90	Fast Lane	1.40	2.38	2.59	1.96	-40%	-37%
	Slow Lane	2.20	3.50	3.09	3.12	-42%	
24.00	Fast Lane	2.20	2.19	3.04	1.71	+22%	-48%
	Slow Lane	2.30	3.23	3.29	3.26	-42%	
24.10	Fast Lane	2.30	2.05	5.34	2.39	-4%	-52%
	Slow Lane	2.30	4.07	3.16	4.99	-117%	

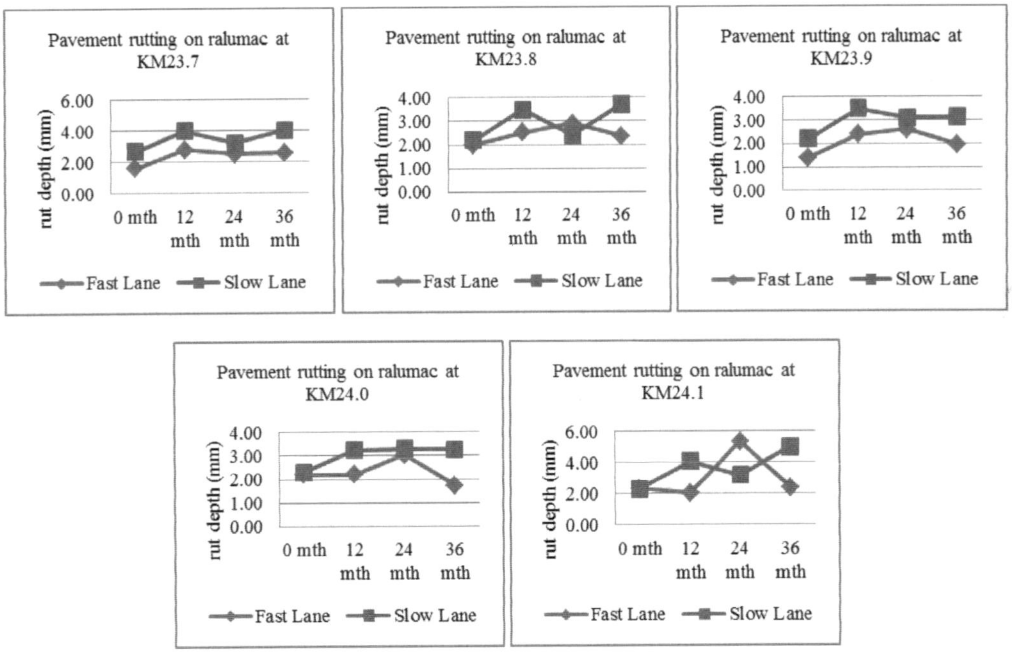

Figure 4. Pavement rutting at fast lane and slow lane.

5.3. Texture Depth

Table 8 and Figure 5 indicate overall texture depth for Ralumac Micro surfacing after 36 months is remain above 0.5mm which are shown as a good rating as per Table 4. The pavement condition with Ralumac Micro surfacing at slow lane is lower up to 57% texture depth compare to fast lane as surveyed at KM23.90. However, the trend of texture depth at both lanes has been improved every year although Annual Average Daily Traffic (AADT) had increased by 24%.

Table 8. Texture depth for Ralumac Micro surfacing.

KM	Lane	0 mth (AADT 56,031)	12 mth (AADT 66,520)	24 mth (AADT 67,616)	36 mth (AADT 69,530)	Performance after 36 mth	Performance at Slow Lane vs Fast Lane
23.70	Fast Lane	1.00	0.60	0.68	0.75	-25%	-34%
	Slow Lane	0.90	0.48	0.50	0.56	-38%	
23.80	Fast Lane	0.90	0.61	0.67	0.72	-20%	-35%
	Slow Lane	0.60	0.48	0.44	0.53	-11%	
23.90	Fast Lane	0.90	0.61	0.62	0.77	-14%	-57%
	Slow Lane	0.50	0.43	0.44	0.49	-1%	
24.00	Fast Lane	0.50	0.57	0.55	0.78	55%	-48%
	Slow Lane	0.50	0.44	0.49	0.53	5%	
24.10	Fast Lane	0.40	0.49	0.27	0.56	40%	-2%
	Slow Lane	0.60	0.42	0.26	0.55	-9%	

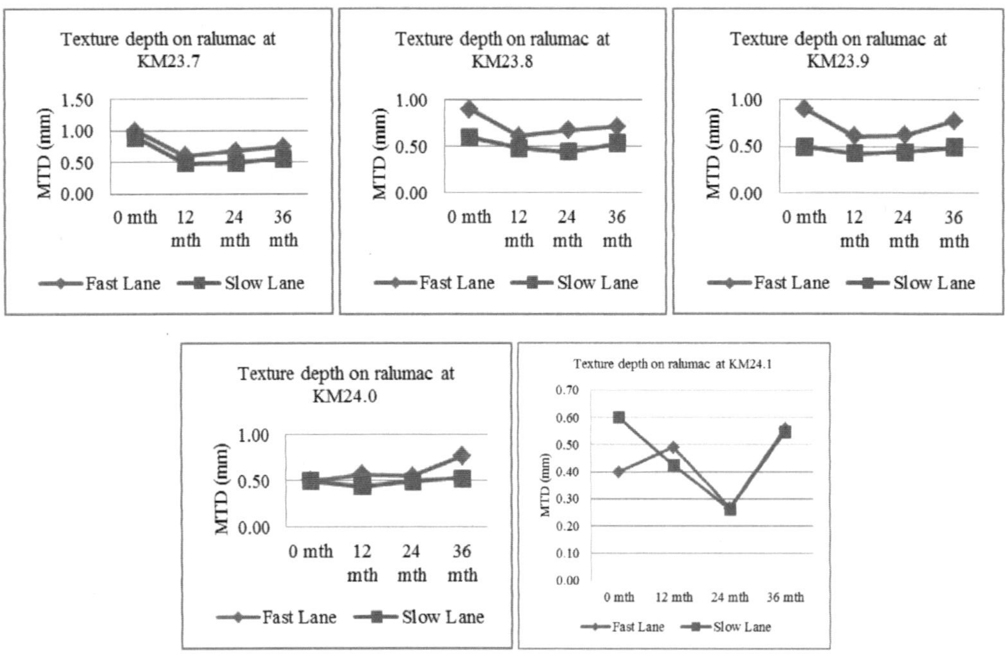

Figure 5. Texture depth at fast lane and slow lane.

5.4. Skid Resistance

Table 9 and Figure 6 show that the overall skid resistance for Ralumac Micro surfacing after 36 months operational at fast lane is still improved up to 24% as surveyed at KM23.70. Meanwhile skid resistance on Ralumac Micro surfacing at slow lane is decreasing from initial pavement condition until 21% as surveyed at KM24.10. Overall skid resistance at slow lane had decreased up to 36% compare to at the fast lane as surveyed at KM23.70. This occurred due to the slow lane receives more heavy vehicles. However, overall skid resistance for the micro surfacing after 36 months operational are still above 0.38 which are shown as a good rating as per Table 5 although Annual Average Daily Traffic (AADT) had increased by 24%.

Table 9. Skid resistance for micro surfacing.

KM	Lane	0 mth (AADT 56,031)	12 mth (AADT 66,520)	24 mth (AADT 67,616)	36 mth (AADT 69,530)	Performance after 36 mth	Performance at Slow Lane vs Fast Lane
23.70	Fast Lane	0.55	0.55	0.62	0.68	+24%	-36%
	Slow Lane	0.51	0.53	0.52	0.5	-3%	
23.80	Fast Lane	0.60	0.55	0.65	0.75	+24%	-29%
	Slow Lane	0.56	0.54	0.56	0.58	+3%	
23.90	Fast Lane	0.62	0.57	0.61	0.64	+3%	-16%
	Slow Lane	0.60	0.53	0.54	0.55	-9%	
24.00	Fast Lane	0.61	0.56	0.59	0.62	+2%	-24%
	Slow Lane	0.61	0.5	0.50	0.5	-18%	
24.10	Fast Lane	0.61	0.53	0.57	0.61	0%	-27%
	Slow Lane	0.60	0.49	0.49	0.48	-21%	

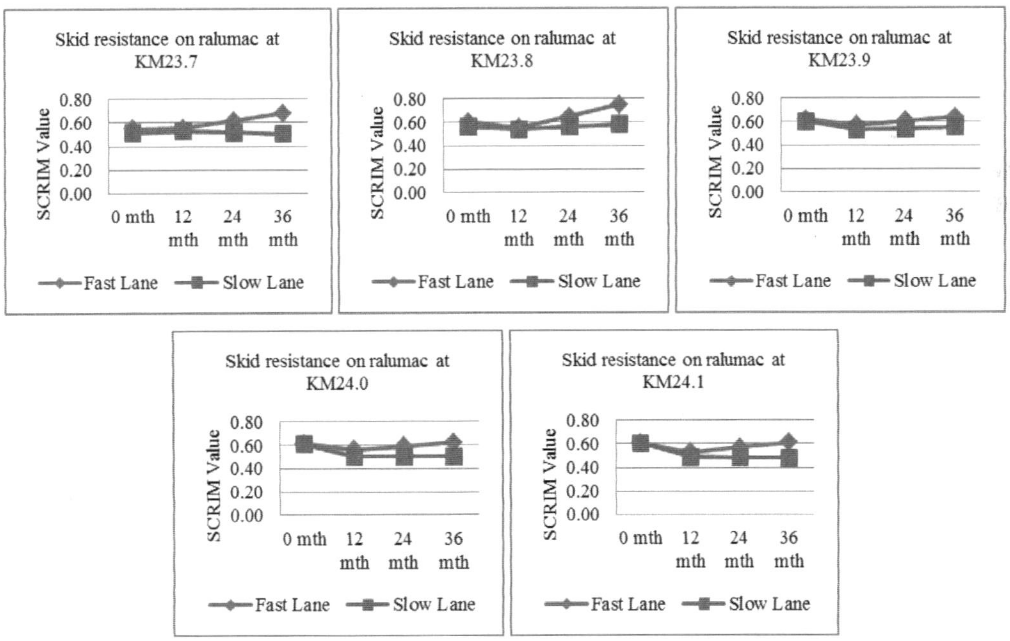

Figure 6. Skid resistance at fast lane and slow lane.

6. Conclusions

Overall performance of Ralumac Micro surfacing on roughness, rutting, texture depth, and skid resistance after 36 months operational are above the requirement by MHA and achieved good rating. The Ralumac Micro surfacing performed differently on fast lane and slow lane as summarized in Table 10.

Table 10. Performance of Ralumac Micro surfacing after 36 months operational.

No	Indicators	Performance
1	Roughness	16% more roughness at slow lane compare to fast lane
2	Rutting	42% more rutting at slow lane compare to fast lane
3	Texture depth	35% less texture depth at slow lane compare to fast lane
4	Skid resistance	26% less skid resistance at slow lane compare to fast lane

This is mainly due to the slow lane of LATAR highway receives more traffic loadings from heavy vehicles. The extra traffic loadings affect condition of pavement structure underneath the micro surfacing subsequently affect it performance. The service life of pavement also is slightly decreased as the traffic volume increased [12]. Currently, the heavy vehicles contribute 2% of AADT and traffic growth is 24% for LATAR highway. This paper recommends continuous monitoring on the performance of Ralumac Micro surfacing at this section to verify the actual service life as it shall extent until 8.5 years [12].

7. References

[1] Al-Mansour A and Sinha KC 1994 Economic analysis of effectiveness of pavement preventive maintenance *Transportation Research Record* 1442

[2] Association ISS 2010 *Recommended performance guidelines for micro surfacing (A143 Revised)*.

[3] Bae A and Stoffels SM 2008 Economic effects of microsurfacing on thermally-cracked pavements *KSCE Journal of Civil Engineering* **12**(3) 177-185

[4] Ben Broughton S J L, Yoo-Jae Kim 2012 "30 Years of Microsurfacing: A Review ISRN Civil Engineering **2012**

[5] Chan S, Lane B, Kazmierowski T and Lee W 2011 Pavement peservation: A solution for sustainability *Transportation Research Record: Journal of the Transportation Research Board* (2235) 36-42

[6] Gransberg D D and Tighe S M 2012 Microsurfacing best practices in North America *Seventh International Conference on Maintenance and Rehabilitation of Pavements and Technological Control*

[7] Takamura K, Rolf Wittlinger K P L and Aktiengesellschaft B A S F 2001 Microsurfacing for preventive maintenance: Eco-efficient strategy *International Slurry Seal Association Annual Meeting* Maui, Hawaii

[8] Dong Q, Stephen B H, Richards H and Yan X 2013 Cost-effectiveness analyses of maintenance treatments for low-and moderate-traffic asphalt pavements in Tennessee *Journal of Transportation Engineering* **139**(8) 797-803

[9] Torres-Machi C, Osorio-Lird A, Chamorro A, Videla C, Tighe SL and Mourgues C 2018 Impact of Environmental assessment and budgetary restrictions in pavement maintenance decisions: Application to an urban network *Transportation Research Part D: Transport and Environment* **59** 192-204

[10] James Wilde W, Thomas L T and Wood J 2014 *Cost-effective Pavement preservation solutions for the real world* (Department of Transportation: Research Services & Library)

[11] Zaniewski J and Mamlouk M 1999 Pavement preventive maintenance: Key to quality highways *Transportation Research Record: Journal of the Transportation Research*

IOP Conf. Series: Materials Science and Engineering **512** (2019) 012049 doi:10.1088/1757-899X/512/1/012049

Board (1680): 26-29
[12] Zhang, Y. and J. P. Mohsen 2018 "A Project-Based Sustainability Rating Tool for Pavement Maintenance." Engineering **4**(2): 200-208.

Capability of chained interlocking plastic beads in enhancing properties of dense bituminous macadam

P W Liew, R Muniandy, F M Jakarni and S Hassim

Department of Civil Engineering, Faculty of Engineering, Universiti Putra Malaysia, 43400 Serdang, Selangor, Malaysia

E-mail: jacqueline.liew@hotmail.com

Abstract. Typically, Dense Bituminous Macadam (DBM) layers on expressway flexible pavements tend to take up the flexural loading more than the top layers with binder and wearing courses. A lot of road failures associated with DBM are reflected as tensile cracks or typically know as Alligator cracks. In view of this mechanism of failure, a new thinking was foreseen in reinforcing the DBM layer with Chained Interlocking Plastic Beads (CIPB) to enhance the fatigue resistance of the layer. The CIPB in the DBM layer greatly reinforced and improved the interlocking mechanism within the pavement layer. The asphaltic binder with 60-70 penetration grade was used and the gradation for this DBM was selected in accordance with Lembaga Lebuhraya Malaysia (LLM) specification. CIPB of two diamond shaped beads are linked with 0.5 mm to 1.0 mm diameter strands. In this study, different percentages of CIPB were added into the bituminous mixture to determine the percentage of CIPB that would optimize the strength of DBM in term of resilient modulus, stability, and density. Several appropriate tests were conducted to determine the volumetric properties and stiffness modulus of the designed mixtures to determine the Optimum Asphalt Content (OAC). Based on the test results obtained, the CIPB did produce better interlocking mechanism within the mixture and greatly improved the stiffness and stability of the mixture. From this study it was observed that, with an adequate amount of CIPB, the performance of the DBM mixture improved the Marshall stability and resilient modulus. The study shows that the newly developed interlocking mechanism with CIPB has great potential to minimize tensile cracks that migrate to the top through the binder and wearing courses and thus extend the life of the pavement. Such long lasting effect may certainly reduce the road rehabilitation costs. The objectives of this study was to determine the Optimum Chained Interlocking Plastic Beads content in Dense Bituminous Macadam and to evaluate the capability of Chained Interlocking Plastic Beads in enhancing properties of Dense Bituminous Macadam.

1. Introduction

Plastic is a cheap and versatile substance that can be recycled from wastes and transformed to CIPB for use in road construction. Therefore, the plastic made products such as water bottles and plastic bags are easily being dumped after single use and this leads to massive amounts of plastic wastes in Malaysia and worldwide. As of 2010, the recycling rate in Malaysian is at 5%. This rate is considered as low compared to other countries such as Singapore (11%), Thailand (14%), Japan (40%), China (13%) and Germany (52.8%). Therefore, numerous recycling campaigns have been conducted by Kementerian Kesejahteraan Bandar, Perumahan dan Kerajaan Tempatan (KPKT) towards achieving the recycling target of 22% by the year 2020 [1].

1.1. Utilization of reclaimed plastics for road construction

Although there are several uses for recycled plastic materials, reutilizing the plastic waste particles as bitumen modifier for flexible pavement construction seem to be very promising. The use of recycled waste plastics in asphalt pavements represents a valuable outlet for such materials. Literature review indicates that the use of modified bitumen with the addition of processed waste plastics of about 5 to10% by weight of bitumen helped in improvement of the Marshall stability, strength, fatigue life and other desirable properties of bituminous concrete mixtures. Such improvements may result in the longevity and pavement performance with saving in bitumen usage [2]. Reclaimed polyethylene derived from low density polythene carry bags collected from domestic waste is also used as an additive in flexible pavement construction. When those polyethylene materials mixed with hot bitumen, they melt to form an oily coat over the aggregate and the mixture when laid on the road surface like a normal asphalt road. The durability of plastic waste modified road was much more as compared with roads of ordinary mix [3].

1.2. The newly developed and designed Chained Interlocking Plastic Beads

Chained Interlocking Plastic Beads (CIPB) is an intellectual propriety of Universiti Putra Malaysia (UPM) that was designed and fabricated to be used as reinforcing structure in asphalt mixtures in general. Based on the previous study by UPM researchers, CIPB was able to reinforce the asphalt concrete through the interlocking mechanism between the aggregates in the mixture. Several tests were carried out such as Marshall stability and resilient modulus on the CIPB reinforced cylindrical sample in order to determine the performance of the CIPB in asphalt mixtures. CIPB is designed and fabricated with diamond shaped terminal beads linked by strands that is able to reinforce Hot Mix Asphalt (HMA), Stone Mastic Asphalt (SMA) and Fiber Mastic Asphalt (FMA). The researchers from Universiti Putra Malaysia looked into the shape, size, dimension and type of plastic materials from new and recyclable plastics that are suitable to be fabricate into CIPB. The newly designed CIPB expected to provide additional fatigue and rutting resistance in most of the asphalt mixtures. Due to the melting point characteristic of above 220°C, the thermoplastic Nylon 66 was found to be ideal in the fabrication of the CIPB. Based on the several shape analyses carried out, the diamond shaped beads turned out to be very effective in rendering increased interlocking between the aggregates in the asphalt mixtures. CIPB produced in small sized beads such as 4 and 6 mm gave better flexibility with the links compared to larger beads and link sizes that posed problems during compaction of asphalt mixtures. Since the beads are linked up with two different sizes of strings or strands of 0.5 mm and 1.0 mm in diameter with the beads spaced at 20 mm, further analysis were carried out to determine their effectiveness. Based on UPM, s research and analysis on the newly developed plastic beads, it shows that the overall performance of the SMA mixtures with the CIPB performed better than the control mix [4].

1.3 Flexible pavement reinforcement in previous study

There are also others methods or technologies used to improve flexible pavements. The application of carbon fiber in asphalt mixture increased the strength, durability and resistance towards creep, fatigue and rutting condition. The amount of fiber that added to the bituminous mix designs, which at the optimum fiber content are resulted an increase in pavement strength [5]. Application of fiber into bituminous mixture slightly increased the Optimum Asphalt Content but it also provided additional tensile strength and fracture energy. It is able to the volumetric properties of hot mix asphalt [6]. The reinforcement mesh is embedded at the bottom of asphalt concrete mixture for the binder course enforced the initiation and propagation of cracks to be bottom-up. This decreased the tensile strain at the bottom of asphaltic mix and increased the number of cycles to initiate the cracks. There is improvement in the number of cycles and deformation rate for reinforced slab samples over the equivalent unreinforced slabs [7]. However, the implementation of CIPB takes the asphalt mixture performance to a different level by providing increased interlocking of aggregates in the mixture.

1.4 The Dense Bituminous Macadam

Bitumen macadam is composed of asphalt binder and three fractions of aggregate which were continuously graded. The three fractions of aggregate for bitumen macadam is consisted of coarse, larger fine and fine fractions of aggregate. Its strength is primarily achieved through friction and mechanical interlock of the coarse aggregates. Air void contents in typical Bituminous Macadam tend to be slightly higher than asphaltic concrete and it is thus relatively permeable to air and water, and consequently not as durable as asphaltic concrete. However, the higher level of air voids ensured the proportions of the mixture are not as critical as for asphaltic concrete, thus easier to produce [8].

Mixture volumetric properties are important to the long-term performance and durability of a pavement. If the required volumetric properties of the mixture are achieved, the minor deviations in gradation and asphalt binder content can be tolerated. The most important factor contributing to pavement performance is the in-place volumetric properties. Reduction of the in-place air voids below 8.0% significantly improved fatigue life, moisture resistance, and ravelling resistance [9].

1.5 Background of study

Generally, the structure of flexible pavement is formed by asphaltic layer, road base layer, sub-base layer and subgrade layer. When there is traffic loading on this pavement structure, each of the layers in the pavement system undergo the vertical compressive strain and also horizontal tensile strain. Figure 1 shows that the horizontal tensile strain and vertical compressive strain in pavement structure during traffic loading. The axle loads on the pavement cause the horizontal tensile strain to develop at the bottom of each layer in the structure, particularly the DBM layer. After certain number of load passes, this horizontal tensile strain could result in the cracking at the bottom of the DBM layer. In order to treat this kind of pavement failures, partial reconstruction may be required depending on its severity.

In order to minimize the severity of the crack propagation, a research was carried out in Universiti Putra Malaysia with a newly developed CIPB pavement reinforcing agent made from reclaimed plastic wastes. CIPB was found to be effective for use in DBM mixtures to improve the tensile strength properties and to provide reinforcement properties of DBM.

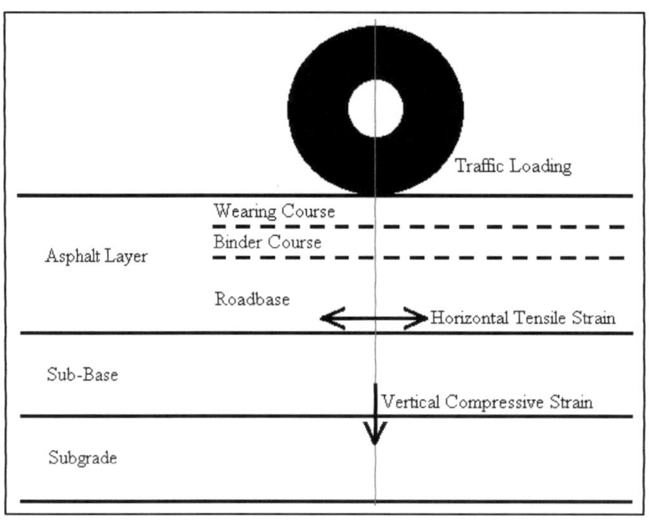

Figure 1. Horizontal tensile and vertical compressive strains in pavement structure.

2. Materials and Methods

2.1 The aggregate, asphalt binder and CIPB

The gradation requirement for DBM in this study was as stated in the Lembaga Lebuhraya Malaysia (LLM) specification: Series 800. The raw materials such as granite aggregates and asphalt binder with penetration grade 60-70 were provided by Kajang Rock quarry while the CIPB materials were supplied by UPM for the DBM study. Several tests were conducted to determine the physical and mechanical properties of the materials prior to formulating the appropriate mix designs. The raw materials that were used in mix design complied with the minimum requirement as stated in the Public Works Department (PWD) standard and specifications. A total of 80 samples with 60 compacted samples and 20 uncompacted loose samples were produced for the Marshall Mix design analysis in accordance with ASTM D1559. The materials properties of the aggregates, asphalt binder and CIPB used in this study are tested in accordance with ASTM and AASHTO standards. The LLM gradation for DBM is as shows in Table 1 and gradation plot is as shows in Figure 2.

Table 1. LLM Gradation Specification for DBM.

Sieve size	% Passing	Median	% Retained
37.5 mm	100	100	0
25 mm	90 - 100	95	5
20 mm	71 - 95	83	12
12.5 mm	56 - 80	68	15
6.3 mm	44 - 60	52	16
3.15 mm	31 - 45	38	14
300 µm	7 -21	14	24
75 µm	2 - 9	5.5	8.5
Pan			5.5
Total			100

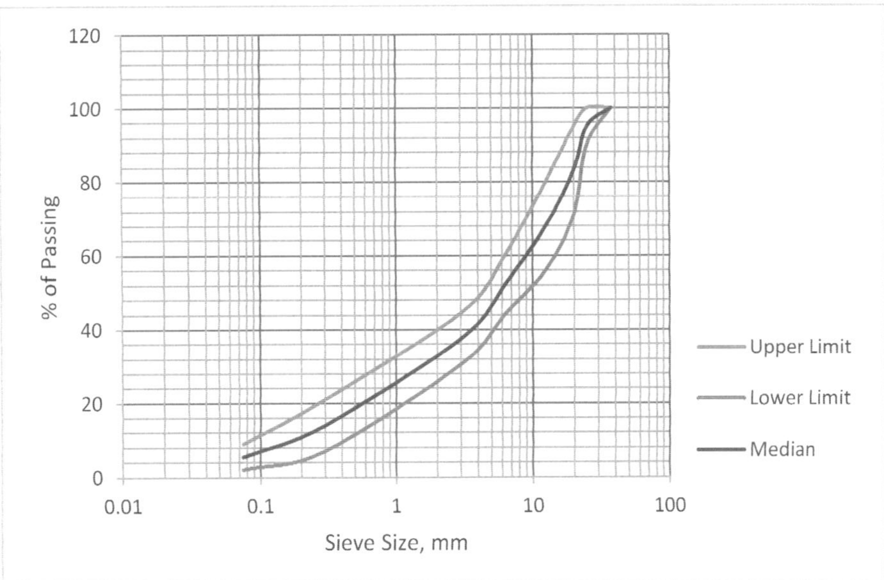

Figure 2. Aggregate Gradation Envelop.

The tensile test on the CIPB was conducted in accordance with ASTM D638 to determine its tensile strength properties. The purpose of the test was to determine the tensile strength of the CIPB and also the effect of heating it at 160°C (mixing temperature) on the tensile strength of CIPB. Therefore, the CIPB that has undergone tensile test was categorized into two set which were the non-heat conditioned CIPB (original) and the 160°C conditioned CIPB. Figure 3 shows that the effect of heating on the CIPB. After one-hour heating in the oven at 160°C, the maximum load bearing capacity of CIPB was increased by 17.91% while the tensile stress at the maximum load was increased by 8.42%. This increased load bearing capacity and the tensile stress showed that the heating process of during mixing would not cause major any damage to the CIPB properties.

Non-conditioned CIPB		160°C conditioned CIPB
Extension 55.50 mm **Maximum Load** 47.35 N **Tensile Stress at Maximum Load** 68.04 Mpa	After 1 hour heating →	**Extension** 47.11mm **Maximum Load** 55.83 N **Tensile Stress at Maximum Load** 80.57 Mpa

Figure 3. Tensile Strength for Non-conditioned and 160°C Conditioned CIPB.

Figure 4 (a) shows that the use of CIPB with a 2 bead and link system. As explained above, the CIPB has a great potential to stiffen at mixing temperatures. A preliminary assessment was carried out to determine whether the CIPB plastic beads with strand link would snap at terminal points upon compaction. The compacted samples were then loosened up to recover the CIPB through rotary extraction process using methylene chloride solvent. Figure 4(b) shows that the CIPB are still intact.

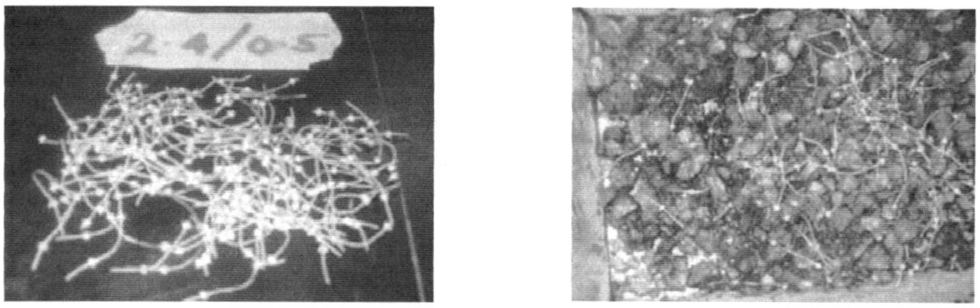

(a) (b)

Figure 4 (a)and (b). CIPB before and after compaction.

2.2 The Marshall mix design analysis

There were four types of mix designs produced for the Marshall mix design analysis namely DBM-0, DBM-5, DBM-10 and DBM-15. DBM-0 was the controlled samples that did not contain any CIPB while DBM with 5 to 15 were the mixtures that contained 5.0 to 15 grams of CIPB for each Marshall specimen. The 5 grams weight of CIPB is also equivalent to 0.43% of CIPB by the weight of aggregate. Similarly, DBM-10 and DBM-15 were equivalent to 0.87% and 1.30% of CIPB by the weight of aggregate respectively.

A total of 20 specimens included 15 compacted specimens and 5 uncompact specimens were prepared for each mix design. Marshall Mix Design Analysis was then conducted on those specimens to determine their Volumetric Properties including VTM (Voids in Total Mix), VMA (Voids in

Mineral Aggregate), VFA (Voids Filled with Asphalt), Bulk Density, Marshall Flow, Marshall Stability and Resilient Modulus.

The targeted air void was selected as the ideal void content of 4% based on research and past performance shown that the value was ideal for asphalt mixtures. Therefore, OAC was determined based on 4% of design VTM. Volumetric Properties Ranking system was carried out to determine the optimum CIPB content. Table 2 shows the OAC for CIPB reinforced DBM Mix Designs. For the determination of Optimum CIPB Content, the mix design with the Optimum CIPB Content was the mix design with the highest Volumetric Properties Ranking.

Table 2. OAC for CIPB reinforced DBM.

Mix Design	Optimum Asphalt Content
DBM-0	5.45%
DBM-5	5.48%
DBM-10	6.45%
DBM-15	6.68%

3. Results and Discussion

Graphs below were plotted based on the data obtained from the Marshall Mix Design Analysis. The specific gravity of CIPB was 1.06 which was much lower than the specific gravity of aggregate. Therefore, introduce the CIPB into mixture reduced the bulk density of the total mixture. Higher percentage of CIPB added into mixture resulted in lower bulk density. Based on Figure 5 plot of Bulk Density against AC, it showed that DBM-15 had the lowest bulk density among the mixes due to DBM-15 contained the highest percentage of CIPB.

Based on Figure 6 plot of Marshall Stability against AC, it showed that mix design with the adequate amount of CIPB, DBM-5, did effectively improved the Marshall Stability of the mix design. By comparing between the controlled sample and DBM-5 (0.43% of CIPB by the weight of aggregate), it was able to improve the stability properties of DBM up to 6.53%. However, from the graph, it was observed that the mix design with increased amount of CIPB resulted in low Marshall Stability due to the higher amount of CIPB in the mixture that resulted in greater influence in the matrix of aggregate in the mix design and may cause segregation. The matrix of aggregate in the mixture seemed to be disturbed by the exceeding amount of CIPB causing the aggregates not slipped into the perfect position during the compaction stage to enable the interlocking mechanism among the aggregate.

From Figure 7 below a plot of Resilient Modulus against AC showed that with introduction of adequate amount of CIPB, it can effectively improve the Resilient Modulus properties of the mix design. But increased amount of CIPB tend to lower the Marshall Stability and also resulted in low Resilient modulus of the asphalt mixtures. Comparing between the controlled sample and DBM-5, by introducing 0.43% of CIPB by the weight of aggregate, it was able to improve the stiffness modulus properties of DBM up to 13.32%.

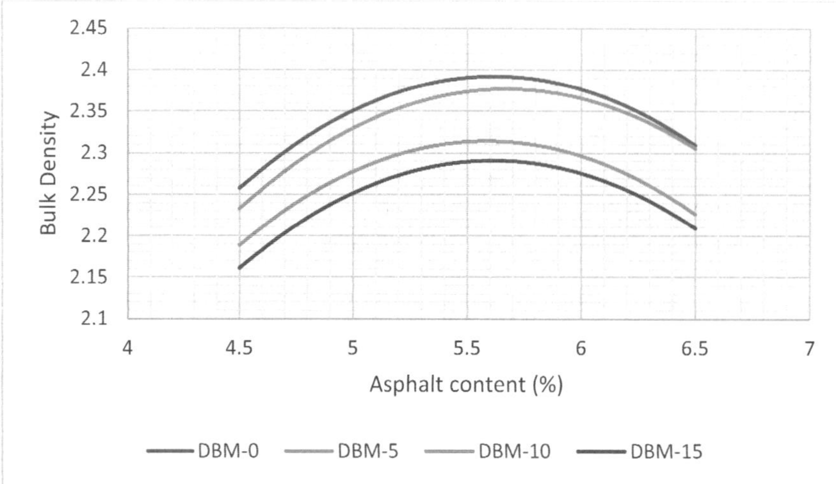

Figure 5. Plot of Bulk Density against Asphalt Content.

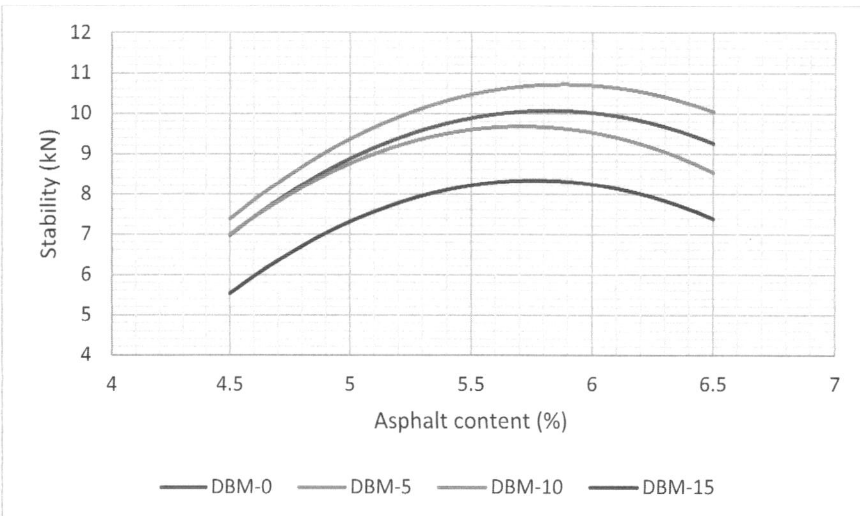

Figure 6. Plot of Marshall Stability against Asphalt Content.

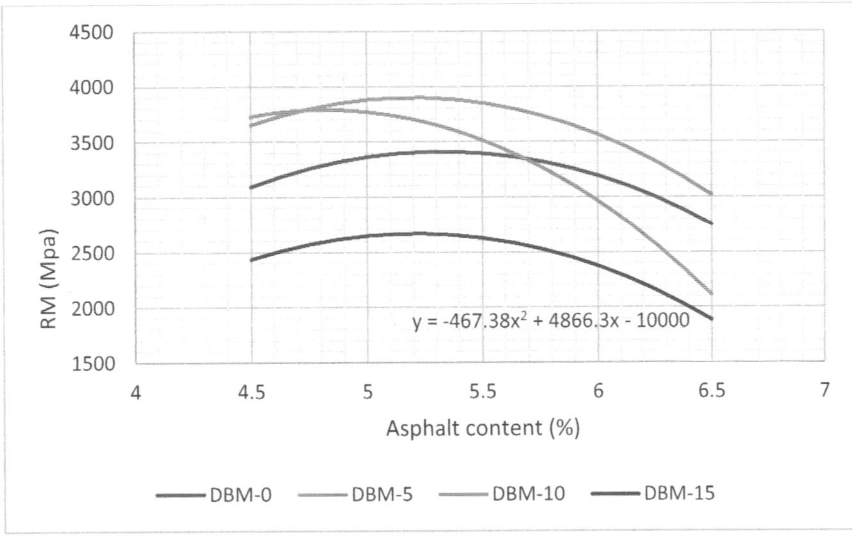

Figure 7. Plot of Resilient Modulus against Asphalt Content.

Table 3 shows that the summarized results of the Volumetric Properties and Resilient Modulus properties for DBM-0, DBM-5, DBM-10 and DBM-15. The results were obtained from Volumetric Properties graphs that plotted based on the Marshall Mix Design Analysis. Table 4 shows that the Volumetric Properties Ranking among the mix designs. Comparisons were made among DBM-0, DBM-5, DBM-10 and DBM-15 to find out the highest ranking mix design and to determine the Optimum CIPB Content. Comparison of Volumetric Properties among the mix design based on results from Table 3 were converted to the Volumetric Properties Ranking table as in Table 4. The ranking was from 1 to 4 (best → least), Therefore, mix design with the smallest number will have higher ranking.

Table 3. Volumetric properties of all the mix design at their OAC.

Mixtures	Resilient Modulus (MPa)	Stability (kN)	VTM (%)	Bulk Density	VMA (%)
DBM-0	3401.7	9.80	4.00	2.388	16.41
DBM-5	3854.9	10.44	4.00	2.374	16.98
DBM-10	2208.9	8.67	4.00	2.235	20.20
DBM-15	1651.3	6.86	4.00	2.170	21.22

Table 4. Volumetric Properties Ranking.

Mixtures	Resilient Modulus (MPa)	Stability (kN)	Bulk Density	VMA	Ranking
DBM-0	2	2	1	2	7
DBM-5	1	1	2	1	5
DBM-10	3	3	3	3	12
DBM-15	4	4	4	4	16

Based on the above results the mix designs with CIPB, DBM-5, DBM-10 and DBM-15, it was observed that DBM with 5 grams of CIPB had the highest ranking between them. Based on the data obtained, the OAC of the DBM-5 mixture was 5.48% and the Optimum CIPB Content was 0.43% by the weight of aggregate. If compared the mix design properties between the DBM-0 and DBM-5, DBM mixture with 5 grams or 0.43% CIPB effectively enhanced the DBM mixtures by improving the properties of stability and stiffness modulus. However, with compare among the mixes with CIPB, DBM-10 and DBM-15 have low stability and stiffness modulus. It was also observed that high percentage of CIPB may cause the mixture homogeneity issue that resulted in greater influence in the matrix of aggregate and may cause segregation. The matrix of aggregate in the mixture been disturbed by the exceeding amount of CIPB caused the aggregate not slipped into the perfect position during the compaction stage to enable the interlocking mechanism among the aggregate.

As shows in Table 4, among the mix designs with CIPB, DBM-5, DBM-10 and DBM-15, DBM-5 had the highest ranking among them. Therefore, DBM-5 with a 0.43% or 5 grams by the weight of aggregate seemed to be the optimum for the selected DBM aggregate gradation. It was also observed that the properties between the controlled sample (DBM-0) and the mix design with Optimum CIPB content (DBM-5), DBM-5 had a higher Resilient Modulus of 3854.9 MPa and a Marshall Stability of 10.44 kN. As such, it can be assured that CIPB did improve the properties of DBM mixture in term of Stiffness Modulus and Stability.

4. Conclusion

Based on the results and analysis above, the following conclusion can be made:

i. From the Volumetric Properties Ranking, DBM-5 had the highest ranking among others. The OAC for DBM-5 was 5.48% while the Optimum CIPB content was 0.43% by the weight of aggregate. The result showed that the Marshall Stability and Resilient Modulus of DBM-5 gave were the highest. It meant that the degree of stiffness of the mixture was higher and that CIPB effectively enhance the DBM mix performance through its interlocking mechanism.
ii. The Volumetric Properties of DBM-5 at its OAC and Optimum CIPB Content: VTM =4%, Bulk density = 2.374, Marshall Stability = 10.44kN, Resilient Modulus = 3854.9 MPa, and VMA = 17% which fulfilled all the specification requirements. Adequate amount of CIPB did improve the properties of the mixture in term of Resilient Modulus and Stability, but it will reduce the bulk density of the mixes as the percentage of added CIPB increases.
iii. Based on the tests results, adequate amount of CIPB can effectively improve the Marshall stability and Resilient Modulus of the mixture.
iv. With the appropriate amount of CIPB, DBM layers on the expressways can be reinforced to resist fatigue cracking. This can minimize cracks migrating to the top of the flexible pavement and prolong rehabilitation intervention measures by the road authorities. Such improvements to DBM layers could save cost on maintenance of roads.

5. References

[1] National Solid Waste Management Department 2011 *A study on plastic management in Peninsular Malaysia*

[2] Ahmad N, Muniandy R, Hassim S and Aburkaba E 2013 Development of interlocking chained plasti-beadsfor use in stone mastic asphalt *Australian Journal of Basic and Applied Sciences* 7(4) 466–74

[3] Amit G, Zamarea G, Rengea V C, Saurabh T and Bharsakale G 2012 An overview on waste plastic utilization in asphalting of roads *Journal of Engineering Research and Studies* (E-ISSN0976-7916)

[4] Donald L C 2013 Roads building fundamentals: Volumetric properties of asphalt mixes *Arizona Pavements and Materials Conference*

[5] Malik S A 2014 Low density polyethylene modified dense graded bituminous macadam *International Journal of Engineering Trends and Technology* **16** 8
[6] Kumar B K, Kumar V R and Rao M K 2016 Experimental behaviour of carbon fiber in dense bituminous macadam *International Journal for Technological Research in Engineering* **3** 6
[7] Kaloush K E 2014 Fiber Reinforced Asphalt Pavements *54th Annual Idaho Asphalt Conference* Idaho
[8] Radzi H M, Harun M H and Hanafi WRW 2010 *Asphalt concrete on roads in Malaysia: Is quality under control?* (Kuala Lumpur: Public Work Department)
[9] Donald L C 2013 Roads building fundamentals: Volumetric properties of asphalt mixes *Arizona Pavements and Materials Conference*

Acknowledgement
The authors would like to thank the Ministry of Higher Education for providing the research grant and Universiti Putra Malaysia for providing the required facilities and support for this study and all those directly or indirectly assisted in the research project.

IOP Conf. Series: Materials Science and Engineering **512** (2019) 012051 doi:10.1088/1757-899X/512/1/012051

Construction of the first Malaysia's hybrid concrete pavement using jointed plain concrete pavement (JPCP) and roller compacted concrete (RCC)

S Sahari, B K Tan, S Shaharudin and M Bakar

Lafarge, Wisma Lafarge, 46050 Petaling Jaya, Selangor, Malaysia

E-mail: sahruzi.sahari@lafargeholcim.com & steven.tan@lafargeholcim.com

Abstract. Lafarge Infrastructure Team had constructed Malaysia's first hybrid concrete pavement using Jointed Plain Concrete Pavement (JPCP) and Roller Compacted Concrete (RCC) at Lafarge's Kanthan Cement Plant, Ipoh, Malaysia. The main purpose of the project is to showcase and study construction challenges in combining JPCP and RCC hybrid product for road solutions. The progress and pavement performance were monitored recorded and detailed through tests and site observation reports. JPCP pavement was constructed using two (2) type of construction methods; slip form and fixed form, while RCC pavement was laid and constructed using an asphalt paver set. The Quality Assurance & Control (QA/QC) monitored comprises of concrete design mix, concrete slump, concrete delivery, joints insertion, concrete curing, compaction, profile levels, saw cut and sequence of construction activities. Throughout the study, we have managed to table out the "Challenges and Best Practices" on the construction methods which mainly involves labours, machinery and material. All the information has been taken into consideration, to prepare a comprehensive report for the project. The study will enhance the knowledge and understanding for road owners, concrete suppliers, consultants and contractors in understanding the right application of JPCP and RCC in the future. We able to share and propose few best practices based on the challenges that we had face.

1. Introduction

Concrete roads have over 30-year history in Malaysia. Starting in 1988, first mega project using concrete pavement was built at the North-South Expressway (NSE) and New Klang Valley Expressway (NKVE). The concrete pavement wa built using two difference method called Continuously Reinforced Concrete Pavement (CRCP) and Jointed Reinforced Concrete Pavement (JPCP). About twenty-two percent (22%) out of 846km Expressway network was constructed using concrete pavement methods [1].

These concrete roads have performed significantly much better than a conventional asphalt pavement in terms of durability. Projek Lebuhraya Utara Selatan (PLUS) concessionaire of the expressway, indicated that in general, the road stretches which were built using a conventional asphalt pavement need regular maintenance for every 7 years. It has been reported that PLUS is spending approximately RM740mil yearly on the maintenance and repair works for the whole stretch [2]. On the other hand, concrete pavement which required minimal maintenance would use only 1% of the budget. It is a proven testimonial on the durability and the superiority on the life cycle cost of concrete roads.

IOP Conf. Series: Materials Science and Engineering **512** (2019) 012051 doi:10.1088/1757-899X/512/1/012051

Despite the benefits of concrete pavement, the adoption rate in Malaysia is still low. Until today, Malaysia has only constructed 220km of concrete road. It consist less than 1% out of the total 144,403 km road networks in Malaysia. The most recent concrete road being built was a dual-lane 10km stretch of CRCP located at Raub, Pahang between Felda Telang to Kg Kechor. The road was part of the Central Spine Road (CSR) Project that had successfully completed in 2015 and had won the Second Runner-up for JKR Project Management Innovation Award 2015 [3].

Malaysia local authorities are facing challenges in increasing the adoption rate of concrete pavement. The benefits of concrete pavement are well recognized by the Kementerian Kerja Raya (KKR) and has set an ambition to build 5% of its new federal road using concrete in the '2013-2015 KKR's Strategy Plan' [4]. However, they face challenges in achieving its ambition to build concrete roads. There are limited roads network in Malaysia other than highway that being constructed using concrete. We reckoned one of the challenges is high initial construction cost; in relation of using a CRCP method is one of the key barriers; resulted in difficulty to justify concrete road adoption for rural roads. However, CRCP is not the only option in constructing concrete roads. Optimizing the design to suit with the needs and identifying the requirements based on different road categories could help in reducing the overall cost on the concrete road construction.

The objective of this paper is to determine the efficiency of alternative concrete pavements. By adopting to different concrete road construction methods, i.e. combination of Jointed Plain Concrete Pavement (JPCP) & Roller Compacted Concrete (RCC) as showcased in the reference project below (hereafter refer to as JPCP-RCC hybrid concrete pavement); can this be a good alternative with better efficiency comparing to CRCP? Meanwhile, at the same time looking at the anticipated challenges, as well as best practices of such hybrid concrete pavement as part of the road alternative and solution in Malaysia.

2. Overview of different concrete road construction methods

2.1. CRCP – Continuously Reinforced Concrete Pavement

CRCP is a continues span of concrete pavement that without any transverse joints. The pavement contains continuous, longitudinal high tensile steel reinforcement. Most of the length of pavement is without any transverse joints, except for end-of-day header joints, transition between bridge approaches and other pavement structures. The highly heavily steel reinforcing in CRCP is to hold the concrete together and controlling all cracking especially transverse cracking caused by cement hydration, thermal effects and contraction/expansion of a concrete pavement [5].

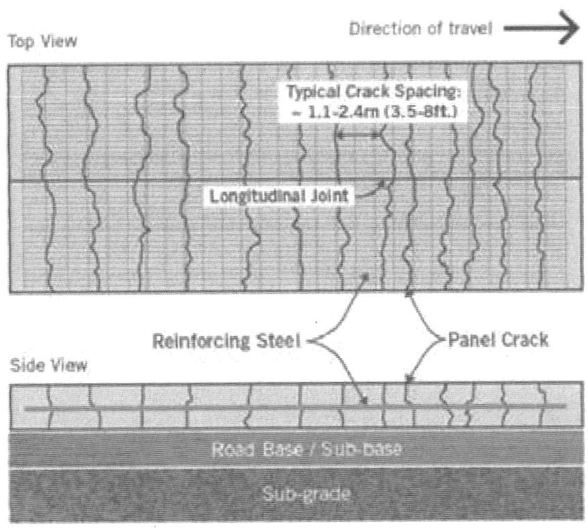

Figure 1. Continuously Reinforced Concrete Pavement (CRCP).

2.2. *JPCP – Jointed Plain Concrete Pavement*

JPCP is a concrete pavement that has transverse joints with panels. The pavement is constructed in a similar manner to CRCP, while the difference is that JPCP only introduces dowel & tie bars as compared to CRCP with fully reinforcing the pavement. The joints with dowel or tie bars have two important functions, i.e. inducing contraction & compression cracks and transferring loads to adjacent slabs. The bars also hold the slabs tightly closed to instill additional stiffness into the concrete slab in-between the panels [5].

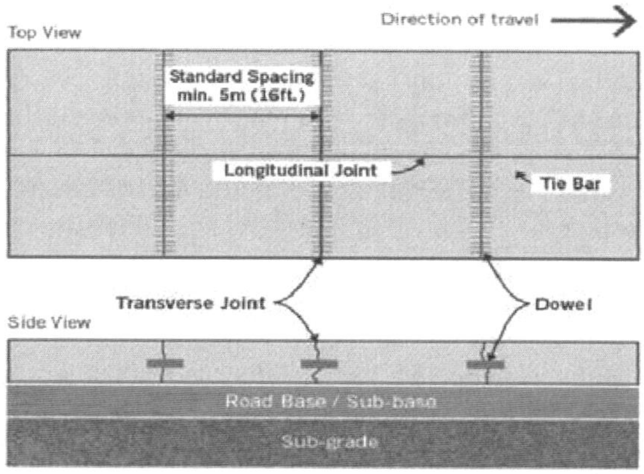

Figure 2. Jointed Plain Concrete Pavement (JPCP).

2.3. *RCC – Roller Compacted Concrete*

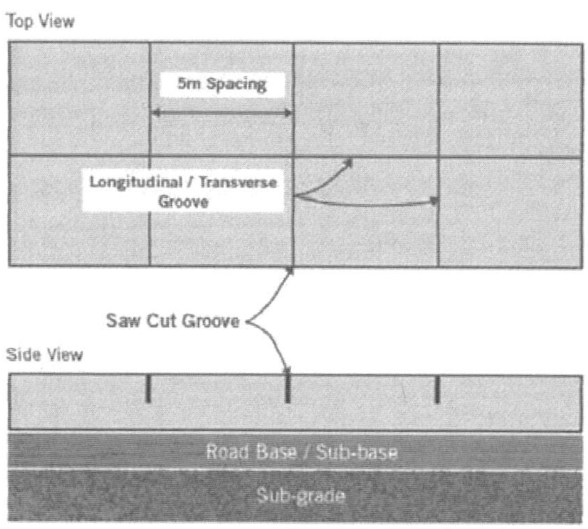

Figure 3. Roller Compacted Concrete (RCC).

RCC is concrete pavement that uses a zero-slump concrete, which can be laid using an asphalt paver. The RCC's material is designed much stiffer than a typical conventional concrete. The design mix has to remain stable under compaction operation using a vibratory roller but at the same time wet enough to permit adequate mixing and distribution of paste of cement mortar without segregation. This product is purposely designed to provide road contractors a solution to construct concrete pavement without the need to invest in slip form concrete pavement machines. RCC is engineered to be place with asphalt-type pavers' followed by compaction by rollers. The RCC method of construction is without the need of any forms/formwork, dowels, or steel reinforcement. The proposed RCC material used in road pavement was a concrete that had a Compressive Strength of 40MPa with a Flexural Strength of 4.5MPa [6].

2.4. JPCP-RCC Hybrid Concept

Figure 4. JPCP-RCC hybrid concept.

JPCP-RCC hybrid pavement is first of its kind in Malaysia. This hybrid design combining both JPCP and RCC construction techniques and is piloted by Lafarge Road Infrastructure team in Malaysia. The first hybrid pavement was successfully constructed and handed-over to Lafarge's Kanthan Cement Plant. The idea of the hybrid pavement is to optimize the pavement in relation to requirement, application and traffic flows to promote cost saving. This pavement is designed to use JPCP as the main driveway and RCC as the parking, lay-by and truck sheeting bay.

Upgrading existing rural asphalt roads to concrete become feasible with concept of JPCP-RCC hybrid pavement. One of the main challenges in upgrading existing rural asphalt roads into concrete roads is the long setting time of a normal concrete. Referring to the industry best practice, the concrete pavement must achieve minimum 20MPa in compressive strength before it can open to traffic. If the pavement is to be constructed with a normal G40 concrete, it will take a minimum 7 days for it to reach the desired strength. Hence, it is not feasible to close the roads for 7 days because it will cause significant nuisance to local communities that depend on it. However, with JPCP-RCC hybrid pavement help in resolving the challenges of long setting time. RCC material could achieve the desired strength and open to light traffic within 8 hours upon the compaction. The concept of the hybrid roads is to construct the RCC as permanent road shoulder but temporarily serve as diversion and access roads. Once the traffic has been diverted to shoulder/temporary access roads, JPCP can then be constructed as the main carriageway. With this approach, it makes upgrading existing rural asphalt roads feasible.

IOP Conf. Series: Materials Science and Engineering **512** (2019) 012051 doi:10.1088/1757-899X/512/1/012051

3. Project Background

3.1. General Overview

Lafarge Kanthan Cement Plant, Chemor, Perak, has planned to upgrade the existing crusher-run base to a reinforced concrete hardstand. Like most of the cement plants, dust pollution is always an environmental issue. By having an un-paved granular hardstand, it makes the issue of dust difficult to control. To reduce the dust, hence resolving environmental issues, the plant manager had decided to upgrade the un-paved hardstand by constructing a new reinforced concrete hardstand. The new concrete pavement is design to serve two main purposes. First, to address the environmental issue by reducing the dust pollution from the existing hardstand to an acceptable level, set by Department of Environment (DOE). Secondly, to address the increase number of traffic (trucks and plant vehicles), by adding in additional areas for driveway, sheeting bay and parking lots; this will allow the plant to have better management on the total traffic movement.

Figure 5. Aerial view of JPCP-RCC.

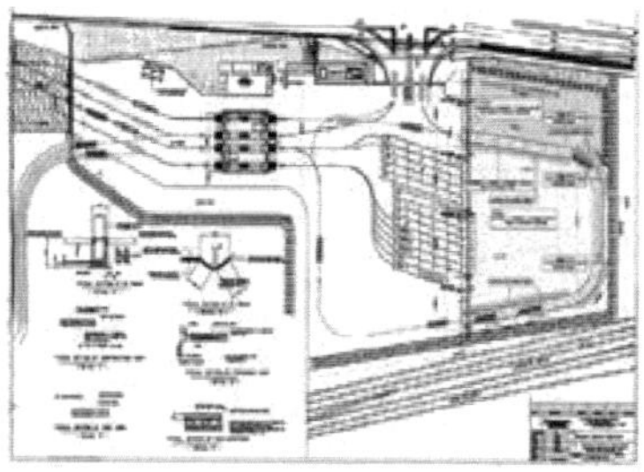

Figure 6. Original design pavement layout plan.

The concrete pavement is designed for 20 years with 20 million of Rigid Equivalent Standard Axle Load (ESAL). The daily traffic volumes for the cement plant can reach up to 200 numbers of cement tankers per day transporting its daily production of 8,000 metric tonne (mt) cement to clients. Each cement tankers can weigh up to 60 mt on full load of delivery. The design ESAL have to be minimum of 0.64 Million per year to support the number of truckload going through the pavement. Hence, for a 20 years design, the pavement has to be designed with 12.8mil flexible ESAL or 19.2mil rigid ESAL [7].

3.2. Project objective

JPCP and RCC hybrid pavement was introduced to reduce the overall construction cost and at the same time shorten the completion period. In optimizing the overall design of the plant hardstand, combination of JPCP and RCC hybrid construction technique has been proposed to the plant. The proposal will help in reducing the overall construction cost and completion period. On the other hand, this proposal will also provide a training ground to gain on-site experiences on the construction of the hybrid concrete road design.

The ultimate goal is to study the concept of constructing the road pavement with JPCP as carriageway and RCC as road shoulders. This project will allow us to understand the best and workable concrete mix design, practical construction method statements, pre, during and post construction testing for the construction of JPCP and RCC hybrid pavement works.

IOP Conf. Series: Materials Science and Engineering **512** (2019) 012051 doi:10.1088/1757-899X/512/1/012051

4. The optimized JPCP-RCC hybrid design

4.1. Original design

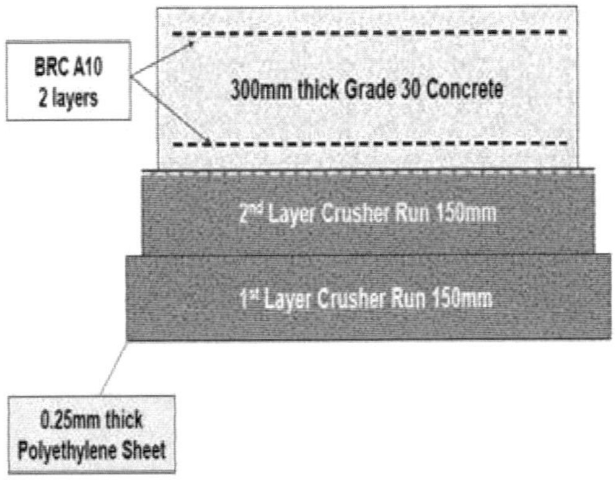

Figure 7. Cross section of original design.

The pavement was originally designed with conventional reinforced concrete pavement. The original design of the concrete pavement is made up of a 300mm thick G30 concrete pavement, reinforced with two layers of BRC A10. The concrete surface course is to be supported with two layers of 150mm thick of crusher run and separated with a layer of 0.25mm thick polyethylene sheet lay under the concrete. The details are referred to Figure 7.

4.2. The optimized design

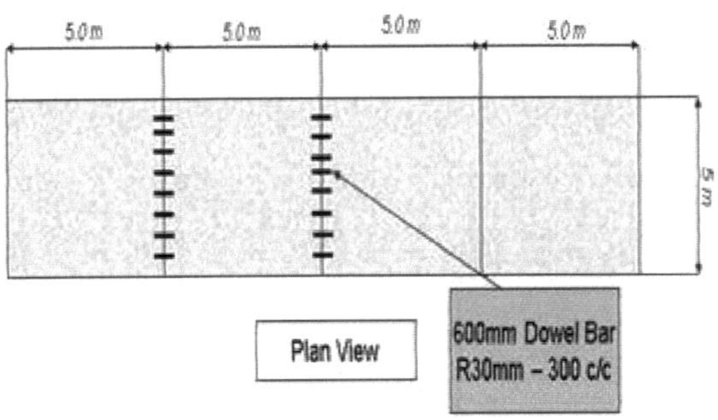

Figure 8. Joints and saw cut for JPCP.

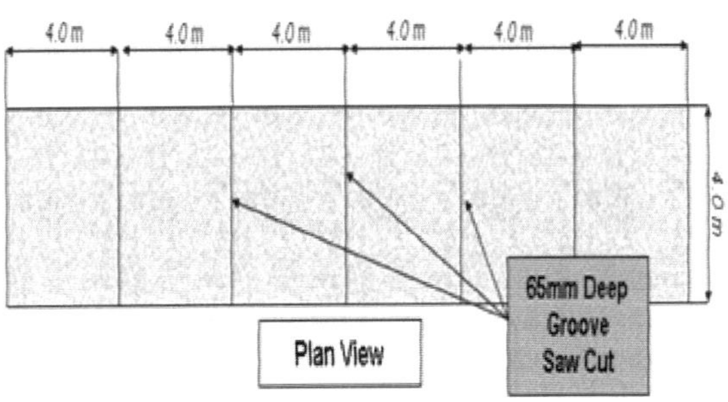

Figure 9. Joints and saw cut for RCC.

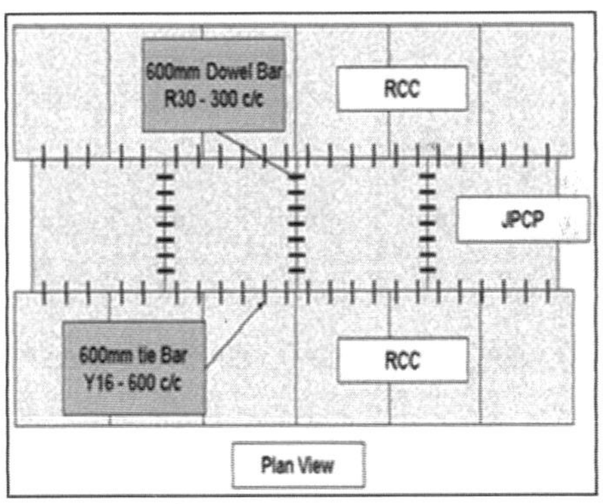

Figure 10. Joints and saw cut for JPCP-RCC hybrid.

The hybrid pavement is designed according to traffic movement to bring better efficiency in saving cost and time. Not all concrete panels are subject to the same loading, the pavement can be optimized by combined JPCP and RCC pavement. By understanding tankers and trucks movement, the pavement can be designed to provide a better cost effective and easy to be constructed by reducing the need to install steel reinforcement and formworks. The optimized design decided to construct the driveway using JPCP with broom finishing texture to cater for cement tankers and trucks movement. While area that have longer vehicles stoppage time and lower movement like parking lot and sheathing bay were design to construct using RCC. The specifications for the pavements are as follow:

- Surface course, 200mm thick G40 Concrete for both JPCP and RCC
- Joints, JPCP pavements are joint with steel basket consist of 600mm R30, 300mm c/c Dowels and 600mm Y16, 600mm c/c tie bars.
- Panel, JPCP is saw cut into a 5m x 5m panel' size while RCC is saw cut into a 4m x 4m panel size.
- Tie bars between JPCP and RCC, 600mm Y16- 600mm c/c tie bars are inserted at conjunction area of JPCP and RCC.

4.3. Benefit of the optimized design
The optimized hybrid pavement is proven in reducing overall construction cost and faster completion. The outcome of the hybrid design is promising, as it has managed to reduce the overall construction cost by 30%. The project has registered saving from concrete, steel reinforcement, formwork and overhead cost. The major material saving came from the reduced quantity of reinforcement steel bar, whereby the hybrid design has managed to reduce 70% reinforcement steels needed to construct the pavement. Beside cost saving, the JPCP-RCC hybrid pavement was completed within 13 days.

5. Challenges in constructing concrete pavement

Table 1. Challenges in JPCP and RCC construction.

No		Challenges	Best Practice
1		Concrete Delivery	
	(i)	Poor mixing Maximum loading of mixer trucks to speed up delivery could result in poor mixing due to limited space inside the mixer drum to thoroughly mix low slump concrete creating a non-homogeneous mix.	To load only 80 % of mixer drum capacity to allow adequate space inside the mixer drum to mix and create a homogeneous mix.
	(ii)	Waiting time Too many trucks being used for concreting will result in extended waiting time and will cause rapid slump loss.	Proper tracking and scheduling of mixer trucks need to be done by the plant and site personnel to ensure on time delivery is maintained during concreting work.
	(iii)	Travel time Concrete being delivered from batching plants located more than 10 km away from the construction site. Due to extended travel time and over mixing during long hauls, excess heat is generated due to friction causing the concrete to lose water rapidly thru evaporation process resulting serious slump drop.	To identify suitable batching plants which are located much nearer or less than 10 km away from the project site. Lesser travel time will enable us to maintain a consistent concrete slump.
2		Inexperience road contractor An incompetent contractor with poor or limited knowledge and experience in RCC paving work will impact the overall operations and product performance.	Competent contractors with the relevant knowledge and experience in paving work must be selected for each job to ascertain the desired results are achieved. Additional class room or on the job trainings can be conducted to further enhance the contractor's knowledge.

Table 1. Challenges in JPCP and RCC construction (continued)

3		Weather	
	(i)	Rain Concreting in bad weather.	Proper canopy need to be in place to protect freshly casted pavement from being exposed to the rain water. Mixer trucks for JPCP need to have proper hopper covers and tipper truck for RCC need to have proper tarpaulin to prevent rain water entry into the concrete drum resulting in high slump.
	(ii)	Hot weather Working in ambient temperatures above 34degree Celsius will lead to concrete drying up much quicker due to rapid evaporation of water. This dry concrete would result in poor pavement finishing.	To provide proper shade and to arrange adequate manpower and equipment to commence pavement finishing work immediately upon paving.
4		Joint/ Connection	
	(i)	Improper installation of bars Improper installation of the tie bar and dowel bar between the JPCP - RCC pavement and JPCP - JPCP pavement will affect the overall levelness of pavement. Besides, it will also result in mid-panel crack or another pavement defect.	Contractor to properly install tie bar and dowel bar into the JPCP pavement according to the design and specification. Installment of the tie bar into the JPCP pavement shall be perpendicular to the concrete edge.
	(ii)	Delay in saw cut Delay in saw-cutting the JPCP/RCC pavement will result in uncontrolled cracking.	Contractor to saw cut the pavement in a straight line and saw cut shall be done within eleven (11) hours of casting. The timing of saw cut is essential to avoid cracks in concrete.

Table 2. Challenges in JPCP construction using slip form.

No		Challenges	Best Practice
1		Slump	
	(i)	Slump non-conformance (too wet) - Slump above 60 mm. When concrete above the specified slump is delivered to the site, it will be difficult to pave platform edges/sides. The concrete basically tend to sag and collapse due to its diluted state.	Compliance to specified slump range. Based on practical experience, it is best to use concrete which is within the specified slump range; 50 mm to 60 mm.
	(ii)	Slump non-conformance (too dry) - Slump below 50 mm. When concrete below the specified slump is delivered to the site, it will have poor workability which will be an issue during placement of concrete. We'll also face difficulty unloading the concrete from the mixer truck onto the road base, etc.	Compliance to specified slump range. Based on practical experience, it is best to use concrete which is within the specified slump range; 50 mm to 60 mm.
2		Improper Concrete Placement Insufficient concrete in the paver during paving resulting in honeycomb or cavity on the paved pavement.	To keep the paver's receiving hopper fully loaded during paving process.
3		Poor Surface Finishing Delay in bull floating, troweling and broom finishing work resulting in poor pavement finishing.	To arrange adequate manpower and equipment to commence finishing work immediately upon paving.

Table 3. Challenges in JPCP construction using fixed form.

No	Challenges	Best Practice
1	Slump Slump non-conformance (too wet) - Slump above 85 mm. When concrete above the specified slump is delivered to the site, slurry loss occurs resulting in work delayed, concrete bleeding and poor finishing work.	Compliance to specified slump range. Based on practical experience, it is best to use concrete which is within the specified slump range; 75 mm to 85 mm.
	Slump non-conformance (too dry) - Slump below 75 mm. When concrete below the specified slump is delivered to the site, manual handling by workers during casting will be difficult and prolonged resulting in honeycomb and cavities.	Compliance to specified slump range. Based on practical experience, it is best to use concrete which is within the specified slump range; 75 mm to 85 mm.
2	Uneven/Dented Formwork Using dented or damaged steel form will cause uneven pavement surface finishing.	To check accuracy and straightness of the steel form using thread alignment method or using dumpy level.

Table 4. Challenges in RCC construction using asphalt paver.

No	Challenges	Best Practice
1	Paver and Equipment	
(i)	Asphalt Paver Paver breakdown during paving and sensor string line not functioning.	Site personnel need to control and monitor site situation and to communicate with plant people consistently on the concrete delivery if any emergency and machinery breakdown happened.
(ii)	Compactor & Water Truck Compaction non-conformance – Compaction value < 90 %	Contractor to use High Density Paver to pave the pavement in order to achieve initial compaction value more than 80 % with an additional tandem & tire roller.

6. Observed performances

6.1. Pre-construction testing and trial

Before the actual site construction work started, preliminary study on the existing platform has been conducted to understand the overall condition of the site. The tests involved for the preliminary study were Dynamic Cone Penetration (DCP) complying to ASTM Code D6951/D6951M-18 [9] and In-situ California Bearing Test (In-situ CBR) complying to BS Code 1377:Part9:1990 [10].

The DCP was conducted to measure the in-situ strength and the thickness of underlying soil layers. Five (5) numbers of tests (Refer Appendix A) have been conducted at different areas, with a total depth of 1.0 meter from the existing surface. Two (2) areas have been found having average CBR value less than 40%. As for these findings, we have decided to excavate these 2 critical areas to a depth of 500mm and replace it with crushed aggregate/crusher run to ensure the concrete pavement have better subgrade to support the pavement.

To further determine the strength of the base, three (3) 'In-situ CBR' using a test method called Load Penetration Test, have been conducted along the alignment of the main road of proposed JPCP. We find out that one area have the lowest CBR value between 19% to 35% at a penetration of 2.50mm and 5.00mm relatively (Refer Appendix B). The others are having a CBR value between 49% and 81%. We found that the area was not properly compacted and additional compaction has resolved the problem area.

The other testing that were done before the construction was designing the JPCP and RCC material at the plant. The plant trial was conducted to ensure that the value of concrete strength conforming to the pavement design. Full scale plant trial mix was conducted at the concrete batching plant that was selected to supply the material to the project. All of the information above is vital to make sure that the alternative proposed design is conforming to the client requirement.

6.2. Testing during construction

Upon confirming the CBR value of the base, profile levels and compaction, trial lay of the JPCP and RCC were conducted at site to verify the concrete performance and to identify the method, labor, machinery and work sequence confirming with the Method Statement submitted by the contractor. Slump tests and Vebe' test were conducted on every delivery of concrete supply to the site. Slump tests were conducted for JPCP and Vebe' test for RCC works for every batch of product produced. The purpose of the tests is to ensure the workability of the material when arrived at the site.

Samples of concrete will also be taken on every batch to prepare cubes to run through compression tests for 7 and 28 days (Refer Appendix C). Although only 75% has achieved the targeted strength for 7 days, it was found that for the overall 28 days we had achieved 130% compression strength more than the design. There is no single cube fail for 28 days compression tests. Upon completion of the work for RCC, Nuclear Density Test (NDT) was conducted at 3 location every 100 meter length of works (Refer Appendix D). Although a small number of the NDT tests failed to achieve 98 % compaction as per target, we managed to achieve a minimum compaction of 95% which are acceptable. Re-compaction of the areas have improved the initial compaction.

6.3. Post construction testing

Upon completion of the overall project, before handing over the site to the client, several subsequent tests have been conducted to measure the performance of the pavement. Tests such as Sand Patch, International Roughness Index (IRI) and Skid Resistance Test were conducted.

The sand patch was conducted, conforming to BS 598 Part 105: ASTM E965, to determine the average texture depth of a selected portion of a concrete pavement surface [11,12]. From the test conducted, the results (Refer Appendix E) were accepted within tolerance, which is, in the range of 1.50mm to 2.91mm in thickness.

Road roughness, or smoothness, was inspected with IRI method. Conforming to ASTM E1364-95(2017), the tests were conducted to monitor the pavement conditions and to evaluate the riding quality of new the pavements. From the tests (Refer Appendix F) we concluded that the range 1.52m/km to 11.01m/km is not smooth enough for traffic traveling above 50km/hr.

The other test that was conducted after the construction was skid resistant. The purpose of the test is to measure the force of skid resistance of a tire is prevented from rotating slides along the pavement surface. Inadequate skid resistance on the pavement will lead to higher skid which could cause accidents. Seventy (70) numbers of tests had been conducted for the whole area (Refer Appendix G). From the results, we found that 46% of the test results fall in category A, 47% in category B and 7% in category C (refer to Appendix Table 1).

7. Conclusion

From the project, we can confirmed that different methods of concrete pavement have its own advantages and disadvantages. By using more than one method of concrete pavement, we could optimize the overall design based on its application and usage. The disadvantages of a concrete pavement, such as initial cost and difficulties to construct, can be overcome by exploiting the advantages at the right application. Understanding the right method for the right application is the key for optimization of design and construction.

In order to construct a concrete pavement, you must do it right for the first time. Experience is very vital to run a concrete pavement project. The right equipment used during the construction is essential to deliver a good concrete pavement job. Observing performance and testing; before, during and after the construction are tools for better understanding to improve the delivering to a good

IOP Conf. Series: Materials Science and Engineering **512** (2019) 012051 doi:10.1088/1757-899X/512/1/012051

concrete pavement job. Finding and experiences should be shared among clients, consultants and contractors to ensure we continue learning the secret of better concrete pavement.

To ensure concrete pavement to become a desired option when building road, bad impression and experiences from the user should be overcome. Delivering a good concrete pavement project that shall be accepted by the user is another challenge. Concrete pavement construction in Malaysia is still young. Road users need to give some time to allow the local road industry to become matured in delivering a superb concrete pavement for Malaysian road network.

8. References

[1] Hoe N G and Currie J V 1997 The north-south expressway – continuously reinforced concrete pavement ITRD 00744376
[2] Corporate News 2018 PLUS spent RM1bil in 2016 to maintain highways The Star Online
[3] JKR 2015 Innovation Day 2015 Programme JKR Malaysia Retrieved from https://www.jkr.gov.my/node/941
[4] KKR 2013 *Pelan Strategik Kementerian Kerja Raya* (KKR Malaysia)
[5] Griffiths G and Thom N 2007 *Concrete Pavement Design Guidance Notes* (London: Taylor & Francis)
[6] The Concrete Society 2003 *Concrete Industrial ground floors: A guide to design and construction* CSTR 3 TR34
[7] Dale H, Fares A, Wayne A and Chetan H 2010 *Guide for roller compacted concrete pavements* NCPTC SR298
[8] Hazaren, Chetan V, Halil C, Peter T, Kasthurirangan G, Kejin W and Fatih B 2010 *Use of chemical admixtures in roller compacted concrete for pavement* PCA R&D 3243
[9] ASTM 2018 *Standard Test Method for Use of the Dynamic Cone Penetrometer in Shallow Pavement Applications (ASTM D6951/D6951M-18)* (ASTM International: West Conshohocken)
[10] Road Engineering Standards Policy Committee 2007 Methods of test for Soils for civil engineering purposes-Part 9: In-situ Tests (BS 1377-9:1990, Part 9)
[11] British Standard Publication 2002 *Sampling and examination of bituminous mixtures for roads and other paved areas. Methods of test for the determination of texture depth BS 598 – Part 105*
[12] ASTM 2015 *Standard Test Method for Measuring Pavement Macrotexture Depth Using a Volumetric Technique (ASTM E965-15)* (ASTM International: West Conshohocken)
[13] ASTM 2017 *Standard Test Method for Measuring Road Roughness by Static Level Method ASTM E1364-95* (ASTM International: West Conshohocken)

A case study of soft ground improvement by dynamic consolidation approach

Lee Eng Choy[1], Qusanssori Noor bin Rusli[2], Muhammad Khudrie Razali[2] and Ramu Andy[1]

[1]Infra Tech Geo Solutions (M) Sdn Bhd, Malaysia
[2]Infra Tech Projects Malaysia Sdn Bhd, Malaysia

E-mail: engchoy.lee@infratechgeo.com & qusanssori@infratechgeo.com

Abstract: There are several ground improvement approaches which can be utilised to deliver results which are sought after in when constructing over poor ground conditions. Where soft grounds, such as soft saturated clays, are encountered, ground improvement utilising consolidation approaches such as, surcharge with PVD, are commonly adopted. It is known that surcharge fill is required to pressurise the porewater within the soft clay, hence creating an excess pore pressure within the pores of the clay matrix. This excess pore pressure drives the pore water towards the PVD and upwards towards the drainage blanket. An alternative method to pressurise the pore water is by applying impact energy such as HIEDYC dynamic compaction. This approach has proven to be capable of developing excess porewater pressures in soft saturated clays. The excess porewater pressure is 'locked-in' due to the low permeability of the clay. The magnitude of this excess pore pressure has been measured as up to 30kPa which is equivalent to the placement of 1.5m surcharge fill height. Hence, this method is coined 'DYCON' or Dynamic Consolidation. This method will require installation of PVD but with reduced surcharge requirements. A case study of application of this approach in ground improvement in soft ground is described in this paper. The case study illustrated that Dycon can effectively improve soft ground and the procedure is recommended for sites with similar soft soil condition.

1. Introduction

It is quite common for designers of embankments on soft ground to specify surcharge preloading to compensate or eliminate post-construction settlements. Frequently this is accompanied by installation of vertical drainage elements such as PVD to assist in dissipating the excess pore pressures generated.
This mechanism of placing surcharge fill is primarily to alter the pore water pressure to exceed the expected future stresses from future loads. This fill also imposes a compression load to collapse the soil matrix once the excess pore water pressures have been dissipated. This will manifest as ground settlements.
An alternative method to pressurise the pore water is by applying impact energy such as utilising dynamic compaction technologies available. This approach has proven to be capable of developing excess pore pressures in soft saturated clays. The excess pore pressure is 'locked-in' due to the low permeability of the clay. Hence, this method is coined 'DYCON' or Dynamic Consolidation. This method will require installation of PVD but with reduced surcharge requirements [1].

Dynamic consolidation adds another dimension to the pre-consolidation treatment of soft clays. This process merges dynamic compaction technologies to conventional consolidation processes. The application of dynamic stresses imposes instantaneous prestress in the pore water and creates instantaneous pore water pressures in the underlying soft clay. This effect is similar to that created by conventional surcharge preload. The provision of vertical drainage elements assists in accelerating the dissipation of the excess pore water pressures.
A case study of application of this approach in ground improvement in soft ground is described in this paper.

2. Conventional 1D Consolidation Model

In a conventional 1D consolidation model, surcharge fill is utilised to increase the stresses within the pore water system. This increase in pore water pressure is termed excess pore water pressure. As the permeability of soft clays is often low, thus the drainage of this excess pore pressure is also a slow process. The objective in soft ground treatment is to accelerate the dissipation of this excess pore water pressure by introducing vertical drains, which reduces the drainage path [2].
This process of consolidation is often explained with an idealized system composed of a spring, a container with a hole in its cover, and water as illustrated in Figure 1. In this system, the spring represents the compressibility or the structure of the soil itself, and the water which fills the container represents the pore water in the soil.

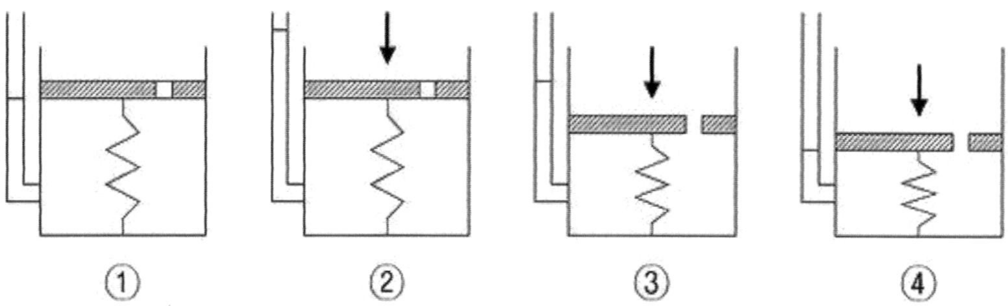

Figure 1. Idealised Concept of 1-Dimensional Consolidation Process.

1. The container is completely filled with water, and the hole is closed. (Fully saturated soil)
2. A load is applied onto the cover, while the hole is still unopened. At this stage, only the water resists the applied load. (Development of excess pore water pressure)
3. As soon as the hole is opened, water starts to drain out through the hole and the spring shortens. (Drainage of excess pore water pressure)
4. After some time, the drainage of water no longer occurs. Now, the spring alone resists the applied load. (Full dissipation of excess pore water pressure. End of consolidation)

The other function of the surcharge fill is to exert an overburden pressure to pre-collapse the soil matrix once the excess pore water pressure has been dissipated through the vertical drains. This process will result in consolidation settlement of the embankment.
This process results in ground settlement. The primary consolidation settlement in normally consolidated clay can be estimated as shown in Equation 1:

$$\delta_c = \frac{C_c}{1+e_o} H_o log\left(\frac{\sigma'_{zf}}{\sigma'_{zo}}\right) \tag{1}$$

Where

δ_c is the settlement due to primary consolidation.
C_c is the compression index.
e_0 is the initial void ratio.
H_o is the thickness of the consolidating soil layer.
σ'_{zf} is the final vertical stress.
σ'_{zo} is the initial vertical stress.

This approach in treatment of soft ground requires adequate and ready supply of fill material to be used as surcharge material. In many instances, this resource is not readily available is adequate quantity. Hence, an alternative approach is proposed in applying this prestress load in place of high surcharge fill. Dynamic consolidation offers an opportunity to re-engineer the ground treatment process. Dynamic consolidation utilises dynamic compaction technology to apply a dynamic impact load to pressurise the pore water.

3. An Overview of Dynamic Compaction Technologies

Dynamic compaction methods have historically involved the use of tall cranes and free-falling weights, imposing limitations on the types of sites that can be treated. These techniques are generally suitable for sands and not suitable for fine grained materials. However, recent developments have shown how dynamic compaction technologies can also be applied for treatment of soft clays, a process called 'dynamic consolidation'. Dynamic consolidation involves applying dynamic energy to pressurise the pore water and to accelerate consolidation of the underlying soft ground using one of the dynamic compaction methods identified below [1, 7].

3.1 Deep Dynamic Compaction (DDC) with drop weights

The Deep Dynamic Compaction technique involves using a crane to drop weights of between 5 to 20 tons, from heights of up to 20m. The number of drops, weights used and the height of the drops depend on the required post-treatment bearing capacity, settlement performance and soil conditions [3]. DDC is commonly used in reclaimed areas, land fill rehabilitation to provide a strong ground with less susceptibility for settlement or differential settlement. The authors experience, however, has shown that DDC can be used in soft clays in combination with Vertical Drains that can withstand the impact of the drop weight.

Figure 2. Example of Deep Dynamic Compaction (ref: Wikipedia).

3.2 Impact Rolling using High Impact Energy Dynamic Compaction (HIEDYC)

HIEDYC ground treatment imparts vertical energy into the ground to depths ranging from 2m to 5m. In view of the near vertical energy input, the spread of energy along the ground surface as surface waves in minimised. With the use of HIEDYC ground treatment process, the levels of vibration are relatively low and the use of HIEDYC ground treatment is suitable close to existing structures.

HIEDYC ground treatment relies upon a towed non-circular module of 3, 4 or 5 sides that compacts as it rotates around a 'corner' and 'falls' to impact onto the ground [1, 4, 5]. Figure 3 illustrates the various modules currently available.

HIEDYC treatment is possible for landfills, sands, silts and clays. it is the author's experience that HIEDYC treatment will be effective when combined with prefabricated vertical drains. This scheme has been adopted in tailing ponds, airports and housing schemes by the authors' company. Figure 4 shows a HIEDYC Tria (3 sided) working in a soft ground condition.

Figure 3. Visuals of Tria (3-sided), Qadra (4-sided) and Penta (5sided) HIEDYC modules (L to R) (courtesy of Infra Tech Group).

Figure 4. A HIEDYC Tria working in a soft ground condition (courtesy of Infra Tech Group).

3.3 Rapid Impact Compactor (RIC) and Controlled Dynamic Compaction (CDYC)

The RIC unit is generally fitted to tracked base excavator of 35 tonne to 70 tonne weight which provides the dual benefit of allowing improved mobility and site accessibility. Having the RIC mounted on a tracked machine gives it the versatility to move about in narrow and limited height spaces, such as within existing warehouses and urban area lots. Also, being mounted on a tracked excavator, transport to site is easier. Figure 5 shows RIC in operation.

CDYC is similar to RIC and is the acronym for Controlled Dynamic Compaction technique. Figure 6 shows CDYC in operation. The authors' experience have been that CDYC has been found to achieve deep compaction in sand and silts greater than 7m depth [4, 6]. Dynamic energy is imparted by a dropping a weight by hydraulics from a controlled height onto a foot plate of $1m^2$ or $1.5m^2$ or a circular plate. The RIC/CDYC impacts the soil at a rate of 10-60 blows per minute using a 5, 7, 9 or 12 Tonne drop weight. The drop height varies from 1m to 2m. Energy is transferred to the ground safely and efficiently as the RIC/CDYC's foot remains in contact with the ground. This is quicker than conventional DDC but the energy input is smaller per blow. Blows at each treatment position create imprints, which are subsequently filled with granular material.

Figure 5. Carrying out dynamic compaction with RIC (ref: Serridge & Synac, 2006).

IOP Conf. Series: Materials Science and Engineering **512** (2019) 012052 doi:10.1088/1757-899X/512/1/012052

Figure 6. Ground Improvement Work using CDYC in progress (courtesy of Infra Tech Group).

The authors' experience is that both DDC and HIEDYC in combination with PVD can be useful for deployment in soft clays.

4. An Overview of Dynamic Consolidation

Dynamic consolidation involves applying dynamic energy to pressurise the pore water [3] and to accelerate consolidation of the underlying soft ground using one of the dynamic compaction methods identified in Section 3 above and in combination of vertical drains.
The authors have carried out a research with application of HIEDYC dynamic compaction on a soft clayey soil where piezometers were installed at depths of 3m, 6m and 12m and the research has shown that the dynamic energy impacted by HIEDYC dynamic compaction creates an instantaneous increase in pore water pressure to a depth exceeding the maximum instrumented depth of 12m. Excess pore water pressures in excess of 30kPa were observed; this is equivalent to the placement of about 1.5m high surcharge fill. These results are indicated in Figure 7. Pore-water pressure measurements were again read a month later and the results show that the pore water pressures were locked in and remained the same. Following these measurements, prefabricated vertical drains (PVD) were installed and these resulted in a quick release of these excess pore water pressures [1].
The fundamental concept of dynamic consolidation utilises these results to demonstrate that dynamic consolidation can work, with installation of PVDs, to accelerate the consolidation of soft soils and to reduce the need for placement of high surcharge fills. The benefits that can be derived from dynamic consolidation are:

i. The reduced requirement for high surcharge fill will greatly reduce the overall project costs, in addition to mitigating need to source suitable fill material in areas where these materials are not readily available or costly [5].

Dynamic consolidation creates instantaneous development of excess pore water pressures, which can be quickly dissipated with the assistance of PVDs. This may result in much reduced requirement for lengthy consolidation or rest period thereby shortening project durations.

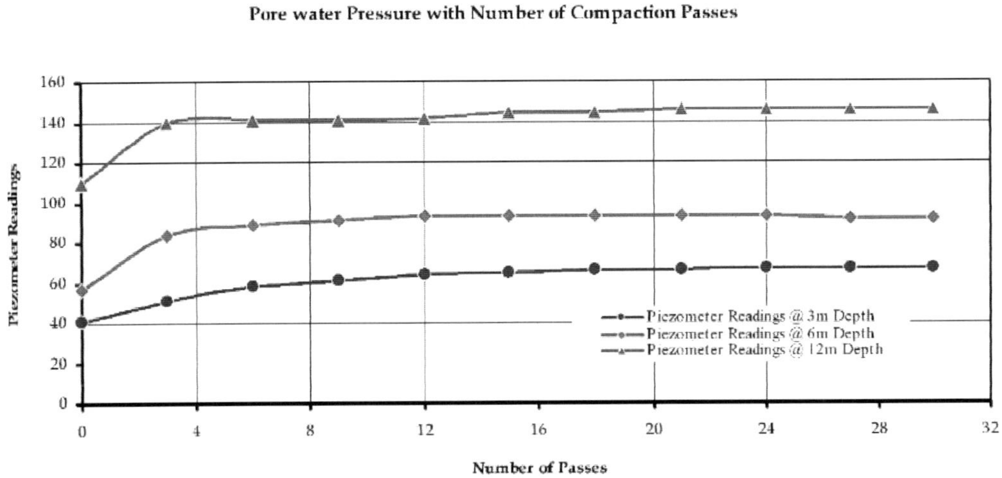

Figure 7. Development of excess pore water pressures with application of HIEDYC dynamic compaction.

5. Case Study

A case study of where the dynamic consolidation approach was utilised in accelerating the consolidation of soft marine clay. The project site is a located in Batu Kawan, Penang where a Sales Gallery and show houses for a housing development are required to be constructed on a fast track basis. The geology map shows that the site is underlain by a quaternary deposit as shown in Figure 8.

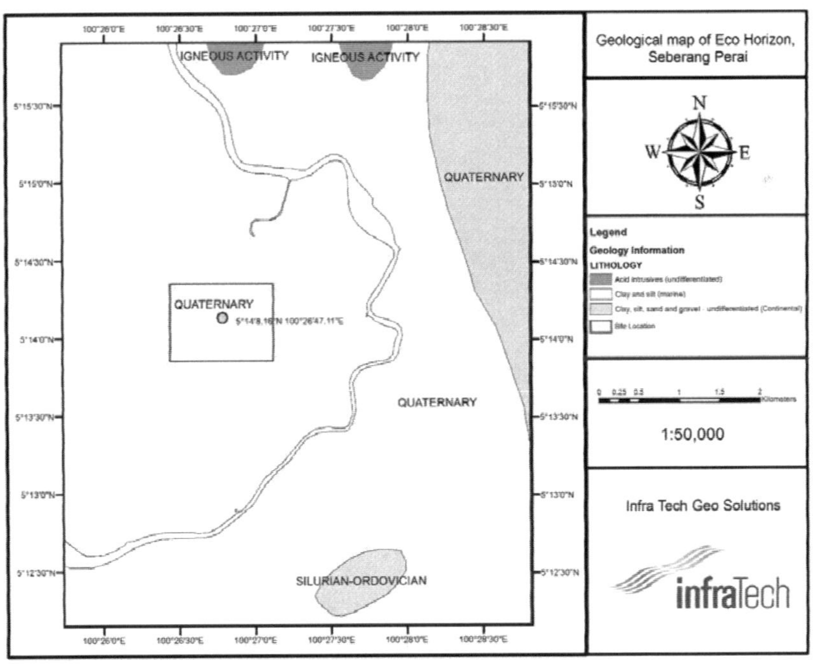

Figure 8. Geology map of the site.

5.1 Subsoil Properties of Site

Typical SPT-value vs depth profile of the subsoil is illustrated in Figure 9.

IOP Conf. Series: Materials Science and Engineering **512** (2019) 012052 doi:10.1088/1757-899X/512/1/012052

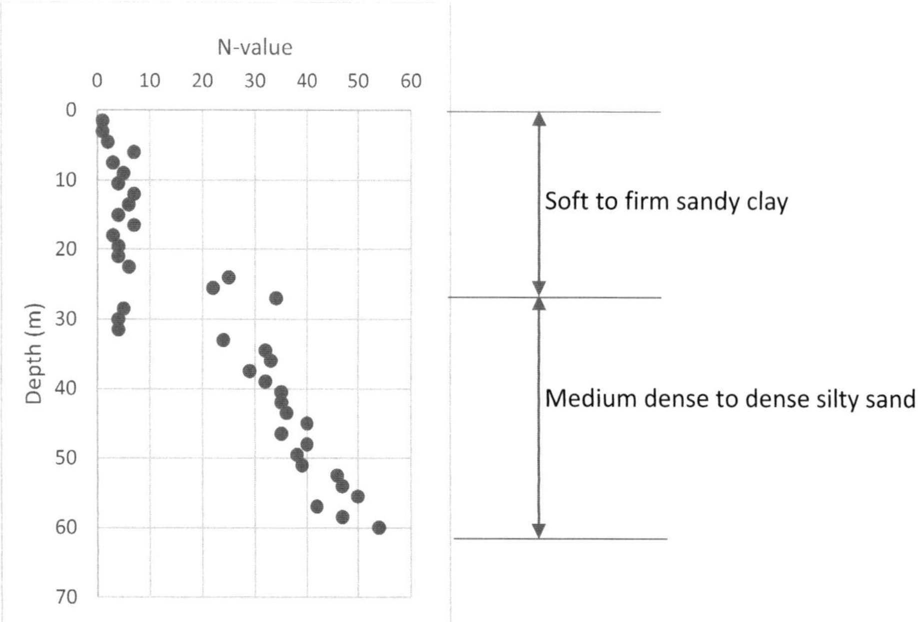

Figure 9. Typical distribution of SPT N-value.

Figures 10, 11 and 12 illustrate the distributions of compression ratio, coefficient of consolidation and over-consolidation ratio with depth below ground surface, respectively.

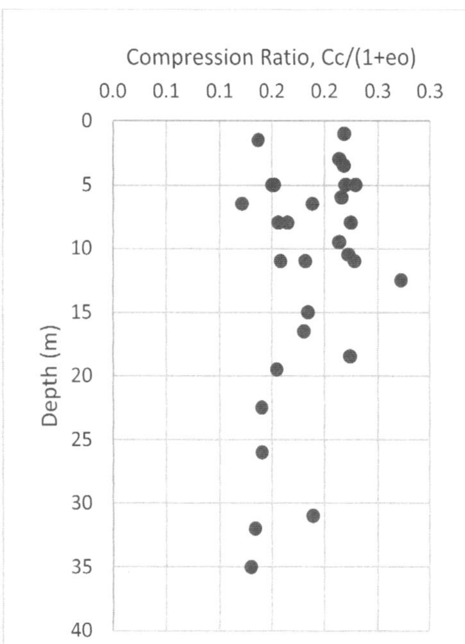

Figure 10. Compression Ratio.

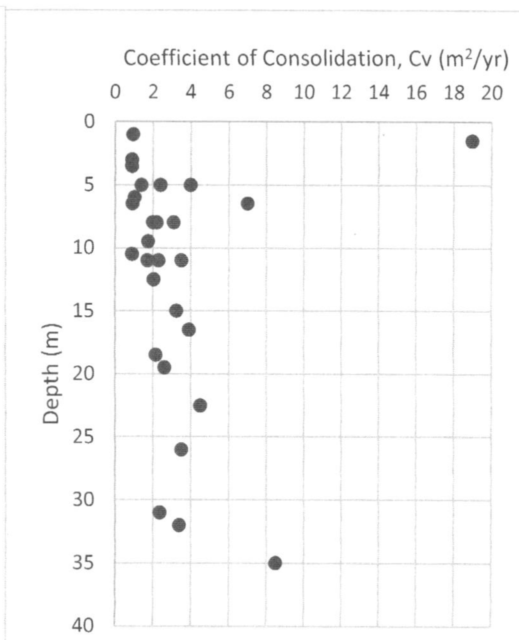

Figure 11. Coefficient of Consolidation.

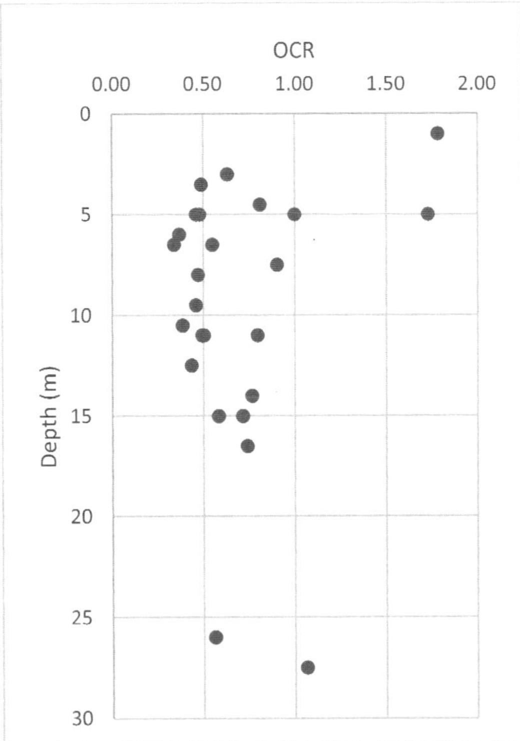

Figure 12. Over-Consolidation Ratios (OCR).

As can be observed, the subsoil is comprised essentially of normally consolidated clay with compression index in the region of 0.2 and a coefficient of consolidation ranging between 1 and 4.

5.2 Earthwork and ground improvement processes

The existing ground surface was for the site was approximately at RL1m. The platform for the proposed Sales Gallery is to be raised to RL4.2m, whereas the platform for the Show Houses is to be raised by 3.0m. The ground improvement design comprised of installation of PVD at 1.0m centres and a surcharge fill of 1.0m. HIEDYC dynamic compaction was carried out at the base of fill as well as at the design platform level. The body of fill and surcharge fill were compacted by conventional vibratory rollers in accordance to the earthworks specification. Geosynthetic strip drains were used as drainage layer in place of a sand layer. The conventional design for the ground improvement was to install PVD at 1.0 centres with a surcharge fill of 2.0m.

Geotechnical instruments installed at the site include rod settlement gauge, piezometers and inclinometers. These were monitored regularly on a weekly schedule during the earthworks and surcharging durations.

5.3 Settlement monitoring results

The settlement monitoring of RSG1 at the Sales Gallery is shown in Figure 13.

IOP Conf. Series: Materials Science and Engineering **512** (2019) 012052 doi:10.1088/1757-899X/512/1/012052

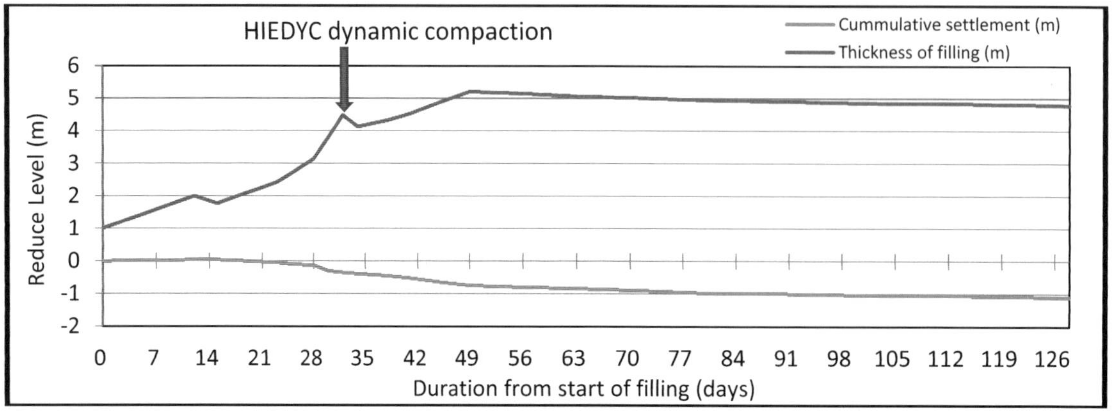

Figure 13. Typical ground settlement profile at the Sales Gallery.

The settlement monitoring of RSG2 at the Show Houses is shown in Figure 14.

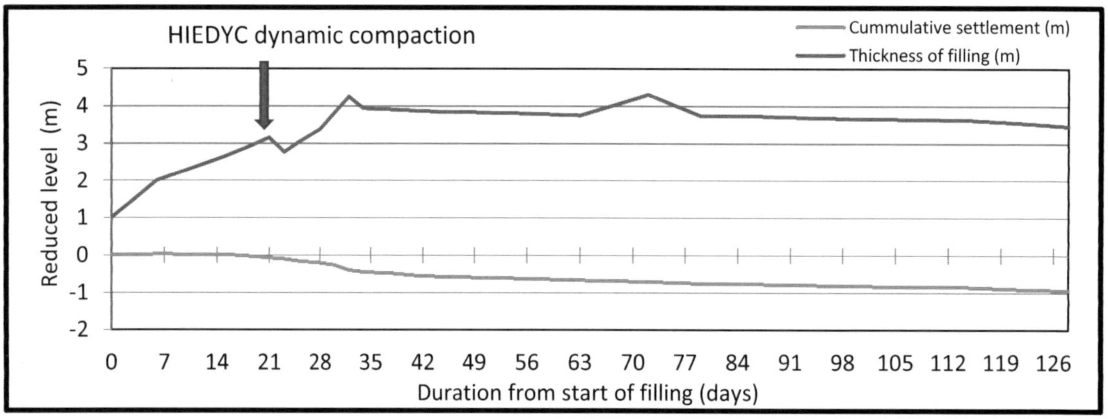

Figure 14. Typical settlement profile at Show Houses.

As can be observed, settlement at both locations levelled off at about 126 days from the start of filling, i.e. approximately 60 days from the completion of fill placement. The bulk of the observed settlement occurred during placement of fill. The predicted total settlement at the Sales Gallery is 1.2m and 1.05m at the Show Unit using lower bound soil properties. The observed settlement appeared to match the predicted settlement fairly accurately.

5.4 Project deliverables and Pictorial illustration of project activities

The project deliverables are an allowable bearing capacity of 200kPa. This was validated via Plate Bearing Tests using 500mm diameter plates. Results show settlements at ultimate load of 200kPa (is less than 3mm and settlements at 400kPa (i.e. twice design load) remain less than 8mm. Typical results from plate bearing tests are shown in Figure 15. With these results the foundation of the Sales Gallery and Show Houses were constructed on Raft Slabs. Figure 16 to 19 illustrate the critical project activities and project deliverables.

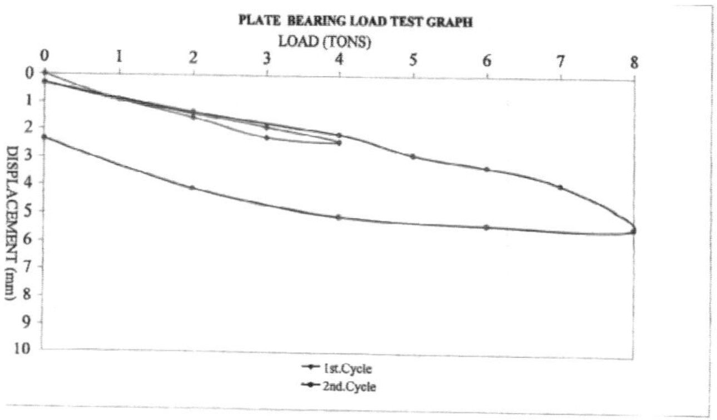

Figure 15. Typical plate bearing test result.

Figure 16. Strip drains provide horizontal drainage.

Figure 17. HIEDYC compaction in operation.

Figure 18. Completed Sales Gallery.

Figure 19. Show Houses nearing completion.

6. Summary

This paper described an overview of conventional consolidation treatment of soft soils, an overview of dynamic compaction technologies currently available and the application of dynamic consolidation processes in soft ground improvement. It illustrated an opportunity to utilise this dynamic consolidation technique for soft ground treatment in situations where road embankments are to be constructed over soft ground.

A case study was described to illustrate the dynamic consolidation approach in treatment of deep soft soil exceeding 25m depth to enable construction of earth embankments and structures over soft ground. The approach illustrated the benefits that can be derived, namely reduction in requirement of surcharge fill, accelerated settlements and increased bearing capacity. The case study also illustrates an innovative use of geosynthetics as horizontal drainage element in place of conventional sand blanket. This geosynthetic strip drain possesses a consistent drainage capacity in comparison with the drainage capacity of sand blanket. It is commonly known that the drainage capacity of sand blanket is highly influenced by its clay or silt content, and it is difficult task to ensure the clay or silt content is kept to a consistent minimum value.

This case study thus illustrates an alternative approach in treatment of soft soil and is recommended for sites with similar soft soil condition. It is particularly suitable where supply of fill for surcharging is sparse or costly.

7. References

[1] Narendranathan N and Lee E C 2015*Ground Improvement Case Histories – Compaction, Grouting and Geosynthetics* Ed B Indranatna, J Chu and C Rujikiatkamjorn (Amsterdam: Elsevier) p 365-98

[2] Hansbo S 1979 Consolidation of clay by band-shaped prefabricated drains *Ground Eng.* **12**(5)

[3] Menard L and Broise Y 1975 Theoretical and practical aspects of dynamic consolidation *Geotechnique* **25** (1) 3–18

[4] Berry A and Narendranathan N 2010 The use of rapid impact compaction for ground improvement prior to sheet pile installation in sandy fill at the Australian Marine Complex *Proceedings of the GI Conference* Perth

[5] Lee E C and Narendranathan N 2011 Performance of ground improvement works for runway extension of Kota Kinabalu Airport *Proceedings of International Conference on Advances in Geotechnical Engineering* Perth, November 7–9

[6] Serridge C J and Synac O 2006 Application of Rapid Impact Compaction (RIC) technique for risk mitigation in problematic soils *Geological Society of London*

[7] Chu J, Yan S W and Zheng Y R 2006 Three soil improvement methods and their applications to road construction *Proceedings of the Institution of Civil Engineers-Ground Improvement* **10**(3) pp103-12

Performance based pavement design and construction

Mohd Azwan Salleh, Nagamuttu Narendranathan, Eng Choy Lee and Qusanssori Noor Rusli

Infra Tech Group, Petaling Jaya, Selangor, Malaysia

E-mail: azwansalleh@infratechgeo.com, nn@infratechgeo.com, engchoy.lee@infratechgeo.com & qusanssori@infratechgeo.com

Abstract. Pavement designs for roads have evolved from empirical approaches whose origins lie in AASHTO road trials in the 1950's. These are embodied in AASHTO guidelines from USA, the British Road note 29 and 31 as well as Arahan Teknik JKR guidelines. This approach constrains the designer and the contractor to use "standard" materials which were assumed in the empirical approach since these were the materials used in the trials. Any materials that are "nonstandard" cannot be utilised in the empirical approach. This leads to expensive pavement designs. With the development of mechanistic pavement design principles along with equipment to measure performance criteria such as stiffness, roughness, surface texture, friction etc performance-based pavement design and construction has become possible. This paper outlines the mechanistic approach with a case history of such pavement along with other important factors to be considered such as drainage of pavements which are essential for better pavement performance. Based on past successful history of performance-based pavement designs and construction there is a strong case that pavement designs can be done by mechanistic designs and the performance criteria be defined in specifications by way of roughness, stiffness, friction, surface texture in all road works contracts.

1. Introduction

The development of pavement engineering is growing from year to year and helps to provide better perspective on present and future practice. Previously, most of the pavement designs were based on empirical approaches whose origins lie in AASHTO road trials in the 1950's. However, the last three decades have seen a shift to mechanistic design methods. Unlike an empirical approach, a mechanistic design approach seeks to explain phenomena by reference to physical causes such as stress, strain and deflections (pavement response) [1]. The mechanistic analysis theory for pavement stresses have developed extensively in Denmark and Australia. Along with this, the ability to measure stiffness by Falling Weight Deflectometer (FWD) has allowed the design of pavements based on stiffness. This approach also allows the stiffness of conventionally designed pavements to be compared to stiffness-based pavement designs. In addition, the stiffness-based approach also allows the use on "non-conforming" materials in pavements and to model them in the mechanistic design approach and later validate the stiffness by FWD. This paper outlines the approach and illustrates with a case history of such pavement design along with other important factors such as drainage of pavements which are essential for pavement performance. Examples of how performance is defined in major highway contracts in Australia are presented along with a case history from Malaysia.

2. History of Pavement Design Development

The Romans have built roads which were initially utilised for military purposes to connect the military camps which were located far apart [2]. With regards to this primary purpose, no basic standard in designing the road was specified. These roads produced high noise levels and were rough. The initial design of Roman road consisted of 4 layers with a total thickness of about 0.9m.
With the extension of masonry knowledge to bridge building, Thomas Telford introduced a new pavement design approach using Tresaguet's theories (use of cubical stone blocks) consisting of 3 layers [3]. The total pavement thickness was reduced to between 350mm to 450mm with the bottom layer comprising of large stones (100mm size). On top of this, the smaller size of stones is placed (65mm size) and followed by a wearing course of gravel about 40 mm in size.

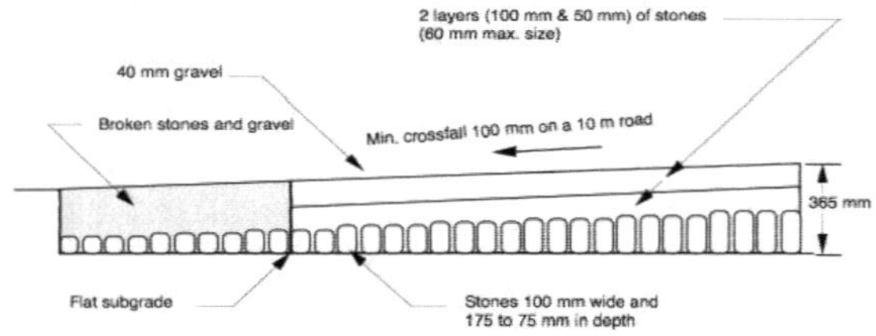

Figure 1. Thomas Telford's pavement cross section (Courtesy of www.pavementinterative.org).

Then, moving on to the Macadam era, the angular aggregates were used over well compacted subgrade as it would be bound together by fines generated by traffic [4]. The subgrade was designed slightly with a gradient unlike Telford pavements which had flat grades. Macadam pavement allowed better drainage and performed substantially better. The typical thickness of Macadam pavement is 250mm and consisted of 3 layers. Two layers with a total thickness of 200mm comprised of angular aggregates (75mm particle size) and wearing course of 50 mm thickness with aggregates size of 25mm were placed at the top.

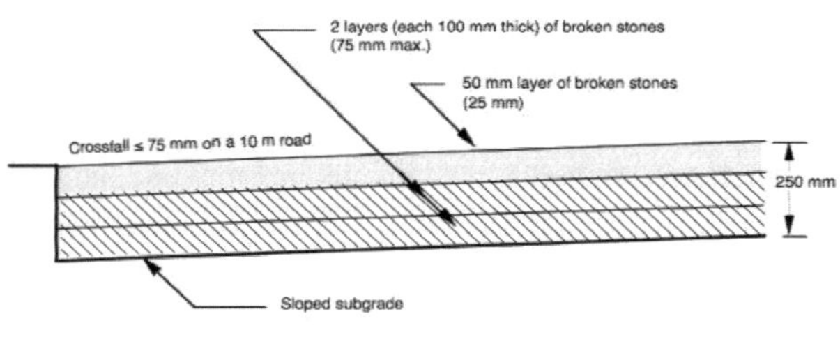

Figure 2. Macadam's pavement cross section (Courtesy of www.pavementinterative.org).

History has showed that pavement designs have gone through many phases of enhancement before asphalt and Portland Cement Concrete (PCC) pavement was introduced. However, as reported by Agg and McCullough [5] to the Iowa State Highway Commission (1916), this PCC pavement has some issue with its performance. Then, extensive experiments were carried out by AASHTO in 1959 and eventually the empirical based guideline was produced.

Figure 3. AASHTO Road Test carried out in Ottawa (Courtesy of www.explorer.acpa.org).

The empirical design approach has been widely used in pavement structural design procedures. The pavement performance is evaluated based on the experimentation and experience or with combination of both. [6]. However, initially this empirical design guideline produced by AASHTO is solely depending on the AASHTO Road Test carried out in Ottawa, IL from 1956 – 1961 [7]. In this guideline, only the specified material (granite) shall be used in pavement design. The California Method is one of many empirical design approaches that were commonly used during the early 1940s. Therefore, for Malaysian conditions the applicability of this method is quite questionable as Malaysia has different climate and soil characteristics resulting in variable soil and rock types and characteristics.

3. Mechanistic Pavement Design Developments

The main consideration in mechanistic-empirical method is the actual response of the pavement when it is subjected to load. Unlike the empirical design method, this mechanistic design method seeks to explain phenomena of only by reference to physical causes such as stress, strain and deflections [1]. Empirical design elements are also used along with mechanistic method to determine the values of the calculated stresses, strains and deflections result in pavement failure. The performance-based pavement design and construction has the following advantages;

- Enables the use of marginal materials with modifications and the ability to model these modified materials in the mechanistic design.
- Saves cost by allowing local "non-standard" materials to be used.
- Reduces the risk of pavement life reduction due to poor quality construction because stiffness can be measured rapidly and economically by FWD testing.
- Better long-term life can be obtained by specifying and ensuring roughness.
- Safer road (especially during rains) can be obtained by measuring Surface texture and Friction as construction acceptance testing.
- This method can be used for both existing pavement rehabilitation and new pavement construction.

The most special value of the mechanistic design method is it allows a rapid analysis of the impact of changes in input items such as changes in traffic and materials. This method also can accurately characterize in situ material using the portable device that is called Falling Weight Deflectometer (FWD). FWD testing is used to evaluate the structural condition of pavement by predicting the layer moduli of the pavement [8].

IOP Conf. Series: Materials Science and Engineering **512** (2019) 012053 doi:10.1088/1757-899X/512/1/012053

The stiffness of the pavement sub grade is denoted by the FWD deflection, D_0. It is widely accepted that the D_0 deflection and the life of the pavement are well correlated. Deflection measurements by FWD and pavement life prediction using deflection measurements have been in practice in Malaysia for more than a decade.

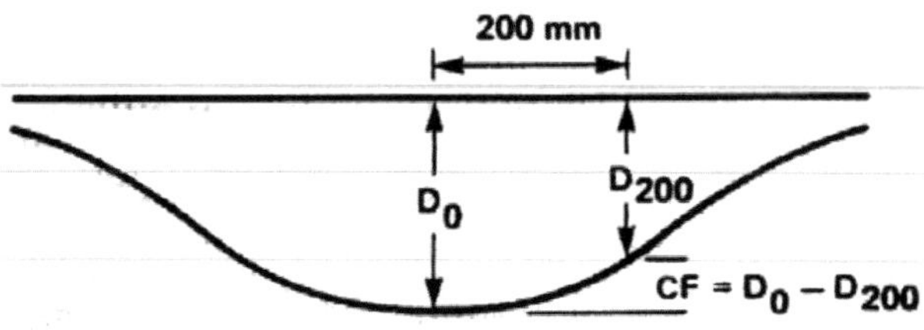

Figure 4. Typical deflection bowl under FWD test plate.

D_0 = maximum deflection for a test point
D_0 - D_{200} = deflection measured where the test load is 200mm from the point of maximum deflection (in the direction of travel).

The deflection (D_0) of the pavement represents the strength, while the curvature (D_0 - D_{200}) represent the asphalt fatigue. The smaller value of the D_0, the stiffer the pavement it can be. The design surface deflection (D_0) and design curvature (D_0 - D_{200}) against design traffic or Equivalent Standard Axles (ESA) can be estimated based on the graph in Figure 5 and Figure 6 (Austroads, 2004) [9]. The Austroads (2004) design deflections for traffic loading less than 10^6 ESAs were derived from research undertaken by Scala in the early 1960s. These design deflections were based on field measurements of the dependence of Benkelman Beam deflections on granular thickness and subgrade CBR, and the relationship between pavement composition and design traffic loading.

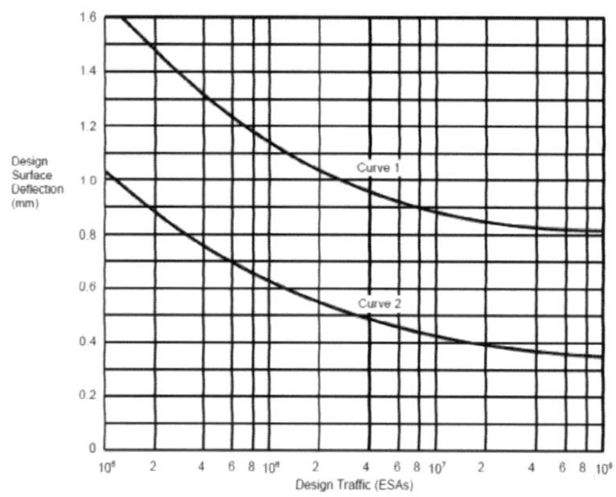

Figure 5. Design deflection levels design traffic (ESA) (Austroads Pavement Design Guide, 2004).

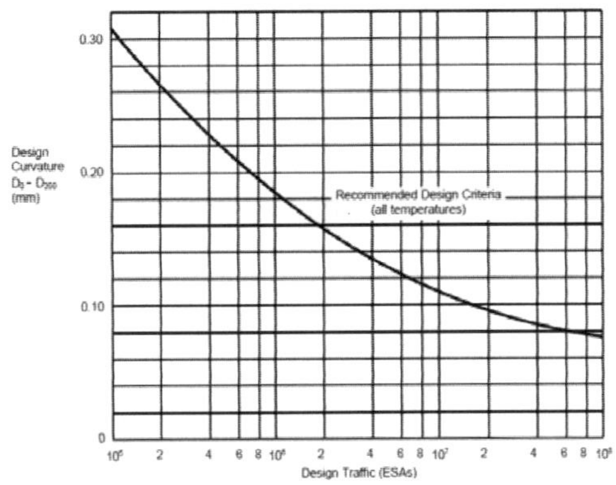

Figure 6. Design curvature function design traffic (ESA) (Austroads Pavement Design Guide, 2004).

4. Pavement performance parameters

Unlike the empirical design method, the mechanistic design method can be evaluated using the in-situ testing. Several performance criteria can be evaluated to determine the pavement performance.

Table 1. Pavement Performance Criteria.

No.	Performance Criteria	Descriptions
1	Rutting	Rutting is when pavement deform due to sub grade strain and mix design issue that cause surface depression in the wheel path.
2	Skid Resistance	The friction force developed between tyre and pavement to prevent the vehicles from sliding - TRRL Pendulum to Laser Profilometer Test.
3	Surface Texture	Pavement surface texture is texture wavelength. Adequate surface texture will provide proper drainage of tyre grooves and reduce water spray when moving at high speed – Sand Patch Test and Laser Profilometer Test
4	Roughness	Pavement roughness is defined as microscopic undulating of the pavement that affect the ride quality of vehicles – Bump Integrator and Lase Profilometer Test.
5	Strength	The maximum deflection (D_0) is noted as the strength of the pavement - Benkelman Beam and Falling Weight Deflectometer Test
6	Stiffness	Deflection ratio (D_{250}/D_0) is used to indicate the stiffness of the pavement structure - Benkelman Beam and Falling Weight Deflectometer Test. ➢ > 0.8 indicates CTB or CTSB bound pavement ➢ 0.6 – 0.8 indicates good quality unbound pavement ➢ < 0.6 indicates a possible weakness in the pavement materials
7	Fatigue	Pavement fatigue is determined from the curvature function ($D_0 - D_{200}$) to predict the fatigue life of an applied asphalt surfacing overlay or an existing asphalt surfacing - Benkelman Beam and Falling Weight Deflectometer Test.

The above performance criteria need to be controlled and monitored during pavement construction. In mechanistic design method, Benkelman Beam and Falling Weight Deflectometer Test are two common tests to determine the stiffness for each layer of the pavement. However, the proper drainage design must be strictly taken into consideration to prevent failure of pavement in long term period.

Based on specifications from Main Road Western Australia – Contract 89/13 [10], the pavement condition is measured based on criteria specified in table below.

Table 2. Performance Measure requirements prior to Practical Completion (Main Roads Western Australia – Contract 89/13).

Pavement Category	Component	Measure	Acceptable Standard
New construction	Structural Capacity	Pavement deflection – FWD (at 700 kPa)	For granular Pavements: mean Segment value ≤ 0.60mm
		Pavement curvature – FWD (at 700 kPa)	For granular Pavements: mean Segment value ≤ 0.23mm
	Functional Capability	Pavement roughness	95th percentile lane value < 40 counts/km and no segment value > 50 counts/km
		Surface shape	3mm maximum
		Texture Index	Segment value not greater than 0
		Texture depth	Measured at any point on the surface must be greater than 1.0 mm

5. Case History of Performance-Based Pavement

The design and construction Senai Desaru Expressway (SDE) pavement for Package 1 and 2 was designed by Infra Tech Projects using the mechanistic design method. Before the actual pavement was constructed, the trial pavements of various thicknesses on subgrade which was compacted by HIEDYC (High Impact Energy Dynamic Compaction) as compared to the original designed where the subgrade was conventionally compacted as per the JKR Design was constructed [11]. Table 3 show the pavement trial sections constructed by Infra Tech.

Figure 7. HIEDYC carried out at SDE.

IOP Conf. Series: Materials Science and Engineering **512** (2019) 012053 doi:10.1088/1757-899X/512/1/012053

Table 3. Trial pavements for SDE.

Pavement Option	Pavement Thickness (Each layer)		Targeted FWD deflection, D_0	Sub-grade Compaction
P1 450mm	ACWC14 ACBC28 BMR28 CRB	40 mm 50 mm 60 mm 300 mm	D_0 - 0.45mm	HIEDYC
P 2 480mm	ACWC14 ACBC28 BMR40 CRB	40 mm 50 mm 90 mm 300 mm	D_0 - 0.460mm	HIEDYC
P 3 530mm	ACWC14 ACBC28 BMR40 CRB	40 mm 50 mm 90 mm 350 mm	D_0 - 0.47mm	HIEDYC
P4 590mm	ACWC14 ACBC28 BMR40 CRB	40 mm 50 mm 100 mm 400 mm	D_0 - 0.58mm	HIEDYC
P5 790mm	ACWC20 ACBC28 DBM40 WMM SB	50 mm 60 mm 100 mm 280 mm 300 mm	D_0 - 1.29mm	Conventional compaction

These trial pavements were used to determine the stiffness and resilience modulus, E_R of each layer of the pavement. The validation test using FWDT was carried by IKRAM and the result as shown in table below.

Table 4. Pavement deflection, D_0 and strength modulus, Mpa.

Pavement Option	D_0 mm 95% (85%)	95% (85%) Modulus, MPa				Sub Grade Compaction
		AC	BMR	CR	SG	
P1 450mm	0.65 (0.62)	4864 (5990)	259 (344)	113 (117)	93 (100)	HIEDYC
P 1 450mm	1.56 (1.45)	2957 (3181)	218 (389)	41 (42)	32 (33)	No HIEDYC
P 2 480mm	0.45 (0.42)	4276 (5211)	413 (527)	141 (181)	142 (149)	HIEDYC
P 3 530mm	0.46 (0.42)	5194 (5648)	401 (480)	186 (189)	144 (156)	HIEDYC
P4 590mm	0.58 (0.54)	3443 (3990)	310 (389)	137 (144)	122 (126)	HIEDYC
P5 790mm	1.28 (1.23)	927 (978)	96 (108)	67 (70)	60 (63)	No HIEDYC

The alternative pavement designed by mechanistic method was constructed using similar materials that was used for pavement designed by Arahan Teknik JKR. From the table above, it shows that the stiffness for the alternative pavement where the subgrade was compacted by HIEDYC is higher than the conventional compaction method and pavement designed using Arahan Teknik JKR. The targeted deflection, D_0 for alternatives is lower than the conventional pavement. In conclusion, empirical design method allows innovation to be adopted and modelled int the design and construction process and reduce the cost as well as long term maintenance. However as explained below other important considerations must be taken into account in pavement designs.

Sub sequent to the trial, the pavements for SDE packages 1 and 2 were quality controlled using the Falling Weight Deflectometer to validate the pavement stiffness of each lane of the highway. The slow lane was tested at 50m intervals while the fast lane was tested at 100m intervals with the 95th percentile deflection targeted to be 0.8mm or less.

6. Important Factors in Pavement Design

In order to ensure the pavement condition can be sustained for long period, several factors need to be taken into account during design and construction stage. Adequate assessment need to be carried out on the subgrade conditions especially when dealing with problematic soil or rock material (variably weathered soil subgrades, swelling soils, collapsing soil, soft soil and etc).

Issues such as surface and subsurface drainage must be looked into at the design stage and then modified where necessary based on site observations of geological features and water seepage when the site is opened up during construction stage. All parties must accept that construction stage modifications will be necessary based on the variabilities observed in subgrade strength, water seepage and such issues that can have an effect on the pavement performance.while it may be possible that surface drainage can be reasonably addressed during design stage it is almost impossible to address sub surface drainage until when the site is opened up during construction. If Clients, Consultants and Contractors do not allow for this, this can be the single highest contributory factor for pavement distress.

Proper compaction needs to be done to ensure the subgrade is highly compacted and no major post-construction settlement will occur after the road is opened for public. It is very important to identify the location and the maximum value critical stress and strain so that the pavement can be designed with adequate strength and thickness to prevent pavement failure under actual loading [12].

Other than that, subsurface drainage is essential for economical, long term performance of roads and highways. Excessive and uncontrolled subsurface water has been cause for many slope failures, pavement failures, and unsatisfactory projects. Subsurface drainage is essential for economical, long term performance of roads and highways. The primary objective of the drainage is to channel out the water along and across the road conveniently without any obstruction in water flow [13]. If groundwater and seepage are not identified and adequately addressed, it can cause severe problem to the pavement in its constructability, pavement performance and slope failure along the construction area.

7. Conclusion

With the use of mechanistic design methods availability of pavement performance measuring equipment such as Laser Profilometers, Falling Weight Deflectometers, Light Weight Deflectometers, it is not only feasible for performance-based specifications but also performance-based construction and quality control of new works as well as maintenance works. Based on past successful history of performance-based pavement designs and construction there is a strong case that pavement designs can be done by mechanistic designs and the performance criteria be defined in specifications by way of roughness, stiffness, friction, surface texture in all road works contracts. However, important factors in pavement design as addressed in Section 6 need to be considered. Other than that, good interaction among the highway agency engineers to identify the proper input parameters for the design is necessary. Even if the designs are done using empirical methods, the pavement construction specifications should be based on roughness, stiffness, friction, surface texture. It is of utmost importance that the sub soil drainage and surface drainage issues be addressed as the project site is opened, since these issues cannot be adequately addressed at the design stage based on limited site investigations. This will lead to better roads and tighter quality control.

8. References

[1] Li Q, Xiao D X, Danny, Wang K C P, Hall K D and Qiu Y 2011 Mechanistic-empirical pavement design guide (MEPDG): a bird's-eye view *Journal of Modern Transportation* **9** 114-33

[2] O'Flaherty CA 2002 *Highways: The location, design, construction and maintenance of road pavements, Fourth Edition* (Butterworth-Heinemann)

[3] Raitz K B and Thompson G F 1996 *The National Road* (Baltimore & London: Johns Hopkins University Press)

[4] Gillette H P 1906 *Economics of Road Construction* (New York: Engineering News Publishing Co.)

[5] Agg T R and McCullough C B 1916 *An Investigation of Concrete Roadways (Technical Report No. 1)* (Ames, Iowa: Iowa State Highway Commission)

[6] Ranadive M S and Tapase A B 2016 Parameter sensitive analysis of flexible pavement *International Journal of Pavement Research and Technology* **9**

[7] Bekele A 2011 *Implementation of the AASHTO Pavement Design Procedures into MULTI-PAVE* (Master Degree Project)

[8] Nega A, Nikraz H and Al-Qadi I L 2016 Dynamic analysis of falling weight deflectometer *Journal of Traffic and Transportation Engineering*

[9] Andy R 2009 *Senai Desaru Expressway (SDE) packages 1 and 2 (Closure report)* (Infra Tech Projects Sdn Bhd)

[10] Austroads 2004 *Austroads Pavement Design Guide* (Sydney, Australia)

[11] Main Road Western Australia *Scope of works and technical criteria (Contract 89/13)* (Bullabuling to Coolgardie)

[12] Arshad A K 2007 Flexible pavement design: Transitioning from empirical to mechanistic-based design methods *JURUTERA*

[13] Toryila T M, Terpase I V and Terlumun I E 2016 The effects of poor drainage system on road pavement: A review *International Journal for Innovative Research In Multidisciplinary Field*

Innovation in earthwork practices

Iradatul Hanis Abd Hamid, Nathan Narendrannathan, Lee Eng Choy and Qusanssori Noor Bin Rusli

Infra Tech Group, Petaling Jaya, Selangor, Malaysia

E-mail: Iradatul.Hanis@infratechgeo.com, nn@infratechgeo.com, engchoy.lee@infratechgeo.com & qusanssori@infratechgeo.com

Abstract Site Earthworks specifications by various authorities have not seen much changes over the last 50 years. However, knowledge and practice relating to earthworks compaction, quality control testing have developed albeit empirically over the last 30 years in different industries such as civil infrastructure, mining and oil and gas, often driven by the need for improvisation or lack of adequate "conforming materials" as per traditional specifications. This paper presents the author's experience and research in earthworks placement and compaction, development of new tools for quality control of earthworks and how earthworks can be engineered to specific situations focus on High Impact Energy Dynamic Compaction (HIEDYC). Case histories from Malaysia and overseas are presented along with discussion of the benefits to owners. These benefits include fast earthworks procedures with quick validation processes.

1. Introduction
Earthworks are engineering works created through the moving of massive quantities of soil or rock. The process of earthwork includes the excavation of the existing land to a suitable level and the compaction of the fill materials to the appropriate level.
The earthworks specifications from various authorities in last 50 years have not changed much. In the meantime, knowledge and practice relating to earthworks compaction, quality control testing has developed albeit empirically over the last 30 years in different industries such as civil infrastructure, mining and oil and gas, which are often driven by the need for improvisation or lack of adequate "conforming materials" as per traditional specifications. Construction industry needs innovation in earthworks practice including fill placement and compaction. The development of in-situ testing and instrumentation has enabled project timelines to be expedited.

2. Observation of Traditional Earthworks and Related Specifications
Earthworks practices and related specifications have changed very little over the last 50 years.

2.1 Site Clearing
The beginning of earthworks is site clearing works. This involves the removal of trees, demolishing buildings, removing all old underground infrastructure, any other obstacles that might affect the construction process in the future or hinder the project to be done.

2.2 Cutting and Filing

Cutting and filling is the process of moving earth from one place to another to prepare platforms to the design levels. A 'cut' is made when earth is cut from above the desired ground height and a 'fill' is when earth is used to fill a low point or valley to desired ground level. Cutting and filling is a common technique used to create an even ground surface.

2.3 Proof Rolling

Proof rolling is carried out for the following purposes;

- Identify loose or unsuitable material pockets before fill is placed
- Assess compaction to check if there are uncompacted areas on finished platforms before placing foundations of pavement layers

It is to be noted that the Proof Rolling equipment must be selected based on the expected CBR of the surface to be rolled. For example, if the subgrade material to be proof rolled is expected to have a CBR of 3% then the proof rolling equipment will need to be lighter than the Proof roller used for a subgrade expected to have a CBR of say 10%. These matters are seldom considered in earthworks specifications. If this is not considered in proof rolling, unnecessary disputes might occur [5].

The Texas Department of Transport (TDT) have issued a Technical Advisory: Proof Rolling and Base Curing Advisory states that the purpose of proof rolling is not to establish soil strength, but to assure that the subgrade soil will provide uniform pavement support [10].

Infra Tech Pty. Ltd. (ITPL) resolved such issues on a project where the plastic clay fill was to be compacted at optimum moisture content (OMC) to 98% Standard Compaction. The fill had OMC in the range of 20% to 25% and Plasticity Index in the range of 40% to 60%. Then the 4 days soaked CBR was in the range of 2% to 3%. The compaction was checked by proof rolling with the Australian Standard 3798 Bulk earthworks Specifications (AS 3798) specified truck and failed. When proof rolling was done as per Table 1, and Figure 1 below the CBR 2% to 3% fill proof rolling was found to be adequate [10 and 12].

An example of suitable Proof rolling requirements by Texas Department of Transport (TDT) is shown below in Table 1 below.

Table 1. Proof Rolling Guide (TDT, 2001).

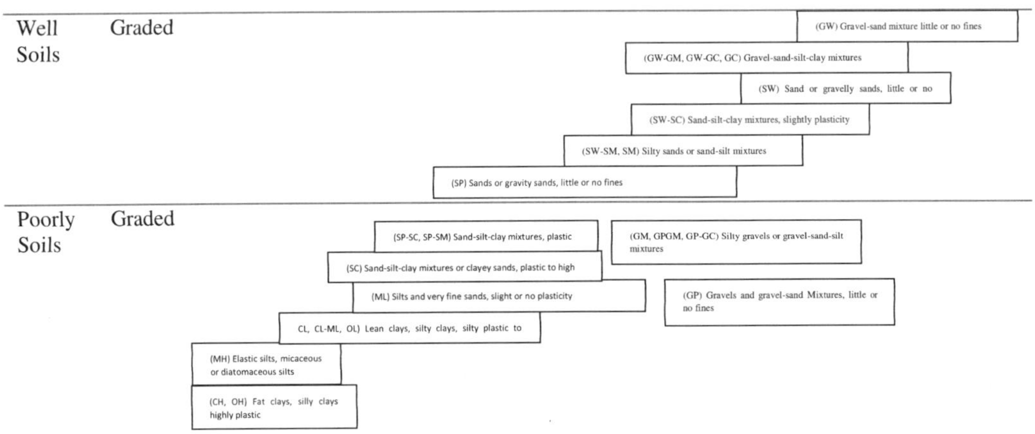

Relative Subgrade Support	Poor	Fair		Good			Excellent
Stress Level	Minimum	1	2	3	4	5	Maximum
Load (tons)	30	34	38	42	46	50	50
Tire Inflation pressure (psi)	40	50	60	70	80	90	130

2.4 Traditional Earthwork and Related Specifications

During conventional compaction, moisture content is an important consideration especially for fine-grained or cohesive soils, as it dictates the response of the soil to compaction and associated movements affected by climate or loading. Optimum Moisture Content (OMC) is defined as the moisture content at which the maximum dry density of soil can be achieved for a given compactify effort. When a fine-grained/cohesive soil is significantly below the OMC, compaction is less effective due to frictional and capillary forces within the soil (ASTM stock number: MNL70, 2011).

Current practice for compaction revolves around the OMC to achieve the best results (BS 6031:1981). However, to ensure that OMC is maintained within the material during compaction, significant volumes of water may be required. Alternatively, in wet weather the fill will need to be dried before achieving optimum compaction. This may require consideration of economic, practical and environmental factors.

Conventional compaction relies on Proctor's (1933) principle (ASTM stock number: MNL70, 2011);

1. Energy used for compaction
2. Water / moisture content of the soil
3. Soil properties

The moisture content at which the maximum dry density is achieved is known as the Optimum Moisture Content (OMC) (see Figure 1). This implies that, to ensure effective compaction, the soils must be kept at the optimum moisture content during compaction. Compaction is generally evaluated by means of degree of compaction, which is calculated as a ratio of the measured field density to the maximum laboratory density. However, if the compactive effort is higher hen the OMC can be lower as illustrated in Figure 1 below.

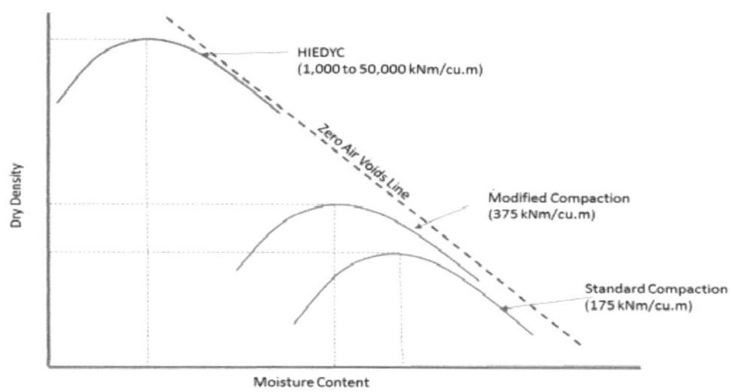

Figure 1. Relationship between compactive effort and optimum moisture content.

The compaction requirements as well as the compaction plant will vary based on the type of soils used for the fill or in the cut subgrade. These are covered in many earthworks specifications and manufacturers hand books (7). Using conventional compaction plant, the layer thickness for fill is normally limited to 300mm loose thickness. In some instances, with heavier vibrating compactor 450mm thick lifts are possible. A maximum lift thickness of 300mm, means that large particles (greater than 37.5mm in size) cannot be accepted in the fill. This will require crushing and screening of fill with large particles and increase the cost of earthworks.

In cohesive soils if over compaction occurs either due to excessive number of passes or a heavier than required compactor, then loss of strength occurs. This means the compacted clays will become weaker. This must be borne in mind by Engineers and Contractors, since soils on a site are expected to be variable it is necessary to have field compaction trials whenever changes in the type of fill material is observed. This phenomenon is not very pronounced in sandy soils. Hence it is important to conduct filed compaction trials for each project or whenever fill type changes in a project.

3. HIEDYC for Deep Lift Compaction in Major Earthworks

HIEDYC plant can be sued for the following;

- Proof rolling of natural ground before fill is placed
- Fill compaction in layers of 1m to 3m (depending on type of fill material and HIEDYC plant used)

3.1 Proof rolling by HIEDYC

High Impact Energy Dynamic Compaction (HIEDYC) can also be used as a proof rolling equipment to identify the presence of loose or uncompacted ground to a greater depth than the conventional smooth drum vibratory roller because of the high energy imparted. The impact of the heavy mass transfers enough energy to achieve medium compaction to a depth of several meters. In addition, HIEDYC can reduce the risk of differential settlement and can achieve proofing to greater depths, typically 2m to 4m, than with conventional methods.

Figure 2. Proof Rolling on soft soil site prior to fill placement.

3.2 Deep Lift Compaction of fill by HIEDYC

Method of earthworks compaction by Deep Lift Compaction allow filling to be done in 1500mm to 3000mm thick layers compared to conventional 300 mm thick layers. The most important is filling can be placed at natural moisture content significantly drier of the optimum moisture content without additional water. This is possible due to the significantly higher (up to 10,000 times) energy that this applied by this compaction method. Therefore, the result is a stronger fill that can be placed more rapidly as compared to conventional compaction methods.

The HIEDYC deep compactor has non-circular compaction modules and is towed by a prime mover shown in Figure 3. The HIEDYC modules impart energy vertically downward as a compressive wave into the ground. The validation testing utilising Plate Bearing Test (PBT), Light Weight Deflectometer Test (LWDT) and Deep Cone Penetrometer (DCP) are used as a quality control to validate the deep lift compaction using HIEDYC.

Conventional compactors are capable of compacting clays in 200mm thick lifts and sands in 300mm thick lifts, with moisture levels maintained at or near OMC while HIEDYC can compact most types of soils in loose thicknesses of 1m to 2.5m at natural moisture content. These thicknesses vary depending on the soil types, compaction conditions and compaction requirements. This process of compaction in thicker (than conventional) layers is termed Deep Lift Compaction [4].an example is shown in Figure 4.

Table 2. HIEDYC™ modules, capabilities and applications.

Module	Number of sides	Mass (tonnes)	Compaction depth limitations (m)	Mining engineering applications	Civil engineering applications
Tria	3	17	5	Pit floors/rock crushing, haul roads, soil dumps, rock dumps, tailings consolidation and strengthening.	Subgrade compaction, earthwork compaction, coarse sand and silt compaction and clay soils in conjunction with Prefabricated Vertical Drains
Qadra	4	14	2	Haul roads, soil dumps, tailings consolidation and strengthening.	Subgrade compaction for road pavements, earthwork compaction and sand and silt compaction not greater than 1.5m
Penta	5	16	3.5	Pit floors/rock crushing, haul roads, soil dumps, rock dumps, tailings consolidation and strengthening.	Subgrade compaction, earthwork compaction, coarse sand and silt compaction and clay soils in conjunction with Prefabricated Vertical Drains

Figure 3. Photographs of HIEDYC™ modules, (L to R) Tria, Qadra and Penta.

The advantages in the use of Deep Lift Compaction are listed below.

- Layer thickness of fill is 1.5m to 2m thick as against 300mm thick layers used in conventional rolling.
- These 1.5m to 2m thick layers are compacted at the natural in place moisture content and therefore, there will be no need for large quantities of compaction water - as would be required for conventional earthworks - which is not only difficult to obtain but can be very expensive. This leads to environmental benefits.
- Because of larger layer thickness it is also possible to use fill containing particle sizes of 300mm where as in conventional earthworks in 300mm layers, the fill cannot have particle sizes more than 37.5mm. this reduces the cost of reprocessing excavated fill to smaller sizes of less than 37.5mm
- Quality control is performed using Light Weight Deflectometer Test (LWDT) which provides on the spot results rather than traditional tests where at least 24 hours of waiting may be required. (alternatively, DCP testing or MP testing can also be performed with or without Plate bearing tests to correlate the readings)
- The earthworks completion time can be slashed by more than half, this would mean a reduction of several months off the construction schedule
- The compacted earth platform will be stronger (can be up to about 3 times the required bearing capacity) which means the foundations cab be (re) designed to be more economical
- The emissions associated with deep lift compaction are less than 20% of emissions from conventional compaction techniques.

Figure 4. Deep Lift Compaction of gravelly clay.

A comparison of productivity between conventional compaction and HIEDYC compaction is summarised in Table 3 below.

Table 3. Comparison of productivity between conventional compaction and deep lift compaction of earthwork fill placement (Nathan 2010).

Technique	Compaction volume per 10-hour shift (m^3)	Mode of compaction	Types of machinery	Number of equipment involved	Number of operators involved	Type of QC testing required	Water usage
HIEDYC deep Lift compaction using HIEDYC TRIA or HIEDYC Penta module in 1.5m to 2m layer thickness a natural moisture content	Varies between 8,000 m^3 to 10,000m^3	Earth spread in 1.5m to 2m loose layers at natural moisture content	HIEDYC	1	1	LWDT, MP, PSP, DCP, EFCPT, Limited PBT for every 1.5m to 2m layers	100 to 200m^3 as surface spray for dust suppression
			Grader to level ground	1	1		
			Water cart (20,000litres capacity)	1	1		
Conventional compaction using a 12tonne vibratory compactor, to match production of Deep Lift compaction by HIEDYC	2,500 m^3 to 3,500 m^3	Earth spread in 200mm to 300mm layers at OMC	Compactor	8 to 10	8 to 10	Laboratory MDD and Field Density testing for every 300mm layer	1,000 to 2,000cu.m depending assuming dry weather and addition of at least 5% by weight of water
			Water cart	6 to 8	6 to 8		

* *Elastic modulus was back calculated from Falling Weight Deflectometer (FWD) results on subgrade.*

4. Rapid Testing Methods

Validation tests including Plate Bearing Test (PBT), in-situ CBR test, and Mackintosh Probe (MP) test can be carried out to validate the performance of the ground improvement works. In addition, seismic refraction or surface wave measurements (Multi Channel Analysis of Surface Waves - MASW) are very useful to assess existing ground conditions as well as ground improvement during HIEDYC.

Figure 6 shows the relationship of seismic wave velocity versus the number of HIEDYC compaction runs for a trial area of the reclamation sand fill at Westport, Malaysia (Infra Tech, 2004) [8].
The data for Figure 6 below was collected from a field trial of HIEDYC compaction to establish optimum number of passes for site sub soils.

The magnitude of the project, type of loads, the structures to be supported and the settlement acceptance limits for structure are influence by the frequency of testing. As an indication, on large projects a test for every 500m^2 interval is considered appropriate for important structures and one test for every 1,000m^2 is appropriate for more settlement tolerant structures.

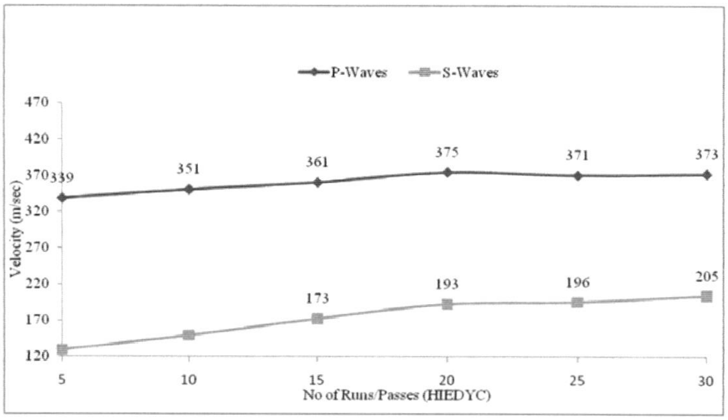

Figure 5. The relationship of seismic wave velocity versus the number of HIEDYC compaction (passes).

Most empirical requirements for compacted densities have been derived from experience in civil projects. There is very little evidence that high levels of bearing capacity or elastic modulus can be obtained with fills compacted conventionally, especially when the fill is supporting houses on shallow foundations or heavy structures such as factory machinery or mine infrastructure. Furthermore, most density testing approaches have been developed for soil size particles which are less than 37.5mm in size. However, in the mining sector oversize and heterogeneous fill are very common. Conventional density testing cannot be applied to such situations. It may be more appropriate to move away from the conventional approach of density testing in specifications. The use of performance-based earthworks specifications allows for flexibility in the fill materials and the earthworks. With the use of PBT, LWDT, DCP and MASW compaction or foundation preparation can be validated and direct strength parameters can be obtained rapidly and economically.

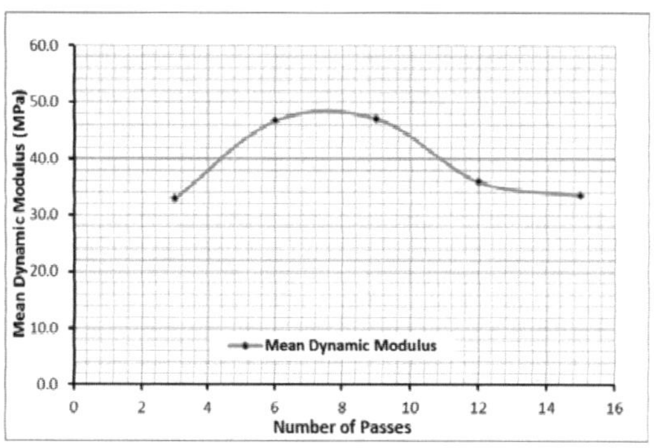

Figure 6. LWDT result after HIEDYC compaction of approx. 1.5m thick general fill (clayey sandy GRAVAL) loosely placed on HIEDYC compacted foundation (ITPL, 2012c).

Figure 6 above shows how LWDT can be used to assess the optimum number of passes during Deep lift Compaction using HIEDYC modules. Figure 7 shows the LWDT equipment in use on sites.

Figure 7. Light Weight Deflectometer Test (LWDT).

5. HIEDYC deep compaction for loose soils in lieu of excavate and replace in layers

The HIEDYC Deep Compaction can be used in many ground conditions, ranging from silts, loose sand, gravel, sites comprising of mixtures of different types soils such as landfills. In addition, for sites with soft clays HIEDYC can be used in combination with prefabricated vertical drains (DYCON method) to improve these sites.

The traditional approach to deal with loose or soft layers has been to remove these layers and replace with suitable fall and compact in 300mm layers until, adequate compaction is achieved. Alternatively, piled foundations can be provided for structures on loose layers. All these approaches are costly. In contrast, in situ improvement of existing loose layers by HIEDYC deep compaction has proven to be a cost-effective solution. Subject to the limitations of improvement depth being in the range of 1m to 5m.

The traditional "cut, replace and compact in 300mm layers" type solutions involve a considerable amount of excavation, handling and re handling, disposing of materials and all other activities associated with cutting and replacing. These activities lead to higher costs, time, and issues arising in relation to occupational health, safety and environmental emissions. These factors are often the catalyst to look into alternative methods of ground improvements that are more cost effective, safer and environmentally friendly. A comparison of HIEDYC deep compaction and "cut, replace and compact in 300mm layers" method is provided in Table 4.

Table 4. Comparison of productivity between HIEDYC compaction ad "excavate, replace and compact in 300mm layers" (Caterpillar performance handbook , 1999).

Technique	Compaction volume per 10 hour shift (m^3)	Mode of compaction	Types of machinery	Number of equipment involved	Number of operators required	Type of QC testing required	Water usage
HIEDYC in situ deep compaction	10,000	Existing materials in situ	HIEDYC	1	1	LWDT, PSP, DCP, EFCPT, Seismic Survey and Limited PBT for every 1.5m to 2m layers	200 m^3 to 300m^3 as surface spray for dust suppression
			Grader to level ground	1	1		
			Water cart (20,000litres capacity)	1	1		
"excavate, replace excavated soil in 300mm layers with conventional compaction at OMC"	3,500	Imported materials in 300mm layers	Trucks	5	5	Laboratory MDD and Field Density testing for every 300mm layer	1,000 m^3 to 2,000m^3 depending assuming dry weather and addition of at least 5% by weight of water
			Excavator	1	1		
			wheel loader	1	1		
			smooth drum or vibratory roller	8	8		
			water cart (20,000litres capacity)	5	5		

6. Summary

Proof rolling is an essential part of earthworks and must be specified by considering the type of soils to be proof rolled. Fill can be compacted in layers thicker than 300mm as has been practiced for more than 40 years. Deep lift Compaction in layers of 1.5m to 3, thickness using HIEDYC can create higher strength, use less diesel, less compaction water. The high strength can be validated rigorously by seismic wave measurements, LWDT, Plate Bearing, Mackintosh Probe or Dynamic Cone Penetrometer. Small areas where HIEDYC plant cannot run at speeds of more than 8km per hour are not suitable for HIEDYC compaction. Field trials need to be done to establish the optimum number of passes to avoid under or over compaction. This is applicable to both conventional earthworks as well as HIEDYC compaction.

7. References

[1] Lambe T W 1958 The structure of compacted clay *Journal of the Soil Mechanical and Foundation Division ASCE* **84** 1-35
[2] Southwell G 2007 Oscillating roller *A New Form of Compaction Edition No.3.*
[3] Department of Transport 1993 *Manual of Contract Documents for Highway Works Volume 1 Specification of Highway Works Series 600 - Earthworks*
[4] Whellens P, Suntharalingham S, Narendranathan N and Samarasinghe BSAM 2010 *High Impact Energy Dynamic Compaction (HIEDYC) applications in ground improvement and earthwork*
[5] Narendranathan N, Archer S and Narendranathan G 2012 Earthwork practice in civil & mining industries and advances in earthworks *International Conference on Ground Improvement and Ground Control*
[6] Alshibli K 2005 Laboratory evaluation of the geogauge and light falling weight deflectometer as construction control tools *Journal of Materials in Civil Engineering* 560-69

[7] Caterpillar 1990 *Caterpillar Performance Handbook Edition 30*
[8] Infra Tech Pty Ltd 2004 *Closure Report for Westport Container Terminal CT4, Malaysia*
[9] Look B 2006 *Earthworks* (Victoria: Engineering Education Australia)
[10] Texas Department of Transport *Technical Advisory: Proof Prolling and Base Curing Advisory.*
[11] Infra Tech Pty. Ltd. (ITPL) 2012a. *Lot 23 (86A) Walters Drive, Osborne Park: Ground Improvement by HIEDYC Compaction*
[12] State of New York Department of Transportation 2015 *Standard Specifications*

IOP Conf. Series: Materials Science and Engineering **512** (2019) 012055 doi:10.1088/1757-899X/512/1/012055

Design and properties of high reclaimed asphalt pavement with RH-WMA

L Gungat[1], M O Hamzah[2], N I M Yusoff[3] and S W Goh[4]

[1]Civil Engineering Program, Faculty of Engineering, Universiti Malaysia Sabah, Kota Kinabalu, Sabah, Malaysia
[2]School of Civil Engineering, Universiti Sains Malaysia, Engineering Campus, Pulau Pinang, Malaysia
[3]Dept. of Civil and Structural Engineering, Faculty of Engineering and Built Environment, Universiti Kebangsaan Malaysia, Bangi, Selangor, Malaysia
[4]Universal Pave Sdn. Bhd., Pulau Pinang, Malaysia

E-mail: lillian@ums.edu.my

Abstract. Economic and environmental benefits gain from the recycling of old pavement promotes the use of reclaimed asphalt pavement (RAP) in the road construction and rehabilitation. However, high production temperature due to RAP stiffness becomes the main concern in the production of RAP. Reduced production temperature with the addition of warm mix asphalt (WMA) may compensate for aging of RAP binder, nevertheless the stiffness of RAP binder is depending on how much it ages and needs localised investigation for a bulk production of RAP. This research investigates the mix design and properties of high RAP content incorporating a WMA additive for a bulk RAP production in Malaysia. Milled RAP obtained from local roads were incorporated with a WMA named RH-WMA. Prior to addition of RAP into mixtures, the milled RAP were processed and characterized as a quality control of RAP from various sources. Mix design of mixture that contains RAP and 3% RH-WMA were performed using the Response Surface Method. Subsequently, the specimens were evaluated for its chemical and mechanical properties. The analysis on mix design showed that compaction temperature and bitumen content were significant factors. Reasonable reduction of optimum binder content and energy consumption were observed with the additions of RH-WMA. Tests on mechanical properties demonstrate increased on stiffness with the addition of RAP and hence reduced the fatigue resistance. However, the addition of RH-WMA on specimens that subjected to combined effects of moisture and aging showed improvement on fatigue resistance. Hence, with proper quality control of RAP material, the incorporation of RH-WMA into RAP indicated potential as an alternative material for green road construction.

1. Introduction

In recent years, awareness on environmental impact particularly on emissions has been growing. The environmental regulation and the need for green construction technology have fostered innovations and renewed interested for old technology in pavement construction. Green construction such as recycling and low energy consumption materials can contribute to achieve the global environment goal. Recycling of used pavement has started since the mid 1970's due to the significant increase in asphalt prices as a result of Arab oil embargo [1]. During resurfacing, rehabilitation and maintenance operations, old asphalt pavement materials are normally removed. The removed and processed materials are defined as reclaimed asphalt pavement (RAP). Asphalt mixture incorporating RAP will produce a comparable performance with conventional hot mix asphalt (HMA) when treated properly

[2]. However, the increasing amount of RAP content requires high production temperatures during production and road construction and thus causes health threat to the asphalt paving workers.

Recycling of HMA can be categorized into cold planning, hot recycling, hot in-place recycling, cold in-place recycling and full depth reclamation. The most commonly implemented method of recycling is hot recycling due to its proven cost effectiveness and environmentally friendly in comparison to other techniques. In the hot recycling process, the old road surfacing is milled, processed and combined with a new/virgin binder, aggregate and/or recycling agent in the asphalt plant. Milling removes one or more layers of an existing pavement surface.

When using RAP in road construction, it should fulfil the requirement as for virgin materials and its performance must be at least equal or better than conventional HMA. The properties of asphalt mixtures with high RAP content or more than 30% RAP content can be improved by modifying the mixture such as using a softer binder grade, incorporating a WMA additive, utilizing rejuvenators, and adding anti-stripping agents [3-4]. WMA additives are widely used in European countries due to its ability to decrease the production temperature. In general, WMA additive reduces production temperature by lowering the temperature to 30-50°C lesser than conventional HMA [5]. The synergy of RAP and WMA aids in the recycling of high RAP content because of the reduced production or mixing temperature. However, the RAP materials need to be characterized properly to ensure the quality of the resultant mixture performance. The RAP needs to be processed and screened into multiple sizes before it is blended with a virgin binder. There are various methods available for processing the RAP aggregates. For high RAP content, fractionation approach is the most suitable for a better quality control of the mixture.

When RAP is mix with virgin materials, the addition of RAP changes the volumetric and mechanistic properties of the resultant mixtures and hence affected the optimum binder content (OBC) [6-7]. Typically, the determination of OBC of mixture containing RAP is similar to the conventional mixture except for the consideration of the RAP binder. In other study, the fabrication of WMA at various compaction temperatures can affect its volumetric and strength properties [8-9]. Hence, the author proposed a simultaneous evaluation of binder content, compaction temperature and WMA additive content for determining the OBC using Response Surface Method (RSM).

Some concerns when utilizing high RAP content are the low temperature and fatigue cracking performance because the stiffness of the asphalt mixture can dramatically increase with the inclusion of RAP [9]. The incorporation of RAP increases the mixture's stiffness and may improve the mixture performance at high service temperatures. Nonetheless, the mixture properties should be monitored to avoid fatigue failure. Previous research findings indicated that the addition of WMA additives reduced stiffening effects and hence produce better resistance to thermal cracking and also improved mixture resistance to fatigue resistance and moisture [10-13]. Nevertheless, the RAP properties are differ as it is influence by level of RAP aging and variability of RAP aggregate gradation. Hence, this study investigates the local RAP in terms of mix design with relate to the effects of RAP content and compaction temperature. Besides that, the effects of RH-WMA on the chemical and mechanical properties are also investigated.

2. Experimental Design

2.1. Materials

The PG64 binder which equivalent to 80/100 pen and commonly used for local road construction in Malaysia was supplied by SHELL Sdn. Bhd for this research. The granite aggregate was supplied by Kuad Kuari Sdn. Bhd and fulfilled the Jabatan Kerja Raya (JKR) Malaysia aggregate gradation specification for AC14. RAP was obtained by milling process from two local roads along North-South Expressway which located at KM 75.4 and KM138. North-South Expressway is the main road networks in Peninsular Malaysia. In this study, 30%, 40% and 50% RAP content based on RAP aggregate were incorporated into virgin materials. A wax WMA additive named RH-WMA was incorporated into the RAP and virgin material to decrease the production temperature. A designation shown in Table 1 was adopted to simplify the identification of the mixture blend. In the following sections, number will be used together with the designation to indicate the RAP content. For example, 30RA1 denotes 30% RAP from RA1 was incorporated in the mixture.

Table 1. Designation of mixture blend.

Source of RAP	Additive (%)	Designation
-	0	PG64
-	3	PG64-RH
RA1	0	RA1
	3	RA1-RH
RA2	0	RA2
	3	RA2-RH

2.2. Specimen preparation and testing

The specimen preparation involves processing of RAP, characterization, preparation of RH-WMA with binder and determination of RH-WMA percentage. Various percentage of RH-WMA (1%, 2%, 3%, 4% and 5%) were added into binder by the total mass of binder and evaluated for its physical and rheological properties. The initial content of RH-WMA was selected based on the review of previous study for the organic WMA additive The RH-WMA was blended with virgin binder at 145°C for 15 minutes using a mechanical mixer. Then, an optimum RH-WMA content was selected to be mixed with RAP. Subsequently, chemical tests were performed on extracted RAP binder blended with virgin binder and RH-WMA.

For mixing the RH-WMA with the RAP, the addition was done by dry mixing whereby required amount of RH-WMA was added into the RAP based on the calculated RAP binder. Mixing was done in accordance to the standard lab procedures and all specimens were compacted using the gyratory machine at 100 gyrations to simulate the high traffic. Optimization using RSM was adopted for the mix design. There were three main parameters included which were the compaction temperature, binder content and RAP content. The responses or inputs for the optimization were the conventional Marshall mix design parameters, energy consumption and cost. Optimization of RAP-WMA mixtures production based on the method suggested by Derringer was performed to propose the optimum binder content and compaction temperature [14]. Energy consumption was calculated by adopting the method proposed by DEFRA and Hamzah as shown in Equation (1) [15-16].

$$Q = \sum_{i=n}^{j=n-1} mc\Delta\theta \tag{1}$$

Where, Q is the sum of heat energy(J), m is the mass of material(kg), c is the specific heat capacity coefficient (J/(kg/°C)), $\Delta\theta$ is the difference between the ambient and mixing temperatures (°C), and i and j indicate different materials types. All specimens were fabricated at 10°C higher than the compaction temperature. Cost of RAP-WMA production was calculated based on the data provided by quarries.

Subsequently, specimens were fabricated based on design mix and examined for cracking potential by performing indirect tensile and fatigue tests. Specimens for the fatigue test were conditioned to simulate the effects moisture and long term aging. All specimens for both tests were kept in the incubator at 15°C for 4 hours prior to testing. The indirect tensile strength and fatigue tests were conducted in accordance with ASTM D7369 and BS EN 12697-24 procedures.

3. Results and discussion

3.1. Processing and characterization of RAP

Prior to incorporating the RAP into the virgin materials, it needs to be processed. The processing of RAP includes heating, crushing and fractionation. For heating the RAP, the temperature and duration are the most important factors in order to avoid further aging of aged binder in the RAP. Based on review, the appropriate temperature for heating high RAP percentage is 110°C to 135°C [4]. The milled RAP used in this research were taken directly from the road maintenance site. Therefore, the moisture content in the RAP was determined to ensure that it does not affect the mixture performance. The changes in moisture content of the two sources of RAP after heated in the oven at 110°C is shown in Figure 1.

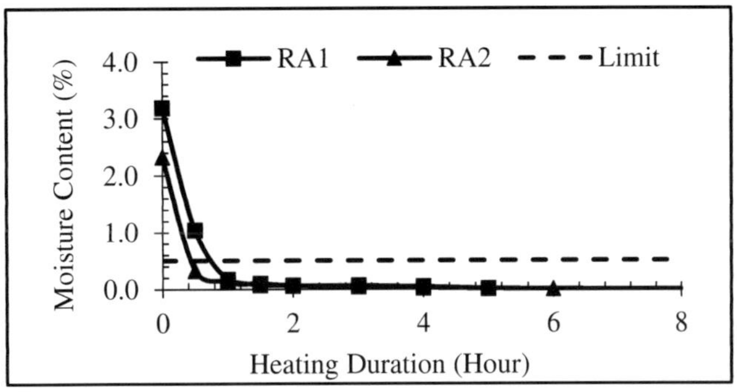

Figure 1. Heating duration of RAP.

All RAP has moisture content that less than 0.5% after one hour of heating. According to Xiao et al., there is no significant influence of the aggregates moisture content in WMA mixtures for values up to 0.5% [17]. Hence, one hour and 30 minutes was chosen to dry the RAP and enable the large RAP lump to be crushed manually. Aggregate gradation of recovered RAP aggregate by ignition method from an average of 5 samples was analysed. The coefficient of variance (COV) is less than 15% for all sources of RAP and considered acceptable. For mixture preparation, RAP from various sources were fractionated or separated into sizes according to AC14. In order to get the acceptable blending within AC14 gradation, a few trials of aggregate blending gradation consists of recovered aggregate and virgin aggregate were conducted.

The physical properties and binder content of reclaimed asphalt is shown in Table 2. The penetration result shows that the recovered RAP binders are stiffer. The RAP binder from RA2 is stiffer than the RAP binder from RA1. When in service, binder progressively ages and becomes stiffer due to various mechanisms. Aging of binder during construction and service are associated with oxidation, volatization, polymeration, air voids in the HMA and position within the road construction [18,9]. Hence, the variations in physical properties of the RAP binder from the different sources are related to the mentioned factors.

Table 2. Physical properties and binder content in RAP.

RAP Source	Physical Properties		RAP Binder Content (%)
	Penetration (dmm)	Softening Point (°C)	
RA1	19	70	5.1
RA2	9	72	4.7

3.2. Effects of RH-WMA on chemical properties

The addition of RAP changes the chemical properties in the binder due to the aging. The Fourier Transform Infrared (FTIR) test was used to identify and quantify the chemical functional group of the binder as the results of the binder modification and effect of aging. The functional group is characterized based on absorbance related to C=O stretching (1700cm-1), C=C stretching (1598 cm-1) and S=O stretching (1030 cm-1). FTIR output for RAP from RA1 is shown in Figure 2.

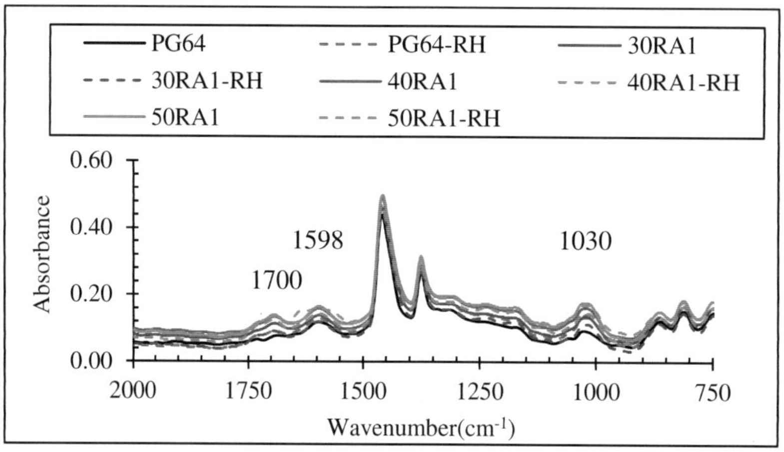

Figure 2. Effects of RAP binder on FTIR.

In general, the binder modified with RAP exhibits higher absorbance for all selected spectral band. Similar trend is observe for RAP from RA2. Since RAP binders in the recycled binder are highly oxidized, RAP binders contain high amount of Sulfoxide and Carbonyl. The Sulfoxide and Carbonyl are related to the spectral band at 1030 cm^{-1} and 1700 cm^{-1}. It can be seen the absorbance increased with RAP binder content. For instance, 50RA1 has higher absorbance than 30RA1. With the addition of RH-WMA additive, the absorbance slightly reduced. This indicates the RH-WMA potential to reduce the oxidation as the results of aging.

3.3. Mix Design of RAP mixture

Mix design of control mixture was carried out in accordance with the job mix formula in JKR specification. The optimum binder content are 5.2% for the control (PG64) and 4.9 % with RH-WMA. For the mix design of RAP mixture, experimental design approach using RSM was adopted in order to reduce the cost, labour and time in mix design. The mix design involves three main tasks that are the aggregate gradation, trial mixing and determination of optimum binder content using RSM. In the mix design of RAP, simultaneous effects of compaction temperature, RAP content and binder content were evaluated by using RSM. The minimum temperature and binder content to enable the properly coating of aggregate were determined prior experimental design in the RSM. Specimen fabrication, experiment and analysis were performed based on the compaction temperature, RAP content and binder content. Analysis of variance on the selected responses according to JKR specification for mix design indicated that the compaction temperature and binder content are the two significant factors affecting the responses. The RAP content does not significantly affect the responses.

Subsequently, optimization on responses were carried out which is according to the JKR design parameter in JKR/SPJ/2008-S4, energy consumption and cost. In line with the steps provided by the Design Expert software, the desired goal of each factor and responses were set. All factors (RAP content, compaction temperature and binder content) were chosen as 'within the range' which refers to JKR specification, while the responses (cost and energy) were selected as minimum. Similarly, the volumetric and strength properties were set according to the JKR specification. The Design Expert software optimized each parameters based on the desired goal. During the optimization, the effects of responses can be illustrated. For examples, Figure 3 demonstrates the relationship between the energy consumption, cost, RAP content and temperature for production of 1 Ton RAP-WMA for RA1.

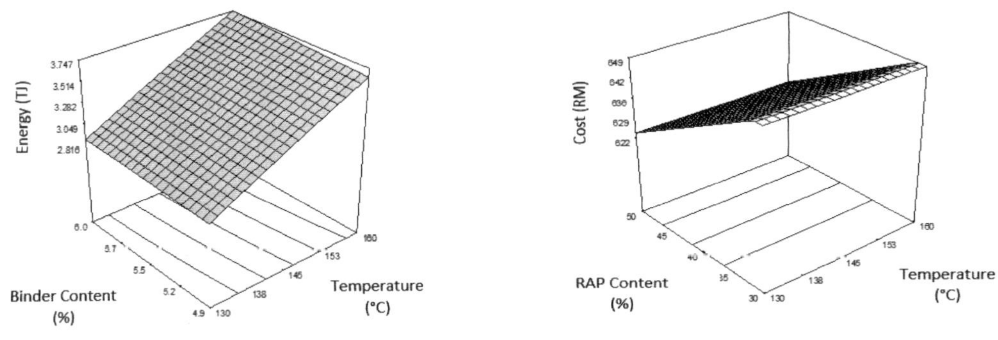

(a) (b)

Figure 3. Relationship between energy consumption, cost, RAP content and temperature for production of 1 Ton RAP-WMA.

As seen in Figure 3, both binder content and mixing temperature increase linearly with energy usage. It appears that mixing temperature at 40% RAP has a significant effect on the energy usage which is indicated by the slope of the graph. Steeper slope relates to the more significant changes in energy usage due to increment in mixing temperature. In terms of cost, increase in RAP content reduces the cost linearly. For example, increasing RAP content from 30% to 50% saves 3.7% production cost when temperature and binder content are constant. Similar pattern is observed for 30% and 50% RAP contents. Overall desirability was calculated to proposed best solution based on the optimized parameters. Based on the optimized parameters, specimens were fabricated in the laboratory at 140°C mixing temperature using the optimized RAP content to verify the optimal condition. The average differences of the experimental and optimized parameters for air voids, bulk specific gravity of mixture (G_{mb}), voids filled with asphalt (VFA) and stability are 10.3%, 0.4%, 2.7% and 5.4 % respectively. Some differences between the optimized and experimental are due to the variation in RAP. The differences are considered acceptable and RAP-WMA mixtures containing 50%, 40% and 30% RAP were fabricated at the suggested condition. Table 3 shows the optimum binder content in mix design using the RSM.

Table 3. Optimum binder content.

RAP	Optimum Binder Content			
(%)	RA1	RA1-RH	RA2	RA2-RH
30	5.6	5.4	5.9	5.7
40	5.7	5.5	5.8	5.6
50	5.7	5.5	5.8	5.6

3.4. Effects of RH-WMA on mechanical properties

In this study, the ITS was conducted to indicate the mixtures' stiffness. As shown in Figure 4, RAP increases the stiffness. At higher RAP content, the increase of ITS produces a stiffer mixture. Previous research reported similar finding on the increased ITS with the addition of RAP but on the porous asphalt. [19]. The stiffness of RAP is affected by the source of RAP and the ITS is consistent with the penetration of RAP binder presented in Table 2.

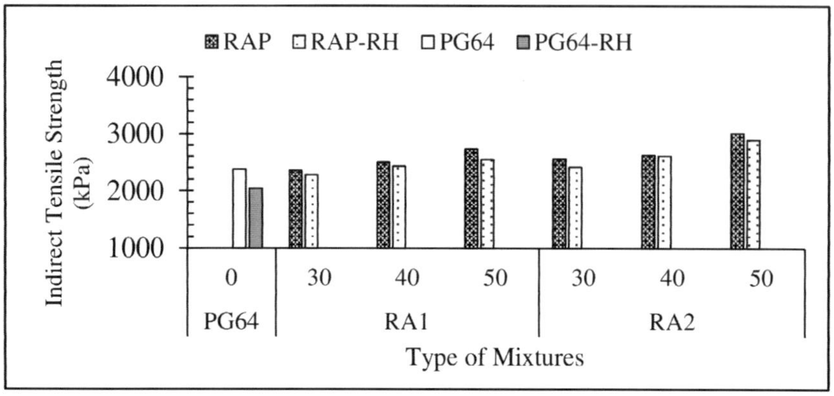

Figure 4. ITS of mixture with RAP.

Table 4. Regression coefficient.

Type of Mixture	Regression Coefficient		R^2
	K	*n*	
PG64	5.00E+08	-2.005	0.9468
PG64-RH	1.00E+08	-1.844	0.9682
RA1	2.00E+07	-1.337	0.9856
RA1-RH	7.00E+07	-1.376	0.8777
RA2	2.00E+06	-0.789	0.9149
RA2-RH	6.00E+06	-0.934	0.9851

The fatigue failure is determined based on the number of cycles that cause fracture to the asphalt specimen. Prior to the determination of fatigue failure, the initial strain and maximum tensile at the center of the specimen are calculated. Figure 5 exhibits the fatigue curve for various RAP mixture. The gradient of the slope (n) can be calculated based on the power relationship. In general, a higher value of n results in a higher fatigue life. The regression parameters obtained from the fatigue relationship with their R2 values are shown in Table 4. Based on Figure 5 and Table 4, RA2-RH results in comparable fatigue resistance with control. This shows that RH-WMA improve the fatigue resistance in RAP mixture.

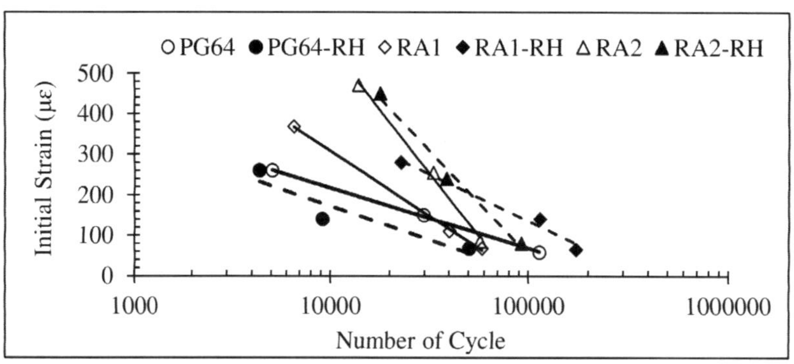

Figure 5. Fatigue curve for mixture with RAP.

In the diametral fatigue test, the applied stress causes horizontal deformation and tensile stress at the center was monitored. The high tensile strain at the bottom of the HMA layer creates cracks that propagate upwards to the pavement surface. The tensile properties of asphalt are essential because of the problem associated with cracking. Specimens subjected to moisture and aging exhibit higher tensile strains indicating the damaging effects of moisture and aging that have weakened the specimen. Moisture and aging are two main components that can significantly increase the potential for fatigue cracking to develop in an asphalt mixture [20]. The variation of fatigue resistance in RAP affected by the binder oxidation and aging, increased binder stiffness and variability of aggregate gradation [21]. In general, the addition of RAP reduces the mixture resistance to fatigue and sometimes the stiffness of the asphalt mixture can dramatically increase with the inclusion of high RAP [9]. When using high RAP content, the fatigue resistance can be reduced due to the addition of aged binder that stiffened the asphalt mixtures.

4. Conclusions

The research findings indicated that proper mix design and localised study on RAP from various sources are necessary due to variation in RAP properties. RAP characterization showed that the RAP met the general quality control characterization. In mix design of 30%, 40% and 50% RAP content, it was found out that temperature and binder content were significant factors. Increased in temperature lead to higher energy consumption which generates negative impact to environment. On the other hand, the addition of high RAP content in RAP-WMA mixtures showed some cost reduction. The lowest compaction temperature of 130°C and 50% RAP content showed the highest cost reduction at lowest energy consumption. In terms of mechanical properties, higher RAP content produced higher stiffness but less resistance to fatigue. However, the addition of RH-WMA in RAP mixture also reduced the stiffness and hence demonstrated better resistance to fatigue compared to RAP mixture only. These findings suggested the potential of recycling the local RAP in bulk production for more than 30%. As a conclusion, study of RAP from local road indicated the possibility for the plant recycling with RH-WMA because it reduced the production temperature, saved cost and produced comparable properties.

5. References

[1] Karlsson R and Isacsson U 2006 Material-related aspects of asphalt recycling—state-of-the art. *Journal of Materials in Civil Engineering* **18**(1) 81-92

[2] Farooq M A and Mir M S 2017 Use of reclaimed asphalt pavement (RAP) in warm mix asphalt (WMA) pavements: a review *Innovative Infrastructure Solutions* **2**(1) 10

[3] Song W Huang B and Shu X 2018 Influence of warm-mix asphalt technology and rejuvenator on performance of asphalt mixtures containing 50% reclaimed asphalt pavement *Journal of Cleaner Production* **192** 191-98

[4] West R C Willis J R and Marasteanu M O 2013 *Improved mix design, evaluation, and materials management practices for hot mix asphalt with high reclaimed asphalt pavement content* (Washington: Transportation Research Board)

[5] Peinado D De Vega M García-Hernando N and Marugán-Cruz C 2011 Energy and exergy analysis in an asphalt plant's rotary dryer *Applied Thermal Engineering* **31**(6-7) 1039-49

[6] Daniel J S and Lachance A 2005 Mechanistic and volumetric properties of asphalt mixtures with recycled asphalt pavement *Transportation Research Record* **1929**(1) 28-36

[7] Hamzah M O Gungat L and Golchin B 2017 Estimation of optimum binder content of recycled asphalt incorporating a wax warm additive using response surface method *International Journal of Pavement Engineering* **18**(8) 682-92

[8] Hamzah M O Golchin B and Tye C T 2013 Determination of the optimum binder content of warm mix asphalt incorporating Rediset using response surface method *Construction and Building Materials* **47** 1328-36

IOP Conf. Series: Materials Science and Engineering **512** (2019) 012055 doi:10.1088/1757-899X/512/1/012055

[9] Zaumanis M and Mallick R B, 2015 Review of very high-content reclaimed asphalt use in plant produced pavements: state of the art *International Journal of Pavement Engineering* **16**(1) 39-55

[10] Evotherm 2009 Warm Mix Technology *Evotherm Pavement Durability: Result of Laboratory and Field Testing*

[11] Kim D Norouzi A Kass S Liske T and Kim Y R 2017 Mechanistic performance evaluation of pavement sections containing RAP and WMA additives in Manitoba *Construction and Building Materials* **133** 39-50

[12] Hill B Behnia B Buttlar W G and Reis H 2012 Evaluation of warm mix asphalt mixtures containing reclaimed asphalt pavement through mechanical performance tests and an acoustic emission approach *Journal of Materials in Civil Engineering* **25**(12) 1887-97

[13] Gungat L Yusoff N I M and Hamzah M O 2016 Effects of RH-WMA additive on rheological properties of high amount reclaimed asphalt binders *Construction and Building Materials* **114** 665-72

[14] Derringer G and Suich R 1980 Simultaneous optimization of several response variables. *Journal of quality technology* **12**(4) 214-19

[15] DERFA 2010 Guidelines to GHGs calculations. *Department for Environment, Food and Rural Affairs. United Kingdom*

[16] Jamshidi A Hamzah M O and Shahadan Z 2012 Selection of reclaimed asphalt pavement sources and contents for asphalt mix production based on asphalt binder rheological properties, fuel requirements and greenhouse gas emissions *Journal of Cleaner Production* **23**(1) 20-27

[17] Xiao F Amirkhanian S N and Putman B J 2010 Evaluation of rutting resistance in warm-mix asphalts containing moist aggregate *Transportation Research Record* **2180**(1) 75-84

[18] Al-Qadi I L Elseifi M and Carpenter S H 2007 Reclaimed asphalt pavement-a literature review *FHWA ICT-07-001 Illinois Center for Transportation Federal Highway Administration*

[19] Goh S W and You Z 2011 Mechanical properties of porous asphalt pavement materials with warm mix asphalt and RAP *Journal of Transportation Engineering* **138**(1) 90-97

[20] Tong Y Luo R and Lytton R L 2015 Moisture and aging damage evaluation of asphalt mixtures using the repeated direct tensional test method *International Journal of Pavement Engineering* **16**(5) 397-410

[21] Bernier A Zofka A and Yut I 2012 Laboratory evaluation of rutting susceptibility of polymer-modified asphalt mixtures containing recycled pavements *Construction and Building Materials* **31** 58-66

The application of CAP-SASW method in determination of sub-surface profile of flexible taxiway in the airport

Norfarah Nadia Ismail[1], Nur Izzi Md Yusoff[2], Nadiah Hussain[1], Jamaludin Mat[3] and Maisarah Ali[1]

[1]Department of Civil Engineering, Faculty of Engineering, International Islamic University Malaysia, Jalan Gombak, 53100 Kuala Lumpur, Malaysia
[2]Civil Engineering Programme, Faculty of Engineering and Built Environment, Universiti Kebangsaan Malaysia 43600 Bangi, Selangor, Malaysia
[3]School of Science and Engineering, Manipal International University, Putra Nilai, 71800. Negeri Sembilan, Malaysia

E-mail: norfarahnadia@iium.edu.my

Abstract. In Malaysia, evaluating and assessing pavement condition always require destructive method such as field-cured cylinders and drilled cores from pavement. Not only these current practices are expensive and labor intensive, they are also may not represent in-place quality of pavement due to the in-situ conditions and site irregularities. In terms of the performance of the flexible pavement, the modulus and thickness of each layer in the pavement are the primary factors that associate with it. In-situ determination of these parameters is crucial, and non-destructive seismic method successfully alleviates the dependency on the destructive method. Utilizing Rayleigh wave propagation as its core element, the technology has proven its ability to obtain important parameters in pavement assessment and delineate the subsurface profile accurately. In this study, a non-destructive seismic method called Common Array Profiling Spectral Analysis of Surface Wave (CAP-SASW) was employed on flexible taxiway at Malaysia airport, as in a new, different configuration compared to the conventional SASW method. This method produces stiffness profile of the flexible pavement at airport taxiway. The inversion procedure is performed to obtain their thickness profile and results were compared with the existing methods such as Heavy Weight Deflectometer (HWD), Ground Penetrating Radar (GPR), Dynamic Cone penetrometer (DCP) and coring data. The pavement moduli are obtained with benefit from equation of Yoder and Witzack, and as for this surface wave technique, the moduli obtained are the small strain modulus. The value of elastic modulus obtained for every layer is also being compared to the modulus obtained from HWD method, which shows good comparison.

1. Introduction

In the management of airport taxiway and runway, the methods or techniques used to measure the pavement condition is very crucial as it will be used to evaluate and monitor the condition of the pavement. Currently the tests used to measure the conditions of the pavements in Malaysian airports are Heavy Weight Deflectometer (HWD), Ground Penetrating Radar (GPR), Single Laser Profile (SLP), coring and Dynamic Cone Penetrometer (DCP).

Nowadays, airports are considered essential in transporting goods and persons, and have been the backbone of economic activities since decades ago. The increasing demand of for air transportation requires greater attention to be given to the overall and continuous assessment of the conditions of runways and taxiways pavement [1,2]. Heavy use of these runways and taxiways has had an impact on the pavement structures, which are either approaching or have exhausted their design life. In view of the deteriorating state of the pavements, airfield management agencies have gradually shifted their emphasis from construction of new pavements to maintenance and rehabilitation (M & R) of existing runways and taxiways [3].

Limited budget is one of the problems in M & R of existing runways and taxiways as monitoring the quality of pavement may take up too much cost, time, and workforce. But some of these limitations might be overcome by adapting geophysical method (such as surface wave method) in civil engineering. These methods are non-destructive and can provide thorough profiling of the subsurface material.

The subsurface profiling was done to evaluate strength and stiffness of the subsurface materials to carry structural load or traffic load. The profiling is also important in the process of quality control, maintenance and rehabilitation so that any distress or failure could be detected at the early stage. The nature of the subsurface materials which varies in terms of characteristics, may make it difficult to obtain the exact parameter value from a test, unless a thorough sampling system is done [4]. However, a thorough sampling system, for example, more drilling hole, will contribute to the increase of cost, time, and workforce.

One of the most important parameters needed for the design of pavement layer system is the shear modulus of the material, which also known as stiffness parameter. Both in-situ measurement and lab measurement could be used to obtain this parameter; however, in-situ measurement give more advantage compare to the later method due to the undisturbed sample material.

The Common Array Profiling of Spectral Analysis of Surface Waves (CAP-SASW), was introduced by Joh et al. [5] and has the advantage in terms of the filtering criteria, to minimize the near-field effect. The measurement technique using CAP-SASW arrangement has been upgraded and allows stiffness evaluation of subsurface being done not only for the whole measurement array, but also at specific subsurface area.

2. Layered System of Airport Taxiway Pavement Design

As for Kuala Lumpur International Airport (KLIA) Sepang, three layers of cement treated base (CTB), each has the thickness of 0.15m were designed and built during the construction. This kind of design may allow crack to happen. To avoid it from happening, the airport pavement was designed and built with another additional layer, called crack relief layer (CRL), with a thickness of 0.10m [6]. The configuration of the layered system is shown in figure 1.

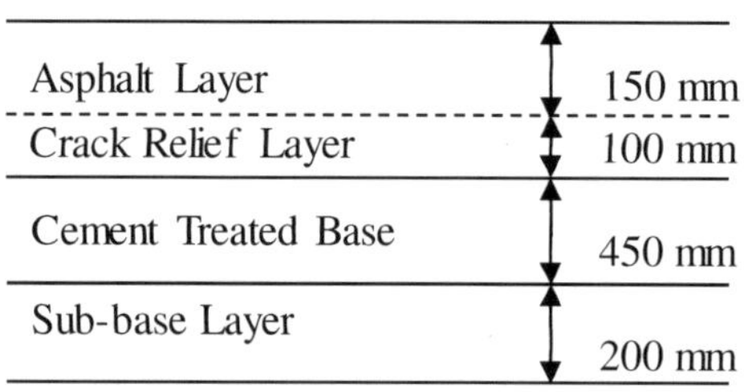

Figure 1. Schematic shows the thickness of pavement design in Kuala Lumpur International Airport.

The asphalt layer in the layered pavement system in Kuala Lumpur International Airport is using polymer modified bitumen as the design material [7]. According to Arahan Teknik Jalan, ATJ 5/85 [8], in the section of Manual of Design of Flexible Pavement, the modulus value of asphalt layer (wearing and binder course) can varies from few hundreds MPa to 3000 MPa. Crack relief layer has lower modulus value compared to typical asphalt mixture, which is around 1380 MPa [9], while cement treated base has the modulus value located in the range of 800 – 3000 MPa [10].

3. Methodology

This paper presents research and data collected at Taxiway B, Kuala Lumpur International Airport (KLIA) Sepang. Measurements were done utilizing CAP-SASW method. Generally, CAP-SASW method consists of three main phases: 1) Experimental data collection, 2) Determination of dispersion curve from the experiment and 3) Inversion process to the dispersion curve to obtain subsurface profile. Typical configuration for CAP-SASW method is by using 2~4 numbers of receivers (sensors) in an array, with a seismic (instrumented) hammer, or impact source acting at the end of the array. Same receiver distance is employed as standard practice during the measurement. An adequate wavelength range is needed to sample the material in the layered pavement system. The configuration for CAP-SASW measurement is shown in figure 2.

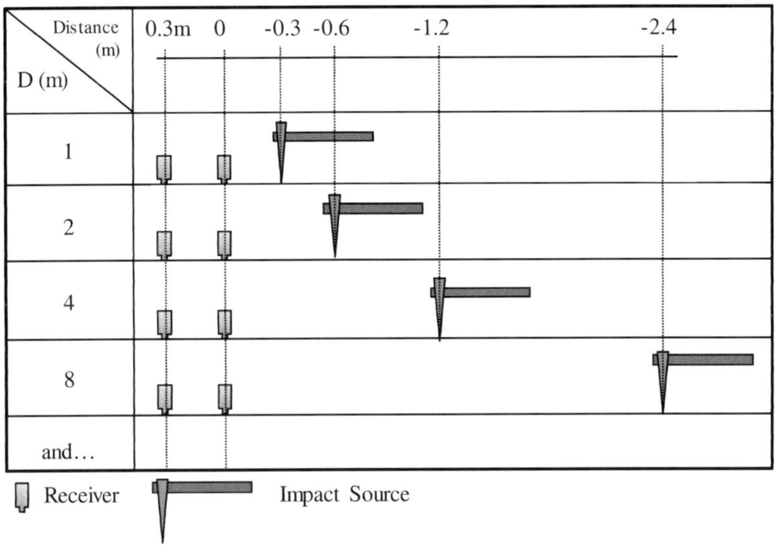

Figure. 2. Configuration of CAP-SASW field measurement.

During field measurement, this research utilizes a dedicated hardware probe called Pavement Integrity Scanner (PiScanner). PiScanner consists of a portable outdoor laptop with customized compact analyzer (POLCCA), and a PiScan probe itself. The PiScan probe is equipped with two receivers (accelerometers), located 0.30m apart and weight to ensure good coupling between the accelerometers and pavement surface. The instrumented hammer is connected to POLCCA so that the impact signal could be recorded. PiScanner is shown in figure 3.

To sample a pavement system with certain depth and evaluation of the stiffness profile, two essential methodologies should be implemented. They are forward modelling process and inversion process. The iteration process for forward modelling is a try-and-error procedure. Based on an initial assumption, a theoretical dispersion curve is computed and then being compared to the experimental dispersion curve. In this analysis, the layer characteristics such as shear wave velocity, layer thickness, Poisson's ratio and density should be assigned. This research is utilizing stiffness matrix method in the forward modelling process. It is better in a way that it creates dispersion curve in superposed mode, in between normal modes and body wave, in which the simulation is more realistic for experimental dispersion curve obtained from field measurement [11].

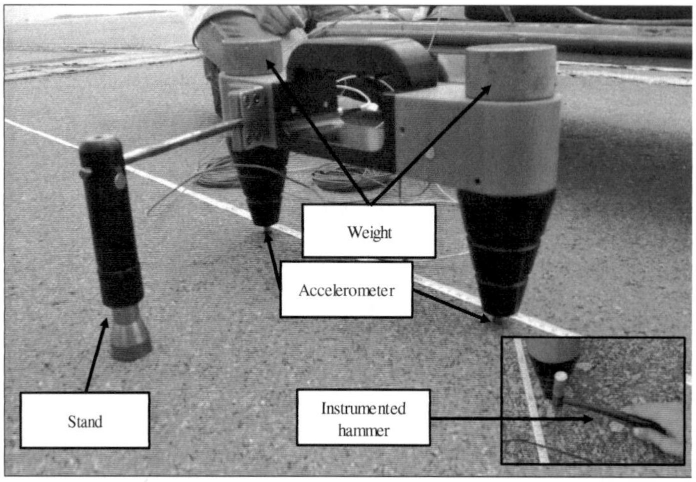

Figure. 3. PiScan probe with weight.

The inversion process on the other hand is the main procedure in characterizing subsurface model from the collected data. It is one of the important parts in using CAP-SASW method in determining the shear wave velocity profile. Simply stated, seismic inversion is a process to determine physical characteristic of rocks or fluids which may be generated from the seismic record that has been obtained. Inversion is the backside of the forward modelling process. The principles of forward modelling and inversion process are illustrated in figure 4.

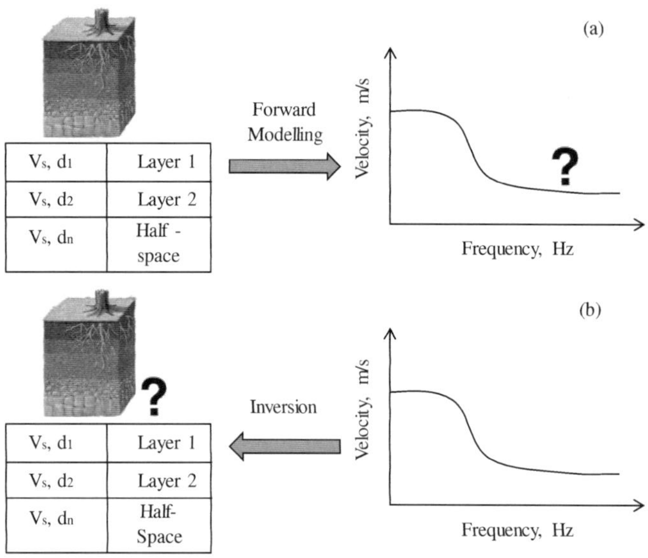

Figure. 4. Principles of: (a) forward modelling process and (b) inversion process.

4. Result and Discussion

The result of this study is presented with comparison to other available methods. This paper presented the outcome of experiment which was done on taxiway B at KLIA Sepang. Taxiway B is located near to the taxiway A, in which both taxiways are connected through series of shorter taxiways. Field measurement using CAP-SASW method has been done to obtain the characteristic of subsurface material, in terms of layer thickness, stiffness (which is represented by the shear wave velocity value), and strength of the layer, which is represented by modulus parameter.

By utilizing CAP-SASW method, there were three measured segments being analyzed. figure 5 shows the outcome of shear wave velocity dispersion curve, plotted against wavelength for all three measured segments. The profiles of pavement layer were also developed and shown in figure 6. These profiles show the stiffness and strength variation of the subsurface materials with depth. figure 5 shows stiffness and strength profile for segment 3 only because this segment is chosen to represent segment 1 and 2 in giving the thickness and modulus of taxiway B.

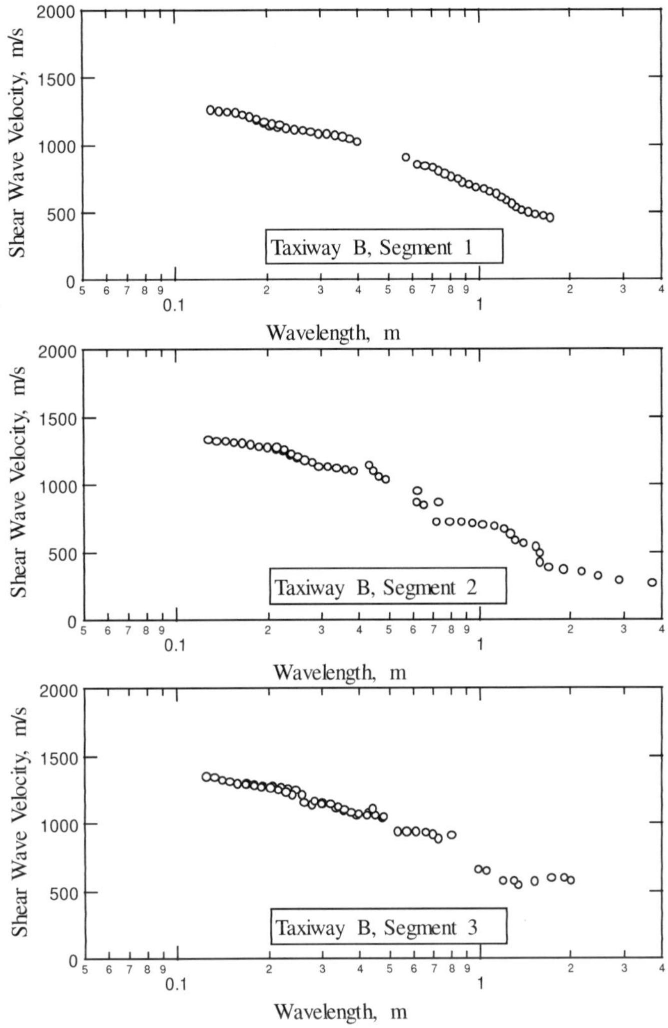

Figure. 5. Shear wave velocity dispersion curve plotted for Taxiway B.

Looking from the general trend of the shear wave velocity dispersion curve (figure 5) and profiles of the pavement structure (figure 6), they are in good terms and reasonable. In terms of value of variation of stiffness and strength with depth, they are also in good and reliable manner. Based on the profiles, it is found that the total depth of asphalt layer and crack relief layer for taxiway B is 0.20m. Based on this CAP-SASW method, the modulus obtained for asphalt layer, crack relief layer and cement treated base for taxiway B are 700 MPa, 950 MPA, and 400 MPa accordingly. These results are reliable up to the depth of 1.0 m.

The output from this study is then used to construct two-dimensional shear wave velocity profile (2-D contour plot). This 2-D contour plot shows shear wave velocity (which represent the material stiffness) variation with depth along the measurement location, 0.30m. The 2-D contour plot for taxiway B is shown in figure 7. The thicker contour line is the boundary between top layer (asphalt layer and crack relief layer) and base layer. The scale on its side shows the value of shear wave velocity which was plotted in this 2-D profile. The higher shear wave velocity represents higher material stiffness.

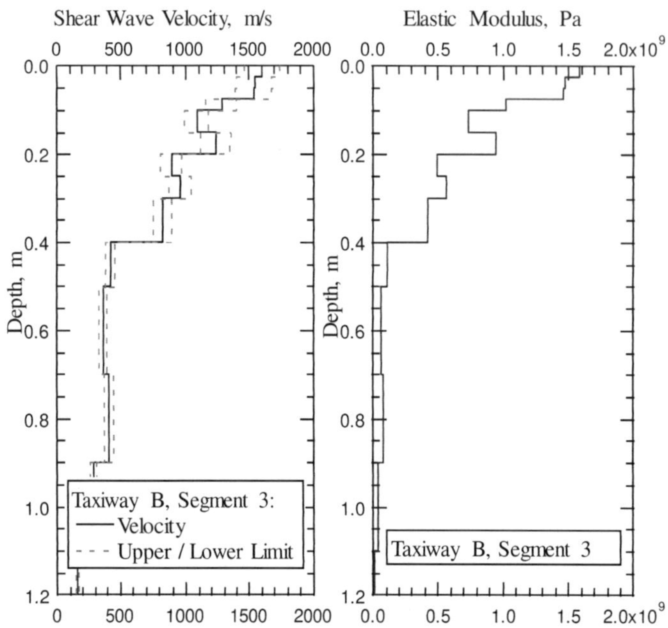

Figure. 6. Stiffness profile and strength profile of subsurface material for taxiway B using CAP- SASW method.

The information synthesized from CAP-SASW method for pavement layer of taxiway B are then being compared to the available information obtained from other measurement methods, as shown in Table 1.

Table 1. Information of pavement layer for Taxiway B.

Method	Thickness (m)			Modulus (MPa)		
	Asphalt	Crack Relief Layer	Cement Treated Base	Asphalt	Crack Relief Layer	Cement Treated Base
CAP-SASW	0.100	0.100	0.300	700	950	400
HWD	0.250	-	0.450	699	6266	259
DCP & Coring	0.140	0.095	0.420	-	-	-
GPR	0.150	0.120	0.451	-	-	-

In the design process of rehabilitation of taxiways at KLIA, asphalt layer and crack relief layer are combined together becoming a top layer with thickness of 0.25m. Seen from Table 1 also, there is no value available for crack relief layer for HWD method. This happened because Taxiway B had undergone rehabilitation process. Extending the research, we found out from Khong (2013) that the thickness of pavement layer obtained from HWD is the design layer thickness, therefore the actual value could not be trusted. It is better to compare the thickness of the layers obtained from DCP & coring with the one obtained from CAP-SASW.

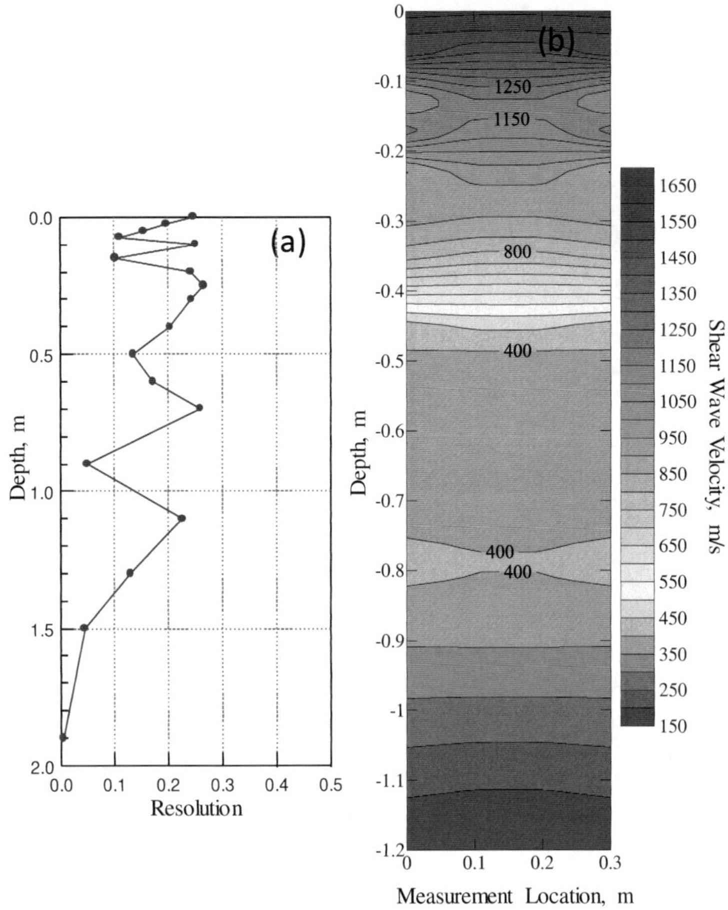

Figure. 7. (a) Parameter model resolution for Segment 3 Taxiway B and (b) 2-D contour plot for Taxiway B.

In terms of strength of the material, the value of asphalt modulus obtained from CAP-SASW method is differ by only 0.1% from the modulus value obtained from HWD method. Therefore, this output is believed to be reasonable and trusted. The modulus value obtained from HWD method for crack relief layer is unfortunately too high and above the assumed value. Hence for crack relief layer, the modulus obtained from CAP-SASW, which is 950 MPa is more trusted. For cement treated base, there is 35% discrepancy in terms of the strength of the materials. Due to the dispute of the layer thickness obtained by HWD method, therefore the modulus obtained from its back-calculation can be doubted and this gives more advantage to CAP-SASW method to provide a more trusted value of modulus.

5. Conclusions

The measurement that has been done on the airport taxiway gave the significant engineering parameters that would be very useful for quality control, maintenance and rehabilitation. The parameters obtained are stiffness which represented by the shear wave velocity, strength of the materials for each layer, represented by the modulus value and lastly is the thickness of each layer. The outcome presented in 2-D contour plot also shows that the thickness values obtained are reliably trusted (0.20m). As for modulus, the CAP-SASW method gives modulus value of 950 MPa, quite in the range of which has been outlined by ATJ 5/85 (2013), which is few hundreds to 3000 MPa.

6. References

[1] Pigozzi F, Portas S, Maltini F and Coni M 2014 Analysis of runway deflectometer campaign for implementation on Airport Pavement Management System *International Journal on Pavement Engineering & Asphalt Technology* **15**(2) 11-26

[2] Abdessemend M, Kenai S and Bali A 2015 Experimental and numerical analysis of the behavior of an airport pavement reinforced by geogrids, *Construction and Building Materials* **94** 547-54

[3] Irfan M, Khurshid M B, Iqbal S and Khan A 2015 Framework for airfield pavements management and approach based on cost-effectiveness analysis *European Transport Research Review* **7**(2) 1-14

[4] Mayne P W, Schneider J A and Martin G K 1999 Small- and large- strain soil properties from seismic flat dilatometer tests *Pre-failure Deformation Characteristics of Geomaterials* 419-26

[5] Joh S H, Jang D W, Kang T H and Lee I W 2005 Evaluation of stiffness profile for a subgrade cross-section by the CAP (Common-Array-Profiling)-SASW technique, *Journal of the Korean Geotechnical Society* **21**(4) 71- 81

[6] Khong K K O 2013 *Proposed full integration of airport pavement management system for Kuala Lumpur International Airport*, *APMS Report* (Kuala Lumpur: Malaysia Airport Holding Berhad)

[7] Nataraj A R and Van de Meer A 2000 Use of asphalt crack relief layer in airport pavements, *Proceedings of the 4th International Rilem Conference: Reflective Cracking in Pavements-Research in Practice*

[8] Arahan Teknik (Jalan) 5/85 2013 Manual for the Structural Design of Flexible Pavement (Kuala Lumpur: JKR)

[9] Bausano J 2007 Reflective crack mitigation using an asphalt concrete interlayer system, Presented at *Pavement Performance Prediction Symposium Laramie*, Wyoming

[10] Jabatan Kerja Raya Malaysia 2008 *Standard Specification for Road Works Section 4: Flexible Pavement* (Kuala Lumpur: JKR)

[11] Joh S H, Kim D S, Kang T H, Kim K S, Ha H S, Chang H S and Jo C H 2006 Comparison of surface-wave techniques in the spatial profiling of subsurface stiffness *American Society of Civil Engineers (ASCE), Geotechnical Special Publication* **149** 188-95

Acknowledgement

The authors would like to thank the Department of Civil & Structural Engineering, Universiti Kebangsaan Malaysia and the Department of Civil Engineering of International Islamic University Malaysia for the financial support and equipment assistance received during conducting this research.

IOP Conf. Series: Materials Science and Engineering **512** (2019) 012057 doi:10.1088/1757-899X/512/1/012057

Review of Malaysian current practice in supply/demand of parking in the hospital

Abdul Razak Ali[1,2] and Sitti Asmah Hassan[1]

[1] School of Civil Engineering, Faculty of Engineering, Universiti Teknologi Malaysia (UTM), Johor Bahru, Malaysia
[2] Prima Reka Konsultan, Jurutera Perunding Awam & Struktur, Ampang Baru, Ipoh, Perak.

E-mail: arazak01@yahoo.com

Abstract. Parking has long been an issue in government hospital, especially the one which are located within urban environment. In view to the fact that private cars are the preferred mode of choice for commuting in Malaysia, the issue could not get any better. This technical paper attempts to evaluate the supply and demand of parking in public hospital. Secondary data are obtained from various parking guidelines. License Plate method is used to collect the parking demand data as a primary data for a period of 12 hours at public hospitals. The actual parking demand is found to be higher than the provided parking space at all public hospitals under investigation. The actual parking demand is found in close trend with parking requirement by JPBD.

1. Introduction

Parking has long been an issue in government hospital, especially the one which are located within urban environment. In view to the fact that private cars are the preferred mode of choice for commuting in Malaysia, the issue could not get any better. Parking facilities are also costly and space intensive. The problems of getting a car park had been a great concern to general public, authorities and policy makers especially at public hospitals. Based on a publication by Ministry of Health, the total annual visit for outpatient attendances and admissions in government hospital summed up to the figure of 23,231,994 visit which obtained all over Malaysia [1]. This 8-digit figure itself shows the substantial turnover to the hospitals for the available parking facilities provided.

The questions of are we providing enough parking space to satisfy the demand has been the topic of discussions among public and transport planners. Excessive number of parking will increase development cost while limited number of parking will cause congestion and increase the risk to public. It is observed that parking is still a big issue in our country that needs an appropriate management strategy. However, documented Transportation Demand Management (TDM) practices are mostly come from the developed world, and contrast markedly with Malaysia, which lacks an organized public transport sector and effective law enforcement and also has a higher car ownership [2-5].

The main reason why parking issues in public hospitals remains unsolved is due to the absence of parking demand data as a basis for analysis in infrastructure management. Therefore, this study attempts to evaluate parking demand and supply in public hospitals in Malaysia. The provided parking spaces are also compared to the various parking generation guidelines to evaluate the current parking demand in government hospitals.

2. Methodology

In this study, the first stage of the work involved putting together the available secondary data from various parking norms and guidelines. In succession, data on the variables were also obtained, which then tabulated and plotted in graph to show the parking generation and requirement trends.

The second stage of the work involved the process on collecting the primary data which is the actual parking demand data. This is the most important part of the study since the are no currently relevant available statistical data in Malaysia. License Plate method was conducted, and the result tabulated and analyzed against the previously analyzed parking trends. Figure 1 below shows the flow chart of the study methodology.

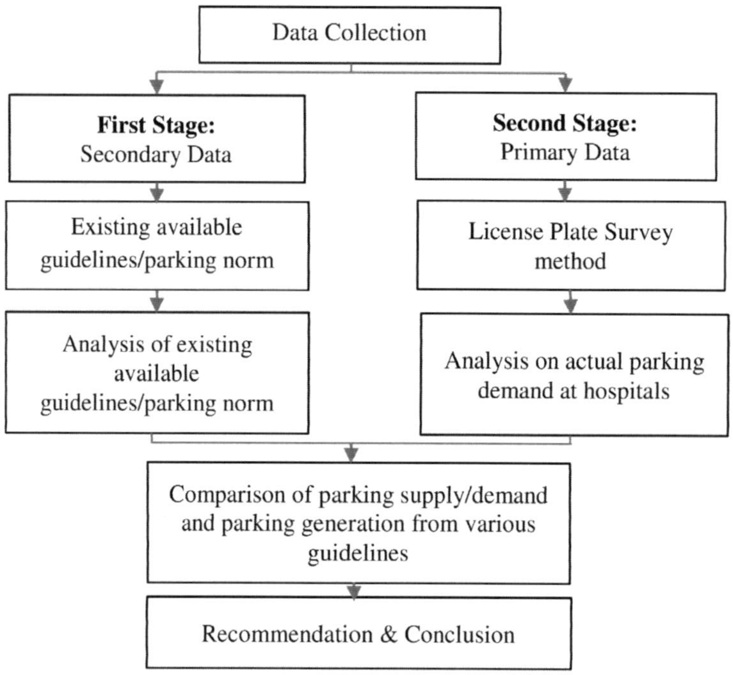

Figure 1. Methodology adopted

2.1 *Parking Requirement*

In order to get to the quantitative derivation for the parking generation, several parking guidelines available in Malaysia, as well as the one established by ITE [6] is referred to. Table 1 below shows the parking requirement based on independent variables such as number of Bed, Number of Professional in the hospitals, as well as number of available Support Staffs for various guidelines.

From Table 1, in the Malaysian very own guideline context, the parking norm by Ministry of Health (MOH Norm) should yield the highest parking generation with the requirement of one parking per one bed. This will be followed by JPBD Norm of one parking per four beds, and lastly the EPU Norm of one parking per five beds. US ITE Norm on the other hand outlined higher requirement with 3.5 parking per one bed. Other variables considered are the number of professional staffs and supporting staff which are relevant to all three Malaysian Requirements.

Table 1. Parking requirement based on available parking guidelines

EPU NORM	JPBD NORM	MOH NORM	ITE NORM
• **1 Parking / 5 Bed** • **1 Parking / 3 Professional Staff**	• 1 Parking / 4 Bed • Additional 1 Parking / 1 Professional Staff • Additional 1 Parking / 3 Support Staff • Additional 20% Parking Space for Motorcycle • 5 Parking Space for Taxi	• 1 Parking / 1 Bed • 1 Parking / 1 Professional Staff • 1 Parking / 3 Support Staff • Additional 20% Parking Space for Motorcycle • 5 Parking Space for Taxi	• 3.5 Parking / 1 Bed

Notes: EPU = Economic Planning Unit
JPBD = Jabatan Perancang Bandar dan Desa (Federal Department of Town and Country Planning)
MOH = Ministry of Health
ITE = Institution of Engineers

Table 2 below shows the obtained information from Engineering Service Department, Ministry of Health Malaysia to be considered as the independent variables [7].

Table 2. Available information from State Hospital in East Malaysia

NO.	HOSPITAL	NUMBER OF BEDS (A)	HOSPITAL STAFF		TOTAL STAFF	TOTAL DAILY OUTPATIENT
			PROFESSIONAL (B)	SUPPORT STAFF (C)		
1.	Hospital Kangar	404	478	1402	1880	572
2.	Hospital Alor Setar	856	1037	2653	3690	1600
3.	Hospital Pulau Pinang	1090	1197	3000	4197	3059
4.	Hospital Ipoh	990	1047	3084	4131	1593
5.	Hospital Klang	1094	1079	2790	3869	2506
6.	Hospital Seremban	1200	1017	2664	3681	1363
7.	Hospital Melaka	901	925	2605	3530	220
8.	Hospital Johor Bahru	989	452	3013	3465	1334
9.	Hospital Kota Bharu	920	1004	2649	3653	1057
10.	Hospital Kuantan	793	838	2419	3257	780
11.	Hospital Kuala Terengganu	821	1046	2633	3679	1251

Based from these two tabulations, the parking generation for each guideline can be substantiated. From the site visit, five out of eleven hospital sites were ruled out from the potential list of suitable study area, due to the undergoing of building construction that in apparent will yield abnormal traffic behavior.

IOP Conf. Series: Materials Science and Engineering **512** (2019) 012057 doi:10.1088/1757-899X/512/1/012057

2.2 License Plate Method

Then, parking study was conducted to obtain the parking demand in the hospitals. License Plate method was carried out for 12 hours (7am – 7pm) with 30-minute interval. For the data collection, the parking facility were subdivided into zones, where team of enumerators were designated at each location, supervised and coordinated by an engineer. The License plate survey method carried out at site by using portable video recorder (Handycam). This apparatus gives advantage over the conventional in-situ plates number note-taking; it gives recorded prove for subsequent validation (if the need arises), as well as expedite the data collection process, where enumerators can cover bigger study area (Figure 2).

Figure 2. License Plate Survey Method, using portable video recording devices

3. Data Analysis and Findings

The Parking Generation for the state government hospitals were tabulated based on the information obtained for each parking requirement (Table 1). The independent variables given out by Ministry of Health were used accordingly (Table 2). As the result, Table 3 below shows the output in term of parking generation, based on the various guidelines considered.

Table 3. Parking generation based on various guidelines

NO.	HOSPITAL	NUMBER OF PARKING TO BE PROVIDED				PROVIDED PARKING IN HOSPITAL
		EPU NORM (A/5) + (B/3)	JPBD NORM (A/4) + B + (C/3)	MOH NORM A + B + (C/3)	ITE NORM (Ax3.5)	
1.	Hospital Kangar (HTF)	240	1046	1349	1414	670
2.	Hospital Kuantan (HTAA)	438	1843	2437	2776	964
3.	Hospital Melaka	489	2019	2694	3154	1401
4.	Hospital Alor Setar (HSB)	517	2135	2777	2996	1736
5.	Hospital Kota Bharu (HRPZII)	519	2117	2807	3220	692
6.	Hospital Seremban (HTJ)	579	2205	3105	4200	1813

To give better graphical representation, the above information was then plotted into graph to show the data variation and how each parking generation trend lines coincide with each other. Figure 3 shows parking generation trend obtained from various parking generation guidelines and the provided parking spaces.

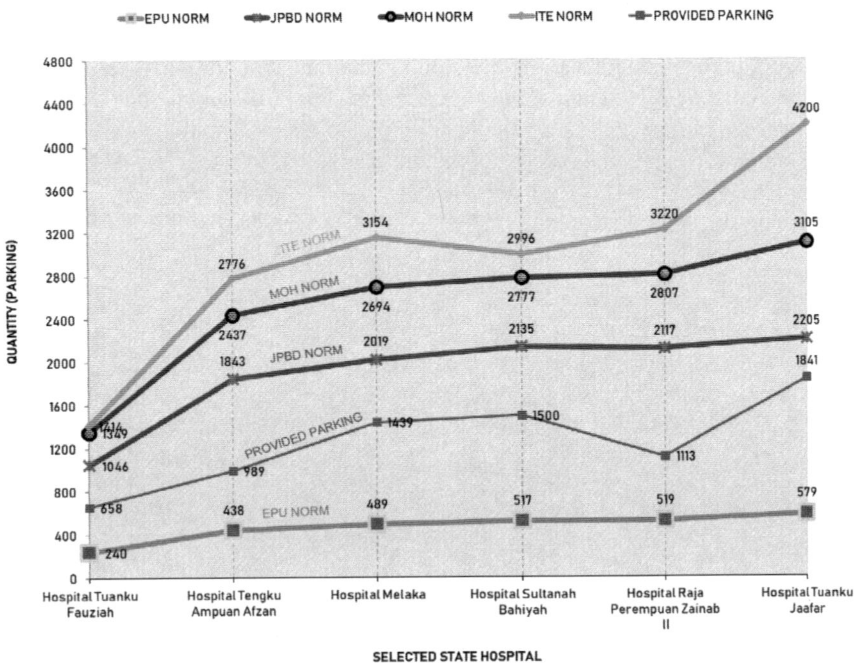

Figure 3. Parking supply and parking generation from various guidelines for each hospitals

The above graph in Figure 3 shows ITE norm leads the trend, followed by MOH norm and JPBD norm, while EPU norm bound to be the lowest trend lines among all. The provided parking at hospitals is shown higher than the numbers required by EPU guideline, but lower than recommended by JPBD.

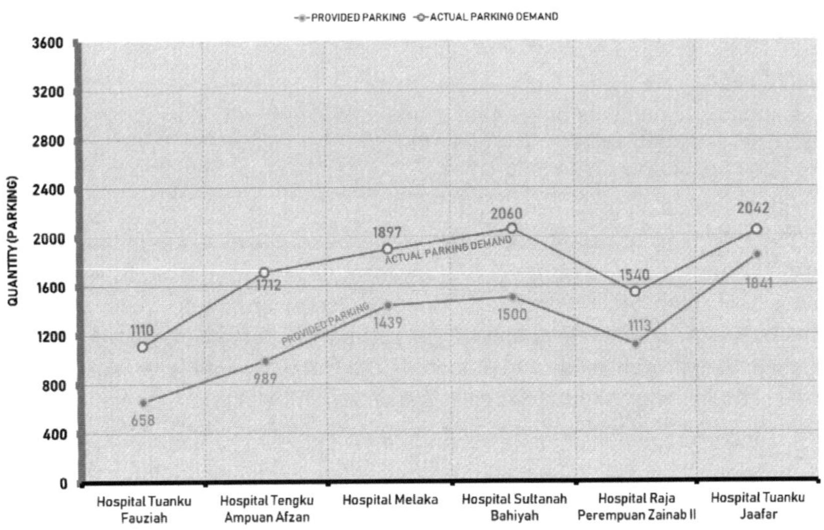

Figure 4. Parking supply and demand at public hospitals

Data collection at each respective hospital were carried out to determine the actual parking demand. This to justify whether the provided parking is sufficient or otherwise. Figure 4 below shows the information on the actual parking demand, overlaid onto the graph of provided parking by hospitals.

With the current situation, parking demand is found to be higher than the provisioned parking, the red line shows the actual demand figure compared to number of parking actually made available

Comparing all the Parking Generation from various guidelines with the provided parking and actual parking demand, the actual parking demand is above the actual parking provided and in close trend with JPBD guideline as shown in Figure 5.

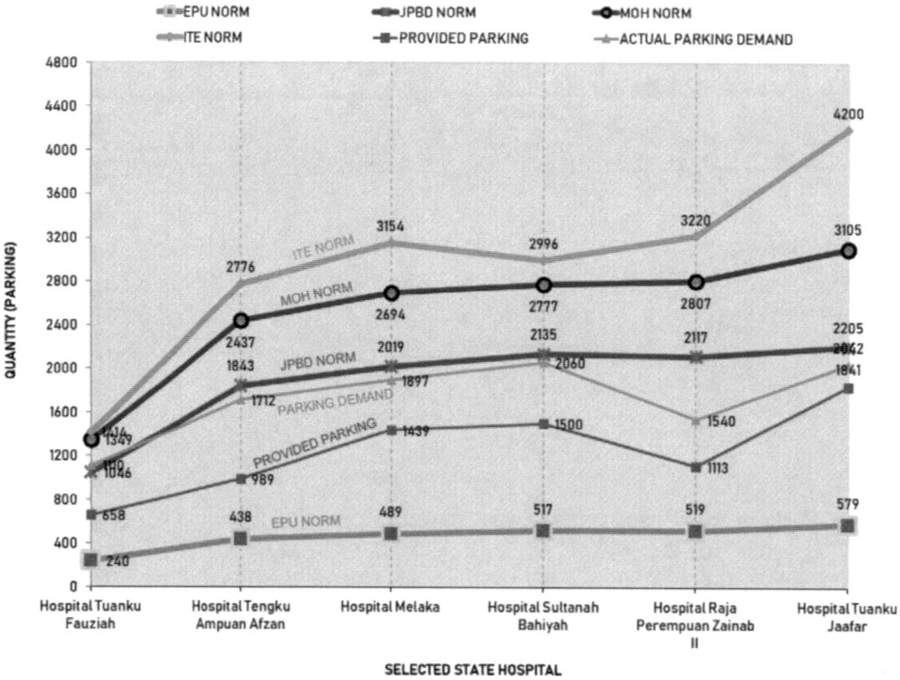

Figure 5. Parking supply-demand trends and parking generation from various guidelines

4. Conclusions and Discussions

As seen in Figure 3, the actual numbers of parking space provided by the hospitals are well below the required amount by JPBD, MOH and ITE. The only Parking generation requirement exceeded over is the EPU norm, which is ranked as the lowest in parking provision criteria. This could be the reason why shortage of parking always been an issue for hospital visitors.

There could be many reasons to the shortcoming; one of the reasons being is the lacked available land to provide the parking spaces. The potential approach is to provide multi-story car parks as introduced in instance of Hospital Pulau Pinang and Hospital Seremban. But to what extent the potential solution really solves the parking deficiency and illegal parking is yet to be validated. Parking demand study is a recommended approach to get to know to the basis of problem issues as well as to quantitatively get the actual parking shortage at each respective study site.

For further study, the impact of public transportation improvement, as well as the e-hailing services to be deliberated and analyzed. Statistical parking model from the correlation of studied variables can also be formed to further enhance the understanding on the parking supply and demand within the study scope.

5. References

[1] Kementerian Kesihatan Malaysia 2017 *Petunjuk Kesihatan 2017* Available from http://www.moh.gov.my/images/gallery/publications/Petunjuk%20Kesihatan%202017.pdf

[2] Weinberger R 2012 Death by a thousand curb-cuts: Evidence on the effect of minimum parking requirements on the choice to drive *Transport Policy* **20** 93–102

[3] Aoun A, Abou-Zeid M, Kaysi I and Myntti C 2013 Reducing parking demand and traffic congestion at the American University of Beirut *Transport Policy* **25** 52–60

[4] Alkheder S A, Al Rajab M M and Alzoubi K 2016 Parking problems in Abu Dhabi, UAE toward an intelligent parking management system "ADIP: Abu Dhabi Intelligent Parking." *Alexandria Engineering Journal* **55**(3) 2679-87

[5] Barter P 2016 On-Street Parking Management: An International Toolkit *Deutche Gesellschaft für Internationale Zusammenarbeit (GIZ)* pp 66-76

[6] Institute of Transportation Engineers (ITE) 2010 *Parking Generation, 4th Edition* (Washington: ITE)

[7] Correspondence letter from Engineering Service Department, Ministry of Health Malaysia, dated 16th April 2016.

Acknowledgement

The authors would like to express their deep gratitude to Universiti Teknologi Malaysia for providing the financial grant (Q.J1300000.2522.11H37) and opportunity to carry out this research.

Establishing a resilient modulus test protocol for miniature cylindrical asphalt mix specimens

Ratnasamy Muniandy, F X Anthoney, Salihudin Hassim and Hussain Hamid

Department of Civil Engineering, Faculty of Engineering, Universiti Putra Malaysia, 43400 UPM Serdang, Selangor, Malaysia

E-mail: ratnas@upm.edu.my

Abstract. The resilient modulus is a very important parameter to be identified and used in pavement design. The indirect tensile test is the most common repeated load test to measure the resilient modulus of asphalt mixtures and the easiest method to test laboratory compacted specimens and field core specimens. The resilient moduli of asphalt mixtures are measured using the indirect tensile procedure (ASTM D4123). A metal loading strip with a concave surface with the radius of the test specimen is required to apply load to the specimen. Typically, specimens are either a nominal 100mm or 150mm in diameter with a minimum thickness over diameter ratio of 0.4. However, typical 100mm diameter core samples with 40mm to 50mm thickness taken from the site most often do not fulfil the minimum ratio of 0.4 after the samples are trimmed for testing. As such, a new procedure was developed to test specimens smaller than 100mm in diameter. This expected to be very useful and minimize the volume requirement from the field. This novel protocol for resilient modulus test using miniature specimens of 37.5mm and 56.3mm in diameter has great potential for practical relevance for the industry.

1. Introduction

The distresses of asphalt pavement, including fatigue, rutting, and low temperature cracking are related to the elastic modulus of the asphalt layer. In addition, the elastic modulus of asphalt concrete is a design variable for asphalt pavement structural design when elastic-layer system theory is employed. However, in the most commonly used asphalt concrete design methods (the Marshall, Hveem, and Superpave methods), the elastic modulus is not used as a control variable. Therefore, these design methods do not ensure that the desired elastic modulus of asphalt concrete will be obtained [1]. Falling Weight Deflectometer (FWD) testing has been used widely to measure the structural conditions and identification of the resilient modulus of flexible pavement layers. The results of resilient modulus values for flexible pavement layers acquired from the back calculation analysis may not be precise, despite the fact that the measured and calculated deflection may remain within acceptable parameters [3]. Wide interpretation is associated with getting the resilient modulus of these flexible pavement layers. FWD interpretation has turned out to be more demanding, since more of our flexible pavements have encountered various milling processes and overlays. The properties of a structure in terms of damaged layers, thickness variety and temperature differences can affect the deflection values, having a much more critical impact than those made by structural layer stiffness.

This paper discusses on the new thinking which was developed in this study to identify a new procedure to carry out resilient modulus testing on cored specimens from field overlay or asphalt concrete wearing course (ACWC) layers. These new findings could allow researchers and practitioners to perform resilient modulus test on specimens smaller than 100 mm in diameter while maintaining equivalence to the standard test procedure as per ASTM D4123.

2. Literature review

2.1. Diameter of specimen

The relative development of the resilient modulus and dynamic modulus of hot mix asphalt as material properties to be used in the flexible pavement design has been carried out [7]. It was found that the size of the sample mathematically influenced the obtained resilient modulus measurement. Resilient modulus measurements acquired for 100mm diameter samples were higher than those obtained for 150mm diameter samples in all variable temperatures used for carrying out the testing [13]. Figure 1 shows the effects of specimen diameter.

Kandhal did a similar assessment on 4-inch (100mm) and 6-inch (150mm) diameter samples and found that the rigidity (resilient modulus) of the 6-inch (150mm) diameter samples was dependably lower than the 4-inch (100mm) diameter samples. Under a similar loading, the strain rate for the 6-inch (150mm) diameter samples was dependably lower than that of the 4-inch (100mm) samples [4].

Likewise, Lim et al. investigated the sample size impacts on the aftereffects of diametrical mechanical testing approaches, in particular the resilient modulus and indirect tension tests. The diameter over height proportion of sample was consistent at 1.6 and it was found that the resilient modulus diminishes as the diameter of the sample increments [12].

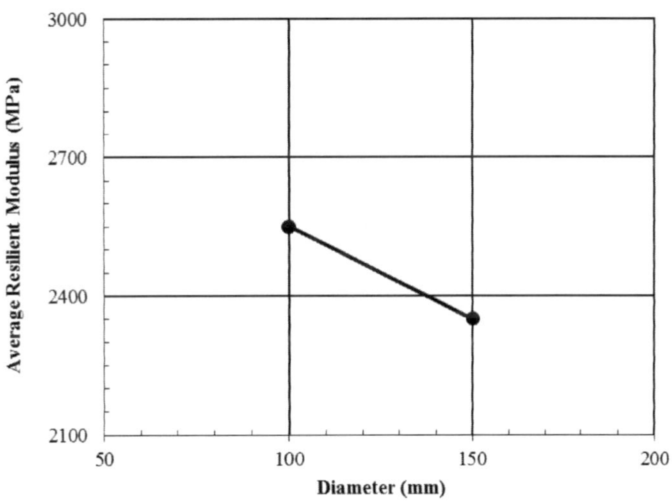

Figure 1. Effect of specimen diameter [2]

2.2. Thickness of specimen

As indicated in the Australian Standard AS 2891.13.1, the thickness of the test samples is to be in the vicinity of 70mm and 35mm thick for 100mm diameter samples and in the vicinity of 90mm and 60mm thick for the 150mm diameter samples [14]. Utilizing static Indirect Tensile Test, Hugo and Schreuder assessed the impact of the sample thickness on the tensile strength and related engineering practices. They found that the indirect tensile strength increments as the sample thickness increments [8]. The sample with thickness more than 20mm will be experiencing stress concentrations at the bottom and top of the contact surface. The stress along the rest of the vertical diameter would be diminished far beneath the average calculated stress level [5]. This could be the reason for the expansion in tensile strength, considering the way that the unequal pressure dissemination makes the sample quality pressure-reliant. This shows the centre part of the samples once the uppermost and the base contact points (very pressured points) outwardly begin to fail [11].

2.3. Maximum nominal aggregate size

In view of the Australian Standard AS 2891.13.1, the element size of up to 40mm can be utilized for the resilient modulus sample test. An examination by Lim et al. on the impact of diameter/utmost nominal aggregate size proportion demonstrated that the resilient modulus diminishes as the proportion upsurges [2].

Research coordinated by Brown and Bassett on the connection between asphalt blend properties and utmost stone size showed a strong relationship between the resilient modulus and the utmost stone size. Consequently, the resilient modulus upsurges as the stone size upsurges [10]. Figure 2 shows the effect of maximum nominal aggregate size on the resilient modulus.

Tongyan et al. carried out research intended to explore the impacts of the material properties of the significant segment on the resilient modulus of asphalt blends with the coarse stone texture considered as the main factor [9]. With modulus experiments carried out at a temperature of 25°C, utilizing coarse stone with more irregular morphologies significantly enhanced the resilient modulus of asphalt blends [6].

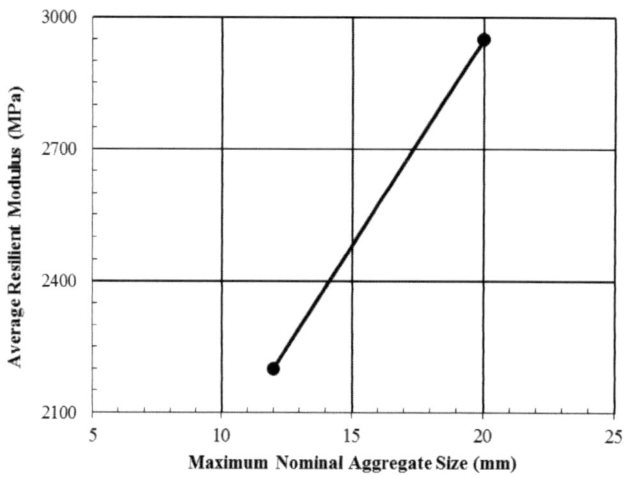

Figure 2. Effect of maximum nominal aggregate size on the resilient modulus [2]

2.4. Resilient modulus test

The five-pulse indirect tensile modulus test is a material stiffness test conforming to ASTM D4123. In the test, a pulsed diametral loading force is applied to a specimen. The resulting total recoverable diametral strain is then measured from the axes 90 degrees from the applied force. Strain in the same axis is not measured; thus, a value of 0.4 for Poisson's ratio is used as a constant. With a fixed level of applied peak force, the test sequence consists of the application of 150 conditioning pulses followed by 5 pulses where data acquisition takes place. The conditioning pulses ensure that the loading plates are seated onto the specimen for consistent results. For controlled temperature testing, the specimen's skin and core temperatures are estimated by transducers inserted in a dummy specimen and located near the specimen under test. The specimens are mounted in the indirect tensile test jigs and the results are recorded in a computer. Resilient modulus tests were performed on laboratory prepared samples.

IOP Conf. Series: Materials Science and Engineering **512** (2019) 012058 doi:10.1088/1757-899X/512/1/012058

3. Design and fabrication of specimen and test assembly

3.1. Establishment of sample thickness and diameter

Performance tests were conducted on 225 samples, with 75 samples with 100mm diameter as control, 75 samples with 56.3mm diameter and 75 samples with 37.5mm diameter. The minimum thickness of samples was determined based on 0.4 ratio of height over diameter for each 100mm diameter sample. Figure 3 shows the calculated minimum thickness for 100mm diameter sample.

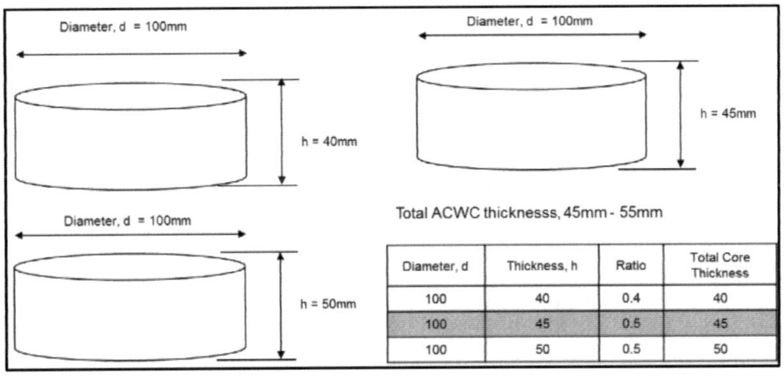

Figure 3. Calculated minimum thickness for 100mm diameter sample

The minimum thickness of samples was determined based on 0.4 ratio of height over diameter for 56.3mm diameter sample. Figure 4 shows the calculated minimum thickness for the 56.3mm diameter sample.

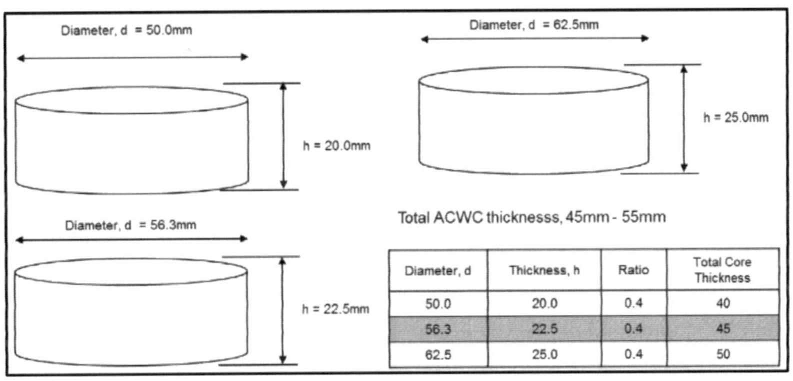

Figure 4. Calculated minimum thickness for 56.3mm diameter sample

The minimum thickness of samples was determined based on 0.4 ratio of height over diameter for 37.5mm diameter sample. Figure 5 shows the calculated minimum thickness for 37.5mm diameter sample.

IOP Conf. Series: Materials Science and Engineering **512** (2019) 012058 doi:10.1088/1757-899X/512/1/012058

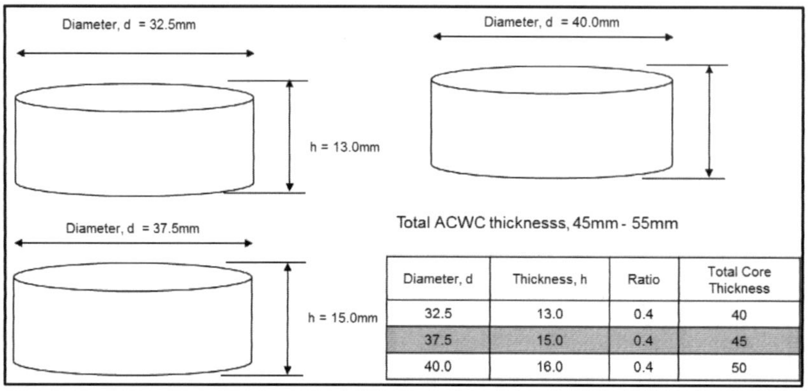

Diameter, d	Thickness, h	Ratio	Total Core Thickness
32.5	13.0	0.4	40
37.5	15.0	0.4	45
40.0	16.0	0.4	50

Figure 5. Calculated minimum thickness for 37.5mm diameter sample

3.2. Coring on cylindrical gyratory compacted sample

The samples extracted from the original 150mm diameter compacted samples are labelled accordingly on top of each extracted samples to ensure that the samples to avoid confusion. Later, the 100mm, 56.3mm and 37.5mm diameter samples are to be sliced to its required thickness. During slicing, the samples have to be again labelled from top to bottom and on the top surface of each sample using a planned labelling method using permanent markers to ensure that the samples are not confused and also to ensure the ability to trace back to the original samples to minimize any discrepancies during the performance test. Figures 6 (a), (b) and (c) below show how the markings have been made on the 150mm samples compacted using gyratory compactor.

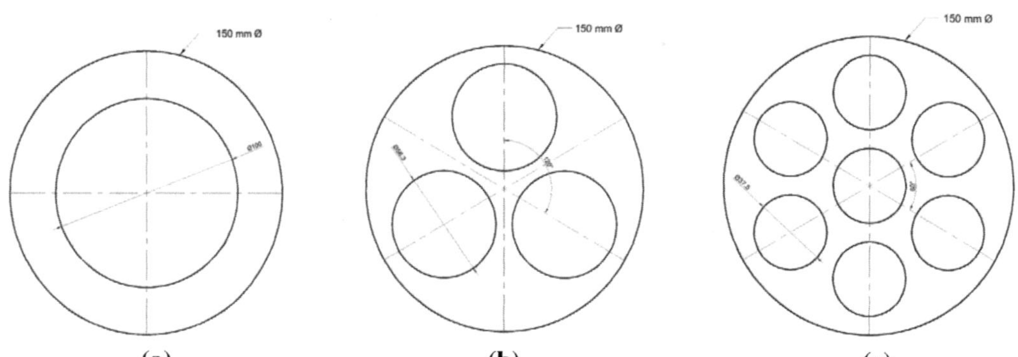

(a) **(b)** **(c)**

Figure 6 (a), (b) and (c). Coring layout to extract 100mm and 56.0mm diameter samples from 150mm diameter

4. Results

4.1. Resilient modulus test on 100mm diameter control sample

Resilient Modulus test has been performed using the Material Testing and Analysis (MATTA) machine in the UPM laboratory in accordance with ASTM D4123 standard. The standard diameters of test samples are 100mm and 150mm. The control sample diameter used in this study is a 100mm diameter sample. The standard load and pulse are 1200N and 3000ms, respectively. The results of the resilient modulus test for various load and pulse for 100mm diameter control samples has been tabulated and shown in Table 1 below.

Table 1. Summary of resilient modulus (Mpa),lLoad (N) and pulse (ms) data for 100mm diameter sample

Load (N)	Pulse (ms)	M_R (Mpa)	Pulse (ms)	M_R (Mpa)	Pulse (ms)	M_R (Mpa)	Pulse (ms)	M_R (Mpa)	Pulse (ms)	M_R (Mpa)
350	2000	2738	2500	2628	3000	2250	3500	2504	4000	2359
550	2000	2928	2500	2888	3000	2291	3500	2723	4000	2802
750	2000	3145	2500	2834	3000	2466	3500	2778	4000	2837
950	2000	3013	2500	2835	3000	2514	3500	2807	4000	3145
1200	2000	2950	2500	2555	3000	2400	3500	2801	4000	2974

The resilient modulus obtained from the test specimen has been plotted as shown in Figure 7 below.

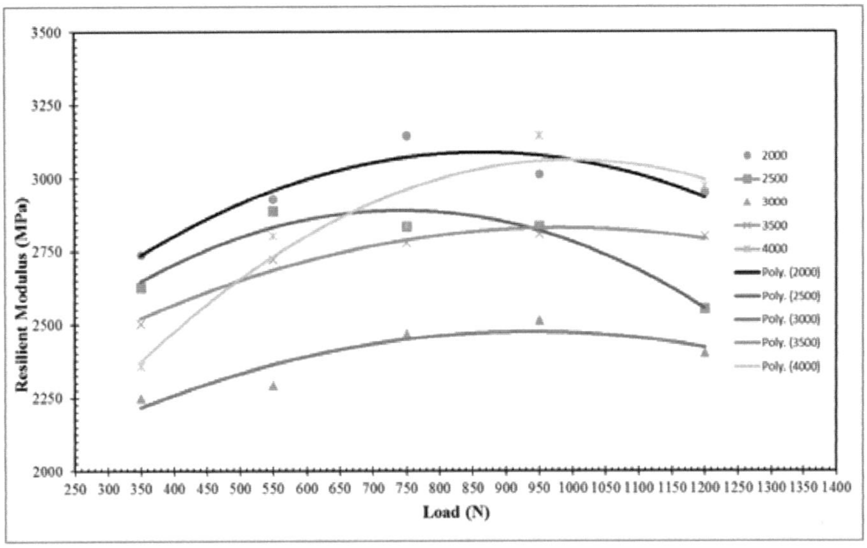

Figure 7. Resilient modulus for 100mm diameter specimens versus load @ a pulse width of 2000ms, 2500ms, 3000ms, 3500ms and 4000ms

4.2. Resilient modulus test on 56.3mm diameter miniature sample

Resilient Modulus test has been performed using the MATTA machine in the laboratory in accordance with the ASTM procedure. The results of the resilient modulus test for 56.3mm diameter samples are tabulated and shown in Table 2 below.

Table 2. Summary of resilient modulus (Mpa), load (N) and pulse (ms) data for 56.3mm diameter sample

Load (N)	Pulse (ms)	M_R (Mpa)	Pulse (ms)	M_R (Mpa)	Pulse (ms)	M_R (Mpa)	Pulse (ms)	M_R (Mpa)	Pulse (ms)	M_R (Mpa)
350	2000	3303	2500	2857	3000	2588	3500	2881	4000	2872
550	2000	3367	2500	3069	3000	2821	3500	3388	4000	2535
750	2000	3543	2500	2927	3000	2930	3500	3753	4000	2953
950	2000	3032	2500	2902	3000	3060	3500	3423	4000	3190
1200	2000	2880	2500	2419	3000	2758	3500	3570	4000	2530

The resilient modulus obtained from the test specimen has been plotted as shown in Figure 8 below.

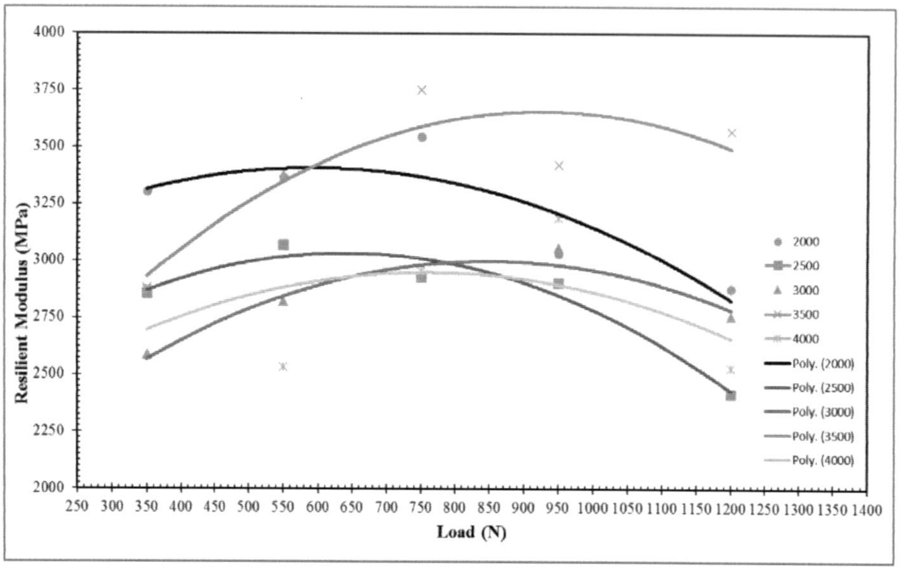

Figure 8. Resilient modulus for 56.3mm diameter specimens versus various load and pulse width

4.3. Resilient modulus test on 37.5mm diameter miniature sample

Resilient Modulus test has been performed using the MATTA machine in the laboratory in accordance with procedure. The results of the resilient modulus test for 37.5mm diameter samples are tabulated and shown in Table 3 below.

Table 3. Summary of resilient modulus (Mpa), load (N) and pulse (ms) data for 37.5mm diameter sample

Load (N)	Pulse (ms)	M_R (Mpa)	Pulse (ms)	M_R (Mpa)	Pulse (ms)	M_R (Mpa)	Pulse (ms)	M_R (Mpa)	Pulse (ms)	M_R (Mpa)
350	2000	2110	2500	2060	3000	3739	3500	2176	4000	2291
550	2000	3389	2500	3855	3000	4250	3500	2554	4000	3456
750	2000	4778	2500	3412	3000	4254	3500	3530	4000	3116
950	2000	3525	2500	5250	3000	5065	3500	4113	4000	3690
1200	2000	3753	2500	4892	3000	4520	3500	3806	4000	3024

The resilient modulus obtained from the test specimen has been plotted as shown in Figure 9.

From the plots as shown in Figures 4.1, 4.2 and 4.3, the optimum resilient modulus and optimum load values for sample diameter 100mm, 56.3mm and 37.5mm are tabulated as shown in Table 4 below. The optimum values are obtained using the quadratic equation.

IOP Conf. Series: Materials Science and Engineering **512** (2019) 012058 doi:10.1088/1757-899X/512/1/012058

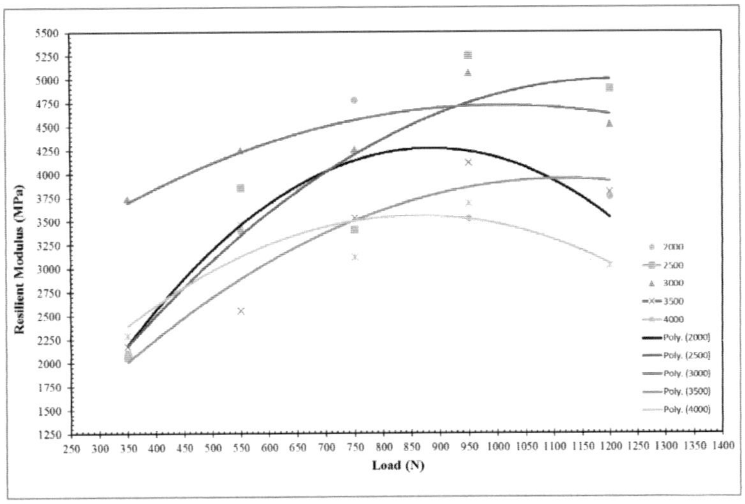

Figure 9. Resilient modulus for 37.5mm diameter specimens versus various load and pulse width

Table 4. Summary of optimum resilient modulus (Mpa), optimum load (N) and pulse (ms) results for the various diameter speciemens

100mm Ø			56.3mm Ø			37.5mm Ø		
Pulse (N)	Load (ms)	M_R (Mpa)	Pulse (N)	Load (ms)	M_R (Mpa)	Pulse (N)	Load (ms)	M_R (Mpa)
2000	888.92	3119	2000	583.75	3398	2000	883.49	4267
2500	733.75	2881	2500	667.11	3027	2500	1190.81	4932
3000	878.44	2434	3000	862.50	3017	3000	1000.71	4711
3500	923.56	2790	3500	920.57	3645	3500	1115.76	3945
4000	1023.53	3103	4000	760.20	2948	4000	869.92	3587

5. Conclusion

The linear equations as shown in Figure 10 below demonstrates a strong relationship between the three (3) type of specimen test results. The 100mm diameter specimen results were correlated with the 56.3mm diameter specimen and 37.5mm diameter specimen. The load values at each pulse width are analysed to identify the shift factor between 100mm diameter specimen against 56.3mm diameter specimen and also 100mm diameter specimen against 37.5mm diameter specimen. From the established linear equation (1), which represents the equation for 56.3mm diameter specimen, a shift factor is introduced for each pulse width as shown in equation 1.1 below. For example, if a sample with 56.3mm diameter is being tested at a pulse width of 3000ms, the equation 1.4 is adopted to identify the appropriate load value to be used to begin the test.

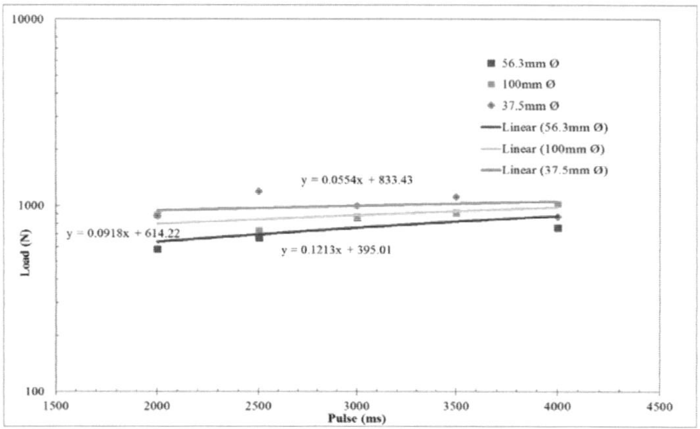

Figure 10. Correlations of load versus pulse for 100mm, 56.3mm & 37.5mm diameter specimens

Equation: $y = 0.1213x + 395.01$ (1)

$L_{56.3,Pulse} = [0.1213P + 395.01] * \text{Shift Factor}$ (1.1)

$L_{56.3,2000} = [0.1213P + 395.01] * 1.251$ (1.2)

$L_{56.3,2500} = [0.1213P + 395.01] * 1.208$ (1.3)

$L_{56.3,3000} = [0.1213P + 395.01] * 1.172$ (1.4)

$L_{56.3,3500} = [0.1213P + 395.01] * 1.141$ (1.5)

$L_{56.3,4000} = [0.1213P + 395.01] * 1.115$ (1.6)

where P is the desired Pulse to be applied and L is the recommended Load.

From linear equation (2) which represents the equation for 56.3mm diameter specimen, a shift factor is introduced for each pulse width as shown in equation 2.1 below. Again, if a sample with 37.5mm diameter being tested at a pulse width of 3000ms, equation 2.4 can be adopted to identify the appropriate load value to be used in the test.

Equation: $y = 0.0554x + 833.43$ (2)

$L_{37.5,Pulse} = [0.0554P + 833.43] * \text{Shift Factor}$ (2.1)

$L_{37.5,2000} = [0.0554P + 833.43] * 0.845$ (2.2)

$L_{37.5,2500} = [0.0554P + 833.43] * 0.868$ (2.3)

$L_{37.5,3000} = [0.0554P + 833.43] * 0.890$ (2.4)

$L_{37.5,3500} = [0.0554P + 833.43] * 0.911$ (2.5)

$L_{37.5,4000} = [0.0554P + 833.43] * 0.930$ (2.6)

where P is the desired Pulse to be applied and L is the recommended Load.

6. References

[1] Pavlovich R D and Goetz W H 1976 Direct tension test results for some asphalt concretes *Journal of the Association of Asphalt Paving Technologist* **45** 400–28

[2] Ghaffarpour J S and Khodaii A 2009 Comparing factors affecting resilient modulus in asphalt mixtures *Transaction A: Civil Engineering* **16**(5) 367–75

[3] Hassim S, Harahap R, Muniandy R, Kadir M and Mahmud A 2005 Cost comparison between stone mastic asphalt and asphalt concrete wearing course *Am J Appl Sci.* **2**(9) 1350-55

[4] Kandhal P S 1991 Design of large-stone asphalt mixes for low volume roads using 6-in. diameter marshall specimens *Transportation Research Record* **1291** 253–62

[5] Loulizi A, Flintsch G W and Mcghee K 2007 Determination of in-place hot-mix asphalt layer modulus for rehabilitation projects by a mechanistic – empirical procedure *Transportation Research Record* **2037** 53–62

[6] Mansour T N, Putman B J and Asce A M 2013 Influence of aggregate gradation on the performance properties of porous asphalt mixtures *Journal of Materials in Civil Engineering* **25**(2) 281–88

[7] Moazami D, Behbahani H and Muniandy R 2011 Pavement rehabilitation and maintenance prioritization of urban roads using fuzzy logic *Expert Systems with Applications* **38**(10) 12869-79.

[8] Muniandy R, Aburkaba E E, Hamid H and Yunus R B T 2009 An initial investigation of the use of local industrial wastes and by-products as mineral fillers in stone mastic asphalt pavements *Journal of Engineering and Applied Sciences* **4**(3) 54–63

[9] Muniandy R, Jakarni F M, Hassim S and Mahmud A R 2008 A newly developed laboratory slab roller compactor (turamesin): an overview *International Journal of Engineering, Transactions A: Basics* **21**(2) 133–42

[10] Arabani M and Ferdowsi B 2009 Evaluating the semi-circular bending test for HMA mixtures *International Journal of Engineering, Transactions A: Basics* **22**(1)

[11] Muniandy R and Huat B B 2006 Laboratory diameteral fatigue performance of stone matrix asphalt with cellulose oil palm fiber *American Journal of Applied Science* **3**(9) 2005-10

[12] Taylor P, Oh J H, Fernando E G, Holzschuher C and Horhota D 2011 Comparison of resilient modulus values for florida flexible mechanistic-empirical pavement design *International Journal of Pavement Engineering* **13**(5) 37–41

[13] Yousif R A, Muniandy R, Hassim S, Jakarni F M and Aburkaba E 2015 An overview of quantification of fatigue resistance of asphalt mixture using pre-aged binder *WALIA Journal* **31** 125–32

Acknowledgement

The Authors wish to thank the Ministry of Higher Education for financing the above research through FRGS.

IOP Conf. Series: Materials Science and Engineering **512** (2019) 012059 doi:10.1088/1757-899X/512/1/012059

Three location referencing method families: a new theoretical framework

R Kenley[1,2], R Yeo[3] and T Harfield[1]

[1]Swinburne Business School, Faculty of Business and Law Engineering, Swinburne University of Technology, John St, Hawthorn VIC 3122, Australia
[2] Faculty of Engineering, Unitec Institute of Technology, Carrington Road, Mount Albert, Auckland, New Zealand
[3] Asset Management Task Force, Austroads, 287 Elizabeth Street, Sydney, New South Wales, Australia

E-mail: tharfield@swin.edu.au

Abstract. This conceptual paper provides a new theoretical framework for location necessary to enable integration of current road asset management systems with digital designs based on IFC standards. Three important concepts are proposed. The first is redefining the meaning of location data taking into account three different types of model view: linear/network location referencing methods, spatial earth coordinates, and digital design modelling. The creation of new families for location referencing methods takes into account these model views. The new families proposed are: Topological (Logical Linear and Network Referencing), Geospatial (Real World Coordinates), and Geometric (Model Coordinates). By expanding the definitions of each of these families, a more complex, and thus, more inclusive concept of location data can be utilised to support the storage and retrieval of both heritage and future road asset location information.

1. Introduction

Globally, road assets are a major responsibility. Approximately 40 million km of roads in 222 countries have been built and now must be maintained. This becomes more difficult because of the increased number and types of vehicles using the roads and ever increasing costs for repair due to the increasing intensity of major weather events. Thus, the escalating costs drives the need for more effective and efficient ways of managing road asset budgets. However, the technology that could be at the centre of the solution is still evolving. To put this into perspective, this paper focuses on future implications based on the importance of location referencing methods for road asset management.

Location Referencing Methods (LRMs) are a fundamental feature of road asset management systems. The location information generated by the LRMs can be collected from a number of sources and in a number of different formats. However, currently there is no theoretical framework to integrate the location data from past, present and future referencing methods. Therefore, the objective of this paper is to fill the theoretical gap concerning types of location referencing methods required for the road asset management transformation in the 21st century.

In order to understand the problem, a desktop critical review explored 120 documents. The extensive review focused on the location referencing requirements of road asset management. This included international standards that underpin current specifications and the expected data for emerging digital technologies. Analysis of the documents showed that, globally, individual jurisdictions have developed a wide variety of LRMs. However, the similarity of properties suggested the concept of clustering the variety of types into families of LRMs [1].

The balance of the paper outlines the findings of the critical review in section two. Section three describes the proposed alternative set of conceptual families; Topological, Geospatial and Geometric.

Details of each of the families is provided in sections four, five and six. The last section provides a summary.

2. Critical literature review

The purpose of the literature review was two-fold; first to find instances of the term 'location' and second to find definitions that could be used to understand the types of applications for use. The first step was to do a content analysis of a sample of 120 documents. The content analysis criteria were:

- The purpose of the document
- Perspective of the writer and the expected reader of the document in relation to location referencing depending life-cycle of asset phase
- Asset type: road networks, transport systems or networks comprised of multiple asset types
- End uses for a location model, each with its own sub-categories. The uses of location included:
  - way finding (specific addresses, co-ordinates, etc.)
  - works & services (network service lines, proximity regions, resources optimisation, maintenance, works & services co-ordination, construction phase planning, etc.)
  - road users (public information, traffic information, haulage, road closures, flood & fire alerts, legal & registration, etc.)
  - co-operation with utilities (telecommunication, water, etc.)
  - future of roads (automated vehicles, GIS, BIM, etc.)

Summaries of each of the content analysis criteria were then considered in terms of location data utilisation, based on end-user perspectives. Four types of perspective were identified; the real world, a dataset, a model, and model view information exchange. These perspectives were used to categorise the documents. One sub-set is focuses specifically on LRMs.

Table 1 provides an example of some of the documents presented as a chronology that provided definitions of location referencing methods.

Table 1. Location referencing methods definitions from a variety of sources.

Source	Year	Definition of Location Referencing Methods (LRMs)
T Ries	2000	A way of describing the location of an object or event relative to some known point in space.
HTC, TNZ Task 2100, Glossary	2001	The technique used to identify the specific point (location) or segment of road, either in the field or in the office.
NCHRP Report 460, Glossary	2001	A mechanism for finding and stating the location of an unknown point by referencing it to a known point.
P Scarponcini	2002	A way to identify a specific location with respect to a known point.
VicRoads, RoadsOne, Glossary	2005	A mechanism (or technique) for finding and stating the location of an unknown point by referencing it to a known point.
ISO 19133	2006	The manner in which measurements are made (and optionally offset from) a curvilinear element
Austroads AP-T190-11	2011	The method used to identify a specific point or event on the road by providing it with a unique address.
ISO 19148	2012	Manner in which measurements are made along (and typically offset from) a linear element. Note 1: A Linear Element is defined as a 1-dimensional object that serves as the axis along which linear referencing is performed.
ISO 17572	2015	Methodology of assigning location references to locations

Throughout this paper the definition of location referencing methods is not specifically given. This is because the critical review found that meanings vary between standards, government or industry reports, and academic papers as shown in table 1. Indeed, variation is the driver of the necessity for the new theoretical framework.

3. Defining a three family theoretical framework for location referencing methods

The proposed theoretical framework has evolved from early 21st century groups of location referencing methods. Ries identified three groups of LRMs: geodetic, geometric, and linear [2].

Table 2. Definitions of Ries' LRM groups

Type	Location Referencing Methods that:
Geodetic (Geographic)	describe locations on the earth's surface
Geometric	represent discrete features on the earth as a map coordinate
Linear	describe locations along discrete network features

Although as table 2 shows Ries proposed three groups and definitions, he also recognized the attributes of a number of sub-sets within each group. The importance of the sub-sets is that they can be conceptualized as families of location referencing methods. Thus, providing a framework for different types of location referencing methods for road asset management [3].

The Ries groups have either a linear or a spatial focus, or a combination of both [2]. Indeed, during the last 20 years road network asset management functions have come to require both linear and spatial LRMs as Ries suggested they would.

However, more recently scholars suggest that emerging digital technologies, such as BIM (3D, information loaded, digital design) should also be incorporated into the future functionality of road asset management.

The belief that these technologies will drive changing requirements for future road asset management performance, means that an expanded framework of location referencing methods is required. That is the purpose of the proposed theoretical framework outlined in this paper.

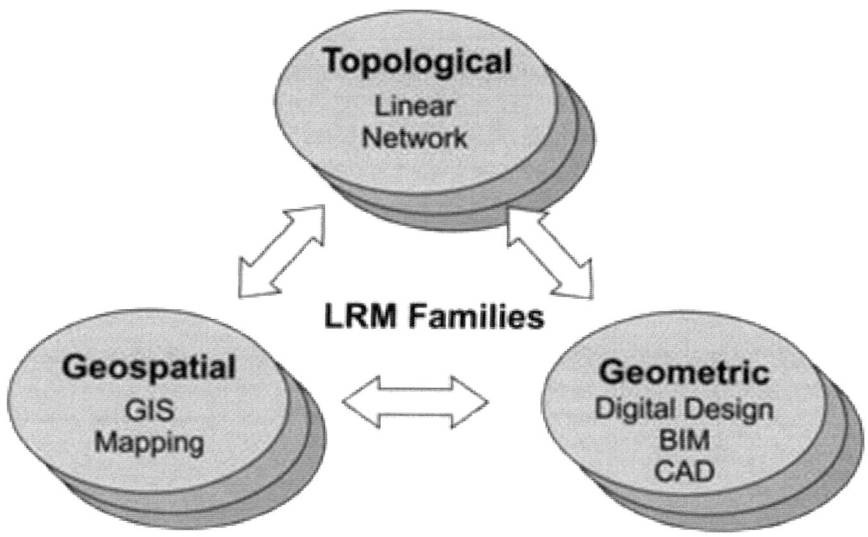

Figure 1. Proposed families of location referencing methods.

IOP Conf. Series: Materials Science and Engineering **512** (2019) 012059 doi:10.1088/1757-899X/512/1/012059

4. Family one: Topological (Linear/Network)

The two groups of LRMs within the Topological family are:

- Linear referencing (discrete linear elements)
- Network referencing (a topologically connected routable network of linear elements).

4.1. Linear referencing

Many studies of linear location referencing methods provide the wide variety of referencing methods for road infrastructure asset management [1, 4, 5]. They are a group of methods that, at the base level, describe locations relative to a one-dimensional object, measured along (and optionally offset from) that object [6]. This group of LRMs does not rely on spatial coordinate information, but often spatial coordinate information is included in their attributes.

An extensive review of locally used linear referencing methods is found in the 2001 report for Transit New Zealand [4]. The report outlines 14 examples from US states and four examples of Australian states. The authors provide the advantages and limitations of the linear reference methods reviewed. This specifically outlines the problem of the wide variety of types of location referencing methods.

The main reason for the variety is LRMs is differing business drivers for the local conditions within which roads are built and maintained [1]. These legacy practices are also embedded in legacy asset management systems. Thus, the benefits for the most commonly implemented LRMs are found in Geographic information: linear referencing, the current international standard ISO19148:2012:

"… a significant amount of information is currently held in huge databases from legacy systems that pre- date Geographic Information Systems (GIS). Many useful applications can and have been built on these data with no understanding of where on the earth's surface the data are located. Knowing where they are located relative to a linear element such as a roadway route or pipeline is sufficient to support these applications and can be used as a means of integrating data from multiple, disparate sources…in some situations, having a linearly referenced location along a known linear element is more advantageous than knowing its spatial position" [6].

4.2. Network referencing

However, it is also possible to hybridize the linear location referencing methods by including geospatial data as attributes (particularly with modern geospatial databases such as Oracle Geospatial [7].

The international standard, ISO 17572-1 requires that location referencing methods should also enable referencing of the relative spatial relationship of objects and, more importantly, not change topological relationships by its own action [8]. For example, the order of points along a line should not get confused.

The network referencing standard ISO19133 explicitly recognizes that the links and nodes of a road system are complex, interconnected and may be routable [9]. Constructing a routable path in a road network may also require significant information other than the information provided by the location referencing methods. For example, addition information could include: grade separation, traffic flow direction, or turn restrictions. These may have both physical and temporal dimensions.

A routable path also requires location attributes such as: road class, road purpose, and environmental conditions. Some networks may also have complex 3D geometry, such as slope gradients for road and rail, or drainage networks. Thus, network referencing has been included in the more recent transport standards [10, 11, 12].

4.3. Importance of Topological Location Referencing

Topological networks provide the same advantages as linear referencing for human cognitive models [13, 14, 15]. However, they also provide a method for solving problems relating to routes or journeys. The obvious use is for journey guidance, such as navigation systems [8]. ISO 19133-2016 defines a navigation as "combination of routing, route transversal [changing] and tracking" [9].

At the same time, for asset management, a more critical use of topological networks is for service delivery management [16]. Network service performance is central to the concept of road user service.

This requires current-time analysis based on routes or journeys providing the possibility of alternative paths in the event of planned or unexpected disruptions, such as road works or accidents provided through all road asset management systems [17, 18].

5. Family 2: Geospatial (Real World Coordinates)

Cartographic systems are included in the geospatial family of location referencing methods. It is framed according to the technology, standards and syntax of geospatial systems [16].

GIS is a system that captures, stores, manipulates, analyzes, manages, and presents all types of geographical data. Traditionally, GIS, based on 2D maps, assigned 2D virtual references to real objects. The most common are longitude and latitude locations [19].

Geospatial referencing methods use a set of spatial coordinates by which location can be interpreted through many forms such as addresses or physical features. However, positioning is the most important locator for road asset management. Positioning consists of coordinates with reference to the Earth. It is supported by survey marks, Global Navigation Satellite Systems, geodetic modelling (coordinate transformations) and the geoid and bathymetric reference surfaces [20].

5.1. Positioning: location data

Geospatial data collection technologies including Total Stations (modern theodolites with built in Global Positioning Systems and laser ranging devices), full 3D static or mobile laser scanners, and photogrammetry cameras.

Traditional road asset management, and current practice for all Australian and New Zealand jurisdictions, limits positioning to two dimensions [1]. However, the traditional 2D location referencing method, is increasingly changing. The state road networks can now be surveyed with terrestrial laser scanners or mobile laser scanning systems to acquire 3D point clouds. These data are processed to extract the relevant location information, such as road centre-lines, location of kerbs, and other roadside furniture [20].

5.2. Importance of Geospatial Referencing

Searching and mapping are two key advantages of geospatial location reference systems. Most governments rely heavily on GIS applications, such as ESRI, for managing geographic data. Thus, they create local standards that provide searching (proximity-based) and modelling abilities [16].

GIS systems are able to utilise much more data than found in the road asset database. For example datasets that provide population, terrain, or ground cover information .The typical application is therefore able to display a huge number of different types of features on a map providing searchable functionality.

Thus, features can be processed according to road user needs, as well as the needs of traditional road asset management systems. It can be argued that no modern road Location Referencing System will be able to avoid having a Geospatial location referencing method based on real world coordinates as part of their asset management system in the future.

6. Family 3: Geometric (Model Coordinates)

de Laat and van Berlo argue that there are two types of modelling worlds that are not able to integrate. The authors indicate this difference by suggesting that "…the 'BIM people' and the 'GIS people' still seem to live in different [modelling] worlds. They use different technology, standards and syntax descriptions." [21].

Understanding this difference is important when considering the potential impact of BIM technology on location referencing methods and systems for road asset management.

Past efforts have paid little attention to either accuracy or discrepancy of location data due to the confusion between how systems use model geometry or cartographic geometry. However, modelling functionality resulting in 3D images continues to be developed. Therefore, road transport agencies are being urged to move towards integration of digital engineering road design data in their asset management systems [1].

6.1. CityGML: linking Geospatial and Geometric location data?

Understanding this difference is important when considering the potential impact of BIM technology on location referencing methods and systems for road asset management.

Past efforts have paid little attention to either accuracy or discrepancy of location data due to the confusion between how systems use model geometry or cartographic geometry. However, modelling functionality resulting in 3D images continues to be developed. Therefore, road transport agencies are being urged to move towards integration of digital engineering road design data in their asset management systems [1].

3D GIS is an emerging technology [19]. Schemas such as KML, COLLADA and GML store 3D attributes of objects in GIS enhancing functionality. Currently in the GIS world, the cross-over from the BIM world for 3D representation uses City Geography Markup Language (CityGML).

CityGML is applicable for large areas and small regions and can represent the terrain and 3D objects in different levels of detail simultaneously integrating model geometry or cartographic geometry. Because of this functionality, both linear and coordinate location referencing datasets can be integrated into road asset management Location Referencing Systems. However, CityCML points do not take into account the curvature of the earth, the continuing problem when using digital platforms [20].

6.2. Designing road infrastructure

The family of Geometric reference systems are those based on geometric models of infrastructure. They are typically created and used during the design phase of road construction. At the simplest level, geometric systems draft 2D representations of the design using relative coordinates (relative to a fixed and known point). Such systems use different methods to calculate the positioning of lines. Importantly, they are generally based on mathematical formula to form vectors [1].

Geometric models of infrastructure include traditional 2D drafting of a project, through to 3D modelling, with the possibility of including modern BIM workflows. These are the tools and methods of 'BIM people' identified by de Laat & van Berlo [19].

Their world-view is based on technology, standards and syntax related to object-based modelling. These types of models make intense use of 3D geometry embedded in Industry Foundation Classes (IFC) [22]. IFC4 provides the rules for using the concepts of constructive solid geometry and boundary representation using Boolean operations to design geometric models of infrastructure, currently this means mainly vertical infrastructure.

However, it should be noted that the inclusion of older 2D and 3D modelling formats for infrastructure design means that Geometic location referencing methods are much more than BIM. At the most advanced level, complex and powerful integrated systems build models of infrastructure [23].

6.3. Alignment, horizontal infrastructure, and modelling

Horizontal Infrastructure is the term that describes built environment and infrastructure assets created through string-based design such as networks. These typically are linear elements with alignment as the principle feature. The location breakdown structure creates continuous centre-lines.

However, BIM modelling has yet to develop effective methods of sending road alignment location information to road asset management systems. This is because an alignment is not composed of coordinate positions with straight lines between, as occurs in most geospatial location reference systems.

Road alignments consists of known points (geo-referenced) and polylines: lines made of multiple segments each being a line segment, a circular arc segment or a clothoidal arc segment [22]. This specification does not force continuity between a point at the end of one segment and the start of the next [1]. This is a common digital modelling dilemma, because of the need for a mathematical trade-off between positions calculated by curves and desired end-points.

An alignment may be seen as the direct equivalent to a Line Location except that dimensionality it is specific and is made of segments. However, some scholars claim that the horizontal alignment [only] is used for linear referencing, which was one of the principles for the development of the IFC Alignment schema [22].

6.4. Importance of Geometric Location Referencing

Clearly geometric modelling using local coordinates is critical for the design of infrastructure. As a coordinate system and with local coordinates of all objects and alignments, this forms an extensive referencing system. Until IFC-Alignment forms the basis of equivalency to topological and geospatial reference systems, CityGML is more likely to be used for road asset management modelling data.

7. Conclusions

Clearly geometric modelling using local coordinates is critical for the design of infrastructure. As a coordinate system and with local coordinates of all objects and alignments, this forms an extensive referencing system. Until IFC-Alignment forms the basis of equivalency to topological and geospatial reference systems, CityGML is more likely to be used for road asset management modelling data.

The Topological, Geospatial and Geometric categories capture the necessity to consider past, present and future location data for road asset management systems. Both the functionality and problems of utilization of each of the families are described to encourage consideration of how each of the families can be adopted.

For many road asset management systems legacy linear/network location referencing methods can be considered mature. The location data from geospatial methods utilising GIS coordinates is becoming more common. However, the inclusion of location information captured in digital models is the most challenging, because it is integrating this category of information integration is at an early stage for road asset management systems.

Therefore, the continuing business drivers for digital information exchange benchmarks suggests the proposed location referencing methods framework will be the foundation of future adoption strategies.

8. References

[1] Kenley R and Harfield T 2018 *Scoping study for a location referencing model to support the BIM environment* (Sydney: Austroads)

[2] Ries T 2000 *Integrating traffic management data via an enterprise LRS* Available from https://ntl.bts.gov/lib/10000/10900/10985/027ppr.pdf

[3] Ebendt R and Touko Tcheumadjeu, L 2017 An approach to geometry-based dynamic location referencing *Eur. Transp. Res. Rev.* **9** 38

[4] HTC Infrastructure Management Ltd 2001 *Task 2100: Principles of location Report for Transit New Zealand* Available from http://www.lpcb.org/index.php/edocman-test/2001-new-zealand-location-referencing-requirements-in-transit/viewdocument

[5] NCHRP 2001 *Report 460-guidelines for the implementation of multimodal transportation location referencing systems* Available from http://onlinepubs.trb.org/onlinepubs/nchrp/nchrp_rpt_460.pdf.

[6] ISO 2012 *Geographic Based Information-Linear Referencing* (Online: International Organization for Standardization)

[7] Oracle 2018 *Spatial Developer's Guide* Available from https://docs.oracle.com/cd/B28359_01/appdev.111/b28400/sdo_intro.htm#SPATL010

[8] ISO 2015 *Intelligent Transport Systems (ITS)-Location Referencing for Geographic Databases: Part 1: General Requirements and Conceptual Model* (Online: International Organization for Standardization)

[9] ISO 2016 *Geographic Information: Location-Based Services: Tracking and Navigation* (Online: International Organization for Standardization)

[10] ISO 2015 *Intelligent Transport Systems (ITS)-Location Referencing for Geographic Databases: Part 2: Pre-Coded Location References (Pre-Coded Profile)* (Online: International Organization for Standardization)
[11] ISO *Intelligent Transport Systems (ITS)-Location Referencing for Geographic Databases: Part 3: Dynamic Location References (Dynamic Profile)* (Online: International Organization for Standardization)
[12] CEN 2011 *Intelligent Transport Systems-DATEX II Data Exchange Specifications for Traffic Management and Information Part 2: Location Referencing* (Online: European Committee for Standardization)
[13] Lobben A 2004 Tasks, strategies, and cognitive processes associated with navigational map reading: a review perspective *Prof. Geogr.* **56** 270-81
[14] Committeri G, Galati G, Paradis A-L, Pizzamiglio L, Berthoz A and LeBihan D 2004 Reference frames for spatial cognition: different brain areas are involved in viewer-object and landmark centred judgements about object location *J. Cognitive. Neurosci.* **16** 1517-35
[15] Choocharukul K, Sinha K C and Mannering F L 2002 User perceptions and engineering definitions of highway level of service: an exploratory statistical comparison *Trans. Res. Part A-Pol* **38** 677-89
[16] South Australia Spatial Services Contractor standards GIS Standard Specifications – GIS-CT001 Online:https://www.dpti.sa.gov.au/__data/assets/pdf_file/0006/393756/GIS-CT001_Geospatial_Services_Specifications.pdf
[17] Kazemi S and Forghani A 2016 Knowledge-based generalisation of road networks *Int. J. Geoinform* **12** 1-13
[18] Roess R P and Prassas E S 2014 *The Highway Capacity Manual: A Conceptual and Research History* (Switzerland: Springer International Publishing)
[19] Deng Y, Cheng J C P and Anumba C 2016 Mapping between BIM and 3D GIS in different levels of detail using schema mediation and instance comparison *Automat.Constr.* **67** 1-21
[20] West GAW and Kenley R 2014 Geographic data and systems in project planning, Position Paper 3 *SBEnrc Project 2.21 New Project Management Models for Productivity Improvement in Infrastructure*
[21] de Laat R and van Berlo L 2011 Integrating BIM & GIS: the development of CityGML GeoBIM extension *Advances in 3D Geo-Information Sciences* ed TH Kolbe, G König and C Nagel pp (Springer Science & Business Media) pp 211-25
[22] Amann A, Borrmann T, Chipman E, Lebegue T, Liebich T and Scarponcini P 2015 *IFC Alignment project, conceptual model, building SMART P6 Project Team* Available from http://www.buildingsmart- tech.org/downloads/ifc/ifc5-extension-projects/ifc-alignment/ifcalignment-conceptualmodel-fs
[23] Chen K, Lu W, Peng Y, Rowlinson S and Huang G Q 2015 Bridging BIM and building: from a literature review to an integrated conceptual framework *Int. J. Prog. Manag.* **33** 1405-16

Acknowledgements
This research has been supported the Australian Research Council Linkage Project 116100524 and the Austroads ASP2112 Asset Management Harmonisation Project

IOP Conf. Series: Materials Science and Engineering **512** (2019) 012060 doi:10.1088/1757-899X/512/1/012060

Leveraging on building information modelling (BIM) for infrastructure project: Pan Borneo Highway Sarawak Phase 1

Zohari Akob[1], Mohd Zaidee Abang Hipni[2], and Mohd Rizal Rosly[3]

[1]Kementerian Kerja Raya Malaysia, Kuala Lumpur, Malaysia
[2]Lebuh Raya Borneo Utara Sdn. Bhd., Kuching, Sarawak, Malaysia.
[3]BIMASIA Sdn. Bhd., Petaling Jaya, Selangor, Malaysia

E-mail: zohari@kkr.gov.my

Abstract. The construction industry in Malaysia has always been confronted with various challenges to enhance productivity, efficiency, infrastructure value, quality and sustainability. The Pan Borneo Highway Sarawak (PBHS) is currently under construction, a large-scale project had to overcome technical and land issues at its outset, as well as manage communication and project management challenges due to its dispersed construction locations and the involvement of multiple regulatory agencies. To address these issues, Building Information Modelling (BIM) offers a better way to address multi-disciplinary inefficiencies and miscommunication throughout the lifecycle of the project. The Project Delivery Partner has adopted BIM to monitor and manage the project to ensure more effective collaboration and communication between all parties, towards delivering on time and within the budgeted cost, as well as to enable a meaningful strategic and tactical asset management once the Highway is completed. Whilst BIM has numerous benefits, implementation has its challenges with regard to people, process, policy and technology (PPPT). PBHS demonstrates how these PPPT challenges were overcome, benefits attained, and what can be improved for future implementation. PBHS is also setting the benchmark for future infrastructure BIM implementation to be used in Malaysia and elsewhere, including for research.

1. Introduction

1.1. Background

The Malaysian construction industry is a key contributing sector towards the Malaysian economy. It contributes approximately 3 to 5 percent of the Gross Domestic Product (GDP) annually [1]. Although it plays a big role in contributing to the growth of Malaysia's economy, the construction industry needs to be improved, especially its productivity. The Malaysian construction industry should upgrade its current construction execution and strategies, whether in terms of practice, management or technology in order to be globally competitive [2].

Research advocates that BIM (Building Information Modelling) is a vital asset for building construction from preconstruction through operation to its end of life [2]. The advantages of BIM for infrastructure construction has started to be recognized and realized. Additionally, advantages of using BIM for infrastructure include the opportunity to record details of every construction action. Such documentation could offer a record for each element based on the history of the construction [3].

Engineers, construction managers, owners, and facility managers are acquiring the benefits of implementation of BIM for infrastructure as being dynamic, decisive, and an organized approach of maintaining their assets [4].

Research has demonstrated the extent of benefits of using BIM in vertical construction (i.e. buildings) and horizontal construction (i.e., infrastructure) [4]. Although there have been many efforts by the Malaysian Government to increase BIM implementation in construction projects, implementation is still slow and less favourable. People, process, policy and technology (PPPT) are the key challenges in the implementation of BIM today in Malaysia [5, 6].

This paper will elaborate how BIM is being used in the Pan Borneo Highway Sarawak Project and how those challenges are solved.

1.2. Building Information Modelling (BIM)

BIM can be viewed as a combination of advanced processes and technology that offers a platform for collaboration between different parties in a construction project by exploiting the use of Information Technology (IT). In the Malaysian construction industry, many construction players regard BIM as a new technology. It is not widely used as yet [11].

Traditionally, 2D design approved for construction are checked manually for discrepancies. This method is time consuming, especially for complex designs. With BIM, the process of creating and using 3D parametric computer-aided-design (CAD) technologies for design allows the exchange of information within a construction project team in a digital format. More importantly, clash detection is possible with the available pool of information [7, 8, 9, 10].

This approach is faster and reduces incidences of human error to a minimum. The models can be passed onto the contractor for estimation and planning of the construction projects.

In general, BIM can be viewed as a single repository system that supplies and receives relevant information in a digital form relevant to the construction project [11].

1.3. BIM for Infrastructure

BIM for infrastructure processes can benefit multiple stakeholders by incorporating project information into a single or multiple 3D models with multiple data sets. For example, owners can use the model for operation and maintenance, while engineers and contractors can use the information for design and building considerations. Various alternatives can be compared to achieve an optimum lifecycle cost [3].

A key benefit is the accurate geometrical representation of parts of the building infrastructure in an integrated data environment. Here, project stakeholders can acquire a greater level of detail at early stages of the project for better decision-making before they are implemented at site [12].

In addition, operation and maintenance histories can be well documented. Transportation infrastructure typically has a lifecycle of decades and generally maintenance is driven by financial considerations. BIM enables this to be done more cost effectively by providing the potential for up-to-date, accurate, and geometric representations of the asset and their sub assets [12].

Overall, the initial cost of constructing and maintaining a BIM model can be minimal in comparison with the benefits gained over the lifecycle of the infrastructure asset [12].

1.4. Value of BIM

BIM potentially allows for such considerations to be assessed and addressed through collaboration using a 3D model. Stakeholders can provide design alternatives in a digital format to address problem areas and apply degradation models to determine the most effective and appropriate means of addressing design and construction issues [12].

The use of BIM can help stakeholders transfer important decisions from the site to the computer where changes are easier to be made and more effective. Additionally, stakeholders can develop a shared understanding of the project through cross-disciplinary collaboration that helps reduce design errors and miscommunication, which in turn reduces risk and liability [13].

Finally, additional value may result through the use of BIM by avoiding data dispersions, and duplication of effort, thereby increasing efficiency and safety, reducing time for routine data collection and recording. All of these translate into cost savings for the owner and increased structural safety of the assets [14].

BIM can also help decision-makers to schedule regular maintenance for infrastructure assets. Research has suggested that BIM implementation can lead to noticeable cost savings. Overall cost diminishes as unplanned maintenance is replaced by planned maintenance, helping infrastructure owners and engineering firms to seek integrated and cost-effective solutions that span the entire project lifecycle [15,16].

2. Case study: PBHS Phase 1 project

This research implemented uses a case study methodology whereby the key tasks were to:

1) Compare and analyse BIM implementation on the PBHS Project with PPPT Challenges with regard to Aryani Ahmad Latifi's [2] research about PPPT;
2) Validate findings through interviews of project representatives.

2.1. Project Overview

The long-awaited Pan Borneo Highway in the Malaysian state of Sarawak on the island of Borneo will span an estimated 1,060 km. Works are on-going to develop and upgrade the current 2-lane single carriageway which connects the entire state into a 4-lane dual carriageway. The project is being undertaken by the Government of Malaysia and State Government of Sarawak through a Project Delivery Partner (PDP). For this, a special purpose vehicle private entity, Lebuhraya Borneo Utara (LBU), has been appointed as the PDP.

Costing RM16.5 billion, Pan Borneo Highway Sarawak is deemed to be the single largest infrastructure project ever awarded by the Government in Sarawak. Implementation of the project involves construction of 786km of a new 4-lane highway in 11 works packages. Works began in January 2016, with the Highway targeted for completion in 2021. Operations and maintenance of the Highway will follow upon completion of construction.

As PDP, LBU is in charge of the entire project lifecycle, from planning to design and construction, to operations and maintenance of the completed highway. As PDP to the Government, LBU is committed to deliver the project on time and within budget. An associated company will then undertake operations and maintenance of the highway.

Phase 1 development of Sarawak's single largest infrastructure project is divided into 11 works package stretches averaging between 60-90 km each. BIM has been implemented for Phase 1 of the PBHS project up to Level 2 of BIM maturity and provided up to LOD 500. Powered by five BIM Modellers and one Coordinator for each package, the total number of dedicated personnel for BIM Implementation on the PBHS project is 55 BIM Modellers, 11 BIM Coordinators, 8 BIM Specialists, 4 BIM Lead Family, 1 BIM Lead Coordinator and 1 BIM Director.

2.2. People, process, policy & technology (PPPT) challenges

With reference to Aryani Ahmad Latifi's [2] research on PPPT, there are several factors and causes which contribute to barriers and challenges of BIM implementation in construction projects. These can be categorised as people, process, policy and technology [17].

2.2.1 People. Many construction players refused to adopt BIM because they are more comfortable with conventional processes in managing construction projects. Lack of knowledge and understanding in BIM, lack of awareness and encouragement to adopt BIM among top management and availability of good BIM resources are the key factors contributing to the slow adoption of BIM [2].

Lack of knowledge and skill in BIM is another factor contributing to the slow implementation of BIM among construction players. A lack of awareness on BIM and little encouragement to implement BIM among clients and top management in construction organizations contribute to the lack of knowledge and skill on BIM among players [17, 18].

2.2.2 Process and Policy. Clear guidelines on BIM are important to assist construction players to implement BIM in construction projects. Without proper guidelines, BIM implementation can be inaccurate, causing construction players more problems and failure to reap the optimum benefits of using such technology [6].

It has been seen that many construction players have developed their own version of BIM implementation guidelines. These guidelines can and in many cases have resulted in confusion among construction players, causing users to be doubtful as to the use of BIM implementation [6, 19].

This has contributed to the slow acceptance and adoption of BIM in the local construction industry due to the reluctance of implementing BIM in future projects [6, 19].

2.2.3 Technology. Respondents felt that the adoption of new technology such as BIM requires high cost, especially for new hardware (computer), software (BIM tools) and BIM training. Costs to implement BIM range from RM15,000 to RM90,000.00. Many felt such costs could only be afforded by large organisations. [2, 8, 19, 20]

The combination of all these factors has resulted in the slow implementation of BIM among Malaysian construction projects in general. Table 1 shows the summary of barriers and challenges of BIM Implementation in construction projects.

Table 1. Barriers and challenges of BIM implementation

No	Factors	Challenges & Problems
1	People	• Comfort with conventional processes used add to refusal of construction players to change. • Lack of knowledge of BIM. • Lack of skill on BIM.
2	Process & Policy	• No BIM guideline and specific model could assist construction players to implement BIM.
3	Technology	• BIM tools are expensive. • New hardware is expensive. • BIM training is expensive.

3. BIM implementation: PBHS Phase 1 project

While People, Process, Policy and Technology are seen to be challenges in BIM Implementation not only in Malaysia but across borders, BIM implementation in the Pan Borneo Highway Sarawak (PBHS) Phase 1 project has proven that these challenges can be mitigated by execution of strategies. Based on the performance and actual site coordination, leveraging on BIM has become part of the daily work of project stakeholders who now understand and reap the benefits of its implementation.

3.1. People

The initial lack of knowledge and skill in BIM was addressed by providing intensive training to employees of the Project Delivery Partner of the PBHS project. BIM Modellers and BIM Coordinators attended a one-month intensive training course for both roles. They were trained on implementing

BIM in real work conditions, understanding the functions of BIM Tools, and familiarising themselves with the workings of a BIM Environment.

For PBHS Phase 1, the BIM Modellers and Coordinators were given basic, intermediate and advance level of BIM Tools usage, such as Autodesk Civil 3D®, Autodesk Revit® and Autodesk Infraworks® for highway modelling, followed by Autodesk Navisworks®, Bentley ProjectWise® and Oracle Primavera P6® for coordination and management.

In addition to training the BIM Modellers and Coordinators, the local construction industry involved on the PBHS project were also introduced to BIM Awareness through seminars, project showcases, and conferences. Socialization and promotion are important activities to introduce BIM to all construction players and carry them along the BIM Process and build acceptance of the technology's usefulness.

Questionnaire is used to determine BIM implementation value based on real approach. Within the duration between March until April 2018 there are 31 respondents are involved in this questionnaire and giving their opinion about benefit of HIM implementation (A.1). In question Have you / your team gained any value / benefit from HIM implementation? the result shown that 80.6% respondents said yes and the rest said no. Most of them answered that BIM help to gain detail information on pre-construct about clashes, better 3d modelling view, give better mitigation and more efficient solution.

3.2. Process and Policy

References for BIM Process and Policy are available extensively to be adopted by projects. For PBHS Phase 1, it established its own process and policy for implementation by adhering to BIM guidelines implemented successfully for other countries such as Singapore and the United Kingdom.

BIM Essentials Guidelines from Building and Construction Authority (BCA) Singapore became the first reference to create a series of BIM Execution Plan templates. For the process of BIM Implementation PBHS Phase 1, the British Standards Institution (BSI) Public Available Specification (PAS) series became a key reference point.

BCA and BSI were able to accommodate the workflow of BIM implementation in PBHS Phase 1 Project. Additional policies and processes may be added if required. Additional workflows need to be documented and researched further if it is to be used as a reference for Malaysian Standards for BIM Infrastructure. The Government of Malaysia is currently preparing this standard to be used as a national guideline.

3.3. Technology

The main software used in PBHS Phase 1 project is Autodesk Infrastructure Design Suite Ultimate® 2016, a subscription base system. This package includes all BIM tools to be used for infrastructure and building projects. Whilst the software may be costly, with proper BIM implementation, the Returns of Investment (ROI) of using such technology has multiple benefits.

Freeware was also being considered for this project, such as Autodesk Navisworks Freedom® for site coordination purposes. Today, users enjoy the benefits of the system. They are able to gain access, view models, and study comparisons between the model and actual site conditions to identify and visualise the clash objects. Based on coordinators clashes report it prove that this tool help to collect all potential clash data (A.2). This report will help to gain all potential clash and inform the site team to justify all the clashes before it constructed. It proves that along the construction stage, site team able to check the clash report and compare with site condition. It helps to reduce time for checking the clash by using visibility check only based on the clash reports.

Usage of Network Attached Storage (NAS) Servers are other cost-saving measures to create your own Common Data Environment (CDE) instead of using the Cloud based systems which are paid monthly and with limited number of users. The NAS Server is also a good inventory investment as it can be used multiple times.

4. Conclusion

BIM Implementation in the PBHS Phase 1 project can be used as a national reference to leverage BIM in infrastructure projects. Among the benefits as seen through the successful use of BIM in implementation of this large-scale Highway project are:

- It has been proven that the challenges of PPPT can be mitigated by using the right strategies.
- BIM has resulted in the more effective use of resources and better coordination.
- Stakeholders benefit with better management and coordination.
- Collaboration between BIM practitioners and the Government plays an important role to drive the implementation of BIM in the local construction industry.
- Cooperation between both parties is also important to develop a standardized BIM guideline to be used by construction players in the country.

The experience of BIM Implementation in the PBHS Phase 1 project can be used as a benchmark for infrastructure projects both in Malaysia and elsewhere.

It is hoped that the information contained in this paper can be useful to construction players in other countries intending to implement BIM. The information can help to expose construction players to the realities of implementing BIM. Lessons learned can be used as guidelines to avoid unexpected circumstances in the journey to implementing BIM. It is anticipated that research will be made richer with the experiences of other types of construction projects. All such resources will be useful to increase BIM implementation in the architecture, engineering and construction (AEC) industry both here in Malaysia and elsewhere.

5. References

[1] Ahmad Latiffi A, Brahim J and Fathi M S 2014 The development of building information modelling (BIM) definition *Appl. Mech. Mater.* **567** 625–30

[2] Ahmad Latiffi A, Suzila M and Syamsul Rakiman U 2015 Potential improvement of Building Information Modeling (BIM) implementation in Malaysian construction projects *IFIP International Conference on Product Lifecycle Management* pp 149-58

[3] Marzouk M and Abdel Aty A 2012 Maintaining subways infrastructures using BIM. *Construction Research Congress 2012* Indiana, 21-23 May 2012

[4] Construction, M H 2012 *The business value of BIM for infrastructure: addressing Americas challenges with collaboration and technology, Smart Market Report* (New York: McGraw-Hill)

[5] Ahmad Latiffi A, Mohd S and Brahim J 2014 Application of building information (BIM) in the malaysian construction industry: A story of the first government project *Appl. Mech. Mater.* **773** 943–8

[6] Zakaria Z, Mohamed Ali N, Tarmizi Haron A, Marshall-Ponting A J and Abd Hamid Z 2013 Exploring the adoption of building information modelling (BIM) in the Malaysian construction industry: A qualitative approach *Int. J. Res. Eng. Technol.* **2**(8)

[7] Revit 2008 *The Five Fallacies of BIM* Available from http://dynamic.ziftsolutions.com/clients/autodesk/assets/emea/whitepapers/The_Five_Fallacies_of_BIM.pdf

[8] Eastman C, Teicholz P, Sacks R and Liston K 2011 *2nd Edition BIM Handbook: A Guide to Building Information Modelling for Owner, Managers, Designers, Engineers, and Contractors* (New Jersey: John Wiley and Sons)

[9] Construction, M H 2008 *Building Information Modeling Trends Smart Market Report.*, (New York: McGraw-Hill)

[10] Taylor J E and Bernstein P G 2008 Paradigm trajectories of building information modelling practice in project networks *ASCE Journal of Management in Engineering* **25**(2) 69-76

[11] Zahrizan Z, Nasly M A, Ahmad Tarmizi H, Amanda M P and Zuhairi A H 2014 Exploring the barriers and driving factors in implementing building information modelling (BIM) in the Malaysian construction industry: a preliminary study *The Institution of Engineers, Malaysia* **75**(1) 1-10
[12] Clevenger C M, Ozbek M E, Mahmoud H and Fanning B 2014 *Impacts and Benefits of Implementing Building Information Modeling on Bridge Infrastructure Projects* (No. MPC 14-272).
[13] Bennett T 2018 *Digitalization of Infrastructure* (Autodesk Inc.)
[14] Lwin M 2006 *The Important Roles of Bridge Maintenance and Management on Transportation Safety and Efficiency* (Washington: Federal Highway Administration)
[15] Cagle R 2003 Infrastructure asset management: An emerging direction *AACE International Transactions* **02**(01) 1-6
[16] Kathy M 2012 *Rapid increase in use of building information modeling (BIM) for infrastructure projects expected* Available from https://www.construction.com/about-us/press/increase-in-building-information-modeling-BIM-for-infrastructure-projects.asp
[17] Haron A T 2013 *Organisational readiness to implement building information modelling: A framework for design consultants in Malaysia* (Doctoral dissertation, University of Salford)
[18] Eastman C, Teicholz P, Sacks R and Liston K 2011 *BIM Handbook: A Guide to Building Information Modelling (BIM) for Owners, Managers, Designers, Engineers and Constructions* (New Jersey: Wiley)
[19] Furneaux C and Kivit R 2008 *BIM: Implications for Government, CRC for Construction Innovation* (Brisbane: Net Pty Ltd)
[20] Forbes L H and Ahmed S M 2011 *Modern Construction Lean Project Delivery and Integrated Practices* (USA: Taylor and Francis)

IOP Conf. Series: Materials Science and Engineering **512** (2019) 012061 doi:10.1088/1757-899X/512/1/012061

Performance of Dense Graded Asphaltic Concrete Using Nanosilica Modified Bitumen

A K Arshad[1], E Shaffie[1], K A Masri[2], W Hashim[3] and Z A Rahman[3]
[1] Institute for Infrastructure Engineering and Sustainability Management (IIESM), Universiti Teknologi MARA, Shah Alam, Selangor, Malaysia
[2] Faculty of Civil Engineering & Earth Resources, Universiti Malaysia Pahang, Malaysia
[3] Faculty of Civil Engineering, Universiti Teknologi MARA, Shah Alam, Selangor, Malaysia

E-mail: drahmadkamil@salam.uitm.edu.my

Abstract. Flexible pavements are constructed to last for its design life, therefore, a good and durable asphalt surface layer is required. However, flexible pavements using dense graded asphaltic concrete have some shortcomings such as prone to rutting when heavy loads are applied at high ambient temperatures. One of the ways to improve the performance of dense graded asphaltic concrete is to modify the bitumen using nanosilica. Among the properties of nanosilica that is advantageous to improve the bitumen is strong absorption, large surface area and excellent stability. The objective of this study is to investigate the use of nanosilica as a bitumen modifier to improve the moisture susceptibility and rutting resistance of dense graded asphaltic concrete. Bitumen PEN 60/70 was modified with nanosilica at 2% by weight of bitumen. The performance of the asphaltic concrete specimens was then evaluated in terms of its moisture susceptibility, resilient modulus and rutting resistance. The results obtained from the testing showed that the addition of nanosilica increased the resilient modulus value and reduced the rutting depth of dense asphaltic concrete mixes, while achieving the required moisture resistance. It concludes that the addition of nanosilica in the bitumen improved the performance of dense asphaltic concrete mix.

1. Introduction

The ever-increasing numbers of heavy vehicles with increased axle loads on the roads today placed high demands on the pavements to withstand the high stresses imposed from these vehicles. Good pavement design and timely maintenance enable flexible pavements to withstand these demands but special attention is required on highly stressed areas particularly the roads and highways with very high percentage of heavy vehicles and loaded with heavy freight. Bitumen modification offers a solution to provide a longer service life on these roads and reduce the frequency of maintenance required [1].

Researchers around the world have investigated various kinds of modifiers to modify the bitumen in order to improve its performance such as polymer. The polymer modifiers used include Styrene-Butadiene-Styrene (SBS), Styrene-Butadiene-Rubber (SBR), Styrene-Ethylene-Butylene-Styrene (SEBS), crumb rubber, and Ethylene Vinyl Acetate (EVA) [1]. These studies found that the use of polymer in asphalt modification can improve the rheological properties in terms of phase angle (δ) and dynamic modulus (G*), resistance to fatigue cracking, rutting and moisture susceptibility. Although these polymers all improved bitumen properties to some extent, there are still some drawbacks limiting their application such as high cost, low ageing resistance and poor storage stability [2]. Therefore, there is a need for further explore the use of other materials such as nano materials to improve the performance of bitumen and at the same time overcome these drawbacks.

Nanotechnology is the creation of new materials, devices, and systems at the molecular level as the phenomena associated with atomic and molecular interaction strongly influence the macroscopic material properties [3]. Nanomaterial/nanoparticle is a material with least one dimension is on the nanometer scale that is one billionth of a meter (10-9 m).

Nanomaterials that have been investigated for use in asphalt mixes include nano-clay, carbon nanotube and carbon nanofibre, nano-TiO2, nano-silica and nano-alumina [4]. Nanoclay was found to improve the viscosity and complex modulus (G*) value of bitumen. Carbon nanotube (CNT) is a one-atom thick sheet of graphite rolled up into a seamless hollow cylinder with a diameter of the order of one nanometer. The addition small percentage (around 1.5%) of CNT helpful in increasing the failure temperature, elastic modulus, complex modulus and rutting resistance of the bitumen after RTFO aging.

Silica is the dioxide form of silicon, SiO2 that is naturally occurring, such as quartz sand, rocks, and clays, and is widely used in industries to produce silica gels, colloidal silica, and fumed silica [5]. Nanosilica (NS), derived from processed silica, has a maximum dimension of about 30 nm. An inorganic material possesses beneficial properties such as large surface area, good dispersal ability, strong adsorption, high chemical clarity and excellent stability [6]. NS has been widely used in polymer, concrete and asphalt binder as inorganic filler to improve the properties of polymeric, mechanical and bituminous materials. In bitumen modification, a study by Yao [7] showed that the addition of NS improved anti-aging property, rutting and fatigue resistance of asphalt mixture. In addition dynamic modulus, flow number and rutting resistance of asphalt mixture also significantly improved.

Since NS promises improvements in properties for bitumen modification, therefore, in this study NS was used as a modifier to modify bitumen. Bitumen PEN 60/70 was modified with different concentrations of nanosilica. The modified bitumen was then used in dense asphaltic concrete and the performance of the asphaltic concrete specimens was then evaluated in terms of its moisture susceptibility, resilient modulus and rutting resistance. The objective of this study is to investigate the use of nanosilica as a bitumen modifier to improve the performance of dense graded asphaltic concrete in terms of moisture resistance, stiffness and rutting resistance.

2. Methodology

The properties of aggregates and bitumen used for the study were first tested to ensure compliance with the specifications. Marshall mix design method was then used for sample preparation using both the unmodified Pen 60/70 bitumen as control sample and the modified bitumen (2% NS by weight of bitumen). The optimum bitumen content was then determined. Finally, moisture susceptibility, resilient modulus and rutting tests were carried out to evaluate the performance of the dense asphaltic concrete mixes.

2.1. Materials

In this study, bitumen of 60/70 penetration grade was modified using 2% NS by weight of the bitumen. The bitumen was supplied from Petronas Berhad. The influence of NS on bitumen was determined by the softening point test, penetration test and ductility test. The aggregate used was obtained from Kajang Rock Quarry, Selangor. The aggregate tests conducted for this study are Los Angeles Abrasion Value, Aggregate Impact Value, Aggregate Crushing Value and Flakiness Index Value. Portland cement was used as filler additive to the mixture. The bitumen and aggregate used complied with the requirements as per PWD Malaysia's Specification for Road Works (JKR/SPJ/2008-S4). The properties of the bitumen, aggregates and liquid, gel type NS is shown in Table 1, Table 2 and Table 3.

Table 1. Physical properties of bitumen.

Test	Test result
Softening point (°C)	48-56
Penetration (0.1 mm)	60-70
Ductility (cm)	>100

Table 2. Properties of aggregates.

Test	Test result
Aggregate crushing value	18.3%
Aggregate Impact Value	20.6%
Los Angeles Abrasion	27.0%

Table 3. Properties of nanosilica.

Parameter	Specification value
Appearance	Slight milky transparent to translucent liquid
SiO2 (wt%)	30±1%
Na2O (wt%)	0.5%
pH (20°C)	8.5-10.5
Density (20°C, g/cm3)	1.19-1.22
Particle size (nm)	10-15

The bitumen was modified using a mechanical mixer. Asphalt binder was placed in a container and heated up to 155oC using a hot plate. NS was gradually added to the asphalt binder while stirring with a mechanical stirrer at a speed of 2000 rpm. The mixing process continued for one hour to achieve uniform dispersion of NS. The optimum amount of NS was determined based on a previous study by the author, after evaluation of the physical properties of the modified bitumen (penetration, softening point, ductility, viscosity and storage stability). The optimum NS content was found to be 2% by weight of bitumen as shown in Table 4. In addition, further evaluation were carried out on the modified bitumen for morphological properties (using atomic force microscopy and x-ray diffraction), rheological properties (using dynamic shear rheometer) and chemical properties (using Fourier transform infrared spectrometer) [8,9].

The AC14 aggregate gradation used was in accordance to PWD Malaysia's Specification for Road Works JKR/SPJ/2008-S4[10]. Table 5 shows the aggregate gradation as per the specification. The middle gradation line between the gradation limits was used for this study.

IOP Conf. Series: Materials Science and Engineering **512** (2019) 012061 doi:10.1088/1757-899X/512/1/012061

Table 4. Physical properties of nanosilica modified bitumen

Physical Properties	Nanosilica Concentration (% by weight of bitumen)	
	0%	2%
Softening point (°C)	52.3	60.3
Penetration (dmm)	65.0	41.1
Ductility (cm)	140.0	70.2
Viscosity (Pa)	0.7	0.9
Storage stability	0	0.2
Penetration Index	0.02	0.62
Penetration Viscosity Number	0.103	-0.054

Table 5. Aggregate Gradation for Samples (PWD Malaysia, 2008).

Mix Designation	AC 14 Gradation (Specification)	AC14 Gradation used (mid-gradation)
B.S Sieve	% Passing By Weight	% Passing By Weight
20 mm	100	100
14 mm	90 -100	95
10 mm	76 – 86	81
5.0 mm	50 - 62	56
3.35 mm	40 - 54	47
1.18 mm	18 - 34	26
425 μm	12 - 24	18
150 μm	6 - 14	10
75 μm	4 - 8	6

2.2. Preparation of samples

The laboratory testing carried out in this study include Marshall mix design to determine the optimum binder content, moisture susceptibility test, resilient modulus test and rutting resistance tests to evaluate the performance of the dense asphaltic concrete mix.

2.2.1. Marshall mix design. Marshall samples were prepared at 4.0%, 4.5%, 5.0%, 5.5% and 6.5% bitumen contents for 0% NS (control) and 2% NS asphalt mixtures. The samples were prepared in accordance to ASTM D6926 to determine the Optimum Bitumen Content (OBC) using the Marshall Test method. Before the mixing process, the bitumen was heated to a temperature of 155°C. The mix was compacted at a temperature of 135 ± 5°C using the standard Marshall compactor with 75 blows for each face. Marshall tests were carried out on the samples and volumetric analysis were carried out on the results obtained from the tests in accordance to ASTM D1559. The average value of bulk specific gravity, stability, flow, VFA and VMA obtained separately were plotted against the bitumen content and smooth curve were drawn through the plotted value. The determination of optimum bitumen content (OBC) for ACW-14 was determined by averaging the optimum bitumen contents as specified in JKR/SPJ/2008-S4.

2.2.2. Moisture susceptibility test. The Modified Lottman test was performed to evaluate the moisture susceptibility of the mixes in accordance with the procedure described in AASHTO T 283. Samples for the test were prepared at 7 + 0.5% air voids. Three samples are selected as a control and tested without moisture conditioning; and another three samples were selected to be conditioned by saturating with water at 70-80 percent followed by immersing in water for 24 hours at 60°C in a water bath. The samples were then tested for indirect tensile strength (ITS) by loading the samples at constant head rate (50 mm/minute vertical deformation at 25°C) and maximum compressive force required to break the samples were recorded. Tensile Strength Ratio (TSR) results were determined by comparing the indirect tensile strength (ITS) of unconditioned samples with the control samples. Retained tensile strength ratio (TSR) was used with 80% as the boundary between mixtures resistant and sensitive to moisture.

2.2.3. Resilient modulus test. Indirect tensile repeated load test was used to determine the resilient modulus. The test used the UTM-5P machine. The resilient modulus test was conducted to measure the stiffness modulus of the design asphalt mixtures. All samples were conditioned for 3 hours at a temperature of 25°C. The test was performed in accordance with ASTM D 4132. Three pulse repetition periods was applied, 1000 ms indicates low traffic volume road (high speed), 2000 ms for medium traffic volume road while 3000 ms is high traffic volume road (slow speed). The compacted cylindrical asphalt mixture was subjected to diametrical repeated loading of 900 N and two points was taken on different axis at each sample.

2.2.4. Rutting resistance test. The Asphalt Pavement Analyzer (APA) device was used to determine the rutting resistance of asphaltic concrete mixes. In this study, asphaltic concrete mixes containing with 0% and 2% NS were prepared with the weight of aggregate of 2965g and compacted at 7± 5% air voids. The specimens were tested at 60°C of temperature with a hose pressure of 100 ± 5 psi and wheel load with 100 ± 5 lb. This test was carried out based on the standard procedure given in AASHTO T340. Figure 1 shows the Asphalt Pavement Analyser.

Figure 1. Asphalt Pavement Analyzer for rutting resistance test.

3. Results and discussion

3.1. Optimum binder content

Marshall Test was carried out on the samples and volumetric analysis was carried out on the results obtained from the tests. The average value of bulk density, stability, flow, VFA and VMA obtained separately were plotted against the bitumen content and smooth curve were drawn through the plotted value. The OBC was determined from the arithmetic mean of the results from the five curves plotted and the value was cross-checked with the limits set in Public Works Department specification. Table 6 shows the OBC for both samples. The OBC obtained for 0%NS samples is 4.92% while the OBC for 2%NS samples 4.96%.

Table 6. Optimum bitumen content determination.

Property	0%NS	2%NS
Peak of curve taken from the graph bulk specific density	5.05	5.20
VIM equals to 4.0% from VIM graph	4.75	4.80
VFB equals to 75% from VFB graph	4.95	4.89
Peak of curve taken from the stability graph	4.70	4.65
Flow equals to 4mm from flow graph	5.15	5.25
Average OBC	4.92	4.96

3.2. Moisture susceptibility

Figure 2 shows the results of Indirect Tensile Strength (ITS) for dry and wet conditioned samples for both 0% and 2% NS asphaltic concrete mixes. For dry conditioned samples, 2% NS mix has higher ITS value (1147 kPa) compared with 0% NSMB mixes (1134 kPa). However, for wet conditioned samples with 0%NS has higher ITS value of 972 kPa compared with 927 kPa for 2% NS mixes.

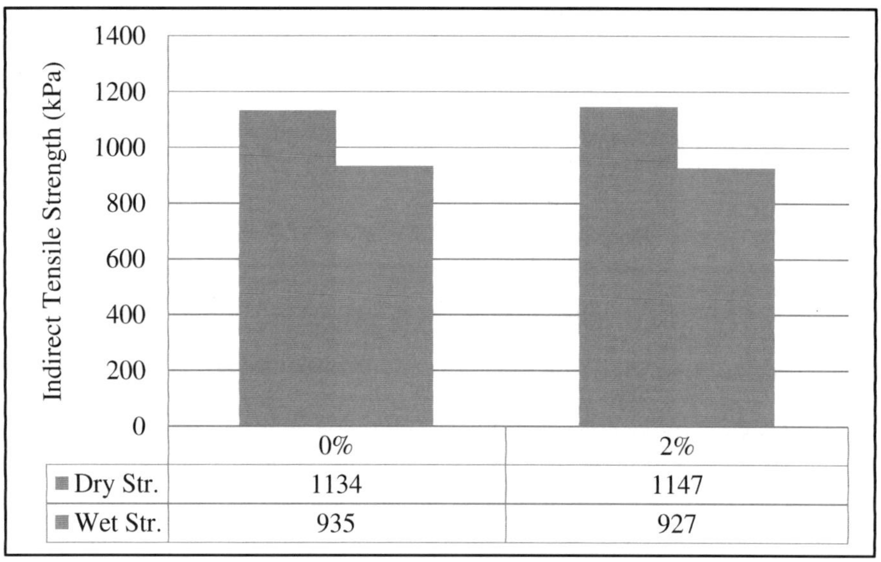

Figure 2. Indirect Tensile Strength (ITS) for dry and wet conditioned samples.

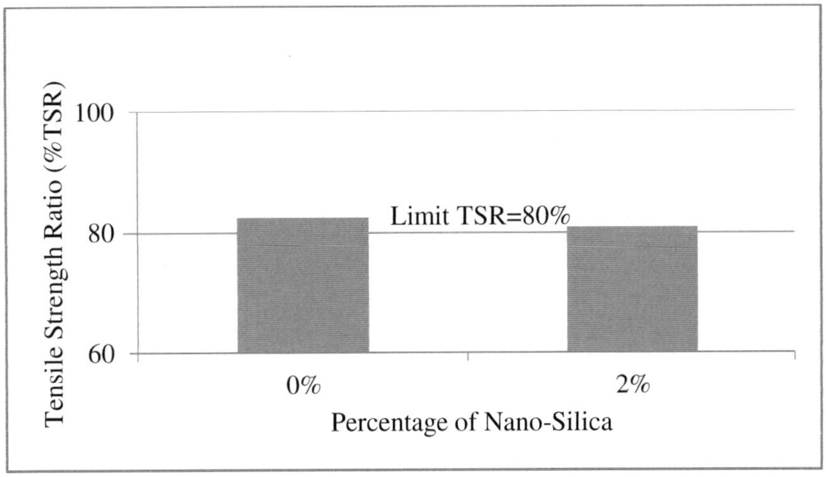

Figure 3. Tensile Strength Ratio (TSR) value for 0% and 2% NSMB.

Figure 3 shows the TSR value for both mixes; the TSR value for mixes with 2% NS (80.8%) was slightly lower than the TSR value for 0%NS mixes (82.5%). However, the result showed that both mixes meet the minimum TSR value of 80% required by AASHTO T283.

3.3. Resilient modulus

Figure 4 shows the resilient modulus graph for pulse repetitive periods of 1000 ms, 2000 ms and 3000 ms. The mixture with 2%NS has higher resilient modulus values for all pulse repetition periods compared to the mix with asphalt 0%NS. When the pulse repetitive period of 1000 ms was applied, the value of resilient modulus for 0% NS mixture was 5,477.5 MPa while for the 2% NS mixture. For the 2000 ms pulse period, the resilient modulus values for 0% and 2% NS mixes were 4,594.75 MPa and 6,060.25 MPa respectively. When the 3000 ms pulse period was applied, the values of resilient modulus for 0%NS and 2%NS mixtures were 4,188.75 MPa and 5,644 MPa. The results showed that the addition of NS can improve the stiffness of asphaltic concrete mixes as indicated with the higher resilient modulus values for the mixes with 2% NS.

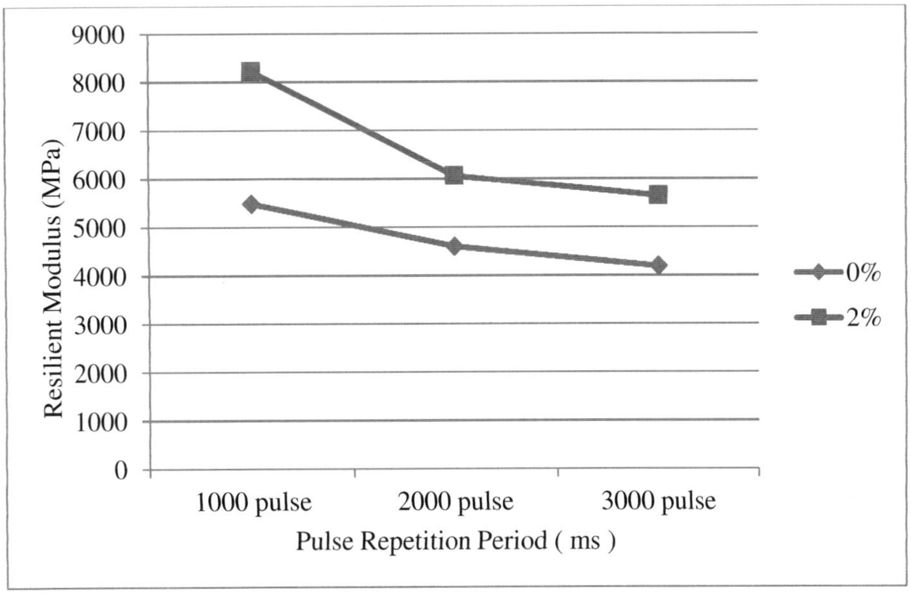

Figure 4. Resilient Modulus versus Pulse Repetition Period.

3.4. Rutting test

Figure 5 shows the graph for rutting depth versus number of cycles. The mixture with 2%NS has lower rutting depth compared to the mix with asphalt 0%NSMB. This result showed that the addition of NS could reduce the rutting depth of asphaltic concrete mixes.

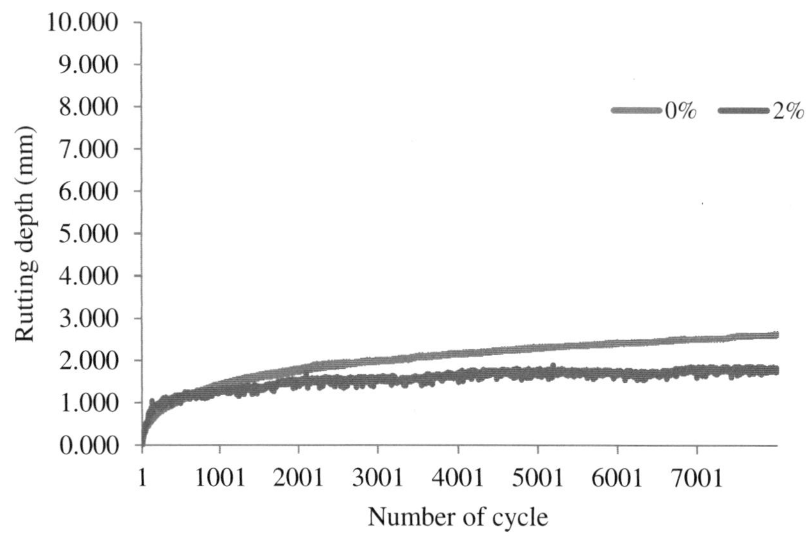

Figure 5. Rutting depth versus number of load repetition cycles.

Table 7 shows the rutting depth for both 0% and 2% NS asphaltic concrete mixes. The result showed that the rutting depth decreased by addition of NS in the asphaltic concrete mixes. The addition of 2%NS reduced about 33.0% of the rutting depth compared with the 0%NS mix. It can be concluded that the addition 2%NS on asphaltic concrete mixes can enhance rutting resistance and produce better quality of asphalt mixt. The rate of rutting for both of 0%NS and 2%NS mixes is also showed in Table 6. This result showed that the addition of 2%NS lowered the rate of rutting compared to 0%NS mix. The rate of rutting is reduced by 67.58% due to the addition of NS. This result showed that asphaltic concrete mix with 2%NS has a better performance for rutting resistance compared to asphaltic concrete mix with 0%NS.

Table 7. Rut depth and rate of rutting.

Type of mix	Rutting depth	Rate of rutting
Mix with 0%NS	2.637 mm	0.438 mm/hr
Mix with 2%NS	1.767 mm	0.142 mm/hr

4. Conclusion

Based on the experimental results on this study, the following conclusion can be made:

1) The moisture susceptibility performance for 2%NS is above the required minimum TSR ratio of 80%.
2) The stiffness of the asphaltic concrete mix increased by addition of 2% of NS in the mixture.
3) The rutting resistance increased by 30% by addition of 2% of NS in the mixture.

It can be concluded that the addition of nanosilica is beneficial to the dense asphaltic concrete mix, as the laboratory evaluation indicate improved performance in stiffness and rutting resistance.

IOP Conf. Series: Materials Science and Engineering **512** (2019) 012061 doi:10.1088/1757-899X/512/1/012061

5. References

[1] Read J and Whiteoak D 2003 *The Shell Bitumen Handbook* (Thomas Telford)

[2] Zhu J, Birgisson B and Kringos N 2014 Polymer modification of bitumen: advances and challenges *Euro. Poly. J.* **54** 18-38

[3] Chong K P 2004 Nanotechnology and information technology in civil engineering *4th Int. Sym. on Info. Tech. in CE* Nashville pp 1-9

[4] Li R, Xiao F, Amirkhanian S, You Z and Huang J 2017 Developments of nano materials and technologies on asphalt materials – A review *Const. and Bldg. Mtrl.* **143** 633-648

[5] Balamurugan M and Saravanan S 2012 producing nanosilica from sorghum vulgare seed heads *Powder Tech.* **224** 345-50

[6] Hsieh T H, Kinloch A J, Taylor A C and Sprenger S 2011 The effect of silica nanoparticles and carbon nanotubes on the toughness of a thermosetting epoxy polymer *J. of App. Poly. Sci.* **119** 2135-42

[7] Yao H, You Z, Li L, Lee C H, Wingard D, Yap Y K, Shi X and Goh S W 2013 Rheological properties and chemical bonding of asphalt modified with nanosilica *ASCE J. of Mtrl. in CE* **25** 1619-30

[8] Arshad A K, Samsudin M S, and Ahmad J 2015 Characterisation of nanosilica modified asphalt binder *Int. Conf. on Adv. in Civil & Envir. Eng.* Penang

[9] Arshad A K, Samsuddin M S, Masri K A, Karim M R and Abdul Halim A G 2017 Multiple stress creep and recovery of nanosilica modified asphalt binder *MATEC Web of Conferences* **103** pp 09005

[10] JKR 2008 *Malaysia Standard Specification for Road Works: Section 4 – Flexible Pavement* (JKR/SPJ/2008-S4) (Kuala Lumpur: JKR)

Acknowledgement

The authors expressed utmost gratitude to FRGS Research Grant: FRGS/1/2015/TK08/UIYM/02/3 from Ministry of Higher Education, Malaysia for financial support, which enables this paper to be written.

Institute of Physics
Dirac House, Temple Back
Bristol BS1 6BE UK

ISSN: 1757-8981
ISBN 978-1-5108-8633-9

CURRAN ASSOCIATES INC.
proceedings.com